AF411402

Muscle Homeostasis and Regeneration

Muscle Homeostasis and Regeneration: From Molecular Mechanisms to Therapeutic Opportunities

Editor

Antonio Musarò

MDPI • Basel • Beijing • Wuhan • Barcelona • Belgrade • Manchester • Tokyo • Cluj • Tianjin

Editor
Antonio Musarò
Sapienza University of Rome
Italy

Editorial Office
MDPI
St. Alban-Anlage 66
4052 Basel, Switzerland

This is a reprint of articles from the Special Issue published online in the open access journal *Cells* (ISSN 2073-4409) (available at: https://www.mdpi.com/journal/cells/special_issues/Muscle_Homeostasis).

For citation purposes, cite each article independently as indicated on the article page online and as indicated below:

LastName, A.A.; LastName, B.B.; LastName, C.C. Article Title. *Journal Name* **Year**, *Article Number*, Page Range.

ISBN 978-3-03943-436-7 (Hbk)
ISBN 978-3-03943-437-4 (PDF)

Cover image courtesy of Antonio Musarò.

Contents

About the Editor

Antonio Musarò (Ph.D.) obtained a doctorate (Ph.D.) in Biotechnological Sciences from Sapienza University of Rome and completed a 4-year postdoctoral term at Harvard University (USA). From 2003 to 2014, Prof. Musarò was Adjunct Associate Professor of the School of Biomedical & Sports Science at Edith Cowan University, Western Australia. Currently, Prof. Musarò is a Full Professor of Medicine and Biotechnology at Sapienza University of Rome, Italy. He has authored over 120 scientific papers published in indexed scientific journals and discussed at international conferences; these research activities have also resulted in different patents. His main research interests are focused on three areas: (1) modulation of the hostile microenvironment to improve stem cell activity and muscle regeneration, (2) characterization of the physiopathologic interplay between muscle and nerve, and (3) establishment of muscle engineered in vitro model to study muscle homeostasis and differentiation.

Editorial

Muscle Homeostasis and Regeneration: From Molecular Mechanisms to Therapeutic Opportunities

Antonio Musarò

Laboratory affiliated to Istituto Pasteur Italia—Fondazione Cenci Bolognetti, DAHFMO-Unit of Histology and Medical Embryology, Sapienza University of Rome, Via Antonio Scarpa, 14, 00161 Rome, Italy; antonio.musaro@uniroma1.it; Tel.: +39-0649766956; Fax: +39-06-4462854

Received: 28 August 2020; Accepted: 3 September 2020; Published: 4 September 2020

Abstract: The capacity of adult muscle to regenerate in response to injury stimuli represents an important homeostatic process. Regeneration is a highly coordinated program that partially recapitulates the embryonic developmental program and involves the activation of the muscle compartment of stem cells, namely satellite cells, as well as other precursor cells, whose activity is strictly dependent on environmental signals. However, muscle regeneration is severely compromised in several pathological conditions due to either the progressive loss of stem cell populations or to missing signals that limit the damaged tissues from efficiently activating a regenerative program. It is, therefore, plausible that the loss of control over these cells' fate might lead to pathological cell differentiation, limiting the ability of a pathological muscle to sustain an efficient regenerative process. This Special Issue aims to bring together a collection of original research and review articles addressing the intriguing field of the cellular and molecular players involved in muscle homeostasis and regeneration and to suggest potential therapeutic approaches for degenerating muscle disease.

Keywords: muscle homeostasis; muscle regeneration; satellite cells; stem cells; FAPs; tissue niche; growth factors; inflammatory response; muscle pathology; aging

---

The regeneration of adult tissues is a highly coordinated process that partially recapitulates the embryonic developmental program. The myogenic program involves an interplay between the intrinsic program of muscle lineage specification and extrinsic influence, such as innervation and growth factor activity. Along with a set of four key transcription factors, namely MyoD, Myf5, Myogenin, MRF4, other factors have been characterized that play a critical role during the different stages of muscle development and regeneration. The transcription factor NF-Y is an evolutionarily conserved heterotrimer formed by the sequence-specific NF-YA and the Histone Fold Domain—HFD—NF-YB/N. Libetti et al., analyzed the role of the two NF-YA isoforms in myogenic program, proposing that NF-YAl, but not NF-YAs, maintains muscle commitment by indirectly regulating Myogenin and MyoD expression in C2C12 cells [1]. The myogenic program is also typically regulated by multiple signaling pathways and by the interaction of several extracellular matrix components with muscle stem cells. Lee et al., demonstrated that Thyroid hormones (THs, thyroxin; T4 and triiodothyronine; T3) regulate the expression of various proteins crucial for muscle development and contractility [2], whereas Torrente et al., discussed the role of insulin-like growth factor axis on fetal and post-natal growth [3]. The characterization of molecular mechanisms involved in the myogenic program also provides a roadmap for recapitulating skeletal myogenesis in vitro from pluripotent stem cells (PSCs). Baci et al., established a reliable protocol to induce the myogenic differentiation of induced pluripotent stem cells (iPSCs), generated from pericytes and fibroblasts, exploiting skeletal muscle-derived extracellular vesicles (EVs), in combination with chemically defined factors [4]. This genetic integration-free approach generated functional skeletal myotubes, maintaining the engraftment ability in vivo. The authors demonstrated that EVs can act as biological "shuttles" to deliver specific bioactive molecules for a successful transgene-free differentiation offering new opportunities for disease modeling and regenerative approaches [4].

The life-long maintenance of muscle tissue has been the objective of numerous studies employing a variety of approaches. It is generally accepted that cumulative failure to repair damage related to an overall decrease in anabolic processes is a primary cause of functional impairment in muscle. Muscle regeneration occurs in five interrelated and time-dependent phases, namely degeneration (necrosis), inflammation, regeneration, remodeling, and maturation/functional repair [5]. Although the phases of muscle regeneration are similar in different organisms (e.g., mouse, rat, human) and after different types of damage/trauma, the kinetics and amplitude of each phase are different in each organism and may depend on the extent of damage and the damage model used. Animals adopt different basic strategies of regeneration that include the activation of adult stem cells, the dedifferentiation of preexisting cells, and/or the proliferation of differentiated cells. This diversity of mechanisms is still widely understudied and underexploited for biomedical applications.

Forcina et al., provided an exhaustive overview about the general aspects of muscle regeneration and discussed the different approaches to study the interrelated and time-dependent phases of muscle healing [5], whereas Zullo et al., discussed, in an interesting review, the insights into the evolutionary aspect of muscle regeneration [6]. In these reviews, the authors integrated the principles of the physiologic muscle regeneration with a technical approach, reporting key experimental methods and markers employed to study cellular and molecular interactors dominating each stage of muscle healing [5] and provided an outline of main animals' clades, muscle types, their development, homeostasis, and regeneration abilities highlighting what is known of their molecular mechanisms [6]. Moreover, the review by Poovathumkadavil and Jagla [7] discussed how the Drosophila model could help understand the mechanisms of muscle homeostasis and regeneration, discussing the genetic control mechanisms of muscle contraction, development, and homeostasis with particular emphasis on the contractile unit of the muscle, the sarcomere.

The maintenance of an efficient regeneration process is guaranteed by both satellite cells' niche environment and satellite cells' pool. By disrupting either one of the two or both, the impairment in muscle regeneration suddenly happens, as likely occurs in many muscular dystrophies. Indeed, one of the critical points that remain to be addressed is why skeletal muscle fails to regenerate under pathological conditions. Either the resident muscle stem cells drastically decrease during aging and in several degenerative diseases or perhaps the pathological muscle is a prohibitive environment for stem cells activation and function. Although we lack definitive answers, several studies suggested that with age or under pathological conditions, the systemic environment impinges the activity of satellite cells and stimulates fibrotic accumulation [8].

Muscle homeostasis and regeneration are indeed severely impinged in several pathologic conditions such as sarcopenia, cachexia, and muscle dystrophies.

In the review by Pratt et al. [9] the authors aimed to identify genetic variants known to be associated with muscle phenotypes relevant to sarcopenia, the progressive deterioration in skeletal muscle mass, strength, and physical function with advancing age. The authors, interrogating PubMed, Embase and Web, using pre-defined search terms such as "aging", "sarcopenia", "skeletal muscle", "muscle strength" and "genetic association", identified the genetic variants associated with muscle phenotypes relevant to sarcopenia in humans; thus, the review might help to further illuminate the genetic basis of sarcopenia. Heterotopic ossification (HO) is another dysregulation process of skeletal muscle homeostasis and regeneration, which results in mature bone formation in atypical locations. Pulik et al., made the survey of cells responsible for heterotopic ossification development in skeletal muscles [10].

Satellite cells (SCs) are the main actors of myofiber regeneration after damage. Nevertheless, successful muscle healing requires the participation of additional cell types that directly or indirectly contribute to this process. SC activity is known to be influenced by signals deriving from the surrounding environment and by interactions with other cellular components of muscle niche. Petrilli et al., provided a comprehensive picture of the dynamics of the major cell populations that sensed and responded to acute damage in wild type mice and in a mouse model of Duchenne muscular dystrophy, a genetic disease characterized by impaired regenerative process and muscle wasting [11].

One of the mechanisms by which cells communicate with each other involves a conserved delivery system based on the generation and release of extracellular vesicles (EVs). These vesicles transfer information between cells through several categories of cargo-enriched biomolecules (i.e., proteins, lipids, nucleic acids, and sugars), each of them selectively influencing different cellular domains. Picca et al., demonstrated that older adults with physical frailty and sarcopenia show increased levels of circulating small extracellular vesicles with a specific mitochondrial signature [12].

Muscle wasting disorders, including sarcopenia and muscular dystrophies, are also characterized by the gradual replacement of muscle fibers by adipose and fibrotic tissue, which represent critical components of a hostile microenvironment/niche. Transforming Growth Factor β (TGF-β) is known for its role in the regulation of skeletal muscle size as well as fibrosis. Hillege et al. [13] assessed the time-dependent effects of TGF-β signaling and downstream signaling on the expression of myogenic, atrophic, and fibrotic genes in both myoblasts and myotubes. The authors showed that TGF-β inhibits myogenic gene expression in both myoblasts and myotubes but does not affect myotube size in vitro. Most importantly, TGF-β regulates the expression of fibrotic genes in both myoblasts and myotubes in a time-dependent manner [13].

Recent studies on the factors involved in skeletal muscle growth, differentiation, homeostasis, and regeneration have provided new insights into the function of these signaling molecules in muscle and suggest promising new avenues for systemic as well as local intervention in the defects associated with many muscle pathologies. Thus, skeletal muscle repair/regeneration may benefit by the treatments with factors that favor regenerative myogenesis and support the robustness of regenerated myofibers.

The inflammatory response of injured skeletal muscle plays an important and critical role in muscle homeostasis and regeneration and involves the recruitment of specific myeloid cell populations within the injured area. Thus, the inflammatory response is a coordinate process that must be finely regulated to obtain an efficient regenerative process, and the perturbed spatial distribution of inflammatory cells, altered identity of the inflammatory infiltrate (cell type and magnitude of influx) and disrupted temporal sequence results in a persistent rather than resolved inflammatory phase. The transcription factor Nfix, a member of the nuclear factor I (Nfi) family, plays a pivotal role during muscle development, regeneration and in the progression of muscular dystrophies. Saclier, et al., showed that *Nfix* is mainly expressed by anti-inflammatory macrophages [14]. Upon acute injury, mice deleted for *Nfix* in myeloid line displayed a significant defect in the process of muscle regeneration. Indeed, Nfix protein is involved in the macrophage phenotypical switch and macrophages lacking *Nfix* failed to adopt an anti-inflammatory phenotype and interact with myogenic cells.

Cappellari et al., taking Duchenne muscular dystrophy as a paradigm of defect in muscle regeneration, reviewed the main effects of drugs on regeneration biomarkers to assess whether targeting pathogenic events can help to protect niche homeostasis and enhance regeneration efficiency other than protecting newly formed fibers from further damage [15]. Squecco et al., reported the beneficial role of Platelet-Rich Plasma (PRP) treatment owing to PRP pro-myogenic and anti-fibrotic effects [16]. The main relevance of this study was the contribution toward defining the molecular and functional mechanisms regulating TGF-β1-induced fibroblast–myofibroblast transition, highlighting the role of Gap Junctions in this process as well as the involvement of voltage-dependent connexin isoform, namely Cx43. Leduc-Gaudet et al., analyzed the effect of Parkin on disease model, demonstrating that Parkin overexpression prevents sepsis-induced skeletal muscle atrophy, likely by improving mitochondrial quality and contents [17].

The study by Soendenbroe et al., supported the role of exercise in maintaining NMJ stability, even in elderly inactive individuals [18], whereas Jin and co-workers demonstrated that the functions of Lysine (Lys), the first limiting essential amino acid for mammals, in muscle mass accumulation are mediated by satellite cells and the mTORC1 pathway [19].

The maintenance of skeletal muscle mass plays a critical role in health and quality of life. Jorgenson et al., discussed how mechanical loads are one of the most potent regulators of muscle mass [20]. The authors have summarized the major structural adaptations that have been implicated in the mechanical load-induced growth of skeletal muscle and have also considered whether each of these adaptations makes a substantive contribution to the overall growth process [20].

Homeostasis represents one of the most important and critical parameters of adult tissues, including skeletal muscle, and it is defined as the capability of a system to maintain a constant state of complexity and order in a dynamic equilibrium. Interestingly, skeletal muscle is well-maintained in hibernators during hibernation, a unique survival strategy exhibited by various mammals in order to cope with adverse environments in winter, during which hibernators not only face the challenge of prolonged skeletal muscle inactivity, but also deal with other stresses, including hypoxia, fasting, and repeated ischemia-reperfusion during the torpor-arousal cycle.

The maintenance of cytoplasmic calcium (Ca^{2+}) homeostasis is important for the preservation of a normal structure and function of skeletal muscle fibers. Skeletal muscle inactivity can trigger Ca^{2+} homeostasis disturbance, often characterized by cytoplasmic Ca^{2+} overload, leading to protein degradation and cell apoptosis, both involved in skeletal muscle loss. Zhang et al., demonstrated that under extreme conditions, such as low temperature, low metabolism, and prolonged hindlimb inactivity during hibernation, hibernating ground squirrels still possess the ability to maintain intracellular Ca^{2+} homeostasis [21]. Therefore, maintaining intracellular Ca^{2+} homeostasis and avoiding skeletal muscle injury caused by its disturbance appear to be priority strategies employed by hibernating squirrels to cope with the various stresses induced during the torpor–arousal cycle. In addition, Malatesta et al., demonstrated that during hibernation, satellite cell nuclei maintain similar transcription and splicing activity as in euthermia, indicating an unmodified status during immobilization and hypometabolism [22]. Skeletal muscle preservation during hibernation is presumably not due to satellite cells activation, but rather to the maintenance of some functional activity in myofibers that is able to counteract muscle wasting

We believe that the papers in this Special Issue, each addressing a specific aspect of muscle homeostasis and regeneration under physiopathologic conditions, will help us to better understand the underlying mechanisms and will help to design more appropriate therapeutic approaches to improve muscle regeneration and to counteract muscle diseases.

Funding: The research in author's lab was supported by ASI, Ricerca Finalizzata (RF-2016-02364503), Ateneo-Sapienza, Fondazione Roma.

Conflicts of Interest: The authors declare no conflict of interest.

References

1. Libetti, D.; Bernardini, A.; Sertic, S.; Messina, G.; Dolfini, D.; Mantovani, R. The Switch from NF-YA1 to NF-YAs Isoform Impairs Myotubes Formation. *Cells* **2020**, *9*, 789. [CrossRef] [PubMed]
2. Lee, E.J.; Shaikh, S.; Choi, D.; Ahmad, K.; Baig, M.H.; Lim, J.H.; Lee, Y.; Park, S.J.; Kim, Y.-W.; Park, S.-Y.; et al. Transthyretin Maintains Muscle Homeostasis through the Novel Shuttle Pathway of Thyroid Hormones during Myoblast Differentiation. *Cells* **2019**, *8*, 1565. [CrossRef] [PubMed]
3. Torrente, Y.; Bella, P.; Tripodi, L.; Villa, C.; Farini, A. Role of Insulin-Like Growth Factor Receptor 2 across Muscle Homeostasis: Implications for Treating Muscular Dystrophy. *Cells* **2020**, *9*, 441. [CrossRef] [PubMed]
4. Baci, D.; Chirivì, M.; Pace, V.; Maiullari, F.; Milan, M.; Rampin, A.; Somma, P.; Presutti, D.; Garavelli, S.; Bruno, A.; et al. Extracellular Vesicles from Skeletal Muscle Cells Efficiently Promote Myogenesis in Induced Pluripotent Stem Cells. *Cells* **2020**, *9*, 1527. [CrossRef] [PubMed]
5. Forcina, L.; Cosentino, M.; Musarò, A. Mechanisms Regulating Muscle Regeneration: Insights into the Interrelated and Time-Dependent Phases of Tissue Healing. *Cells* **2020**, *9*, 1297. [CrossRef]
6. Zullo, L.; Bozzo, M.; Daya, A.; Di Clemente, A.; Mancini, F.P.; Megighian, A.; Nesher, N.; Röttinger, E.; Shomrat, T.; Tiozzo, S.; et al. The Diversity of Muscles and Their Regenerative Potential across Animals. *Cells* **2020**, *9*, 1925. [CrossRef]

7.   Poovathumkadavil, P.; Jagla, K. Genetic Control of Muscle Diversification and Homeostasis: Insights from Drosophila. *Cells* **2020**, *9*, 1543. [CrossRef]

8.   Scicchitano, B.M.; Sica, G.; Musarò, A. Stem Cells and Tissue Niche: Two Faces of the Same Coin of Muscle Regeneration. *Eur. J. Transl. Myol.* **2016**, *26*, 6125. [CrossRef]

9.   Pratt, J.; Boreham, C.; Ennis, S.; Ryan, A.W.; De Vito, G. Genetic Associations with Aging Muscle: A Systematic Review. *Cells* **2019**, *9*, 12. [CrossRef]

10.   Pulik, Ł.; Mierzejewski, B.; Ciemerych, M.A.; Brzóska, E.; Łęgosz, P. The Survey of Cells Responsible for Heterotopic Ossification Development in Skeletal Muscles-Human and Mouse Models. *Cells* **2020**, *9*, 1324. [CrossRef]

11.   Petrilli, L.L.; Spada, F.; Palma, A.; Reggio, A.; Rosina, M.; Gargioli, C.; Castagnoli, L.; Fuoco, C.; Cesareni, G. High-Dimensional Single-Cell Quantitative Profiling of Skeletal Muscle Cell Population Dynamics during Regeneration. *Cells* **2020**, *9*, 1723. [CrossRef] [PubMed]

12.   Picca, A.; Beli, R.; Calvani, R.; Coelho-Júnior, H.J.; Landi, F.; Bernabei, R.; Bucci, C.; Guerra, F.; Marzetti, E. Older Adults with Physical Frailty and Sarcopenia Show Increased Levels of Circulating Small Extracellular Vesicles with a Specific Mitochondrial Signature. *Cells* **2020**, *9*, 973. [CrossRef] [PubMed]

13.   Hillege, M.M.G.; Galli Caro, R.A.; Offringa, C.; de Wit, G.M.J.; Jaspers, R.T.; Hoogaars, W.M.H. TGF-β Regulates Collagen Type I Expression in Myoblasts and Myotubes via Transient Ctgf and Fgf-2 Expression. *Cells* **2020**, *9*, 375. [CrossRef] [PubMed]

14.   Saclier, M.; Lapi, M.; Bonfanti, C.; Rossi, G.; Antonini, S.; Messina, G. The Transcription Factor Nfix Requires RhoA-ROCK1 Dependent Phagocytosis to Mediate Macrophage Skewing during Skeletal Muscle Regeneration. *Cells* **2020**, *9*, 708. [CrossRef] [PubMed]

15.   Cappellari, O.; Mantuano, P.; De Luca, A. "The Social Network" and Muscular Dystrophies: The Lesson Learnt about the Niche Environment as a Target for Therapeutic Strategies. *Cells* **2020**, *9*, 1659. [CrossRef]

16.   Squecco, R.; Chellini, F.; Idrizaj, E.; Tani, A.; Garella, R.; Pancani, S.; Pavan, P.; Bambi, F.; Zecchi-Orlandini, S.; Sassoli, C. Platelet-Rich Plasma Modulates Gap Junction Functionality and Connexin 43 and 26 Expression During TGF-β1-Induced Fibroblast to Myofibroblast Transition: Clues for Counteracting Fibrosis. *Cells* **2020**, *9*, 1199. [CrossRef]

17.   Leduc-Gaudet, J.P.; Mayaki, D.; Reynaud, O.; Broering, F.E.; Chaffer, T.J.; Hussain, S.N.A.; Gouspillou, G. Parkin Overexpression Attenuates Sepsis-Induced Muscle Wasting. *Cells* **2020**, *9*, 1454. [CrossRef]

18.   Soendenbroe, C.; Bechshøft, C.J.L.; Heisterberg, M.F.; Jensen, S.M.; Bomme, E.; Schjerling, P.; Karlsen, A.; Kjaer, M.; Andersen, J.L.; Mackey, A.L. Key Components of Human Myofibre Denervation and Neuromuscular Junction Stability are Modulated by Age and Exercise. *Cells* **2020**, *9*, 893. [CrossRef]

19.   Jin, C.L.; Ye, J.L.; Yang, J.; Gao, C.Q.; Yan, H.C.; Li, H.C.; Wang, X.Q. mTORC1 Mediates Lysine-Induced Satellite Cell Activation to Promote Skeletal Muscle Growth. *Cells* **2019**, *8*, 1549. [CrossRef]

20.   Jorgenson, K.W.; Phillips, S.M.; Hornberger, T.A. Identifying the Structural Adaptations that Drive the Mechanical Load-Induced Growth of Skeletal Muscle: A Scoping Review. *Cells* **2020**, *9*, 1658. [CrossRef]

21.   Zhang, J.; Li, X.; Ismail, F.; Xu, S.; Wang, Z.; Peng, X.; Yang, C.; Chang, H.; Wang, H.; Gao, Y. Priority Strategy of Intracellular Ca2+ Homeostasis in Skeletal Muscle Fibers During the Multiple Stresses of Hibernation. *Cells* **2019**, *9*, 42. [CrossRef] [PubMed]

22.   Malatesta, M.; Costanzo, M.; Cisterna, B.; Zancanaro, C. Satellite Cells in Skeletal Muscle of the Hibernating Dormouse, a Natural Model of Quiescence and Re-Activation: Focus on the Cell Nucleus. *Cells* **2020**, *9*, 1050. [CrossRef] [PubMed]

Article

The Switch from NF-YAl to NF-YAs Isoform Impairs Myotubes Formation

Debora Libetti, Andrea Bernardini, Sarah Sertic, Graziella Messina, Diletta Dolfini and Roberto Mantovani *

Dipartimento di Bioscienze, Università degli Studi di Milano, Via Celoria 26, 20133 Milano, Italy; debora.libetti@unimi.it (D.L.); andrea.bernardini@unimi.it (A.B.); sarah.sertic@unimi.it (S.S.); graziella.messina@unimi.it (G.M.); diletta.dolfini@unimi.it (D.D.)
* Correspondence: mantor@unimi.it

Received: 14 February 2020; Accepted: 21 March 2020; Published: 24 March 2020

Abstract: NF-YA, the regulatory subunit of the trimeric transcription factor (TF) NF-Y, is regulated by alternative splicing (AS) generating two major isoforms, "long" (NF-YAl) and "short" (NF-YAs). Muscle cells express NF-YAl. We ablated exon 3 in mouse C2C12 cells by a four-guide CRISPR/Cas9n strategy, obtaining clones expressing exclusively NF-YAs (C2-YAl-KO). C2-YAl-KO cells grow normally, but are unable to differentiate. Myogenin and—to a lesser extent, MyoD— levels are substantially lower in C2-YAl-KO, before and after differentiation. Expression of the fusogenic Myomaker and Myomixer genes, crucial for the early phases of the process, is not induced. Myomaker and Myomixer promoters are bound by MyoD and Myogenin, and Myogenin overexpression induces their expression in C2-YAl-KO. NF-Y inactivation reduces MyoD and Myogenin, but not directly: the Myogenin promoter is CCAAT-less, and the canonical CCAAT of the MyoD promoter is not bound by NF-Y in vivo. We propose that NF-YAl, but not NF-YAs, maintains muscle commitment by indirectly regulating Myogenin and MyoD expression in C2C12 cells. These experiments are the first genetic evidence that the two NF-YA isoforms have functionally distinct roles.

Keywords: splicing isoforms; CRISPR-Cas9; exon deletion; NF-Y; muscle differentiation; C2C12 cells

1. Introduction

Cell specification and differentiation during development of multicellular organisms is a complex set of events resulting in the formation of organs, whose physiology is maintained by a balance of cell proliferation and differentiation. A paradigmatic example of these phenomena is formation of skeletal muscle. In the case of mammals—mouse in particular—the process begins at early developmental stages, proceeding through embryonic, fetal and adult stages [1,2]. Sequence-specific transcription factors—TFs—play a central role in specifying the identities of myoblasts, their migration to different body locations, organization and the capacity to self-renew and differentiate into myotubes. These properties are key to guarantee maintenance and functionality of the different muscles throughout the lifespan of the organism, including repair after injury in adult life. A set of four key TFs —MyoD, Myf5, Myogenin, MRF4, termed myogenic regulatory factors (MRFs)—have been identified and thoroughly studied by genetic and biochemical means for their capacity to specify myoblasts identity [3,4]. During development, PAX3/7 are located upstream of MRFs [5]; downstream are many TFs, such as the MADS box MEF2A/C/D [6,7], the bHLH ID1/3 [8–10] and SNAI1 [11], the HOX SIX1/4/5 [12–15], STAT3 [16], NFIX [17,18] and the ZNF KLF2/4/5 [19,20]. Unlike MRFs, most of these TFs are not expressed predominantly in muscle cells and are equally important for development and differentiation of other tissues and organs [21–25].

NF-Y is an evolutionarily conserved heterotrimer formed by the sequence-specific NF-YA and the Histone Fold Domain—HFD—NF-YB/NF-YC [26]. The sequence recognized by NF-Y is the CCAAT box, which plays an important role in the activation of 25%–30% of mammalian genes. NF-Y has been classified as "pioneer" TF, in mammals and plants [27–31]. NF-YA is the regulatory subunit; it is alternatively spliced, generating two major isoforms "short" (NF-YAs) and "long" (NF-YAl), differing in 28 amino acids coded by exon 3 [32]. This stretch is located at the N-terminus of the protein, in the Gln-rich transactivation domain (TAD). NF-YAs and NF-YAl have identical subunits-interactions and DNA-binding properties *in vitro*; ChIP-seq from cells harboring predominantly either one of the two isoforms showed recovery of peaks enriched in CCAAT. The isoforms are expressed at various levels in different tissues and cell lines [32,33]. Importantly, no cell line has been so far described lacking NF-YA—nor the HFDs—and NF-YA inactivation was reported to be fatal to cells [28,34]. NF-YAl is the predominant isoform in muscle C2C12 cells: it is abundant in proliferating cells, but it drops to low levels following terminal differentiation to myotubes, unlike the HFD partners [35–37]. Highly reduced NF-YA protein was found in myotubes of adult mice [38]. This suggested that genes up-regulated in the terminal phases of muscle differentiation are either CCAAT-less or not NF-Y-dependent, whereas the trimer activates cell-cycle and growth-promoting genes required during the proliferative state. However, overexpression of NF-YAl led to improved differentiation of C2C12 [39], suggesting that NF-YAl does take part in the differentiation process.

For decades, C2C12 myoblast cells have represented an informative tool to identify genes involved in muscle differentiation [40]. Ablation of the whole NF-YA gene is early embryonic lethal [41], and KO in stable cell lines could not be generated so far. We investigated the role of NF-YAl by genetically ablating exon 3, leading to the production of an intact NF-YAs. We successfully generated homozygous C2C12 lines expressing only NF-YAs and went on to study differentiation properties.

2. Materials and Methods

2.1. Cell Culture and Proliferation Assay

Mouse myoblast cells (C2C12, ATCC) were cultured in Dulbecco's modified Eagle's medium (DMEM) supplemented with 10% Fetal Bovine Serum (FBS, Gibco-Thermo Fisher Scientific), 4 mM L-Glutamine, 100 units/mL penicillin and 100 µg/mL streptomycin (GM, growth medium), in a humidified 5% CO_2 atmosphere at 37 °C. C2C12 cells differentiation was induced plating cells in DMEM with 2% horse serum (Gibco-Thermo Fisher Scientific), 4 mM L-Glutamine, 100 units/mL penicillin and 100 µg/mL streptomycin (DM, differentiation medium). Proliferation assay was performed by plating 1.5×10^5 cells into a 12-well plate and counting every 24 h for 3 days, using the Trypan Blue dye exclusion test. All data were gathered from at least three independent biological replicates. Multiple comparisons were performed using the One-way ANOVA test.

2.2. Derivation of C2-YAl-KO Clones

To delete the exon 3 of NF-YA gene in C2C12 cells, four guide RNAs (gRNAs) were designed to simultaneously target the two flanking introns by using the online tool https://zlab.bio/guide-design-resources. Potential off-target sites were monitored by the online tool https://crispr.cos.uni-heidelberg.de: Table S1 shows the results of such analysis for the four guides. The selected gRNAs had no common off-target sites and were cloned in the two plasmids pX330A_D10A-1x2_ac and pX330A_ D10A-1x2_bd, following the Multiplex CRISPR/Cas9n Assembly System Kit protocol [42]. 1×10^6 C2C12 cells were transfected with 3 µg of the two gRNAs/CRISPR/Cas9n plasmids by electroporation and plated at low density. 72 h after transfection, single clones were picked, expanded and screened.

For DNA extraction, cells from the individual clones were washed with PBS, collected by scraping, lysed in 100 µL ice-cold lysis buffer (40 mM Tris-HCl, 2 mM EDTA, 0.08% SDS, 80 mM NaCl, 0.5 µg/µL Proteinase K) and incubated overnight at 37 °C in agitation. To precipitate DNA, 100 µL of ice-cold 2-propanol and 0.3 M NaAc were added, samples were mixed and centrifuged at 13,000 rpm for 30 min at 4 °C. The pellet was washed with 150 µL of 70% ethanol, centrifuged at 13,000 rpm for 30 min at

4 °C. Supernatant was discarded, the pellet was dried and resuspended in 30 µL H_2O. The resulting DNAs were then screened for positive exon 3 deletion by PCR.

We screened a total of 335 individual clones and obtained 2 independent homozygously edited clones.

2.3. Protein Extraction and Western Blot Analysis

For Whole Cell Extracts preparation, cells were pelleted by centrifugation, resuspended in ice-cold RIPA buffer (10 mM TrisHCl pH 8.0, 1 mM EDTA, 0.5 mM EGTA, 0.1% SDS, 0.1% sodium deoxycholate, 140 mM NaCl, 1% Triton X-100, 1 mM PMSF, Protease inhibitor cocktail) and incubated for 30 min on ice, with occasional shaking. Samples were centrifuged at 13,000 rpm for 10 min at 4°C and the supernatant recovered and quantified using the Bradford protein assay.

20 µg of extracts were loaded on a 4–10% SDS-polyacrylamide gel and analyzed by Western blot using primary antibodies and a peroxidase-conjugate secondary antibody (Sigma-Aldrich). Primary antibodies: anti-NF-YA (G-2, Santa Cruz), anti-NF-YB (GeneSpin), anti-NF-YC (home-made) anti-Vinculin (H-10, Santa Cruz), anti-MyHCs (MF20, DHSB), anti-Myogenin (IF5D, DHSB), anti-MyoD (C-20, Santa Cruz), anti-Myf5 (C-20, Santa Cruz), anti-Pax3 (DHSB), anti-Snai1 (C15D3, Cell Signaling). Western blot experiments were performed on three independent biological replicates.

2.4. Reverse Transcriptase PCR and Real-Time PCR

RNA was isolated by the Tri Reagent (Sigma-Aldrich) protocol according to the manufacturer's instruction. The cDNA was produced starting from 1 µg of total RNA using the MMLV Reverse Transcription Mix (GeneSpin) and used for real-time PCR (SYBR® Green Master Mix, Bio-rad Laboratories) analysis. Real-time PCRs were performed with oligonucleotides designed to amplify 100–200 bp fragments (Table S2). The housekeeping gene Rsp15a was used to normalize expression data. The relative sample enrichment was calculated with the formula $2^{-(\Delta\Delta Ct)}$, where $\Delta\Delta Ct$ = [(Ct sample – Ct Rps15a)$_x$ – (Ct sample – Ct Rps15a)$_y$], x = sample and y = sample control. RT-qPCR analyses were performed on three independent biological replicates. For ChIP experiments, we figured out the percentage of input immunoprecipitated by NF-YB and nc (negative control) antibodies. Results of three independent experiments were represented as Fold change (Fc) between NF-YB sample and nc sample as: %Input NF-YB/%Input nc.

2.5. Flow Cytometry Analyses

Cells were harvested by trypsinization and washed in PBS, fixed in ice-cold 70% ethanol and stored at 4 °C at least 24 hours. Cells were then washed with 1% BSA in PBS and resuspended in 500 µL of PI-staining solution (20 µg/mL Propidium Iodide, 10 µg/mL RNaseA, 1X PBS) at room temperature for 30 minutes, light protected. FACS analyses were performed using the BD FACSCantoII, analyzed with FACSDiva software and quantified with FlowJo. A total of 10^4 events were acquired for each sample. Three independent FACS experiments were performed.

2.6. Immunofluorescence

For immunofluorescence analyses, cells were washed three times with PBS and fixed 10 min with ice-cold acetone-methanol (1:1) at room temperature. After three washes, cells were permeabilized with 0.25% Triton X-100 in PBS for 5 min and incubated 1 h with the primary antibody anti-sarcomeric MyHCs (MF20, DHSB) at room temperature. Cells were washed three times, permeabilized 5 min with 0.25% Triton X-100 in PBS and incubated with secondary FITC anti-mouse antibody (1:500, Sigma-Aldrich) plus DAPI (2 µg/mL) for 40 min at room temperature, light protected. The acquisition was performed by using the inverted microscope Leica DMI6000 B. Three independent immunofluorescence experiments were performed.

2.7. Overexpression and RNA Interference Experiments

Myogenin overexpression was performed by electroporating 1×10^6 C2C12 cells with 3 µg of plasmid (pEMSV-Empty/pEMSV-Myog) and plating them in DM for 96 h. Cells were then collected and analyzed. Three independent biological replicates were performed.

For small interfering RNA (siRNA)-mediated knockdown of NF-YB [29], 2×10^6 C2C12 cells were transfected by electroporation with 100 nM of NF-YB [29] or scrambled control siRNA (Qiagen, SI01327193) and plated into a 10 cm plate in GM condition. 72 h after transfection, cells were collected by scraping for total protein preparation and RNA extraction. Gene expression was analyzed performing real-time PCR. Two independent biological replicates of siRNA interference were performed.

2.8. Chromatin Immunoprecipitation Assay (ChIP)

ChIPs were performed as described previously [43] with the following modifications. Briefly, 2×10^7 cells were crosslinked using 1% formaldehyde for 7 min, the reaction was quenched with 125 mM glycine and cells were collected by scraping. After lysis, nuclei were resuspended in Sonication buffer (50 mM Tris-HCl pH 8, 10 mM EDTA, 0.1% SDS, 0.5% sodium deoxycholate, protease Inhibitor cocktail) and sonicated (Bioruptor, Diagenode) to obtain fragments of approximately 150–300 bp, verified on agarose gel electrophoresis. Samples were centrifuged at 13,000 rpm for 10 min at 4 °C and supernatants recovered and quantified by Bradford assay. One hundred micrograms of chromatin were immunoprecipitated with 5 µg of anti-NF-YB (GeneSpin) and anti-FLAG (Sigma-Aldrich) antibodies. Protein-G beads (KPL) were used for recovery of antibody-bound proteins. Crosslinking was reversed by incubation at 65 °C overnight. Reactions were digested with RNase A and Proteinase K and DNA purified using the DNA purification kit (PCR clean Up, GeneSpin). The DNA was eluted in TE (10 mM Tris-HCl pH 8, 1 mM EDTA) and used in real-time PCR. Three independent biological replicates of ChIP experiments were performed.

3. Results

3.1. Ablation of NF-YA Exon 3 in C2C12 Cells by a Four Guides CRISPR/Cas9n Strategy

Mouse C2C12 cells mostly express NF-YAl [35–38]. To study the role of this isoform in maintenance and differentiation of C2C12, we set out a strategy to selectively eliminate exon 3, coding for the 28 extra amino acids present in NF-YAl and absent in NF-YAs. We figured that the use of four guides flanking precisely the exon 3 regions and of the single strand-cutting Cas9-nickase (Cas9n) would minimize off-target effects, which potentially affect the outcome of this technology [44]. Figure 1A shows the design of the four guide oligonucleotides, two couples targeting the 5' and 3' intronic DNA flaking exon 3, respectively. The two couples of oligos were first checked for absence of common genomic targets (Table S1) and cloned unpaired in the final pX330A_D10A-1x2_ad and pX330A_D10A-1x2_cb (Figure S1A), also expressing the Cas9n gene. The two plasmids were transfected in growing C2C12 cells by electroporation. Individual clones were isolated, expanded and analyzed by PCR, employing the amplicons shown in Figure 1B. As expected, the strategy was less efficient if compared to the standard use of two guides plus wt Cas9: 335 clones were individually screened and two were positive for correct ablation in homozygosity, as shown in Figure 1C. The results of PCRs of the two positive clones, #83 and #117, show the expected bands for the A, B and C amplicons, absent in the DNA of the parental C2C12 cells. The regions surrounding exon 3 in both clones were then amplified and sequenced: Figure S1B confirms the deletion of coding sequences of exon 3, with somewhat different ends in the two clones. In summary, we successfully ablated NF-YA exon 3, deriving two clones termed C2-YAl-KO. To the best of our knowledge, this is the second system of genome editing describing a clean deletion of an individual exon [45] and the first one employing the Cas9 nickase system coupled with four gRNAs.

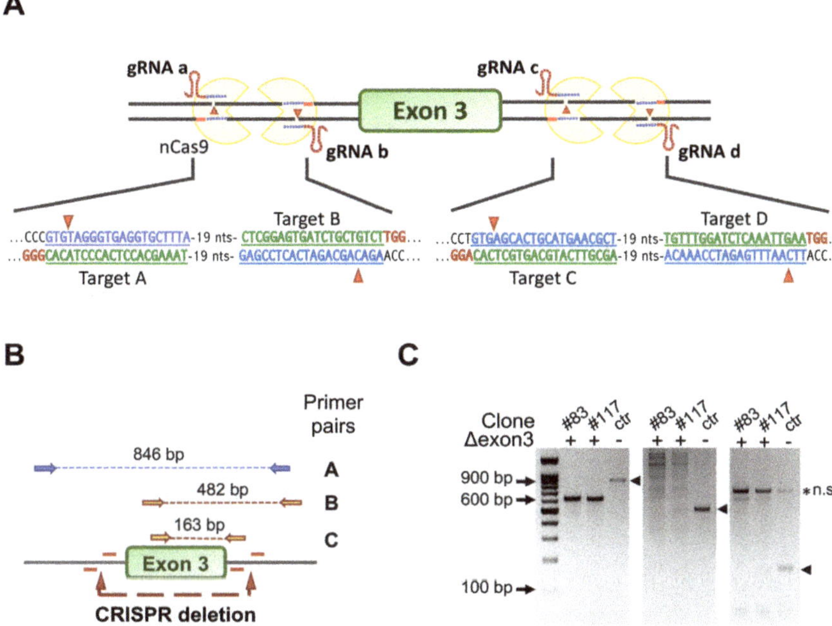

Figure 1. Strategy for ablation of NF-YA exon 3 in C2C12 cells using CRISPR/Cas9n and four gRNAs. (**A**) Gene editing strategy for NF-YA exon 3 deletion using the Cas9-nickase (Cas9n) and four guide RNAs. The targeted sequence by each guide RNA and the deletion sites are shown. Note that Cas9n cuts only the DNA strand that is complementary to and recognized by the gRNA, making necessary the simultaneous presence of two gRNAs/Cas9n complexes to induce a double-strand break (DSB). (**B**) The three primer pairs used to check for positive C2-YAl-KO clones are shown with the specific amplification products highlighted by the dashed lines. (**C**) Example of PCR products run into a 1.2% agarose gel. The expected bands in control cells (ctr) are marked with arrowheads; clones #83 and #117 represent positive C2-YAl-KO clones.

3.2. Characterization of C2-YAl-KO Cells

The two C2-YAl-KO clones were characterized first for expression of NF-YA. We performed qRT-PCR analysis with oligos specific for the individual isoforms [46]; Figure 2A shows that the NF-YAl mRNA is absent in the C2-YAl-KO clones. Extracts were prepared and Western blots performed: as expected, the parental C2C12 cells show expression of NF-YAl (Figure 2B). Instead, the clones express uniquely the NF-YAs isoform. We exposed the blots for long times to verify that no NF-YAl is visible in the two KO clones. Note that the levels of the two isoforms in parental cells—NF-YAl—and edited clones—NF-YAs—are essentially identical, as are the levels of NF-YB and NF-YC: since there is an important level of autoregulation among NF-Y subunits [47], this result indicates that HFD subunits are available for trimer formation and DNA-binding in C2C12 and C2-YAl-KO cells. In summary, genetic ablation of exon 3 in C2C12 was effective, leading to generation of clones that express uniquely the short isoform of NF-YA at physiological levels.

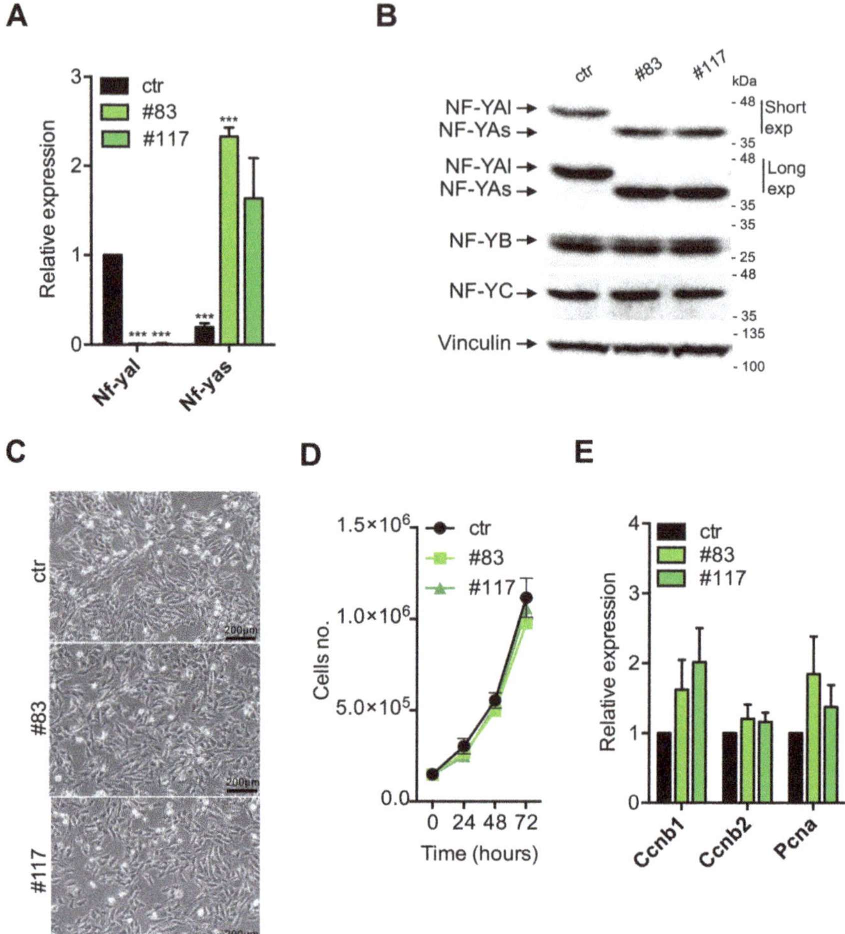

Figure 2. C2-YAl-KO clone characterization. (**A**) Gene expression analysis of NF-YA short and long levels in ctr and C2-YAl-KO clones (#83 and #117) in growth medium (GM) condition. Error bars represent the SD of three independent experiments. P-values were calculated using the one-sample t-test. (**B**) Western blot analysis of NF-Y protein subunits (NF-YA, NF-YB, NF-YC) in ctr cells and C2-YAl-KO clones (#83 and #117) in GM condition. For NF-YA isoforms analysis, short and long exposures are shown. Vinculin was used as loading control. (**C**) Phase contrast analysis of myoblast cells (ctr and C2-YAl-KO clones) morphology in GM condition. Scale bar 200 μm. (**D**) Proliferation assay performed in GM condition counting every 24 h for 3 days using the Trypan Blue dye exclusion test. Error bars represent the SD of three independent experiments. P-values were calculated using the one-way ANOVA test. (**E**) Gene expression analysis of key cell-cycle regulators in ctr and C2-YAl-KO clones (#83 and #117) in GM condition. Error bars represent the SD of three independent experiments. P-values were calculated using the one-sample t-test.

Next, we started to analyze the phenotype of the KO clones: they are stable upon repeated cycles of freezing and thawing and their morphology looks apparently similar to the parental C2C12 cells (Figure 2C). In mouse embryonic stem cells, expression of NF-YAs is associated with growth, and NF-YAl to differentiation [43]: in theory, NF-YAs-expressing C2C12 clones could be enhanced in proliferation. Cells were compared for growth under standard conditions: Figure 2D shows that growth curves are similar, with the two edited clones being marginally slower. In FACS analysis, we did notice some differences: a higher number of S-phase and G2/M cells in the two clones (Figure S2, 21% and 28%, with respect to 18% in C2C12). We checked the mRNA levels of PCNA, Cyclin B1/B2: a slight increase of Cyclin B1 and PCNA in the KO clones is observed (Figure 2E); although not statistically significant, this is consistent with the FACS data. The most noticeable difference, however, was the lower number of sub-G1 cells: 6%–7% in the two clones compared to 12% in the parental C2C12 cells (Figure S2): such non cycling cells are possibly undergoing cell death, suggesting that the switch to NF-YAs is not provoking negative effects on cellular vitality, and, if anything, the opposite. In summary, C2-YAl-KO clones expressing NF-YAs have an apparently normal morphology, grow well, but not faster, with the expected partitioning in cell cycle phases, bar slightly elevated G2/M and decreased sub-G1 populations.

3.3. C2-YAl-KO Cells Fail to Differentiate and Fuse into Myotubes

The levels of NF-YAl drop following terminal differentiation of C2C12 cells and myotubes of mouse muscles show low-to-nil levels of NF-YAl [35–39]. To ascertain whether NF-YAs-expressing cells could form myotubes, we switched the parental C2C12 and the two C2-YAl-KO clones at 70%–80% confluence to a differentiation medium. Before and after 72 h, we monitored cell morphology, performed Immunofluorescence experiments and derived whole cell extracts. Figure 3A shows that parental C2C12 form well organized, multinucleated myotubes, as expected (Upper Panels). The average number of nuclei per fiber is 15, in keeping with an efficient process (Figure 3B). On the other hand, the two edited clones showed a dramatic lack of myotubes formation: cells did not fuse; they were disorganized (Figure 3A, lower panels). We reasoned that the process could be simply slower in these cells and prolonged differentiation up to 5 days: this did not lead to formation of myotubes, nor cell fusion in the C2-YAl-KO clones (not shown). Immunofluorescence and Western blot data are consistent: the MyHCs marker is clearly visible in IFs (Figure 3A, right panels) and WB (Figure 3C) in C2C12 cells after differentiation, but not in the two edited clones. Interestingly, the levels of Myogenin and MyoD were substantially lower both in growing cells and at these late stages of differentiation in C2-YAl-KO clones. As previously reported, NF-YAl, in C2C12, and NF-YAs, in the edited clones, are down-regulated after 72 h of differentiation; NF-YB remained unchanged (Figure 3C). In summary, we conclude that terminal differentiation is completely blocked in C2C12 cells expressing NF-YAs instead of NF-YAl.

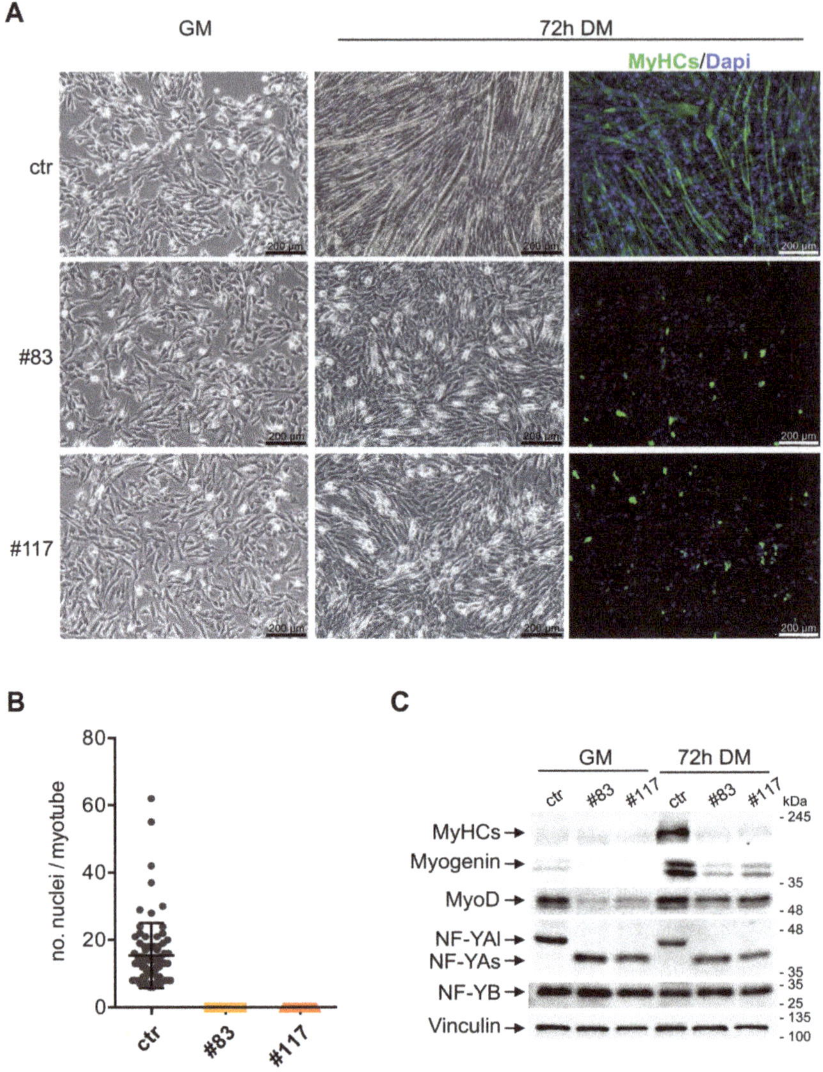

Figure 3. C2-YAl-KO clones fail to differentiate into myotubes. (**A**) Phase contrast analysis of myoblast cells (ctr and C2-YAl-KO clones) before and after 72 h of differentiation (differentiation medium (DM) condition) and immunofluorescence analyses after 72 h of differentiation. Antibody against all sarcomeric MyHCs and DAPI were used. (**B**) Fusion index was calculated as the number of nuclei in each myotube (with three or more nuclei). (**C**) Western blot analysis of key muscle differentiation regulators (MyHCs, MyoD), NF-YA isoforms (NF-YAl, NF-YAs) and NF-YB proteins, before (GM) and after 72 h of differentiation (72 h DM). Vinculin was used as loading control. The experiment was performed three times.

3.4. Expression of TFs in C2-YAl-KO

We analyzed expression of MRFs and TFs with a proven role in differentiation, in the parental and in the C2-YAl-KO cells under growing conditions and 24 h after differentiation. Profiling experiments established this as an early time point to detect significant changes in gene expression [48]. Note that most of the TFs analyzed have CCAAT in promoters and some formally shown to be under NF-Y control. First, we verified expression levels of MRFs in parental C2C12 (Figure S3): Myogenin is robustly induced; MyoD is modestly increased; Myf5 is modestly decreased after differentiation; Mef2C, but not Mef2D, is robustly increased. These changes are in agreement with expectations [49]. At the same time, we analyzed other TFs shown to be important for muscle differentiation: Six1/4/5, Snai1, Stat3 and Klf5 are all increased upon C2C12 differentiation, Id1/3 are modestly decreased, Pax3 is unchanged (Figure S3). These results are also in agreement with published data. Having established that our differentiation program runs normally in C2C12 cells, we monitored expression of these genes in the C2-YAl-KO clones. The results are shown in Figure 4A for growing conditions and Figure 4B for differentiation. MRFs show the most conspicuous differences: Myogenin is almost undetectable in growing C2-YAl-KO clones and marginally increased upon differentiation. MyoD basal levels are normal, but induction is reduced upon differentiation, compared to parental C2C12. Myf5 expression is basally similar in the edited clones and higher after differentiation (Figure 4A,B). Mef2C levels are similar in growing conditions, but lower after differentiation: note that the levels are very low basally and differences with parental C2C12 cells are not statistically significant. Mef2D expression is identical in C2C12 and edited clones. As for the other TFs, Six1/4/5, Klf5 and Pax3 show similar expression patterns (Figure 4A,B). Minor changes are observed in growing conditions for Snai1, Stat3 and Id1 (one clone only) and for Id1 (same clone) after differentiation. Finally, Id3 shows somewhat higher levels before and after differentiation, but again, these changes are variable in the three experiments and thus not statistically significant.

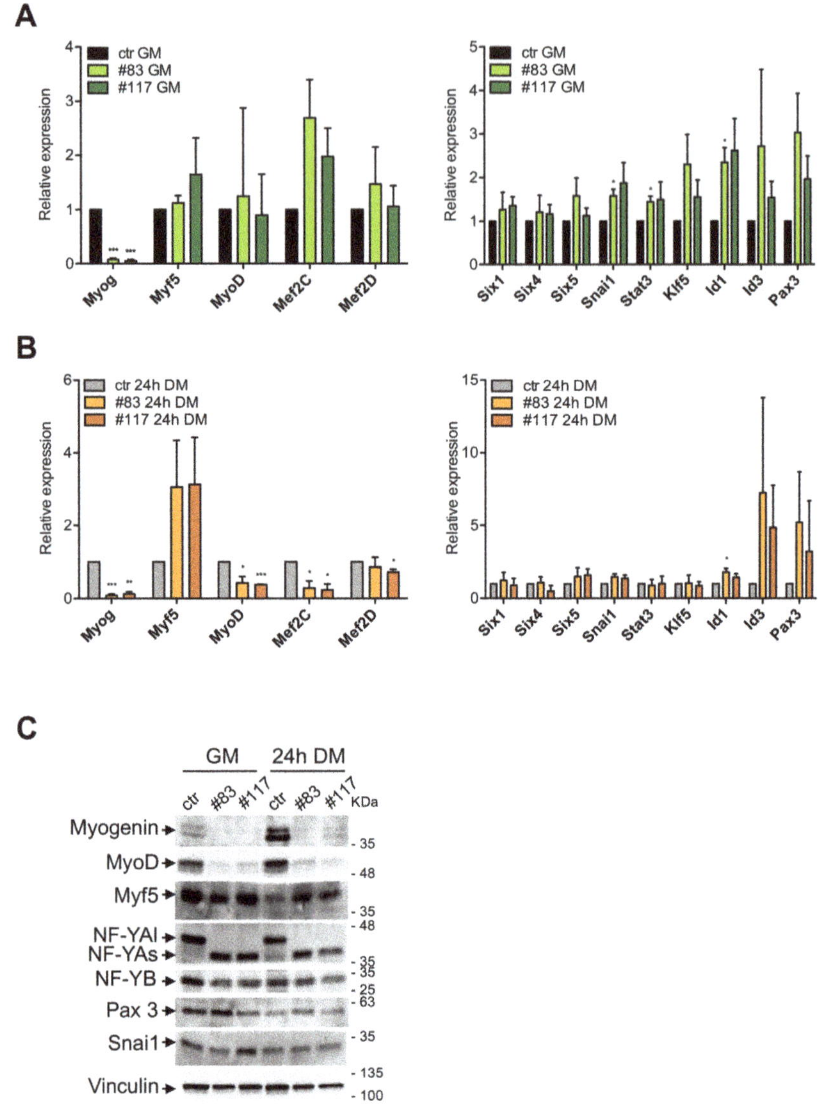

Figure 4. MRFs are downregulated in C2-YAl-KO clones. (**A**) Gene expression analysis of key muscle differentiation regulators (left panel) and other TFs shown to be important for muscle differentiation (right panel) in GM condition. Error bars represent the SD of three independent experiments. P-values were calculated using the one-sample t-test. (**B**) Gene expression analysis of key muscle differentiation regulators (left panel) and other TFs shown to be important for muscle differentiation (right panel) 24 h after differentiation (24 h DM). Error bars represent the SD of three independent experiments. P-values were calculated using the one-sample t-test. (**C**) Western blot analysis of key muscle differentiation regulators (Myogenin, MyoD, Myf5), NF-YA isoforms (NF-YAl, NF-YAs) and NF-YB proteins and other TFs shown to be important for muscle differentiation (Pax3, Snai1), in GM and 24 h DM. Vinculin was used as loading control.

To substantiate these results, protein expression of selected TFs was monitored by Western Blot analysis. Figure 4C shows that Myogenin levels are consistent with the mRNA data, being much lower in C2-YAl-KO clones than in parental cells, both in growing cells and after 24 h of differentiation. MyoD is substantially reduced in growing and differentiating clones, compared to parental C2C12. Note that protein levels were far lower than expected based on the mRNA levels, especially under growing conditions: this calls for post-transcriptional control in edited clones. Myf5 protein is downregulated in C2C12 after differentiation, as expected; in edited clones, it shows lower levels in growing cells, but higher after induction. NF-YA and NF-YB show the expected patterns; Pax3 is very modestly increased in C2-YAl-KO clones and Snai1 is unchanged. In summary, C2-YAl-KO cells have substantial differences in MRFs levels with respect to C2C12 cells, both before and after differentiation, whereas the other TFs showed rather minor changes.

3.5. Expression of Myomaker and Myomixer Is Activated by Myogenin and It Is Impaired in C2-YAl-KO

We were intrigued by the lack of cell fusion of the C2-YAl-KO clones after induction of differentiation. Myomaker—Mymk—and Myomixer—Mymx—are genes induced transcriptionally during muscle terminal differentiation, including in the C2C12 system [50,51]. Specifically, their expression is essential for the process of myocytes fusion [52]. We checked expression by qRT-PCR in parental C2C12 and in the two edited clones 24 h after differentiation. Figure 5A shows a strong induction—20-fold—of both Myomaker and Myomixer in C2C12 cells. C2-YAl-KO have much lower levels in growing cells (Figure 5B) and even more after differentiation (Figure 5C).

The obvious hypothesis was that these genes are under direct NF-Y control. We surveyed their promoter sequences and verified that no bona fide CCAAT box is present, notably within the evolutionary conserved areas: given the specificity of NF-Y CCAAT recognition, we considered unlikely that it acts directly on their expression. Genetic experiments in zebrafish have recently shown that Myomaker and Myomixer are directly activated by Myogenin [53]. We analyzed ENCODE datasets of C2C12 cells and found that Myogenin and MyoD target both promoters. Myomixer has apparently one promoter, Myomaker has two promoters, some 4 kb distant from each other: Figure 5D shows the overlapping peaks of Myogenin and MyoD. Myogenin binds exclusively after 24 h of differentiation, in accordance with its induced expression. One MyoD peak is visible already under growing conditions on Myomaker, and two additional peaks are found at 24 h. Importantly, the regions bound by MyoD and Myogenin in these two promoters are conserved across vertebrates, as shown by PhastCons data in Figure S4A: this corroborates the functional relevance proven in zebrafish [53]. To verify whether Myogenin activates Myomaker and Myomixer, we overexpressed it in parental C2C12 and in one of the C2-YAl-KO clones (#83) and induced to differentiate: Western blot of Figure 5E shows the increased levels of Myogenin compared to cells transfected with an Empty vector control; q-RT-PCR of Figure 5F shows that Myogenin overexpression has negligible effects on expression of the endogenous Myomaker and Myomixer in parental C2C12, but it increases expression of both genes in the C2-YAl-KO cells. Finally, morphological observation of the edited cells shows —incomplete—improvement in differentiation (Figure S4B).

In essence, we find that the marginal levels of Myogenin in C2-YAl-KO cells could result in lack of induction of the Myomaker and Myomixer targeted genes, entailing lack of cell fusion in NF-YAs-expressing clones.

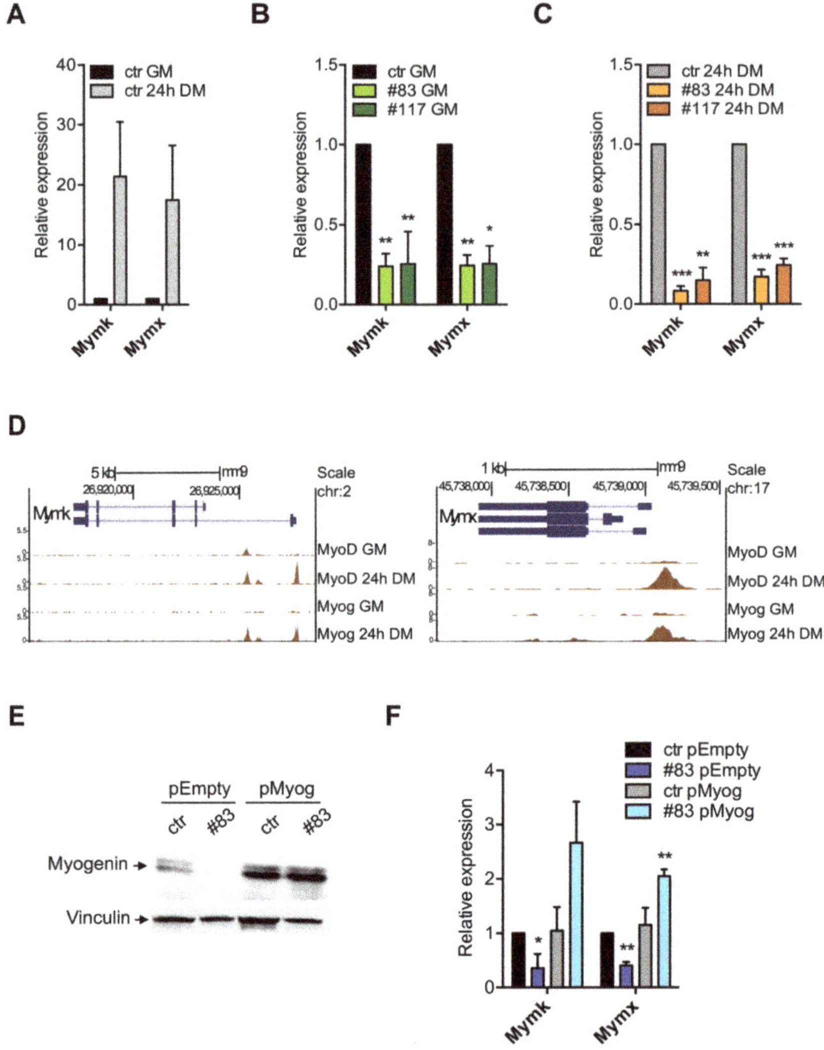

Figure 5. Myogenin directly regulates Myomaker and Myomixer expression. (**A**) Relative expression levels of Myomaker (Mymk) and Myomixer (Mymx) in C2C12 cells before and after 24 h of differentiation (24 h DM). Error bars represent the SD of three independent experiments. P-values were calculated using the one-sample t-test. (**B,C**) Relative expression levels of Myomaker (Mymk) and Myomixer (Mymx) in C2C12 cells before (**B**) and after 24 h of differentiation (**C**) in ctr and the two C2-YAl-KO clones. Error bars represent the SD of three independent experiments. P-values were calculated using the one-sample t-test. (**D**) ChIP-seq peaks of MyoD and Myogenin on Mymk and Mymx promoters in GM and after 24 h of differentiation (24 h DM) (UCSC-genome browser available tracks). Vertical viewing range Mymk: min 0, max 5.5. Vertical viewing range Mymx: min 0, max 8. (**E**) Western blot analysis of Myogenin protein levels in C2C12 cells transfected with a control plasmid (pEmpty) and the Myogenin-overexpressing plasmid (pMyog) 96 h after differentiation induction. Vinculin was used as loading control. (**F**) Relative expression levels of Mymk and Mymx in C2C12 Myog-overexpressing cells after 96 h of differentiation. Error bars represent the SD of three independent experiments. *p*-values were calculated using the one-sample t-test.

3.6. Myogenin and MyoD Are—Indirectly—Regulated by NF-Y

The results shown above beg the question as to whether NF-Y directly regulates MRFs. Myogenin and Myf5 promoters do not contain CCAAT boxes, MyoD does [54]. To verify the NF-Y dependence of these genes, we transitorily inactivated NF-Y activity. In our hands, NF-YA inactivations by shRNA or siRNA were rather inefficient in C2C12 cells (not shown). We thus turned to NF-YB by treating C2C12 cells with an siRNA previously shown to be active and very specific, including in profiling experiments [29]. NF-YB is a necessary component of the DNA-binding trimer: this allows us to inhibit CCAAT-binding activity, upon siRNA treatment. Most importantly, unlike NF-YA, NF-YB inactivation does not trigger apoptosis [29,34], making this a suitable choice for long differentiation processes. Figure 6 shows the results of experiment 1, Figure S5 those of experiment 2: in both, RT-qPCR (Figure 6A and Figure S5A) and Western blots (Figure 6B and Figure S5B) show far lower expression of NF-YB in C2C12 cells treated with NF-YB siRNA, with respect to the control siRNA. In mRNA analysis, Myogenin, MyoD and Mef2C, but not Myf5 nor Mef2D, are substantially downregulated upon NF-Y inactivation; Myomaker and Myomixer are also reduced. Six1/4/5 are reduced: for Six4, this in keeping with an NF-Y dependence predicted from previous data on NF-Y binding to a canonical promoter CCAAT [39]. As for Id1 and Id3, they are somewhat reduced, but the results are borderline significant: Id1 in experiment 2 and Id3 in experiment 1. We conclude that NF-Y removal entails a reduction of MRFs, which, in turn, could explain the observed drop of Myomaker and Myomixer. We also show that members of the Six family are NF-Y targets. Analysis of proteins levels in extracts of siRNA-inactivated cells by Western blots confirmed these results: the levels of NF-YB were lower (although not to the extent of the mRNA) and paralleled by somewhat lower levels of NF-YA. Myogenin is substantially decreased and MyoD is also affected, to a lesser extent (Figure 6B and Figure S5B). We conclude that NF-Y regulates the expression of MyoD and Myogenin in C2C12 cells.

The Myogenin promoter is CCAAT-less and was not bound by NF-Y in C2C12 cells [39] and, despite the presence of a canonical CCAAT, the MyoD promoter was also not bound [39]. To understand whether the positive effect of NF-Y on MyoD is direct, we checked the parental C2C12 cells for the presence of NF-Y in ChIP experiments. Three independent experiments are shown in Figure 6C and Figure S5C. The absence of enrichment of NF-Y on MyoD is indeed confirmed, whereas the Stard4 positive control promoter is clearly bound. Equally positive was the promoter of Id1, but not that of Id3. Note that there is variability in the fold-enrichments in the three experiments: as this is high (from 60 to 800-folds), we consider quantitative changes difficult to interpret, especially when compared to completely negative promoters such as MyoD and Id3. Therefore, we conclude that NF-Y does not regulate MyoD directly—and despite promoter binding—NF-Y has modest effects on Id1 transcription in C2C12 cells.

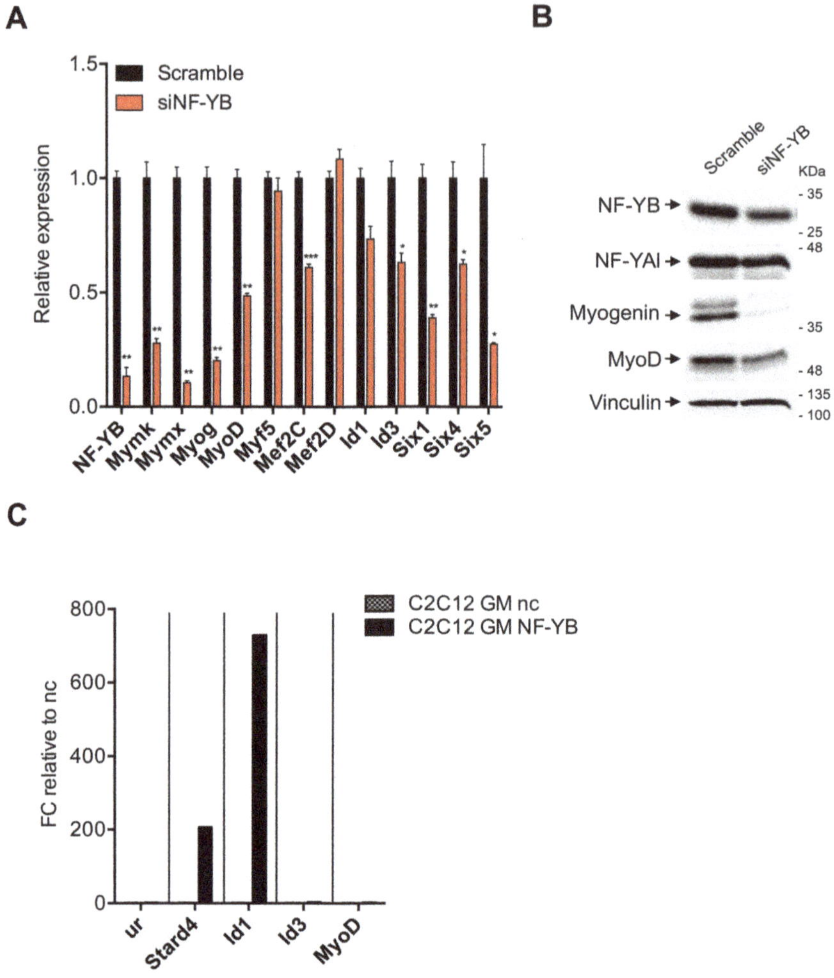

Figure 6. Analysis of NF-Y involvement in muscle specific genes expression. (**A**) Gene expression analysis of NF-YB and key muscle differentiation regulators in C2C12 cells 72 h after NF-YB silencing (siNF-YB) and scrambled siRNA control. Error bars represent the SD of two different RT-qPCR replicates. P-values were calculated using the one-sample t-test. (**B**) Western blot analysis of NF-YB, NF-YA and key muscle differentiation regulators (Myogenin, MyoD) protein levels 72 h after NF-YB silencing (siNF-YB) and the scrambled siRNA control. Vinculin was used as loading control. (**C**) ChIP experiment performed on C2C12 ctr cells in GM condition using NF-YB and negative control (nc) antibodies. The unrelated region (ur) and Stard4 were used as negative and positive control, respectively. Results are represented as the input percentage of each sample normalized to the input percentage of the nc antibody.

4. Discussion

By genome editing, we derived clones of C2C12 cells that express NF-YAs instead of NF-YAl. We verified that NF-YAs—and companion HFDs—are expressed at comparable levels and that it decreases after differentiation. The edited C2C12 clones are stable, grow normally, yet they are completely deficient in differentiation. We report defects of basal and induced expression of Myomaker and Myomixer, early response-genes likely responsible for lack of cell fusion. Their promoters are targeted by MyoD and Myogenin. In turn, we find low—basal and induced—levels of MyoD and Myogenin in the NF-YAs-expressing clones. Finally, expression of both MRFs are indirectly controlled by NF-Y.

4.1. Role of NF-YA Alternative Splicing in Muscle Cells

Specific isoforms of TFs have long been known to impact heavily on transcriptional regulation. Paradigmatic examples are the members of the p53/p63/p73 families, whose isoforms, produced by multiple promoters and alternative splicing, have different targets and often opposing transcriptional effects [55]. The muscle system is no exception [56,57]. Mef2C and Mef2D undergo alternative splicing during muscle differentiation [57,58]: a muscle-specific isoform of Mef2D contains exon α2 rather than α1, both expressed in muscle cells. Growing and early differentiating cells harbors MEF2Dα1; the switch to MEF2Dα2 occurs in terminal stages of C2C12 differentiation, leading to activation of late genes. MEF2Dα1 is phosphorylated at two serines by PKA [59], which mediate association with HDACs, resulting in repression. MEF2Dα2 lacks these residues, functioning as a transcriptional activator. Parallel molecular mechanisms appear to be operating for the related MEF2Cα1/α2 alternative splicing isoforms [58]. The key issue in Mef2 splicing regulation is involvement in late stages of differentiation. Alternative splicing was reported for the master TFs of muscle commitment PAX3 and PAX7, but the functional roles of the single isoforms are less well characterized [60–64].

We show here that a switch from NF-YAl to NF-YAs causes a major difference in the differentiation properties of C2C12 cells. The major NF-YA isoforms, originally reported decades ago [32], are only recently attracting the attention they deserve. In part, this was due to the elusive logic of their expression patterns: in some systems, cells have NF-YAs before—and NF-YAl after—differentiation; in others, such as in muscle cells, NF-YAl is mostly found. In part, it was because of the rather unimpressive nature of the exon 3 amino acids incorporated into NF-YAl: a short stretch rich in glutamines and hydrophobic residues amid the larger transactivation domain. Overexpression experiments suggested differences in gene activation [39,65], but these experiments are to be taken with a grain of salt, because of the large amount of proteins produced, targeting the large number of potential NF-Y sites in the genome. NF-YA AS is likely more complex than what is shown here. First, NF-YAx is another alternatively spliced isoform, recently reported in glioblastomas, devoid of exons 3 and 5: this greatly reduces the activation domain, with important functional consequences [66]. Expression of NF-YAx will have to be monitored in normal cells, to verify whether it is specific for glioblastomas. Second, there are micro differences—6 amino acids—produced in many cell types within the acceptor site of exon 5. Third, some cells show the inclusion of an additional Gln residue at the acceptor splicing site of exon 3, producing a 29 amino acids insertion [32]. Note that a similar situation was reported for PAX3, in which an extra Gln causes differences in DNA-binding affinity [59]. Precise editing techniques, as we have started to use here in C2C12 cells, could sort out the functionality of the various isoforms.

4.2. NF-Y Does Not Target Directly Genes Involved in C2C12 Differentiation

Sequence-specific TFs target specific genomic sites, driven by the discriminatory power of their DNA-binding Domains. However, they are also known to be binding indirectly, being tethered by other TFs or complexes: analysis of genomic locations by ENCODE has shown that this latter mechanism is far from marginal [67]. In addition to ENCODE, several independent ChIP-seq of TFs—and cofactors—identified binding to CCAAT locations [68]. One such example regards the orphan receptor Rev-Erb, important for muscle regeneration, targeting NF-Y sites in C2C12 cells [69]. The reverse, namely NF-Y being tethered to CCAAT-less locations by other TFs, has yet to be described. The issue could theoretically be relevant, since the genes down-regulated after NF-Y removal, or by switching from NF-YAl to NF-YAs, have generally no CCCAT in promoters. The effects appear to be largely indirect, but we do not favor the promoter tethering hypothesis. Rather, we report binding of Myogenin and MyoD to the promoters of Myomaker and Myomixer and show that Myogenin overexpression leads to recovery of their expression in C2-YAl-KO cells. This extends to mouse cells genetic experiments made in zebrafish [53]. It also indicates that NF-Y does not regulate other TFs essential for expression of these two genes. In summary, NF-Y/CCAAT interactions in promoters, which are structurally identical for NF-YAl and NF-YAs, are likely not crucial for genes induced during myotubes formation: rather, the focus is shifted to the control of MRFs, or other TFs.

We have analyzed expression of TFs involved in myoblast/C2C12 differentiation. The majority are not dramatically altered in edited clones. Mef2C induction is impaired, but previous studies indicated that NF-Y is bound to the Mef2D, not to Mef2C promoter [39]. We find that Mef2C, not Mef2D, is regulated by NF-YB RNAi interference. Note that these TFs are also targeted by MyoD and Myogenin, as they play a role in the final stages of differentiation [7,59]. This suggests indirect regulation by NF-Y via MRFs. Id1/Id3 do have bona fide functional CCAAT in promoters [70], bound in cancer cells as per ENCODE data (M. Ronzio, A.B., D.D., R.M., in preparation) and in NTera2 cells [71]: Id1, but not Id3, is bound in vivo by NF-Y in C2C12, parental cells and edited clones. The levels are decreased in C2-YAl-KO upon differentiation, but NF-Y-inactivation brings very marginal decrease in Id1 expression. PAX3, which acts upstream of MyoD, shows variable, somewhat increased mRNA levels in the edited clones, but this is not supported by analysis of protein levels. In summary, there is no clear CCAAT-driven TF that could explain the phenotype: instead, we propose that the decrease of Myogenin and MyoD expression entails a cascade of transcriptional events leading to failure of differentiation (Figure 7).

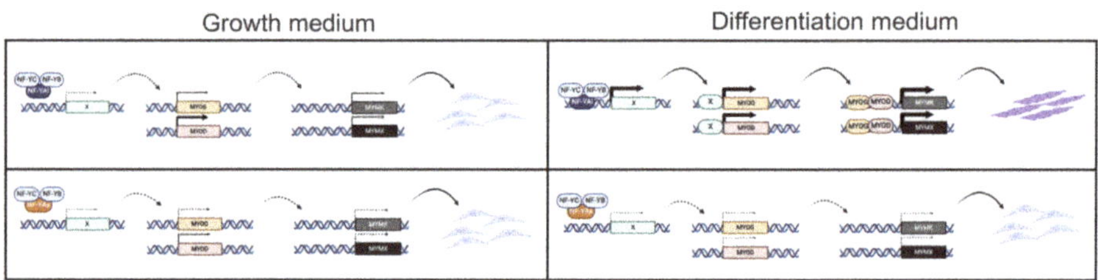

Figure 7. NF-YA isoforms involvement in regulation of expression of muscle genes. Model for NF-YA isoforms mediated regulation of expression of muscle genes in growth condition (left panel) and differentiation condition (right panel).

4.3. NF-Y Regulates MRFs Expression Indirectly

Switching from NF-YAl to NF-YAs—and NF-YB inactivation—negatively affects MRFs expression. Myf5 is moderately down in growing cells, remaining somewhat higher after differentiation. NF-Y-inactivation leads to a severe drop in Myogenin expression and a decrease of MyoD, which indicates an impact of NF-Y on their expression. The regulation appears to be transcriptional for Myogenin, not for MyoD, whose mRNA levels are variable, but overall similar. The Myogenin promoter is CCAAT-less and an indirect effect of NF-Y must be invoked. As for MyoD, the promoter harbors a high affinity NF-Y site, extremely conserved in evolution [54] and at the expected position (at -70 from TSS). Yet, NF-Y is not bound in vivo (Figure 6C). This is the only such example in nearly 200 promoters for which genetic analysis was reported [72]. The combination of an evolutionarily conserved, canonical CCAAT in a standard promoter position might function through NF-Y somewhen during the physiological activation of MyoD in development, while it has become expendable in the C2C12 system. Thus, down-regulation of MyoD in NF-YAs-expressing cells is also an indirect effect. It was proposed that MyoD serves as "pioneer" TF predisposing chromatin configurations for Myogenin to act as powerful activator of terminal differentiation genes and repressor of cell-cycle genes [73]. The latter function might be robustly counteracted by NF-YAs, but we have no evidence of that (Figure 2). It is now clear that the focus is set on transcriptional regulation of the MyoD and Myogenin units and on which activator TF(s)—or cofactor(s)—is under NF-YAl—but not NF-YAs—direct control.

For the time being, the "candidate" TFs approach used here failed to offer a plausible explanation on how NF-YAl regulates MRFs expression, thereby muscle differentiation. We must resolve to more systematic analysis, such as RNA-seq, to identify potential NF-Y-mediated regulators in C2C12. In light of the low intrinsic levels of muscle-commitment by MRFs in C2-YAl-KO clones, such analysis could also shed light on the actual identity of these cells.

Supplementary Materials: The following are available online at http://www.mdpi.com/2073-4409/9/3/789/s1, Figure S1: CRISPR/Cas9n system and ablation of NF-YA exon 3 in C2C12 cells. (A) Schematic representation of plasmids construction, following the Multiplex CRISPR/Cas9n Assembly System Kit protocol (Yamamoto lab) [40]. (B) Sequencing of the two C2-YAl-KO clones (#83, #117) compared to the control (ctr). Deleted sequence, targeted sequence and exon 3 sequence are highlighted. Figure S2: Cell-cycle analysis of C2-YAl-KO clones. Flow cytometry analysis of ctr C2C12 cells and the two C2-YAl-KO clones in GM condition. The analysis of three independent experiments and the average of percentage of cells in each cell-cycle phase are shown. Figure S3: Gene expression analysis of TFs in growing and differentiated C2C12 cells. Gene expression analysis by RT-qPCR of key muscle differentiation regulators (left panel) and other TFs shown to be important for muscle differentiation (right panel) in GM condition and 24 h after differentiation in C2C12 ctr cells. Error bars represent the SD of three different experiments. P-values were calculated using the one-sample t-test. Figure S4: Myomaker and Myomixer expression are regulated by MyoD and Myogenin. (A) UCSC view of Mymk and Mymk *loci* showing alignment of ChIP-Seq data and DNA regulatory motifs conserved across Vertebrates by PhastCons. (B) Phase-contrast analysis of C2C12 cells (ctr and #83) morphology transfected with pEmpty or pMyog, 96 h after differentiation. Figure S5: Analysis of NF-Y involvement in muscle specific genes expression. (A) Gene expression analysis by RT-qPCR of NF-YB and key muscle differentiation regulators in C2C12 cells 72 h after NF-YB silencing (siNF-YB) and the scrambled siRNA control (II° experiment). Error bars represent the SD of two different q-PCR replicates. *p*-values were calculated using the one-sample t-test. (B) Western blot analysis of NF-YB, NF-YA and key muscle differentiation regulators (Myogenin, MyoD) protein levels 72 h after NF-YB silencing (siNF-YB) and the scrambled control. Vinculin was used as loading control. (C) Analysis of II° and III° ChIP experiments performed on C2C12 ctr cells in GM condition using NF-YB antibody and the negative control (nc). The unrelated region (ur) and Stard4 were used as negative and positive control, respectively. Results are represented as the input percentage of sample normalized to the nc. Table S1: Off-targets analysis. Analysis of possible off-target sites of each gRNA using the online tool https://crispr.cos.uni-heidelberg.de. For each gRNA the off-target gene name, gene id, position (intronic, intergenic, exonic), mismatches (MM) and the PAM sequence are reported. Table S2: Primers used. The specific sequence of each primer (forward and reverse) used for RT-qPCR and ChIP analysis are reported.

Author Contributions: D.L.: investigation, formal analysis and visualization; A.B.: data curation and methodology; S.S.: investigation; G.M.: resources, writing (review and editing); D.D.: methodology, data curation, writing (review and editing); R.M.: supervision, funding acquisition, writing (Original draft and editing). All authors have read and agreed to the published version of the manuscript.

Funding: The work was supported by AFM Téléthon Trampoline Grant (n.16408), AIRC Grant (n.19050) and PRIN 2015 to R.M. and AFM Téléthon Research Grant (n. 18364) to G.M.

Acknowledgments: The authors would like to thank prof. Carol Imbriano (Università degli Studi di Modena e Reggio Emilia) for providing antibodies against Myod and Myf5.

Conflicts of Interest: The authors declare no conflicts of interest.

References

1. Buckingham, M. Skeletal muscle formation in vertebrates. *Curr. Opin. Genet. Dev.* **2001**, *11*, 440–448. [CrossRef]

2. Buckingham, M.; Rigby, P.W. Gene Regulatory Networks and Transcriptional Mechanisms that Control Myogenesis. *Dev. Cell* **2014**, *28*, 225–238. [CrossRef]

3. Hernández-Hernández, J.M.; García-González, E.G.; Brun, C.E.; Rudnicki, M.A. The myogenic regulatory factors, determinants of muscle development, cell identity and regeneration. *Semin. Cell Dev. Boil.* **2017**, *72*, 10–18. [CrossRef] [PubMed]

4. Zammit, P.S. Function of the myogenic regulatory factors Myf5, MyoD, Myogenin and MRF4 in skeletal muscle, satellite cells and regenerative myogenesis. *Semin. Cell Dev. Boil.* **2017**, *72*, 19–32. [CrossRef] [PubMed]

5. Buckingham, M.; Relaix, F. PAX3 and PAX7 as upstream regulators of myogenesis. *Semin. Cell Dev. Boil.* **2015**, *44*, 115–125. [CrossRef] [PubMed]

6. Black, B.L.; Olson, E.N. Transcriptional Control Of Muscle Development by Myocyte Enhancer Factor-2 (MEF2) Proteins. *Annu. Rev. Cell Dev. Boil.* **1998**, *14*, 167–196. [CrossRef]

7.   Taylor, M.V.; Hughes, S.M. Mef2 and the skeletal muscle differentiation program. *Semin. Cell Dev. Boil.* **2017**, *72*, 33–44. [CrossRef]

8.   Kumar, D.; Shadrach, J.L.; Wagers, A.J.; Lassar, A.B. Id3 Is a Direct Transcriptional Target of Pax7 in Quiescent Satellite Cells. *Mol. Boil. Cell* **2009**, *20*, 3170–3177. [CrossRef]

9.   Wu, J.; Lim, R.W. Regulation of inhibitor of differentiation gene 3 (Id3) expression by Sp2-motif binding factor in myogenic C2C12 cells: Downregulation of DNA binding activity following skeletal muscle differentiation. *Biochim. et Biophys. Acta (BBA) Gene Struct. Expr.* **2005**, *1731*, 13–22. [CrossRef]

10.   Atherton, G.T.; Travers, H.; Deed, R.; Norton, J.D. Regulation of cell differentiation in C2C12 myoblasts by the Id3 helix-loop-helix protein. *Cell Growth Differ. Mol. Boil. J. Am. Assoc. Cancer Res.* **1996**, *7*, 1059–1066.

11.   Soleimani, V.D.; Yin, H.; Jahani-Asl, A.; Ming, H.; Kockx, C.; Van Ijcken, W.F.J.; Grosveld, F.; Rudnicki, M.A. Snail regulates MyoD binding-site occupancy to direct enhancer switching and differentiation-specific transcription in myogenesis. *Mol. Cell* **2012**, *47*, 457–468. [CrossRef] [PubMed]

12.   Grifone, R.; Demignon, J.; Houbron, C.; Souil, E.; Niro, C.; Seller, M.J.; Hamard, G.; Maire, P. Six1 and Six4 homeoproteins are required for Pax3 and Mrf expression during myogenesis in the mouse embryo. *Development* **2005**, *132*, 2235–2249. [CrossRef] [PubMed]

13.   Yajima, H.; Motohashi, N.; Ono, Y.; Sato, S.; Ikeda, K.; Masuda, S.; Yada, E.; Kanesaki, H.; Miyagoe-Suzuki, Y.; Takeda, S.; et al. Six family genes control the proliferation and differentiation of muscle satellite cells. *Exp. Cell Res.* **2010**, *316*, 2932–2944. [CrossRef]

14.   Santolini, M.; Sakakibara, I.; Gauthier, M.; Aulinas, F.R.; Takahashi, H.; Sawasaki, T.; Mouly, V.; Concordet, J.-P.; Defossez, P.-A.; Hakim, V.; et al. MyoD reprogramming requires Six1 and Six4 homeoproteins: genome-wide cis-regulatory module analysis. *Nucleic Acids Res.* **2016**, *44*, 8621–8640. [CrossRef] [PubMed]

15.   Yajima, H.; Kawakami, K. Low Six4 and Six5 gene dosage improves dystrophic phenotype and prolongs life span of mdx mice. *Dev. Growth Differ.* **2016**, *58*, 546–561. [CrossRef]

16.   Guadagnin, E.; Mázala, D.; Chen, Y.-W. STAT3 in Skeletal Muscle Function and Disorders. *Int. J. Mol. Sci.* **2018**, *19*, 2265. [CrossRef] [PubMed]

17.   Messina, G.; Biressi, S.A.M.; Monteverde, S.; Magli, A.; Cassano, M.; Perani, L.; Roncaglia, E.; Tagliafico, E.; Starnes, L.; Campbell, C.E.; et al. Nfix Regulates Fetal-Specific Transcription in Developing Skeletal Muscle. *Cell* **2010**, *140*, 554–566. [CrossRef]

18.   Rossi, G.; Antonini, S.; Bonfanti, C.; Monteverde, S.; Vezzali, C.; Tajbakhsh, S.; Cossu, G.; Messina, G. Nfix Regulates Temporal Progression of Muscle Regeneration through Modulation of Myostatin Expression. *Cell Rep.* **2016**, *14*, 2238–2249. [CrossRef]

19.   Hayashi, S.; Manabe, I.; Suzuki, Y.; Relaix, F.; Oishi, Y. Klf5 regulates muscle differentiation by directly targeting muscle-specific genes in cooperation with MyoD in mice. *eLife* **2016**, *5*, 37798. [CrossRef]

20.   Sunadome, K.; Yamamoto, T.; Ebisuya, M.; Kondoh, K.; Sehara-Fujisawa, A.; Nishida, E. ERK5 Regulates Muscle Cell Fusion through Klf Transcription Factors. *Dev. Cell* **2011**, *20*, 192–205. [CrossRef]

21.   Potthoff, M.J.; Olson, E.N. MEF2: a central regulator of diverse developmental programs. *Dev.* **2007**, *134*, 4131–4140. [CrossRef] [PubMed]

22.   Ling, F.; Kang, B.; Sun, X.H. Id proteins: small molecules, mighty regulators. *Curr. Top Dev. Biol.* **2014**, *110*, 189–216. [PubMed]

23.   Christensen, K.L.; Patrick, A.N.; McCoy, E.L.; Ford, H.L. Chapter 5 The Six Family of Homeobox Genes in Development and Cancer. *Advances in Cancer Research* **2008**, *101*, 93–126. [PubMed]

24.   Bialkowska, A.; Yang, V.W.; Mallipattu, S.K. Krüppel-like factors in mammalian stem cells and development. *Development* **2017**, *144*, 737–754. [CrossRef]

25.   Piper, M.; Gronostajski, R.; Messina, G. Nuclear Factor One X in Development and Disease. *Trends Cell Boil.* **2019**, *29*, 20–30. [CrossRef]

26.   Dolfini, D.; Gatta, R.; Mantovani, R. NF-Y and the transcriptional activation of CCAAT promoters. *Crit. Rev. Biochem. Mol. Boil.* **2011**, *47*, 29–49. [CrossRef]

27.   Fleming, J.D.; Pavesi, G.; Benatti, P.; Imbriano, C.; Mantovani, R.; Struhl, K. NF-Y coassociates with FOS at promoters, enhancers, repetitive elements, and inactive chromatin regions, and is stereo-positioned with growth-controlling transcription factors. *Genome Res.* **2013**, *23*, 1195–1209. [CrossRef]

28.   Sherwood, R.I.; Hashimoto, T.; O'Donnell, C.P.; Lewis, S.; A Barkal, A.; Van Hoff, J.P.; Karun, V.; Jaakkola, T.; Gifford, D.K. Discovery of directional and nondirectional pioneer transcription factors by modeling DNase profile magnitude and shape. *Nat. Biotechnol.* **2014**, *32*, 171–178. [CrossRef]

29. Oldfield, A.; Yang, P.; Conway, A.E.; Cinghu, S.; Freudenberg, J.; Yellaboina, S.; Jothi, R. Histone-fold domain protein NF-Y promotes chromatin accessibility for cell type-specific master transcription factors. *Mol. Cell* **2014**, *55*, 708–722. [CrossRef]

30. Oldfield, A.; Henriques, T.; Kumar, D.; Burkholder, A.B.; Cinghu, S.; Paulet, D.; Bennett, B.D.; Yang, P.; Scruggs, B.S.; Lavender, C.A.; et al. NF-Y controls fidelity of transcription initiation at gene promoters through maintenance of the nucleosome-depleted region. *Nat. Commun.* **2019**, *10*, 3072. [CrossRef]

31. Lu, F.; Liu, Y.; Inoue, A.; Suzuki, T.; Zhao, K.; Zhang, Y. Establishing Chromatin Regulatory Landscape during Mouse Preimplantation Development. *Cell* **2016**, *165*, 1375–1388. [CrossRef] [PubMed]

32. Li, X.Y.; Van Huijsduijnen, R.H.; Mantovani, R.; Benoist, C.; Mathis, D. Intron-exon organization of the NF-Y genes. Tissue-specific splicing modifies an activation domain. *J. Boil. Chem.* **1992**, *267*, 8984–8990.

33. Ceribelli, M.; Benatti, P.; Imbriano, C.; Mantovani, R. NF-YC Complexity Is Generated by Dual Promoters and Alternative Splicing. *J. Biol. Chem.* **2009**, *284*, 34189–34200. [CrossRef] [PubMed]

34. Benatti, P.; Dolfini, D.; Vigano, M.A.; Ravo, M.; Weisz, A.; Imbriano, C. Specific inhibition of NF-Y subunits triggers different cell proliferation defects. *Nucleic Acids Res.* **2011**, *39*, 5356–5368. [CrossRef] [PubMed]

35. Farina, A.; Manni, I.; Fontemaggi, G.; Tiainen, M.; Cenciarelli, C.; Bellorini, M.; Mantovani, R.; Sacchi, A.; Piaggio, G. Down-regulation of cyclin B1 gene transcription in terminally differentiated skeletal muscle cells is associated with loss of functional CCAAT-binding NF-Y complex. *Oncogene* **1999**, *18*, 2818–2827. [CrossRef] [PubMed]

36. Gurtner, A.; Manni, I.; Fuschi, P.; Mantovani, R.; Guadagni, F.; Sacchi, A.; Piaggio, G. Requirement for Down-Regulation of the CCAAT-binding Activity of the NF-Y Transcription Factor during Skeletal Muscle Differentiation. *Mol. Boil. Cell* **2003**, *14*, 2706–2715. [CrossRef] [PubMed]

37. Gurtner, A.; Fuschi, P.; Magi, F.; Colussi, C.; Gaetano, C.; Dobbelstein, M.; Sacchi, A.; Piaggio, G. NF-Y Dependent Epigenetic Modifications Discriminate between Proliferating and Postmitotic Tissue. *PLOS ONE* **2008**, *3*, e2047. [CrossRef]

38. Goeman, F.; Manni, I.; Artuso, S.; Ramachandran, B.; Toietta, G.; Bossi, G.; Rando, G.; Cencioni, C.; Germoni, S.; Straino, S.; et al. Molecular imaging of nuclear factor-Y transcriptional activity maps proliferation sites in live animals. *Mol. Boil. Cell* **2012**, *23*, 1467–1474. [CrossRef]

39. Basile, V.; Baruffaldi, F.; Dolfini, D.; Belluti, S.; Benatti, P.; Ricci, L.; Artusi, V.; Tagliafico, E.; Mantovani, R.; Molinari, S.; et al. NF-YA splice variants have different roles on muscle differentiation. *Biochim. et Biophys. Acta (BBA) - Gene Regul. Mech.* **2016**, *1859*, 627–638. [CrossRef]

40. Mauro, A. Satellite cell of skeletal muscle fibers. *J. Cell Boil.* **1961**, *9*, 493–495. [CrossRef]

41. Maity, S.N. NF-Y (CBF) regulation in specific cell types and mouse models. *Biochim. et Biophys. Acta (BBA) Bioenerg.* **2016**, *1860*, 598–603. [CrossRef] [PubMed]

42. Sakuma, T.; Nishikawa, A.; Kume, S.; Chayama, K.; Yamamoto, T. Multiplex genome engineering in human cells using all-in-one CRISPR/Cas9 vector system. *Sci. Rep.* **2014**, *4*, 5400. [CrossRef] [PubMed]

43. Dolfini, D.; Minuzzo, M.; Pavesi, G.; Mantovani, R. The Short Isoform of NF-YA Belongs to the Embryonic Stem Cell Transcription Factor Circuitry. *STEM CELLS* **2012**, *30*, 2450–2459. [CrossRef] [PubMed]

44. Cullot, G.; Boutin, J.; Toutain, J.; Prat, F.; Pennamen, P.; Rooryck, C.; Teichmann, M.; Rousseau, E.; Lamrissi-Garcia, I.; Guyonnet-Duperat, V.; et al. CRISPR-Cas9 genome editing induces megabase-scale chromosomal truncations. *Nat. Commun.* **2019**, *10*, 1136. [CrossRef] [PubMed]

45. Min, Y.-L.; Bassel-Duby, R.; Olson, E.N. CRISPR Correction of Duchenne Muscular Dystrophy. *Annu. Rev. Med.* **2018**, *70*, 239–255. [CrossRef] [PubMed]

46. Bungartz, G.; Land, H.; Scadden, D.T.; Emerson, S.G. NF-Y is necessary for hematopoietic stem cell proliferation and survival. *Blood* **2012**, *119*, 1380–1389. [CrossRef] [PubMed]

47. Belluti, S.; Semeghini, V.; Basile, V.; Rigillo, G.; Salsi, V.; Genovese, F.; Dolfini, D.; Imbriano, C. An autoregulatory loop controls the expression of the transcription factor NF-Y. *Biochim. et Biophys. Acta (BBA) Bioenerg.* **2018**, *1861*, 509–518. [CrossRef]

48. Moran, J.; Li, Y.; Hill, A.A.; Mounts, W.M.; Miller, C.P. Gene expression changes during mouse skeletal myoblast differentiation revealed by transcriptional profiling. *Physiol. Genom.* **2002**, *10*, 103–111. [CrossRef]

49. Clever, J.L.; Sakai, Y.; Wang, R.A.; Schneider, D.B. Inefficient skeletal muscle repair in inhibitor of differentiation knockout mice suggests a crucial role for BMP signaling during adult muscle regeneration. *Am. J. Physiol. Physiol.* **2010**, *298*, C1087–C1099. [CrossRef]

50. Salizzato, V.; Zanin, S.; Borgo, C.; Lidron, E.; Salvi, M.; Rizzuto, R.; Pallafacchina, G.; Donella-Deana, A. Protein kinase CK2 subunits exert specific and coordinated functions in skeletal muscle differentiation and fusogenic activity. *FASEB J.* **2019**, *33*, 10648–10667. [CrossRef]

51. Millay, D.P.; Gamage, D.G.; Quinn, M.E.; Min, Y.-L.; Mitani, Y.; Bassel-Duby, R.; Olson, E.N. Structure–function analysis of myomaker domains required for myoblast fusion. *Proc. Natl. Acad. Sci.* **2016**, *113*, 2116–2121. [CrossRef] [PubMed]

52. Petrany, M.J.; Millay, D.P. Cell Fusion: Merging Membranes and Making Muscle. *Trends Cell Boil.* **2019**, *29*, 964–973. [CrossRef] [PubMed]

53. Ganassi, M.; Badodi, S.; Quiroga, H.P.O.; Zammit, P.S.; Hinits, Y.; Hughes, S.M. Myogenin promotes myocyte fusion to balance fibre number and size. *Nat. Commun.* **2018**, *9*, 4232. [CrossRef] [PubMed]

54. Pedraza-Alva, G.; Zingg, J.M.; Jost, J.P. AP-1 binds to a putative cAMP response element of the MyoD1 promoter and negatively modulates MyoD1 expression in dividing myoblasts. *J. Boil. Chem.* **1994**, *269*, 6978–6985.

55. Murray-Zmijewski, F.; Lane, D.P.; Bourdon, J.C. p53/p63/p73 isoforms: an orchestra of isoforms to harmonise cell differentiation and response to stress. *Cell Death Differ.* **2006**, *13*, 962–972. [CrossRef]

56. Imbriano, C.; Molinari, S. Alternative Splicing of Transcription Factors Genes in Muscle Physiology and Pathology. *Genes* **2018**, *9*, 107. [CrossRef]

57. Nakka, K.; Ghigna, C.; Gabellini, D.; Dilworth, F.J. Diversification of the muscle proteome through alternative splicing. *Skelet. Muscle* **2018**, *8*, 8. [CrossRef]

58. Zhang, M.; Zhu, B.; Davie, J. Alternative Splicing of MEF2C pre-mRNA Controls Its Activity in Normal Myogenesis and Promotes Tumorigenicity in Rhabdomyosarcoma Cells*. *J. Boil. Chem.* **2014**, *290*, 310–324. [CrossRef]

59. Sebastian, S.; Faralli, H.; Yao, Z.; Rakopoulos, P.; Palii, C.; Cao, Y.; Singh, K.; Liu, Q.-C.; Chu, A.; Aziz, A.; et al. Tissue-specific splicing of a ubiquitously expressed transcription factor is essential for muscle differentiation. *Genome Res.* **2013**, *27*, 1247–1259. [CrossRef]

60. Vogan, K.; Underhill, D.A.; Gros, P. An alternative splicing event in the Pax-3 paired domain identifies the linker region as a key determinant of paired domain DNA-binding activity. *Mol. Cell. Boil.* **1996**, *16*, 6677–6686. [CrossRef]

61. Barber, T.D.; Barber, M.C.; Cloutier, T.E.; Friedman, T.B. PAX3 gene structure, alternative splicing and evolution. *Gene* **1999**, *237*, 311–319. [CrossRef]

62. Pritchard, C.; Grosveld, G.; Hollenbach, A.D. Alternative splicing of Pax3 produces a transcriptionally inactive protein. *Gene* **2003**, *305*, 61–69. [CrossRef]

63. Charytonowicz, E.; Matushansky, I.; Castillo-Martin, M.; Hricik, T.; Cordon-Cardo, C.; Ziman, M. Alternate PAX3 and PAX7 C-terminal isoforms in myogenic differentiation and sarcomagenesis. *Clin. Transl. Oncol.* **2011**, *13*, 194–203. [CrossRef] [PubMed]

64. Vorobyov, E.; Horst, J. Expression of two protein isoforms of PAX7 is controlled by competing cleavage-polyadenylation and splicing. *Gene* **2004**, *342*, 107–112. [CrossRef] [PubMed]

65. LiBetti, D.; Bernardini, A.; Chiaramonte, M.L.; Minuzzo, M.; Gnesutta, N.; Messina, G.; Dolfini, D.; Mantovani, R. NF-YA enters cells through cell penetrating peptides. *Biochim. et Biophys. Acta (BBA) Bioenerg.* **2019**, *1866*, 430–440. [CrossRef] [PubMed]

66. Cappabianca, L.; Farina, A.R.; Di Marcotullio, L.; Infante, P.; De Simone, D.; Sebastiano, M.; Mackay, A. Discovery, characterization and potential roles of a novel NF-YAx splice variant in human neuroblastoma. *J. Exp. Clin. Cancer Res.* **2019**, *38*, 1–25. [CrossRef]

67. Wang, J.; Zhuang, J.; Iyer, S.; Lin, X.; Whitfield, T.W.; Greven, M.C.; Pierce, B.G.; Dong, X.; Kundaje, A.; Cheng, Y.; et al. Sequence features and chromatin structure around the genomic regions bound by 119 human transcription factors. *Genome Res.* **2012**, *22*, 1798–1812. [CrossRef]

68. Zambelli, F.; Pavesi, G. Genome wide features, distribution and correlations of NF-Y binding sites. *Biochim. et Biophys. Acta (BBA) Bioenerg.* **2017**, *1860*, 581–589. [CrossRef]

69. Welch, R.D.; Guo, C.; Sengupta, M.; Carpenter, K.J.; Stephens, N.A.; Arnett, S.A.; Meyers, M.J.; Sparks, L.M.; Smith, S.R.; Zhang, J.; et al. Rev-Erb co-regulates muscle regeneration via tethered interaction with the NF-Y cistrome. *Mol. Metab.* **2017**, *6*, 703–714. [CrossRef]

70. Van Wageningen, S.; Ridder, M.C.B.-D.; Nigten, J.; Nikoloski, G.; Erpelinck-Verschueren, C.A.J.; Löwenberg, B.; De Witte, T.; Tenen, D.G.; Van Der Reijden, B.A.; Jansen, J.H. Gene transactivation without direct DNA binding defines a novel gain-of-function for PML-RARα. *Blood* **2008**, *111*, 1634–1643. [CrossRef]

71. Moeinvaziri, F.; Shahhosseini, M. Epigenetic role of CCAAT box-binding transcription factor NF-Y onIDgene family in human embryonic carcinoma cells. *IUBMB Life* **2015**, *67*, 880–887. [CrossRef] [PubMed]
72. Dolfini, D.; Mantovani, R.; Zambelli, F.; Pavesi, G. A perspective of promoter architecture from the CCAAT box. *Cell Cycle* **2009**, *8*, 4127–4137. [CrossRef] [PubMed]
73. Singh, K.; Dilworth, F.J. Differential modulation of cell cycle progression distinguishes members of the myogenic regulatory factor family of transcription factors. *FEBS J.* **2013**, *280*, 3991–4003. [CrossRef]

Article

Transthyretin Maintains Muscle Homeostasis through the Novel Shuttle Pathway of Thyroid Hormones during Myoblast Differentiation

Eun Ju Lee [1,†], Sibhghatulla Shaikh [1,†], Dukhwan Choi [1], Khurshid Ahmad [1], Mohammad Hassan Baig [1], Jeong Ho Lim [1], Yong-Ho Lee [2], Sang Joon Park [3], Yong-Woon Kim [4], So-Young Park [4] and Inho Choi [1,*]

[1] Department of Medical Biotechnology, Yeungnam University, Gyeongsan 38541, Korea; gorapadoc0315@hanmail.net (E.J.L.); sibhghat.88@gmail.com (S.S.); apdltkd@naver.com (D.C.); ahmadkhursheed2008@gmail.com (K.A.); mohdhassanbaig@gmail.com (M.H.B.); lim2249@kitech.re.kr (J.H.L.)
[2] Department of Biomedical Science, Daegu Catholic University, Gyeongsan 38430, Korea; ylee325@cu.ac.kr
[3] College of Veterinary Medicine, Kyungpook National University, Daegu 41566, Korea; psj26@knu.ac.kr
[4] Department of Physiology, College of Medicine, Yeungnam University, Daegu 42415, Korea; ywkim@yumail.ac.kr (Y.-W.K.); sypark@med.yu.ac.kr (S.-Y.P.)
* Correspondence: inhochoi@ynu.ac.kr; Fax: +82-53-810
† These two authors contributed equally to this work.

Received: 8 October 2019; Accepted: 2 December 2019; Published: 4 December 2019

Abstract: Skeletal muscle, the largest part of the total body mass, influences energy and protein metabolism as well as maintaining homeostasis. Herein, we demonstrate that during murine muscle satellite cell and myoblast differentiation, transthyretin (TTR) can exocytose via exosomes and enter cells as TTR- thyroxine (T_4) complex, which consecutively induces the intracellular triiodothyronine (T_3) level, followed by T_3 secretion out of the cell through the exosomes. The decrease in T_3 with the TTR level in 26-week-old mouse muscle, compared to that in 16-week-old muscle, suggests an association of TTR with old muscle. Subsequent studies, including microarray analysis, demonstrated that T_3-regulated genes, such as FNDC5 (Fibronectin type III domain containing 5, irisin) and RXRγ (Retinoid X receptor gamma), are influenced by TTR knockdown, implying that thyroid hormones and TTR coordinate with each other with respect to muscle growth and development. These results suggest that, in addition to utilizing T_4, skeletal muscle also distributes generated T_3 to other tissues and has a vital role in sensing the intracellular T_4 level. Furthermore, the results of TTR function with T_4 in differentiation will be highly useful in the strategic development of novel therapeutics related to muscle homeostasis and regeneration.

Keywords: muscle satellite cell; transthyretin; thyroid hormone; myogenesis; exosomes; skeletal muscle

---

1. Introduction

Skeletal muscle is comprised of multinucleated myofibers and has excellent regeneration capability, which deteriorates progressively with age, restraining the voluntary functions of daily life. The regenerative capacity is mostly facilitated by muscle satellite or stem cells (MSCs) that reside between the basal lamina and sarcolemma, a distinct 'niche' in the muscle fibers [1,2]. MSCs vigorously regulate myofiber growth, and MSC progression is typically regulated by the expression of myogenic transcription factors (Pax3, Pax7, myoblast determination protein; MYOD, and myogenin; MYOG) [3]. After injury, quiescent Pax7[+] MSCs are triggered to undergo sequential activation, proliferation and differentiation involving MYOD, Myf5 and MYOG to generate multinucleated myotubes [4].

MSC differentiation is indispensable in the regeneration of skeletal muscle and is typically regulated by multiple signaling pathways and by the interaction of several extracellular matrix components with MSCs. Fibromodulin was reported to have a robust role in muscle regeneration by enhancing the recruitment of MSCs to injury sites [5].

Thyroid hormones (THs, thyroxin; T_4 and triiodothyronine; T_3) have vital roles in the development of various tissues, as well as in postnatal life, by modulating gene expressions [6,7]. THs regulate the expression of various proteins crucial for muscle development and contractility [8–10]. Indeed, the foremost targets of THs are muscles, as they regulate the expression of several genes at the transcriptional level [11,12]. The effects of TH signaling in the development and function of skeletal muscle are the result of a remarkably complex mechanism [11]. Generally, to retain homeostasis, regeneration capability, and development, binding of T_3 to thyroid hormone receptors (TR) is essential [13]. TRs are encoded by two genes (THRA and THRB), and alternate splicing of each gene produces TRα1, TRβ1, and TRβ2 receptor subtypes. TRα is the predominant subtype in cardiac and skeletal muscle [14]. TRα has a key role in regulation of heart rate and basal metabolism [15]. Transcription of MYOD is directly regulated by T_3 [16]. Therefore, TH signaling can control several events during myogenesis via direct and/or indirect regulation of myogenic gene expression.

Retinoids (synthetic vitamin A derivatives) can influence development and metabolism through nuclear hormone receptors (retinoic acid receptor and retinoid X receptor, RXR). RXR forms heterodimers with retinoic acid, TH, and vitamin D receptors, enhancing transcriptional function on their respective response elements [17]. Three different RXR isoforms (RXRα, β and γ) have been characterized. RXRγ is the dominant isoform in adult heart and skeletal muscle [18].

Exosomes are small (40–100 nm) membrane vesicles of endocytic origin that are released from most cell types into the extracellular environment [19]. Exosomes were first defined in 1983, and interest in these vesicles increased markedly after finding that they contain mRNA and microRNA [20]. Exosomes have been shown to facilitate cellular communication by transporting proteins, cytokines, and nucleic acids and to sustain the normal physiological function of cells [21].

Transthyretin (TTR) is a 55-kDa homotetrameric transporter protein for T_4 and retinol-binding protein in the blood [22,23]. The liver is the main contributing organ for TTR synthesis in plasma. TTR null ($TTR^{-/-}$) mice exhibit a delayed suckling-to-weaning transition, delayed growth, reduced muscle mass, and stunted longitudinal bone growth [24]. Among the transporters existing in blood, thyroxine-binding globulin (TBG) has the highest affinity for T_4 and T_3 (1.0×10^{10} and $4.6 \times 10^8 \ M^{-1}$, respectively), followed by TTR (7.0×10^7 and $1.4 \times 10^7 \ M^{-1}$) and albumin (7.0×10^5 and $1.0 \times 10^5 \ M^{-1}$) [25]. The binding efficacy of TH distributor proteins determines the transportation times for distribution of THs to tissues, thus, TTR (with transitional affinity), more than TBG, is responsible for instant delivery of THs to tissues [6,25].

Though it is known that human placenta, trophoblasts, JEG-3 and HepG2 cells secrete and internalize TTR [26–28], its cellular uptake in skeletal muscle has not been fully described. We have demonstrated that TTR initiates myoblast differentiation by inducing the expression of myogenic genes involved in the early phase of myogenesis and the associated calcium channels [6], and we have elucidated its functional role in maintaining the cellular T_4 level. Furthermore, we reported that TTR enhances recruitment of MSCs to the site of injury, thereby regulating muscle regeneration [29]. However, the detailed mechanism of TTR with T_4 in MSCs differentiation into muscle cells is unclear. In the current work, we have confirmed TTR secretion and internalization in myoblast cell. We found that TTR uptake and internalization by myoblast cells is increased by T_4. By using microarray analysis and other studies, we have elucidated that TTR and TH coordinate with each other to modulate gene expression in muscle growth, development, and homeostasis.

2. Materials and Methods

2.1. Animal Experiments

C57BL/6 male mice were obtained from Daehan Biolink (Dae-Jeon, South Korea) and housed at four per cage in a temperature controlled room under a 12 h light/12 h dark cycle. In the period mice (six weeks) were fed a normal diet containing 4.0% (*w/w*) total fat (Rodent NIH-31 Open Formula Auto; Zeigler Bros., Inc., Gardners, PA, USA). Gastrocnemius muscle tissues were collected after 10 or 20 weeks. After collection, muscle tissues were fixed and stored at −80 °C until required for RNA and protein extraction or fixed overnight at 4 °C for paraffin-embedded tissue blocks to be used in immunohistochemistry. All experimental were done by following the guidelines issued by the Institutional Animal Care and Use Committees of the Catholic University of Daegu (IACUC-2017-051).

2.2. C2C12 Cell Culture

C2C12 cells (murine myoblast, Korean Cell Line Bank, Seoul, Korea) were cultured in DMEM (HyClone Laboratories, Logan, UT, USA) supplemented with 10% FBS (fetal bovine serum, HyClone Laboratories) and 1% P/S (penicillin/streptomycin, Thermo Fisher Scientific, Waltham, MA, USA) in a humidified 5% CO_2 incubator at 37 °C. For differentiation, cells were cultured for two or three days in DMEM + 2% FBS + 1% P/S (serum (+) differentiation media) or DMEM + 1% P/S (serum (−) differentiation media). T_4 (50 ng/mL, Sigma Aldrich, St. Louis, MO, USA), I-850 (5 ug/mL, Sigma Aldrich), TTR (0.1 ug/mL, Sigma Aldrich) or bovine serum albumin (BSA,1 mg/mL, Sigma Aldrich) was added to the indicated differentiation medium after two or three days.

2.3. Mouse MSCs Culture

Gastrocnemius and cranial thigh muscles were collected from C57BL male mice (six weeks) and minced, digested with 1% pronase (Roche, Mannheim, Germany) for 1 h at 37 °C, and then centrifuged at 1000× *g* for 3 min followed by passage of the digested tissue phase through a 100 mm syringe filter (Millipore, Darmstadt, Germany). After centrifugation of the filtrate at 1000× *g* for 5 min, the pellets were suspended in DMEM + 20% FBS + 1% P/S + 5 ng/mL FGF2 (fibroblast growth factor 2, Miltenyi Biotec GmbH, Auburn, CA, USA), seeded on collagen-coated plates (Corning, Brooklyn, NY, USA), and incubated in a humidified 5% CO_2 atmosphere at 37 °C. The medium was changed every day. For induction of MSC differentiation into muscle cells, media were switched to DMEM + 2% FBS + 1% P/S or DMEM + 1% P/S followed by incubation for two days. MSC purity was confirmed with Pax7 protein expression (Santa Cruz Biotechnology, Paso Robles, CA, USA) using immunocytochemistry.

2.4. MTT Assay

C2C12 cells were cultured with DMEM + 10% FBS + 1% P/S for two days for analysis of cell viability. The cells were washed with DMEM and then incubated with 0.5 mg/mL MTT reagent (Sigma Aldrich) for 1 h. After dissolving the formazan crystals with DMSO (Sigma Aldrich), absorbance was measured at 540 nm (Tecan Group Ltd., Männedorf, Switzerland).

2.5. Immunoneutralization

TTR protein neutralization was carried out with TTR-specific antibodies (5 µg/mL, Santa Cruz Biotechnology) for two or three days in DMEM + 2% FBS + 1% P/S or DMEM + 1% P/S differentiation media.

2.6. Exosomes Isolation

Cells were cultured with DMEM + 1% P/S differentiation media. The cells were incubated for two or three days and the media were then collected, centrifuged at 2000× g for 30 min, and the upper phase collected for exosomes isolation. Using a total exosomes isolation reagent (Thermo Fisher Scientific, MA, USA), the exosomes from the upper phase were isolated according to the manufacturer's protocol. In brief, the media were incubated with the total exosomes isolation reagent at 4 °C overnight and centrifuged at 10,000× g for 60 min. After discarding the supernatant, the pellet was dried at room temperature and suspended in PBS.

Mouse plasma (4 mL) was filtered with a 0.8 um syringe filter (Sartorius, Goettingen, Germany), and the exosomes were then isolated according to the manufacturer's protocol (exoEasy Maxi Kit, Qiagen, Germantown, MD, USA).

2.7. T_4 and T_3 Concentration Measurement

An ELISA kit (DRG International, Marburg, Germany) was used to measure the concentration of T_4 or T_3 hormones. In brief, cell lysates or cultured media with T_4 or T_3 enzyme conjugate reagent were homogenized and added to specific antibody-coated microtiter plates and then incubated for 60 min at room temperature. After discarding the mixtures, the unbound materials were removed by washing the plates. Substrate solution was added followed by incubation for 20 min. Stop solution was then applied to terminate the reaction. Color intensities were then measured at 450 nm by using a spectrophotometer (Tecan Group Ltd., Switzerland).

2.8. Gene Knockdown

When C2C12 cells confluency reached 30%, 1 ng TTR, TR-α, RXRγ, or fibronectin type III domain containing 5 (FNDC5) shRNA vector (Santa Cruz Biotechnology) and scrambled vector (empty vector as negative control, Santa Cruz Biotechnology) were transfected using plasmid transfection reagent and transfection medium according to the manufacturer's protocol (Santa Cruz Biotechnology). After three days, transfected cells were selected with puromycin (2 ug/mL, shRNA or scrambled vector is a puromycin selection vector, Santa Cruz Biotechnology). Selected cells were grown to 70% confluence before switching to differentiation media. Knockdown efficiencies were determined by analyzing the expressions of control (scrambled vector transfected cell) and knockdown cells. Supplementary Table S1 shows the sequences of the shRNA constructs.

2.9. RNA Isolation, cDNA Synthesis and RealTime RT-PCR

Trizol reagent (Thermo Fisher Scientific) was used following the manufacturer's instructions to extract total RNA from cells. Two micrograms of RNA in 20 µL of reaction mixture was employed for the synthesis of 1st strand cDNA with random hexamer and reverse transcriptase at 25 °C for 10 min, 37 °C for 120 min, and 85 °C for 5 min. The cDNA product (2 µL) and gene-specific primers (10 pmole, 2 µL) were used for analysis of real-time RT-PCR (40 cycles), which was performed using a 7500 real-time PCR system with power SYBR Green PCR Master Mix (Thermo Fisher Scientific) as the fluorescence source. Glyceraldehyde 3-phosphate dehydrogenase (GAPDH) was used as the reference gene. Primer information is presented in Supplementary Table S2.

2.10. RT-PCR

Exosomes RNA was synthesized into cDNA and 2 μL cDNA and gene-specific primers (10 pmole, 2 μL) were used for PCR, which was performed using a 2720 Thermal Cycler PCR machine with PCR Master mix (Genetbio, Daejeon, Korea). The PCR conditions were as follow; denaturation 95 °C for 30 s, annealing at 59 °C for 30 s, extension at 72 °C, post-extension 72 °C for 5 min followed by holding (40 cycles). The PCR product was examined by performing electrophoresis on agarose gel.

2.11. Protein Isolation from Culture Media

The cells were cultured with DMEM + 1% P/S differentiation media for two days, centrifuged at 5000× g for 5 min, and the supernatant was then incubated with 1.3% potassium acetate (Sigma Aldrich) for 1 h at 4 °C. The mixture was centrifuged at 1500× g for 10 min, the supernatant was discarded, and the pellet washed with 100% acetone (Merck, Darmstadt, Germany). The isolated proteins were then dried, and Western blot analysis was performed by adding buffer with protease inhibitor cocktail (Thermo Fisher Scientific).

2.12. Western Blot

After washing the cells with PBS, they were lysed with RIPA buffer supplemented with protease inhibitor cocktail (Thermo Fisher Scientific). The Bradford assay was used to estimate the total protein concentration. Proteins (60 μg) were electrophoresed in 10% or 12% SDS-polyacrylamide gel and then transferred to PVDF membrane (EMS–Millipore, Billerica, MA, USA). The blots were then blocked with 3% skim milk or BSA in Tris-buffered saline (TBS) Tween 20 for 1 h, incubated overnight with protein-specific primary antibodies [TTR (1:400), MYOD (1:500), MYOG (1:500), D2 (iodothyronine deiodinase type 2; 1:500), RXRγ (1:500) (Santa Cruz Biotechnology) or β-actin (1:2000) antibody (Santa Cruz Biotechnology), TR-α (1:500, Thermo Fisher Scientific), MYL2 (myosin light chain 2, 1:1000, Abcam, Cambridge, MA, USA) or FNDC5 (1:500, Bioss Antibodies, Woburn, MA, USA)] in 1% skim milk or BSA in TBS at 4 °C. The blots were then washed and incubated with horse radish peroxidase (HRP)-conjugated secondary antibody (Santa Cruz Biotechnology) for 1 h at room temperature and then developed with Super Signal West Pico Chemiluminescent Substrate (Thermo Fisher Scientific). Supplementary Table S3 shows the molecular weight of protein.

2.13. Fusion Index

After washing with PBS, cells were fixed with methanol and then stained with 0.04% Giemsa G250 (Sigma Aldrich). Images were taken randomly of three different sections per dish. The number of nuclei in myotubes and the total number of nuclei in the cells were counted in each field. Fusion indices were calculated by expressing the number of nuclei in the myotubes as percentages of the total numbers of nuclei.

2.14. TTR Protein Labeling with Fluorescence

TTR protein and BSA were labeled with the Alexa Fluor 594 protein labeling kit (Thermo Fisher Scientific) following the manufacturer's instructions. Briefly, 100 μL TTR proteins and BSA (0.1 μg/μL) were incubated with 4.7 μL Alexa Fluor 594 succinimidyl ester (12.2 nmole/μL) for 15 min at room temperature, and the conjugated reaction mixture was then purified with resin gel-spin filter. Labeled TTR proteins (0.2 μg) and BSA were added to the cells and detected by fluorescence microscope (Nikon, Tokyo, Japan).

2.15. TTR Overexpression Vector

The region corresponding to the TTR gene open reading frame (ORF) was PCR amplified with TTR ORF primer (Forward: 5′-ATGGCTTCCCTTCGACTCTTCC-3′, Reverse: 5′-GATTCTGGGGGTTGCTGACGA-3′) and ligated into the pcDNA 3.1/CT-GFP-TOPO vector (Invitrogen, Waltham, MA, USA). The ligated sequence was confirmed by sequencing analysis. The construct (2.5 µg) was transfected using 10 µL lipofectamine (1 mg/mL) and Opti MEM medium (Invitrogen) into C2C12 cells following the manufacturer's directions and positive cells were selected using G418 antibiotics (2 µg/mL, AppliChem GmbH, Darmstadt, Germany).

2.16. Immunocytochemistry

The cells were fixed with 4% formaldehyde (Sigma Aldrich) and permeabilized with 0.2% Triton X 100 (Sigma Aldrich). After blocking with 1% normal goat serum (SeraCare Life Sciences, Milford, MA, USA) for 30 min in a humid environment, cells were incubated with primary antibodies [TTR (1:50), MYOD (1:50), MYOG (1:50), MYL2 (1:50), D2 (1:50), RXRγ (1:50), TRα (1:50), or FNDC5 (1:50)] at 4 °C in a humid environment overnight. Secondary antibody (1: 100; Alexa Fluor 594 goat anti-rabbit or anti-mouse; Thermo Fisher Scientific) was applied for 1 h at room temperature. DAPI was used to stain the cells (Sigma-Aldrich) and imaged using a fluorescence microscope equipped with a digital camera (Nikon, Tokyo, Japan).

2.17. Immunohistochemistry

The sections of paraffin-embedded muscle tissue were deparaffinized and hydrated with xylene (Junsei, Tokyo, Japan) and ethanol (Merck), respectively, and endogenous peroxidase activity was quenched in 0.3% H_2O_2/methanol. The sections were then either stained with hematoxylin and eosin (Thermo Fisher Scientific) for morphological observation or blocked with 1% normal goat serum (SeraCare Life Sciences), incubated with primary antibodies [TTR (1:50), D2 (1:50), or FNDC5 (1:50)] overnight at 4 °C, and then incubated with horse radish peroxidase–conjugated secondary antibody (1:100). Positive signals were visualized by adding horse radish peroxidase-conjugated streptavidin (Vector, CA, USA). Nuclei of stained sections were stained with hematoxylin and then dehydrated, mounted, and observed by a light microscope (Leica, Wetzlar, Germany).

2.18. Microarray Analysis

Microarray analysis was conducted with the Agilent Technologies mouse GE4X 44K (V2) chip to determine the differentially expressed genes in wild-type (TTR$_{wt}$) and TTR$_{kd}$ (TTR knockdown) cells as described previously [30]. Briefly, TTR$_{wt}$ and TTR$_{kd}$ cells were grown in serum (+) differentiation media for two days and RNAs were extracted, synthesized into cDNA with fluorescence using a Low RNA Input Linear Amplification kit (Agilent Technologies, CA, USA) according to the manufacturer's instructions. A total of three hybridizations were performed, and the statistical relevance of gene expression differences was confirmed by SAM (Standard University, Palo Alto, CA, USA). The significance cut-off was a median false discovery rate ≤5% for the SAM analysis.

2.19. DAVID Analysis

DAVID was performed as described previously [5]. In brief, enriched biological themes in up- and down-regulated gene lists ($p \leq 0.05$ and 2 fold≤) were categorized by employing the Gene Ontology (GO) terms of cellular component, molecular function, and biological process in DAVID.

2.20. Statistical Analysis

Mean values of normalized expressions were evaluated by Tukey's Studentized range test to categorize expressional differences of genes, considering $p \leq 0.05$ statistically significant. The real-time RT-PCR data was normalized using glyceraldehyde 3-phosphate dehydrogenase (GAPDH) as the internal standard and was analyzed by one-way ANOVA using PROC GLM in SAS 9.0 (SAS Institute, Cary, NC, USA).

3. Results

3.1. TTR Secretion During Myoblast Differentiation

To investigate TTR secretion from cells during C2C12 myoblast differentiation, normal and TTR knockdown cells were cultured in serum-free media for two days, after which the isolated protein level from cultured media was analyzed by Western blotting. The appearance of more TTR protein in cultured media compared to that in cell lysate indicates that TTR is secreted during myoblast differentiation (Figure 1A). Furthermore, TTR mRNA and protein were decreased in TTR_{kd} cells and cultured media, respectively, compared to those in TTR_{wt} cells (Figure 1A). Next, the TTR mRNA level was analyzed in normal cells and exosomes from mouse plasma and media of cultured cells (CM): T_4-treated cells, TTR_{wt}, and TTR_{kd}. TTR mRNA was evident in exosomes isolated from culture media and mouse plasma and was increased by T_4 treatment but decreased by TTR_{kd} (Figure 1B). TTR immunoneutralization using TTR antibody was performed during differentiation. Myotube formation and the expression of the myogenic genes were decreased by TTR neutralization. However, TTR expression was significantly enhanced in neutralized cells (Figure 1C,D). Interestingly, when the T_4 concentration was measured in cells, it was higher in non-neutralized cells than in neutralized cells supplemented with T_4 (Figure 1E). Taken together, these results show that TTR secreted from cells transported T_4 into the cells during myoblast differentiation.

3.2. Enhancement of Myoblast Viability and Differentiation by TTR with T_4

To assess the role of TTR and T_4 on myoblast viability and differentiation, C2C12 cells were grown with T_4 or T_4 + TTR protein for two or three days. Cell viability was increased in T_4 + TTR protein treated cells compared to that in only T_4 treated cells (Figure 2A). The T_4 and T_3 concentrations were measured in CM and cells. A lower T_4 concentration in media with a consequent higher concentration in cells was observed with T_4 + TTR treatment than in those with only T_4 treatment. The results indicate that TTR outside the cell enhances the transport of T_4 to the cell interior in myoblast viability (Figure 2A). Further, cells were cultured in serum-free media with added T_4 or T_4 + TTR protein for three days to induce differentiation. The T_4 + TTR treatment significantly induced myotube formation with elevated mRNA (MYL2) and protein expression of myogenic factors (MYOD and MYL2), RXRγ, and TRα. However, TTR mRNA and protein expression were decreased by TTR + T_4 treatment and their expression in exosomes was also reduced from that of only T_4-treated cells (Figure 2B). T_4 and T_3 concentrations were increased by TTR + T_4 treatment (Figure 2C). TTR in mouse MSCs was assessed to determine its expression during differentiation. For this, MSCs were incubated with differentiation media for zero or two days. Expression of TTR and myogenic genes or proteins were increased on day 2 compared to that on day 0 (Figure 2D). Next, MSCs were cultured in serum-free conditions supplemented with T_4 or T_4 + TTR protein for two days to induce differentiation. Similar to the results with C2C12 cells, MSCs exhibited increased myotube formation with elevated thyroid hormone concentration under T_4 + TTR treatment (Figure 2E). Interestingly, decreased TTR mRNA was observed in the exosomes following T_4 + TTR treatment (Figure 2E). Furthermore, T_3 was present in exosomes isolated from serum-free MSCs culture media supplemented with T_4 (Figure 2F). These data showed that TTR protein with T_4 not only enhanced myoblast proliferation and myogenic differentiation, but also increased MSC differentiation into muscle cells.

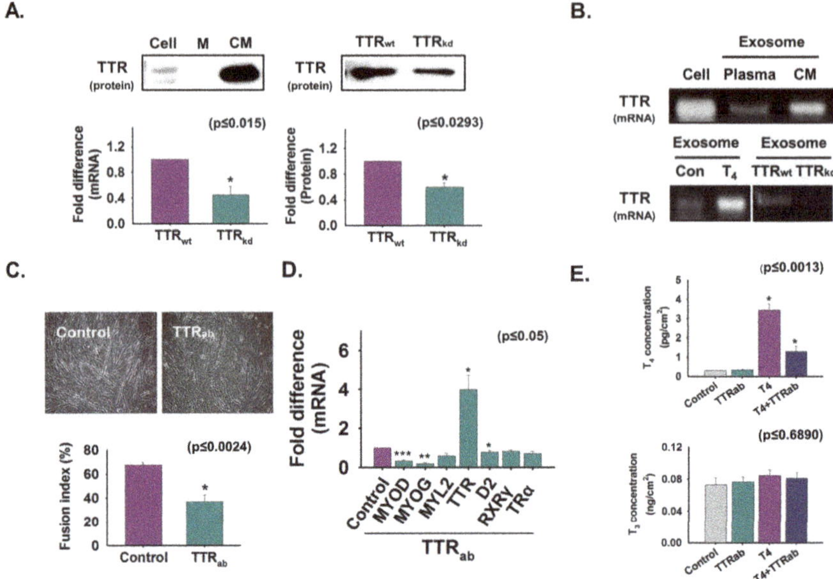

Figure 1. The role of secreted TTR from cells during myogenic differentiation. Normal and TTR knockdown cells were cultured with serum-free media for two days (**A**,**B**). (**A**) Proteins were isolated from cells, DMEM (control) and cultured media (CM). TTR protein level was analyzed by Western blot. TTR mRNA level in cells by real-time RT-PCR, and protein level in cell culture media of TTR$_{wt}$ and TTR$_{kd}$ by Western blot. Band intensity was measured by using ImageJ. (**B**) TTR mRNA levels in normal cell, exosomes isolated from mouse plasma, media of cultured C2C12 cells (CM) with or without T$_4$ treatment, and TTR$_{wt}$ and TTR$_{kd}$ by RT-PCR. Cells were cultured in 2% FBS or serum-free media supplemented with TTR antibody for two (**C**) or three days (**D**,**E**) for immunoneutralization. (**C**) Myotube formation and fusion index was observed by Giemsa staining. (**D**) Gene expression was observed by real-time RT-PCR. (**E**) T$_4$ and T$_3$ concentration in cells was observed by ELISA. TTR$_{wt}$ indicates cells transfected with scrambled vector. Means ± SD (n = 3). * $p \leq 0.05$, ** $p \leq 0.001$, *** $p \leq 0.0001$.

3.3. Reduction of T$_4$ Concentration Inside Cells and Myoblast Differentiation by Bovine Albumin Serum (BSA) Treatment

For comparative assessment of T$_4$ transport through TTR to the cell interior, C2C12 cells were cultured in serum-free media supplemented with T$_4$ or T$_4$ + BSA protein for two days. Myotube formation and MYOG expression were decreased in BSA-treated cells, while TTR and D2 expressions were increased at the translational level (Figure 3A). Interestingly, elevated TTR in both exosomes (mRNA) and CM (protein) was also observed in BSA-treated cells (Figure 3B). High T$_4$ and T$_3$ concentrations in T$_4$ + BSA supplemented media with subsequent low levels in both hormone concentrations in the cell, under the same conditions, indicated that BSA reduced the transport of T$_4$ to the cell interior (Figure 3C). Furthermore, elevated T$_4$ concentration was observed in T$_4$ + BSA + TTR supplemented cells relative to that in T$_4$ + BSA treated cells (Figure 3D). Interestingly, decreased TTR mRNA was found in exosomes of T$_4$ + BSA + TTR treated cells (Figure 3E). Additionally, T$_3$ was present in exosomes, and there was no difference in T$_3$ concentration in exosomes supplemented with T$_4$, T$_4$ + BSA, or T$_4$ + TTR (Figure 3F). We observed that BSA reduces myotube formation by decreasing T$_4$ transport.

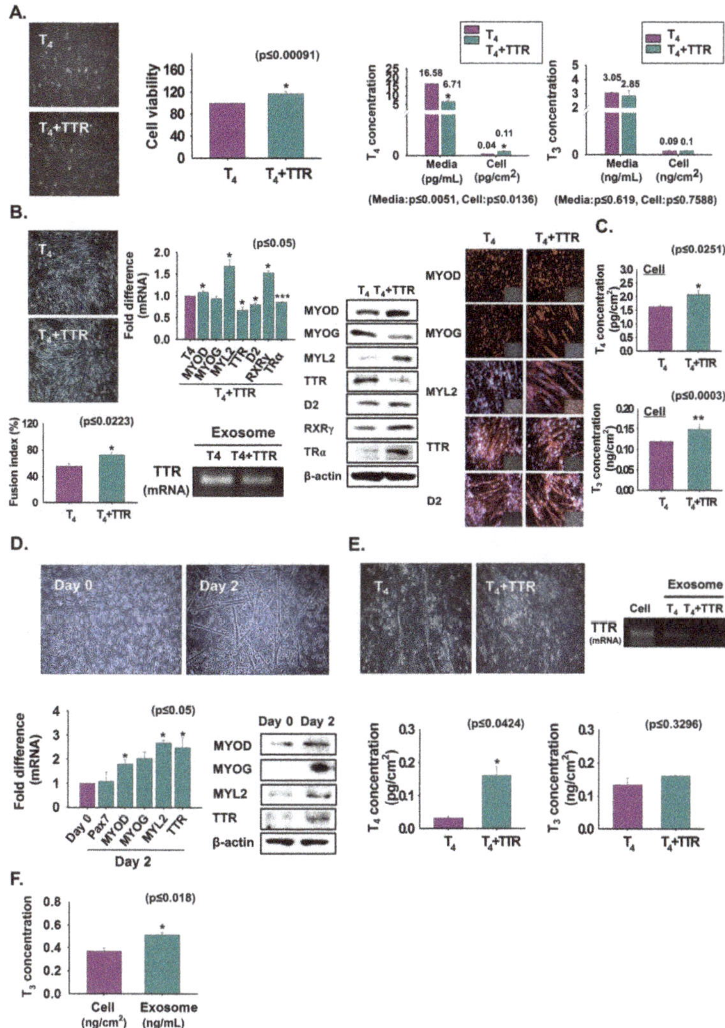

Figure 2. Myoblast viability and differentiation by treatment with TTR proteins. (**A**) C2C12 cells were cultured in 10% FBS media supplemented with T_4 or T_4 + TTR protein for two days. Cell viability was observed by MTT assay. T_4 or T_3 concentration in cultured media and cells were observed by ELISA. Cells were cultured in serum-free media supplemented with T_4 or T_4 + TTR protein for three days (**B**,**C**). (**B**) Myotube formation and fusion index by Giemsa staining, mRNA level in cells by real-time RT-PCR, exosomes by RT-PCR, protein expression by Western blot and immunocytochemistry. (**C**) T_4 or T_3 concentration in cells was observed by ELISA. (**D**) When mouse MSCs reached 100% confluency, media were switched to 2% FBS and cultured for zero and two days. MSC differentiation, TTR mRNA level by real-time RT-PCR and protein expression by Western blot. (**E**) MSCs were cultured in serum-free media supplemented with T_4 or T_4 + TTR protein for two days. T_4 or T_3 concentration in cells was observed by ELISA. (**F**) MSCs were cultured with serum-free media supplemented with T_4 for two days and exosomes were isolated from cultured media. T_3 concentration in cell and exosomes was observed by ELISA. Means ± SD (n = 3). * $p \leq 0.05$, ** $p \leq 0.001$, *** $p \leq 0.0001$.

3.4. TTR Internalization Into Myoblast

To elucidate TTR internalization to the cell interior, TTR protein or BSA was fluorescently labeled and C2C12 cells were cultured under serum-free conditions supplemented with labeled TTR protein or BSA for one day. Higher fluorescence of labeled TTR protein was evident in the cells treated with labeled TTR than with BSA or in non-treated cells (Figure 4A). TTR overexpression was achieved by transfection with the TTR ORF plasmid and cultured with 10% FBS for two days. Increased cell viability was observed in TTR-overexpressing cells (Figure 4B). Next, TTR-overexpressing cells were cultured with serum-free media for two days. Increased myotube formation with enhanced TTR mRNA/protein expression was observed in TTR-overexpressing cells (Figure 4C). Additionally, elevated concentrations of THs were observed in TTR-overexpressing cells supplemented with T_4 (Figure 4C).

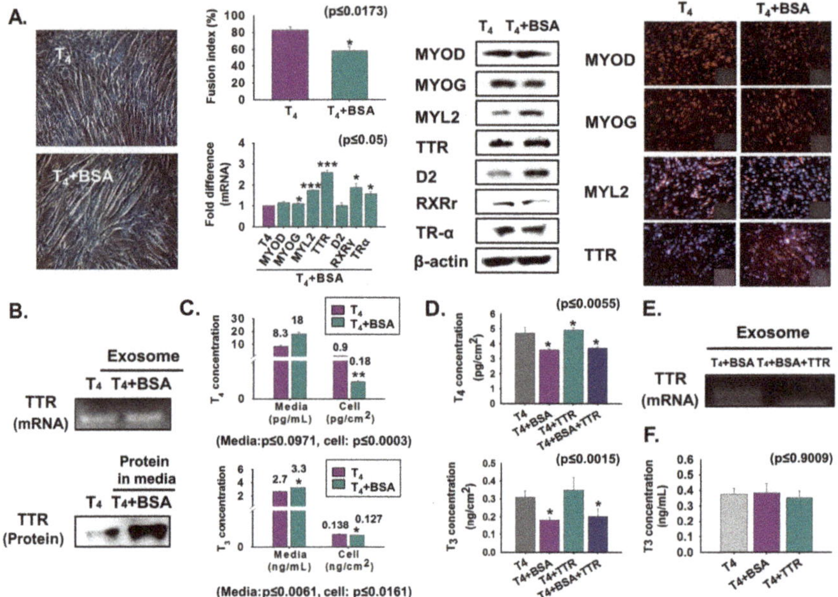

Figure 3. Myoblast differentiation following BSA treatment. Cells were cultured in serum-free media supplemented with T_4 or T_4 + BSA for two days (**A–E**). (**A**) Myotube formation and fusion index were observed by Giemsa staining. mRNA level was observed by real-time RT-PCR and protein expressions by Western blot and immunocytochemistry. (**B**) TTR mRNA in exosomes of cultured media using RT-PCR and protein level in cultured media by Western blot. (**C**) T_4 or T_3 concentration in cultured media or cells was observed by ELISA. (**D,E**) Cells were cultured with serum-free media supplemented with T_4, T_4 + BSA, T_4 + TTR or T_4 + BSA + TTR for two days. T_4 or T_3 concentration in T_4 + BSA or T_4 + BSA + TTR treated cells. TTR mRNA in exosomes of cultured media (in T_4 + BSA or T_4 + BSA + TTR treated cells) using RT-PCR. (**F**) Cells were cultured in serum-free media supplemented with T_4 or T_4 + BSA or T_4 + TTR for two days and exosomes were isolated from each cultured medium. T_3 concentration in exosomes. Means ± SD (n = 3). * $p \leq 0.05$, ** $p \leq 0.001$, *** $p \leq 0.0001$.

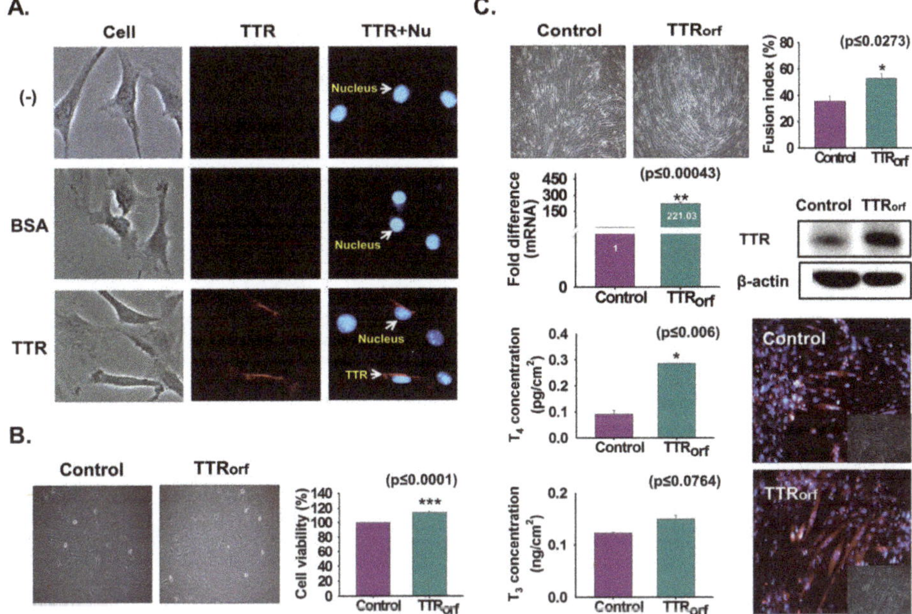

Figure 4. Endocytosis of TTR protein and TTR overexpression effects. (**A**) TTR protein or BSA were labeled with fluorescence and cells were cultured with serum-free media supplemented with labeled TTR protein or BSA for 1 day. Detection of labeled TTR protein and BSA in cells (Red: TTR, Blue: Nucleus). (**B**) TTR overexpression was performed by transfecting with TTR ORF plasmid followed by incubation with 10% FBS for two days. Cell viability was analyzed by MTT assay. (**C**) TTR overexpressing cells were incubated with serum-free media for two days. Myotube formation and fusion index were observed by Giemsa staining, TTR mRNA level by real-time RT-PCR, and protein expression by Western blot and immunocytochemistry. Control or TTR-overexpressing cells were incubated with serum-free media supplemented with T_4 for two days. T_4 or T_3 concentration was measured by ELISA. Means ± SD ($n = 3$). * $p \leq 0.05$, ** $p \leq 0.001$, *** $p \leq 0.0001$.

3.5. Regulation of RXRγ and TRα Expression by TTR During Myoblast Differentiation

To determine the role of T_4 or TTR on RXRγ and TRα expression, C2C12 cells were grown with or without serum in normal or TTR knockdown cells, and the effects were studied during myoblast differentiation. Increases in mRNA and protein expression of RXRγ and TRα were evident on day 2 compared to the levels on day 0 (Figure 5A). Next, T_4 treatment under serum-free conditions stimulated RXRγ expression at both the transcriptional and translational level. However, TRα protein expression was decreased by T_4 treatment (Figure 5B). Interestingly, TTR knockdown reduced expression of RXRγ and TRα (Figure 5C). Further, RXRγ and TRα knockdown were performed and followed by culturing with 2% FBS for two days. Myotube formation, myogenic genes and D2 expression were decreased by RXRγ or TRα knockdown, whereas TTR and TRα expressions were increased in RXRγ$_{kd}$ cells. Most gene or protein expressions were decreased in TRα knockdown cells (Figure 5D,E). Overall, the above results indicate that expression of RXRγ is controlled by TTR via T_4 transportation into the cell during myoblast differentiation.

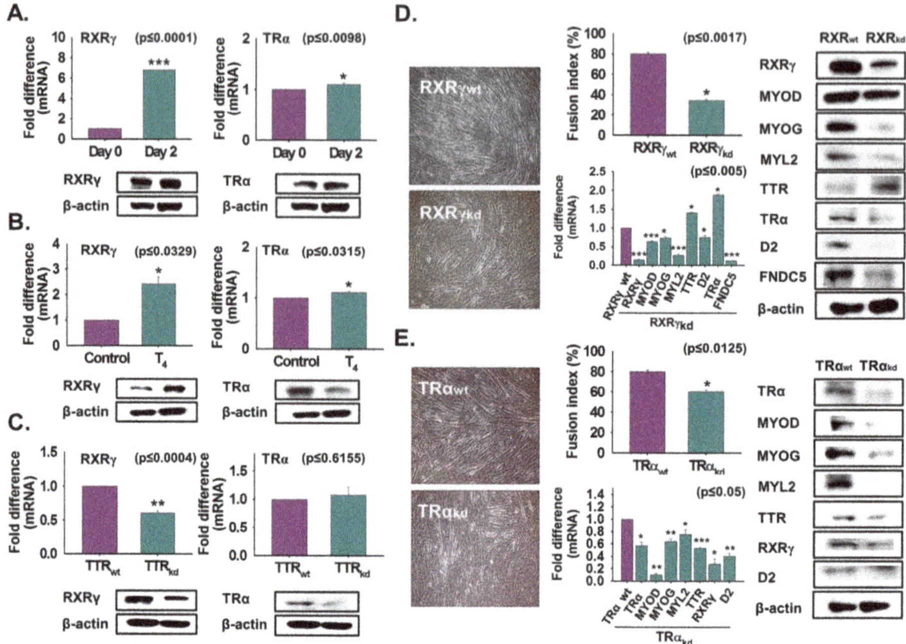

Figure 5. RXRγ and TRα expression during myoblast differentiation. (**A**) Cells were cultured with 2% FBS for two days. RXRγ and TRα expressions using real-time RT-PCR or Western blot. (**B**) Cells were cultured in serum-free media supplemented with T_4 for two days. RXRγ and TRα expression by real-time RT-PCR or Western blot. (**C**) RXRγ and TRα expression in TTR_{kd} and TTR_{wt} cells using real-time RT-PCR or Western blot. (**D**) RXRγ knockdown was performed and followed by culture with 2% FBS for two days. Myotube formation and fusion index were observed by Giemsa staining, mRNA expression by real-time RT-PCR, and protein expression by Western blot in $RXR\gamma_{kd}$ and $RXR\gamma_{wt}$ cells. (**E**) TRα knockdown was performed and followed by culture with 2% FBS for two days. Myotube formation and fusion index were observed by Giemsa staining, mRNA expression by real-time RT-PCR, and protein expression by Western blot in $TR\alpha_{kd}$ and $TR\alpha_{wt}$ cells. TTR_{wt}, $RXR\gamma_{wt}$, or $TR\alpha_{wt}$ indicate cells transfected with the scrambled vector. Means ± SD (n = 3). * $p \leq 0.05$, ** $p \leq 0.001$, *** $p \leq 0.0001$.

3.6. Relationship between TTR and D2 According to Muscle Age

To determine the effect of muscle age on TTR and D2 expression, mouse muscle at 16- and 26-weeks were collected. Myofiber size (width) and expression of TTR and D2 were decreased in 26-week muscle compared with 16-week muscle (Figure 6A). Interestingly, a decreased TTR level was observed in exosomes isolated from 26-week plasma (Figure 6A). The T_3 concentration in 16-week muscle was higher than that in 26-week muscle (Figure 6B). Further, a significant increase in the T_4 concentration in the plasma of 26-week mice the was observed, whereas there was no difference in the T_3 concentration in the plasma of either age group (Figure 6B). The above findings suggest that expressions of TTR and D2 correlate with the age-dependent differences of muscle.

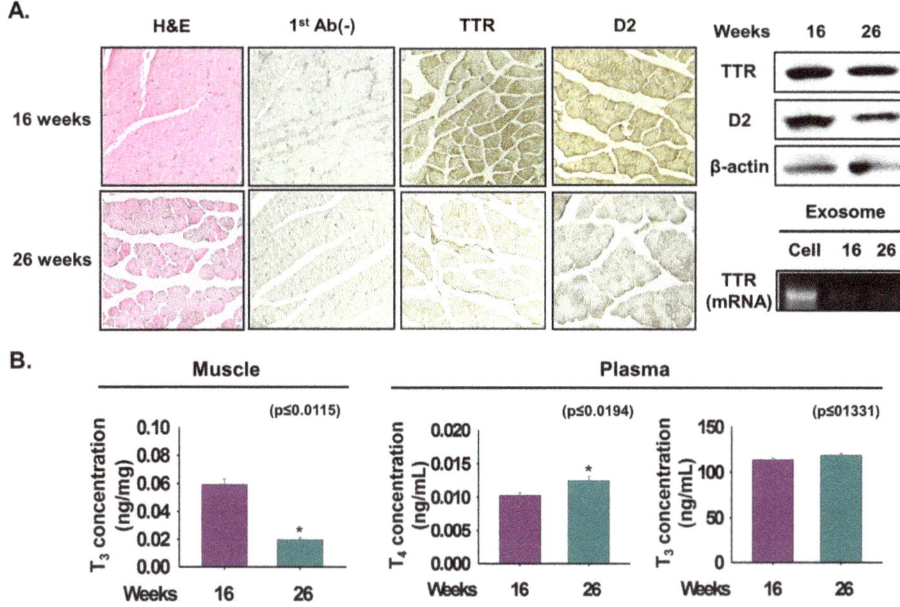

Figure 6. TTR expression and T_3 concentration in age-dependent differences of muscle.. Expression of TTR and D2 proteins were analyzed in 16- and 26-week mouse muscles. (**A**) TTR and D2 proteins expression by Immunohistochemistry and Western blot. Exosomes were isolated from 16- and 26-week plasma. TTR mRNA level in cell and exosomes of 16- or 26-week plasma by RT-PCR. (**B**) T_3 or T_4 concentration in 16- or 26-week muscles or plasma was observed by ELISA. Means ± SD (n = 3). * $p \leq 0.05$, ** $p \leq 0.001$, *** $p \leq 0.0001$.

3.7. Microarray Assessment of Gene Expression in TTR$_{kd}$ Cells and Effect of T$_4$ on Gene Expression

To explore TTR function in myoblast differentiation, TTR$_{kd}$ and TTR$_{wt}$ C2C12 cells were cultured with 2% FBS for two days. TTR/MYOG expression and myotube formation were decreased by TTR$_{kd}$ (Supplementary Figure S1A,B). Microarray analysis was performed with TTR$_{wt}$ and TTR$_{kd}$ cells. After applying two-fold cut-offs for down- and up-regulated genes, analysis of the effects of knocking down TTR on myoblasts revealed that, among the genes involved in sarcomere formation, specific genes are actively up- or down-regulated, and some novel genes that are not involved in sarcomere formation functioned at the onset of myogenesis. Among the identified genes, 29 and 7 genes were down- or up-regulated, respectively, by greater than two-fold in TTR$_{kd}$ cells (Table 1A,B; Supplementary Figure S1C,D). Many genes were previously reported to be involved in MSC maintenance (Heyl, Sox8), myogenesis (Fgf21, Ankrd2, Sox8, Asb2), proliferation (Ankrd2), myokine secretion (Fndc5), neuromuscular junction (Dok7), and Ca^{2+} release of sarcoplasmic reticulum (Asph). Even though some of these genes have identified roles in myogenesis, many novel genes were also affected by TTR$_{kd}$ (R3hdml, Inpp4b, Igf2as, Btbd17, Sema6b and Ddc) (Table 1A; Supplementary Table S4A). However, the upregulated genes were mostly involved in the cell cycle, cell proliferation, and transcription regulation. Interestingly, there was little information indicating that those genes were related to muscle differentiation. Moreover, most of the up-regulated genes were novel genes, but their main functions have been studied in other tissues or organs (Table 1B; Supplementary Table S4B).

Table 1. Microarray analysis of TTR knockdown.

							A.
Gene	**Set1**	**Set2**	**Set3**	**Set4**	**Average**	***p* Value**	**Description**
Myh1	0.16	0.31	0.13	0.06	0.16	0.0001	Mus musculus myosin, heavy polypeptide 1, skeletal muscle, adult (Myh1)
Heyl	0.25	0.19	0.15	0.1	0.17	0.0001	Mus musculus hairy/enhancer-of-split related with YRPW motif-like (Heyl)
Myo18b	0.17	0.4	0.13	0.05	0.19	0.0001	Mus musculus myosin XVIIIb (Myo18b)
Myh8	0.2	0.39	0.12	0.08	0.2	0.0001	Mus musculus myosin, heavy polypeptide 8, skeletal muscle, perinatal (Myh8)
Nmrk2	0.16	0.37	0.06	0.22	0.2	0.0001	Mus musculus nicotinamide riboside kinase 2 (Nmrk2)
Fgf21	0.14	0.35	0.06	0.32	0.22	0.0001	Mus musculus fibroblast growth factor 21 (Fgf21)
Ankrd2	0.17	0.36	0.06	0.28	0.22	0.0001	Mus musculusankyrin repeat domain 2 (stretch responsive muscle) (Ankrd2)
Myom3	0.18	0.49	0.11	0.11	0.22	0.0001	Mus musculusmyomesin family, member 3 (Myom3)
Myh3	0.18	0.39	0.14	0.21	0.23	0.0001	Mus musculus myosin, heavy polypeptide 3, skeletal muscle, embryonic (Myh3)
Myh7	0.3	0.33	0.12	0.2	0.24	0.0001	Mus musculus myosin, heavy polypeptide 7, cardiac muscle, beta (Myh7)
Myh3	0.21	0.4	0.15	0.21	0.24	0.0001	Mus musculus myosin, heavy polypeptide 3, skeletal muscle, embryonic (Myh3)
Fndc5	0.44	0.26	0.18	0.1	0.25	0.0001	Mus musculus fibronectin type III domain containing 5 (Fndc5)
R3hdml	0.35	0.4	0.09	0.15	0.25	0.0001	Mus musculus R3H domain containing-like (R3hdml)
Myh7b	0.31	0.39	0.07	0.22	0.25	0.0001	Mus musculus myosin, heavy chain 7B, cardiac muscle, beta (Myh7b)
Rbm24	0.2	0.41	0.15	0.23	0.25	0.0001	Mus musculus RNA binding motif protein 24 (Rbm24)
Sox8	0.45	0.21	0.13	0.22	0.25	0.0001	Mus musculus SRY (sex determining region Y)-box 8 (Sox8)
Dok7	0.21	0.5	0.08	0.29	0.27	0.0002	Mus musculus docking protein 7 (Dok7)
Tnnt1	0.31	0.37	0.12	0.29	0.27	0.0001	Mus musculus troponin T1, skeletal, slow (Tnnt1), transcript variant 1
Ttn	0.17	0.48	0.24	0.2	0.27	0.0001	Mus musculus titin (Ttn), transcript variant N2-B
Asph	0.28	0.46	0.18	0.16	0.27	0.0001	Mus musculus aspartate-beta-hydroxylase (Asph), transcript variant 8
Inpp4b	0.2	0.41	0.33	0.16	0.27	0.0001	Mus musculus inositol polyphosphate-4-phosphatase, type II (Inpp4b)
Igf2os	0.37	0.45	0.15	0.16	0.28	0.0001	Mus musculus insulin-like growth factor 2, opposite strand (Igf2os), antisense RNA
Myh6	0.42	0.23	0.18	0.32	0.28	0.0001	Mus musculus myosin, heavy polypeptide 6, cardiac muscle, alpha (Myh6)
Asb2	0.48	0.39	0.23	0.17	0.32	0.0001	Mus musculusankyrin repeat and SOCS box-containing 2 (Asb2)
Mybpc1	0.45	0.44	0.25	0.13	0.32	0.0001	Mus musculus myosin binding protein C, slow-type (Mybpc1)
Btbd17	0.47	0.45	0.1	0.29	0.33	0.0002	Mus musculus BTB (POZ) domain containing 17 (Btbd17)
Sema6b	0.46	0.39	0.09	0.38	0.33	0.0002	Mus musculussema domain, transmembrane domain (TM), and cytoplasmic domain
Actc1	0.38	0.39	0.33	0.25	0.34	0.0001	Mus musculus actin, alpha, cardiac muscle 1 (Actc1)
Ddc	0.48	0.48	0.22	0.39	0.39	0.0001	Mus musculusdopa decarboxylase (Ddc), transcript variant 1

							B.
Gene	**Set1**	**Set2**	**Set3**	**Set4**	**Average**	***p*Value**	**Description**
Gm10536	4.95	7.99	11.97	13.93	9.71	0.0049	Mus musculus predicted gene 10536 (Gm10536), long non-coding RNA
Iws1	3.92	4.7	2.15	10.28	5.26	0.5130	Mus musculus IWS1 homolog (S. cerevisiae) (Iws1)
Dkk2	4.02	3.54	4.95	2.98	3.87	0.0005	Mus musculusdickkopf homolog 2 (Xenopuslaevis) (Dkk2)
Cdc45	3.44	2.22	4.92	3.16	3.43	0.0048	Mus musculus cell division cycle 45 (Cdc45), transcript variant 1
Suv420h1	3.25	2.86	2.71	3.45	3.07	0.0001	Mus musculus suppressor of variegation 4–20 homolog 1 (Drosophila) (Suv420h1)
Cdc42bpa	2.28	2.11	3.11	4.32	2.95	0.0082	Mus musculus CDC42 binding protein kinase alpha (Cdc42bpa)
Zfp318	2.02	2.89	2.23	4.18	2.83	0.0094	Mus musculus zinc finger protein 318 (Zfp318), transcript variant 2

Table 1. *Cont.*

C.			
Term	**Count**	**%**	***p* Value**
Transcription regulation	3	42.9	7.5×10^{-2}
Nucleus	4	57.1	9.8×10^{-2}

D.			
Term	**Count**	**%**	***p* Value**
Muscle protein	8	28.6	1.40×10^{-13}
Thick filament	6	21.4	1.00×10^{-12}
Myosin	7	25	1.40×10^{-11}
Motor protein	7	25	5.90×10^{-9}
Actin-binding	6	21.4	8.50×10^{-6}
ATP-binding	10	35.7	1.20×10^{-5}
Calmodulin-binding	5	17.9	1.80×10^{-5}
Methylation	6	21.4	6.60×10^{-5}
Nucleotide-binding	10	35.7	8.90×10^{-5}
Coiled coil	11	39.3	1.40×10^{-3}
Cytoplasm	10	35.7	4.80×10^{-2}
Isopeptide bond	4	14.3	9.40×10^{-2}

TTR_{wt} or TTR_{kd} were cultured with 2% FBS for two days and microarray analysis was performed on TTR_{wt} or TTR_{kd}. (**A** and **B**) List of down- or up-regulated genes in TTR_{kd} (2-fold≤). (**C** and **D**) Functional analysis by DAVID (2-fold≤). TTR_{wt} indicates cells transfected with scrambled vector. Means ± SD (n = 3).

Down-regulated genes were analyzed at different myogenic times (0, 2, 4 and 6 days). Interestingly, most gene expressions were increased under myogenic conditions than that at the proliferating stage (Day 0). Similar to MYOG (the myogenic marker gene) the expression of 15 genes increased greatly during myogenic differentiation (Supplementary Figure S2). DAVID analysis was performed using the up- and down-regulated genes. More than half of the up-regulated genes were classified as transcription regulators (Table 1C), especially cell-cycle regulators. Although some of the down-regulated genes were identified as being involved in Ca^{2+}-mediated signal transduction and were reported to regulate transcription, most down-regulated genes were classified as components or regulators of the sarcomere motor unit or the ATPase-related group, which are the main structural components of the sarcomere (Table 1D).

Even though genes were selected based on their high statistical significance among all differentially expressed genes, the genes were also cross-examined by performing real-time RT-PCR with TTR_{kd} and comparing the results to those of TTR_{wt} (Figure 7A). To investigate the effect of T_4 on TR expression, cells were grown under serum-free conditions with added T_4 and/or TR-specific antagonist 1-850 and examined both for morphological appearance and for changes in mRNA expression levels of certain genes. Myotube formation and mRNA levels of the myogenic marker genes (MYOD, MYOG and MYL2) were decreased by T_4 + 1-850 treatment (Figure 7B). In contrast, T_4 treatment increased myofibril diameter. The T_4 treatment elevated most of the gene mRNA levels, whereas T_4 + 1-850 treatment had opposite effects. However, T_4 treatment reduced the suppressing effect of 1-850 on mRNA expression of Nmrk2 (40% rescue), Sox8 (40%), Myh1 (20%), and Myh8 (30%) (Figure 7C).

To determine whether the TTR_{kd} effects were produced by TH and its specific receptor, the TR binding site was scanned in genome portions containing the 5′ flanking region and the first intron of each gene. For precise analysis, two nuclear receptor scanning software programs, NHR-scan and NUBI-scan, were utilized. All binding site candidates were predicted by using the AGGTCA sequence arranged by the DR0, DR4, IR0, IR4, ER4, and ER6 patterns, as was used in the in silico thyroid hormone response elements (TRE) prediction models. Consequently, most of the genes contained more than one TRE at the 5′ flanking region. However, some genes such as Fgf21 did not have a suitable TRE. In addition, Myh1 did not possess a TRE upstream of the first exon (Supplementary Figures S3 and S4). Altogether, these results showed that T_4 transported to the cell interior activated TR to induce gene expression and modulated novel and major transcription regulating genes that markedly increased during myogenic differentiation in a TH-dependent manner.

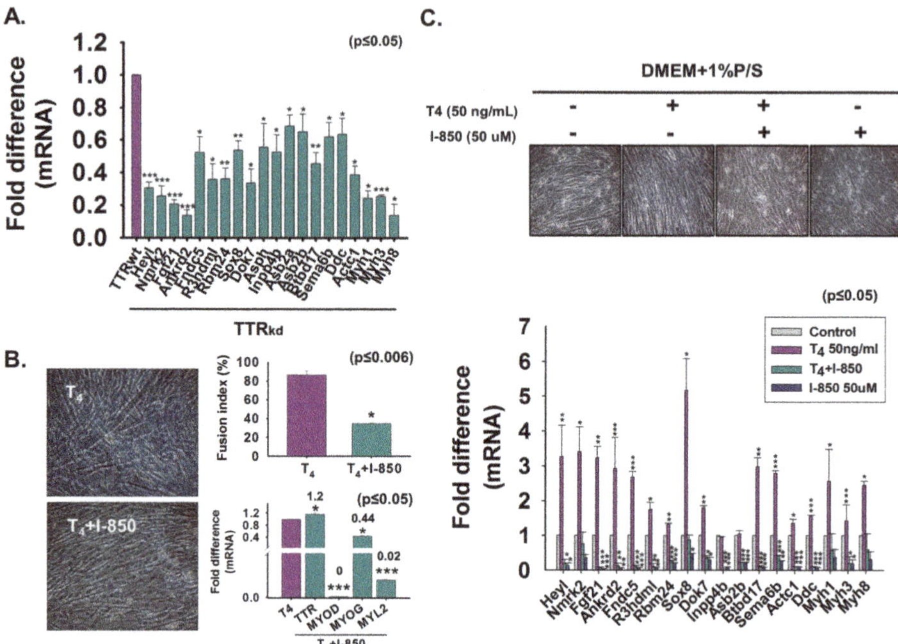

Figure 7. Expression of down-regulated genes in TTR knock-down cells and effect of T_4 treatment on down-regulated genes. (**A**) TTR$_{wt}$ or TTR$_{kd}$ were cultured with 2% FBS for two days. Down-regulated gene expression was assessed by real-time RT-PCR in TTR$_{wt}$ or TTR$_{kd}$. (**B**) Cells were cultured with serum-free media supplemented with T_4 or T_4 + 1-850 and incubated for two days. Myotube formation and fusion index were observed by Giemsa staining and mRNA expression by real-time RT-PCR. (**C**) Cells were incubated without or with T_4, T_4 + 1-850 or 1-850 for two days. Expression of down-regulated genes without or with T_4, T_4 + 1-850 or 1-850 by real-time RT-PCR. Control indicates non-treated cells. Means ± SD (n = 3). * $p \leq 0.05$, ** $p \leq 0.001$, *** $p \leq 0.0001$.

3.8. FNDC5 Expression During Myoblast Differentiation

To confirm the function of the genes that were down-regulated by TTR$_{kd}$, myokine FNDC5 was selected. FNDC5 knockdown was performed followed by culture with 2% FBS for two days. Myotube formation and myogenic gene expression were decreased in FNDC5$_{kd}$ cells, whereas TTR and TRα expressions were increased at both the transcriptional and translational levels (Figure 8A). Next, cells were grown in serum-free media or supplemented with T_4 for two days, and the FNDC5 mRNA level was analyzed in normal cells and exosomes from plasma and media of cultured cells (FNDC5$_{kd}$ and FNDC5$_{wt}$). FNDC5 mRNA was evident in exosomes from culture media and plasma and decreased in FNDC5$_{kd}$ cells (Figure 8B). Additionally, decreased FNDC5 mRNA was observed in T_4 + TTR treatment in MSCs exosomes (Figure 8B). Expression of FNDC5 was decreased in 26-week muscle compared with that in 16-week muscle (Figure 8C). These results show that FNDC5 positively regulates myoblast differentiation.

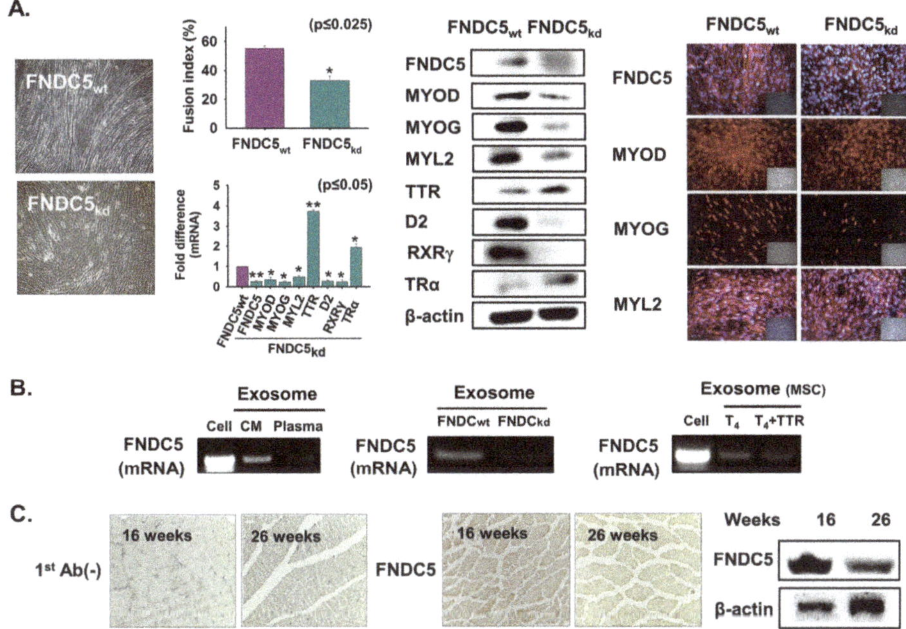

Figure 8. FNDC5 expression during myoblast differentiation. (**A**) FNDC5 knockdown was performed and cells were incubated with 2% FBS for two days. Myotube formation and fusion index were observed by Giemsa staining, mRNA expression using real-time RT-PCR and protein expression were observed by Western blot and immunocytochemistry. (**B**) Cells were cultured with only serum-free media for two days and exosomes were isolated from cultured media. FNDC5 mRNA level in normal cells, exosomes isolated from plasma, and media of cultured cells (FNDC5$_{wt}$ and FNDC5$_{kd}$). MSCs were cultured with only serum-free media or supplemented with T_4 for two days. FNDC5 mRNA level in exosomes from cell, media of cultured cells with T_4 or T_4 + TTR. (**C**) FNDC5 protein expression in 16- or 26-week muscle by immunohistochemistry and Western blot. FNDC5$_{wt}$ indicates cells transfected with scrambled vector. Means ± SD (n = 3). * $p \leq 0.05$, ** $p \leq 0.001$, *** $p \leq 0.0001$.

4. Discussion

Skeletal muscle accounts for nearly half of the body mass and represents the largest protein reservoir in the human body [31]. Although the importance of TH signaling in muscle physiology has been documented for several years, its precise mechanism in skeletal muscle during postnatal myogenesis remains unclear. Initially, we demonstrated the role of TTR in sustaining the cellular T_4 level during myoblast proliferation and differentiation [6,29]. In this study, we give the first direct evidence of TTR secretion and uptake in C2C12 mouse myoblast cells. We also identify TTR mRNA in exosomes and its increased expression following T_4 treatment, which may act as a mediator in this process. In addition, we studied the role of TTR in T_4 transport into C2C12 cells and murine MSCs during the assessment of cell viability and differentiation. The appearance of TTR in cultured serum-free media from myoblasts strongly suggests that TTR synthesized by C2C12 cell is secreted. This suggestion was confirmed by TTR immunoneutralization using TTR antibody, which demonstrated a reduction in myotube formation and mRNA level of some myogenic marker genes (especially, MYOD and MYOG), and T_4 uptake into cells, along with an increase in TTR retained in the cells. Further, it is important to emphasize the presence of T_3 in exosomes, which indicates that T_3 produced in cells is secreted out of the cell through exosomes. These results imply that muscle may not only utilize T_4 but also act as a reservoir of T_3 in order to distribute it to other tissues or to more distant sites.

Cosmo et al. reported that TH uptake by skeletal muscle can occur independently of monocarboxylate transporter 8 (Mct8). However, they found enhanced TH action, T_3 content, and glucose metabolism in Mct8 knockout mice [32]. We speculate that TTR might maintain the TH content in Mct8 knockout mice and, hence, normal muscle metabolism and development. The binding affinity of TTR for T_4 is high, hence, it serves as a primary distributor protein in muscle. We showed that TTR with T_4 treatment significantly increased cell viability and differentiation compared to that of only T_4 treated cells. This was consistent with our previous finding that TTR expression increases myoblast differentiation by increasing T_4 transport into the cell [29]. Similar to what we observed in the C2C12 cell line, the T_4 + TTR protein treated mouse MSCs also showed increased myotube formation with elevated T_4 concentration. Additionally, a progressive increase in TTR expression was observed during differentiation (day 2) in primary MSC cultures. These data confirm that TTR promotes myogenesis by enhancing the transport of T_4.

Kassem et al. showed that the availability of TTR in cerebrospinal fluid (CSF) was associated with enhanced T_4 uptake into the choroid plexus and brain and this uptake was increased in the presence of TTR [33]. Accordingly, in the present study, low T_4 concentration in media with its consequent high concentration in cells supplemented with T_4 + TTR indicated that TTR enhanced the transport of T_4 to the cell interior during myoblast viability. The enhanced cell uptake may be a simple consequence of the increased T_4 level in serum, providing a concentration gradient that promotes TTR secretion and subsequent cell uptake. Furthermore, increased uptake of TH in cells treated with both T_4 and TTR probably involves a T_4 complex with TTR, as well as passive diffusion of T_4, allowing for greater cell uptake than can be accomplished by diffusion only, which is consistent with observations in human ependymoma cells [34]. Although TTR has been reported to be the main component in maintaining high TH levels in CSF and brain [33,35], in this study we observed that TTR also sustained the TH concentration in skeletal muscle and, hence, promoted myogenesis.

TTR is one of three proteins required for T_4 transport: TBG is the major transporter and albumin has the lowest affinity, acting as the third T_4 binding protein in human plasma [25,36]. Consistent with this theory, we found that BSA reduced T_4 transport to the cells, which was also increased with TTR treatment as it has high efficiency for TH. Additionally, BSA treatment decreased myotube formation and myogenic protein expression, while TTR and D2 expressions were increased at the translational level, which might reflect the drop in the T_4 level in the cells. TH regulates several genes that are responsible for muscle development and homeostasis. Among those genes, MYOD, MYOG and contractility-determining proteins are transcriptionally regulated by TH and are important for regeneration and myogenesis [37]. MYOD expression regulated by TH is involved in the fast muscle fiber phenotype, with transcriptional stimulation of the myosin-1, myosin-2 and myosin-4 isoforms [38]. TH metabolizing enzyme D2 can activate TH by outer-ring deiodination and can influence local tissue TH levels [39]. Collectively, our findings suggest that TTR acts to maintain the TH level in myoblast cells.

Evidence of high fluorescence-labeled TTR protein levels in cells reveals that TTR was internalized into the myoblast cells. This supports previous results showing endocytosis of fluorescence-labeled TTR in ependymoma cells [34]. In other reports, ^{125}I-TTR and digoxigenin labeled TTR were internalized by an endocytic process in rat yolk sac and β-cells, respectively [40,41]. Furthermore, increased uptake of T_4 in TTR-overexpressing cells supplemented with T_4 implies that an even distribution of T_4 within the cell is not only dependent on the free fraction of T_4 in serum but also on the T_4 bound to TTR. The presence of T_4 or T_3 significantly enhanced TTR internalization in JEG-3 cells, with TTR entering the cells as a TTR-T_4 complex [27]. In addition, Divino and Schussler [42] reported increased TTR internalization in HepG2 cells with increasing amounts of T_4 and suggested that a T_4-stimulated conformational alteration in TTR somehow enhanced the uptake of TTR.

Reduced T_4 serum concentrations have been reported in old rats [43,44], though their serum T_3 level remains more controversial [43]. We show that TTR and D2 expressions with T_3 concentration have a correlation with muscle age. The reduced D2 activity is suggestive of impaired T_4 conversion in 26-week muscle. Silvestri et al. observed reduced D1 activity in 26-month-old rats relative to that in young (6- and 12-month-old male) rats [44]. Interestingly, decreased TTR expression in 26-week muscle was consistent with the decreased TH transporter Mct8 protein level in liver of 24-month-old rats [44]. Furthermore, higher plasma T_3 or T_4 concentration in 26-week muscle could be associated with the reduced free T_4 concentration in the 16-week muscle, probably due to a higher TBG expression, as described elsewhere [45]. Additionally, decreased T_3 concentration in the 26-week muscle at the cellular level was consistent with the findings of Silvestri et al. [44] in which decreased T_3 concentration was observed in 24-month-old rats. Nevertheless, T_3 generation has been observed in 11-month-old rats relative to that in seven-month-old rats [46], indicating that the mechanisms of T_3 production from T_4 in old muscle remain poorly understood.

TH is the main endocrine regulator that acts by binding to TRs and imposing a signature type of gene expression [11]. TH primarily functions either via nuclear receptor-mediated stimulation that is T_3 dependent or by switching off the gene transcription machinery [13]. In muscle, this signaling pathway is regulated by the THRA1 isoform of TR [47]. The heterodimer complex formed by the TR with RXR- binds to a TRE, leading to activation or suppression of gene transcription [13]. Accordingly, we showed that T_4 treatment induced RXRγ expression. However, myotube formation and myogenic factors were decreased in RXRγ and TRα knockdown cells. Interestingly, RXRγ knockout mice are unable to increase their mass in response to high-fat feeding, suggesting a specific effect of RXRγ in skeletal muscle [48]. In muscle, the proteins whose expression are transcriptionally controlled by T_3 are SERCA1a [12], SERCA2a [49], uncoupling protein 3 (UCP3) [50], GLUT4 [51], cytosolic malic enzyme (ME1) [52], muscle glycerol-3-phosphate dehydrogenase (mGPDH) [53], and myosin-7 [54]. Furthermore, we found that TTR and D2 expression were decreased in $TR\alpha_{kd}$ cells, which explains the retarded transport of TH into the cell. The selective functions of TRs are controlled by local ligand availability [39,55] or by TH transport to the cell interior via Mct8 or other associated transporters [56]. The TH metabolizing enzymes D2 and D3, as well as transporters Mct8 and Mct10, are expressed in both rodent and human skeletal muscle [57,58].

The TTR-affected genes identified by TTR_{kd}-based microarray analysis included important transcription factors or mediators that have the potential to control several other genes. For example, Rbm24 is reported to regulate MYOG expression [59] and mediate skeletal muscle-specific splicing events [60]. In contrast, Sox8, a negative regulator of myogenesis [61], has increased expression during myogenesis of C2C12 cells. In addition, Sox8 and Heyl genes are marker genes of MSCs [61,62]; however, the Heyl gene showed increased expression during myogenic differentiation. Interestingly, those opposing results were also observed for Nmrk2 [63]. Another research group reported that Ddc is not produced by myotubes [64], but in the present study, it was induced by suitable myogenic differentiation. Altogether, some genes that have been reported to be negatively correlated with myogenesis were markedly increased in expression during myogenic differentiation in this study.

Another interesting observation from the time-course expression study is that several novel genes that show increased expression during myogenesis responded to T_4 as they did to TTR. However, Inpp4b and Asb2, genes that contain TREs in proximity to the transcription start site (TSS), did not show any change with T_4 treatment. In the case of Inpp4b, TREs in the proximity of the promoter were only downstream of the TSS, and the first intron was approximately 130 kb. This characteristic indicates a rare aspect of the TTR_{kd}-affected genes. Moreover, Fgf21 and Myh1, which do not seem to contain TREs, showed increased expression levels. The various TRE elements have only been predicted by a one-dimensional arrangement, moreover, a proper, precise, and complete nucleotide matrix for this one-dimensional arrangement is not present in public databases. Due to these limitations, many other researchers [65,66] have reported different nucleotide matrices for TREs and different reactivity of each.

Interaction and cooperation between TR and the mammalian insulator CCCTC-binding factor have already been reported [67,68]. An insulator can mediate multi-dimensional chromosomal changes [69]. In addition, based on the results of the TTR_{kd} microarray analysis, the T_4 affected sarcomere genes Myh1, Myh3 and Myh8 may be suitable candidates for TR-insulator mediated transcriptional regulation. In the case of the Myh1 gene, no TREs were present in its promoter region.

In contrast, the FNDC5 gene, downregulated by TTR_{kd}, also showed a high expression level during myogenic differentiation and after T_4 treatment. The FNDC5 gene encodes the irisin protein, which is considered as a circulating myokine. The most remarkable feature of the FNDC5/irisin protein is that it generates brown fat from white fat [70,71]. Recently, it has been shown that irisin injection stimulated muscle hypertrophy and increased regeneration in injured skeletal muscle [72]. Additionally, enhanced irisin levels have been found during myogenic differentiation and the additional irisin enhances the expression of p-Erk, which has a vital role in the protein synthesis pathway [73]. Thus, knockdown of the FNDC5 gene was undertaken. We showed that interruption of the FNDC5 gene produced a low level of myotube formation. In humans, FNDC5 protein is cleaved to provide detectable irisin levels in circulation. Additionally, increased irisin concentrations occur in response to exercise in humans [74]. Therefore, based on the pro-myogenic role of FNDC5 in the present study, we suggest that FNDC5 may be a potential curative target for the intrusion of muscle dystrophy. Thus, we conclude that one control pathway within TTR myogenesis is mediated by the protein FNDC5.

5. Conclusions

In conclusion, these results suggest that: (1) a portion of the extracellular T_4 enters myoblasts or myocytes via MCT via passive diffusion and is converted to T_3 by the D2 enzyme which, in turn, induces the expression of several genes including TTR; (2) synthesized TTR exocytoses the cell through exosomes; (3) TTR brings T_4 inside the cells as a TTR-T_4 complex through an endocytic mechanism; (4) intracellularly synthesized T_3 can exocytose via exosomes (Figure 9A); and (5) TTR, through the action of T_3 converted from T_4, regulates gene expression of TTR intermediates, such as RXRγ and FNDC5 (irisin), which ultimately induces myogenesis (Figure 9B). In this study, we have shown that muscle cells use a much more active mechanism than previously thought to bring T_4 into cells. Moreover, intracellularly-generated T_3, besides being used in the target muscle cells, also moves out of the cell and affects adjacent cells as well as probably other tissues. Herein, we propose a novel mechanism for the uptake and release of T_4 and T_3 in myoblasts and for TTR to act as a sensor for intracellular T_4 during myogenesis. However, this study has presented a most rudimentary picture of T_4 and T_3 transport into and out of muscle cells, and further studies will undoubtedly reveal more detailed mechanisms.

A.

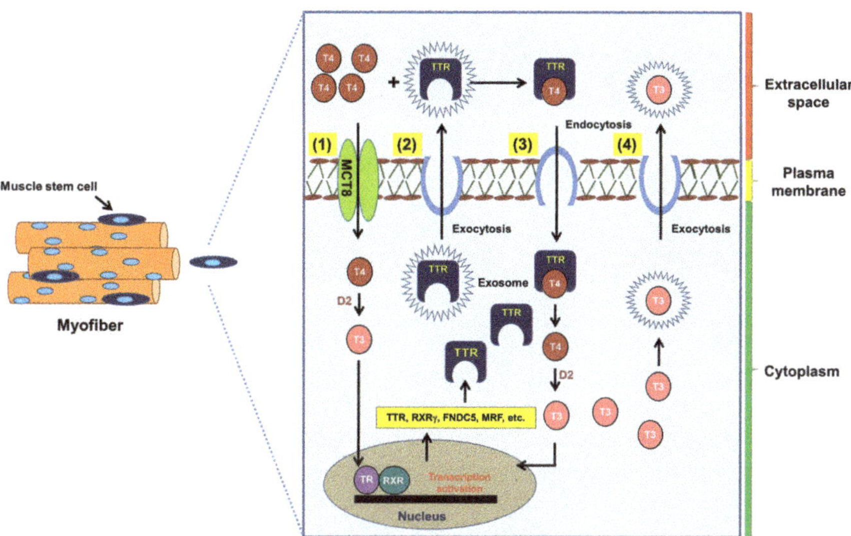

B.

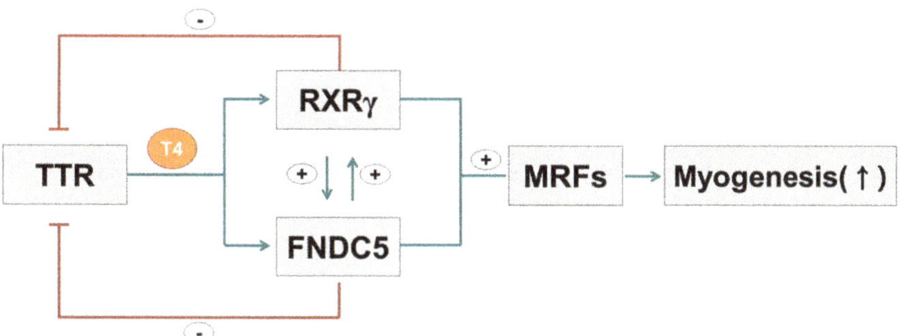

Figure 9. Hypothesis for the role of TTR with T_4 during myoblast differentiation. (**A**) Hypothetical figure depicting role of TTR with T_4 during myoblast differentiation. (1) T_4 enters cells via Mct8 by passive diffusion and is converted to T_3 by D2 enzyme, which in turn triggers the expression of several genes including TTR. (2) Synthesized TTR is exocytosed through exosomes, and (3) subsequently enters the cells as TTR-T_4 complex via an endocytic mechanism. (4) T_3 produced in the cells can exocytose via exosomes. (**B**) TTR positively regulates RXRγ and FNDC5 and triggers myogenic regulatory factors, hence promoting myogenesis. RXRγ and FNDC5 negatively regulate TTR while RXRγ and FNDC5 regulate each other.

Supplementary Materials: The following are available online at http://www.mdpi.com/2073-4409/8/12/1565/s1, Figure S1: Microarray analysis of TTR_{kd} cells, Figure S2: Time-course study of down-regulated genes during myoblast differentiation, Figure S3: Promoter of down-regulated genes was analyzed to predict TRE binding site, Figure S4: Promoter of down-regulated genes was analyzed to predict TRE binding site, Table S1: shRNA information, Table S2: Primer information, Table S3: Molecular weight of protein, Table S4: Functional analysis of up- or down-regulated genes affected by TTR_{kd}.

Author Contributions: Conceptualization: E.J.L. and I.C.; formal analysis: Y.-W.K. and I.C.; funding acquisition: E.J.L. and I.C.; investigation: E.J.L. and D.C.; methodology: J.H.L., Y.-H.L. and S.-Y.P.; resources: S.J.P. and S.-Y.P.; writing—original draft: E.J.L., S.S. and I.C.; writing—review and editing: E.J.L., S.S., K.A. and M.H.B.

Funding: This research was supported by the National Research Foundation of Korea (NRF) funded by the Korean government (MSIP: grant no. NRF-2018R1A2B6001020) and a grant from the Next-Generation BioGreen 21 Program (project no. PJ01324701), Rural Development Administration, Republic of Korea.

Conflicts of Interest: The authors declare that they have no conflict of interest.

Abbreviations

TTR	Transthyretin
RXR	Retinoid X receptor
MSCs	Muscle satellite cells
MYOG	Myogenin
T_4	Thyroxin
T_3	Triiodothyronine
THs	Thyroid hormones
TR	Thyroid hormone receptors
MYOD	Myoblast determination protein
TBG	Thyroxine binding globulin
FBS	Fetal bovine serum
P/S	Penicillin/Streptomycin
D2	Iodothyronine deiodinase type 2
TRE	Thyroid hormone response elements
Mct8	Monocarboxylate transporter 8
TSS	Transcription start site
CSF	Cerebrospinal fluid
BSA	Bovine serum albumin
FNDC5	Fibronectin type III domain containing 5
CM	Culture media MYL2 (Myosin light chain 2)

References

1. Blau, H.M.; Cosgrove, B.D.; Ho, A.T. The central role of muscle stem cells in regenerative failure with aging. *Nat. Med.* **2015**, *21*, 854–862. [CrossRef] [PubMed]

2. Ahmad, K.; Lee, E.J.; Moon, J.S.; Park, S.Y.; Choi, I. Multifaceted Interweaving between Extracellular Matrix, Insulin Resistance, and Skeletal Muscle. *Cells* **2018**, *8*, 332. [CrossRef] [PubMed]

3. Baig, M.H.; Jan, A.T.; Rabbani, G.; Ahmad, K.; Ashraf, J.M.; Kim, T.; Min, H.S.; Lee, Y.H.; Cho, W.K.; Ma, J.Y.; et al. Methylglyoxal and Advanced Glycation End products: Insight of the regulatory machinery affecting the myogenic program and of its modulation by natural compounds. *Sci. Rep.* **2017**, *7*, 5916. [CrossRef] [PubMed]

4. Zhang, K.; Zhang, Y.; Gu, L.; Lan, M.; Liu, C.; Wang, M.; Su, Y.; Ge, M.; Wang, T.; Yu, Y.; et al. Islr regulates canonical Wnt signaling-mediated skeletal muscle regeneration by stabilizing Dishevelled-2 and preventing autophagy. *Nat. Commun.* **2018**, *9*, 5129. [CrossRef]

5. Lee, E.J.; Jan, A.T.; Baig, M.H.; Ashraf, J.M.; Nahm, S.S.; Kim, Y.W.; Park, S.Y.; Choi, I. Fibromodulin: A master regulator of myostatin controlling progression of satellite cells through a myogenic program. *FASEB J.* **2016**, *30*, 2708–2719. [CrossRef]

6. Lee, E.J.; Bhat, A.R.; Kamli, M.R.; Pokharel, S.; Chun, T.; Lee, Y.H.; Nahm, S.S.; Nam, J.H.; Hong, S.K.; Yang, B.; et al. Transthyretin is a key regulator of myoblast differentiation. *PLoS ONE* **2013**, *8*, e63627. [CrossRef]

7. Mishra, A.; Zhu, X.G.; Ge, K.; Cheng, S.Y. Adipogenesis is differentially impaired by thyroid hormone receptor mutant isoforms. *J. Mol. Endocrinol.* **2010**, *44*, 247–255. [CrossRef]

8. Ambrosio, R.; De Stefano, M.A.; Di Girolamo, D.; Salvatore, D. Thyroid hormone signaling and deiodinase actions in muscle stem/progenitor cells. *Mol. Cell Endocrinol.* **2017**, *459*, 79–83. [CrossRef]

9. Milanesi, A.; Lee, J.W.; Yang, A.; Liu, Y.Y.; Sedrakyan, S.; Cheng, S.Y.; Perin, L.; Brent, G.A. Thyroid Hormone Receptor Alpha is Essential to Maintain the Satellite Cell Niche During Skeletal Muscle Injury and Sarcopenia of Aging. *Thyroid* **2017**, *27*, 1316–1322. [CrossRef]

10. Soukup, T.; Smerdu, V. Effect of altered innervation and thyroid hormones on myosin heavy chain expression and fiber type transitions: A mini-review. *Histochem. Cell Biol.* **2015**, *143*, 123–130. [CrossRef]

11. Salvatore, D.; Simonides, W.S.; Dentice, M.; Zavacki, A.M.; Larsen, P.R. Thyroid hormones and skeletal muscle—new insights and potential implications. *Nat. Rev. Endocrinol.* **2014**, *10*, 206–214. [CrossRef] [PubMed]

12. Simonides, W.S.; Brent, G.A.; Thelen, M.H.; van der Linden, C.G.; Larsen, P.R.; van Hardeveld, C. Characterization of the promoter of the rat sarcoplasmic endoplasmic reticulum Ca2+-ATPase 1 gene and analysis of thyroid hormone responsiveness. *J. Biol. Chem.* **1996**, *271*, 32048–32056. [CrossRef] [PubMed]

13. Brent, G.A. Mechanisms of thyroid hormone action. *J. Clin. Investig.* **2012**, *122*, 3035–3043. [CrossRef] [PubMed]

14. Lazar, M.A. Thyroid hormone receptors: Multiple forms, multiple possibilities. *Endocr. Rev.* **1993**, *14*, 184–193. [PubMed]

15. Weiss, R.E.; Murata, Y.; Cua, K.; Hayashi, Y.; Seo, H.; Refetoff, S. Thyroid hormone action on liver, heart, and energy expenditure in thyroid hormone receptor beta-deficient mice. *Endocrinology* **1998**, *139*, 4945–4952. [CrossRef] [PubMed]

16. Muscat, G.E.; Mynett-Johnson, L.; Dowhan, D.; Downes, M.; Griggs, R. Activation of myoD gene transcription by 3,5,3'-triiodo-L-thyronine: A direct role for the thyroid hormone and retinoid X receptors. *Nucleic Acids Res.* **1994**, *22*, 583–591. [CrossRef]

17. Leid, M.; Kastner, P.; Lyons, R.; Nakshatri, H.; Saunders, M.; Zacharewski, T.; Chen, J.Y.; Staub, A.; Garnier, J.A.; Mader, S.; et al. Purification, cloning, and RXR identity of the HeLa cell factor with which RAR or TR heterodimerizes to bind target sequences efficiently. *Cell* **1992**, *68*, 377–395. [CrossRef]

18. Mangelsdorf, D.J.; Borgmeyer, U.; Heyman, R.A.; Zhou, J.Y.; Ong, E.S.; Oro, A.E.; Kakizuka, A.; Evans, R.M. Characterization of three RXR genes that mediate the action of 9-cis retinoic acid. *Genes Dev.* **1992**, *6*, 329–344. [CrossRef]

19. Simpson, R.J.; Jensen, S.S.; Lim, J.W. Proteomic profiling of exosomes: Current perspectives. *Proteomics* **2008**, *8*, 4083–4099. [CrossRef]

20. Valadi, H.; Ekström, K.; Bossios, A.; Sjöstrand, M.; Lee, J.J.; Lötvall, J.O. Exosome-mediated transfer of mRNAs and microRNAs is a novel mechanism of genetic exchange between cells. *Nat. Cell Biol.* **2007**, *9*, 654–659. [CrossRef]

21. Jan, A.T.; Malik, M.A.; Rahman, S.; Yeo, H.R.; Lee, E.J.; Abdullah, T.S.; Choi, I. Perspective Insights of Exosomes in Neurodegenerative Diseases: A Critical Appraisal. *Front. Aging Neurosci.* **2017**, *9*, 317. [CrossRef] [PubMed]

22. Johnson, S.M.; Connelly, S.; Fearns, C.; Powers, E.T.; Kelly, J.W. The transthyretin amyloidoses: From delineating the molecular mechanism of aggregation linked to pathology to a regulatory-agency-approved drug. *J. Mol. Biol.* **2012**, *421*, 185–203. [CrossRef] [PubMed]

23. Richardson, S.J. Cell and molecular biology of transthyretin and thyroid hormones. *Int. Rev. Cytol.* **2007**, *258*, 137–193. [PubMed]

24. Monk, J.A.; Sims, N.A.; Dziegielewska, K.M.; Weiss, R.E.; Ramsay, R.G.; Richardson, S.J. Delayed development of specific thyroid hormone-regulated events in transthyretin null mice. *Am. J. Physiol. Endocrinol. Metab.* **2013**, *304*, E23–E31. [CrossRef] [PubMed]

25. Alshehri, B.; D'Souza, D.G.; Lee, J.Y.; Petratos, S.; Richardson, S.J. The diversity of mechanisms influenced by transthyretin in neurobiology: Development, disease and endocrine disruption. *J. Neuroendocrinol.* **2015**, *27*, 303–323. [CrossRef] [PubMed]

26. Blaner, W.S.; Bonifacio, M.J.; Feldman, H.D.; Piantedosi, R.; Saraiva, M.J. Studies on the synthesis and secretion of transthyretin by the human hepatoma cell line Hep G2. *FEBS Lett.* **1991**, *287*, 193–196. [CrossRef]

27. Landers, K.A.; McKinnon, B.D.; Li, H.; Subramaniam, V.N.; Mortimer, R.H.; Richard, K. Carrier-mediated thyroid hormone transport into placenta by placental transthyretin. *J. Clin. Endocrinol. Metab.* **2009**, *94*, 2610–2616. [CrossRef] [PubMed]

28. McKinnon, B.; Li, H.; Richard, K.; Mortimer, R. Synthesis of thyroid hormone binding proteins transthyretin and albumin by human trophoblast. *J. Clin. Endocrinol. Metab.* **2005**, *90*, 6714–6720. [CrossRef]

29. Lee, E.J.; Pokharel, S.; Jan, A.T.; Huh, S.; Galope, R.; Lim, J.H.; Lee, D.M.; Choi, S.W.; Nahm, S.S.; Kim, Y.W.; et al. Transthyretin: A Transporter Protein Essential for Proliferation of Myoblast in the Myogenic Program. *Int. J. Mol. Sci.* **2017**, *18*, 115. [CrossRef]

30. Lee, E.J.; Jan, A.T.; Baig, M.H.; Ahmad, K.; Malik, A.; Rabbani, G.; Kim, T.; Lee, I.K.; Lee, Y.H.; Park, S.Y.; et al. Fibromodulin and regulation of the intricate balance between myoblast differentiation to myocytes or adipocyte-like cells. *FASEB J.* **2018**, *32*, 768–781. [CrossRef]

31. Gonzalez-Freire, M.; Semba, R.D.; Ubaida-Mohien, C.; Fabbri, E.; Scalzo, P.; Hojlund, K.; Dufresne, C.; Lyashkov, A.; Ferrucci, L. The Human Skeletal Muscle Proteome Project: A reappraisal of the current literature. *J. Cachexia Sarcopenia Muscle* **2017**, *8*, 5–18. [CrossRef] [PubMed]

32. Di Cosmo, C.; Liao, X.H.; Ye, H.; Ferrara, A.M.; Weiss, R.E.; Refetoff, S.; Dumitrescu, A.M. Mct8-deficient mice have increased energy expenditure and reduced fat mass that is abrogated by normalization of serum T3 levels. *Endocrinology* **2013**, *154*, 4885–4895. [CrossRef] [PubMed]

33. Kassem, N.A.; Deane, R.; Segal, M.B.; Preston, J.E. Role of transthyretin in thyroxine transfer from cerebrospinal fluid to brain and choroid plexus. *Am. J. Physiol. Regul. Integr. Comp. Physiol.* **2006**, *291*, R1310–R1315. [CrossRef] [PubMed]

34. Kuchler-Bopp, S.; Dietrich, J.B.; Zaepfel, M.; Delaunoy, J.P. Receptor-mediated endocytosis of transthyretin by ependymoma cells. *Brain Res.* **2000**, *870*, 185–194. [CrossRef]

35. Chen, R.L.; Kassem, N.A.; Preston, J.E. Dose-dependent transthyretin inhibition of T4 uptake from cerebrospinal fluid in sheep. *Neurosci. Lett.* **2006**, *396*, 7–11. [CrossRef]

36. Palha, J.A. Transthyretin as a thyroid hormone carrier: Function revisited. *Clin. Chem. Lab. Med.* **2002**, *40*, 1292–1300. [CrossRef]

37. Bentzinger, C.F.; Wang, Y.X.; Dumont, N.A.; Rudnicki, M.A. Cellular dynamics in the muscle satellite cell niche. *EMBO Rep.* **2013**, *14*, 1062–1072. [CrossRef]

38. Allen, D.L.; Sartorius, C.A.; Sycuro, L.K.; Leinwand, L.A. Different pathways regulate expression of the skeletal myosin heavy chain genes. *J. Biol. Chem.* **2001**, *276*, 43524–43533. [CrossRef]

39. Bianco, A.C.; Salvatore, D.; Gereben, B.; Berry, M.J.; Larsen, P.R. Biochemistry, cellular and molecular biology, and physiological roles of the iodothyronine selenodeiodinases. *Endocr. Rev.* **2002**, *23*, 38–89. [CrossRef]

40. Dekki, N.; Refai, E.; Holmberg, R.; Kohler, M.; Jornvall, H.; Berggren, P.O.; Juntti-Berggren, L. Transthyretin binds to glucose-regulated proteins and is subjected to endocytosis by the pancreatic beta-cell. *Cell Mol. Life Sci.* **2012**, *69*, 1733–1743. [CrossRef]

41. Sousa, M.M.; Norden, A.G.; Jacobsen, C.; Willnow, T.E.; Christensen, E.I.; Thakker, R.V.; Verroust, P.J.; Moestrup, S.K.; Saraiva, M.J. Evidence for the role of megalin in renal uptake of transthyretin. *J. Biol. Chem.* **2000**, *275*, 38176–38181. [CrossRef] [PubMed]

42. Divino, C.M.; Schussler, G.C. Transthyretin receptors on human astrocytoma cells. *J. Clin. Endocrinol. Metab.* **1990**, *71*, 1265–1268. [CrossRef] [PubMed]

43. Mariotti, S.; Franceschi, C.; Cossarizza, A.; Pinchera, A. The aging thyroid. *Endocr. Rev.* **1995**, *16*, 686–715. [CrossRef] [PubMed]

44. Silvestri, E.; Lombardi, A.; de Lange, P.; Schiavo, L.; Lanni, A.; Goglia, F.; Visser, T.J.; Moreno, M. Age-related changes in renal and hepatic cellular mechanisms associated with variations in rat serum thyroid hormone levels. *Am. J. Physiol. Endocrinol. Metab.* **2008**, *294*, E1160–E1168. [CrossRef]

45. Savu, L.; Vranckx, R.; Rouaze-Romet, M.; Maya, M.; Nunez, E.A.; Treton, J.; Flink, I.L. A senescence up-regulated protein: The rat thyroxine-binding globulin (TBG). *Biochim. Biophys. Acta* **1991**, *1097*, 19–22. [CrossRef]

46. Jang, M.; DiStefano, J.J., 3rd. Some quantitative changes in iodothyronine distribution and metabolism in mild obesity and aging. *Endocrinology* **1985**, *116*, 457–468. [CrossRef]

47. Van Mullem, A.; van Heerebeek, R.; Chrysis, D.; Visser, E.; Medici, M.; Andrikoula, M.; Tsatsoulis, A.; Peeters, R.; Visser, T.J. Clinical phenotype and mutant TRalpha1. *N. Engl. J. Med.* **2012**, *366*, 1451–1453. [CrossRef]

48. Haugen, B.R.; Jensen, D.R.; Sharma, V.; Pulawa, L.K.; Hays, W.R.; Krezel, W.; Chambon, P.; Eckel, R.H. Retinoid X receptor gamma-deficient mice have increased skeletal muscle lipoprotein lipase activity and less weight gain when fed a high-fat diet. *Endocrinology* **2004**, *145*, 3679–3685. [CrossRef]

49. Hartong, R.; Wang, N.; Kurokawa, R.; Lazar, M.A.; Glass, C.K.; Apriletti, J.W.; Dillmann, W.H. Delineation of three different thyroid hormone-response elements in promoter of rat sarcoplasmic reticulum Ca2+ATPase gene. Demonstration that retinoid X receptor binds 5′ to thyroid hormone receptor in response element 1. *J. Biol. Chem.* **1994**, *269*, 13021–13029.

50. Solanes, G.; Pedraza, N.; Calvo, V.; Vidal-Puig, A.; Lowell, B.B.; Villarroya, F. Thyroid hormones directly activate the expression of the human and mouse uncoupling protein-3 genes through a thyroid response element in the proximal promoter region. *Biochem. J.* **2005**, *386*, 505–513. [CrossRef]

51. Zorzano, A.; Palacin, M.; Guma, A. Mechanisms regulating GLUT4 glucose transporter expression and glucose transport in skeletal muscle. *Acta Physiol. Scand.* **2005**, *183*, 43–58. [CrossRef] [PubMed]

52. Desvergne, B.; Petty, K.J.; Nikodem, V.M. Functional characterization and receptor binding studies of the malic enzyme thyroid hormone response element. *J. Biol. Chem.* **1991**, *266*, 1008–1013. [PubMed]

53. Dummler, K.; Muller, S.; Seitz, H.J. Regulation of adenine nucleotide translocase and glycerol 3-phosphate dehydrogenase expression by thyroid hormones in different rat tissues. *Biochem. J.* **1996**, *317*, 913–918. [CrossRef] [PubMed]

54. Morkin, E. Control of cardiac myosin heavy chain gene expression. *Microsc. Res. Tech.* **2000**, *50*, 522–531. [CrossRef]

55. Gereben, B.; Zavacki, A.M.; Ribich, S.; Kim, B.W.; Huang, S.A.; Simonides, W.S.; Zeold, A.; Bianco, A.C. Cellular and molecular basis of deiodinase-regulated thyroid hormone signaling. *Endocr. Rev.* **2008**, *29*, 898–938. [CrossRef] [PubMed]

56. Visser, W.E.; Friesema, E.C.; Visser, T.J. Minireview: Thyroid hormone transporters: The knowns and the unknowns. *Mol. Endocrinol.* **2011**, *25*, 1–14. [CrossRef] [PubMed]

57. Friesema, E.C.; Jansen, J.; Jachtenberg, J.W.; Visser, W.E.; Kester, M.H.; Visser, T.J. Effective cellular uptake and efflux of thyroid hormone by human monocarboxylate transporter 10. *Mol. Endocrinol.* **2008**, *22*, 1357–1369. [CrossRef]

58. Marsili, A.; Ramadan, W.; Harney, J.W.; Mulcahey, M.; Castroneves, L.A.; Goemann, I.M.; Wajner, S.M.; Huang, S.A.; Zavacki, A.M.; Maia, A.L.; et al. Type 2 iodothyronine deiodinase levels are higher in slow-twitch than fast-twitch mouse skeletal muscle and are increased in hypothyroidism. *Endocrinology* **2010**, *151*, 5952–5960. [CrossRef]

59. Jin, D.; Hidaka, K.; Shirai, M.; Morisaki, T. RNA-binding motif protein 24 regulates myogenin expression and promotes myogenic differentiation. *Genes Cells* **2010**, *15*, 1158–1167. [CrossRef]

60. Cardinali, B.; Cappella, M.; Provenzano, C.; Garcia-Manteiga, J.M.; Lazarevic, D.; Cittaro, D.; Martelli, F.; Falcone, G. MicroRNA-222 regulates muscle alternative splicing through Rbm24 during differentiation of skeletal muscle cells. *Cell Death Dis.* **2016**, *7*, e2086. [CrossRef]

61. Schmidt, K.; Glaser, G.; Wernig, A.; Wegner, M.; Rosorius, O. Sox8 is a specific marker for muscle satellite cells and inhibits myogenesis. *J. Biol. Chem.* **2003**, *278*, 29769–29775. [CrossRef] [PubMed]

62. Yamaguchi, M.; Murakami, S.; Yoneda, T.; Nakamura, M.; Zhang, L.; Uezumi, A.; Fukuda, S.; Kokubo, H.; Tsujikawa, K.; Fukada, S. Evidence of Notch-Hesr-Nrf2 Axis in Muscle Stem Cells, but Absence of Nrf2 Has No Effect on Their Quiescent and Undifferentiated State. *PLoS ONE* **2015**, *10*, e0138517. [CrossRef] [PubMed]

63. Li, J.; Mayne, R.; Wu, C. A novel muscle-specific beta 1 integrin binding protein (MIBP) that modulates myogenic differentiation. *J. Cell. Biol.* **1999**, *147*, 1391–1398. [CrossRef] [PubMed]

64. Smith, J.L.; Patil, P.B.; Minteer, S.D.; Lipsitz, J.R.; Fisher, J.S. Possibility of autocrine beta-adrenergic signaling in C2C12 myotubes. *Exp. Biol. Med.* **2005**, *230*, 845–852. [CrossRef] [PubMed]

65. Harbers, M.; Wahlstrom, G.M.; Vennstrom, B. Transactivation by the thyroid hormone receptor is dependent on the spacer sequence in hormone response elements containing directly repeated half-sites. *Nucleic Acids Res.* **1996**, *24*, 2252–2259. [CrossRef] [PubMed]

66. Weth, O.; Weth, C.; Bartkuhn, M.; Leers, J.; Uhle, F.; Renkawitz, R. Modular insulators: Genome wide search for composite CTCF/thyroid hormone receptor binding sites. *PLoS ONE* **2010**, *5*, e10119. [CrossRef]

67. Ali, T.; Renkawitz, R.; Bartkuhn, M. Insulators and domains of gene expression. *Curr. Opin. Genet. Dev.* **2016**, *37*, 17–26. [CrossRef]

68. Lutz, M.; Burke, L.J.; LeFevre, P.; Myers, F.A.; Thorne, A.W.; Crane-Robinson, C.; Bonifer, C.; Filippova, G.N.; Lobanenkov, V.; Renkawitz, R. Thyroid hormone-regulated enhancer blocking: Cooperation of CTCF and thyroid hormone receptor. *EMBO J.* **2003**, *22*, 1579–1587. [CrossRef]

69. Hou, C.; Zhao, H.; Tanimoto, K.; Dean, A. CTCF-dependent enhancer-blocking by alternative chromatin loop formation. *Proc. Natl. Acad. Sci. USA* **2008**, *105*, 20398–20403. [CrossRef]
70. Bostrom, P.; Wu, J.; Jedrychowski, M.P.; Korde, A.; Ye, L.; Lo, J.C.; Rasbach, K.A.; Bostrom, E.A.; Choi, J.H.; Long, J.Z.; et al. A PGC1-alpha-dependent myokine that drives brown-fat-like development of white fat and thermogenesis. *Nature* **2012**, *481*, 463–468. [CrossRef]
71. Roca-Rivada, A.; Castelao, C.; Senin, L.L.; Landrove, M.O.; Baltar, J.; Belen Crujeiras, A.; Seoane, L.M.; Casanueva, F.F.; Pardo, M. FNDC5/irisin is not only a myokine but also an adipokine. *PLoS ONE* **2013**, *8*, e60563. [CrossRef] [PubMed]
72. Reza, M.M.; Subramaniyam, N.; Sim, C.M.; Ge, X.; Sathiakumar, D.; McFarlane, C.; Sharma, M.; Kambadur, R. Irisin is a pro-myogenic factor that induces skeletal muscle hypertrophy and rescues denervation-induced atrophy. *Nat. Commun.* **2017**, *8*, 1104. [CrossRef] [PubMed]
73. Huh, J.Y.; Dincer, F.; Mesfum, E.; Mantzoros, C.S. Irisin stimulates muscle growth-related genes and regulates adipocyte differentiation and metabolism in humans. *Int. J. Obes.* **2014**, *38*, 1538–1544. [CrossRef] [PubMed]
74. Jedrychowski, M.P.; Wrann, C.D.; Paulo, J.A.; Gerber, K.K.; Szpyt, J.; Robinson, M.M.; Nair, K.S.; Gygi, S.P.; Spiegelman, B.M. Detection and Quantitation of Circulating Human Irisin by Tandem Mass Spectrometry. *Cell Metab.* **2015**, *22*, 734–740. [CrossRef] [PubMed]

Review

Role of Insulin-Like Growth Factor Receptor 2 across Muscle Homeostasis: Implications for Treating Muscular Dystrophy

Yvan Torrente *, Pamela Bella, Luana Tripodi, Chiara Villa and Andrea Farini *

Stem Cell Laboratory, Department of Pathophysiology and Transplantation, University of Milan, Unit of Neurology, Fondazione IRCCS Cà Granda Ospedale Maggiore Policlinico, Dino Ferrari Center, 20122 Milan, Italy; pamelabella@hotmail.it (P.B.); tripodiluana@libero.it (L.T.); kiaravilla@gmail.com (C.V.)

* Correspondence: yvan.torrente@unimi.it (Y.T.); farini.andrea@gmail.com (A.F.); Tel.: +39-0255033874 (Y.T.); +39-0255033852 (A.F.)

Received: 20 January 2020; Accepted: 11 February 2020; Published: 14 February 2020

Abstract: The insulin-like growth factor 2 receptor (IGF2R) plays a major role in binding and regulating the circulating and tissue levels of the mitogenic peptide insulin-like growth factor 2 (IGF2). IGF2/IGF2R interaction influences cell growth, survival, and migration in normal tissue development, and the deregulation of IGF2R expression has been associated with growth-related disease and cancer. IGF2R overexpression has been implicated in heart and muscle disease progression. Recent research findings suggest novel approaches to target IGF2R action. This review highlights recent advances in the understanding of the IGF2R structure and pathways related to muscle homeostasis.

Keywords: IGF2R; muscle homeostasis; inflammation; muscular dystrophy; pericytes

---

1. Introduction

The cation-independent mannose 6-phosphate/insulin-like growth factor 2 receptor (CI-M6P/IGF2R, hereafter IGF2R) is a type-1 transmembrane glycoprotein consisting of a large N-terminal extracytoplasmic domain, which allows it to bind to a wide variety of ligands [1,2]. The IGF2 and M6P ligands [3–5] of IGF2R have distinct but important roles in normal development and mesoderm differentiation [6]. Many studies have demonstrated the suppression action of IGF2R on insulin-like growth factor 1 receptor (IGF1R) signaling by scavenging extracellular IGF2 [7]. Furthermore, several lines of evidence demonstrate that IGF2 is highly expressed in rodent embryos, where it functions as an embryonic growth factor, while its amount is diminished at birth [8]. Smith et al. recently showed that a transgene-induced overexpression of IGF2 blocked programmed cell death, one of the main pathological features of cancer [9]. Furthermore, in some cancers such as mammary tumors, IGF2R behaves as a tumor suppressor gene [10], whereas in other cancers such as cervical tumors or glioblastomas, IGF2R acts as an oncogene [11,12]. Thus, these two traits of IGF2R might depend on cell type. Interestingly, cervical tumors and glioblastomas have common mesenchymal founders, namely myofibroblasts [13], which are also involved in muscle disease. Muscle repair is a complex and tightly regulated event that recruits different cell types, starting from macrophage and lymphocyte consecutive involvement and terminating with satellite cell (SC) activation and differentiation [14]. Among the common hallmarks of muscular dystrophy are the infiltration of immune cells into skeletal muscle fibers, and fibrotic cell proliferation [15–18].

Impaired muscle regeneration with SC pool exhaustion is considered an additional pathological feature of Duchenne muscular dystrophy (DMD) [19]. The main biological function of IGF2R is the suppression of IGF1R signaling via the deprivation of extracellular IGF2 ligands. Some studies have explained the tumor suppressive functions of IGF2R by its negative regulation of the oncogenic IGF2–IGF1R signal axis [2]. However, in muscle tissues, the IGFs bind to the IGF1R leading to conformational changes and activation of its tyrosine kinase activity, promoting muscle regeneration [20]. In injured tissues, IGF1 is secreted by SCs and mediates muscle-derived stem cell proliferation and differentiation into myoblasts, which contribute to myofiber formation for restoring normal tissue structures [21]. It has been demonstrated that prolonged expression of IGF1 causes an exaggerated protein synthesis and is responsible for muscular hypertrophy, by increasing myofiber diameter [22–25], and also preventing muscle atrophy in cases of cachexia or chronic inflammation [26]. Thus, tissue-specific IGF1 upregulation rises to the challenge of counteracting the development of muscular dystrophy.

Similar to IGF1, autocrine IGF2 is fundamental to mediate the differentiation of SCs in vitro, but little is known about its role in skeletal muscle development and regeneration in vivo [27]. The expression profile of *IGF2* is quite complicated as it depends on multiple-promoter activation, alternative translation initiation and messenger RNA (mRNA) stability. IGF2R functions as a negative regulator of IGF2 in embryonic skeletal muscles and modulates the amount of systemic IGF2 by inducing its degradation through lysosomes and clearance from the circulation [28,29].

Even if the signaling cascade that regulates the activation of IGF2 at muscular level is not determined, it has been shown that the phosphatidylinositol 3-kinase (PI3K)–the serine/threonine protein kinase B (AKT) pathway is the IGF2 downstream pathway contributing to mammalian target of rapamycin (mTOR) functions [30]. A study by Ge et al. [31] showed that the synergic activity of mTOR with miR-125b regulated *IGF2* production both transcriptionally and post-transcriptionally, and that these events positively influence myogenesis.

In muscle pathology and ageing contexts, where there is a predominant switch of the fiber phenotype from fast to slow [32,33], IGF2 was also able to orchestrate the development of fast myofibers by acting as a twitch motor unit during secondary myogenesis. The modulation of IGF2 expression had a dramatic impact on the amount of fast myofibers in the respiratory (intercostal and diaphragmatic) muscles, likely lessening cardio-respiratory dysfunction related to DMD. IGF2 targeting was suggested as a feasible therapeutic strategy, since IGF2 has a small size and consequently it could be easily distributed to a skeletal muscle target [34]. However, the IGF2R signaling pathways involved in muscle repair and disease remain to be identified.

2. Structure, Genomic Organization and Gene Imprinting of IGF2R

Imprinting genes are those whose expression is determined by one's parents. They occur in discrete clusters that are regulated by DNA elements called imprinting control regions (ICR). The two copies of one imprinted gene are characterized by methylation of cytosine–guanine base pairs. This modification originates in the paternal germ cells—after adding a methyl group, the chromatin becomes inaccessible to transcription machinery, so the gene is silenced. The IGF2/H19 locus is one of the imprinted gene clusters in human chromosome 11p15.5 or mouse distal chromosome 7 and plays a primary role in muscle cell development [35]. The expression of this cluster is regulated by a distant enhancer downstream of the H19 coding region. A recent study presents paxillin (PXN), a focal adhesion protein, as a transcriptional regulator of the IGF2 and H19 genes; in particular, it has the opposite effect on the activity of the IGF2 and H19 promoters [36]. The knockdown of PXN in human HepG2 cells allows for an increase in the activity of the H19 promoter and at the same time a decrease in the activity of the IGF2 promoter [36]. In a recent study, it was demonstrated that the loss of imprinting (LOI) in a mouse model of Beckwith–Wiedemann syndrome (BWS) results in impaired muscle differentiation and hypertrophy. It was also proposed that there is a signaling pathway in which IGF2 overexpression allows for an overactivation of mitogen-activated protein kinase (MAPK) signaling, while a loss of H19 long non-coding RNA (lncRNA) prevents the regulation of p53 levels, resulting in reduced AKT/mTOR signaling [35].

The *IGF2R* is an example of differential imprinting in the human and mouse genomes; the *IGF2R* is repressed on the paternal chromosome in the murine genome, but the same gene is expressed from both alleles in humans. For humans, the study was conducted on several tissues—adult liver, placenta, fetus, kidney, adrenal, brain, intestine, heart, tongue, skin and muscle: in all these samples both IGF2R alleles were expressed more or less at the same level. Accordingly, it was established that this character is subject to Mendelian segregation, escaping imprinting. As a plausible explanation for this phenomenon, Kalscheuer et al. suggested that the stages of initiation and maintenance of imprinting could be under the control of trans-acting factors that could act differently in mice compared to in humans [37]. They also hypothesized an alternative explanation based on the structural difference of the mouse and human *IGF2R* genes, as an "imprinting box" was previously discovered that could be modified by methylation in the female gamete and allowed maternal expression [37]. The *IGF2R* gene is located on mouse chromosome 17: it is composed of 48 exons and encoded for a 2482-amino acid protein. Exons 1–46 comprise the extracellular part of the receptor, while the transmembrane portion and the cytoplasmic region are located, respectively, on exon 46 and 46–48 [38].

3. IGF2R-Dependent Pathway

M6P/IGF2-R lacks intrinsic kinase activity, and the role of G-proteins in its downstream pathway has been investigated. Functionally, G-proteins are a class of proteins that interact with multi-spanning receptors (seven transmembrane receptors or heptahelical receptors). Some studies have speculated that M6P/IGF2-R, although characterized by a single-spanning structure, might initiate signaling cascades through G-proteins in a direct or indirect manner. It is well known that the pertussis toxin exerts its activity by binding and blocking the activation of the α subunit of the Gq/11-protein [39]. El-Shewy et al. showed that the use of this toxin is able to inhibit the function of M6P/IGF2-R, therefore suggesting the involvement of G-proteins in the downstream pathway of M6P/IGF2-R. Pre-treatment of Human embryonic kidney 293 cells (HEK-293) cells with the pertussis toxin can significantly reduce the level of extracellular signal-regulated kinase 1/2 (ERK1/2) phosphorylation resulting from the interaction of IGFs with their receptors. The indirect activity of these receptors is carried on through parallel activation of G-protein-coupled receptors (GPCR). They also observed that the administration of IGF1 and IGF2 ligands activates sphingosine kinase (SK) 4, which is translocated from the cytosol to the plasma membrane. There, it promotes an increase in extra- and intracytoplasmic levels of sphingosine-1-phosphate (S1P). S1P's interaction with its G-protein-coupled receptor represents a general mechanism for indirect G-protein-dependent signaling of M6P/IGF2-R resulting in ERK1/2 phosphorylation [40]. Conversely, the direct activity of IGF2R and G-proteins was hypothesized by Nishimoto et al. [41,42]: based on their observations, a region of the cytoplasmatic domain of M6P/IGF2-R may contain aminoacidic residues (2410–2423) that directly bind and activate G-proteins, in particular, Gi-2. This is supported by evidence that the use of both human and rat antibodies targeting aminoacidic residues is able to inhibit the activation of Gi-2 resulting from M6P/IGF2-R stimulation.

4. Functions of IGF2R

4.1. IGF2R Expression Levels Regulate Cardiac Development and Remodeling

The expression of the *IGF2R* gene is particularly abundant in embryo hearts. IGF2R-deficient mice display dramatic cardiac dysfunction and heart failure development. In adults, low levels of IGF2R expression lead to heart disease and apoptosis in cardiac myocytes [43], while upregulation determines myocardial infarction, remodeling [44] and hypertrophy [45]. In particular, the IGF2R activates phospholipase C (PLC) through the heterotrimeric G-protein-coupled receptor: this interaction is mediated by αq G subunits (Gαq) that in turn allow the function of different enzymes such as the protein kinase C-α (PKC-α), Ca^{2+}-calmodulin-dependent protein kinase II (CaMKII) and ERK kinase—all upregulated in cardiac hypertrophy [46]. Alternatively, the modulation of IGF2R can enhance cardiomyocyte apoptosis and cardiac contractility by inhibiting protein kinase A (PKA) phosphorylation [47]. A recent study showed that IGF2R expression is negatively controlled by the cardioprotective heat shock transcription factor 1 (HSF1). Antitumor drugs such as doxorubicin (DOX), meanwhile, lead to high levels of IGF2R expression in cardiomyocytes, through a decrease in HSF1, and trigger cardiac apoptotic processes [48]. IGF2R expression in the heart may also mediate increased sizes of cardiomyocytes, in a manner dependent on PKA activation, and mediate atrial natriuretic peptide (ANP), calcium-dependent channels (SERCA) and phospho-troponin I signaling, as described recently by Wang et al. [49].

4.2. IGF2R Modulates Vascular Remodeling and Skeletal Muscle Growth

IGF proteins play an essential role in skeletal muscle homeostasis and vascular mechanisms involving smooth muscle cells (SMCs). The latter are driven by IGF2 in the development and migration processes occurring during vascular growth or in response to vascular damage. A study has verified the role of IGF2 in SMC migration by studying the interaction between the cellular repressor of E1A-stimulated gene (CREG) factor and M6P/IGF2-R. Specifically, when the CREG factor binds M6P/IGF2-R, it is able to inhibit the SMC proliferation process and the migration process. Further studies showed that CREG knockdown leads to an increased release of IGF2, mitigating its internalization and partly restoring the migration process of the SMCs. Accordingly, the use of an anti-human IGF2-neutralizing antibody on a CREG knockdown population promotes the inhibition (in a concentration-dependent manner) of the SMCs' migration process [50]. Despite the shortage of detailed articles, an indirect point of view of the solid connection between vascular remodeling and IGF2R is offered by Ca^{2+} signaling. A comprehensive body of literature has already demonstrated that IGF2R triggers several intracellular signaling pathways aimed at Ca^{2+} mobilization [41]. This cascade may occur through an increase in PKC-α phosphorylation or in a Gαq-dependent manner [51], such as for controlling hypertrophy in cardiac cells, or through PLC-induced interactions between IGF2R and G(i) protein, as in endothelial progenitor cells (EPCs) [52]. In the latter case, the upstream role of IGF2R possibly affects the migration, adhesion and invasion of EPCs in the neovascular zone, therefore raising the importance of IGF2R for vessel network formation, both in embryonic and in post-ischemic vasculo-genesis. Vascular SMCs, composing the medial layer of blood vessels, are also essential for the maintenance of post-natal vascular homeostasis, and their correct functionality is subordinated to Ca^{2+} signaling. Intracellular calcium is likely to regulate the mechanical properties of SMCs through a tight modulation of $\alpha5\beta1$ integrin, α-SMA and cell–ECM interactions. The Ca^{2+} dynamic across cells may activate the elasticity and the adhesion properties of vascular SMCs, with physiologically important consequences on vascular tone and resistance, and blood flow and pressure [53].

A more complex signaling pathway underlying Ca^{2+} entry and exit from cells [54] is tuned by ATP-dependent pumps, which counterbalance the calcium ion levels and the electrolyte homeostasis. Among these pumps, sarcoplasmic/endoplasmic reticulum (SR) Ca^{2+}ATPase (SERCA) is the one in charge of the Ca^{2+} homeostasis within the reticulum, with a role susceptible to the type of cells [55,56]. There are three isoforms of SERCA characterized by tissue-specific expression. Briefly, the fast twitch skeletal muscle isoform SERCA1 has been found in the heart, and, to a lesser extent, in the liver, kidney, brain and pancreas [57]. Variants of SERCA2 have been detected preferentially in cardiac, skeletal and vascular smooth muscle [58]. SERCA3 isoforms are heterogeneously expressed through tissues and, conversely to the others, are characterized by a low affinity to Ca^{2+} [59]. Altered levels of SERCA proteins lead to aberrant calcium flux dynamics, which are responsible for the reduced contractility and dysfunction of SMCs observed in numerous diseases including cardiomyopathies, atherosclerosis, metabolic syndromes, and diabetes [60,61]. Although this evidence implies a connection between SERCA, IGF2R and calcium flux, a thorough explanation of the causal relationships is yet to be provided. Recently, we have identified a possible pathway in the context of muscular dystrophies, which are often associated with Ca^{2+} dysfunction: the contractile function of muscle fibers is dependent on the expression of many proteins involved in the calcium cycle between the cytosol and SR. Ca^{2+} signaling includes the ryanodine receptor, which is the SR Ca^{2+} release channel; the troponin protein complex, which leads to contraction; the extracellular Ca^{2+} reuptake pump, which mediates the flux of Ca^{2+} into the SR by a mechanism called store-operated Ca^{2+} entry (SOCE); and calsequestrin, the Ca^{2+}

storage protein in the SR. In addition, several Ca^{2+}-binding proteins are present in muscle tissue such as calmodulin, annexins, myosin, calcineurin and calpain [62]. In our study, we found that IGF2R expression is increased in dystrophic muscles, and IGF2R and the store-operated Ca^{2+} channel CD20 share a common hydrophobic binding motif stabilizing their association. We verified that the intravenous administration of an anti-IGF2R-neutralizing antibody facilitates IGF2/IGF1R interactions, while the occurrence of CD20 phosphorylation activation promotes the entrance of Ca^{2+} ions into the sarcoplasm [63]. Based on this evidence, we proposed a signaling pathway to explain the activation of SERCA and the Ca^{2+} flux through cells. Among the pathway proteins, STIM1 and ORAI1 are engaged in SOCE regulation and activation. STIM1 acts as a sensor of Ca^{2+} concentration in cellular stores and undergoes a horizontal movement in the SR membrane when the ER is calcium-depleted. Due to this shifting, STIM1 interacts with the membrane channel protein ORAI1 and causes calcium to enter the cell. After the replenishment of calcium stores, the ORAI1/STIM1 interaction dissolves and the Ca^{2+} influx stops. CD20 phosphorylation decreases the interaction between CD20 and ORAI1 in store-depleted myoblasts, largely in anti-IGF2R-treated myoblasts.

In dystrophic muscle, SERCA activity is reduced [64,65] leading to higher cytoplasmic levels of calcium ions. After anti-IGF2R treatment, calcium uptake is activated by CaMKII-dependent regulation of SERCA1: the blockade of IGF2R allows the activation of calcineurin, which dephosphorylates the nuclear factor of activated T cells (NFAT), which, consequently, shuttles into the nucleus, promoting activation of the genes involved in myogenic differentiation. Moreover, the binding of anti-IGF2R to domain 11 of IGF2R activates IGF2R/Gαi2 interactions, preventing the interplay with IGF2. This event increases the bioavailability of IGF2 for IGF1R and promotes the activation of PI3-K/AKT/mTOR signaling involved in expression of myogenic genes. Our results demonstrated that the blockade of IGF2R in mdx muscles rescued the murine dystrophic phenotype and increased force production. The in vivo experiments also revealed a marked vascular remodeling consisting of structural modifications resulting in higher linearization and wrapping of muscle fibers. It was conceivable that EPCs and pericyte cells accounted for the amelioration of the microvasculature and the blood supply. Adhesion and migration of EPCs could be affected directly by the IGF2R blockade, while pericytes cells, which present a contractile activity, could sense the calcium uptake activation [66]. Initially, pericytes had been discovered as mural cells able to provide capillary stability by interacting with endothelial cells. This classification was exceeded by anatomical and morphological evidence [67] demonstrating that pericytes could not only have a contractile activity, but they could regulate blood flow, capillary diameters and vascular tone [66,68,69]. In turn, pericytes' behavior in microcirculation can be regulated by upstream and downstream signaling, depending on the tissues and cell types. For instance, in healthy conditions, pericytes' coverage of the retina is essential for maintaining the contact with the endothelium and the integrity of the vascular barrier. Diabetic mice exhibit altered retinal vasal permeability caused by high levels of macrophage-secreted cathepsin D (CD). CD has been shown to disrupt the pericyte–endothelial junction either by increasing Rho/ROCK-dependent cell contractility [70] or by binding to IGF2R via 2M6P binding sites and changing PKC-α–CaMKII signaling [71].

Additional demonstrations of the association of IGF2R with insulin resistance and glucose homeostasis are based on the discovery of IGF2R genetic variants and soluble circulating IGF2R [72] in both type 1 (T1DM) and type 2 (T2DM) diabetes mellitus [73,74]. In T2DM, hyperglycemia seems to severely affect the islet capillary pericytes in terms of reduced numbers, improper islet coverage, altered calcium flux sensitivity, and relaxation phenotype shift [75]. As a response, diabetic capillaries dilate, blood pressure increases, and islets lose the ability to adapt and control their blood flow. Finally, high glucose concentrations and streptozotocin-induced diabetic conditions lead to abnormal activation of the IGF2R pathway and a downstream signaling for the expression of proteins related to hypertrophy in cardiac tissues, and to apoptosis in cardiomyocytes [76].

Taken together, this evidence suggests the importance of IGF2R in the pathogenesis or treatment of disorders related to energy metabolism, vascular remodeling and muscle homeostasis.

However, unraveling the signaling of the whole process within different tissues requires further and extensive investigation.

4.3. IGF2R Is Involved in Carcinogenesis

As described above, the proteins that constitute the IGF system—IGF1/IGF2, the surface receptors, and the IGF-binding proteins—regulate a plethora of functions related to growth and development operating through AKT1, mitogen-activated protein kinase (MAPK) and the phosphatidylinositol 3-kinase (PI3K) [77]. Consequently, dysfunction in this complex system is often associated with cancer. The upregulation of IGF2 was described in colorectal cancers (CRC) due to epigenetic mechanisms, and it was associated with poor survival [78]. In particular, IGF2 was dramatically overexpressed in tumorigenic clones related to IGF1, leading to constitutive activation of IGF1R and AKT [79] and to malignant modulation of apoptosis and stemness [80]. In addition, the loss of IGF2 imprinting was associated with esophageal adenocarcinoma [81], while the hypomethylation of one of the IGF2 promoters caused dysfunction in the transcriptional regulator Kruppel-like factor 4 (KLF4) in humane prostate cancer [82]. Genetic mutations in IGF2R can modify the bioavailability of IGF2 so that cancer cells can dramatically proliferate. Different studies demonstrated that IGF2R expression could be involved in the development of hepatocarcinoma, breast and ovarian human cancers by encoding for a tumor suppressor gene [83]. In particular, the loss of heterozygosity (LOH) at the M6P/IGF2R gene locus on 6q26–27 chromosome seemed to be associated with the invasiveness of breast cancers, while M6P/IGF2R point mutations were identified in hepatoma, gastrointestinal (mainly associated with microsatellite instability) and prostate tumors [84]. This condition likely led to uncontrolled IGF2 upregulation that enhanced cancer growth and survival by binding to IGF1R. In particular, the work of Delaine et al. [85] showed a new hydrophobic patch on the domain 11 of the IGF2R that is fundamental for the high binding affinity of IGF2/IGF2R. The first direct demonstration of IGF2R in tumor growth came from the work of Chen et al., in which they described how the downregulation of M6P/IGF2R expression in adenocarcinoma cells led to increased cell proliferation and decreased susceptibility to apoptosis, according to the bioavailability of TNFα and activated TGF-β. In addition, this condition was probably hampered by the action of IGF2 and cathepsins B and D [86]. Interestingly, ligands other than IGF2 can bind to IGF2R: among them are urokinase-type plasminogen activator receptor (uPAR) and retinoic acid, whose activity allows the internalization of IGF2 [7]. All the IGF2R-dependent pathways are summarized in Figure 1.

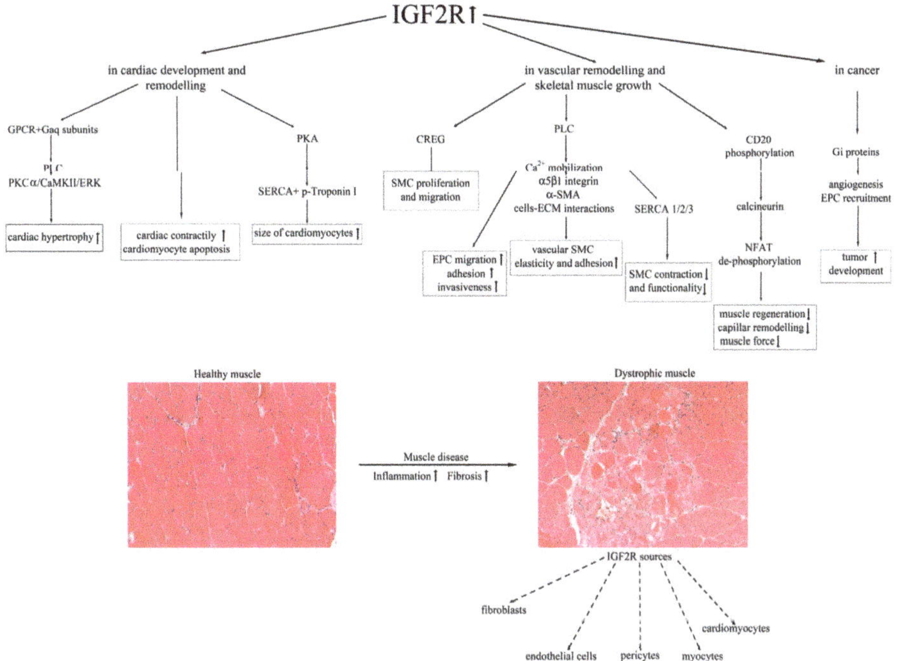

Figure 1. A schematic model of the IGF2R-dependent mechanisms leading to cardiac and skeletal muscle impairment, and cell sources of IGF2R expression in dystrophic muscle.

5. Conclusions

Fetal growth and post-natal growth are closely regulated by the insulin-like growth factor axis: alterations in the IGF signaling pathways could cause severe dysfunction in somatic growth and development and be responsible for tumor proliferation. We have recently demonstrated in mdx mice that intravenous administration of an anti-IGF2R neutralizing antibody significantly upregulated muscle regeneration and decreased fibrosis, leading to the rescue of the pathological phenotype. The inhibition of IGF2R resulted in an increase in intracellular Ca^{2+} in myoblasts and increased SERCA1 activity, possibly operating through CD20. This condition allowed NFAT dephosphorylation and its translocation into the nucleus. The dystrophic phenotype rescue activated in vivo by the anti-IGF2R antibody was further corroborated by higher numbers of structurally more linear microvessels enveloping myofibers. It is likely that anti-IGF2R acts on pericyte function, determining normalization of the vascular wall and consequent amelioration of the oxygenation of the dystrophic muscle. As contractile cells, pericytes can regulate their tone and contraction depending on the intracellular calcium concentration and consequently tune the capillary diameter and blood flow [69].

This is an exciting time in our understanding of muscular dystrophies. Increasing knowledge of IGF2R's role in muscle disease is starting to suggest new therapeutic approaches. A challenge for the future will be to understand how IGF2R interacts with other components of the muscle system to influence muscular dystrophy progression. Similarly, genetic and epigenetic changes affecting *IGF2R* need to be considered. Therefore, targeting IGF2R may be a potential therapeutic strategy for muscular dystrophies.

Author Contributions: A.F., Y.T. and C.V. wrote the paper. P.B. and L.T. prepared the original draft of the manuscript. All the authors stated were involved in the critical revision of the manuscript and approved the final version of the article, including the authorship list. The corresponding authors had final responsibility for the decision to submit for publication. All authors have read and agreed to the published version of the manuscript.

Funding: This study was supported by the Associazione Centro Dino Ferrari.

Acknowledgments: Funders of the study had no role in manuscript design.

Conflicts of Interest: The authors declare that no conflicts of interest exist.

Abbreviations

α-SMA	alpha smooth muscle actin
ANP	atrial natriuretic peptide
CaMKII	Ca^{2+}-calmodulin-dependent protein kinase II
CREG	Cellular Repressor of E1A-stimulated Gene
ECM	extracellular matrix
EPC	endothelial progenitor cell
Gαq	αq G subunits
GPCR	G-protein-coupled receptors
NFAT	nuclear factor of activated T cells
PKA	protein kinase A
PKC-α	protein kinase C-α
PLC	phospholipase C
SERCA	sarcoplasmic/endoplasmic reticulum (SR) Ca^{2+} ATPase
SMC	smooth muscle cell

References

1. Brown, J.; Jones, E.Y.; Forbes, B.E. Keeping IGF-II under control: Lessons from the IGF-II-IGF2R crystal structure. *Trends Biochem. Sci.* **2009**, *34*, 612–619. [CrossRef]
2. Livingstone, C. IGF2 and cancer. *Endocr. Relat. Cancer* **2013**, *20*, R321–R339. [CrossRef]
3. Dahms, N.M.; Brzycki-Wessell, M.A.; Ramanujam, K.S.; Seetharam, B. Characterization of mannose 6-phosphate receptors (MPRs) from opossum liver: Opossum cation-independent MPR binds insulin-like growth factor-II. *Endocrinology* **1993**, *133*, 440–446. [CrossRef] [PubMed]
4. Reddy, S.T.; Chai, W.; Childs, R.A.; Page, J.D.; Feizi, T.; Dahms, N.M. Identification of a low affinity mannose 6-phosphate-binding site in domain 5 of the cation-independent mannose 6-phosphate receptor. *J. Biol. Chem.* **2004**, *279*, 38658–38667. [CrossRef] [PubMed]
5. Williams, C.; Rezgui, D.; Prince, S.N.; Zaccheo, O.J.; Foulstone, E.J.; Forbes, B.E.; Norton, R.S.; Crosby, J.; Hassan, A.B.; Crump, M.P. Structural insights into the interaction of insulin-like growth factor 2 with IGF2R domain 11. *Structure* **2007**, *15*, 1065–1078. [CrossRef] [PubMed]
6. Morali, O.G.; Jouneau, A.; McLaughlin, K.J.; Thiery, J.P.; Larue, L. IGF-II promotes mesoderm formation. *Dev. Biol.* **2000**, *227*, 133–145. [CrossRef] [PubMed]
7. Martin-Kleiner, I.; Gall Troselj, K. Mannose-6-phosphate/insulin-like growth factor 2 receptor (M6P/IGF2R) in carcinogenesis. *Cancer Lett.* **2010**, *289*, 11–22. [CrossRef] [PubMed]
8. Rotwein, P.; Pollock, K.M.; Watson, M.; Milbrandt, J.D. Insulin-like growth factor gene expression during rat embryonic development. *Endocrinology* **1987**, *121*, 2141–2144. [CrossRef] [PubMed]
9. Smith, J.; Goldsmith, C.; Ward, A.; LeDieu, R. IGF-II ameliorates the dystrophic phenotype and coordinately down-regulates programmed cell death. *Cell Death Differ.* **2000**, *7*, 1109–1118. [CrossRef]
10. Wise, T.L.; Pravtcheva, D.D. Delayed onset of Igf2-induced mammary tumors in Igf2r transgenic mice. *Cancer Res.* **2006**, *66*, 1327–1336. [CrossRef]
11. Takeda, T.; Komatsu, M.; Chiwaki, F.; Komatsuzaki, R.; Nakamura, K.; Tsuji, K.; Kobayashi, Y.; Tominaga, E.; Ono, M.; Banno, K.; et al. Upregulation of IGF2R evades lysosomal dysfunction-induced apoptosis of cervical cancer cells via transport of cathepsins. *Cell Death Dis.* **2019**, *10*, 876. [CrossRef] [PubMed]
12. Varghese, R.T.; Liang, Y.; Guan, T.; Franck, C.T.; Kelly, D.F.; Sheng, Z. Survival kinase genes present prognostic significance in glioblastoma. *Oncotarget* **2016**, *7*, 20140–20151. [CrossRef] [PubMed]
13. Kast, R.E.; Skuli, N.; Karpel-Massler, G.; Frosina, G.; Ryken, T.; Halatsch, M.E. Blocking epithelial-to-mesenchymal transition in glioblastoma with a sextet of repurposed drugs: The EIS regimen. *Oncotarget* **2017**, *8*, 60727–60749. [CrossRef]

14. Guiraud, S.; Davies, K.E. Regenerative biomarkers for Duchenne muscular dystrophy. *Neural Regen. Res.* **2019**, *14*, 1317–1320. [CrossRef]

15. Villalta, S.A.; Nguyen, H.X.; Deng, B.; Gotoh, T.; Tidball, J.G. Shifts in macrophage phenotypes and macrophage competition for arginine metabolism affect the severity of muscle pathology in muscular dystrophy. *Hum. Mol. Genet.* **2009**, *18*, 482–496. [CrossRef]

16. Villalta, S.A.; Rosenberg, A.S.; Bluestone, J.A. The immune system in Duchenne muscular dystrophy: Friend or foe. *Rare Dis.* **2015**, *3*, e1010966. [CrossRef]

17. Kobayashi, Y.M.; Rader, E.P.; Crawford, R.W.; Campbell, K.P. Endpoint measures in the mdx mouse relevant for muscular dystrophy pre-clinical studies. *Neuromuscul. Disord.* **2012**, *22*, 34–42. [CrossRef]

18. Angelini, C.; Nardetto, L.; Borsato, C.; Padoan, R.; Fanin, M.; Nascimbeni, A.C.; Tasca, E. The clinical course of calpainopathy (LGMD2A) and dysferlinopathy (LGMD2B). *Neurol. Res.* **2010**, *32*, 41–46. [CrossRef]

19. Emery, A.E. The muscular dystrophies. *Lancet* **2002**, *359*, 687–695. [CrossRef]

20. Lawrence, M.C.; McKern, N.M.; Ward, C.W. Insulin receptor structure and its implications for the IGF-1 receptor. *Curr. Opin. Struct. Biol.* **2007**, *17*, 699–705. [CrossRef]

21. Barton-Davis, E.R.; Shoturma, D.I.; Sweeney, H.L. Contribution of satellite cells to IGF-I induced hypertrophy of skeletal muscle. *Acta Physiol. Scand.* **1999**, *167*, 301–305. [CrossRef]

22. Hennebry, A.; Oldham, J.; Shavlakadze, T.; Grounds, M.D.; Sheard, P.; Fiorotto, M.L.; Falconer, S.; Smith, H.K.; Berry, C.; Jeanplong, F.; et al. IGF1 stimulates greater muscle hypertrophy in the absence of myostatin in male mice. *J. Endocrinol.* **2017**, *234*, 187–200. [CrossRef]

23. Shavlakadze, T.; Chai, J.; Maley, K.; Cozens, G.; Grounds, G.; Winn, N.; Rosenthal, N.; Grounds, M.D. A growth stimulus is needed for IGF-1 to induce skeletal muscle hypertrophy in vivo. *J. Cell Sci.* **2010**, *123*, 960–971. [CrossRef]

24. Slusher, A.L.; Huang, C.J.; Acevedo, E.O. The Potential Role of Aerobic Exercise-Induced Pentraxin 3 on Obesity-Related Inflammation and Metabolic Dysregulation. *Mediators Inflamm.* **2017**, *2017*, 1092738. [CrossRef]

25. Yang, S.; Alnaqeeb, M.; Simpson, H.; Goldspink, G. Cloning and characterization of an IGF-1 isoform expressed in skeletal muscle subjected to stretch. *J. Muscle Res. Cell. Motil.* **1996**, *17*, 487–495. [CrossRef]

26. Clemmons, D.R. Role of IGF-I in skeletal muscle mass maintenance. *Trends Endocrinol. Metab.* **2009**, *20*, 349–356. [CrossRef]

27. Florini, J.R.; Ewton, D.Z.; Coolican, S.A. Growth hormone and the insulin-like growth factor system in myogenesis. *Endocr. Rev.* **1996**, *17*, 481–517. [CrossRef]

28. Fargeas, C.A.; Florek, M.; Huttner, W.B.; Corbeil, D. Characterization of prominin-2, a new member of the prominin family of pentaspan membrane glycoproteins. *J. Biol. Chem.* **2003**, *278*, 8586–8596. [CrossRef]

29. Spicer, L.J.; Aad, P.Y. Insulin-like growth factor (IGF) 2 stimulates steroidogenesis and mitosis of bovine granulosa cells through the IGF1 receptor: Role of follicle-stimulating hormone and IGF2 receptor. *Biol. Reprod.* **2007**, *77*, 18–27. [CrossRef]

30. Erbay, E.; Park, I.H.; Nuzzi, P.D.; Schoenherr, C.J.; Chen, J. IGF-II transcription in skeletal myogenesis is controlled by mTOR and nutrients. *J. Cell Biol.* **2003**, *163*, 931–936. [CrossRef]

31. Ge, Y.; Sun, Y.; Chen, J. IGF-II is regulated by microRNA-125b in skeletal myogenesis. *J. Cell Biol.* **2011**, *192*, 69–81. [CrossRef]

32. Deschenes, M.R. Effects of aging on muscle fibre type and size. *Sports Med.* **2004**, *34*, 809–824. [CrossRef]

33. Pedemonte, M.; Sandri, C.; Schiaffino, S.; Minetti, C. Early decrease of IIx myosin heavy chain transcripts in Duchenne muscular dystrophy. *Biochem. Biophys. Res. Commun.* **1999**, *255*, 466–469. [CrossRef]

34. Merrick, D.; Ting, T.; Stadler, L.K.; Smith, J. A role for Insulin-like growth factor 2 in specification of the fast skeletal muscle fibre. *BMC Dev. Biol.* **2007**, *7*, 65. [CrossRef]

35. Park, K.S.; Mitra, A.; Rahat, B.; Kim, K.; Pfeifer, K. Loss of imprinting mutations define both distinct and overlapping roles for misexpression of IGF2 and of H19 lncRNA. *Nucleic Acids Res.* **2017**, *45*, 12766–12779. [CrossRef]

36. Marasek, P.; Dzijak, R.; Studenyak, I.; Fiserova, J.; Ulicna, L.; Novak, P.; Hozak, P. Paxillin-dependent regulation of IGF2 and H19 gene cluster expression. *J. Cell Sci.* **2015**, *128*, 3106–3116. [CrossRef]

37. Kalscheuer, V.M.; Mariman, E.C.; Schepens, M.T.; Rehder, H.; Ropers, H.H. The insulin-like growth factor type-2 receptor gene is imprinted in the mouse but not in humans. *Nat. Genet.* **1993**, *5*, 74–78. [CrossRef]

38. Szebenyi, G.; Rotwein, P. The mouse insulin-like growth factor II/cation-independent mannose 6-phosphate (IGF-II/MPR) receptor gene: Molecular cloning and genomic organization. *Genomics* **1994**, *19*, 120–129. [CrossRef]

39. Lemamy, G.J.; Sahla, M.E.; Berthe, M.L.; Roger, P. Is the mannose-6-phosphate/insulin-like growth factor 2 receptor coded by a breast cancer suppressor gene? *Adv. Exp. Med. Biol.* **2008**, *617*, 305–310. [CrossRef]

40. El-Shewy, H.M.; Johnson, K.R.; Lee, M.H.; Jaffa, A.A.; Obeid, L.M.; Luttrell, L.M. Insulin-like growth factors mediate heterotrimeric G protein-dependent ERK1/2 activation by transactivating sphingosine 1-phosphate receptors. *J. Biol. Chem.* **2006**, *281*, 31399–31407. [CrossRef]

41. Nishimoto, I.; Hata, Y.; Ogata, E.; Kojima, I. Insulin-like growth factor II stimulates calcium influx in competent BALB/c 3T3 cells primed with epidermal growth factor. Characteristics of calcium influx and involvement of GTP-binding protein. *J. Biol. Chem.* **1987**, *262*, 12120–12126. [PubMed]

42. Nishimoto, I.; Murayama, Y.; Katada, T.; Ui, M.; Ogata, E. Possible direct linkage of insulin-like growth factor-II receptor with guanine nucleotide-binding proteins. *J. Biol. Chem.* **1989**, *264*, 14029–14038.

43. Chu, C.H.; Lo, J.F.; Hu, W.S.; Lu, R.B.; Chang, M.H.; Tsai, F.J.; Tsai, C.H.; Weng, Y.S.; Tzang, B.S.; Huang, C.Y. Histone acetylation is essential for ANG-II-induced IGF-IIR gene expression in H9c2 cardiomyoblast cells and pathologically hypertensive rat heart. *J. Cell. Physiol.* **2012**, *227*, 259–268. [CrossRef] [PubMed]

44. Chang, M.H.; Kuo, W.W.; Chen, R.J.; Lu, M.C.; Tsai, F.J.; Kuo, W.H.; Chen, L.Y.; Wu, W.J.; Huang, C.Y.; Chu, C.H. IGF-II/mannose 6-phosphate receptor activation induces metalloproteinase-9 matrix activity and increases plasminogen activator expression in H9c2 cardiomyoblast cells. *J. Mol. Endocrinol.* **2008**, *41*, 65–74. [CrossRef]

45. Chen, R.J.; Wu, H.C.; Chang, M.H.; Lai, C.H.; Tien, Y.C.; Hwang, J.M.; Kuo, W.H.; Tsai, F.J.; Tsai, C.H.; Chen, L.M.; et al. Leu27IGF2 plays an opposite role to IGF1 to induce H9c2 cardiomyoblast cell apoptosis via Galphaq signaling. *J. Mol. Endocrinol.* **2009**, *43*, 221–230. [CrossRef] [PubMed]

46. Wang, K.C.; Brooks, D.A.; Thornburg, K.L.; Morrison, J.L. Activation of IGF-2R stimulates cardiomyocyte hypertrophy in the late gestation sheep fetus. *J. Physiol.* **2012**, *590*, 5425–5437. [CrossRef]

47. Chu, C.H.; Huang, C.Y.; Lu, M.C.; Lin, J.A.; Tsai, F.J.; Tsai, C.H.; Chu, C.Y.; Kuo, W.H.; Chen, L.M.; Chen, L.Y. Enhancement of AG1024-induced H9c2 cardiomyoblast cell apoptosis via the interaction of IGF2R with Galpha proteins and its downstream PKA and PLC-beta modulators by IGF-II. *CHINESE J. Physiol.* **2009**, *52*, 31–37. [CrossRef]

48. Huang, C.Y.; Kuo, W.W.; Lo, J.F.; Ho, T.J.; Pai, P.Y.; Chiang, S.F.; Chen, P.Y.; Tsai, F.J.; Tsai, C.H.; Huang, C.Y. Doxorubicin attenuates CHIP-guarded HSF1 nuclear translocation and protein stability to trigger IGF-IIR-dependent cardiomyocyte death. *Cell Death Dis.* **2016**, *7*, e2455. [CrossRef]

49. Wang, K.C.; Brooks, D.A.; Botting, K.J.; Morrison, J.L. IGF-2R-mediated signaling results in hypertrophy of cultured cardiomyocytes from fetal sheep. *Biol. Reprod.* **2012**, *86*, 183. [CrossRef]

50. Han, Y.; Cui, J.; Tao, J.; Guo, L.; Guo, P.; Sun, M.; Kang, J.; Zhang, X.; Yan, C.; Li, S. CREG inhibits migration of human vascular smooth muscle cells by mediating IGF-II endocytosis. *Exp. Cell Res.* **2009**, *315*, 3301–3311. [CrossRef]

51. Chu, C.H.; Tzang, B.S.; Chen, L.M.; Kuo, C.H.; Cheng, Y.C.; Chen, L.Y.; Tsai, F.J.; Tsai, C.H.; Kuo, W.W.; Huang, C.Y. IGF-II/mannose-6-phosphate receptor signaling induced cell hypertrophy and atrial natriuretic peptide/BNP expression via Galphaq interaction and protein kinase C-alpha/CaMKII activation in H9c2 cardiomyoblast cells. *J. Endocrinol.* **2008**, *197*, 381–390. [CrossRef] [PubMed]

52. Maeng, Y.S.; Choi, H.J.; Kwon, J.Y.; Park, Y.W.; Choi, K.S.; Min, J.K.; Kim, Y.H.; Suh, P.G.; Kang, K.S.; Won, M.H.; et al. Endothelial progenitor cell homing: Prominent role of the IGF2-IGF2R-PLCbeta2 axis. *Blood* **2009**, *113*, 233–243. [CrossRef] [PubMed]

53. Zhu, Y.; Qu, J.; He, L.; Zhang, F.; Zhou, Z.; Yang, S.; Zhou, Y. Calcium in Vascular Smooth Muscle Cell Elasticity and Adhesion: Novel Insights Into the Mechanism of Action. *Front. Physiol.* **2019**, *10*, 852. [CrossRef] [PubMed]

54. Nance, M.E.; Whitfield, J.T.; Zhu, Y.; Gibson, A.K.; Hanft, L.M.; Campbell, K.S.; Meininger, G.A.; McDonald, K.S.; Segal, S.S.; Domeier, T.L. Attenuated sarcomere lengthening of the aged murine left ventricle observed using two-photon fluorescence microscopy. *Am. J. Physiol. Heart Circ. Physiol.* **2015**, *309*, H918–H925. [CrossRef]

55. Bers, D.M. Calcium cycling and signaling in cardiac myocytes. *Annu. Rev. Physiol.* **2008**, *70*, 23–49. [CrossRef]

56. Lipskaia, L.; Hulot, J.S.; Lompre, A.M. Role of sarco/endoplasmic reticulum calcium content and calcium ATPase activity in the control of cell growth and proliferation. *Pflug. Arch. Eur. J. Phy.* **2009**, *457*, 673–685. [CrossRef]

57. Chami, M.; Gozuacik, D.; Lagorce, D.; Brini, M.; Falson, P.; Peaucellier, G.; Pinton, P.; Lecoeur, H.; Gougeon, M.L.; le Maire, M.; et al. SERCA1 truncated proteins unable to pump calcium reduce the endoplasmic reticulum calcium concentration and induce apoptosis. *J. Cell Biol.* **2001**, *153*, 1301–1314. [CrossRef]

58. Gelebart, P.; Martin, V.; Enouf, J.; Papp, B. Identification of a new SERCA2 splice variant regulated during monocytic differentiation. *Biochem. Biophys. Res. Commun.* **2003**, *303*, 676–684. [CrossRef]

59. Bobe, R.; Bredoux, R.; Corvazier, E.; Andersen, J.P.; Clausen, J.D.; Dode, L.; Kovacs, T.; Enouf, J. Identification, expression, function, and localization of a novel (sixth) isoform of the human sarco/endoplasmic reticulum Ca2+ATPase 3 gene. *J. Biol. Chem.* **2004**, *279*, 24297–24306. [CrossRef]

60. Davies, M.G. New Insights on the Role of SERCA During Vessel Remodeling in Metabolic Syndrome. *Diabetes* **2015**, *64*, 3066–3068. [CrossRef]

61. Johny, J.P.; Plank, M.J.; David, T. Importance of Altered Levels of SERCA, IP3R, and RyR in Vascular Smooth Muscle Cell. *Biophys. J.* **2017**, *112*, 265–287. [CrossRef] [PubMed]

62. Harisseh, R.; Chatelier, A.; Magaud, C.; Deliot, N.; Constantin, B. Involvement of TRPV2 and SOCE in calcium influx disorder in DMD primary human myotubes with a specific contribution of alpha1-syntrophin and PLC/PKC in SOCE regulation. *Am. J. Physiol. Cell Physiol.* **2013**, *304*, C881–C894. [CrossRef] [PubMed]

63. Bella, P.; Farini, A.; Banfi, S.; Parolini, D.; Tonna, N.; Meregalli, M.; Belicchi, M.; Erratico, S.; D'Ursi, P.; Bianco, F.; et al. Blockade of IGF2R improves muscle regeneration and ameliorates Duchenne muscular dystrophy. *EMBO Mol. Med.* **2020**, *12*, e11019. [CrossRef] [PubMed]

64. Divet, A.; Huchet-Cadiou, C. Sarcoplasmic reticulum function in slow- and fast-twitch skeletal muscles from mdx mice. *Pflug. Arch. Eur. J. Phy.* **2002**, *444*, 634–643. [CrossRef]

65. Kargacin, M.E.; Kargacin, G.J. The sarcoplasmic reticulum calcium pump is functionally altered in dystrophic muscle. *Biochim. Biophys. Acta* **1996**, *1290*, 4–8. [CrossRef]

66. Burdyga, T.; Borysova, L. Calcium signalling in pericytes. *J. Vasc. Res.* **2014**, *51*, 190–199. [CrossRef]

67. Toribatake, Y.; Tomita, K.; Kawahara, N.; Baba, H.; Ohnari, H.; Tanaka, S. Regulation of vasomotion of arterioles and capillaries in the cat spinal cord: Role of alpha actin and endothelin-1. *Spinal cord* **1997**, *35*, 26–32. [CrossRef]

68. Borysova, L.; Wray, S.; Eisner, D.A.; Burdyga, T. How calcium signals in myocytes and pericytes are integrated across in situ microvascular networks and control microvascular tone. *Cell Calcium.* **2013**, *54*, 163–174. [CrossRef]

69. Hamilton, N.B.; Attwell, D.; Hall, C.N. Pericyte-mediated regulation of capillary diameter: A component of neurovascular coupling in health and disease. *Front. Neuroenerg.* **2010**, *2*. [CrossRef]

70. Monickaraj, F.; McGuire, P.G.; Nitta, C.F.; Ghosh, K.; Das, A. Cathepsin D: An Mvarphi-derived factor mediating increased endothelial cell permeability with implications for alteration of the blood-retinal barrier in diabetic retinopathy. *FASEB J.* **2016**, *30*, 1670–1682. [CrossRef]

71. Monickaraj, F.; McGuire, P.; Das, A. Cathepsin D plays a role in endothelial-pericyte interactions during alteration of the blood-retinal barrier in diabetic retinopathy. *FASEB J.* **2018**, *32*, 2539–2548. [CrossRef] [PubMed]

72. Chanprasertyothin, S.; Jongjaroenprasert, W.; Ongphiphadhanakul, B. The association of soluble IGF2R and IGF2R gene polymorphism with type 2 diabetes. *J. Diabetes Res.* **2015**, *2015*, 216383. [CrossRef] [PubMed]

73. McCann, J.A.; Xu, Y.Q.; Frechette, R.; Guazzarotti, L.; Polychronakos, C. The insulin-like growth factor-II receptor gene is associated with type 1 diabetes: Evidence of a maternal effect. *J. Clin. Endocrinol. Metab.* **2004**, *89*, 5700–5706. [CrossRef] [PubMed]

74. Villuendas, G.; Botella-Carretero, J.I.; Lopez-Bermejo, A.; Gubern, C.; Ricart, W.; Fernandez-Real, J.M.; San Millan, J.L.; Escobar-Morreale, H.F. The ACAA-insertion/deletion polymorphism at the 3' UTR of the IGF-II receptor gene is associated with type 2 diabetes and surrogate markers of insulin resistance. *Eur. J. Endocrinol.* **2006**, *155*, 331–336. [CrossRef] [PubMed]

75. Almaca, J.; Weitz, J.; Rodriguez-Diaz, R.; Pereira, E.; Caicedo, A. The Pericyte of the Pancreatic Islet Regulates Capillary Diameter and Local Blood Flow. *Cell Metab.* **2018**, *27*, 630–644. [CrossRef] [PubMed]

76. Feng, C.C.; Pandey, S.; Lin, C.Y.; Shen, C.Y.; Chang, R.L.; Chang, T.T.; Chen, R.J.; Viswanadha, V.P.; Lin, Y.M.; Huang, C.Y. Cardiac apoptosis induced under high glucose condition involves activation of IGF2R signaling in H9c2 cardiomyoblasts and streptozotocin-induced diabetic rat hearts. *Biomed. Pharmacother.* **2018**, *97*, 880–885. [CrossRef]

77. Kasprzak, A.; Adamek, A. Insulin-Like Growth Factor 2 (IGF2) Signaling in Colorectal Cancer-From Basic Research to Potential Clinical Applications. *Int. J. Mol. Sci.* **2019**, *20*, 4915. [CrossRef]

78. Unger, C.; Kramer, N.; Unterleuthner, D.; Scherzer, M.; Burian, A.; Rudisch, A.; Stadler, M.; Schlederer, M.; Lenhardt, D.; Riedl, A.; et al. Stromal-derived IGF2 promotes colon cancer progression via paracrine and autocrine mechanisms. *Oncogene* **2017**, *36*, 5341–5355. [CrossRef]

79. Sanderson, M.P.; Hofmann, M.H.; Garin-Chesa, P.; Schweifer, N.; Wernitznig, A.; Fischer, S.; Jeschko, A.; Meyer, R.; Moll, J.; Pecina, T.; et al. The IGF1R/INSR Inhibitor BI 885578 Selectively Inhibits Growth of IGF2-Overexpressing Colorectal Cancer Tumors and Potentiates the Efficacy of Anti-VEGF Therapy. *Mol. Cancer Ther.* **2017**, *16*, 2223–2233. [CrossRef]

80. Kessler, S.M.; Haybaeck, J.; Kiemer, A.K. Insulin-Like Growth Factor 2 - The Oncogene and its Accomplices. *Curr. Pharm. Des.* **2016**, *22*, 5948–5961. [CrossRef]

81. Zhao, R.; DeCoteau, J.F.; Geyer, C.R.; Gao, M.; Cui, H.; Casson, A.G. Loss of imprinting of the insulin-like growth factor II (IGF2) gene in esophageal normal and adenocarcinoma tissues. *Carcinogenesis* **2009**, *30*, 2117–2122. [CrossRef] [PubMed]

82. Schagdarsurengin, U.; Lammert, A.; Schunk, N.; Sheridan, D.; Gattenloehner, S.; Steger, K.; Wagenlehner, F.; Dansranjavin, T. Impairment of IGF2 gene expression in prostate cancer is triggered by epigenetic dysregulation of IGF2-DMR0 and its interaction with KLF4. *Cell Commun. Signal.* **2017**, *15*, 40. [CrossRef] [PubMed]

83. Lemamy, G.J.; Roger, P.; Mani, J.C.; Robert, M.; Rochefort, H.; Brouillet, J.P. High-affinity antibodies from hen's-egg yolks against human mannose-6-phosphate/insulin-like growth-factor-II receptor (M6P/IGFII-R): Characterization and potential use in clinical cancer studies. *Int. J. Cancer* **1999**, *80*, 896–902. [CrossRef]

84. Oates, A.J.; Schumaker, L.M.; Jenkins, S.B.; Pearce, A.A.; DaCosta, S.A.; Arun, B.; Ellis, M.J. The mannose 6-phosphate/insulin-like growth factor 2 receptor (M6P/IGF2R), a putative breast tumor suppressor gene. *Breast Cancer Res. Treat.* **1998**, *47*, 269–281. [CrossRef] [PubMed]

85. Delaine, C.; Alvino, C.L.; McNeil, K.A.; Mulhern, T.D.; Gauguin, L.; De Meyts, P.; Jones, E.Y.; Brown, J.; Wallace, J.C.; Forbes, B.E. A novel binding site for the human insulin-like growth factor-II (IGF-II)/mannose 6-phosphate receptor on IGF-II. *J. Biol. Chem.* **2007**, *282*, 18886–18894. [CrossRef]

86. Chen, Z.; Ge, Y.; Landman, N.; Kang, J.X. Decreased expression of the mannose 6-phosphate/insulin-like growth factor-II receptor promotes growth of human breast cancer cells. *BMC Cancer* **2002**, *2*, 18. [CrossRef]

Article

Extracellular Vesicles from Skeletal Muscle Cells Efficiently Promote Myogenesis in Induced Pluripotent Stem Cells

Denisa Baci [1,2], Maila Chirivì [1], Valentina Pace [1], Fabio Maiullari [3], Marika Milan [1], Andrea Rampin [1], Paolo Somma [4], Dario Presutti [1], Silvia Garavelli [5], Antonino Bruno [6], Stefano Cannata [7], Chiara Lanzuolo [8,9], Cesare Gargioli [7], Roberto Rizzi [8,9,*] and Claudia Bearzi [1,9,*]

[1] Institute of Biochemistry and Cell Biology, National Research Council, 00015 Rome, Italy; denisa.baci@uninsubria.it (D.B.); maila.chirivi@yahoo.it (M.C.); va.pace3@gmail.com (V.P.); marika.milan@ibbc.cnr.it (M.M.); rampin88@gmail.com (A.R.); presuttidario@gmail.com (D.P.)
[2] Department of Biotechnology and Life Sciences, University of Insubria, 21100 Varese, Italy
[3] Gemelli Molise Hospital, 86100 Campobasso, Italy; fabio.maiullari3d@gmail.com
[4] Flow Cytometry Core, Humanitas Clinical and Research Center, 20089 Milan, Italy; paolo.somma@humanitasresearch.it
[5] Institute for Endocrinology and Oncology "Gaetano Salvatore", National Research Council, 80131 Naples, Italy; silvia.garavelli@gmail.com
[6] IRCCS MultiMedica, 20138 Milan, Italy; antonino.bruno@multimedica.it
[7] Department of Biology, University of Rome Tor Vergata, 00133 Rome, Italy; cannata@uniroma2.it (S.C.); cesare.gargioli@uniroma2.it (C.G.)
[8] Institute of Biomedical Technologies, National Research Council, 20090 Milan, Italy; chiara.lanzuolo@cnr.it
[9] Fondazione Istituto Nazionale di Genetica Molecolare, 20122 Milan, Italy
*   Correspondence: roberto.rizzi@cnr.it (R.R.); claudia.bearzi@cnr.it (C.B.); Tel.: +39-02-0066-0230 (R.R.); +39-02-0066-0230 (C.B.)

Received: 19 April 2020; Accepted: 15 June 2020; Published: 23 June 2020

Abstract: The recent advances, offered by cell therapy in the regenerative medicine field, offer a revolutionary potential for the development of innovative cures to restore compromised physiological functions or organs. Adult myogenic precursors, such as myoblasts or satellite cells, possess a marked regenerative capacity, but the exploitation of this potential still encounters significant challenges in clinical application, due to low rate of proliferation in vitro, as well as a reduced self-renewal capacity. In this scenario, induced pluripotent stem cells (iPSCs) can offer not only an inexhaustible source of cells for regenerative therapeutic approaches, but also a valuable alternative for in vitro modeling of patient-specific diseases. In this study we established a reliable protocol to induce the myogenic differentiation of iPSCs, generated from pericytes and fibroblasts, exploiting skeletal muscle-derived extracellular vesicles (EVs), in combination with chemically defined factors. This genetic integration-free approach generates functional skeletal myotubes maintaining the engraftment ability in vivo. Our results demonstrate evidence that EVs can act as biological "shuttles" to deliver specific bioactive molecules for a successful transgene-free differentiation offering new opportunities for disease modeling and regenerative approaches.

Keywords: iPSC; extracellular vesicles; pericytes; skeletal muscle

---

1. Introduction

Skeletal muscle is a dynamic tissue with remarkable features for endogenous regeneration provided by muscle progenitors, such as satellite cells. However, in the presence of progressive muscle loss or degeneration, such as muscular dystrophies or aging, satellite cell function is largely affected due to an incorrect asymmetric division or aberrant transcriptional regulation [1–4]. Other adult progenitor cells with myogenic properties, including mesangioblasts [5], pericytes [6,7], muscle side cell population [8], interstitial cells PW1$^+$/PAX7$^-$ [9], or stem cells derived from bone marrow [10] would be considered promising candidates for muscle repair therapy. Despite this, the reduction of proliferative capacity after isolation and the progressive loss of self-renewal potential strongly limit the use of adult progenitor cells for clinical application [11].

On the other hand, induced pluripotent stem cells (iPSCs) represent a valuable source of myogenic progenitors (MPs), essential for cell-based therapy. Indeed, iPSCs not only would allow autologous transplantation but they can also be produced in large quantities, with an unlimited replication ability in vitro. Furthermore, differentiated iPSCs can be used as individual-specific tissue modeling for the validation of innovative therapies, limiting the toxic effects for the patient and providing early indications on the efficacy [12,13].

Many efforts have been made to establish efficient methods for obtaining MPs from iPSCs, mostly relying on the transgenic expression of major myogenesis regulators, such as myoblast determination protein 1 (MyoD) and Pax7 (key myogenic transcription factors) [14–17]. The main disadvantage of these approaches is that forced expression of the MyoD protein leads to cell cycle arrest along with the consequent loss of the in vitro muscle progenitor generation. The risk of unwanted genetic recombination is a widespread limiting issue for future clinical application.

The use of chemical modulators to activate relevant myogenic pathways represents a promising approach to enhance the efficiency of myogenic iPSC differentiation [18–20]. In particular, a myogenic differentiation improvement of human ESC/iPSC through the treatment with a homologous wingless and Int-1 (Wnt) agonist, the glycogen synthase kinase-3 inhibitor (GSK-3, CHIR9902), has been reported [18,19]. Early inhibition of GSK3β is mandatory for the induction of paraxial mesoderm and activation of the myogenic program [21].

In this study we explored the possibility to exploit the content of extracellular vesicles (EVs), released from differentiated myotubes (MTs), in combination with GSK3 inhibitor, in order to synergistically enhance myogenic differentiation.

To date, the scientific interest regarding the role of EVs in cell-to-cell communication, both in physiological and pathological conditions, is rapidly increasing. EVs are similar in composition to their cell of origin, and their cargo can activate signaling pathways in target cells, thus modulating their activities. In particular, the content of EVs derived from skeletal muscle plays a fundamental role for skeletal muscle homeostasis and development [22,23]. Several studies have shown that skeletal muscle cells release protein/nucleic acid complexes within microvesicles, which promote myogenesis and muscle regeneration [24–27]. EVs derived from MTs (MT-derived EVs) were found to be able to promote the differentiation of myoblasts by altering the expression of cyclin-D1 and myogenin [27]. Another study reported that exosomes, a subclass of EVs measuring approximately 100 nm in diameter, secreted during myotube differentiation, contribute significantly to the myogenic differentiation of stem cells derived from human adipose tissue [28].

Previous researches have shown that iPSCs retain molecular characteristics of the cell from which they originate, named 'epigenetic memory', which is able to strengthen the propensity for re-differentiation in the same tissue [29,30]. On the basis of this, in order to enhance muscle differentiation and exploit myogenic predisposition, muscle-derived pericytes (PCs) and skin fibroblasts (FBs) derived from the same donor were employed as cell sources for iPSC generation. PCs surround the endothelial layer of small/medium vessels that reside beneath the microvascular basement membrane. Despite their role in regulating blood flow, angiogenesis, and maintenance of vascular tissue homeostasis [31], not much is known about pericytes as a source of muscle progenitor cells [6,7]. However, several studies have shown that pericytes are strongly predisposed to differentiate into myogenic lineage and repair muscle damage [6,7,32].

In this study, we established a defined transgene-free protocol, which allows iPSCs, derived either from muscular pericytes or skin fibroblasts, to differentiate into MT-like cells when exposed to GSK-3 inhibitor and EV cargo. This combination improved the differentiation yield into muscle cells up to 70% and the fusion index. After 30 days, evidence of an enhanced muscle differentiation was further revealed by an increased expression of myogenic markers. Furthermore, we found a propensity in pericyte-derived iPSCs to re-differentiate toward the skeletal muscular fate compared to fibroblast-derived counterpart.

Finally, in a pilot study, differentiated iPSCs were injected intramuscularly into anterior tibialis (TA) muscle of immunodeficient alpha-sarcoglycan knockout (KO) mice. The differentiated cells were able to integrate into the host regenerating myofibers, revealing a possible application of the proposed method in regenerative medicine.

2. Materials and Methods

2.1. Cell Isolation

Skin and muscle specimens were obtained from 3 healthy donors, aged between 20 and 40, upon informed consent in line with the Declaration of Helsinki. Tissues were digested and muscular cell suspension was cultured in alpha Minimum Essential Medium (αMEM; Thermo Fisher Scientific, Waltham, MA, USA), 20% fetal bovine serum (FBS; Thermo Fisher Scientific), and penicillin (100 U/mL; Thermo Fisher Scientific) and streptomycin (100 µg/mL; Thermo Fisher Scientific). Pericytes were then selected by their ability to grow on plastic at low confluence ($0.1–1 \times 10^4$ cell/cm^2). Skin cells were plated in Dulbecco's Modified Eagle Medium (DMEM; Thermo Fisher Scientific) supplemented with 10% FBS, 0.5 mM β-mercaptoethanol (Thermo Fisher Scientific), 100 U/mL penicillin, and 100 µg/mL streptomycin (Thermo Fisher Scientific) at a density of 5×10^4 cells/cm^2.

2.2. Tube-Formation Assay

Pericytes and human umbilical vein endothelial cells (HUVEC; Lonza, Basel, Switzerland) were seeded in 8-well Permanox chamber slides coated with Matrigel (Becton Dickinson Franklin Lakes, NJ, USA), either separately (3.75×10^4 cells per well) or co-cultured together at a 1:4 ratio, in Endothelial Cell Growth Basal Medium-2 (EBM-2; Lonza). Cells were incubated for 5 h to allow tube formation. Images of newly formed networks were captured at 10X magnification.

All experiments were performed in duplicates. Analysis was achieved using the Angiogenesis Analyzer tool (ImageJ Software, https://imagej.nih.gov/ij/).

2.3. Pericyte and C2C12 Myogenic Differentiation

Spontaneous skeletal myogenic differentiation of human pericytes and C2C12 cells was induced by plating 10^4 cells/cm^2 in αMEM, 20% FBS and penicillin/streptomycin. After the cells reached confluence, we replaced the medium with low-serum medium (2% horse serum, HS, Thermo Fisher Scientific) for about 10 (pericyte differentiation) and 5 (C2C12 differentiation) days.

2.4. Lentiviral Vector Generation

The lentiviral vector employed for the induction of reprogramming was composed of a single excisable polycistronic lentiviral stem cell cassette (STEMCCA), encoding the Yamanaka factors [33]. Low passage 293T cells (Cell Biolabs, San Diego, CA, USA) were used to produce lentiviruses, employing the psPAX2 and vesicular stomatitis virus G protein (VSV-G) packaging constructs and a calcium phosphate transfection protocol. Supernatants containing STEMCCA lentiviruses were collected 48 h later, filtered, and used immediately right after preparation. The lentiviral vector used to introduce a Green Fluorescent Protein (GFP) transgene for the isolation of GFP$^+$-EVs was produced employing the calcium phosphate method into 293FT packaging cells.

2.5. iPSC Generation

To induce reprogramming, we exposed pericytes and fibroblasts, at early passages, to fresh lentiviral medium 3 times at 12 h intervals. Lentiviral medium was then replaced with fresh medium. After a further 5 days, 1×10^3 transduced cells/cm^2 were plated on a feeder layer constituted of inactivated mouse embryonic fibroblast (iMEF). Cells were then cultured in iPSC medium composed of knockout DMEM (Life Technologies, Carlsbad, CA, USA), supplemented with 20% knockout Serum Replacement (Life Technologies), 20 ng/mL of basic fibroblast growth factor (bFGF; Life Technologies), 1% N-2 (Life Technologies), 2% B27 (Life Technologies), 2 mM Glutamax (Life Technologies), 100 µM Eagle's minimum essential medium non-essential amino acid solution (MEM-NEAA, Life Technologies), 100 µM β-mercaptoethanol, 100 U/mL penicillin and 100 µg/mL streptomycin. After iPSC line expansion and characterization was carried out, cells were adapted to feeder-free condition, by seeding them on Geltrex matrix (Thermo Fisher Scientific) in Essential 8 medium (Life Technologies).

2.6. iPSC Multilineage Differentiation

Cardiomyocyte differentiation was performed using STEMdiff Cardiomyocyte Differentiation Kit (StemCell Technologies, Vancouver, BC, Canada) according to the manufacturer's instructions. Briefly, uniform undifferentiated iPSC colonies were harvested and seeded as single cells at 3.5×10^5 cells per well in a 12-well format. After 48 h, the iPSC medium was replaced with Medium A to induce the cells toward a cardiomyocyte fate. On day 2, a full medium change was performed with fresh Medium B. On days 4 and 6, medium B was replaced with fresh Medium C. On day 8, medium was switched to cardiomyocyte Maintenance Medium with full medium changes on days 10, 12 and 14, to promote further differentiation into cardiomyocyte cells.

Neural differentiation was promoted plating 1×10^6 cells/mL in Neural Induction Medium (Thermo Fisher Scientific) for 7 days. On day 8, iPSC-derived neural stem cells were harvested and expanded in Neural Expansion Medium (Thermo Fisher Scientific).

For endothelial differentiation, human iPSC cells were cultured in Roswell Park Memorial Institute medium (RPMI; Sigma-Aldrich, St. Louis, MO, USA) plus B27 medium with 6 µM CHIR99021 (CHIR; Sigma-Aldrich). On day 2, we replaced the medium with fresh RPMI supplemented with B27 and 2 µM CHIR. After 48 h, the medium was changed with EGM-2 medium supplied with vascular endothelial growth factor (VEGF; PeproTech, London, UK), bFGF, and SB431542 (Merck, Darmstadt, Germany). Every other day, the medium was changed with fresh EGM-2 medium supplied with VEGF, bFGF, and SB431542.

2.7. Isolation of MT-derived EVs

EVs were isolated from conditioned medium of C2C12 myoblasts, differentiated into myotubes, using HS, previously centrifuged at 100,000× g for 16 h at 4 °C for EV depletion. After 48 h of incubation in fresh medium, EVs were harvested and purified by differential centrifugation—cell debris and organelles were eliminated at 500× g for 20 min followed by another centrifugation at 3500× g for 15 min at 4 °C. EVs were pelleted by ultracentrifugation at 100,000× g for 70 min at 4 °C by L-80-XP ultracentrifuge (Beckman-Coulter, Brea, CA, USA). Finally, the pellet was washed with cold PBS (Phosphate Buffered Saline) in order to minimize sticking and trapping of non-vesicular materials. Purified EVs were used immediately after isolation.

2.8. Myogenic Differentiation by MT-Derived EVs

Human iPSCs with no differentiated colonies, expressing pluripotency markers were used for the differentiation process. The iPSCs were cultured under feeder-free conditions using Essential 8 medium on Geltrex matrix. A critical variable for the generation of robust myotube culture was the relative confluence at the onset of differentiation that it should be approximately 30%. After they were seeded for about 48 h, iPSCs were induced toward mesodermal commitment in Essential 6 medium (Life Technologies) and 1% ITS (insulin-transferrin-selenium) supplemented with 10 uM GSK3 inhibitor CHIR (Sigma-Aldrich). After 2 days, we withdrew CHIR from the culture medium. The mesodermal induction medium was replaced with fresh expansion medium composed of Essential 6 medium enriched with 1% ITS, 5 mM LiCl, 10 ng bFGF, 10 ng insulin-like growth factor 1 (IGF-1; Thermo Fisher Scientific) and 50 ug/mL MT-derived EVs. After further 4 days, LiCl was removed from the medium. During this period, cells underwent enhanced proliferation. Between days 8–10, cells reached confluence and were expanded using TryplE (Thermo Fisher Scientific) and Collagen Type I matrix coating (BD Biosciences). The final differentiation and maturation phase into myotubes took additional 2 weeks: by day 20, muscular progenitors were seeded on Collagen type I dishes; after cells reached confluence, growth factors and MT-derived EVs were removed from the medium, and cells were cultured only in Essential 6 medium supplemented with 1% ITS.

2.9. Flow Cytometry and Cell Sorting

Fluorescence-activated cell sorting (FACS) analysis on physical parameters (forward and side light scatter, FSC and SSC, respectively), was first performed in order to exclude small debris, while the LIVE/DEAD Fixable Dead Cell Stain (Invitrogen, Carlsbad, CA, USA) allowed for the discrimination between live and dead cells. Muscle pericytes were labelled with the following conjugated antibodies: anti-alkaline phosphatase-Cy5 (BD Pharmingen), anti-CD45-FITC/CD14-PE (BD Biosciences, San Jose, CA, USA), anti-NG2-PE (BD Pharmingen), anti-CD56-APC (NCAM; BD Biosciences), anti-CD146-Cy5 (MCAM; R&D Systems, Minneapolis, MN, USA), anti-PDGF-R-beta-FITC (R&D Systems), and anti-CD44-APC (BD Pharmingen). Skin fibroblasts were characterized by staining with anti-CD90-FITC (BD Pharmingen). iPSC-derived skeletal muscle progenitor cells were stained with primary antibodies: PAX3 (Thermo Fisher Scientific), MyoD1 (Abcam, Cambridge, UK), PAX7 (DHSB), MyoG (Clone F5D, eBioscience, San Diego, CA, USA), and myosin heavy chain (Clone MF20; R&D Systems) (Abcam), followed by staining with the FITC-conjugated secondary antibody (R&D System). All antibodies were diluted in accordance with the manufacturers' instructions. Fluorescence intensity for surface antigens and intracellular cytokines was detected by flow cytometry using a BD FACS Canto II analyzer. Flow data were analyzed with the FACSDiva 6.1.2 software (Becton Dickinson, Franklin Lakes, NJ, USA) and the FlowLogic software (Miltenyi Biotec, Bergisch Gladbach, Germany).

The ALP$^+$/CD56$^-$ subpopulation was sorted by FACSAria II Cell Sorter (Becton Dickinson) and subsequently characterized by FACS analysis for the expression of pericyte markers (as listed above) following 2 passages in vitro.

To detect and analyze surface EVs markers by FACS analysis, we bound them to 4 μm aldehyde sulphate latex beads (Thermo Fisher Scientific) overnight at 4 °C in rotation. EV-coated beads were then incubated with fluorochrome-conjugated antibodies CD63-APC (eBioscience) and CD81-PE (Invitrogen), and diluted in accordance with the manufacturers' instructions. A "beads only" control sample was used to set gating parameters.

For EV internalization, we labelled the purified vesicles isolated from C2C12 with 5 μg/mL CellMask Deep Red plasma membrane stain (Molecular Probes, Eugene, OR, USA) and 5 mM CellTrace Violet (Invitrogen) at 37 °C for 30 min. The labeled EVs were washed in PBS and ultra-centrifuged at 100,000× g at 4 °C for 90 min, suspended in differentiating medium and used to treat the cells. After 48 h, we detected fluorescence on differentiating cells by flow cytometry.

2.10. Gene Expression Analysis

Total RNA was extracted using small RNA miRNeasy Mini Kit (Qiagen, Hilden, Germany). A total of 1 μg of total RNA was reverse-transcribed to cDNA using SuperScript VILO cDNA synthesis kit (Life Technologies). qRT-PCR were performed using SYBR Green Master Mix (Applied Biosystems, Foster City, CA, USA). Each sample was analyzed in triplicate using QuantStudio 6 Flex Real-Time PCR System Software (Applied Biosystems). The relative gene expression for pluripotency markers was expressed relative to a certified Episomal iPSC lineage (EpiPSC, Thermo Fisher Scientific), and normalized to Glyceraldehyde-3-Phosphate Dehydrogenase (GAPDH) (Table S1).

The expression of the lentiviral vector was assessed by qualitative RT-PCR according to standard procedure. Amplified products were separated by electrophoresis on a 1% agarose gel. Primers were designed to identify cMyc (one of the four human transcription factors included in the polycistronic lentiviral backbone—forward oligonucleotide) and WPRE (woodchuck hepatitis virus post-transcriptional regulatory element, a lentiviral-specific transgene—reverse oligonucleotide) (Table S1).

2.11. Immunohistochemistry and Immunocytochemistry

Cells were fixed in 4% paraformaldehyde for 20 min at room temperature, washed twice in PBS and blocked with 10% donkey serum (Sigma-Aldrich) for 30 min at room temperature. For intracellular immunostaining, we permeabilized cells for 10 min in 0.1% Triton X-100 (Sigma-Aldrich). Immunofluorescence assays were carried out with the following primary antibodies as follows: pericytes were stained with anti-PDGFRβ (1:100 Abcam), anti-α-smooth muscle actin (1:400, SMA; Dako, Santa Clara, CA, USA) and anti-NG2 (1:200, Millipore, Burlington, MA, USA); fibroblasts were labelled with anti-vimentin (1:50, Sigma-Aldrich); HUVEC were incubated with anti-von Willebrand factor (1:100, Abcam); iPSC pluripotency was verified by anti-OCT4, anti-SSEA4, anti-SOX2 and anti-TRA-1-60 (1:100, all from Invitrogen); iPSC-derived MT were identified by anti-myosin heavy chain (MHC; 1:100; R&D).Incubations with the secondary antibodies (1:100, Jackson ImmunoResearch Laboratories, West Grove, PA, USA) were performed for 1 hour at 37 °C. Cells were then counterstained using Vectashield Mounting Medium with DAPI (4′,6-diamidino-2-phenylindole). Fluorescence was detected by microscope (Axio Observer A1, Zeiss, Oberkochen, Germany).

Differentiation evaluation was assessed by immunofluorescence for MHC and fusion index scoring, defined as the ratio between the number of myosin heavy chain expressing myotubes with greater than 2 nuclei with respect to the total number of nuclei.

2.12. Western Blotting Analysis

EVs were also characterized by Western blotting for the expression of specific markers. EVs were lysed in radioimmunoprecipitation assay buffer (RIPA buffer), and supplemented with protease and phosphatase inhibitor cocktails (Roche Diagnostics GmbH, Mannheim, Germany). Proteins (30 μg) were separated on the Nupage Novex on 4–12% Bis-Tris Gel (Life Technologies) and transferred to a Polyvinylidene fluoride (PVDF) membrane Amersham Hybond (GE Healthcare Biosciences, Piscataway, NJ, USA). Membranes were incubated overnight at 4 °C with primary antibodies anti-CD81, anti-CD63, anti-αHSP70 (ExoAb Antibody Kit, System Biosciences, Palo Alto, CA, USA), anti-TSG-101 (Thermo Fisher Scientific), anti-MyoD1 (Abcam), anti-MHC (R&D) followed by peroxidase-linked anti-rabbit IgG or anti-mouse IgG secondary antibodies (GE Healthcare Life science) for 1 h at room temperature. Specific protein bands were detected with Pierce ECL Western Blotting Substrate (Thermo Fisher Scientific).

2.13. In Vivo Studies

Two-month-old male αSGKO/SCIDbg mice (n = 5) were anesthetized with an intramuscular injection of physiologic saline (10 mL/kg) containing ketamine (5 mg/mL) and xylazine (1 mg/mL) and then 5×10^5 PC-derived iPSCs were injected into the Tibialis Anterior muscle (TA), according to standardized procedures [5]. Mice were sacrificed 20 days after implantation for morphological analysis. Experiments on animals were conducted according to the rules of good animal experimentation I.A.C.U.C. no 432 of 12 March 2006 and under Italian Health Ministry approval no. 228/2015-PR. In vivo experiments were conducted in accordance with the principles of the 3Rs (replacement, reduction and refinement).

2.14. Engrafted Human Muscular Cell Identification

Human differentiated iPSCs were identified by immunofluorescence for anti-human lamin A/C (1:100, SIGMA). Anti-laminin (1:100, SIGMA) was used to identify the fibers. The images were obtained by confocal laser scanning microscope.

2.15. Statistical Analysis

Statistical significance of the differences between means was assessed by one-way analysis of variance (ANOVA), followed by the Student-Newman-Keuls test, to determine which groups were significantly different from the others. When only two groups had to be compared, we used the unpaired Student's t-test. $p < 0.05$ was considered significant. Values are expressed as means ± standard deviation (SD). All the analyses were performed using Graph-Pad PRISM 7 and 8.

3. Results

3.1. Pericyte and Fibroblast Isolation and Characterization

Pericytes were isolated from three healthy human skeletal muscle biopsies and characterized by flow cytometry using a specific panel of markers according to previous studies [6]. The harvested cells highly expressed well-known pericyte markers, such as ALP (alkaline phosphatase), PDGFRβ (platelet derived growth factor receptor-beta), CD146 (MCAM, melanoma cell adhesion molecule), NG2 (Neuron/glial antigen 2), and CD44 (HCAM, homing cell adhesion molecule) (Figure 1A) as well as CD56 (NCAM, neural-cell adhesion molecule), a glycoprotein specifically expressed in muscle by human satellite cells [6,34].

Skeletal muscle resident PCs, expressing ALP, represent a myogenic cell compartment, distinct from satellite cells, capable of promoting myofiber regenerating [6]. Therefore, we selected pericytes with myogenic potential by fluorescence-activated cell sorting (FACS), combining the cell surface markers ALP and CD56. We enriched the pericyte population selecting the fraction ALP$^+$CD56$^-$, which represented the 28% of the total population (Figure 1B).

After expansion and before reprogramming, ALP$^+$CD56$^-$ subpopulation was analyzed for the expression of the canonic muscular pericyte markers—almost the totality of the tested cells expressed ALP, PDGFRβ, CD146, NG2, and CD44, while the expression of CD56 was dramatically reduced (Figure 1C), suggesting that ALP$^+$CD56$^-$ fraction retains pericyte features.

These results are in line with previous studies in which pericyte identification was performed through the combination of NG2, PDGFβ, and CD146 markers [35,36]. PC phenotype was further confirmed by the ALP colorimetric assay (Figure 1D) and immunofluorescence positivity for NG2, PDGFRβ, and αSMA (Figure 1E).

We further examined the myogenic potential of ALP$^+$CD56$^-$ cells by measuring myosin heavy chain (MHC) expression, upon skeletal muscle differentiation, induced by cellular confluence and serum depletion [37]. After two weeks, ALP$^+$CD56$^-$ cells spontaneously differentiated into myosin positive multinucleated myotubes, as confirmed by the expression of MHC (Figure 1F). These results indicate that pericytes possess myogenic potential, along with supporting vessel formation and angiogenesis.

PCs are crucial in several phases during angiogenesis and vascular homeostasis, regulating the germination of the capillaries and the stabilization of the vessels. We therefore evaluated the ability of ALP$^+$CD56$^-$ cells to generate networks and to cooperate with endothelial cells (HUVECs) to form capillary-like structures. For this purpose, we transduced cells with a lentivirus expressing GFP, co-cultured with HUVECs (GFP$^+$ ALP$^+$CD56$^-$ cells/HUVECs in a 1:4 ratio), and assembled on Matrigel for 6 h. We found that the ALP$^+$CD56$^-$ cells significantly enhanced capillary-like structure formation of HUVECs. Indeed, PCs co-cultivated with HUVECs displayed higher segment total length, total mesh area and total branch length compared to HUVECs cultured alone (Figure 1G). These results demonstrate that pericytes isolated from skeletal muscle maintain their ability to support vessel formation and myogenic potential after isolation, sorting and expansion procedures.

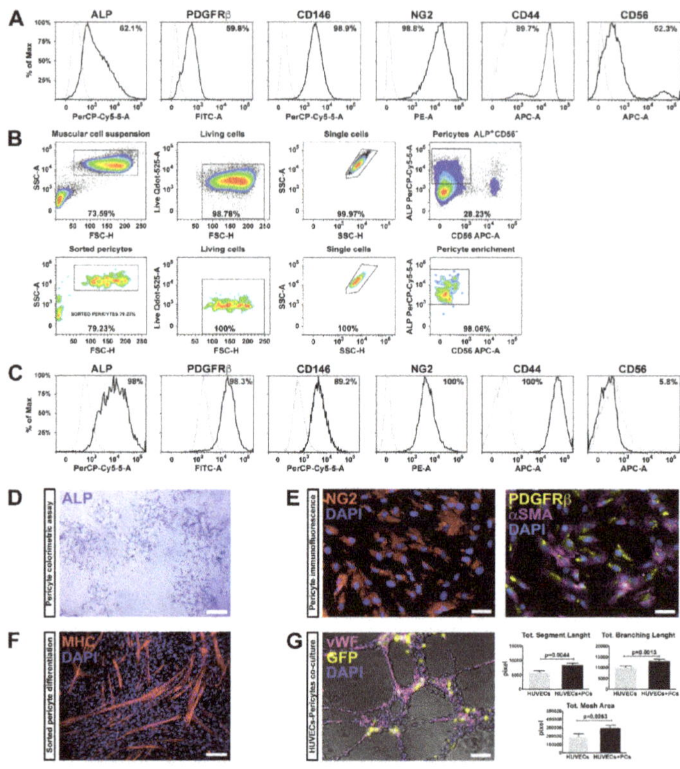

Figure 1. Pericyte characterization. (**A**) Representative histograms indicating the percentage of alkaline phosphatase (ALP⁺), platelet derived growth factor receptor-beta (PDGFRβ⁺), MCAM, melanoma cell adhesion molecule (CD146⁺), (αSMA⁺), Neuron/glial antigen 2 (NG2⁺), HCAM, homing cell adhesion molecule (CD44⁺) and NCAM, neural-cell adhesion molecule (CD56⁺) positivity (black peaks) determined by flow cytometry in pre-sorted cells isolated from muscular biopsy (n = 4). Matched isotypes were used as negative controls (grey peaks). (**B**) Representative gating strategy for ALP⁺ and CD56⁻ cell sorting (n = 3). Cells were first gated for cell size (side light scatter SSC-A vs. forward light scatter FSC-H) and vitality (Live Qdot-525-A). The muscular cell gate was further analyzed for singlets (SSC-A vs. SSC-H) and their expression for ALP and CD56. Pericytes, ALP⁺ and CD56⁻ were then sorted from this gated population. The lower set of four plots confirmed the efficiency of the sorting. (**C**) Representative post-sorting histograms for key pericyte markers after two passages in vitro, indicating an enhanced expression of ALP, NG2, PDGFRβ, CD146, and CD44. (**D**) Sorted pericytes stained for ALP showing fibroblast colony-forming units (CFU-F) when seeded at low confluence. Scale bar represents 300 μm. (**E**) Immunofluorescence labeling for NG2 (red) and the co-staining for PDGFRβ (green) and αSMA (magenta) on sorted ALP⁺CD56⁻ cells. Nuclei were stained with DAPI. Scale bar represents 50 μm. (**F**) Representative fluorescence image for myosin heavy chain (MHC) (red), validating the differentiation of sorted pericytes toward skeletal muscle phenotype. Scale bar represents 100 μm. (**G**) Illustrative images of human umbilical vein endothelial cells (HUVEC) in co-culture with pericytes displaying the formation of capillary-like networks with HUVEC labeled for von Willebrand factor (vWF; magenta), and GFP⁺ pericytes. Nuclei were identified by DAPI (blue). Scale bar represents 100 μm. Tubular structures were photographed at 5× magnification and quantified by the angiogenesis analyzer ImageJ tool. Total segment length, total mesh area and total branching length exhibited significant differences between HUVEC alone and in co-culture with pericytes, as shown in the graphs.

Epigenetic memory inherited from their original tissue have been demonstrated to influence the iPSC differentiation potential [29], suggesting that pericyte myogenic and angiogenic potential could be advantageous in tissue regeneration. Hence, we isolated human adult skin fibroblasts from the same donor of pericytes in order to compare the capability of iPSCs derived from pericytes (PC-derived iPSCs, ALP$^+$CD56$^-$ subpopulation) and fibroblasts (FB-derived iPSCs) to re-differentiate into muscle cells. Skin fibroblasts, isolated by enzymatic digestion, were characterized by immunofluorescence (Figure S1) and FACS analysis (Figure S1) for the expression of vimentin and CD90.

3.2. iPSC Generation and Characterization

We generated human muscular PC-derived iPSCs and skin FB-derived iPSCs from the same donor (n = 3 donors) using a polycistronic vector harboring the four Yamanaka factors (OCT4, Sox2, Klf4, cMyc) [33,38]. Colonies were initially expanded on a feeder layer of inactivated mouse embryonic fibroblasts (iMEF) and then adapted to feeder free conditions replacing iMEF with geltrex matrix (Figure 2A, left panels). iPSC colonies expressed typical pluripotent markers, including OCT4, SSEA4, SOX2, and TRA-1-60, as assessed by immunofluorescence staining (Figure 2A, middle and right panels). These results were confirmed by quantitative real-time PCR (qRT-PCR) for the expression of OCT4, SOX2, NANOG, LIN28, and TERT genes. Certified Episomal iPSC lineage (EpiPSC, Thermo Fisher Scientific) was used as control (Figure 2B).

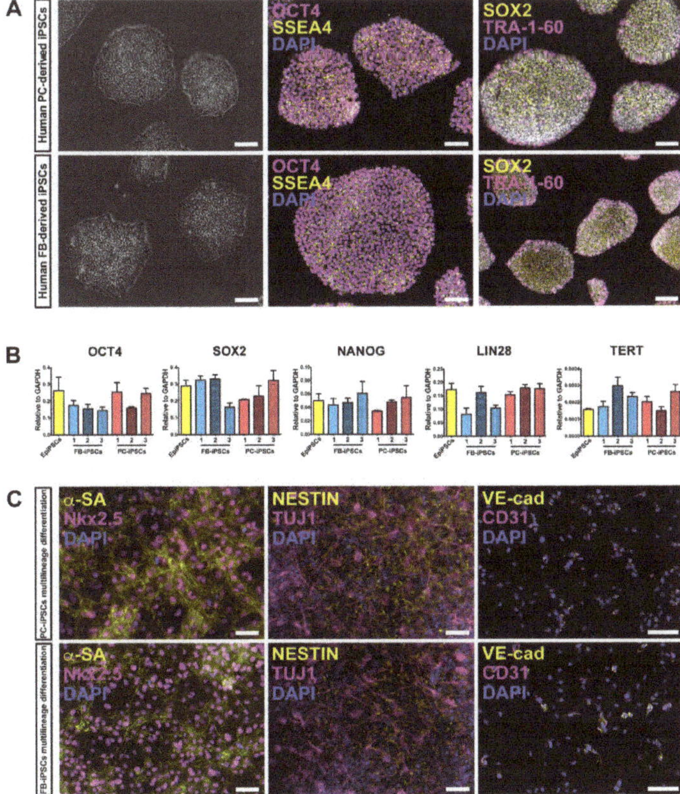

Figure 2. Characterization of human pericyte (PC)- and fibroblast (FB)-derived iPSCs. (**A**) Morphology of PC-derived iPSCs (left upper panel) and FB-derived iPSCs (left lower panel) cultured in feeder-free conditions.

Scale bar represents 100 µm. Immunofluorescence labeling for the expression of the pluripotent markers SSEA4 (yellow), OCT4 (magenta; middle panels), SOX2 (yellow), and TRA-1-60 (magenta; right panels) on PC- and FB-derived iPSCs (PC-iPSCs and FB-iPSCs, upper and lower panels respectively). Nuclei were stained with DAPI. Scale bar represents 100 µm and, for the lower right image, 200 µm. (**B**) Quantitative RT-PCR analysis for the expression of the embryonic genes, OCT4, SOX2, NANOG, LIN28, and TERT—in PC- and FB-derived iPSC lines, derived from three donors (named FB-/PC-derived iPSCs 1, 2, 3), compared to the certified Episomal iPSC lineage (EpiPSC), used as control. Results were normalized to GAPDH. (**C**) Representative fluorescence images demonstrating multilineage differentiation capacity of PC- and FB-derived iPSCs toward cardiomyocyte (alpha-sarcomeric actin–α-SA, Nkx2.5), neuronal (Nestin, TUJ1) and endothelial (CD31, VE-cadherin–VE-cad) lineages. Nuclei were stained with DAPI (blue). Scale bars represent 50 µm (left and middle panels) and 100 µm (right panels).

Qualitative RT-PCR was conducted to verify the silencing of the exogenous transgenes in PC- and FB-derived iPSCs (Figure S2).

In addition, both derived iPSC lines retained the ability to differentiate toward multiple lineages, including cardiomyocytes, neuronal precursors, and endothelial cells (Figure 2C).

3.3. MT-derived EV Characterization

EVs represent an important vector of intercellular communication, acting as vehicles for the transfer of cytosolic factors, proteins, lipids and RNA [39]. EV cargo is cell-type specific, and the molecular composition reflects specific functions of the donor cells.

EV secretion and extracellular signaling occurs during muscle differentiation, repair and regeneration [25,40]. Both myoblasts and myotubes release EVs, but their contribution in muscle physiology and specific biological functions on recipient cells have not been fully elucidated.

Proteomic analysis of muscle-derived EVs revealed that, in addition to proteins involved in their biogenesis, EVs also contain functionally relevant proteins such as myogenic growth factors and contractile proteins [26,40,41].

On the basis of these premises, we have exploited the capacity of EVs, secreted during myotube formation of skeletal myoblasts (C2C12), to promote the differentiation of iPSCs into the myogenic lineage.

EVs were purified from conditioned media of C2C12-derived MTs cultured in 2% EV-depleted horse serum. EV isolation was performed by differential centrifugations according to well-established protocols [42]. EVs were further analyzed by FACS analysis for the expression of CD81 and CD63. We found that MT-derived EVs were enriched in membrane-bound tetraspanins CD63 and CD81, which are common markers for EV subsets released from most cell types (Figure 3A,B). The expression of CD81 is observed on vesicles of various sizes indicating that multivesicular endosomes in muscle cells contain intraluminal vesicles of heterogeneous sizes [27,41]. The expression of proteins associated with EVs, such as tumor susceptibility gene 101 (TSG101), α heat shock 70 kDa protein 4 (HSP70), CD81 and CD63 was further confirmed by Western blot analysis.

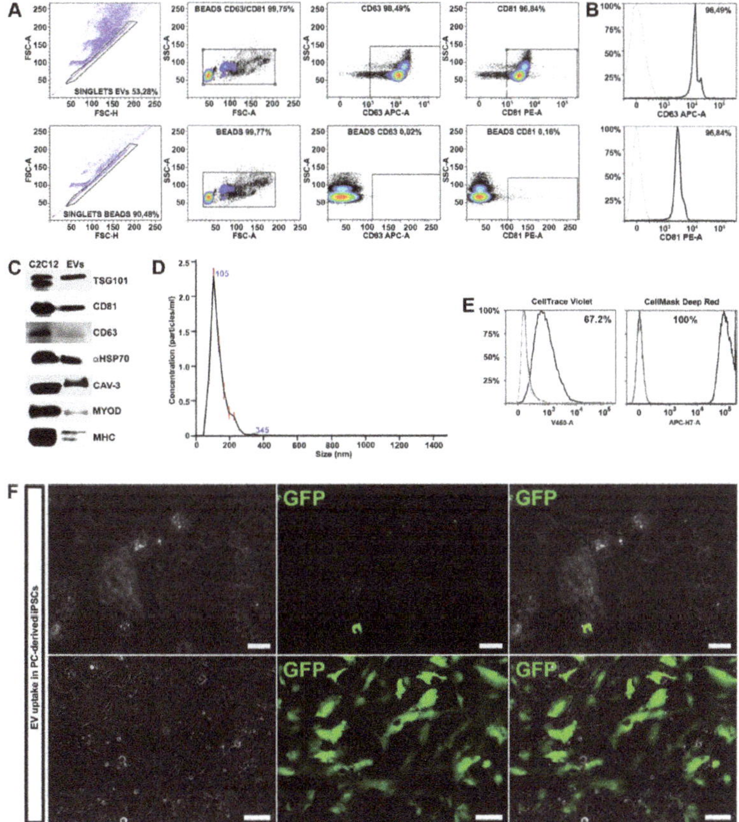

Figure 3. Extracellular vesicle (EV) characterization and uptake. (**A**) Sample gating strategy indicating the percentage of CD63$^+$ and CD81$^+$ in myotube (MT)-derived EVs coated with beads (n = 3). In the upper plots, singlets and subsequently EVs were selected according to physical parameters (FSC-A vs. FSC-H and SSC-A vs. FSC-A, respectively) are shown. Fluorescent intensity signal for CD63 and CD81 was detected on gated EVs (SSC-A vs. APC-A and SSC-A vs. PE-A, respectively). Beads alone were used as control (lower four scatter plots). (**B**) Representative histograms displaying the percentage of EV specific markers, such as CD81 and CD63, determined by flow cytometry in purified MT-derived EVs. Matched isotypes were used as negative controls (grey peaks) (**C**) Western blot for specific the expression of EV markers, such as TSG101, CD81, CD63, aHSP70, and skeletal muscle markers, such as CAV-3, MYOD, MHC, in C2C12-derived myotubes and C2C12 myotube-derived EV lysates. (**D**) Size distribution profile of MT-derived EVs (n = 5). (**E**) Representative histograms, determined by flow cytometry, displaying the percentage of positive cells (black) after the treatment with EVs stained for cell trace violet or cell mask deep red. Non-treated cells (light grey peaks) and cells treated with unstained EVs (dark grey peaks) were used as controls. (**F**) Green spots indicate the presence of fluorescent (GFP$^+$) EVs, derived from GFP transduced myotubes, in the cytoplasm of the recipient differentiating iPSC after 48 h of exposure (upper panels; scale bars represent 20 µm); GFP$^+$ cells 10 days after GFP$^+$ MT-derived EV exposure (lower panels; scale bars represent 100 µm), demonstrating that GFP was transferred through the EVs into the recipient cells.

We found that MT-derived EVs exhibited specific membrane proteins associated with mature muscle tissue, such as caveolin 3 (Cav3), expressed only during the late stage of differentiation, and MHC, a differentiated myotubes marker (Figure 3C). Interestingly, MyoD, a transcription factor implicated in myogenesis, was also detected within the EV cargo.

Finally, the purity, size and concentration of the MT-derived EVs were determined by nanoparticle tracking analysis (NTA), using instrument-optimized analysis settings in NTA 3.1 build 54 software. The results showed a mean size distribution, consistent with what is expected from a sample enriched in microvesicles and exosomes (130.3 ± 0.3 nm), and free from contamination by apoptotic bodies (>1 µm) (Figure 3D).

3.4. Detection of the MT-derived EV Uptake

To exert their functional influence, EV cargo must be internalized within the cell in adequate concentrations. As first step, in order to verify if EVs can transfer their contents to differentiating iPSCs, we performed EV uptake assays. For this experiment, we employed two different methods: FACS analysis and immunofluorescence microscopy. Upon isolation, MT-derived EVs were marked either with cell mask deep red or with cell trace violet. As controls we used untreated cells and cells exposed to unstained EVs. After 48 h, FACS analysis revealed an increased fluorescence of both tracers in recipient cells, indicating a successful uptake (Figure 3E).

In order to demonstrate the EV cargo delivery, we transduced C2C12 cells with a lentivirus expressing the GFP protein, and subsequently differentiated into myotubes. The EVs released by GFP$^+$ myotubes were collected and used to treat the differentiating iPSCs. After 48 h, a fluorescent signal was detected in the cytoplasm of cells undergoing differentiation (Figure 3F, upper panels).

Cells were treated with EVs released by GFP$^+$ myotubes every other day and, after 10 days, approximately 40% of cells expressed GFP in the cytoplasm (Figure 3F, lower panels) indicating a high functional cargo delivery to the recipient cells. EVs may enter into a cell via more than one route, depending on proteins and glycoproteins found on the surface of both the vesicle and the target cell [43]. The mechanisms responsible for MT-derived EVs delivery have not yet been determined highlighting the need for further research [44].

Considering these results, we have shown that EVs target iPSCs via delivery of effector molecules directly affecting their phenotype and functions.

3.5. Myogenic Differentiation by MT-derived EVs

MT-derived EVs express specific cell-adhesion molecules on their surfaces (ITGB1, CD9, CD81, CD44, Myoferlin) that are involved in the recognition and adhesion during the process of myoblast fusion [45–47].

Further, MT-derived EVs contain functionally active proteins of the G-protein family, which are involved in many cellular processes including myogenesis [48].

Wnt signaling and its modulation via GSK3 inhibitors, such as CHIR99021 (CHIR), is essential in the mesoderm induction and to obtain a reliable, reproducible and efficient myogenic differentiation protocol [18,19,49].

On the basis of these findings, we developed a method to induce a robust differentiation of iPSC toward skeletal muscle phenotype combining MT-derived EV cargo and the chemical modulator CHIR for paraxial mesoderm-like muscle progenitor commitment (Figure 4A).

PC- and FB-derived iPSCs, derived from three donors, were divided in four groups and treated as follows: (i) group 1, PC-/FB-derived iPSCs cultured in differentiation medium; (ii) group 2, PC-/FB-derived iPSCs cultured in differentiation medium augmented with EVs; (iii) group 3, PC-/FB-derived iPSCs cultured in differentiation medium enriched with CHIR; and (iv) group 4, PC-/FB-derived iPSCs cultured in differentiation medium supplemented with EVs and CHIR.

Cells were treated with 10 uM CHIR for 48 h to induce the expression of paraxial mesoderm genes. Consistent with other studies [18,19], iPSCs from both sources presented evident morphological changes losing the typical ES-like morphology after 24 h of treatment (Figure 4B). Following replacement of CHIR with FGF2, cells underwent proliferation and reached full confluence approximately at day 8–10.

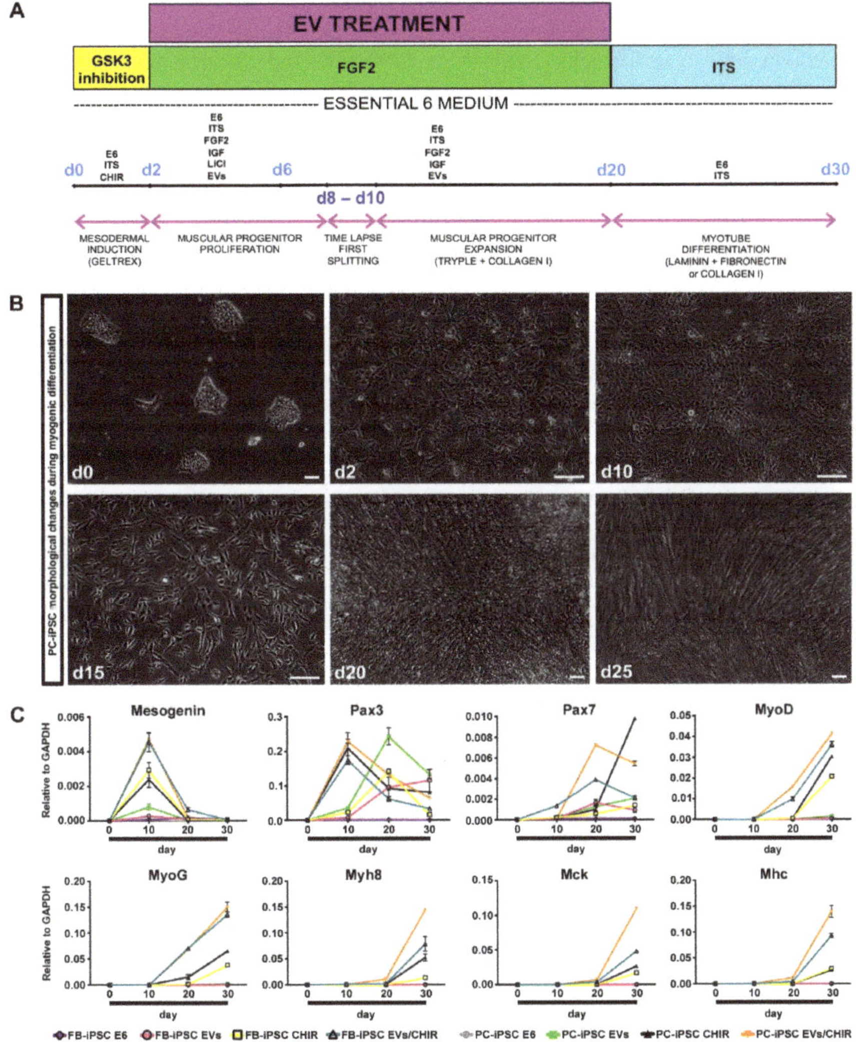

Figure 4. Myogenic differentiation by MT-derived EVs. (**A**) Schematic diagram of the myogenic differentiation procedure: 30% confluent iPSCs were differentiated to early mesoderm using CHIR for 48 h, subsequently proliferation and expansion of muscular progenitors were stimulated utilizing EVs, FGF2, IGF, and the myotube-like cell maturation was induced, removing growth factors from the culture medium. (**B**) Morphological changes of PC-derived iPSC colonies during skeletal muscle differentiation. Scale bars represent 100 μm. (**C**) Representative qRT-PCR for the expression of early (Mesogenin, Pax3, Pax7, MyoD, MyoG) and late (MyH8, MCK, MHC) skeletal muscle genes. The analysis was performed at different time points (d0, d10, d20, d30) in the following conditions: FB- and PC-derived iPSCs not treated with EVs or CHIR (FB-iPSCs E6 and PC-iPSCs E6), exposed to CHIR or EVs only (FB-iPSC EVs, FB-iPSC CHIR, PC-iPSC EVs, PC-iPSC CHIR) and treated with both, EVs and CHIR (FB-iPSC EVs/CHIR and PC-iPSC EVs/CHIR) (n = 3 donors). Results were normalized to GAPDH.

Following 48 h of treatment, CHIR was removed and, in order to enhance the myogenic differentiation, the media were enriched with freshly MT-derived EVs (50 μg/mL) and replenished every 2 days with fresh vesicles until the formation of myoblast-like cells. Between days 15 and

25 paraxial mesoderm-like muscle progenitors can be either expanded or terminally differentiated upon withdrawal of all the growth factors from the medium. The outcome of myotube formation was greater when myoblast-like cells were passaged 2–3 times and consequently exposed longer to MT-derived EVs.

We have analyzed the expression of genes related to mesoderm induction and myogenic differentiation at different time points through qRT-PCR experiments (Figure 4C).

In PC- and FB-derived iPSCs treated with the CHIR/EV combination, pre-myogenic mesoderm genes, such as Mesogenin, Pax3, and Pax7, were upregulated between day 10 and day 20. The myogenic regulatory factor MyoD and MyoG started to be expressed around day 10 and increased up to day 30. Finally, the mature myocyte genes, Myh8 (myosin heavy chain 8), MCK (muscle creatine kinase) and MHC, were detected after day 20 and augmented until the end of the differentiation confirming the increase in differentiation and maturation of the iPSCs treated with both factors.

Cells cultured in differentiation medium were negative for the expression of myogenic genes, while iPSCs treated with EVs exhibited only pre-myogenic genes (Pax3 and Pax7), indicating that the MT-derived EVs are not sufficient, at least within 30 days, to induce myotube differentiation. Lastly, cells exposed to CHIR presented a myogenic inclination similar, but less efficient, to PC-/FB-derived iPSCs treated with the CHIR–EVs mishmash.

Our method proved that the combination CHIR-EVs induces a higher differentiation compared to CHIR treatment (Figure 5A).

To verify whether PC-derived iPSCs, generated from the three donors, possessed a greater propensity to differentiate into myotubes compared to FB-derived iPSCs, we matched the expression of Myh8, MCK, and MHC by qRT-PCR analysis at day 30 of differentiation (Figure 5A). The expression of mature myogenic genes resulted as being higher in differentiated PC-derived iPSCs.

To further determine the role of the EVs/CHIR combination we calculated the fusion index, which was significantly greater compared to the index obtained with the separate exposure of EVs or CHIR (Figure 5B). Consistently, also the cell number expressing MHC was also significantly greater in those generated using the CHIR–EVs cocktail compared to the other treatments (Figure 5C).

The ability of PC-derived iPSCs to generate differentiated cells presenting a characteristic mark of myogenesis, was further confirmed by FACS analysis. Myogenic regulatory factors, Pax3, Pax7, CD56, MYOD, MYOG, and MF20, were used to characterize the expression of myogenic proteins (Figure 5D). At day 30 roughly 80% of the cells were positive for the myocyte mature marker MF20.

Finally, after 25 days of CHIR–EV differentiation, PC-iPSC-derived muscular cells, were implanted in vivo in a limb girdle muscular dystrophy murine model, namely, SCID-Beige α–Sarcoglycan null mice (αSGKO/SCIDbg) [5] in order to evaluate their engraftment capabilities. Three different PC-derived iPSC clones were intramuscularly injected (5×10^5 cells/injection) into the anterior tibialis (TA) muscle of αSGKO/SCIDbg, showing a sufficient engraftment and integration into regenerating muscle fibers revealed by the labelling of human derived cells by immunofluorescence against human specific lamin A/C antibody (Figure 5E).

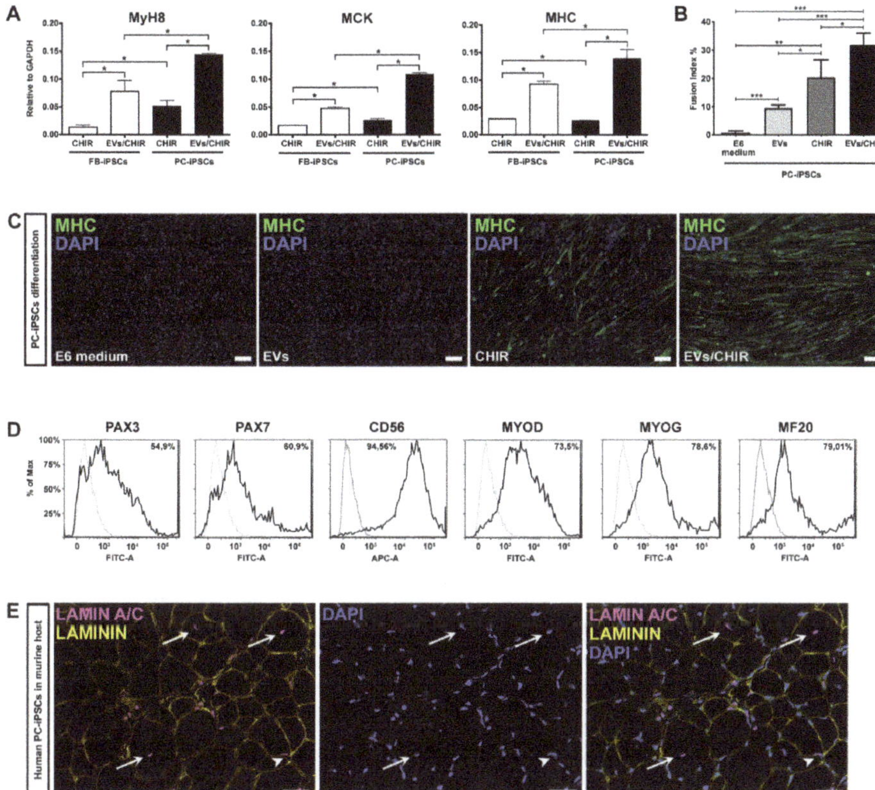

Figure 5. iPSC-derived MT characterization. (**A**) Representative qRT-PCR for the expression of Myh8, MCK and MHC after 30 days of differentiation. The combination CHIR-EVs induced a greater differentiation compared to CHIR treatment and PC-derived iPSCs possessed a greater propensity to differentiate into myotubes compared to FB-derived iPSCs. Results were normalized to GAPDH. (**B**) Fusion index was calculated as the percentage of nuclei within myosin heavy chain–positive myotubes (≥2 nuclei) divided by the total number of nuclei. A minimum of five random fields at 10X magnification were counted. (**C**) Immunofluorescence labeling for MHC (green), validating the differentiation capacity of PC-derived iPSCs toward skeletal muscle phenotype after 30 days of differentiation. The images represent sequentially: PC-derived iPSCs without treatment, exposed to EVs alone, to CHIR only and to the EV–CHIR cocktail. CHIR in combination with MT-derived EVs induced a higher expression of the skeletal marker (n = 3 donors). Nuclei were stained with DAPI. Scale bar represents 100 μm. (**D**) Representative histograms indicating the percentage of PAX3, PAX7, CD56, MYOD, MYOG, and MF20 (black peaks) determined by flow cytometry in PC-iPSC-derived MTs exposed to CHIR and MT-derived EVs after 30 days of differentiation. Matched isotypes were used as negative controls (grey peaks) (n = 3 donors). (**E**) Representative immunofluorescence against human lamin A/C (magenta) and laminin (yellow) on mouse anterior tibialis (TA) injected with human PC-derived iPSC exposed to CHIR-EVs (n = 3 donors). Arrows indicate center-nucleated human derived myofibers, arrowhead labelling integrated human nuclei into mature host muscle fiber. Nuclei were stained with DAPI. Scale bars represent 25 μm.

4. Discussion

In the last few years, there has been a growing interest in the use of iPSCs as a source of myogenic progenitors for cell-based treatment in muscle degenerative diseases. iPSCs overcome several of the limitations related to the use of adult myoblast therapy, such as, non-invasive biopsy for cell isolation, unlimited proliferative capacity in vitro, and a tool for the in vitro correction of genetically mutated somatic cells derived from patients affected by dystrophies [14–16,19,50–52]. Moreover, it is well known that iPSCs maintain the epigenetic memory of the original cell source, influencing the re-differentiation toward the same lineage [30,53]. Pericytes resident in adult skeletal muscle have shown remarkable angiogenic and myogenic differentiation capacities in vitro and in vivo [7]. Hence, in this study, exploiting the advantage of the "epigenetic retention", we proposed the isolation of human derived muscular pericytes to verify whether PC-derived iPSCs are more prone to differentiate into muscular cells compared to FB-derived iPSCs.

To date, the strategies used for iPSC myogenic differentiation are subdivided into two approaches: transgenic (by forced expression of Pax7 or MyoD) or non-transgenic (co-culture, EBs, small molecules) [15,16,51]. The former consists in the differentiation via overexpression of myogenic transcription factors, obtaining efficient iPSC differentiation into myogenic lineage in a relatively short amount of time [15,16,51]. However, forced expression of skeletal master genes drastically reduces the proliferative capacity of skeletal muscle progenitor cells, hindering the molecular mechanisms leading to myogenic differentiation (important for disease modeling in vitro), as well as their in vitro and in vivo self-renewal. Another point to be considered is the potential of insertional mutagenesis events due to the random integration, since these genes are commonly delivered using lentiviral vectors. Integrated virus genomes are frequently associated with chromosomal damage, rearrangements and deletions [54], making these approaches not suitable in terms of clinical applications. On the other hand, the non-transgenic methods use different sequential culture conditions, including cell sorting. These protocols are successful at producing myogenic progenitors capable of engrafting in vivo but lack reproducibility, besides being inefficient and time-consuming [55–57].

First-generation of transgene-free protocols employing different sequential culture conditions were not satisfactory and usually required post-differentiation selection to isolate expandable skeletal muscle progenitor cells [57]. In addition, in some described methods, iPSCs failed to express mature myogenic markers or form myotubes in vitro, underlying the poor efficiency of spontaneous myogenic differentiation method [55,56].

Improvements in muscle differentiation protocols were recently performed by employing a small molecule, CHIR99021, a GSK-3b inhibitor that activates Wnt signaling cascade, which plays a crucial role during early somite induction. These approaches have been proven to be efficient at boosting iPSC towards a myogenic pathway, leading to an ameliorated myotube differentiation in vitro [19,58,59].

Quite recently, considerable attention has been dedicated to extracellular vesicles (EVs) as important mediators of cell-to-cell communication. EVs can influence the behavior of recipient cells and are involved in many processes, including immune signaling, angiogenesis, stress response, proliferation, and cell differentiation [60–62]. Furthermore, EVs, through their paracrine signaling can be used in tissue engineering to modulate cell recruitment, differentiation, and proliferation [63]. Skeletal muscle cells secrete a large number of myokines and EVs that influence the growth, function and development of muscle tissue [26,27,41]. Transcriptome and proteome studies reported that EVs from C2C12 or human myoblasts are enriched in miRNAs and proteins that are implicated in myogenesis [22,27,40,41,64]. EVs isolated from differentiated C2C12 cells express specific cell-adhesion molecules on their surfaces (ITGB1, annexins CD56, CD9, CD81, CD44, Myoferlin), and other myokines (HGF, IGFI/II, FGF2, PDGF, TGFβ, myostatin) involved in myoblast fusion and muscle regeneration. [40,41,45,47,65]. In addition, differentially expressed miRNAs also contained in EVs during C2C12 myotube differentiation, such as miR-1, miR-133a, miR-133b, and miR-206, have been linked with muscle differentiation [64,66]. Interestingly, the Wnt signaling and Sirt 1 that are involved in muscle gene expression and differentiation, were predicted as the most significant targeted pathway by skeletal muscle EVs–miRNAs [64]. In the present study, we found that MT-derived EVs harbored myogenic factors that are crucial in enhancing the iPSC myogenic commitment. Despite several scientific papers reporting transcriptome and proteome profiling of EVs from skeletal muscle, the mechanisms and the key factors playing a major role in promoting skeletal muscle differentiation remain unclear. Evidence suggests that instead of single factors, the diverse EVs myogenic factors such as miRNAs or proteins, synergistically trigger skeletal muscle differentiation in recipient cells [22,26,27,40,41,64]. Further studies are required to define the biological role/function of single myogenic factors in MT-derived EVs.

Here, we have developed an efficient method for iPSC skeletal muscle differentiation combining defined factors with myogenic elements carried by EVs. We were able to generate a consistent number of iPSC-derived positive MHC skeletal muscle cells, evidencing differences in epigenetic signatures between PC-derived iPSCs and their counterpart FB-derived iPSCs. Indeed, quantitative PCR analyses performed at different timepoints and at the end of the differentiation, showed that PC-derived iPSCs are more prone to differentiate in skeletal muscle cells when compared to FB-derived iPSCs. Importantly, the myoblasts obtained with this approach could be expanded, and cryopreserved at all steps during expansion, without losing fusion competence. The obtained MT-like cells were positive for MyoD, MYOG, and MHC, and were able to actively participate in myogenesis in vivo. Transplanted myogenic progenitors demonstrated their ability to successfully engraft and integrate into TA muscle of αSGKO/SCIDbg, identified by the detection of centrally located human lamin A/C positive nuclei, a characteristic perceived only in fusion-competent myoblasts.

In this paper, we explored for the first time EVs as "physiological liposomes" enriched with myogenic factors that were able to trigger skeletal myogenesis in iPSC, raising exciting possibilities for therapeutic use. The use of EVs represent novel tools to deliver signaling molecules for treating muscle-wasting diseases that are better tolerated by the immune system. Additional studies are now required to investigate the role of EVs' skeletal muscle content that may allow the development of better therapeutic approaches to promote skeletal muscle growth, differentiation and regeneration.

Our method may better recapitulate human developmental myogenesis providing inroads for patient-specific drug testing, disease modeling and regenerative approaches.

Supplementary Materials: The following are available online at http://www.mdpi.com/2073-4409/9/6/1527/s1, Figure S1: Fibroblast characterization, Figure S2: Silencing of the exogenous factors, Table S1: List of human primers used for qRT-PCT analysis.

Author Contributions: Conceptualization, D.B., R.R., and C.B.; data curation, D.B.; formal analysis, D.B., M.C., F.M., S.G., and A.B.; funding acquisition, R.R., and C.B.; investigation, D.B., M.C., V.P., F.M., M.M., A.R., P.S., D.P., S.G., and A.B.; methodology, D.B., M.C., R.R., and C.B.; project administration, D.B., R.R., and C.B.; resources, S.C., C.L., C.G., R.R., and C.B.; supervision, S.C., C.G., R.R., and C.B.; validation, D.B., M.C., V.P., F.M., S.G., and A.B.; writing—original draft preparation, D.B.; writing—review and editing, C.L., C.G., R.R., and C.B. All authors have read and agreed to the published version of the manuscript.

Funding: This research was funded by the Italian Ministry of Health—grant Giovani Ricercatori 2009 (GR-2009-1606636), by the Ministry of Education, University and Research—grant PRIN 2010-2011 (2010B5B2NL), by Regione Lazio, LAZIO INNOVA (85-2017-15095) and by the Italian Regenerative Medicine Infrastructure (IRMI), Cluster ALISEI (CTN01_00177_888744).

Acknowledgments: We acknowledge Gustavo Mostoslavsky from Boston University School of Medicine, for providing the STEMCCA plasmid. We thank Massimiliano Gubello for the English editing and proofreading. We want also to thank the reviewers for their insightful comments in previous version of the manuscript.

Conflicts of Interest: The authors declare no conflict of interest.

References

1. Bianchi, A.; Mozzetta, C.; Pegoli, G.; Lucini, F.; Valsoni, S.; Rosti, V.; Petrini, C.; Cortesi, A.; Gregoretti, F.; Antonelli, L.; et al. Dysfunctional polycomb transcriptional repression contributes to lamin A/C-dependent muscular dystrophy. *J. Clin. Investig.* **2020**, *130*, 2408–2421. [CrossRef] [PubMed]

2. Cossu, G.; Tajbakhsh, S. Oriented cell divisions and muscle satellite cell heterogeneity. *Cell* **2007**, *129*, 859–861. [CrossRef] [PubMed]

3. Kuang, S.; Kuroda, K.; Le Grand, F.; Rudnicki, M.A. Asymmetric self-renewal and commitment of satellite stem cells in muscle. *Cell* **2007**, *129*, 999–1010. [CrossRef]

4. Lanzuolo, C. Epigenetic alterations in muscular disorders. *Comp. Funct. Genom.* **2012**, *2012*, 256892. [CrossRef]

5. Fuoco, C.; Salvatori, M.L.; Biondo, A.; Shapira-Schweitzer, K.; Santoleri, S.; Antonini, S.; Bernardini, S.; Tedesco, F.S.; Cannata, S.; Seliktar, D.; et al. Injectable polyethylene glycol-fibrinogen hydrogel adjuvant improves survival and differentiation of transplanted mesoangioblasts in acute and chronic skeletal-muscle degeneration. *Skelet. Muscle* **2012**, *2*, 24. [CrossRef]

6. Dellavalle, A.; Sampaolesi, M.; Tonlorenzi, R.; Tagliafico, E.; Sacchetti, B.; Perani, L.; Innocenzi, A.; Galvez, B.G.; Messina, G.; Morosetti, R.; et al. Pericytes of human skeletal muscle are myogenic precursors distinct from satellite cells. *Nat. Cell Biol.* **2007**, *9*, 255–267. [CrossRef]

7. Fuoco, C.; Sangalli, E.; Vono, R.; Testa, S.; Sacchetti, B.; Latronico, M.V.; Bernardini, S.; Madeddu, P.; Cesareni, G.; Seliktar, D.; et al. 3D hydrogel environment rejuvenates aged pericytes for skeletal muscle tissue engineering. *Front. Physiol.* **2014**, *5*, 203. [CrossRef]

8. Asakura, A.; Seale, P.; Girgis-Gabardo, A.; Rudnicki, M.A. Myogenic specification of side population cells in skeletal muscle. *J. Cell Biol.* **2002**, *159*, 123–134. [CrossRef]

9. Mitchell, K.J.; Pannerec, A.; Cadot, B.; Parlakian, A.; Besson, V.; Gomes, E.R.; Marazzi, G.; Sassoon, D.A. Identification and characterization of a non-satellite cell muscle resident progenitor during postnatal development. *Nat. Cell Biol.* **2010**, *12*, 257–266. [CrossRef]

10. Ferrari, G. Muscle regeneration by bone marrow-derived myogenic progenitors. *Science* **1998**, *279*, 1528–1530. [CrossRef]

11. Negroni, E.; Bigot, A.; Butler-Browne, G.S.; Trollet, C.; Mouly, V. Cellular Therapies for Muscular Dystrophies: Frustrations and Clinical Successes. *Hum. Gene Ther.* **2016**, *27*, 117–126. [CrossRef] [PubMed]

12. Shi, Y.; Inoue, H.; Wu, J.C.; Yamanaka, S. Induced pluripotent stem cell technology: A decade of progress. *Nat. Rev. Drug Discov.* **2017**, *16*, 115–130. [CrossRef] [PubMed]

13. Bearzi, C.; Gargioli, C.; Baci, D.; Fortunato, O.; Shapira-Schweitzer, K.; Kossover, O.; Latronico, M.V.; Seliktar, D.; Condorelli, G.; Rizzi, R. PlGF-MMP9-engineered iPS cells supported on a PEG-fibrinogen hydrogel scaffold possess an enhanced capacity to repair damaged myocardium. *Cell Death Dis.* **2014**, *5*, e1053. [CrossRef] [PubMed]

14. Darabi, R.; Arpke, R.W.; Irion, S.; Dimos, J.T.; Grskovic, M.; Kyba, M.; Perlingeiro, R.C. Human ES- and iPS-derived myogenic progenitors restore DYSTROPHIN and improve contractility upon transplantation in dystrophic mice. *Cell Stem Cell* **2012**, *10*, 610–619. [CrossRef]

15. Goudenege, S.; Lebel, C.; Huot, N.B.; Dufour, C.; Fujii, I.; Gekas, J.; Rousseau, J.; Tremblay, J.P. Myoblasts derived from normal hESCs and dystrophic hiPSCs efficiently fuse with existing muscle fibers following transplantation. *Mol. Ther. J. Am. Soc. Gene Ther.* **2012**, *20*, 2153–2167. [CrossRef] [PubMed]

16. Tedesco, F.S.; Gerli, M.F.; Perani, L.; Benedetti, S.; Ungaro, F.; Cassano, M.; Antonini, S.; Tagliafico, E.; Artusi, V.; Longa, E.; et al. Transplantation of genetically corrected human iPSC-derived progenitors in mice with limb-girdle muscular dystrophy. *Sci. Transl. Med.* **2012**, *4*, 140ra189. [CrossRef]

17. Uchimura, T.; Otomo, J.; Sato, M.; Sakurai, H. A human iPS cell myogenic differentiation system permitting high-throughput drug screening. *Stem Cell Res.* **2017**, *25*, 98–106. [CrossRef]

18. Borchin, B.; Chen, J.; Barberi, T. Derivation and FACS-mediated purification of PAX3+/PAX7+ skeletal muscle precursors from human pluripotent stem cells. *Stem Cell Rep.* **2013**, *1*, 620–631. [CrossRef]

19. Shelton, M.; Metz, J.; Liu, J.; Carpenedo, R.L.; Demers, S.P.; Stanford, W.L.; Skerjanc, I.S. Derivation and expansion of PAX7-positive muscle progenitors from human and mouse embryonic stem cells. *Stem Cell Rep.* **2014**, *3*, 516–529. [CrossRef]

20. Xu, C.; Tabebordbar, M.; Iovino, S.; Ciarlo, C.; Liu, J.; Castiglioni, A.; Price, E.; Liu, M.; Barton, E.R.; Kahn, C.R.; et al. A zebrafish embryo culture system defines factors that promote vertebrate myogenesis across species. *Cell* **2013**, *155*, 909–921. [CrossRef]

21. von Maltzahn, J.; Chang, N.C.; Bentzinger, C.F.; Rudnicki, M.A. Wnt signaling in myogenesis. *Trends Cell Biol.* **2012**, *22*, 602–609. [CrossRef] [PubMed]

22. Rome, S.; Forterre, A.; Mizgier, M.L.; Bouzakri, K. Skeletal Muscle-Released Extracellular Vesicles: State of the Art. *Front. Physiol.* **2019**, *10*, 929. [CrossRef] [PubMed]

23. Trovato, E.; Di Felice, V.; Barone, R. Extracellular Vesicles: Delivery Vehicles of Myokines. *Front. Physiol.* **2019**, *10*, 522. [CrossRef] [PubMed]

24. Guescini, M.; Maggio, S.; Ceccaroli, P.; Battistelli, M.; Annibalini, G.; Piccoli, G.; Sestili, P.; Stocchi, V. Extracellular Vesicles Released by Oxidatively Injured or Intact C_2C1_2 Myotubes Promote Distinct Responses Converging toward Myogenesis. *Int. J. Mol. Sci.* **2017**, *18*, 2488. [CrossRef] [PubMed]

25. Murphy, C.; Withrow, J.; Hunter, M.; Liu, Y.; Tang, Y.L.; Fulzele, S.; Hamrick, M.W. Emerging role of extracellular vesicles in musculoskeletal diseases. *Mol. Asp. Med.* **2018**, *60*, 123–128. [CrossRef] [PubMed]

26. Romancino, D.P.; Paterniti, G.; Campos, Y.; De Luca, A.; Di Felice, V.; d'Azzo, A.; Bongiovanni, A. Identification and characterization of the nano-sized vesicles released by muscle cells. *FEBS Lett.* **2013**, *587*, 1379–1384. [CrossRef]

27. Forterre, A.; Jalabert, A.; Berger, E.; Baudet, M.; Chikh, K.; Errazuriz, E.; De Larichaudy, J.; Chanon, S.; Weiss-Gayet, M.; Hesse, A.M.; et al. Proteomic analysis of C_2C1_2 myoblast and myotube exosome-like vesicles: A new paradigm for myoblast-myotube cross talk? *PLoS ONE* **2014**, *9*, e84153. [CrossRef]

28. Choi, J.S.; Yoon, H.I.; Lee, K.S.; Choi, Y.C.; Yang, S.H.; Kim, I.S.; Cho, Y.W. Exosomes from differentiating human skeletal muscle cells trigger myogenesis of stem cells and provide biochemical cues for skeletal muscle regeneration. *J. Control. Release* **2016**, *222*, 107–115. [CrossRef]

29. Polo, J.M.; Liu, S.; Figueroa, M.E.; Kulalert, W.; Eminli, S.; Tan, K.Y.; Apostolou, E.; Stadtfeld, M.; Li, Y.; Shioda, T.; et al. Cell type of origin influences the molecular and functional properties of mouse induced pluripotent stem cells. *Nat. Biotechnol.* **2010**, *28*, 848–855. [CrossRef]

30. Rizzi, R.; Di Pasquale, E.; Portararo, P.; Papait, R.; Cattaneo, P.; Latronico, M.V.; Altomare, C.; Sala, L.; Zaza, A.; Hirsch, E.; et al. Post-natal cardiomyocytes can generate iPS cells with an enhanced capacity toward cardiomyogenic re-differentation. *Cell Death Differ.* **2012**, *19*, 1162–1174. [CrossRef]

31. Chappell, J.C.; Bautch, V.L. Vascular development: Genetic mechanisms and links to vascular disease. *Curr. Top. Dev. Biol.* **2010**, *90*, 43–72. [CrossRef] [PubMed]

32. Birbrair, A.; Zhang, T.; Wang, Z.M.; Messi, M.L.; Enikolopov, G.N.; Mintz, A.; Delbono, O. Skeletal muscle pericyte subtypes differ in their differentiation potential. *Stem Cell Res.* **2013**, *10*, 67–84. [CrossRef]

33. Somers, A.; Jean, J.C.; Sommer, C.A.; Omari, A.; Ford, C.C.; Mills, J.A.; Ying, L.; Sommer, A.G.; Jean, J.M.; Smith, B.W.; et al. Generation of transgene-free lung disease-specific human induced pluripotent stem cells using a single excisable lentiviral stem cell cassette. *Stem Cells (Dayt. Ohio)* **2010**, *28*, 1728–1740. [CrossRef] [PubMed]

34. Cappellari, O.; Cossu, G. Pericytes in development and pathology of skeletal muscle. *Circ. Res.* **2013**, *113*, 341–347. [CrossRef] [PubMed]

35. Winkler, E.A.; Bell, R.D.; Zlokovic, B.V. Pericyte-specific expression of PDGF beta receptor in mouse models with normal and deficient PDGF beta receptor signaling. *Mol. Neurodegener.* **2010**, *5*, 32. [CrossRef]

36. Huang, F.J.; You, W.K.; Bonaldo, P.; Seyfried, T.N.; Pasquale, E.B.; Stallcup, W.B. Pericyte deficiencies lead to aberrant tumor vascularizaton in the brain of the NG2 null mouse. *Dev. Biol.* **2010**, *344*, 1035–1046. [CrossRef]

37. Peault, B.; Rudnicki, M.; Torrente, Y.; Cossu, G.; Tremblay, J.P.; Partridge, T.; Gussoni, E.; Kunkel, L.M.; Huard, J. Stem and Progenitor Cells in Skeletal Muscle Development, Maintenance, and Therapy. *Mol. Ther. J. Am. Soc. Gene Ther.* **2007**, *15*, 867–877. [CrossRef]

38. Sommer, A.G.; Rozelle, S.S.; Sullivan, S.; Mills, J.A.; Park, S.M.; Smith, B.W.; Iyer, A.M.; French, D.L.; Kotton, D.N.; Gadue, P.; et al. Generation of human induced pluripotent stem cells from peripheral blood using the STEMCCA lentiviral vector. *J. Vis. Exp. JoVE* **2012**, *68*, e4327. [CrossRef]

39. Raposo, G.; Stoorvogel, W. Extracellular vesicles: Exosomes, microvesicles, and friends. *J. Cell Biol.* **2013**, *200*, 373–383. [CrossRef]

40. Demonbreun, A.R.; McNally, E.M. Muscle cell communication in development and repair. *Curr. Opin. Pharm.* **2017**, *34*, 7–14. [CrossRef]

41. Le Bihan, M.-C.; Bigot, A.; Jensen, S.S.; Dennis, J.L.; Rogowska-Wrzesinska, A.; Lainé, J.; Gache, V.; Furling, D.; Jensen, O.N.; Voit, T.; et al. In-depth analysis of the secretome identifies three major independent secretory pathways in differentiating human myoblasts. *J. Proteom.* **2012**, *77*, 344–356. [CrossRef] [PubMed]

42. Thery, C.; Amigorena, S.; Raposo, G.; Clayton, A. Isolation and characterization of exosomes from cell culture supernatants and biological fluids. *Curr. Protoc. Cell Biol.* **2006**, *30*, 3–22. [CrossRef]

43. Mulcahy, L.A.; Pink, R.C.; Carter, D.R. Routes and mechanisms of extracellular vesicle uptake. *J. Extracell. Vesicles* **2014**, *3*, 24641. [CrossRef] [PubMed]

44. Murphy, D.E.; de Jong, O.G.; Brouwer, M.; Wood, M.J.; Lavieu, G.; Schiffelers, R.M.; Vader, P. Extracellular vesicle-based therapeutics: Natural versus engineered targeting and trafficking. *Exp. Mol. Med.* **2019**, *51*, 32. [CrossRef] [PubMed]

45. Grabowska, I.; Szeliga, A.; Moraczewski, J.; Czaplicka, I.; Brzoska, E. Comparison of satellite cell-derived myoblasts and C_2C1_2 differentiation in two- and three-dimensional cultures: Changes in adhesion protein expression. *Cell Biol. Int.* **2011**, *35*, 125–133. [CrossRef] [PubMed]

46. Guescini, M.; Guidolin, D.; Vallorani, L.; Casadei, L.; Gioacchini, A.M.; Tibollo, P.; Battistelli, M.; Falcieri, E.; Battistin, L.; Agnati, L.F.; et al. C_2C1_2 myoblasts release micro-vesicles containing mtDNA and proteins involved in signal transduction. *Exp. Cell Res.* **2010**, *316*, 1977–1984. [CrossRef] [PubMed]

47. Mylona, E.; Jones, K.A.; Mills, S.T.; Pavlath, G.K. CD44 regulates myoblast migration and differentiation. *J. Cell. Physiol.* **2006**, *209*, 314–321. [CrossRef]

48. Estelles, A.; Sperinde, J.; Roulon, T.; Aguilar, B.; Bonner, C.; LePecq, J.B.; Delcayre, A. Exosome nanovesicles displaying G protein-coupled receptors for drug discovery. *Int. J. Nanomed.* **2007**, *2*, 751–760.

49. Tan, J.Y.; Sriram, G.; Rufaihah, A.J.; Neoh, K.G.; Cao, T. Efficient derivation of lateral plate and paraxial mesoderm subtypes from human embryonic stem cells through GSKi-mediated differentiation. *Stem Cells Dev.* **2013**, *22*, 1893–1906. [CrossRef]

50. Roca, I.; Requena, J.; Edel, M.; Alvarez-Palomo, A. Myogenic Precursors from iPS Cells for Skeletal Muscle Cell Replacement Therapy. *J. Clin. Med.* **2015**, *4*, 243–259. [CrossRef]

51. Shoji, E.; Woltjen, K.; Sakurai, H. Directed Myogenic Differentiation of Human Induced Pluripotent Stem Cells. *Methods Mol. Biol.* **2016**, *1353*, 89–99. [CrossRef] [PubMed]

52. Tanaka, A.; Woltjen, K.; Miyake, K.; Hotta, A.; Ikeya, M.; Yamamoto, T.; Nishino, T.; Shoji, E.; Sehara-Fujisawa, A.; Manabe, Y.; et al. Efficient and reproducible myogenic differentiation from human iPS cells: Prospects for modeling Miyoshi Myopathy in vitro. *PLoS ONE* **2013**, *8*, e61540. [CrossRef]

53. Kim, K.; Zhao, R.; Doi, A.; Ng, K.; Unternaehrer, J.; Cahan, P.; Huo, H.; Loh, Y.H.; Aryee, M.J.; Lensch, M.W.; et al. Donor cell type can influence the epigenome and differentiation potential of human induced pluripotent stem cells. *Nat. Biotechnol.* **2011**, *29*, 1117–1119. [CrossRef] [PubMed]

54. Thomas, C.E.; Ehrhardt, A.; Kay, M.A. Progress and problems with the use of viral vectors for gene therapy. *Nat. Rev. Genet.* **2003**, *4*, 346–358. [CrossRef]

55. Zheng, J.K.; Wang, Y.; Karandikar, A.; Wang, Q.; Gai, H.; Liu, A.L.; Peng, C.; Sheng, H.Z. Skeletal myogenesis by human embryonic stem cells. *Cell Res.* **2006**, *16*, 713–722. [CrossRef]

56. Awaya, T.; Kato, T.; Mizuno, Y.; Chang, H.; Niwa, A.; Umeda, K.; Nakahata, T.; Heike, T. Selective development of myogenic mesenchymal cells from human embryonic and induced pluripotent stem cells. *PLoS ONE* **2012**, *7*, e51638. [CrossRef] [PubMed]

57. Zhu, S.; Wurdak, H.; Wang, J.; Lyssiotis, C.A.; Peters, E.C.; Cho, C.Y.; Wu, X.; Schultz, P.G. A small molecule primes embryonic stem cells for differentiation. *Cell Stem Cell* **2009**, *4*, 416–426. [CrossRef]

58. Choi, I.Y.; Lim, H.; Estrellas, K.; Mula, J.; Cohen, T.V.; Zhang, Y.; Donnelly, C.J.; Richard, J.P.; Kim, Y.J.; Kim, H.; et al. Concordant but Varied Phenotypes among Duchenne Muscular Dystrophy Patient-Specific Myoblasts Derived using a Human iPSC-Based Model. *Cell Rep.* **2016**, *15*, 2301–2312. [CrossRef]

59. Chal, J.; Oginuma, M.; Al Tanoury, Z.; Gobert, B.; Sumara, O.; Hick, A.; Bousson, F.; Zidouni, Y.; Mursch, C.; Moncuquet, P.; et al. Differentiation of pluripotent stem cells to muscle fiber to model Duchenne muscular dystrophy. *Nat. Biotechnol.* **2015**, *33*, 962–969. [CrossRef]

60. Valadi, H.; Ekstrom, K.; Bossios, A.; Sjostrand, M.; Lee, J.J.; Lotvall, J.O. Exosome-mediated transfer of mRNAs and microRNAs is a novel mechanism of genetic exchange between cells. *Nat. Cell Biol.* **2007**, *9*, 654–659. [CrossRef] [PubMed]

61. Xu, D.; Tahara, H. The role of exosomes and microRNAs in senescence and aging. *Adv. Drug Deliv. Rev.* **2013**, *65*, 368–375. [CrossRef] [PubMed]

62. Gutzeit, C.; Nagy, N.; Gentile, M.; Lyberg, K.; Gumz, J.; Vallhov, H.; Puga, I.; Klein, E.; Gabrielsson, S.; Cerutti, A.; et al. Exosomes Derived from Burkitt's Lymphoma Cell Lines Induce Proliferation, Differentiation, and Class-Switch Recombination in B Cells. *J. Immunol. (Baltim. Md. 1950)* **2014**, *192*, 5852–5862. [CrossRef] [PubMed]

63. Muylaert, D.E.; Fledderus, J.O.; Bouten, C.V.; Dankers, P.Y.; Verhaar, M.C. Combining tissue repair and tissue engineering; bioactivating implantable cell-free vascular scaffolds. *Heart (Br. Card. Soc.)* **2014**, *100*, 1825–1830. [CrossRef] [PubMed]

64. Forterre, A.; Jalabert, A.; Chikh, K.; Pesenti, S.; Euthine, V.; Granjon, A.; Errazuriz, E.; Lefai, E.; Vidal, H.; Rome, S. Myotube-derived exosomal miRNAs downregulate Sirtuin1 in myoblasts during muscle cell differentiation. *Cell Cycle* **2014**, *13*, 78–89. [CrossRef]
65. Charlton, C.A.; Mohler, W.A.; Blau, H.M. Neural cell adhesion molecule (NCAM) and myoblast fusion. *Dev. Biol.* **2000**, *221*, 112–119. [CrossRef]
66. Townley-Tilson, W.H.; Callis, T.E.; Wang, D. MicroRNAs 1, 133, and 206: Critical factors of skeletal and cardiac muscle development, function, and disease. *Int. J. Biochem. Cell Biol.* **2010**, *42*, 1252–1255. [CrossRef]

Review

Mechanisms Regulating Muscle Regeneration: Insights into the Interrelated and Time-Dependent Phases of Tissue Healing

Laura Forcina, Marianna Cosentino and Antonio Musarò *

Laboratory affiliated to Istituto Pasteur Italia—Fondazione Cenci Bolognetti, DAHFMO-Unit of Histology and Medical Embryology, Sapienza University of Rome, Via Antonio Scarpa, 14, 00161 Rome, Italy; laura.forcina@uniroma1.it (L.F.); marianna.cosentino@uniroma1.it (M.C.)
* Correspondence: antonio.musaro@uniroma1.it; Tel.: +39-06-4976-69-56; Fax: +39-06-4462-854

Received: 26 April 2020; Accepted: 20 May 2020; Published: 22 May 2020

Abstract: Despite a massive body of knowledge which has been produced related to the mechanisms guiding muscle regeneration, great interest still moves the scientific community toward the study of different aspects of skeletal muscle homeostasis, plasticity, and regeneration. Indeed, the lack of effective therapies for several physiopathologic conditions suggests that a comprehensive knowledge of the different aspects of cellular behavior and molecular pathways, regulating each regenerative stage, has to be still devised. Hence, it is important to perform even more focused studies, taking the advantage of robust markers, reliable techniques, and reproducible protocols. Here, we provide an overview about the general aspects of muscle regeneration and discuss the different approaches to study the interrelated and time-dependent phases of muscle healing.

Keywords: muscle regeneration; inflammatory response; satellite cells; cell precursors; experimental methods; stem cell markers; muscle homeostasis

---

1. Introduction

Muscle regeneration represents an important homeostatic process of adult skeletal muscle, which retains, after development, the ability to regenerate in response to different injured stimuli, restoring damaged myofibers [1–3]. This property of the adult muscle tissue has drawn great scientific attention over time, since the impairment of skeletal muscle regenerative potential characterizes a suite of physiopathologic conditions severely affecting human health. A significant contribution to regenerative studies is derived from the development of experimental protocols to induce controlled muscle damage and from the validation of cellular, molecular, and histological analysis to reveal, monitor, and characterize each step of tissue repair. Several models of muscle injury have been developed in rodents; however, the complex dynamic of events following different types of muscle injury has still to be clarified. Confounding interpretations can derive from the indiscriminate use of experimental damaging techniques, since an increasing body of evidence suggests that skeletal muscle can differentially respond to injuries which affect, at various degree, the distinct cellular and structural components.

In this review, we integrated the principles of the physiologic muscle regeneration with a technical approach, reporting key experimental methods and markers employed to study cellular and molecular interactors dominating each stage of muscle healing.

2. From Tissue Destruction to Recovery: Highlighting the Stages of Muscle Regeneration

The dynamic response of skeletal muscle to damaging events can be roughly divided into two main stages: tissue destruction and the stage of reconstruction. However, a suite of cellular and molecular events has been identified in these stages, leading to a more refined classification of the regenerative process. Indeed, muscle regeneration occurs in five interrelated and time-dependent phases, namely degeneration-necrosis, inflammation, regeneration, maturation/remodelling, and functional recovery, reflecting the hierarchy of the overall process dominating the tissue (Figure 1). Although the kinetics and amplitude of each phase can vary among organisms and may depend on the characteristic and intensity of the damaging agent, the overall dynamic of the phases of muscle healing is similar in different mammals (e.g., mouse, rat, and human) and can be monitored at morphologic, molecular, and functional levels.

2.1. Muscle Degeneration

Muscle necrosis occurs when the integrity of myofibers is severely compromised, and the irreversible damage generally involves alteration of plasmalemma permeability, associated with the uncontrolled ionic flux, organelle disfunction, and the loss of a proper architecture. Although necrotic fibers can be histologically identified as pale and enlarged, reflecting internal abnormalities, other methods can be used to rigorously evaluate and quantify the degree of muscle damage upon injury.

Evans Blue Dye (EBD) has been described as a necrosis-avid agent in mammal muscles since it showed the ability to penetrate only the damaged, necrotic myofibers [4–10]. EBD, also called T-1824 or Direct Blue 53, is a synthetic bis-azo dye characterized by a high-water solubility, a strong affinity for serum albumin, and a slow excretion. When injected intravenously or intraperitoneally in living animals, EBD can bind serum albumin, remaining stable and confined in the blood, and can be distributed throughout the entire body. However, at the site of the lesion, the dye can permeate altered cell membranes, accumulating in the cytoplasm of damaged cells [6]. A satisfactory labelling of permeable myofibers can be obtained in mice with a single intraperitoneal injection of a 1% EBD solution, injected at 1% volume relative to body mass and administered between 16 and 24 h prior to tissue sampling [10,11]. Moreover, EBD presents a double advantage for its visualization. It can be both easily identified macroscopically, by the striking blue color within tissue, or revealed through fluorescent microscopy in tissue sections or even in a whole muscle [12,13]. Indeed, EBD can emit a bright red fluorescence (620 nm excitation/680 nm emission) and the amount of biological dye penetrating in a damaged tissue can be quantified as a total intensity of fluorescence in a tissue sample by using confocal microscopy [11,14]. Since it is well known that serum proteins can cross into damaged fibers, sharing the same basic principle of the EBD, the presence of necrotic fibers in skeletal muscle sections can be histologically highlighted by immunofluorescence analysis for the intracellular accumulation of albumin or immunoglobulin G (IgG). For instance, in the mouse, IgG uptake has been recognized as a marker for necrosis in muscle tissue (Figure 1) [5,15].

Markers of tissue damage can be also detected in serum, since skeletal muscle proteins such as creatine kinase (CK), lactate dehydrogenase (LDH), and troponin, when systemically distributed, are well-recognized indexes of muscle tissue alterations, the intensity of which can vary under different physiopathologic conditions (Table 1) [16]. The most commonly used serum marker of myocellular damage is serum CK, a globular protein catalyzing the exchange of high-energy phosphate bonds between phosphocreatine and ADP produced during contraction [16,17]. Based on the critical role of CK in the maintenance of the energy homeostasis of muscle tissue, a specific isoform of the enzyme CK3 (CK-MM) is highly abundant in myofibers and it is released in the extracellular space when the sarcolemma loses the physiologic integrity.

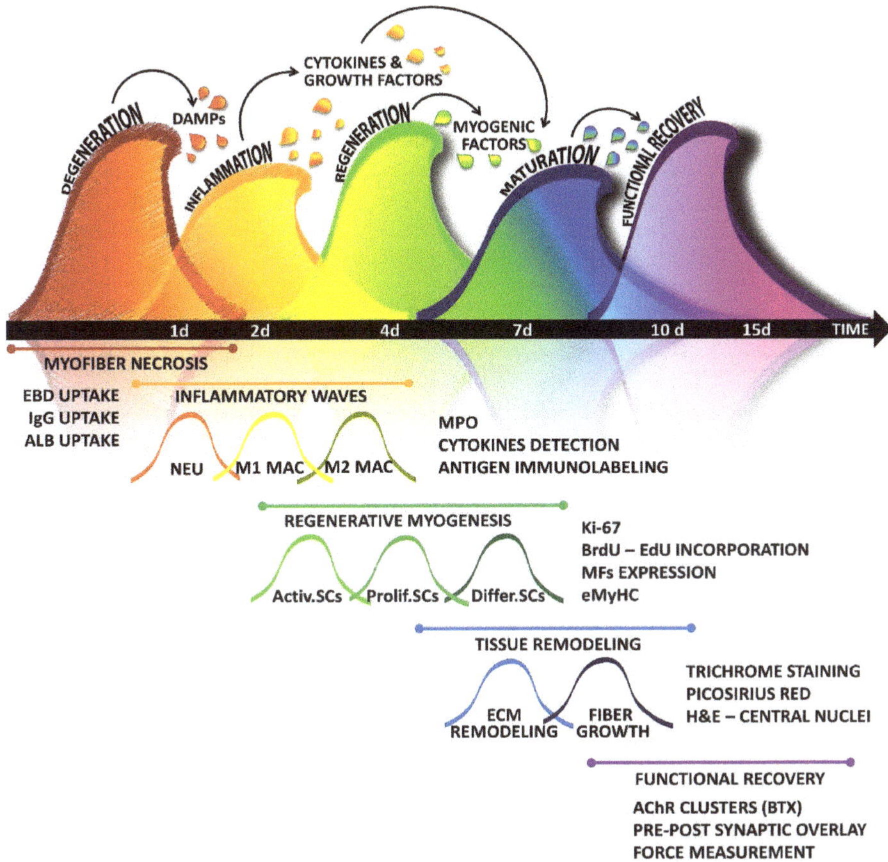

Figure 1. A simplified "wave on wave" model of skeletal muscle healing: The regenerative program activated by muscle tissue in response to damage can be outlined in five interrelated and time-dependent waves, namely degeneration, inflammation, regeneration, maturation-remodelling, and functional recovery, which can be highlighted by using different methodologies. Tissue injury leads to myofiber degeneration/necrosis. Damage stimuli activate the so-called sterile inflammation, characterized by the infiltration of different immune cells dominating in succession the lesion. Inflammation triggers also the regenerative stage, in which satellite cells, along with the support of other stem cells and precursors, undergo activation, expansion, and differentiation. The maturation of myofibers is accompanied by the fine remodelling of tissue architecture, with matrix rearrangement and angiogenesis. The last step of the healing process is characterized by the reconstitution of neuromuscular connections, necessary to regain tissue functionality. DAMPs: Damage-associated molecular patterns; EBD: Evans Blue Dye; IgG: Immunoglobulin G; ALB: Albumin; NEU: neutrophils; MAC: macrophages; MPO: myeloperoxidase; SCs: satellite cells; Activ.SCs: activated SCs; Prolif.SCs: proliferating SCs; Diff.SCs: differentiating SCs; BrdU: 5-bromo-2′-deoxyuridine; EdU: 5-ethynyl-2′-deoxyuridine; MFs: myogenic factors; eMyHC: embryonal myosin heavy chain; H&E: Haematoxylin and Eosin; AchR: Acetylcholine receptor; BTX: Bungarotoxin.

Table 1. Relevant markers of pivotal cellular and molecular actors in the different stages of muscle healing.

Stage	Markers	Recognition	References
Degeneration	Serum CK, LDH, troponin, miR-378a-3p, miR-434-3p	Muscle damage	[16,18]
	Albumin, IgG fiber uptake	Myofiber permeability	[5,15]
Inflammation	$CD11b^{pos.}/Ly6G^{pos.}/Ly6C^{neg.}$	Neutrophils	[19,20]
	$Ly6C^{high}/CCR2^{pos.}/CX3CR1^{low}$	Pro-inflammatory monocytes	[21–26]
	$Ly6C^{low}/CCR2^{neg.}/CX3CR1^{high}$	Patrolling monocytes	
	CD11b, Ly6C, F4/80, CD68, CD38, Gpr18, Fpr2	M1 Macrophages	[27–29]
	CD206, CD11c, CD163, Arginase1, Egr2, c-Myc	M2 Macrophages	
Regeneration	Pax3, Pax7, CD34, NCAM, VCAM-1, Cav1, Mcad, Syndecan 3-4, Sox8-15, Integrin α7-β1, CTR, Emerin, Hey1, Heyl	Quiescent SCs	[1,12,30–39]
	$Pax7^{high}/MyoD^{low}$, DGC, p38γ	Proliferating/Self renewing SCs	
	$Pax7^{low}/MyoD^{high}$, Myf-5, p38α-β	Committed SCs	[1,12,40–43]
	MyoD, Myogenin, Mrf4, miR206, miR486	Differentiating SCs	
	$CD45^{neg.}/CD31^{neg.}/\alpha7$ $int^{neg.}/Sca^{pos.}/PDGFR\ \alpha^{pos.}$	FAPs	[1,11,44,45]
Remodeling, Maturation and Functional retrieval	Collagen I–III–IV, laminin, fibronectin, proteoglicans	ECM	[46–50]
	eMyHC	Regenerating Myofibers	[12]
	AchRs/Synaptohysin/ Neurofilament markers	NMJs	[51]

CK: creatine kinase; LDH: lactate dehydrogenase; IgG: immunoglobulin G; CD: cluster of differentiation; Ly6C, Ly6G: lymphocyte antigen 6 complex, locus C, locus G; CCR2: C-C chemokine receptor type 2; CX3CR1: C-X3-C Motif Chemokine Receptor 1; Gpr18: G-protein coupled receptor 18; Fpr2: formyl peptide receptor 2; Egr2: early growth response protein 2; Pax3, Pax7: paired box transcription factor 3, 7; NCAM: neural cell adhesion molecule; VCAM: Vascular Cell Adhesion protein; Cav1: caveolin 1; Mcad: M-cadherin; Sox 8, 15: SRY-Box transcription Factor 8, 15; CTR: calcitonin receptor; SCs: Satellite cells; Hey1, Heyl: hairy/enhancer-of-split related with YRPW motif proteins; MyoD: myoblast determination protein; DGC: dystrophin-associated glycoprotein complex; Myf-5: myogenic factor 5; Mrf4: myogenic regulatory factor 4; Int: integrin; Sca: stem cell antigen; PDGFRα: platelet derived growth factor receptor alpha; FAPs: fibroadipogenic progenitors; eMyHC: embryonal myosin heavy chain; AchRs: acetylcholine receptors; NMJs: neuromuscular junctions.

Among biochemical markers of muscle damage, the serum levels of muscle-specific or muscle-enriched microRNAs (miRNAs) has been proposed [18]. Indeed, a number of miRNAs, including miR-1, miR-133, and miR-206 (myomiRs), have been involved in the regulation of critical myocellular processes such as satellite cell activity, skeletal muscle growth, adaptation, and regeneration [42,43,52,53]. Furthermore, in a recent profiling study, performed on notexin-injured rats, Siracusa and colleagues identified circulating miR-378a-3p and miR-434-3p as reliable biomarkers of acute muscle damage [18] (Table 1).

2.2. Inflammatory Waves

Tissue necrosis is known to stimulate a host inflammatory response named sterile inflammation because no exogenous infectious agents participate in the immune process. Necrotic cell death is mainly characterized by the swelling of organelles, increased cell volume, and the disruption of the plasma membrane, which leads to the release of the intracellular content. When intracellular components are dispersed throughout the extracellular space, they can act as signals, which have been termed as damage-associated molecular patterns (DAMPs), triggering inflammatory reactions [54,55]. Although it has been recognized as a contributor to pathologic changes, inflammation represents an important physiologic process playing a critical role in muscle homeostasis and regeneration. Indeed, the sequential recruitment of specific myeloid cell populations at the site of the lesion is considered the second of five interrelated phases of muscle regeneration (Figure 1) [12,56–60]. The first sensor of the innate immunity to be activated early after injury is the complement system, which allows the immediate immune response against damaged tissue and leads to the infiltration of inflammatory cells at the site of the lesion [21,61]. Neutrophils, along with mast cells, represent the first inflammatory myeloid cells that invade the site of muscle injury [21]. In particular, resident mast cell degranulate in response to muscle injury and release pro-inflammatory factors such as TNF-α (tumor necrosis factor alpha), IFN-γ (Interferon-γ), and IL-1β (interleukin-1β) [21], which stimulate the recruitment of peripheral neutrophils to the lesion site [12,62–66]. Furthermore, it has been recently demonstrated that ADAM8, a member of a disintegrin and metalloprotease (ADAM) family, contributes to the invasiveness of neutrophils into injured muscle fibers by reducing their adhesiveness to blood vessels after the infiltration into interstitial tissues [19]. The pro-inflammatory action of neutrophils is necessary to allow the removal of myofiber debris and to stimulate the homing of other pivotal inflammatory cells, facilitating the progress of muscle regeneration. The phagocytic activity of neutrophils involves the release of high concentrations of free radicals and proteases as well as the secretion of pro-inflammatory cytokines such as IL-1, IL-8, IL-6, and the soluble interleukin-6 receptor alpha (sIL6R). In particular, sIL6R can stimulate, within 24 h after damage, the homing of other inflammatory cell populations, namely monocytes and macrophages [21,59,67,68].

Macrophages becomes the predominant inflammatory cell type 2 days after injury, while neutrophils decline [21,59,67]. Although macrophages are generally recognized as highly specialized cells with phagocytic activity, responsible for tissue debris removal, this inflammatory population cannot be unequivocally labelled because of its heterogeneity, which still lack a comprehensive classification. A differential phagocytic activity has been described in resident macrophages, with ED1[pos.] cells highly participating in tissue response to acute damage and ED2[pos.]/ED3[pos.] cells showing no phagocytic activity and abundantly present in uninjured muscles [69]. Furthermore, macrophages found at the lesion can also derive from blood monocytes. Circulating monocytes derive from bone marrow and can be classed into at least two populations, based on the variable expression levels of specific markers, namely lymphocyte antigen 6 complex locus C (Ly6C) and chemokine receptors (CCR2 and CX3CR1) (Table 1) [21–23].

It has been proposed that Ly6C[high] monocytes can be recruited to the lesion thanks to the elevated expression of C-C chemokine receptor type 2 (CCR2) that then differentiate into pro-inflammatory macrophages M1 [24,26]. In contrast, patrolling Ly6C[low] monocytes are characterized by a low expression of CCR2 and can enter the damaged tissue in a CX3CR1-dependent manner, participating in tissue repair during the third wave of regenerative inflammation, as M2 pro-regenerative macrophages [25]. Other studies support the hypothesis that only inflammatory monocytes are recruited in injured skeletal muscle and then switch to anti-inflammatory subtype to support myogenesis [67,70]. Thus, the origin, the distinction, and even the existence of M1 and M2 populations are still controversial. However, it has been widely accepted that a first outbreak of macrophages works initially to remove the muscle debris and to secrete pro-inflammatory cytokines, while a subsequent appearance of non-phagocytic macrophages contributes to the shift of the inflammatory response toward resolutive events.

Thus, the enhanced expression of inflammatory mediators, mainly TNF-α, IL-6, and IL-1β, can clearly indicate an ongoing inflammatory response; however, these factors can be secreted by a wealth of cellular agents in a damaged muscle, being unspecific markers of cellular interactors in the pro-inflammatory stage of muscle regeneration. On the other hand, the detection and identification of different inflammatory population at the site of the lesion and thus the temporary collocation of the regenerative event can be obtained through the expression of specific markers.

Histological analysis is frequently performed to reveal the presence of inflammatory cells in regenerative studies. Immunofluorescence analysis for lymphocyte antigen 6 complex locus G

(Ly6G) and F4/80 expression in damaged murine muscle sections has been extensively used to detect neutrophils and macrophages, respectively. Additionally, other histological methods, such as the cytochemical myeloperoxidase (MPO) staining, can be used to detect the extent of infiltrating myeloid cells in damaged muscles. Indeed, MPO is a lysosomal enzyme contained in cytoplasmic primary granules of myeloid cells and can be detected through the oxidation of benzidine or the reaction of p-phenylenediamine and catechol in the presence of hydrogen peroxide (H_2O_2). This staining has been widely used to reveal the presence of neutrophils. However, although MPO mainly characterize azurophil neutrophilic granules, the assay can detect mammal monocytes without guaranteeing a fine discrimination of single cell types [71]. Of note, primary granules are absent in lymphocytes; thus, the MPO biochemical assay can be used as a marker for discerning myeloid from lymphoid cells.

Cytofluorimetric analysis can be useful to evaluate the quality of inflammation and to obtain an accurate quantification of inflammatory cell populations. Neutrophils have been identified as $CD11b^{pos.}/Ly6G^{pos.}/Ly6C^{neg.}$, whereas CD11b positive cells expressing Ly6C but not Ly6G have been identified as monocytes [19,29].

Of note, novel technologies are contributing to expanding the current knowledge about inflammatory cell function and fate. Intravital microscopy, high specific markers, along with the generation of novel transgenic animals allowed the visualization of fast-moving cells, providing promising tools to unravelling inflammatory-associated processes [20,72,73]. In a recent study, Wang and colleagues [74] marked $Ly6G^{pos.}$ cells with a photoactivatable green fluorescent protein (Ly6G-PA-GFP). Using this advanced technique, they combined intravital imaging and photoactivation methods to demonstrate that murine neutrophils do not die at the site of the lesion as previously thought [74,75]. Conversely, it has been shown that neutrophils, fulfilling their inflammatory tasks, are able to perform reverse migration from the local lesion, moving back to circulation and eventually home back to the bone marrow [20,74].

Macrophages (Mac) are a heterogeneous population of cells and their distinction often require the setup of a panel of markers, for which the combination specifically identifies a Mac subset. A marker panel for the detection of macrophages in skeletal muscle can be comprised of Siglec-F, CD11b, Ly6C, F4/80, and CD206 (Table 1) [28,29]. Other markers are required to detect M1 or M2 macrophages. For instance, it has been reported that M1 phenotype expresses CD68 whereas M2 macrophages express CD163. Furthermore, Jablonski and colleagues identified genes common or exclusive to either subset [27]. They report also a validated M1-exclusive pattern of expression for CD38, G-protein coupled receptor 18 (Gpr18), and Formyl peptide receptor 2 (Fpr2), whereas Early growth response protein 2 (Egr2) and c-Myc were recognized as M2 exclusive. Interestingly, they observed that Egr2, rather than the canonical M2 macrophage marker Arginase-1, labelled preferentially M2 macrophages (~70%), indicating that the unambiguous identification of macrophages still deserves further research. Of note, Insulin-like growth factor 1 (IGF-1) is a potent enhancer of tissue regeneration hastening the resolution of the inflammatory phase. It has been demonstrated that local macrophage-derived IGF-1 represents a key factor in inflammation resolution and macrophage polarization during muscle regeneration [76].

2.3. Regeneration

2.3.1. The Role of Satellite Cells

The reconstruction of injured muscle relies on the muscle stem cells, known as satellite cells (SCs), which reside between the basal lamina and sarcolemma of myofibers and are mitotically quiescent until required for growth or repair [77].

Although satellite cells can be easily recognized in healthy skeletal muscle tissue, in light of their sublaminar position, a wealth of markers has been identified to characterize the biology of these myogenic progenitors and to study their behavior during regenerative events. Quiescent satellite cells are characterized by the expression of Paired box transcription factors (Pax3 and Pax7), Neural cell adhesion molecule (NCAM), M-cadherin (Mcad), Forkhead box protein K (FoxK), tyrosine-protein kinase Met (c-Met), Vascular Cell Adhesion protein 1 (VCAM-1), CD34, Syndecan 3 and 4, Sox 8, Sox 15, Integrins (α7 and β1), Caveolin-1, Calcitonin receptor (CTR), Lamin A/C, Emerin, and hairy/enhancer-of-split related with YRPW motif proteins Hey1 and Heyl (Table 1) [1,12,30–39,41]. However, the transition of SCs from the quiescent state toward activation, commitment, and differentiation involves the genetic and epigenetic adaptation to novel biologic functions, entailing dynamic changes in the protein expression profile. Indeed, activated SCs retain the expression of Pax7, Mcad, VCAM1, Caveolin 1, and Integrin α7 along with the induction of proliferative and myogenic markers such as desmin, Myogenic factor 5 (Myf-5), and Myoblast determination protein (MyoD) [78–80].

Proliferating satellite cells can be also effectively identified by using non cell-specific markers of proliferation such as the Ki-67 and Proliferating cell nuclear antigen (PCNA). Ki-67 protein has been detected during all active phases of the cell cycle, namely G(1), S, G(2), and mitosis, but not in resting cells (G(0)), making its expression an excellent marker for determining the cycling fraction of a cell population [81]. Other cell proliferation assays involved the use of the thymidine analog 5-bromo-2'-deoxyuridine (BrdU) or 5-ethynyl-2'-deoxyuridine (EdU) and are based on the de novo synthesis of DNA occurring during cell duplication, which will be labelled by the incorporated nucleosides [82].

It is worth to report that the proliferation of satellite cells has a dual role: the generation of committed cells participating in regenerative processes and the replenishment of the stem cell pool after the exploitation. To achieve this activity, SCs are able to both symmetrically and asymmetrically divide. The symmetric division gives rise to an identical progeny with stem cell properties. Otherwise, through the asymmetric process, a single SC can generate a self-renewing daughter cell, retaining the expression of Pax7 and repressing MyoD (Pax7^{high}/MyoD^{low}) and a committed cell which downmodulates Pax7 and expresses MyoD (Pax7^{low}/MyoD^{high}). When the fine balance between self-renewal and commitment is altered, muscle homeostasis is impaired, leading to failure of the regenerative process and/or to the exhaustion of the stem cell pool. These conditions have been observed in Pax7$^{CreER/+}$:p38$\gamma^{fl/fl}$ mice and in dystrophin-deficient mice, respectively lacking p38γ and dystrophin expression [40]. This is because both dystrophin, as a pivotal member of the dystrophin-associated glycoprotein complex (DGC), and the γ isoform of the p38 MAP kinases are determinants which regulate SC asymmetric division, through the polarized restriction of factors involved in the cell fate decision. Indeed, it has been described that, during the asymmetric division, the apical daughter cell, retaining the expression of the DGC and presenting the phosphorylated p38γ isoform, sequestrates in the cytoplasm molecular mediators of the progression of the myogenic program, undergoing self-renewing. In contrast, other members of the p38 family, such as the α and β isoforms, participate to the commitment and differentiation of satellite cells [40].

The specificity of surface antigens can be used to quantify and isolate satellite cells by Fluorescence Activated Cell Sorting (FACS analysis). This method that has been described as robust and reliable for the isolation of SCs has been widely used, and different panels of antigen detection have been reported [11,32]. Among them, two panels for mouse skeletal muscle analysis, designed to exclude hematopoietic and stromal cells (CD45, CD11b, Ter119, CD31, and Sca-1) and to recognize surface markers present on satellite cells (β1-integrin/CXCR4, α7-integrin/CD34, and VCAM1) have been recently reported and validated [32]. Since it has been well established that satellite cells represent about 2–5% of the total nuclei in skeletal muscle tissue, an accurate evaluation of the muscle stem cell pool can provide indications about the physiopathologic state of the muscle. An abnormal number of SCs can be considered an index of ongoing regenerative events.

Besides the specific analysis of satellite cell activity and fate, overall signs of muscle regeneration can be histologically highlighted by the presence of central nuclei and cytoplasm basophilia. Both characteristics are easily evaluable through Hematoxylin and Eosin (H&E) staining, a standard staining for microscope examination of tissues (Figure 2) [74]. Hematoxylin presents a deep blue-purple color and stains nucleic acids, whereas eosin is pink and stains proteins. Although unspecific, this common staining can allow the visualization of both central nuclei and basophilic small fibers, readily identifying regenerating myofibers [83].

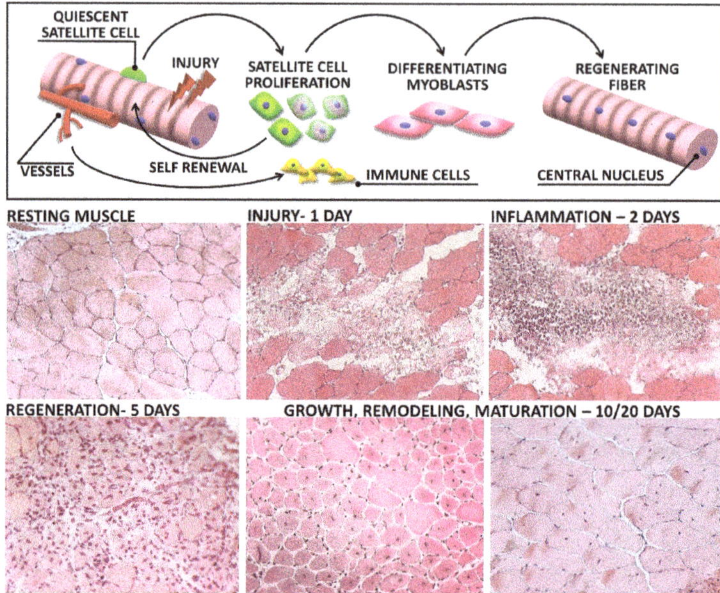

Figure 2. Skeletal muscle regeneration upon acute injury: The upper panel shows a schematic representation of relevant biological responses activated in muscle tissue following damage. Lower panel reports haematoxylin and eosin images of muscle sections, representative of each step of muscle degeneration and regeneration after cardiotoxin (CTX) injection. Early, after the injection (1 day), necrotic myofibers are evident in damaged muscle. During the second day after damage, the lesion is dominated by inflammatory infiltrated cells. Activated satellite cells undergo active proliferation, and newly regenerating fibers appears within the first week. Ten days after injection, the overall tissue architecture is restored and most of myofibers display centrally located nuclei. Regenerated myofibers then undergo progressive growth and maturation, highlighted by the increasing cross-sectional area and the nuclear relocation towards the periphery.

The downmodulation of proliferative genes ratifies the exit of satellite cells from the cell cycle. Committed and differentiating cells, based on the expression levels of Pax7 and MyoD, have been recognized as Pax7^{low}/MyoD^{high} and are characterized by the activation of late markers of the myogenic program, such as myogenin and the myogenic regulatory factor 4 (Mrf4).

The committed population of myoblasts can either fuse with existing myofibers, repair damaged muscle fibers, or alternatively fuse to each other to form new myofibers. This is a complex mechanism not yet fully elucidated which involves tightly regulated events of cell migration, recognition, and adhesion, resulting in an efficacious fusion process [84,85]. In addition to the recognized role of transforming growth factor beta (TGFβ) and IL-4 in myoblast fusion, a crucial role in muscle differentiation is also played by actin cytoskeleton and by components of the contractile apparatus [84–88]. Of note, regenerating myofibers can be also identified by the immunohistochemical detection of the embryonal myosin heavy chain (eMyHC) (Table 1). Indeed, it is well known that the embryonal isoform of the cytoskeletal protein, expressed during muscle development, can be transiently re-expressed in adult muscle upon injury. Newly formed myofibers express eMyHC within 2–3 days after damage and the embryonic protein can be detected for 2–3 weeks, being a robust marker of muscle regeneration [89].

The mature phenotype, which is successively finalized during the regenerative phase of maturation, can be then highlighted by the presence of markers including adult myosin heavy chain (MyHC) isoforms, conferring various contractile and metabolic properties to myofibers, enolase 3 (ENO3), and the muscle creatine kinase (MCK), pivotal components of terminally differentiated fibers [1,84,87].

2.3.2. The Role of "Non-Muscle" Stem Cells in Muscle Regeneration

It has been suggested that other stem cells and precursors, other than satellite cells, such as endothelial-associated cells [90], interstitial cells [91,92], bone marrow-derived side population [93,94], and fibroadipogenic progenitors (FAPs), can participate in muscle regeneration exerting a supportive role for SC activity [95]. These stem cell populations could either reside within muscle or be recruited via the circulation in response to homing signals emanating from the injured skeletal muscle. Among them FAPs, recognized as $CD45^{neg.}/CD31^{neg.}/\alpha7$-integrin$^{neg.}$ interstitial cells highly expressing Sca-1 expression [44] and PDGF receptor alpha (PDGF-R-alpha) [45], excited great interest, being involved in muscle regeneration and degeneration. Indeed, these mesenchymal progenitors are known to persist in an undifferentiated state in resting muscles, while under physiologic regenerative stimuli, FAPs undergo a transient expansion and produces paracrine factors promoting satellite cell-mediated regeneration [96]. A suite of recent findings clearly indicated that the cooperative activity of FAPs is required for muscle homeostasis and regeneration [97–99]. It has been reported that the inducible depletion of FAPs as well as the pharmacologic inhibition of their expansion in murine muscles resulted in a significative impairment of the healing process, affecting regenerative fibrogenesis and SC activity [97,98]. However, the physiologic action of FAPs is transient and finely regulated. These observations suggest that a qualitative microenvironment, generated by the balanced action of cellular and molecular players, is necessary to instruct stem cells to efficiently regenerate the injured tissue.

2.4. Tissue Remodelling and Maturation

In order to rebuild a functional muscle tissue, satellite cells and differentiating myoblasts need the structural and functional support of other cellular and molecular components. From a myogenic-centric point of view, an efficient muscle regeneration can be ratified by the formation and maturation of novel myofibers and/or the complete repair of damaged ones. This mark can be easily highlighted by the peripheralization of nuclei in mature myofibers (Figure 2). However, skeletal muscle is a multifaceted tissue with a complex cellular and molecular architecture, necessary for its functionality. Indeed, a complete muscle retrieval after injury requires the proper reconstitution of all the inner workings of the muscular machinery, namely extracellular matrix, vessels, and re-innervation. It is worth remembering that, during the degenerative and inflammatory phases following muscle injury, extracellular matrix (ECM), vascular network, and innervation undergo extensive degradation. The traumatic event, per se, can alter the ECM structure also damaging vasculature and nerves. Furthermore, several cell actors of muscle healing, including inflammatory cells and stem cells, can degrade matricellular proteins by secreting degrading agents such as metalloproteinases and elastase [100–102]. Since ECM is known to function as a scaffold to guide the formation of novel myofibers and neuromuscular junctions, the active deposition of matricellular components closely accompany muscle healing, and the remodelling of connective tissue, along with angiogenesis, defines the fourth stage of the regenerative process (Figure 1) [12,103,104]. The process starts with matrix deposition within a week post-injury, primarily due to the activity of fibroblasts in response to locally produced mediators such as TGF-β1 [105]. Although the fibrosis formation in case of self-healing injuries represents a beneficial response leading often to the efficient retrieval of muscle architecture, the overproduction of collagens within the injured area can lead to heavy scarring and the loss of muscular function.

The ECM is composed of specialized layers characterized by a variable composition of proteins, proteoglycans, and glycoproteins playing an integral role in structural support, force transmission, and regulation of the stem cell niche [46,106]. Thus, different collagen types can be labelled to evaluate the matrix composition and can be used as markers of connective tissue deposition. Indeed, although collagen I is the predominant type in the perimysium, the basement membrane is mainly comprised of laminin and collagen IV [47,48], whereas collagen I, III, and VI along with fibronectin in a proteoglycan-rich gel constitute the reticular lamina below the basement membrane (Table 1) [46–50]. The use of quantitative and qualitative high-magnification electron microscopy allowed the detailed description of the structure and composition of wild-type and fibrotic ECM. In particular, Gillies and colleagues not only clarified that collagen in the ECM is organized into large bundles of fibrils or cables but also reported that the number of the collagen cables were increased in fibrotic muscles [107]. Interestingly, since the increased number of cables but not the size was associated with an enhanced muscle stiffness, they suggested that alterations in fibrotic muscles can be related to the deregulated organization of ECM components and not only to the altered collagen content [107]. Despite the valuable and accurate results that can be obtained by using specific markers or imaging modalities, the restoration of the matrix or the excessive deposition of connective tissue can be revealed through a suite of standard histological techniques.

Trichrome staining has been frequently used to efficiently visualize connective tissue in muscle sections. The staining procedure is based on the combination of different dyes in a sequential manner. Acid fuchsin dye is used to stain muscle tissue, and although the dye can indiscriminately stain collagen, it can be removed from connective tissue by a polyacid of large molecular size such as phosphomolybdic acid. In addition, aniline blue can be used to stain collagens. Thus, in a standard Masson's Trichrome staining of muscle tissue, collagen appears blue, muscle tissue is stained red, and nuclei are stained dark brown thanks to the employment of the decolorization-resistant Weigert's hematoxylin [108].

Another sensitive method to perform qualitative and quantitative analysis of collagen network is Picrosirius red (F3BA) staining, developed by Junqueira and colleagues at the end of the 1970s [109]. The staining is based on the anionic properties of F3BA structure, which comprise sulfonate groups able to bind cationic collagen fibers, enhancing their natural birefringence under cross-polarized light [109–111]. Thus, under polarized light, collagen bundles stand out from the background appearing as green, red, or yellow. In particular, yellow-red birefringence has been associated with collagen type I bundles, whereas collagen type III has shown a weak birefringence and a green color [109–111]. Although the specificity of Picrosirius red for collagen types is controversial, this staining procedure is still considered one of the most powerful method to study and quantify collagen network [111,112].

2.5. Re-Innervation and Functional Recovery

The healing process is completed when regenerated myofibers rescue their functional performance and contractile apparatus (Figure 1). Thus, the regeneration of damaged muscles is only beneficial if the regenerated muscles become effectively innervated. Of note, this final stage of muscle regeneration must be also finely regulated. Interestingly, it has been demonstrated that, in addition to their specific role in the formation and/or repair of injured myofibers, satellite cells play a critical role in controlling myofiber innervation by upregulating the chemorepulsive semaphorin 3A expression [113]. Indeed, semaphorin 3A would prevent neuritogenesis when the regeneration of myofibers has not yet completed [113].

The first sign of a functional retrieval is the appearance of newly formed neuromuscular junctions (NMJs) between the surviving axons and the regenerated muscle fibers. The muscular terminal of NMJs can be visualized by labelling nicotinic acetylcholine receptor (nAChR) clusters on myofibers by using modified neurotoxins as probes (Table 1). In particular, α-Bungarotoxin (BTX), deriving from the venom of the banded krait, *Bungarus multicinctus*, showed the ability to bind the nAChR at the acetylcholine binding sites. Since the binding event occurs with high affinity and in a relatively irreversible manner, fluorescent α-bungarotoxin conjugates are valuable tools to localize and morphometrically analyze

NMJs [51]. Furthermore, α-BTX staining can be combined with immunolabeling of presynaptic vesicle proteins such as synaptophysin and syntaxin. Indeed, the exact overlay of BTX-derived fluorescence with the signal derived from the staining of presynaptic proteins can be considered an index of NMJ innervation [51,114]. Furthermore, it has been reported that the NMJ functionality can be indirectly evaluated through dual ex vivo electrical stimulation. In particular, we recently described an experimental protocol combining the direct electrical stimulation of muscle membrane and the stimulation through the nerve. Although the technique cannot be used to reveal morphological changes or biochemical changes in NMJs, the comparison of the muscle response to the two different stimulations can provide sensible indications about alterations in the NMJ functionality [115].

Electrical stimulations can be applied to freshly isolated muscles to evaluate the isometric contractile properties of the regenerated tissue [11,115–117]. Indeed, the recovery of the physiologic force-generating capability represents the most robust indicator of the effective muscle recovery after damage.

3. The Dynamic and the Regulation of Regenerative Phases are Altered in Pathologic Conditions: The Case of Muscular Dystrophy

The physiologic sequence of reparative phases, upon muscle injury, generally leads to the complete rescue of tissue morpho-functional properties. Unfortunately, the endogenous regenerative potential of skeletal muscle is not always sufficient to guarantee tissue restoration and/or maintenance. Pathologic conditions, including muscular dystrophies, are known to raise alterations in the dynamic and efficiency of regenerative steps. For this reason, complementary to physiologic regeneration, valuable information to clarify regenerative mechanisms can derive from models in which tissue healing is compromised because of cell-intrinsic and extrinsic defects. A well-characterized model of muscle wasting and regenerative impairment is the dystrophic mdx mouse, a classical model of Duchenne Muscular Dystrophy (DMD) pathology [118]. DMD is a degenerative disease in which the absence of the dystrophin protein leads to sarcolemma instability and fragility. The genetic defect is associated with the extensive damage of myofibers upon contraction which cannot be rescued by newly regenerated myotubes, being itself dystrophin deficient. This means that the degenerative stage of muscle healing, which is generally restricted to the first day after injury in wild-type mouse models, persists in dystrophic muscles throughout the necrotic stage of the pathology (mainly from 3 to 6 weeks of age) [119]. In this stage, EBD-injected mdx mice show high muscle permeability and dye uptake within damaged myofibers [11,120,121]. This is also associated with a sensible increase of serum CK and circulating myomiRs, further confirming the intense muscle damage in dystrophic muscle [11,120,122–128].

The continuous degeneration of dystrophin-deficient fibers represents a persistent stimulus to both inflammation and regeneration, thus inducing the alteration of the dynamic of both the inflammatory and regenerative stages. Indeed, it has been extensively described that dystrophic muscles are chronically dominated by inflammation, which can induce muscle fiber death through NO-mediated and perforin-dependent/independent mechanisms, respectively [68,119,129,130]. In accordance, Wehling and colleagues observed an improved membrane integrity upon antibody depletion of macrophages [131]. On the other hand, it has been recently reported that the local and transient depletion of macrophages in dystrophic muscle affected the balance between SC proliferation and differentiation, associated with defects in the formation of mature myofibers, inducing an exacerbation of the dystrophic phenotype [132]. Conflicting results can be associated with the technical approaches used in the studies, with reference to the persistence of the depletion and the stage of pathology in which the intervention acted. Indeed, Wehling and colleagues treated mdx mice with an anti-F4/80 antibody beginning at 1 week of age and continuing to 4 weeks of age, whereas Madaro and coworkers acted during the regenerative stage of the disease, that peaks between 9 to 12 weeks of age in mdx mice [131–133]. These observations, conflicting at first glance, provided intriguing insights about the complex impact of inflammatory events on the different stages of muscle regeneration. Thus, a better understanding of the inflammatory process in the dystrophic muscle and of the mediators involved might open novel therapeutic perspectives.

Inflammatory cells are responsible for the secretion not only of trophic factors but also of elevated levels of inflammatory mediators, influencing SC behavior. Among them, the enhanced expression of IL-6 is thought be involved in the alteration of the muscle stem cell pool by promoting the proliferation of SCs and the impairment of myoblast differentiation [11,134,135]. Interestingly, blockade of IL6 activity, using a neutralizing antibody against the IL6 receptor, conferred robustness to dystrophic muscle, impeded the activation of a chronic inflammatory response, significantly reduced necrosis, and activated the circuitry of muscle differentiation and maturation. This resulted in a functional homeostatic maintenance of dystrophic muscle. [121,136].

It is also worth to report that, in addition to maladaptive environmental signals, the altered SC behavior can be intrinsically dictated by the absence of dystrophin protein, since a defective compartmentalization of factors during asymmetric division in dystrophin deficient SCs can alter the daughter cell fate [40].

In addition, the delicate interaction between FAPs and SCs is altered in dystrophic muscles. FAPs desist from their supportive role and turn into fibro-adipocytes, which mediate fat deposition and fibrosis, contributing to the exacerbation of the dystrophic hostile microenvironment [44].

A recent study also uncovers the Wingless-related integration site (WNT)/GSK3/β-catenin axis as a new and previously unexplored pathway contributing to control FAP adipogenesis and muscle fatty degeneration, thus contributing to develop strategy to counteract intramuscular fat infiltrations in myopathies [137].

The persistence of degeneration, chronic inflammation, and defective myogenesis contribute to the alteration of the final phases of muscle healing in dystrophic muscles, resulting in the continuous attempt of defective regeneration. Altogether, these alterations lead to the progressive exhaustion of the SC pool, to accumulation of fatty/fibrotic tissue, and thus to the loss of muscle mass and functionality.

4. Technical Approaches to Induce Experimental Muscle Damage and Regeneration

The adaptative response of skeletal muscle to damage can differ in relation with the type of insult, and this is exactly coherent with the elevated plasticity of skeletal muscle tissue. Indeed, although the phases of muscle regeneration are closely interrelated and their time-dependent sequence is highly conserved in different vertebrates, the kinetic and amplitude of these procedural steps can vary depending on the organism or the extent/quality of damaging events. Indeed, a traumatic event can lead to the lesion of a single myofiber or of a localized segment of a fascicle. Furthermore, an insult can induce the degeneration of an entire fiber or can pertain to a number of myofibers dispersed throughout uninjured tissue [138,139]. Furthermore, an increasing body of evidence suggests that the events starting early after the injury profoundly influence the dynamic of tissue regeneration, affecting at various degree the different muscle components [140]. These heterogeneous events can contribute to the occurrence of doubts and difficulties in the field of clinical and experimental pathology [138]. Thus, a comprehensive understanding of the mechanisms underlining skeletal muscle adaptation to different insults can contribute to extending the current knowledge about muscle physiology and can allow the development of specific pro-regenerative therapies. To this aim, animal models of muscle injury represent a valuable and powerful tool to monitor and study muscle response to damage.

4.1. Models of Physical Injury

Murine models of acute muscle injury have been extensively studied to investigate molecular mechanisms underlining regenerative events characterizing each phase of muscle regeneration. Since the induction of an injurious assault is a prerequisite for muscle regeneration, with the necrotic phase being the first step of muscle restoration, the choice of a proper model of injury is critical for a correct interpretation of data and for the dissection of molecular mechanisms taking place in damaged muscles. A wealth of experimental procedures, adopted to induce muscle damage, have been developed and described over time, and qualitative/quantitative differences in the tissue response have been reported. Among them, the most commonly used methods are physical and chemical procedures. Most of the protocols of physical injury, which include freeze injury and crush models, are highly invasive and present technical complexities.

4.1.1. Freeze Injury

The freeze injury (FI) method mainly consists of a skin incision to expose the target muscle, and a single or repetitive action of freeze-thawing by applying for a prefixed time (10–15 s) a liquid nitrogen or dry ice cooled metallic rod [140–142]. Using this protocol, the operator can induce a diffuse necrosis in the treated muscle but the extent of muscle damage can vary not only depending on the number of freezing cycles but also with the pressure applied to the tissue with the cooled probe. The operator-dependent variability can limit the reproducibility of the data and the homogeneous use of the protocol in different laboratories. However, Hardy and colleagues, in a comparative study of muscle damage and repair, highlighted how the freeze-injury method induced a severe necrosis at the site of the lesion, destroying muscle cell components such as myofibers and satellite cells, along with basal lamina and vascular bed [140]. This can allow the study of the dynamics of infiltrating cells. Indeed, the region of damage, the so-called "dead zone", is well marked by the absence of viable cells, and the activation of regenerative events particularly requires the migration inside the lesion not only of inflammatory cells and non-myogenic supportive cells but also of myogenic progenitors [142]. Thus, viable cells infiltrating the lesion are easily identifiable since they appeared as directionally displaced from the spared tissue into the dead zone, allowing the dissection of regenerative events. In particular, the first invading population 18 h after FI is comprised of neutrophils and their presence is accompanied by an early increase of monocyte chemoattractant protein 1 (MCP-1) IL-6. The peak of expression of MCP-1 and IL-6 forewarns the second wave of cellular infiltration, characterized by F4/80[pos.] macrophages and myogenic cells, since both macrophages and muscle precursor cells express the CCR2 and are thus responsive to MCP-1 [141]. IL-6 is a pro-inflammatory cytokine with important regulatory actions on muscle stem cell functions; thus, the heightened expression of this pleiotropic factor can participate in both inflammatory and regenerative processes. This is consistent with the observation of a regenerative front of myoblasts at the periphery of the death zone at a few days after the neutrophilic peak since neutrophils are also a pivotal source of the soluble isoform of the IL-6 receptor alpha (IL6R) necessary to amplify IL-6 signal transduction [1,120,140,143]. Of note, the levels and activity of IL-6 must be finely tuned, since circulating levels of the proinflammatory cytokine IL-6 can also perturb the physiologic redox balance in skeletal muscle and can contribute to exacerbating muscle disease [143,144].

It is worth to report that the extent of tissue damage in FI model is highly elevated, and this event can profoundly affect the behavior of satellite cells from early after trauma. Indeed, it has been reported that there is a dramatic delay of satellite cells early after freeze injury (18 h), which has been quantified as about 90% of devoid cells compared to uninjured muscles. The regeneration of the damaged site is largely accomplished by progenitors deriving from outside the lesion; thus, the proliferation of SCs occurs days after damage and cycling SCs have been reported until one month after FI [140–142]. Although the histological retrieval of muscle architecture appeared complete one month after damage, the total number of SCs in FI mice returned to control levels three months after freezing.

The features of skeletal muscle response to FI were useful to study cellular actors of muscle regeneration and to clarify the involvement of the different cell populations. For instance, in 1986, Shultz and colleagues used the FI technique on the entire extensor digitorum longus (EDL) muscle to verify whether myogenic cells could migrate from adjacent muscles or could be delivered through the bloodstream [145]. Their results highlighted how muscle regeneration mainly depends on the activity of the local population of satellite cells [145]. Although extrinsic myogenic cells, such as migrating myoblasts and CD133[pos.] mononucleated cells identified in adult peripheral blood, can potentially reach the lesion, they would not be sufficient to regenerate the entire muscle [145,146].

4.1.2. Crush Injury and Ischemia-Reperfusion Damage

A muscle-crush injury occurs when high pressure is applied to skeletal muscle, which undergoes blood flow interruption, inducing the damage of myofibers. The combination of mechanical force and ischemia is known to cause an acute rhabdomyonecrosis since the profound alteration of the pressure balance can impair the volume regulation of myocytes, along with their permeability, leading to cell swelling [147].

Several experimental procedures have been developed over time to induce muscle-crush injury in rodents as a model of common trauma in humans and to study acute muscle inflammation and regeneration [148–153]. Among them, one of the most used is the opened model in which the muscle of the animal, generally the pelvic limb muscle, is surgically exposed and a force is applied by using a clamp.

Considering the invasiveness of the methodology and the needs of technical skills to perform the experiments guaranteeing the reproducibility of the data, closed noninvasive protocols have been tested. However, most of them involved dropping of weights upon the interested muscle region and such procedures can result in unwanted bone fractures. The fine-tuning of the procedure is still ongoing in order to minimize additional tissue damage induced by surgical interventions in opened models and to reduce the incidence of fractures due to dropping weights in closed models. For instance, Dobek and colleagues [148] proposed a sustained-force model of lower-extremity crush injury able to induce an acute inflammatory response, thereby reducing the extent of bone fractures. The proposed method, which has been described as a refinement of previous models, involved the use of a crush injury device platform. An air compressor activated a piston, situated in direct contact with the area to be injured, providing a contained force to the selected muscle. They reported that, although the force imposed was smaller than that applied in other studies (about 30 N in comparison with about 250 N), it was sufficient to induce muscle damage and the expected acute inflammatory response. Accordingly, it has been reported that a violent crush can destroy muscle tissue; however, even when the force is insufficient to directly wreck myofibers, the combination of mechanical force and ischemia will rapidly induce tissue degeneration [147]. Indeed, as a physiologic response to tissue injury, it has been reported that neutrophils, identified as 1A8, 7/4, and granulocyte antigen 1 (Gr1)-positive cells, rapidly invade the crushed muscle at the site of the lesion and then decreased from 24 to 48 h after injury [148]. CD68[pos.] and F4/80[pos.] macrophages followed the neutrophilic invasion increasing from 24 to 48 h after injury. However, it is plausible that the controlled force applied to the muscle would induce a mild tissue degeneration, useful to study the kinetic of inflammatory cell infiltration but probably with limitations regarding the study of regenerative events under critical conditions.

On the other hand, Criswell and colleagues [154] proposed and described a procedure of muscle crush in rats, which was able to induce tissue degeneration and to mimic the compartment syndrome (CS), a severe consequence of intense crush injuries frequently occurring in humans. Of note, the compartment syndrome occurs when the pressure within muscle fascicles dramatically increase due to posttraumatic ischemic swelling. This results in both ischemic and reperfusion insults, destroying the vasculature and the neural network and inducing the extensive necrosis of muscle tissue. In this protocol, a controlled compression of the rat hindlimb, proximal to the EDL muscle, has been cleverly obtained by using a neonatal blood pressure cuff in order to constantly maintain the pressure in a range of 120–140 mmHg for 3 h [154]. The persistent compression resulted in a composite injury of the muscular, vascular, and neural compartments. Tissue edema and disorganization were early observed 24 h after injury, and within the first 4 days, 50% of the muscle fibres underwent degeneration. Immune cell infiltration, as described in other models, occurred within 2 days after damage and persisted throughout the first week, with a peak at day four [154–156]. Following the acute inflammatory response, fibroblast and myofibroblast growth resulted in enhanced collagen deposition, also supporting the formation of newly regenerating fibers. Indeed, although early markers of satellite cell activation such as Pax7 and MyoD were observed early after damage (2 days), regenerating myofibers were detected 7 days after injury in correspondence with the phase of collagen deposition

(Figure 1) [154]. The extensive damage induced by muscle compression was also highlighted by signs of denervation, such as the dispersed localization of acetylcholine receptors around crushed myofibers and of vasculature alterations, including neo-angiogenesis preceded by the presence of enlarged vessels and hemorrhagic areas [154,156,157].

4.2. Chemical Damage Induced by Myotoxic Agents

The injection of myotoxic agents, such as the snake venom-derived toxins notexin or cardiotoxin, is one of the most frequently used methods to experimentally induce muscle damage and to study the subsequent regeneration. This is because the degeneration induced by these agents has been described as rapid, vigorous, and reproducible [139,158]. Moreover, venom toxins have been recognised as quite specific toxic agents on muscle fibers, without undermining blood vessels, basal lamina, and thus the activity of satellite cells [159–164]. This quite specific action can allow the dissection of regenerative events in a simplified model of muscle injury and study of the behavior of cellular actors and the profile of molecular players in muscle regeneration.

4.2.1. Cardiotoxin Injection

Cardiotoxins (CTX) are small polypeptides made of 60–63 amino acid residues acting as protein kinase C-specific inhibitor. Over 40 homologous cardiotoxins have been isolated and sequenced over time [165]. The purified toxin derived from the venom of the Indian cobra snake *Naja naja* or from the *Naja mossambica mossambica* is the most widely used myotoxic agent in protocols of experimental injury [140]. Although protocols can vary among laboratories, the method mainly consists of an intramuscular injection of about 20–50 µL of a 10 µM CTX working solution in sterile phosphate buffered saline (PBS). The most frequently treated muscles in murine models are hindlimb muscles such as the tibialis anterior muscle (TA) or gastrocnemius [140,166,167]. The myolytic activity of CTX involves the alteration of ion fluxes, induced by membrane depolarization, and is accompanied by the loss of protein content and organelle breakdown. Muscle degeneration occurs early after the injection, and the injured tissue is rapidly invaded by inflammatory mononucleated cells (Figure 2). Enlarged necrotic fibers in CTX-treated muscles are reached firstly by neutrophils and, later on, by macrophages which have been also described as penetrating swollen fibers [168]. Hardy and colleagues, in a benchmark work on different models of muscle injury, reported that the inflammatory response stimulated by CTX-induced necrosis was not exuberant, and it has been described that, although the kinetic of infiltrating cells is maintained, the phases are more defined and staggered if compared to other models of injury [140]. Furthermore, pro- and anti-inflammatory mediators, except for IL-6 levels showing a significant heightening, undergo a weak induction at early stages to then return to basal levels. The persistence of elevated levels of IL-6 have been detected one month after damage, possibly explaining the highly increased number of satellite cells in completely regenerated CTX-injected muscles [140]. It is worth to report that a significant myofiber hypertrophy and an increased muscle weight has been observed in muscle regenerated after CTX injection [167,169]. On the other hand, it has been postulated that CTX itself may have chemotactic properties enhancing both macrophages and satellite cell activity, thus inducing an early and efficient tissue regeneration [139,167]. Accordingly, newly regenerating myotubes with central nuclei can be observed 4 days after damage; 7 days after injury, the inflammatory response declines and the diameter of centrally nucleated fibers considerably increase [168]. Although it has been extensively described that CTX toxic action does not directly influence microvasculature, a complete destruction of the capillary network has been reported in a 3D study on CTX-injected muscles derived from Flk1$^{GFP/+}$ mice [140]. However, the initial vasculature breakdown was followed by active angiogenesis, and 1 month after the injection, the vessel network was restored [140]. The retrieval of a proper blood supply in injured muscle contributes to the efficient regeneration, without the occurrence of tissue fibrosis [164].

The ability of CTX to induce myofiber degeneration sparing the integrity of both basal lamina and satellite cells and thus inducing a controlled and rapid process of tissue reconstruction has been extensively used over time to study the role of molecular and cellular interactors in muscle regeneration. For instance, the specific action of cardiotoxin in inducing myofiber necrosis but not satellite cells (SCs) death in combination with a model of local Pax7[pos.] cell depletion contributed to clarifying the role of Pax7[pos.] cells in adult myogenesis [170]. Sambasivan and colleagues, inducing muscle damage in the presence or absence of satellite cells and monitoring the subsequent regenerative events, not only reported evidence about the essential role of SCs in skeletal muscle regeneration but also suggested the intriguing possibility that a threshold number of Pax7[pos.] cells would be required to obtain an efficient tissue reconstruction. This action would be associated not only with the proliferative rate of activated SCs but also with the potential ability of satellite cell to orchestrate the pro-regenerative action of non-myogenic cells in damaged muscle tissue. Furthermore, the cardiotoxin method was largely employed to dissect the action of inflammatory mediators in the stem cell niche after injury and to study the behaviour of satellite cell under pathologic conditions in which regeneration is known to be impaired [99,166,170–173].

4.2.2. Notexin Injection

Notexin (NTX) is a myotoxic agent contained in the venom of the *Notechis scutatus*, the Australian tiger snake, and has been described as having a more toxic effect than cardiotoxin (four times more toxic than CTX) [174]. This phospholipase A_2 presents an elevated myolytic impact, through the hydrolyzation of sarcolemma lipids, inducing the alteration of ionic fluxes, hypercontraction, and thus the degeneration of myofibers. Tissue insult provoked by notexin has been described as highly degenerative, causing myofiber breakdown and the loss of skeletal muscle functionality three days after the injection [175]. Moreover, a comparative study performed by Plant et al. reported that the maximum force of contraction recovered by notexin-injected muscles was reduced in comparison with other experimental models of muscle injury, with a percentage of force retrieval of only 10% seven days after damage and 39% after 10 days [175]. These data, along with the results from others benchmark studies, suggested that the extent and the modality of damage can influence the entity and the velocity of muscle recovery and thus the study of regenerative events [140,175–177]. For instance, myofiber regeneration, highlighted by the presence of centrally positioned nuclei, has been observed in the entire muscle injected with the toxin only 7 days after injury in contrast with the earlier observation reported in other models such as CTX-injury [140,167]. Despite the recognised action of notexin in inducing a generalized muscle damage, it is worth to report that fibers have a differential sensitivity to phospholipase A depending on the metabolism, with oxidative fibers showing an elevated susceptibility to NTX-induced damage [139].

Another important difference in the muscle response to notexin injection is the inflammatory response. Indeed, it has been described that there is a granulomatous inflammatory reaction to NTX-induced degeneration. Although the extensive necrosis produced by the myotoxic action of notexin did not activate an immediate inflammatory cell invasion, the immune response occurs with cell infiltration 4 days after damage. Instead of a typical kinetic of acute inflammatory response directed to the resolution, 12 days after injury in notexin-injected muscle, it was possible to observe multifocal calcium deposits, remains of necrotic myofibers, as a midpoint of a granulomatous reaction [140].

Interestingly, these chronically inflamed foci persist even when muscle tissue is quite completely regenerated 3–6 months after the experimental injury, potentially contributing to the establishment of an altered immune milieu. Furthermore, it has been reported that NTX can exert a neurotoxic action by blocking acetylcholine release, thereby altering the neuromuscular junctions, which must be restored for a functional tissue retrieval [178].

4.2.3. Bupivacaine Administration

Another agent used to induce reversible muscle damage with the purpose to study regenerative events is Bupivacaine (BPVC). Bupivacaine is a local anesthetic that, thanks to its highly lipophilic properties, can efficiently penetrate the sarcolemma. Although the precise mechanisms have to be fully clarified, it has been described that Bupivacaine can induce muscle degeneration by perturbing the homeostasis of mitochondria and sarcoplasmic reticulum (SR), producing a dramatic calcium efflux and a simultaneous block of calcium reuptake by the SR. This action results in the hypercontraction and rapid death of myofibers along with the mitochondria membrane depolarization and sarcoplasmic reticulum alterations, which can further induce muscle degeneration [116,179,180]. Despite the intense impact recognized on rat skeletal muscle which has been reported, Bupivacaine can induce only a faint degeneration when injected in murine muscles. Indeed, the degenerative potential of this anesthetic has been described as limited, if compared to notexin or cardiotoxin, since its injection causes the degeneration only of a 45 percentage of fibers [175]. In accordance with the low degree of muscle degeneration induced by Bupivacaine, it has been reported that the force-generating capability of injected muscle was reduced to 42% of control muscles three days after damage and that this impairment was quite completely restored ten days after the injection [175]. However, the analysis of injected muscle cross sections performed by Plant and colleagues revealed that bupivacaine can spread throughout the muscle, equally affecting both inner and peripheral regions of the muscle. In accordance with the low degree of muscle damage induced by BPVC in murine muscle, a rapid inflammatory response and regenerative phase has been described. Three days after the injection of 50–100 µL of 0.5% BPVC, a robust inflammatory infiltrate can be observed surrounding necrotic fibers, whereas small regenerating fibers appearing on day 5 seem to quite completely regenerate the lesion by day 14, when inflammatory cells are significantly reduced [181].

5. Conclusions

Muscle regeneration is one of the most important homeostatic processes of adult tissue and, as such, must be finely regulated to guarantee functional recovery and to avoid muscle alteration and diseases [182]. Skeletal muscle regeneration is a coordinate process in which several factors are sequentially activated to maintain and/or restore a proper muscle structure and function. Although the main actors of the entire process are satellite cells, a heterogenous group of other cells cooperate to reestablish muscle homeostasis after damage. Indeed, each stage of the stepwise muscle healing is dominated by a peculiar combination of cell agents and molecular signals playing a specific role in the complex framework of regeneration. However, the multifaceted nature of the regenerative process has to be still completely unveiled, and a number of pathologic conditions impairing muscle regeneration still lack an effective therapy. Thus, the comprehensive understanding of healing mechanisms still deserves further research to identify novel reliable biomarkers and to develop advanced techniques supporting the future innovation of regenerative studies.

Author Contributions: A.M. conceptualized the study; A.M. and L.F. wrote the original draft; L.F. and M.C. wrote/reviewed and edited the text and figures. All authors have read and agreed to the published version of the manuscript.

Funding: This work was supported by ASI, Ricerca Finalizzata (RF-2016-02364503), Ateneo-Sapienza.

Acknowledgments: We apologize to colleagues whose studies were not cited. The authors are thankful to Carmine Nicoletti and Carmen Miano for technical support.

Conflicts of Interest: The authors declare no conflict of interest.

Abbreviations

AchR	Acetylcholine receptor
Activ.SCs	Activated SCs
ADAM	A Disintegrin and Metalloproteinase
ADAM8	A Disintegrin and Metalloproteinase Domain-Containing Protein 8
ADP	Adenosine diphosphate
ALB	Albumin
BPVC	Bupivacaine
BrdU	5-bromo-2′-deoxyuridine

BTX	α-Bungarotoxin
Cav1	Caveolin 1
CCR2	Chemokine Receptor type 2
CD11b	Cluster of Differentiation 11 b also known as Integrin Alpha M
CD133	Cluster of Differentiation 133
CD163	Cluster of Differentiation 163
CD206	Cluster of Differentiation 206 also known as Mannose receptor C-type 1
CD31	Cluster of Differentiation 31
CD34	Cluster of Differentiation 34
CD38	Cluster of Differentiation 38
CD45	Cluster of Differentiation 45
CD68	Cluster of Differentiation 68
CK	Creatine kinase
CK3 (CK-MM)	Creatine Kinase MM isoform
c-Met	Tyrosine-protein Kinase Met
c-Myc	MYC proto-oncogene, bHLH transcription factor
CS	Compartment Syndrome
CTR	Calcitonin Receptor
CTX	Cardiotoxin
CX3CR1	C-X3-C Motif Chemokine Receptor 1
CXCR4	C-X-C Chemokine Receptor type 4
DAMPs	Damage-Associated Molecular Patterns
DGC	Dystrophin-associated Glycoprotein Complex
Diff.SCs	Differentiating SCs
DMD	Duchenne Muscular Dystrophy
EBD	Evans Blue Dye
ECM	Extracellular Matrix
ED1	Monoclonal antibody staining a single chain glycoprotein of 110 kDa on the lysosomal membrane of myeloid cells, i.e., the majority of tissue macrophages (being the rat homologue of human CD68)
ED2	Monoclonal antibody reacting with a membrane antigen (175, 160, and 95 kDa) on resident rat macrophages such as monocytes and dendritic cells. ED2 discriminates between thymic cortical (positive for ED2) and medullary (negative for ED2) macrophages. The antigen is identical with CD163.
ED3	Monoclonal antibody recognizing the rat CD169 cell surface antigen, a 185 kDa molecule expressed by macrophages in lymphoid organs (no monocytes or granulocytes). In the thymus, the antigen is expressed on clusters of dendritic cells (thymic nurse cells or TNC's) in the (outer) cortex.
EDL	Extensor Digitorum Longus
EdU	5-ethynyl-2′-deoxyuridine
Egr2	Early growth response protein 2
eMyHC	embryonal Myosin Heavy Chain
ENO3	Enolase 3
F3BA	Picrosirius red
F4/80	Mouse macrophage marker. Also known as Ly71 and EMR1, the F4/80 antigen is part of the EGF-TM7 family.
FACS	Fluorescence-Activated Cell Sorting
FAPs	Fibroadipogenic Progenitors
FI	Freeze Injury
FoxK	Forkhead box protein K
Fpr2	Formyl peptide receptor 2
Gpr18	G-protein coupled receptor 18
Gr1	Granulocyte antigen 1
GSK3	Glycogen Synthase Kinase 3
H&E	Haematoxylin and Eosin staining

H2O2	Hydrogen peroxide
Hey1	Hairy/enhancer-of-split related with YRPW motif protein 1
Heyl	Hairy/enhancer-of-split related with YRPW motif protein l
IFN-β	Interferon-β
IFN-γ	Interferon-γ
IGF-1	Insulin-like Growth Factor 1
IgG	Immunoglobulin G
IL-1	Interleukin 1
IL-1β	Interleukin-1β
IL-4	Interleukin 4
IL-6	Interleukin 6
IL6R	IL-6 receptor alpha
IL-8	Interleukin 8;
Ki-67 protein (also known as MKI67)	marker of proliferation KI-67
LDH	Lactate Dehydrogenase
Ly6C	Lymphocyte antigen 6 complex, locus C
Ly6G	Lymphocyte antigen 6 complex, locus G
Ly6G-PA-GFP	Ly6Gpos. cells with a photoactivatable GFP
MAC	Macrophages
Mcad	M-cadherin
MCK	Muscle Creatine Kinase
MCP-1	Monocyte Chemoattractant Protein 1
MFs	Myogenic Factors
MyHC	Myosin Heavy Chain
miRNAs	microRNAs
MPO	Myeloperoxidase
Mrf4	Myogenic regulatory factor 4
Myf-5	Myogenic factor 5
MyoD	Myoblast determination protein
myomiRs	microRNAs involved in the regulation of myocellular processes
nAChR	nicotinic Acetylcholine Receptor
NCAM	Neural Cell Adhesion Molecule
NEU	Neutrophils
NMJs	Neuromuscular Junctions
NTX	Notexin
P38 MAP kinases	P38 mitogen-activated protein kinases
Pax3	Paired box transcription factor 3
Pax7	Paired box transcription factor 7
PBS	Phosphate Buffered Saline
PCNA	Proliferating Cell Nuclear Antigen
PDGF-R-alpha	Platelet derived growth factor receptor alpha
Prolif.SCs	Proliferating SCs
Sca-1	Stem cell antigen-1
SCs	Satellite Cells
sIL6R	soluble Interleukin-6 Receptor alpha
Sox 15	SRY-Box Transcription Factor 15
Sox 8	SRY-Box Transcription Factor 8
TA	Tibialis Anterior muscle
Ter 119 or Ly76	Lymphocyte antigen-76
TGFβ	Transforming Growth Factor beta
TNF-α	Tumor Necrosis Factor alpha
VCAM1	Vascular Cell Adhesion protein 1
WNT	Wingless-related integration site

References

1. Forcina, L.; Miano, C.; Pelosi, L.; Musarò, A. An Overview About the Biology of Skeletal Muscle Satellite Cells. *Curr. Genom.* **2019**, *20*, 24–37. [CrossRef]
2. Seale, P.; Sabourin, L.A.; Girgis-Gabardo, A.; Mansouri, A.; Gruss, P.; Rudnicki, M.A. Pax7 is required for the specification of myogenic satellite cells. *Cell* **2000**, *102*, 777–786. [CrossRef]
3. Blaauw, B.; Reggiani, C. The role of satellite cells in muscle hypertrophy. *J. Muscle Res. Cell Motil.* **2014**, *35*, 3–10. [CrossRef] [PubMed]
4. Yao, L.; Xue, X.; Yu, P.; Ni, Y.; Chen, F. Evans Blue Dye: A Revisit of Its Applications in Biomedicine. *Contrast Media Mol. Imaging* **2018**, *2018*, 18–24. [CrossRef] [PubMed]
5. Straub, V.; Rafael, J.A.; Chamberlain, J.S.; Campbell, K.P. Animal models for muscular dystrophy show different patterns of sarcolemmal disruption. *J. Cell Biol.* **1997**, *139*, 375–385. [CrossRef] [PubMed]
6. Feng, Y.; Chen, F.; Ma, Z.; Dekeyzer, F.; Yu, J.; Xie, Y.; Cona, M.M.; Oyen, R.; Ni, Y. Towards stratifying ischemic components by cardiac MRI and multifunctional stainings in a rabbit model of myocardial infarction. *Theranostics* **2014**, *4*, 24–35. [CrossRef]
7. Cona, M.M.; Koole, M.; Feng, Y.; Liu, Y.; Verbruggen, A.; Oyen, R.; Ni, Y. Biodistribution and radiation dosimetry of radioiodinated hypericin as a cancer therapeutic. *Int. J. Oncol.* **2014**, *44*, 819–829. [CrossRef] [PubMed]
8. Klyen, B.R.; Shavlakadze, T.; Radley-Crabb, H.G.; Grounds, M.D.; Sampson, D.D. Identification of muscle necrosis in the mdx mouse model of Duchenne muscular dystrophy using three-dimensional optical coherence tomography. *J. Biomed. Opt.* **2011**, *16*, 076013. [CrossRef]
9. Wooddell, C.I.; Zhang, G.; Griffin, J.B.; Hegge, J.O.; Huss, T.; Wolff, J.A. Use of Evans blue dye to compare limb muscles in exercised young and old mdx mice. *Muscle Nerve* **2010**, *41*, 487–499. [CrossRef]
10. Hamer, P.W.; McGeachie, J.M.; Davies, M.J.; Grounds, M.D. Evans Blue Dye as an in vivo marker of myofibre damage: Optimising parameters for detecting initial myofibre membrane permeability. *J. Anat.* **2002**, *200*, 69–79. [CrossRef]
11. Pelosi, L.; Berardinelli, M.G.; Forcina, L.; Spelta, E.; Rizzuto, E.; Nicoletti, C.; Camilli, C.; Testa, E.; Catizone, A.; De Benedetti, F.; et al. Increased levels of interleukin-6 exacerbate the dystrophic phenotype in mdx mice. *Hum. Mol. Genet.* **2015**, *24*, 6041–6053. [CrossRef] [PubMed]
12. Musarò, A. The Basis of Muscle Regeneration. *Adv. Biol.* **2014**, *2014*, 1–16. [CrossRef]
13. Matsuda, R.; Nishikawa, A.; Tanaka, H. Visualization of dystrophic muscle fibers in mdx mouse by vital staining with evans blue: Evidence of apoptosis in dystrophin-deficient muscle. *J. Biochem.* **1995**, *118*, 959–963. [CrossRef] [PubMed]
14. Wang, H.L.; Lai, T.W. Optimization of Evans blue quantitation in limited rat tissue samples. *Sci. Rep.* **2014**, *4*, 1–7. [CrossRef] [PubMed]
15. Morgan, J.E.; Prola, A.; Mariot, V.; Pini, V.; Meng, J.; Hourde, C.; Dumonceaux, J.; Conti, F.; Relaix, F.; Authier, F.J.; et al. Necroptosis mediates myofibre death in dystrophin-deficient mice. *Nat. Commun.* **2018**, *9*, 3655. [CrossRef]
16. Brancaccio, P.; Lippi, G.; Maffulli, N. Biochemical markers of muscular damage. *Clin. Chem. Lab. Med.* **2010**, *48*, 757–767. [CrossRef]
17. Komulainen, J.; Kytola, J.; Vihko, V. Running-induced muscle injury and myocellular enzyme release in rats. *J. Appl. Physiol.* **1994**, *77*, 2299–2304. [CrossRef]
18. Siracusa, J.; Koulmann, N.; Bourdon, S.; Goriot, M.E.; Banzet, S. Circulating miRNAs as Biomarkers of Acute Muscle Damage in Rats. *Am. J. Pathol.* **2016**, *186*, 1313–1327. [CrossRef]
19. Nishimura, D.; Sakai, H.; Sato, T.; Sato, F.; Nishimura, S.; Toyama-Sorimachi, N.; Bartsch, J.W.; Sehara-Fujisawa, A. Roles of ADAM8 in elimination of injured muscle fibers prior to skeletal muscle regeneration. *Mech. Dev.* **2015**, *135*, 58–67. [CrossRef]
20. Wang, J. Neutrophils in tissue injury and repair. *Cell Tissue Res.* **2018**, *371*, 531–539. [CrossRef]
21. Yang, W.; Hu, P. Skeletal muscle regeneration is modulated by inflammation. *J. Orthop. Transl.* **2018**, *13*, 25–32. [CrossRef] [PubMed]
22. Geissmann, F.; Jung, S.; Littman, D.R. Blood monocytes consist of two principal subsets with distinct migratory properties. *Immunity* **2003**, *19*, 71–82. [CrossRef]

23. Nahrendorf, M.; Swirski, F.K.; Aikawa, E.; Stangenberg, L.; Wurdinger, T.; Figueiredo, J.L.; Libby, P.; Weissleder, R.; Pittet, M.J. The healing myocardium sequentially mobilizes two monocyte subsets with divergent and complementary functions. *J. Exp. Med.* **2007**, *204*, 3037–3047. [CrossRef] [PubMed]

24. Kratofil, R.M.; Kubes, P.; Deniset, J.F. Monocyte Conversion During Inflammation and Injury. *Arterioscler. Thromb. Vasc. Biol.* **2017**, *37*, 35–42. [CrossRef] [PubMed]

25. Orekhov, A.N.; Orekhova, V.A.; Nikiforov, N.G.; Myasoedova, V.A.; Grechko, A.V.; Romanenko, E.B.; Zhang, D.; Chistiakov, D.A. Monocyte differentiation and macrophage polarization. *Vessel Plus* **2019**, *3*, 10. [CrossRef]

26. Jetten, N.; Verbruggen, S.; Gijbels, M.J.; Post, M.J.; De Winther, M.P.J.; Donners, M.M.P.C. Anti-inflammatory M2, but not pro-inflammatory M1 macrophages promote angiogenesis in vivo. *Angiogenesis* **2014**, *17*, 109–118. [CrossRef]

27. Jablonski, K.A.; Amici, S.A.; Webb, L.M.; de Dios Ruiz-Rosado, J.; Popovich, P.G.; Partida-Sanchez, S.; Guerau-de-Arellano, M. Novel Markers to Delineate Murine M1 and M2 Macrophages. *PLoS ONE* **2015**, *10*, e0145342. [CrossRef] [PubMed]

28. Kastenschmidt, J.M.; Avetyan, I.; Armando Villalta, S. Characterization of the inflammatory response in dystrophic muscle using flow cytometry. In *Methods in Molecular Biology*; Humana Press: New York, NY, USA, 2018; pp. 43–56. [CrossRef]

29. Crane, M.J.; Daley, J.M.; van Houtte, O.; Brancato, S.K.; Henry, W.L.; Albina, J.E. The Monocyte to Macrophage Transition in the Murine Sterile Wound. *PLoS ONE* **2014**, *9*, e86660. [CrossRef]

30. Gnocchi, V.F.; White, R.B.; Ono, Y.; Ellis, J.A.; Zammit, P.S. Further Characterisation of the Molecular Signature of Quiescent and Activated Mouse Muscle Satellite Cells. *PLoS ONE* **2009**, *4*, e5205. [CrossRef]

31. Fukada, S.I.; Yamaguchi, M.; Kokubo, H.; Ogawa, R.; Uezumi, A.; Yoneda, T.; Matev, M.M.; Motohashi, N.; Ito, T.; Zolkiewska, A.; et al. Hesr1 and Hesr3 are essential to generate undifferentiated quiescent satellite cells and to maintain satellite cell numbers. *Development* **2011**, *138*, 4609–4619. [CrossRef]

32. Maesner, C.C.; Almada, A.E.; Wagers, A.J. Established cell surface markers efficiently isolate highly overlapping populations of skeletal muscle satellite cells by fluorescence-activated cell sorting. *Skelet. Muscle* **2016**, *6*, 35. [CrossRef] [PubMed]

33. Relaix, F.; Montarras, D.; Zaffran, S.; Gayraud-Morel, B.; Rocancourt, D.; Tajbakhsh, S.; Mansouri, A.; Cumano, A.; Buckingham, M. Pax3 and Pax7 have distinct and overlapping functions in adult muscle progenitor cells. *J. Cell Biol.* **2006**, *172*, 91–102. [CrossRef] [PubMed]

34. Mechtersheimer, G.; Staudter, M.; Möller, P. Expression of the natural killer cell-associated antigens CD56 and CD57 in human neural and striated muscle cells and in their tumors. *Cancer Res.* **1991**, *51*, 1300–1307. [PubMed]

35. Tatsumi, R.; Anderson, J.E.; Nevoret, C.J.; Halevy, O.; Allen, R.E. HGF/SF is present in normal adult skeletal muscle and is capable of activating satellite cells. *Dev. Biol.* **1998**, *194*, 114–128. [CrossRef]

36. Jesse, T.L.; LaChance, R.; Iademarco, M.F.; Dean, D.C. Interferon regulatory factor-2 is a transcriptional activator in muscle where it regulates expression of vascular cell adhesion molecule-1. *J. Cell Biol.* **1998**, *140*, 1265–1276. [CrossRef]

37. Cornelison, D.D.W.; Filla, M.S.; Stanley, H.M.; Rapraeger, A.C.; Olwin, B.B. Syndecan-3 and syndecan-4 specifically mark skeletal muscle satellite cells and are implicated in satellite cell maintenance and muscle regeneration. *Dev. Biol.* **2001**, *239*, 79–94. [CrossRef]

38. Sherwood, R.I.; Christensen, J.L.; Conboy, I.M.; Conboy, M.J.; Rando, T.A.; Weissman, I.L.; Wagers, A.J. Isolation of adult mouse myogenic progenitors: Functional heterogeneity of cells within and engrafting skeletal muscle. *Cell* **2004**, *119*, 543–554. [CrossRef]

39. Volonte, D.; Liu, Y.; Galbiati, F. The modulation of caveolin-1 expression controls satellite cell activation during muscle repair. *FASEB J.* **2005**, *19*, 1–36. [CrossRef]

40. Chang, N.C.; Sincennes, M.C.; Chevalier, F.P.; Brun, C.E.; Lacaria, M.; Segalés, J.; Muñoz-Cánoves, P.; Ming, H.; Rudnicki, M.A. The Dystrophin Glycoprotein Complex Regulates the Epigenetic Activation of Muscle Stem Cell Commitment. *Cell Stem Cell* **2018**, *22*, 755.e6–768.e6. [CrossRef]

41. Relaix, F.; Zammit, P.S. Satellite cells are essential for skeletal muscle regeneration: The cell on the edge returns centre stage. *Development* **2012**, *139*, 2845–2856. [CrossRef]

42. Chen, J.F.; Mandel, E.M.; Thomson, J.M.; Wu, Q.; Callis, T.E.; Hammond, S.M.; Conlon, F.L.; Wang, D.Z. The role of microRNA-1 and microRNA-133 in skeletal muscle proliferation and differentiation. *Nat. Genet.* **2006**, *38*, 228–233. [CrossRef] [PubMed]

43. Hak, K.K.; Yong, S.L.; Sivaprasad, U.; Malhotra, A.; Dutta, A. Muscle-specific microRNA miR-206 promotes muscle differentiation. *J. Cell Biol.* **2006**, *174*, 677–687. [CrossRef]

44. Uezumi, A.; Fukada, S.I.; Yamamoto, N.; Takeda, S.; Tsuchida, K. Mesenchymal progenitors distinct from satellite cells contribute to ectopic fat cell formation in skeletal muscle. *Nat. Cell Biol.* **2010**, *12*, 143–152. [CrossRef] [PubMed]

45. Mutsaers, S.E.; Bishop, J.E.; McGrouther, G.; Laurent, G.J. Mechanisms of tissue repair: From wound healing to fibrosis. *Int. J. Biochem. Cell Biol.* **1997**, *29*, 5–17. [CrossRef]

46. Dunn, A.; Marcinczyk, M.; Talovic, M.; Patel, K.; Haas, G.; Garg, K. Role of Stem Cells and Extracellular Matrix in the Regeneration of Skeletal Muscle. In *Muscle Cell and Tissue—Current Status of Research Field*; Sakuma, P.K., Ed.; InTechOpen: London, UK, 2018.

47. Gillies, A.R.; Lieber, R.L. Structure and function of the skeletal muscle extracellular matrix. *Muscle Nerve* **2011**, *44*, 318–331. [CrossRef]

48. Kjær, M. Role of Extracellular Matrix in Adaptation of Tendon and Skeletal Muscle to Mechanical Loading. *Physiol. Rev.* **2004**, *84*, 649–698. [CrossRef]

49. Garg, K.; Boppart, M.D. Influence of exercise and aging on extracellular matrix composition in the skeletal muscle stem cell niche. *J. Appl. Physiol.* **2016**, *121*, 1053–1058. [CrossRef]

50. Sanes, J.R. The basement membrane/basal lamina of skeletal muscle. *J. Biol. Chem.* **2003**, *278*, 12601–12604. [CrossRef]

51. Tu, H.; Zhang, D.; Corrick, R.M.; Muelleman, R.L.; Wadman, M.C.; Li, Y.L. Morphological regeneration and functional recovery of neuromuscular junctions after tourniquet-induced injuries in mouse hindlimb. *Front. Physiol.* **2017**, *8*, 207. [CrossRef]

52. Kirby, T.J.; Chaillou, T.; McCarthy, J.J. The role of microRNAs in skeletal muscle health and disease. *Front. Biosci. (Landmark Ed.)* **2016**, *20*, 37–77.

53. Pelosi, L.; Coggi, A.; Forcina, L.; Musarò, A. MicroRNAs modulated by local mIGF-1 expression in mdx dystrophic mice. *Front. Aging Neurosci.* **2015**, *7*, 69. [CrossRef] [PubMed]

54. Yang, Y.; Jiang, G.; Zhang, P.; Fan, J. Programmed cell death and its role in inflammation. *Mil. Med. Res.* **2015**, *2*, 12. [CrossRef] [PubMed]

55. Roh, J.S.; Sohn, D.H. Damage-associated molecular patterns in inflammatory diseases. *Immune Netw.* **2018**, *18*, e27. [CrossRef] [PubMed]

56. Grounds, M. Phagocytosis of necrotic muscle in muscle isografts is influenced by the strain age and sex of host mice. *J. Pathol.* **1987**, *153*, 71–82. [CrossRef]

57. Tidball, J.G.; Wehling-Henricks, M. Macrophages promote muscle membrane repair and muscle fibre growth and regeneration during modified muscle loading in mice in vivo. *J. Physiol.* **2007**, *578*, 327–336. [CrossRef]

58. Summan, M.; Warren, G.L.; Mercer, R.R.; Chapman, R.; Hulderman, T.; Van Rooijen, N.; Simeonova, P.P. Macrophages and skeletal muscle regeneration: A clodronate-containing liposome depletion study. *Am. J. Physiol. Regul. Integr. Comp. Physiol.* **2006**, *290*, R1488–R1495. [CrossRef]

59. Tidball, J.G. Inflammatory processes in muscle injury and repair. *Am. J. Physiol. Regul. Integr. Comp. Physiol.* **2005**, *288*, R345–R353. [CrossRef]

60. Teixeira, C.F.P.; Zamunér, S.R.; Zuliani, J.P.; Fernandes, C.M.; Cruz-Hofling, M.A.; Fernandes, I.; Chaves, F.; Gutiérrez, J.M. Neutrophils do not contribute to local tissue damage, but play a key role in skeletal muscle regeneration, in mice injected with Bothrops asper snake venom. *Muscle Nerve* **2003**, *28*, 449–459. [CrossRef]

61. Frenette, J.; Cai, B.; Tidball, J.G. Complement activation promotes muscle inflammation during modified muscle use. *Am. J. Pathol.* **2000**, *156*, 2103–2110. [CrossRef]

62. Kishimoto, T.K.; Rothlein, R. Integrins, ICAMs, and Selectins: Role and Regulation of Adhesion Molecules in Neutrophil Recruitment to Inflammatory Sites. *Adv. Pharmacol.* **1994**, *25*, 117–169. [CrossRef]

63. Muller, W.A. Leukocyte-endothelial-cell interactions in leukocyte transmigration and the inflammatory response. *Trends Immunol.* **2003**, *24*, 326–333. [CrossRef]

64. Walzog, B.; Gaehtgens, P. Adhesion Molecules: The Path to a New Understanding of Acute Inflammation. *News Physiol. Sci.* **2000**, *15*, 107–113. [CrossRef] [PubMed]

65. Sixt, M.; Hallmann, R.; Wendler, O.; Scharffetter-Kochanek, K.; Sorokin, L.M. Cell adhesion and migration properties of β2-integrin negative polymorphonuclear granulocytes on defined extracellular matrix molecules: Relevance for leukocyte extravasation. *J. Biol. Chem.* **2001**, *276*, 18878–18887. [CrossRef] [PubMed]

66. Pizza, F.X.; Peterson, J.M.; Baas, J.H.; Koh, T.J. Neutrophils contribute to muscle injury and impair its resolution after lengthening contractions in mice. *J. Physiol.* **2005**, *562*, 899–913. [CrossRef]

67. Chazaud, B.; Brigitte, M.; Yacoub-Youssef, H.; Arnold, L.; Gherardi, R.; Sonnet, C.; Lafuste, P.; Chretien, F. Dual and beneficial roles of macrophages during skeletal muscle regeneration. *Exerc. Sport Sci. Rev.* **2009**, *37*, 18–22. [CrossRef]

68. Villalta, S.A.; Nguyen, H.X.; Deng, B.; Gotoh, T.; Tidball, J.G. Shifts in macrophage phenotypes and macrophage competition for arginine metabolism affect the severity of muscle pathology in muscular dystrophy. *Hum. Mol. Genet.* **2009**, *18*, 482–496. [CrossRef]

69. McLennan, I.S. Resident macrophages (ED2- and ED3-positive) do not phagocytose degenerating rat skeletal muscle fibres. *Cell Tissue Res.* **1993**, *272*, 193–196. [CrossRef] [PubMed]

70. Arnold, L.; Henry, A.; Poron, F.; Baba-Amer, Y.; Van Rooijen, N.; Plonquet, A.; Gherardi, R.K.; Chazaud, B. Inflammatory monocytes recruited after skeletal muscle injury switch into antiinflammatory macrophages to support myogenesis. *J. Exp. Med.* **2007**, *204*, 1057–1069. [CrossRef]

71. Silva, M.T. When two is better than one: Macrophages and neutrophils work in concert in innate immunity as complementary and cooperative partners of a myeloid phagocyte system. *J. Leukoc. Biol.* **2010**, *87*, 93–106. [CrossRef]

72. Kolaczkowska, E.; Kubes, P. Neutrophil recruitment and function in health and inflammation. *Nat. Rev. Immunol.* **2013**, *13*, 159–175. [CrossRef]

73. Nourshargh, S.; Alon, R. Leukocyte Migration into Inflamed Tissues. *Immunity* **2014**, *41*, 694–707. [CrossRef] [PubMed]

74. Wang, C.; Yue, F.; Kuang, S. Muscle Histology Characterization Using H&E Staining and Muscle Fiber Type Classification Using Immunofluorescence Staining. *Bio Protoc.* **2017**, *7*, e2279. [CrossRef]

75. Soehnlein, O.; Lindbom, L. Phagocyte partnership during the onset and resolution of inflammation. *Nat. Rev. Immunol.* **2010**, *10*, 427–439. [CrossRef] [PubMed]

76. Tonkin, J.; Temmerman, L.; Sampson, R.D.; Gallego-Colon, E.; Barberi, L.; Bilbao, D.; Schneider, M.D.; Musarò, A.; Rosenthal, N. Monocyte/macrophage-derived IGF-1 orchestrates murine skeletal muscle regeneration and modulates autocrine polarization. *Mol. Ther.* **2015**, *23*, 1189–1200. [CrossRef] [PubMed]

77. Mauro, A. Satellite cell of skeletal muscle fibers. *J. Biophys. Biochem. Cytol.* **1961**, *9*, 493–495. [CrossRef] [PubMed]

78. Scharner, J.; Zammit, P.S. The muscle satellite cell at 50: The formative years. *Skelet. Muscle* **2011**, *1*, 28. [CrossRef] [PubMed]

79. Creuzet, S.; Lescaudron, L.; Li, Z.; Fontaine-Pérus, J. MyoD, myogenin, and desmin-nls-lacZ transgene emphasize the distinct patterns of satellite cell activation in growth and regeneration. *Exp. Cell Res.* **1998**, *243*, 241–253. [CrossRef] [PubMed]

80. Yablonka-Reuveni, Z.; Rivera, A.J. Temporal expression of regulatory and structural muscle proteins during myogenesis of satellite cells on isolated adult rat fibers. *Dev. Biol.* **1994**, *164*, 588–603. [CrossRef]

81. Scholzen, T.; Gerdes, J. The Ki-67 protein: From the known and the unknown. *J. Cell. Physiol.* **2000**, *182*, 311–322. [CrossRef]

82. Mead, T.J.; Lefebvre, V. Proliferation assays (BrdU and EdU) on skeletal tissue sections. *Methods Mol. Biol.* **2014**, *1130*, 233–243. [CrossRef]

83. Ikeda, T.; Ichii, O.; Otsuka-Kanazawa, S.; Nakamura, T.; Elewa, Y.H.A.; Kon, Y. Degenerative and regenerative features of myofibers differ among skeletal muscles in a murine model of muscular dystrophy. *J. Muscle Res. Cell Motil.* **2016**, *37*, 153–164. [CrossRef] [PubMed]

84. Abmayr, S.M.; Pavlath, G.K. Myoblast fusion: Lessons from flies and mice. *Development* **2012**, *139*, 641–656. [CrossRef] [PubMed]

85. Chal, J.; Pourquié, O. Making muscle: Skeletal myogenesis in vivo and in vitro. *Development* **2017**, *144*, 2104–2122. [CrossRef] [PubMed]

86. Mizuno, Y.; Suzuki, M.; Nakagawa, H.; Ninagawa, N.; Torihashi, S. Switching of actin isoforms in skeletal muscle differentiation using mouse ES cells. *Histochem. Cell Biol.* **2009**, *132*, 669–672. [CrossRef] [PubMed]

87. Sharma, A.; Agarwal, M.; Kumar, A.; Kumar, P.; Saini, M.; Kardon, G.; Mathew, S.J. Myosin Heavy Chain-embryonic is a crucial regulator of skeletal muscle development and differentiation. *bioRxiv* **2018**, 261685. [CrossRef]

88. Chal, J.; Oginuma, M.; Al Tanoury, Z.; Gobert, B.; Sumara, O.; Hick, A.; Bousson, F.; Zidouni, Y.; Mursch, C.; Moncuquet, P.; et al. Differentiation of pluripotent stem cells to muscle fiber to model Duchenne muscular dystrophy. *Nat. Biotechnol.* **2015**, *33*, 962–969. [CrossRef]

89. Guiraud, S.; Edwards, B.; Squire, S.E.; Moir, L.; Berg, A.; Babbs, A.; Ramadan, N.; Wood, M.J.; Davies, K.E. Embryonic myosin is a regeneration marker to monitor utrophin-based therapies for DMD. *Hum. Mol. Genet.* **2019**, *28*, 307–319. [CrossRef]

90. De Angelis, L.; Berghella, L.; Coletta, M.; Lattanzi, L.; Zanchi, M.; Cusella-De Angelis, M.G.; Ponzetto, C.; Cossu, G. Skeletal myogenic progenitors originating from embryonic dorsal aorta coexpress endothelial and myogenic markers and contribute to postnatal muscle growth and regeneration. *J. Cell Biol.* **1999**, *147*, 869–877. [CrossRef]

91. Kuang, S.; Chargé, S.B.; Seale, P.; Huh, M.; Rudnicki, M.A. Distinct roles for Pax7 and Pax3 in adult regenerative myogenesis. *J. Cell Biol.* **2006**, *172*, 103–113. [CrossRef]

92. Wosczyna, M.N.; Biswas, A.A.; Cogswell, C.A.; Goldhamer, D.J. Multipotent progenitors resident in the skeletal muscle interstitium exhibit robust BMP-dependent osteogenic activity and mediate heterotopic ossification. *J. Bone Miner. Res.* **2012**, *27*, 1004–1017. [CrossRef]

93. Asakura, A.; Seale, P.; Girgis-Gabardo, A.; Rudnicki, M.A. Myogenic specification of side population cells in skeletal muscle. *J. Cell Biol.* **2002**, *159*, 123–134. [CrossRef] [PubMed]

94. Gussoni, E.; Soneoka, Y.; Strickland, C.D.; Buzney, E.A.; Khan, M.K.; Flint, A.F.; Kunkel, L.M.; Mulligan, R.C. Dystrophin expression in the mdx mouse restored by stem cell transplantation. *Nature* **1999**, *401*, 390–394. [CrossRef] [PubMed]

95. Messina, G.; Biressi, S.; Cossu, G. Non Muscle Stem Cells and Muscle Regeneration. In *Skeletal Muscle Repair and Regeneration*; Springer: Dordrecht, The Netherlands, 2008; pp. 65–84.

96. Joe, A.W.B.; Yi, L.; Natarajan, A.; Le Grand, F.; So, L.; Wang, J.; Rudnicki, M.A.; Rossi, F.M.V. Muscle injury activates resident fibro/adipogenic progenitors that facilitate myogenesis. *Nat. Cell Biol.* **2010**, *12*, 153–163. [CrossRef] [PubMed]

97. Wosczyna, M.N.; Konishi, C.T.; Perez, E.E.; Gan, Q.; Wagner, M.W.; Rando, T.A.; Perez Carbajal, E.E.; Wang, T.T.; Walsh, R.A. Mesenchymal Stromal Cells Are Required for Regeneration and Homeostatic Maintenance of Skeletal Muscle In Brief Article Mesenchymal Stromal Cells Are Required for Regeneration and Homeostatic Maintenance of Skeletal Muscle. *Cell Rep.* **2019**, *27*, 2029.e5–2035.e5. [CrossRef] [PubMed]

98. Fiore, D.; Judson, R.N.; Low, M.; Lee, S.; Zhang, E.; Hopkins, C.; Xu, P.; Lenzi, A.; Rossi, F.M.V.; Lemos, D.R. Pharmacological blockage of fibro/adipogenic progenitor expansion and suppression of regenerative fibrogenesis is associated with impaired skeletal muscle regeneration. *Stem Cell Res.* **2016**, *17*, 161–169. [CrossRef] [PubMed]

99. Lukjanenko, L.; Karaz, S.; Stuelsatz, P.; Gurriaran-Rodriguez, U.; Michaud, J.; Dammone, G.; Sizzano, F.; Mashinchian, O.; Ancel, S.; Migliavacca, E.; et al. Aging Disrupts Muscle Stem Cell Function by Impairing Matricellular WISP1 Secretion from Fibro-Adipogenic Progenitors. *Cell Stem Cell* **2019**, *24*, 433.e7–446.e7. [CrossRef]

100. Davis, M.E.; Gumucio, J.P.; Sugg, K.B.; Bedi, A.; Mendias, C.L. MMP inhibition as a potential method to augment the healing of skeletal muscle and tendon extracellular matrix. *J. Appl. Physiol.* **2013**, *115*, 884–891. [CrossRef]

101. Kim, J.; Lee, J. Matrix metalloproteinase and tissue inhibitor of metalloproteinase responses to muscle damage after eccentric exercise. *J. Exerc. Rehabil.* **2016**, *12*, 260–265. [CrossRef]

102. Arecco, N.; Clarke, C.J.; Jones, F.K.; Simpson, D.M.; Mason, D.; Beynon, R.J.; Pisconti, A. Elastase levels and activity are increased in dystrophic muscle and impair myoblast cell survival, proliferation and differentiation. *Sci. Rep.* **2016**, *6*, 1–20. [CrossRef]

103. Grounds, M.D. Complexity of Extracellular Matrix and Skeletal Muscle Regeneration. In *Skeletal Muscle Repair and Regeneration*; Springer: Amsterdam, The Netherlands, 2008; pp. 269–302.

104. Mann, C.J.; Perdiguero, E.; Kharraz, Y.; Aguilar, S.; Pessina, P.; Serrano, A.L.; Muñoz-Cánoves, P. Aberrant repair and fibrosis development in skeletal muscle. *Skelet. Muscle* **2011**, *1*, 21. [CrossRef]

105. Garg, K.; Corona, B.T.; Walters, T.J. Therapeutic strategies for preventing skeletal muscle fibrosis after injury. *Front. Pharmacol.* **2015**, *6*, 87. [CrossRef] [PubMed]

106. Lu, P.; Takai, K.; Weaver, V.M.; Werb, Z. Extracellular Matrix degradation and remodeling in development and disease. *Cold Spring Harb. Perspect. Biol.* **2011**, *3*, a005058. [CrossRef] [PubMed]

107. Gillies, A.R.; Chapman, M.A.; Bushong, E.A.; Deerinck, T.J.; Ellisman, M.H.; Lieber, R.L. High resolution three-dimensional reconstruction of fibrotic skeletal muscle extracellular matrix. *J. Physiol.* **2017**, *595*, 1159–1171. [CrossRef] [PubMed]

108. Dey, P. *Basic and Advanced Laboratory Techniques in Histopathology and Cytology*; Springer: Singapore, 2018.

109. Junqueira, L.C.U.; Bignolas, G.; Brentani, R.R. Picrosirius staining plus polarization microscopy, a specific method for collagen detection in tissue sections. *Histochem. J.* **1979**, *11*, 447–455. [CrossRef]

110. Montes, G.S.; Junqueira, L.C. The use of the Picrosirius-polarization method for the study of the biopathology of collagen. *Mem. Inst. Oswaldo Cruz* **1991**, *86*, 1–11. [CrossRef]

111. Lattouf, R.; Younes, R.; Lutomski, D.; Naaman, N.; Godeau, G.; Senni, K.; Changotade, S. Picrosirius Red Staining: A Useful Tool to Appraise Collagen Networks in Normal and Pathological Tissues. *J. Histochem. Cytochem.* **2014**, *62*, 751–758. [CrossRef]

112. Séguier, S.; Godeau, G.; Brousse, N. Collagen Fibers and Inflammatory Cells in Healthy and Diseased Human Gingival Tissues: A Comparative and Quantitative Study by Immunohistochemistry and Automated Image Analysis. *J. Periodontol.* **2000**, *71*, 1079–1085. [CrossRef]

113. Tatsumi, R.; Sankoda, Y.; Anderson, J.E.; Sato, Y.; Mizunoya, W.; Shimizu, N.; Suzuki, T.; Yamada, M.; Rhoads, R.P.; Ikeuchi, Y.; et al. Possible implication of satellite cells in regenerative motoneuritogenesis: HGF upregulates neural chemorepellent Sema3A during myogenic differentiation Possible implication of satellite cells in regenerative motoneuritogenesis: HGF upregulates neural che-morepellent Sema3A during myogenic differentiation. *Am. J. Physiol. Cell Physiol.* **2009**, *297*, 238–252. [CrossRef]

114. Ling, S.C.; Dastidar, S.G.; Tokunaga, S.; Ho, W.Y.; Lim, K.; Ilieva, H.; Parone, P.A.; Tyan, S.H.; Tse, T.M.; Chang, J.C.; et al. Overriding FUS autoregulation in mice triggers gain-of-toxic dysfunctions in RNA metabolism and autophagy-lysosome axis. *Elife* **2019**, *8*, e40811. [CrossRef]

115. Rizzuto, E.; Pisu, S.; Nicoletti, C.; Del Prete, Z.; Musarò, A. Measuring neuromuscular junction functionality. *J. Vis. Exp.* **2017**, *2017*, 55227. [CrossRef]

116. Rizzuto, E.; Pisu, S.; Musarò, A.; Del Prete, Z. Measuring Neuromuscular Junction Functionality in the SOD1G93A Animal Model of Amyotrophic Lateral Sclerosis. *Ann. Biomed. Eng.* **2015**, *43*, 2196–2206. [CrossRef] [PubMed]

117. Del Prete, Z.; Musarò, A.; Rizzuto, E. Measuring mechanical properties, including isotonic fatigue, of fast and slow MLC/mIgf-1 transgenic skeletal muscle. *Ann. Biomed. Eng.* **2008**, *36*, 1281–1290. [CrossRef] [PubMed]

118. Grounds, M.D.; Radley, H.G.; Lynch, G.S.; Nagaraju, K.; De Luca, A. Towards developing standard operating procedures for pre-clinical testing in the mdx mouse model of Duchenne muscular dystrophy. *Neurobiol. Dis.* **2008**, *31*, 1–19. [CrossRef] [PubMed]

119. Lagrota-Candido, J.; Vasconcellos, R.; Cavalcanti, M.; Bozza, M.; Savino, W.; Quirico-Santos, T. Resolution of skeletal muscle inflammation in mdx dystrophic mouse is accompanied by increased immunoglobulin and interferon-γ production. *Int. J. Exp. Pathol.* **2002**, *83*, 121–132. [CrossRef] [PubMed]

120. Forcina, L.; Pelosi, L.; Miano, C.; Musarò, A. Insights into the Pathogenic Secondary Symptoms Caused by the Primary Loss of Dystrophin. *J. Funct. Morphol. Kinesiol.* **2017**, *2*, 44. [CrossRef]

121. Pelosi, L.; Berardinelli, M.G.; De Pasquale, L.; Nicoletti, C.; D'Amico, A.; Carvello, F.; Moneta, G.M.; Catizone, A.; Bertini, E.; De Benedetti, F.; et al. Functional and Morphological Improvement of Dystrophic Muscle by Interleukin 6 Receptor Blockade. *EBioMedicine* **2015**, *2*, 285–293. [CrossRef]

122. Klein, S.M.; Prantl, L.; Geis, S.; Felthaus, O.; Dolderer, J.; Anker, A.M.; Zeitler, K.; Alt, E.; Vykoukal, J. Circulating serum CK level vs. muscle impairment for in situ monitoring burden of disease in Mdx-mice. *Clin. Hemorheol. Microcirc.* **2017**, *65*, 327–334. [CrossRef]

123. Liu, N.; Williams, A.H.; Maxeiner, J.M.; Bezprozvannaya, S.; Shelton, J.M.; Richardson, J.A.; Bassel-Duby, R.; Olson, E.N. MicroRNA-206 promotes skeletal muscle regeneration and delays progression of Duchenne muscular dystrophy in mice. *J. Clin. Investig.* **2012**, *122*, 2054–2065. [CrossRef]

124. Cacchiarelli, D.; Legnini, I.; Martone, J.; Cazzella, V.; D'Amico, A.; Bertini, E.; Bozzoni, I. miRNAs as serum biomarkers for Duchenne muscular dystrophy. *EMBO Mol. Med.* **2011**, *3*, 258–265. [CrossRef]

125. Vignier, N.; Amor, F.; Fogel, P.; Duvallet, A.; Poupiot, J.; Charrier, S.; Arock, M.; Montus, M.; Nelson, I.; Richard, I.; et al. Distinctive Serum miRNA Profile in Mouse Models of Striated Muscular Pathologies. *PLoS ONE* **2013**, *8*, e55281. [CrossRef]

126. Jeanson-Leh, L.; Lameth, J.; Krimi, S.; Buisset, J.; Amor, F.; Le Guiner, C.; Barthélémy, I.; Servais, L.; Blot, S.; Voit, T.; et al. Serum profiling identifies novel muscle miRNA and cardiomyopathy-related miRNA biomarkers in golden retriever muscular dystrophy dogs and duchenne muscular dystrophy patients. *Am. J. Pathol.* **2014**, *184*, 2885–2898. [CrossRef] [PubMed]

127. Zaharieva, I.T.; Calissano, M.; Scoto, M.; Preston, M.; Cirak, S.; Feng, L.; Collins, J.; Kole, R.; Guglieri, M.; Straub, V.; et al. Dystromirs as Serum Biomarkers for Monitoring the Disease Severity in Duchenne Muscular Dystrophy. *PLoS ONE* **2013**, *8*, e80263. [CrossRef] [PubMed]

128. Perfetti, A.; Greco, S.; Bugiardini, E.; Cardani, R.; Gaia, P.; Gaetano, C.; Meola, G.; Martelli, F. Plasma microRNAs as biomarkers for myotonic dystrophy type 1. *Neuromuscul. Disord.* **2014**, *24*, 509–515. [CrossRef] [PubMed]

129. McDouall, R.M.; Dunn, M.J.; Dubowitz, V. Nature of the mononuclear infiltrate and the mechanism of muscle damage in juvenile dermatomyositis and Duchenne muscular dystrophy. *J. Neurol. Sci.* **1990**, *99*, 199–217. [CrossRef]

130. Spencer, M.J.; Walsh, C.M.; Dorshkind, K.A.; Rodriguez, E.M.; Tidball, J.G. Myonuclear apoptosis in dystrophic mdx muscle occurs by perforin- mediated cytotoxicity. *J. Clin. Investig.* **1997**, *99*, 2745–2751. [CrossRef] [PubMed]

131. Wehling, M.; Spencer, M.J.; Tidball, J.G. A nitric oxide synthase transgene ameliorates muscular dystrophy in mdx mice. *J. Cell Biol.* **2001**, *155*, 123–131. [CrossRef]

132. Madaro, L.; Torcinaro, A.; De Bardi, M.; Contino, F.F.; Pelizzola, M.; Diaferia, G.R.; Imeneo, G.; Bouchè, M.; Puri, P.L.; De Santa, F. Macrophages fine tune satellite cell fate in dystrophic skeletal muscle of mdx mice. *PLoS Genet.* **2019**, *15*, e1008408. [CrossRef]

133. Turk, R.; Sterrenburg, E.; de Meijer, E.J.; van Ommen, G.J.B.; den Dunnen, J.T.; 't Hoen, P.A.C. Muscle regeneration in dystrophin-deficient mdx mice studied by gene expression profiling. *BMC Genom.* **2005**, *6*, 98. [CrossRef]

134. Pelosi, M.; De Rossi, M.; Barberi, L.; Musarò, A. IL-6 Impairs Myogenic Differentiation by Downmodulation of p90RSK/eEF2 and mTOR/p70S6K Axes, without Affecting AKT Activity. *Biomed Res. Int.* **2014**, *2014*, 206026. [CrossRef]

135. Kurosaka, M.; Machida, S. Interleukin-6-induced satellite cell proliferation is regulated by induction of the JAK2/STAT3 signalling pathway through cyclin D1 targeting. *Cell Prolif.* **2013**, *46*, 365–373. [CrossRef]

136. Wada, E.; Tanihata, J.; Iwamura, A.; Takeda, S.; Hayashi, Y.K.; Matsuda, R. Treatment with the anti-IL-6 receptor antibody attenuates muscular dystrophy via promoting skeletal muscle regeneration in dystrophin-/utrophin-deficient mice. *Skelet. Muscle* **2017**, *7*, 23. [CrossRef] [PubMed]

137. Reggio, A.; Rosina, M.; Palma, A.; Cerquone Perpetuini, A.; Lisa Petrilli, L.; Gargioli, C.; Fuoco, C.; Micarelli, E.; Giuliani, G.; Cerretani, M.; et al. Adipogenesis of skeletal muscle fibro/adipogenic progenitors is affected by the WNT5a/GSK3/β-catenin axis. *Cell Death Differ.* **2020**. [CrossRef] [PubMed]

138. Cullen, M.; Mastaglia, F. Pathological reactions of skeletal muscle. In *Skeletal Muscle Pathology*; Mastaglia, F.L., Walton, J., Eds.; Churchill Livingstone: London, UK, 1982; pp. 88–139.

139. Harris, J.B. Myotoxic phospholipases A2 and the regeneration of skeletal muscles. *Toxicon* **2003**, *42*, 933–945. [CrossRef] [PubMed]

140. Hardy, D.; Besnard, A.; Latil, M.; Jouvion, G.; Briand, D.; Thépenier, C.; Pascal, Q.; Guguin, A.; Gayraud-Morel, B.; Cavaillon, J.M.; et al. Comparative Study of Injury Models for Studying Muscle Regeneration in Mice. *PLoS ONE* **2016**, *11*, e0147198. [CrossRef]

141. Warren, G.L.; Hulderman, T.; Mishra, D.; Gao, X.; Millecchia, L.; O'Farrell, L.; Kuziel, W.A.; Simeonova, P.P. Chemokine receptor CCR2 involvement in skeletal muscle regeneration. *FASEB J.* **2005**, *19*, 1–23. [CrossRef]

142. Le, G.; Lowe, D.A.; Kyba, M. Freeze injury of the tibialis anterior muscle. *Methods Mol. Biol.* **2016**, *1460*, 33–41. [CrossRef]

143. Forcina, L.; Miano, C.; Musarò, A. The physiopathologic interplay between stem cells and tissue niche in muscle regeneration and the role of IL-6 on muscle homeostasis and diseases. *Cytokine Growth Factor Rev.* **2018**, *41*, 1–9. [CrossRef]

144. Forcina, L.; Miano, C.; Scicchitano, B.M.; Rizzuto, E.; Berardinelli, M.G.; De Benedetti, F.; Pelosi, L.; Musarò, A. Increased Circulating Levels of Interleukin-6 Affect the Redox Balance in Skeletal Muscle. *Oxid. Med. Cell. Longev.* **2019**, *2019*, 3018584. [CrossRef]

145. Schultz, E.; Jaryszak, D.L.; Gibson, M.C.; Albright, D.J. Absence of exogenous satellite cell contribution to regeneration of frozen skeletal muscle. *J. Muscle Res. Cell Motil.* **1986**, *7*, 361–367. [CrossRef]

146. Torrente, Y.; Belicchi, M.; Sampaolesi, M.; Pisati, F.; Meregalli, M.; D'Antona, G.; Tonlorenzi, R.; Porretti, L.; Gavina, M.; Mamchaoui, K.; et al. Human circulating AC133+ stem cells restore dystrophin expression and ameliorate function in dystrophic skeletal muscle. *J. Clin. Investig.* **2004**, *114*, 182–195. [CrossRef]

147. Reis, N.D.; Better, O.S. Mechanical muscle-crush injury and acute muscle-crush compartment syndrome. *J. Bone Joint Surg. Br.* **2005**, *87*, 450–453. [CrossRef] [PubMed]

148. Dobek, G.L.; Fulkerson, N.D.; Nicholas, J.; Schneider, B.S.P. Mouse model of muscle crush injury of the legs. *Comp. Med.* **2013**, *63*, 227–232. [PubMed]

149. Crisco, J.J.; Jokl, P.; Heinen, G.T.; Connell, M.D.; Panjabi, M.M. A Muscle Contusion Injury Model: Biomechanics, Physiology, and Histology. *Am. J. Sports Med.* **1994**, *22*, 702–710. [CrossRef] [PubMed]

150. Kerkweg, U.; Schmitz, D.; de Groot, H. Screening for the Formation of Reactive Oxygen Species and of NO in Muscle Tissue and Remote Organs upon Mechanical Trauma to the Mouse Hind Limb. *Eur. Surg. Res.* **2006**, *38*, 83–89. [CrossRef]

151. McBrier, N.M.; Neuberger, T.; Denegar, C.R.; Sharkey, N.A.; Webb, A.G. Magnetic resonance imaging of acute injury in rats and the effects of buprenorphine on limb volume. *J. Am. Assoc. Lab. Anim. Sci.* **2009**, *48*, 147–151.

152. Stratos, I.; Graff, J.; Rotter, R.; Mittlmeier, T.; Vollmar, B. Open blunt crush injury of different severity determines nature and extent of local tissue regeneration and repair. *J. Orthop. Res.* **2010**, *28*, 950–957. [CrossRef]

153. Takagi, R.; Fujita, N.; Arakawa, T.; Kawada, S.; Ishii, N.; Miki, A. Influence of icing on muscle regeneration after crush injury to skeletal muscles in rats. *J. Appl. Physiol.* **2011**, *110*, 382–388. [CrossRef]

154. Criswell, T.L.; Corona, B.T.; Ward, C.L.; Miller, M.; Patel, M.; Wang, Z.; Christ, G.J.; Soker, S. Compression-induced muscle injury in rats that mimics compartment syndrome in humans. *Am. J. Pathol.* **2012**, *180*, 787–797. [CrossRef]

155. Ciciliot, S.; Schiaffino, S. Regeneration of Mammalian Skeletal Muscle: Basic Mechanisms and Clinical Implications. *Curr. Pharm. Des.* **2010**, *16*, 906–914. [CrossRef]

156. Oyster, N.; Witt, M.; Gharaibeh, B.; Poddar, M.; Schneppendahl, J.; Huard, J. Characterization of a compartment syndrome-like injury model. *Muscle Nerve* **2015**, *51*, 750–758. [CrossRef]

157. Lømo, T. What controls the position, number, size, and distribution of neuromuscular junctions on rat muscle fibers? *J. Neurocytol.* **2003**, *32*, 835–848. [CrossRef] [PubMed]

158. Klein-Ogus, C.; Harris, J.B. Preliminary observations of satellite cells in undamaged fibres of the rat soleus muscle assaulted by a snake-venom toxin. *Cell Tissue Res.* **1983**, *230*, 671–776. [CrossRef] [PubMed]

159. Benoit, P.W.; Belt, W.D. Destruction and regeneration of skeletal muscle after treatment with a local anaesthetic, bupivacaine (Marcaine). *J. Anat.* **1970**, *107*, 547–556.

160. Gutiérrez, J.M.; Ownby, C.L. Skeletal muscle degeneration induced by venom phospholipases A 2: Insights into the mechanisms of local and systemic myotoxicity. *Toxicon* **2003**, *42*, 915–931. [CrossRef] [PubMed]

161. Harris, J.B.; Grubb, B.D.; Maltin, C.A.; Dixon, R. The neurotoxicity of the venom phospholipases A2, notexin and taipoxin. *Exp. Neurol.* **2000**, *161*, 517–526. [CrossRef]

162. Harris, J.B.; Vater, R.; Wilson, M.; Cullen, M.J. Muscle fibre breakdown in venom-induced muscle degeneration. *J. Anat.* **2003**, *202*, 363–372. [CrossRef]

163. Rosenblatt, J.D. A time course study of the isometric contractile properties of rat extensor digitorum longus muscle injected with bupivacaine. *Comp. Biochem. Physiol. Comp. Physiol.* **1992**, *101*, 361–367. [CrossRef]

164. Gutiérrez, J.M.; Escalante, T.; Hernández, R.; Gastaldello, S.; Saravia-Otten, P.; Rucavado, A. Why is skeletal muscle regeneration impaired after myonecrosis induced by viperid snake venoms? *Toxins (Basel)* **2018**, *10*, 182. [CrossRef]

165. Hodges, S.J.; Agbaji, A.S.; Harvey, A.L.; Hider, R.C. Cobra cardiotoxins. Purification, effects on skeletal muscle and structure/activity relationships. *Eur. J. Biochem.* **1987**, *165*, 373–383. [CrossRef]

166. Guardiola, O.; Andolfi, G.; Tirone, M.; Iavarone, F.; Brunelli, S.; Minchiotti, G. Induction of acute skeletal muscle regeneration by cardiotoxin injection. *J. Vis. Exp.* **2017**, *2017*, 54515. [CrossRef]

167. Czerwinska, A.M.; Streminska, W.; Ciemerych, M.A.; Grabowska, I. Mouse gastrocnemius muscle regeneration after mechanical or cardiotoxin injury. *Folia Histochem. Cytobiol.* **2012**, *50*, 144–153. [CrossRef] [PubMed]

168. Hirata, A.; Masuda, S.; Tamura, T.; Kai, K.; Ojima, K.; Fukase, A.; Motoyoshi, K.; Kamakura, K.; Miyagoe-Suzuki, Y.; Takeda, S. Expression profiling of cytokines and related genes in regenerating skeletal muscle after cardiotoxin injection: A role for osteopontin. *Am. J. Pathol.* **2003**, *163*, 203–215. [CrossRef]

169. El Andalousi, R.B.; Daussin, P.A.; Micallef, J.P.; Roux, C.; Nougues, J.; Chammas, M.; Reyne, Y.; Bacou, F. Changes in mass and performance in rabbit muscles after muscle damage with or without transplantation of primary satellite cells. *Cell Transplant.* **2002**, *11*, 169–180. [CrossRef]

170. Sambasivan, R.; Yao, R.; Kissenpfennig, A.; van Wittenberghe, L.; Paldi, A.; Gayraud-Morel, B.; Guenou, H.; Malissen, B.; Tajbakhsh, S.; Galy, A. Pax7-expressing satellite cells are indispensable for adult skeletal muscle regeneration. *Development* **2011**, *138*, 3647–3656. [CrossRef]

171. Ho, A.T.V.; Palla, A.R.; Blake, M.R.; Yucel, N.D.; Wang, Y.X.; Magnusson, K.E.G.; Holbrook, C.A.; Kraft, P.E.; Delp, S.L.; Blau, H.M. Prostaglandin E2 is essential for efficacious skeletal muscle stem-cell function, augmenting regeneration and strength. *Proc. Natl. Acad. Sci. USA* **2017**, *114*, 6675–6684. [CrossRef] [PubMed]

172. Siles, L.; Ninfali, C.; Cortés, M.; Darling, D.S.; Postigo, A. ZEB1 protects skeletal muscle from damage and is required for its regeneration. *Nat. Commun.* **2019**, *10*, 1–18. [CrossRef] [PubMed]

173. Sun, K.-T.; Cheung, K.-K.; Au, S.W.N.; Yeung, S.S.; Yeung, E.W. Overexpression of Mechano-Growth Factor Modulates Inflammatory Cytokine Expression and Macrophage Resolution in Skeletal Muscle Injury. *Front. Physiol.* **2018**, *9*, 999. [CrossRef] [PubMed]

174. Mirtschin, P.; Davis, R. *Dangerous Snakes of Australia. An Illustrated Guide to Australia's Most Venomous Snakes*; Rigby Publishers: Adelaide, Australia, 1982.

175. Plant, D.R.; Colarossi, F.E.; Lynch, G.S. Notexin causes greater myotoxic damage and slower functional repair in mouse skeletal muscles than bupivacaine. *Muscle Nerve* **2006**, *34*, 577–585. [CrossRef]

176. Vignaud, A.; Hourdé, C.; Butler-Browne, G.; Ferry, A. Differential recovery of neuromuscular function after nerve/muscle injury induced by crude venom from Notechis scutatus, cardiotoxin from Naja atra and bupivacaine treatments in mice. *Neurosci. Res.* **2007**, *58*, 317–323. [CrossRef]

177. Gayraud-Morel, B.; Chrétien, F.; Flamant, P.; Gomès, D.; Zammit, P.S.; Tajbakhsh, S. A role for the myogenic determination gene Myf5 in adult regenerative myogenesis. *Dev. Biol.* **2007**, *312*, 13–28. [CrossRef]

178. Baghdadi, M.B.; Tajbakhsh, S. Regulation and phylogeny of skeletal muscle regeneration. *Dev. Biol.* **2018**, *433*, 200–209. [CrossRef] [PubMed]

179. Zink, W.; Seif, C.; Bohl, J.R.E.; Hacke, N.; Braun, P.M.; Sinner, B.; Martin, E.; Fink, R.H.A.; Graf, B.M. The acute myotoxic effects of bupivacaine and ropivacaine after continuous peripheral nerve blockades. *Anesth. Analg.* **2003**, *97*, 1173–1179. [CrossRef] [PubMed]

180. Zink, W.; Graf, B.M.; Sinner, B.; Martin, E.; Fink, R.H.A.; Kunst, G. Differential effects of bupivacaine on intracellular Ca 2+ regulation: Potential mechanisms of its myotoxicity. *Anesthesiology* **2002**, *97*, 710–716. [CrossRef] [PubMed]

181. Kimura, N.; Hirata, S.; Miyasaka, N.; Kawahata, K.; Kohsaka, H. Injury and subsequent regeneration of muscles for activation of local innate immunity to facilitate the development and relapse of autoimmune myositis in C57BL/6 mice. *Arthritis Rheumatol.* **2015**, *67*, 1107–1116. [CrossRef] [PubMed]

182. Scicchitano, B.M.; Dobrowolny, G.; Sica, G.; Musaro, A. Molecular Insights into Muscle Homeostasis, Atrophy and Wasting. *Curr. Genom.* **2018**, *19*, 356–369. [CrossRef]

Review

The Diversity of Muscles and Their Regenerative Potential across Animals

Letizia Zullo [1,2,*], Matteo Bozzo [3], Alon Daya [4], Alessio Di Clemente [1,5],
Francesco Paolo Mancini [6], Aram Megighian [7,8], Nir Nesher [4], Eric Röttinger [9], Tal Shomrat [4],
Stefano Tiozzo [10], Alberto Zullo [6,*] and Simona Candiani [3]

[1]   Istituto Italiano di Tecnologia, Center for Micro-BioRobotics & Center for Synaptic Neuroscience and
      Technology (NSYN), 16132 Genova, Italy; Alessio.DiClemente@iit.it
[2]   IRCCS Ospedale Policlinico San Martino, 16132 Genova, Italy
[3]   Laboratory of Developmental Neurobiology, Department of Earth, Environment and Life Sciences,
      University of Genova, Viale Benedetto XV 5, 16132 Genova, Italy; matteo.bozzo@edu.unige.it (M.B.);
      candiani@unige.it (S.C.)
[4]   Faculty of Marine Sciences, Ruppin Academic Center, Michmoret 40297, Israel; alond@ruppin.ac.il (A.D.);
      nirn@ruppin.ac.il (N.N.); talsh@ruppin.ac.il (T.S.)
[5]   Department of Experimental Medicine, University of Genova, Viale Benedetto XV, 3, 16132 Genova, Italy
[6]   Department of Science and Technology, University of Sannio, 82100 Benevento, Italy; mancini@unisannio.it
[7]   Department of Biomedical Sciences, University of Padova, 35131 Padova, Italy; aram.megighian@unipd.it
[8]   Padova Neuroscience Center, University of Padova, 35131 Padova, Italy
[9]   Institute for Research on Cancer and Aging (IRCAN), Université Côte d'Azur, CNRS, INSERM, 06107 Nice,
      France; eric.rottinger@univ-cotedazur.fr
[10] Laboratoire de Biologie du Développement de Villefranche-sur-Mer (LBDV), Sorbonne Université, CNRS,
      06230 Paris, France; tiozzo@obs-vlfr.fr
*   Correspondence: letizia.zullo@iit.it (L.Z.); albzullo@unisannio.it (A.Z.)

Received: 23 July 2020; Accepted: 17 August 2020; Published: 19 August 2020

Abstract: Cells with contractile functions are present in almost all metazoans, and so are the related processes of muscle homeostasis and regeneration. Regeneration itself is a complex process unevenly spread across metazoans that ranges from full-body regeneration to partial reconstruction of damaged organs or body tissues, including muscles. The cellular and molecular mechanisms involved in regenerative processes can be homologous, co-opted, and/or evolved independently. By comparing the mechanisms of muscle homeostasis and regeneration throughout the diversity of animal body-plans and life cycles, it is possible to identify conserved and divergent cellular and molecular mechanisms underlying muscle plasticity. In this review we aim at providing an overview of muscle regeneration studies in metazoans, highlighting the major regenerative strategies and molecular pathways involved. By gathering these findings, we wish to advocate a comparative and evolutionary approach to prompt a wider use of "non-canonical" animal models for molecular and even pharmacological studies in the field of muscle regeneration.

Keywords: myogenesis; evolution; metazoans; differentiation; transdifferentiation; muscle precursors; regenerative medicine

---

1. Introduction

One particular challenge in regenerative biology concerns the development of reconstructive strategies after muscle-related injuries, but also the treatments of degenerative myopathies for which no reliable clinical strategy exists such as muscle dystrophy, sarcopenia, cachexia, to mention just a few [1,2]. In mammals regenerative capacities are restricted to only a small number of organs [3], yet, in other metazoans, the ability to respond to environmental injuries ranges from "simple" wound healing to complete anatomical and functional restoration of the lost or damaged part of the body, including muscles [4]. The musculature is a tissue specialized in contraction and cells with contractile functions are present in almost all metazoans but, despite their structural similarity, the origin of muscles is considered to be the outcome of a process of convergent evolution [5]. Indeed, typical muscle protein core sets are present even in unicellular organisms and in early diverged organisms like sponges, which lacks a proper tissue organization and therefore "true" muscles, and in cnidarians, where muscle-like cells are present but lack almost all molecular hallmarks of bilaterian striated muscles thus suggesting evolution from cells with ancient contractile machinery [5]. The processes of myogenesis and muscle homeostasis have also various degrees of conservation among different clades, and so is the extent of muscle regenerative capabilities [5–10].

Animals adopt different basic strategies of regeneration that include the activation of adult stem cells, the dedifferentiation of preexisting cells, and/or the proliferation of differentiated cells. This diversity of mechanisms is still widely understudied and underexploited for biomedical applications.

In this review, we provide an outline of main animals' clades (see Figure 1), muscle types, their development, homeostasis, and regeneration abilities highlighting what is known of their molecular mechanisms. We emphasize some potential contributions of comparative studies into the biomedical fields, therefore advocating deeper employment of 'non-canonical' animals as models for muscle regeneration studies.

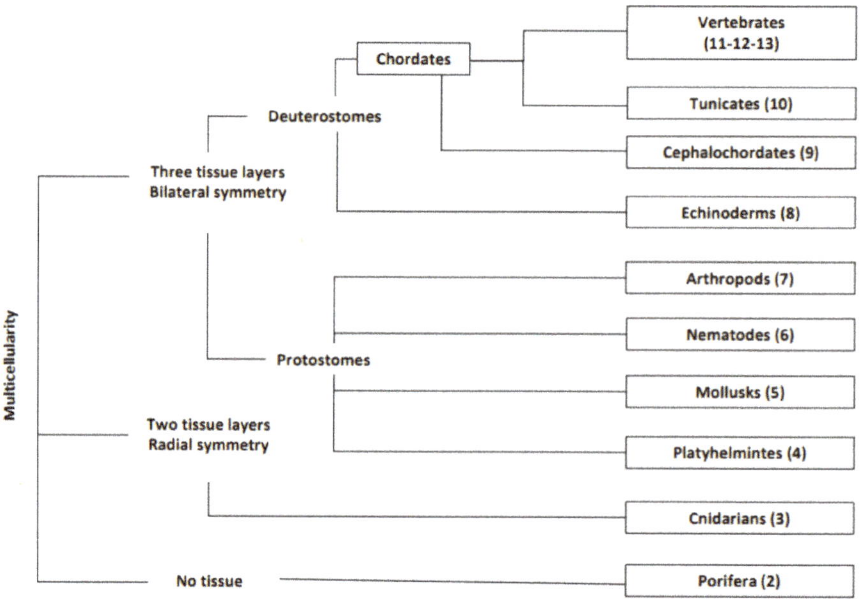

Figure 1. Tree of the animals' clades treated in this review (in brackets the corresponding section numbers).

Understanding the molecular pathways and mechanistic underlying regenerative events may offer insights into potential methods to unlock regeneration in animals where the regenerating capabilities are more restricted, e.g., mammals. Indeed, regeneration is greatly attenuated in mammals although portions of major organs, such as the liver, retain event-triggered regenerative potentials for the entire animal's life [2]. Interestingly, recent pieces of evidence suggest that regeneration can be induced even in non-regenerating species by altering specific signaling pathways [11–13]. This might be also the case for mammals and thus the principles underlying the induction of regeneration in non-regenerating species may be transferred to humans to trigger regeneration [3,14].

Elucidating muscle regeneration in metazoans also provides opportunities to 'model' a complex biological process relevant to human health and offers a window into fundamental principles underpinning this important response.

2. Porifera: Low Body Complexity with High Regenerative Capabilities

The phylum Porifera includes mainly sponges, aquatic multicellular organisms with relatively simple anatomy, lacking an organization of tissues and organs.

They have a very simple functioning relying on water circulating throughout a system of canals and chambers, called a water-current system. Circulatory, digestive, nervous, and muscular systems are completely absent. Their body is composed of a few types of cells [15,16]. For instance, in demosponges we found pinacocytes (the skin cells), mesenchymal cells, choanocytes (lining in the interior body walls), archaeocytes (totipotent cells), sclerocytes, myocytes, and porocytes (surrounding canal openings). Two types of contractile cells can be identified: the pinacocytes, and the myocytes. Pinacocytes form a functional contractile epithelium. They are composed of actin networks and actin-dense plaques allowing a coordinated contraction in adjacent pinacocytes but their mechanism of contractility remains to be further elucidated [17].

Most sponge species have an extraordinary capacity to regenerate lost body parts. Four cell types have been identified as stem-cell-like in sponge: choanocytes and archaeocytes, also referred to as adult stem cells (ASCs), pinacocytes, and particular ameboid vacuolar cells [18,19].

Muscle-Like Cells

Myocytes are spindle-shaped, smooth muscle-like cells containing microtubules lying parallel to peripheral microfilaments. They contract similarly to muscle cells thanks to a non-muscle myosin type II with high homology to that found in bilaterians and vertebrates. They allow the sponge to change shape and expel sediments even without the presence of a nervous system as their contraction is entirely controlled at a cellular level through variation of calcium (Ca^{2+}) concentration [20,21]. In particular, channels located at the plasma membrane allow the control of intracellular Ca^{2+} concentration that, in turn, regulates cell contractility. This mechanism is believed to be rely on the activation of type II myosin by Ca^{2+}-dependent protein kinases [20,22]. Myocytes allow only movement internal to the sponge but these animals remain essentially sessile. Predation and physical injuries are events very common during the entire adult life of sessile organisms that had to develop efficient strategies of repair and replacement of lost structures to survive [23,24].

Very limited information is currently available on the molecular pathways involved in body regeneration. The activity of ADP-ribosyl cyclase (ADPRC) is related to physiological activities in sponges, such as stem cell duplication and regeneration events [25]. Sponges regenerate using diverse and complex morphogenetic mechanisms involving different cell sources depending on the species. Regeneration can occur through epimorphosis and morphallaxis. The first process involves the formation of a mass of undifferentiated cells (blastema) at the wound site. Pluripotent cells such as archaeocytes and choanocytes from sites adjacent to the injury, undergo a process of epithelial-to-mesenchymal transition (EMT) and migrate to the injured area. Here they actively proliferate and form a typical blastema with dedifferentiated cells. Thereafter a process of mesenchymal-to-epithelial transition (MET) re-establish the differentiated cell identity. Different members of the transforming growth factor (TGF) superfamily are also involved in these processes [26]. In morphallaxis, of particular importance is cell transdifferentiation, the conversion of a differentiated cell to another type of differentiated cell [18]. During this process, spreading and fusion of the epithelia surrounding the wound is accompanied by the transdifferentiation of the choanocytes and exopinacocytes without the formation of a blastema. This supports the hypothesis that these cells combine properties of somatic and stem cells.

Taken together, Porifera represents, for their exceptional regenerative capacities and low body complexity, a promising model for investigating mechanisms of cell recognition, adhesion, migration, and cell type transition during regeneration.

3. Cnidarians: The Starlet Sea Anemone, *Nematostella vectensis* (Anthozoa)

Cnidarians (*Hydra*, jellyfish, corals, sea anemones) are aquatic animals that are the sister group to the bilaterian clade [27] and hold a key phylogenetic position for understanding the evolution of common biological processes and mechanisms [28]. The two main groups of the phylum Cnidaria are Medusozoa (jellyfish, hydroids, *Hydra*) and Anthozoa (corals, sea anemones) (Figure 2A). Cnidarians are structurally simple animals with remarkable regeneration capacity. They can regenerate amputated head, foot, and intact animals can even regenerate grouping single dissociated cells [29].

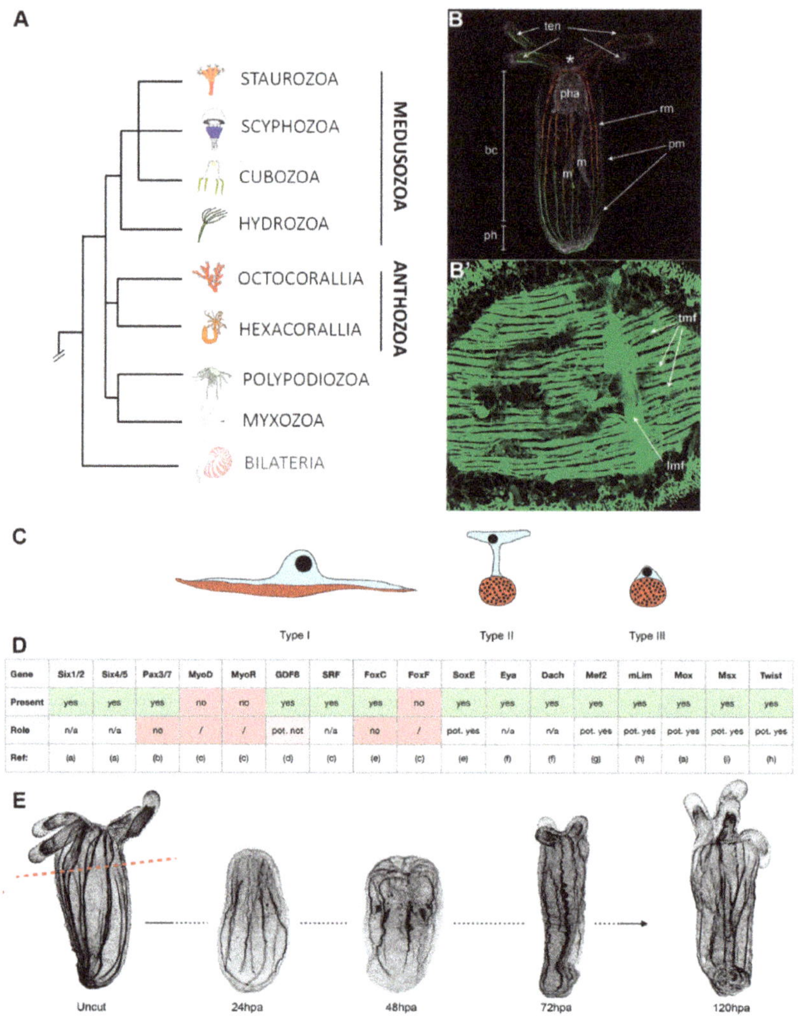

Gene	Six1/2	Six4/5	Pax3/7	MyoD	MyoR	GDF8	SRF	FoxC	FoxF	SoxE	Eya	Dach	Mef2	mLim	Mox	Msx	Twist
Present	yes	yes	yes	no	no	yes	yes	yes	no	yes	yes	yes	yes	yes	yes	yes	yes
Role	n/a	n/a	no	/	/	pot. not	n/a	no	/	pot. yes	n/a	n/a	pot. yes	pot. yes	pot. yes	pot. yes	pot. yes
Ref:	(a)	(a)	(b)	(c)	(c)	(d)	(c)	(e)	(c)	(e)	(f)	(f)	(g)	(h)	(a)	(i)	(h)

Figure 2. The epitheliomuscular system and regenerative capacity of the anthozoan cnidarian *Nematostella vectensis*. (**A**) Schematic representation of the relationship between the main cnidarian lineages and the phylogenetic position of *Nematostella vectensis* (Anthozoa, Hexacorallia). (**B**) The upper

panel shows the muscle network of nematostella in a fixed MyHC1::mCherry transgene [30] labeling the retractor muscles, co-stained with phalloidin thus showing the entire muscle network in green. (ten) tentacles, (*) mouth, (pha) pharynx, (bc) body column, (ph) physa, (m) mesenteries, (rm) retractor muscles, (pm) parietal muscles. (**B′**) Magnification of a body column region to highlight the orientation of the muscle fibers. (tmf) transverse muscle fibers, (lmf) longitudinal muscle fibers. (**C**) Three epitheliomuscular cell types have been identified in nematostella; they vary in their apical and basal cell junctions as well as their localizations within the body [31]. (**D**) Overview of the known bilateral myogenic factors identified in nematostella. (Present) indicates that the gene has been identified in the genome, (Role) indicates a myogenic role (or not) of this gene in nematostella; (pot. yes), indicates evidence of a myogenic role based on functional experiments or gene expression. (pot. not), indicates evidence of a non-myogenic role based on functional experiments or gene expression. (n/a) data not available. References cited: (a) [32], (b) [33], (c) [34], (d) [35], (e) [36], (f) [37], (g) [38], (h) [39], (i) [40]. (**E**) Oral regeneration of lost body parts after sub-pharyngeal amputation (red dashed line) is completed after 120 h post amputation and reforms a fully functional organism. Animals were fixed at various time points during regeneration and stained with phalloidin to show f-actin filaments (black). Elements of the figure are extracted from [28,41].

The present section focuses on the sea anemone *Nematostella vectensis* that belongs to the Anthozoa, mostly sessile cnidarians that are represented by individual or colony-forming polyps.

The sea anemone *Nematostella vectensis* (Anthozoa, Figure 2B), was initially employed for studying the evolution of embryonic developmental mechanisms [42] and is now emerging as a novel complementary whole-body regeneration model [41]. Nematostella possesses a range of fundamental advantages, such as the access to biological material, a relatively short life-cycle, an annotated genome that revealed astonishing similarities with the one from vertebrates [34], a wealth of -omics data [43,44] and well developed functional genomics and genome editing approaches [45–47].

Nematostella is a rather small sea anemone (juveniles ~0.5 mm, adults ~3 cm), translucent, and well suited for imaging purposes (Figure 2B). It is a diploblastic animal formed by a bifunctional internal endomesoderm and an outer ectoderm. On the oral extremity are the tentacles that surround the mouth and the so-called physa on the opposite. Food caught by the tentacles is ingested via a muscular and neuron-rich pharynx and digested within the body cavity. While most of the digestive enzymes are secreted by the mesenteries that also store nutrients [37], these internal structures play another role as they harbor the gonads that are crucial for sexual reproduction [42] and for inducing a regenerative response [48].

3.1. Muscle Types, Organization, and Myogenic Genes

Cnidarians display a large diversity of muscle types and organizations that are involved in multiple crucial physiological functions such as feeding, locomotion, or defense [28]. Although this group of marine invertebrates lacks a large part of the molecular hallmarks of striated muscles [5], jellyfish present some ultrastructural and functional features (such as striated myonemes, thick and thin myofilaments, desmosomes as well as a mechanism of excitation–contraction coupling based on intracellular calcium stores [49]) resembling the structure and function of striated muscles [50–53].

For a global overview of cnidarian muscle diversity, their development, and regeneration, please refer to [28]. Most anthozoan muscle cells, and nematostella is no exception, are epitheliomuscular; they contain smooth myofilaments [28] forming a transverse and longitudinal muscle fiber network clearly visible using a MyHC1::mCherry transgenic line [30] and phalloidin/actin staining (Figure 2B′). The epitheliomuscular cells, whose actin fibers form more or less condensed muscle fibers are responsible for various functions of the animal such as feeding or locomotion.

A recent study has characterized three epitheliomuscular cell types (Figure 2C); two types (I and II) with elongated cytoplasmic bridges present in the endodermal parietal and retractor muscles (Figure 2B) and one type III corresponding to basiepithelial muscle cells encountered in the ectoderm of the tentacles [31]. While the known bilaterian myogenic regulatory factors (MRFs [54]) are missing in nematostella (e.g., MyoD, MyoR), a large part of the conserved myogenic gene regulatory network (e.g., Pax3, Pax7, Six1 [55]) has been identified (Figure 2D, reviewed in [28]). However, their exact roles in the formation of the epitheliomuscular cells in nematostella are yet to be understood.

3.2. Muscle Regeneration and Role of Epitheliomuscular Cells in the Regenerative Process

Like cnidarians in general, nematostella possesses proliferation-dependent whole-body regenerative capacities and can regrow fully functional animals from isolated body pieces within less than a week (Figure 2E) [56–59]. In addition, nematostella is very well suited to compare embryonic development and whole-body regeneration within the same organism [43,57,60], one of the long-lasting questions in regenerative biology [61].

By assessing the MyHC1::mCherry transgenic gene expression after bisection, Renfer and colleagues have observed that the retractor muscles retract from the wound site immediately after amputation. In later steps of the regenerative process, mCherry positive cells accumulate in the regenerating body part suggesting active cellular re-organization and differentiation to reform the retractor muscles [30]. However, the cellular and molecular mechanisms underlying retractor muscle regeneration remain unknown.

On the other hand, there are shreds of evidence suggesting an active role of muscle fibers in the regenerative process. Bossert and colleagues have shown that muscular contractions are involved in reducing the epithelia of the wound site and potentially favoring the wound-healing process [62]. In addition, a stereotyped tissue dynamics that may reflect the above-mentioned observation of MyHc1::mCherry positive cells retracting from the amputation site, supports the idea that muscle contractions play also a crucial role during various phases of the regenerative process [59]. A recent study has shown that the retractor-muscle containing mesenteries are fundamental in inducing regeneration in nematostella [48]. Based on data from planarian [63] and mammalian myoepithelial cells [64], one could thus speculate that the epitheliomuscular cells that form the retractor muscle are also involved in the regeneration process via contraction-independent biochemical signals. However, additional work is required to further support those evidences and to determine the cellular and molecular mechanisms involved.

4. Platyhelminthes: The Freshwater Planarian

Since the beginning of the 21st century, the freshwater, free-living, flatworm planarian, has become a leading model for the study of development and regeneration mechanisms [65]. As a model organism, it possesses a set of clear advantages. 1. Reductionism: although its relative simplicity, planarians exhibits much of the "complexity" of vertebrate systems, including a well-differentiated nervous system, simple eyes, central brain, triploblastic organization, and bilateral symmetry. 2. Planarians are inexpensive and very easy to rear and maintain in the lab, therefore ideal for primary high-throughput screening processes [66]. 3. Planarians are molecularly-tractable model organisms, easily manipulated by RNAi interference [67] and their thin and somewhat transparent body allows whole-mount in situ hybridization in an intact worm [68].

However, there is no doubt that the most astounding feature is its regeneration capability. Planarians are considered as the "Masters of Regeneration". Adult pluripotent stem cells that are called neoblasts and are the only proliferating cells, account for 25–30% of all cells distributed in the planarian body, and give them remarkable regenerative abilities. Whole worms can regenerate from only a small proportion of the adult worm, within 1–2 weeks. Consequently, full results from regeneration experiments are revealed in a relatively short time. For a broad review of planarian as a model system for regeneration see Ivankovic et al. [69].

4.1. Muscle Types

The planarian's muscle cells combine features common to both skeletal and smooth muscles [63]. The planarian contains two main muscular systems, somatic and pharyngeal, that differ in their myosin heavy-chain (MHC) muscle isoforms along with their function and location possibly due to their different biological functions [63]. Without a supportive skeleton, the maintenance of body shape, posture, movements, and defense (strength for their soft bodies) depends on the somatic muscular system. Locomotion is mainly executed by ciliary gliding. The muscular body wall, organized beneath the epithelium, is arranged in a grid work of 4 layers of fibers lying in different orientations and linked to an extracellular network of filaments associated with the body's organs [63]. Moreover, recent works [70–72] revealed that in planarian, the muscles also provide patterning signals essential to regeneration and guidance of tissue turnover and regrowth after injury. Interestingly, this resembles what has been suggested to be a function of the connective tissues in vertebrates [73]. The somatic muscular system provides regeneration guidance through the expression of position control genes (PCGs) differing over time, body region, and types of genes expressed.

In addition to the somatic muscular system, planarians possess a separate pharyngeal musculature system. Planarians have an incomplete digestive system with pharynx (proboscis and anus) connected to the intestine duct by its anterior thus providing a single opening that functions as both, anus and mouth. The pharynx is composed of a muscular tube and demonstrates repertoires of movement capabilities. It is extruded from the body center during feeding and can direct itself toward the food by bending and stretching till it reaches the food; it thus swallows the food and transfers it to the intestine by peristaltic movements. The pharynx does not contain neoblasts [74] and therefore is incapable to regenerate the rest of the worm when amputated. However, a worm losing its pharynx can regenerate it in a few days [75]. The pharynx can thus serve as a module of organ regeneration where stem cells differentiate into distinct cell types to form an organ that integrates within the rest of the body [76].

4.2. Muscle Regeneration and Homeostasis

Upon regeneration, planarian muscle cells, as all other tissue components, arise from the large reservoir of the existent neoblasts population that migrate to the wound area and start proliferate, thus creating a blastema where they differentiate to form the missing body parts. Irradiation protocols applied to the whole body or to specific areas allows neoblasts ablation [77]. Further transplantation of a single pluripotent neoblast can restore regenerative ability and the whole process can be monitored from scratch [78]. Therefore, planarian is an ideal model for deciphering the mystery of stem cell differentiation [79], allowing experimental approaches that are unavailable in any other model organism. Research on planarian muscle regeneration is still limited but provides some interesting perspectives.

One other unique feature of the planarian model (e.g., *Dugesia japonica* and *Schmidtea mediterranea*) is the ability to shift from growing (up to few centimeters) to de-growing (down to few millimeters) by food deprivation and vice versa. The process depends on the balance between cell proliferation and cell death and by keeping stable body shape and proportions through constant remodeling mechanisms [80]. Therefore, it is a perfect model system for the study of tissue homeostasis (for a broad overview of the subject see [81]).

In spite of their exceptional features and their growing popularity as a model for basic research on regeneration, planarians are not yet considered as a conventional organism for studying human pathologies and diseases, maybe it is time to rethink [82].

5. Mollusks: The Cephalopods

Cephalopods represent one the main and most evolved mollusk class. They are the most intelligent, mobile, and the largest of all mollusks and include very diverse species such as squid, octopus, cuttlefish and the chambered nautilus.

Regeneration is a frequent event occurring during cephalopods' lifetime. Wild animals often lose body parts such as portions of arms and fins and as a consequence, it is common to find signs of traumatic events on their bodies [83,84]. These events can dramatically impair their capacity to swim, capture, and manipulate preys [85], and therefore they can seriously impact their survival in the natural environment. Indeed, cephalopods can regenerate their cornea, peripheral nerves, and body limb (arms and tentacles) [86]. Cephalopod limbs are complex organs composed of a tightly packed three-dimensional array of muscle fibers controlled by a sophisticated peripheral nervous system (PNS). The arm PNS is composed of three distinct parts: the arm nerve cord (composed of axial nerve cords and the ganglionic core), the sucker ganglia, and the intramuscular nerves. This assembly allows the transmission of a large amount of sensory and motor information to and from the brain [87–90]. All of these structures are fully and functionally recovered during regeneration.

5.1. Muscle Types

Similar to vertebrates, muscle cells in cephalopods can be found in a variety of different organs that differ dramatically in structure and function. Indeed, muscle cells are in the mantle and appendages (arms and tentacles) but also in eyes, hearts, viscera and chromatophores. Such diversity is paralleled by specific adaptations in muscular organization and physiology.

The majority of the musculature of arms and tentacles is composed by uninucleate transverse or obliquely striated muscle fibers with shared morphological and physiological characteristics. When oblique striation is present, this pattern is uniform and continuous among adjacent cells. Generally, these muscle cells do not exceed 8–20 µm in diameter and 0.8–1 mm in length. The nucleus is in the central portion of the cell whose transverse section is usually round or polygonal, with a mitochondria-rich core and a contractile apparatus in the cortical zone. The contractile apparatus lies along the main axis of the fiber and is organized in sarcomeres with identifiable acto-myosin striations. Cephalopod muscle actin and myosin heavy chain, show strong sequence identity to other invertebrates and vertebrate gene orthologs suggesting a similar contraction mechanism [91,92]. On the contrary, regulatory proteins are very cephalopod-specific [93] suggesting that specific control kinetics and cross-bridge cycle regulation might be developed in cephalopods (for a review see [94,95]). Different from typical skeletal muscles, cephalopod arm muscle cells do not possess a proper T-tubules system, but smaller sarcoplasmic structures named "terminal cisternae" that take contacts with plasma membrane invaginations thus forming "dyads" at the level of the Z-disks. In contrast to the muscle cells of other invertebrates, they are isopotential, and thus each synaptic input can control the membrane potential of the entire muscle cell (for a review see [94]).

Among cephalopods, muscle cells can differ in their activation properties. As an example, in octopus arm, muscle action potentials rely on Ca^{2+} spikes [96,97] generating a massive entrance of Ca^{2+} that activates a calcium-induced calcium release (CICR) process from the internal stores [98]. An intriguing analogy here can be found with vertebrate cardiac muscle cells that represent an important target of regeneration medicine [99,100].

In contrast to the octopus arm muscles, transverse muscles of squid tentacles show 'graded' Na^+ based action potentials different from the typical 'all or nothing' action potentials of squid giant axon or vertebrate muscle fibers. Interestingly, the transverse muscle of the squid arms lacks Na^+-based action potentials [101]. All the above-mentioned characteristics co-evolved with the complex brain to body adaptation and limb specialization [102–104] whose integrity is essential to the animal survival.

Several myogenic genes have been identified in some (but not all) cephalopod species. As an example myoblast-specific Myf5 and MyoD proteins have been identified in *Sepia officinalis* tentacles during late stage development and NK4 is found to be involved in cephalopod striated muscle formation just as in vertebrate cardiac cells [94,105,106]. In addition, an hh-homolog signaling molecule and its receptor Patched (Ptc) have been found to be expressed during myoblast differentiation in *Sepia officinalis* [107].

5.2. Muscle Regeneration

Cephalopod mollusks are a powerful model of limb regeneration due to their similarities in early arm development to vertebrate models and their fast and efficient regenerative abilities (for a review see [94]) and, among regeneration studies of other body parts, rather ample literature is currently available on the regeneration of their limbs (for a review see [84]). However, very little is known about the molecular pathways controlling the regenerative process.

Hereafter, we will employ the octopus arm as a template to describe the step of a regeneration process. Morphologically, a sequence of events can be identified during arm regeneration: (1) wound healing; (2) formation of a knob at the stump tip; (3) elongation of the knob and formation of a hook-like structure; and (4) elongation of the regenerating arm till complete restoration of a functional structure [85,94,108,109].

At early steps of regeneration, a mixture of extracellular matrix (ECM), vesicles, and mucus are present at the plug region, and only subsequently the connective tissue is deposited by fibrocytes migrating to this region [109]. The presence of ECM and connective fibers might be relevant for the correct reorganization of the regenerating structure, a role that has been also suggested in octopus pallial nerve regeneration [84]. Cephalopods might have evolved fine mechanisms of regulating ECM composition and organization during regeneration that favor the tissue competency to regrow. Interestingly, similar fibrillary elements are the ones limiting vertebrates skeletal muscle regeneration as their accumulation at the injury site negatively interferes with regeneration and drives instead scarring and fibrosis of the tissues [94].

At a cellular level, cells composing the stump are first characterized by a layer of undifferentiated cells together with diffuse vascular components forming a typical blastemal region. This structure then disappears, and cells start differentiating [110]. Cell proliferation remains active throughout the entire regeneration process, but while at an early step is primarily localized at the blastema, at later stages it is present within differentiating tissues such as the axial nerve cord and the musculature.

Unfortunately, no study reported so far could reveal the molecular identity of muscle cell precursor during regeneration. It has been speculated that new muscles and nerve cells can originate from dedifferentiated cells of the same type (for a review see [84]) but due to the lack of species-specific molecular markers, we are currently not able to assess the existence of pluripotent vs. lineage-committed progenitor cells, as well as vertebrate satellite-like cells associated with adult muscles. From a mechanicistic viewpoint it has been shown that after an arm lesion, muscles close to the injury site degenerate fast, and large cells containing little protoplasm and a large nucleus appear within the same area. These cells are supposed to be sarcoblasts that later migrate to the most distal part of the wound and undergo active proliferation. Sarcoblasts will then differentiate into the arm and sucker muscle fibers in different time intervals [108]. This process is possibly paralleled by the recovery of the arm functional capacity.

Few data are available on the molecular pathways underlying muscle formation during regeneration in cephalopods. It is known that cephalopods muscle development rely on MRFs, however, still, no data are available on their expression during muscle regeneration in octopus. Several studies suggested that acetylcholinesterase (AChE), a conserved molecule between vertebrates and invertebrates, may orchestrate the formation of the octopus arm during regeneration [110,111] similarly to what happens in regeneration phenomena occurring in other animal phyla such as Platyhelminthes, Mollusca, Arthropoda, and even Chordata (for a review see [94]).

6. Nematodes: The *Caenorabditis elegans* Model

Nematodes are one of the most diverse animal phyla. They occupy a large variety of environments, and many species are parasitic. Nematodes are relatively small animals (~1 mm long adults), and, given their size, a heart and a closed circulatory system are not required.

C. elegans has been employed as a model to study extrinsic and intrinsic factors crucial for axon regeneration and wound healing. In particular it have disclosed important aspects of the mechanisms of wound healing and cellular plasticity, axon regeneration and transdifferentiation in vivo [112].

Muscle Type and Homeostasis

The majority of muscles of the animal body wall are used for the animal's locomotion [113]. *C. elegans* body wall muscle cells are spindle-shaped mononuclear cells with multiple sarcomeres per cell [114]. Muscle cells are obliquely striated and form body-wall muscles running along the length of the body underneath the epidermis [115,116]. Unlike most other animals, their innervation is unusual in that the nerves do not branch out into the muscles but the muscle cells send extensions (muscle arms) to the nerve cord to receive *en passant* synapses from the motor neurons [117,118].

Embryonic development of body wall muscle is controlled by maternally expressed genes initially, but then there is a switch to control by zygotically expressed genes. Several molecular players (e.g., Wnt/Mitogen-Activated Protein (MAP) kinase signaling, Myogenic regulatory factors (MRFs) as many others) act during muscle development and differentiation. For a detailed description please refer to [119,120].

C. elegans muscles lack satellite cells (muscle stem cells) and therefore muscles cannot regenerate. Adult worms only carry post-mitotic body wall muscles [119]. It is interesting to notice that, although lacking an open circulatory system, proteins and structures composing the body wall muscles manifest a high homology with that of human heart muscle. In addition, many molecules involved in sarcomere assembly and maintenance are in common with other animals. A dystrophin ortholog, dys-1 gene, has been identified in *C. elegans* with a key role in the sarcomere structural regulation [121]. The mechanism of assembly of sarcomeres into functional muscles have been extensively investigated in *C. elegans* within the context of repair following activity-induced muscle stress and muscle degeneration [117].

For the reasons listed above, *C. elegans* has become a model study for muscle diseases such as Duchenne's Muscular Dystrophy (DMD) [122] and cardiomyopathies [113]. A more explanatory and detailed list of advantages and disadvantages of this animal as a model of human heart pathologies can be found in [113].

Interestingly, the lack of regeneration capacity of *C. elegans* muscles has been key to the use of this animal as a model of DMD. Indeed, as *C. elegans* adult muscle cells are mono-nucleated and post-mitotic, they can be individually tracked during the process of muscle degeneration and do not undergo fibrosis and chronic inflammation, processes that are common in vertebrate models [121].

7. Artropods: The Insect *Drosophila melanogaster*

Insects have a reduced lifespan and events related to degeneration/regeneration processes following physical, pathological or aging damage are less frequent. Hence, the establishment of a real physiological regenerative mechanism have been under a lower evolutionary pressure. However insects manifest adult muscle hypertrophy, which can be viewed as a degeneration/regeneration-like process, in response to particular hormones as well as to environmental factors, population density, food availability, or mating [123,124].

Drosophila melanogaster is a model organism in which genetic and molecular techniques, coupled with physiological and structural approaches, have been used to unravel specific issues of invertebrates and vertebrates muscle biology, including regeneration processes.

Muscle Type and Homeostasis

In *D. melanogaster* larvae and adults, three types of muscles can be recognized: (1) Tubular muscles, including most of the adult skeletal muscles. They are striated with a centrally located nucleus and are synchronous muscles because each nerve stimulation evokes calcium release from internal stores which triggers a mechanical contraction of the muscle similarly to vertebrate skeletal muscles [125]. (2) Adult indirect flight muscles, or "fibrillar" muscles. In these muscles individual myofibrils can be identified by light microscopy; they are striated and asynchronous muscles as their mechanical response is activated both by calcium following nerve impulse and by stretch-activation due to the elastic recoil of thorax cuticle [125,126]. (3) Supercontractile striated visceral and heart muscles, and larval body wall muscles. Supercontractile muscles are called "supercontractile" because they can contract to a length well below 50% of their resting length [127–131]. They contract in response to caffeine also in the absence of external calcium, showing that a functional store of calcium is present in the sarcoplasmic reticulum and that it is sufficient for muscle contraction [132]. Interestingly, a similar activation property has been also found in the octopus arm muscles [98].

Larval and adult muscle cells and fibers derive from progenitor cells of the embryonic mesoderm. Signaling crosstalk between ectoderm and mesoderm (for instance Decantaplegic and Wingless) and gene (e.g., *twist*, *even-skipped*, and *floppy-paired*) dynamic temporal expression, regulate the muscle cell fate of these cells [133,134]. Cardiac and visceral muscle cell progenitors are formed from these generic muscle cell progenitors by their compartmentalization in segmental regions with low Twist high Even-skipped domains. High Twist high Sloppy-paired domains are, on the other hand, a key point for the development of somatic cell progenitors [134,135]. Both these muscle cell progenitors undergo then asymmetric cell divisions, which generate low Twist cells that fuse to form embryonic myoblasts and subsequently embryonic muscles. Some, but not all, asymmetric division give rise also to a single founder cell and to an adult muscle precursor (AMP), an adult muscle stem cell that remains in quiescence.

Larval muscles degenerate throughout metamorphosis. In some cases (indirect flight muscles in the thorax) larval muscles are utilized as templates for the formation of adult muscles. In other cases, peripheral nerve fibers and the space between larval muscle fibers drive adult muscle fibers development and differentiation. Adult muscle fibers origin from AMP precursors which proliferate, differentiate, and fuse to form myotubes and then adult fibers. This process was deeply studied in indirect flight muscles that are the main power source for flight. From these studies, however, it was found that only a small number of AMPs descendent stem cells remain associated with the adult differentiated indirect flight muscle fibers. These cells resemble mammalian satellite cells which are associated with the adult skeletal muscle fibers and retain the competence to proliferate and differentiate in myoblasts and then adult myofibers when stimulated (for example following skeletal muscle fiber degeneration). It is interesting to note also that these "fly satellite cells" undergo a proliferation/differentiation program leading to the generation of myoblasts which fuse with a damaged indirect flight muscle fiber. This process, similarly to what was observed in vertebrate satellite cells of skeletal muscle fibers, points to repair the damaged fibers, and it is activated by Notch-Delta signaling [136]. In the absence of tissue damage, satellite cells are maintained, not differentiated, and "quiescent" probably by the transcription factor Zhf1 [137].

Other authors claim that there are probably no "satellite cells" in adult flies' muscles. Indeed, almost all of these studies were investigating the indirect flight muscles (IFMs) that are considered the most similar to mammalian skeletal muscles. In these, regeneration processes are triggered by damage consequent to physical or pathological injuries as well as damages related to aging. Considering fly lifespan, these events are less frequent and therefore they should have exerted a minor pressure from the evolutionary point of view to establish a real regenerative mechanism in adult muscles. Moreover, a regeneration-like process is considered for adult skeletal muscles as regarded as muscle hypertrophy. Again, in flies, differently from other invertebrates, the small dimension of IFMs could have been a factor against a "pressure" from an evolutionary perspective.

8. Echinoderms: A Compendium of Regeneration Strategies

Echinodermata is a phylum consisting of radially symmetrical marine animals. All larval and adult echinoderms exhibit high regenerative capacities of entire lost parts following predation or traumatic events [138]. Echinoderms manifest all the regenerative strategies identified in other animals, such as epimorphosis and morphallaxis, and have an impressive high genetic homology with Chordates. They can show epimorphic processes, by which a blastemal is formed through active proliferation of migratory undifferentiated cells. They can also show morphallaxis, where cells derive from differentiation, transdifferentiation, or migration of existing tissues. Most classical and bio-molecular tools currently available have been successfully employed in this animal species giving rise to a large body of literature on Echinoderms regeneration from molecular, cellular, and tissue level. These features make them interesting models in translational research [139].

8.1. Echinoderm Muscles

Movements in echinoderms are assured by a muscular and a water vascular system; two main muscular systems, the visceral and the somatic, are present. Similar to what happens in nematodes and amphioxus, echinoderm visceral muscles may extend cytoplasmic prolongations towards the nerve fibers that they make contact with. Echinoderm muscles retain some epithelial features. Indeed, the epithelial cell of coelom can give rise to peritoneocytes, myoepithelial cells, and myocytes through successive stages of specialization. Despite differences in anatomical location, echinoderm muscles share a similar structure. They are made up of numerous contractile bundles and each of them is composed of several myocytes containing myofilaments of variable thickness [140].

Two main types of muscle fibers can be identified, the first (and most common) in which individual bundles are composed of myocytes of fusiform shape and resemble vertebrate smooth muscle fibers. These fibers are embedded in the extracellular matrix of connective tissue composed of a network of thick striated (collagenous) and thin unstriated fibers and an amorphous component; it also comprises fibroblasts, nerve cells, and different coelomocytes (the immune effector cells of sea urchins). A second muscle type, typical of crinoid arms, consists of obliquely striated fibers with each muscle bundle composed of 8–20 myocytes and surrounded by a basal lamina.

8.2. Echinoderm Muscle Regeneration

Several signaling pathways are involved in regeneration. These include the bone morphogenetic protein/transforming growth factor (BMP/TGFB)-signaling pathway, the homeobox (HOX) signaling pathway and the Ependymin pathway. Nonetheless, it is not directly possible to associate these pathways with the specific process of muscle regeneration. To provide this information, it would be necessary to screen for genes expressed by muscle cells such as cytoskeletal genes, actin, and myosin genes. Interestingly, their expressions are known to be modulated during different stages of the regeneration process [141–143].

Upon injury, echinoderm muscles undergo processes of de-differentiation and myogenesis. In the wound region damaged myocytes degenerate and muscle bundles disintegrate. De-differentiation of the coelomic epithelial cells represents an early regeneration event occurring already during wound healing and continuing at different rates during the regenerative period.

These cells dedifferentiate and start migrating toward the region occupied by the injured muscle, here they form clusters of muscle bundle rudiments. Then, they increase in number and start developing the contractile filaments of future muscle cells. The process of myogenesis goes on in parallel with many other regenerative events and brings to the restoration of a functional muscular tissue [142,144].

In conclusion, echinoderms represent an interesting model with a high potential for muscle regeneration studies (for extended reviews see also [138,142,144]). Their close phylogenetic relation to vertebrates makes them attractive models to determine what cellular and molecular processes are required for successful muscle regeneration to occur.

9. Cephalochordates, the Basal Chordates: Amphioxus

The cephalochordate amphioxus is the sister group of the tunicate–vertebrate clade [145] (Figure 1) and represents a new emerging model for studies on cell and tissue regeneration and in particular for muscle regeneration. Its phylogenetic position also offers insights into the evolution of regenerative capacity in vertebrates.

As demonstrated in other organisms, amphioxus regeneration can vary among body parts and several variables affect the speed of healing such as species, animal age, and body size [146–149].

In amphioxus, two districts show the highest regenerative capacity: the oral cirri, skeletal structures surrounding the mouth, and the post-anal tail.

9.1. Structure of Amphioxus Muscles

The adult amphioxus possesses almost exclusively striated muscles, the most prominent of which are the segmental axial muscles providing force for burrowing and swimming, the notochord and the pterygial muscle. The axial musculature is composed of the myomeres thatare segmentally repeated in dozens of pairs throughout the entire length of the body. Myomeres are composed of flattened striated muscle cells 0.8 μm thick [150], 100 μm wide and at least 500 μm long. They are similar to vertebrate skeletal muscle cells in banding and in the arrangement of the myofilaments [151]. Moreover, the sarcoplasmic reticulum, although not arranged to form the typical T-tubule system, is present and, as in vertebrate striated muscles, might serve as calcium storage [152]. Despite these similarities, amphioxus axial muscle cells are mononucleated, in contrast to the fused myofibers of vertebrate skeletal muscles. The notochord, a modified muscle structure extending from the tail to the most anterior tip of the rostrum, consists of a row of coin-shaped striated muscle cells plus other non-muscle cells known as Müller cells. The pterygial muscle is constituted by striated fibers [153] responsible for the contraction of the branchial cavity, which results in gamete emission from the atriopore.

9.2. Muscle Regeneration in the Amphioxus Tail

Due to its ability to regenerate skin, nerve cord, and muscles after amputation, the post-anal tail is the most studied system for understanding regeneration in amphioxus (see Figure 3 for timing and principal steps of regeneration in *B. lanceolatum* and *B. japonicum*). Indeed, the tail contains the two most prominent striated muscle structures: the myomeres and the notochord.

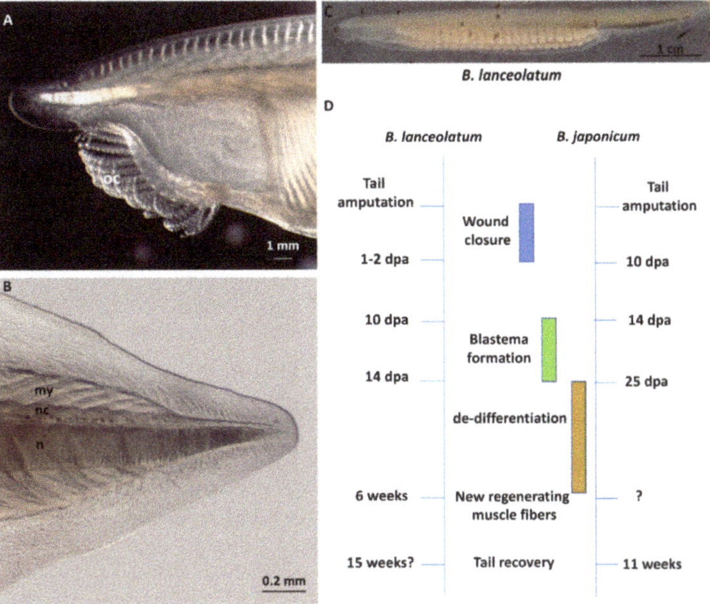

Figure 3. Timing and principal steps of regeneration in *B. lanceolatum* and *B. japonicum*. (**A**) Enlargement of the most anterior end of a *Branchiostoma lanceolatum* adult showing the oral cirri (oc). (**B**) Post-anal tail of the same animal. my, myomeres; nc, nerve cord; n, notochord. (**C**) *B. lanceolatum* individual collected in Banyuls-sur-Mer, France. (**D**) Schematic overview of the series of events occurring during tail regeneration in the *B. lanceolatum* and *B. japonicum*. dpa, days post-amputation.

While little data are available on the role of myogenic factors during tail regeneration in amphioxus, much more is known on muscle development formation in the embryo. Like vertebrates, amphioxus axial muscles derive from the myotomal portion of the somites. Here, the Six1/2-Pax3/7 myogenic program is activated [154] and regulates the differential expression of members of the myogenic regulatory factors (MRFs) gene family, which underwent independent expansion in the amphioxus lineage [155]. Amphioxus somitic mesoderm, which extends for the whole length of the body, is divided into three portions, each induced by a unique combination of transcription factors [154]. The most anterior somites depend on fibroblast growth factor (FGF) signaling [156]. The signals regulating the other two populations are yet to be identified but it has been shown that the most posterior somites, arising from the tail bud as the embryo elongates, do not require FGF nor retinoic acid signaling [156] and that Notch is required for correct separation of contiguous somites [157].

From a mechanistic viewpoint, Somorjai and coworkers [146,147] described a blastema-like structure in the amputated tail of *B. lanceolatum*, with proliferating cells from notochord, myotomes and nerve cord positive for phospho-histone H3. Subsequently, Liang and coworkers [148] confirmed cell proliferation in blastemal cells of the regenerating tail of *B. japonicum* by Bromodeoxyuridine (BrdU) labeling. Conversely, Kaneto and Wada [149] identified in the amputated oral cirri of *Branchiostoma belcheri* a large number of mesenchymal cells able to reform the skeleton, but most likely without the influence of a proliferative cell population, as phospho-histone H3 was not detected. Thus, oral cirri seem to undergo tissue remodeling by morphallaxis, whereas the tail can respond to amputation injury by epimorphic regeneration. In addition, de-differentiation of existing structures but not trans-differentiation or lineage reprogramming seems to occur during amphioxus tail regeneration, as seen in amphibians [146,147].

10. Tunicates: The Sister Group of Vertebrates

Tunicates (Phylum Chordata) encompass a large group of ubiquitous and diverse animals that occupy a wide variety of marine habitats and ecosystems around the world [158]. Despite their appearance, these animals are the sister group of vertebrates, with whom they share a common ancestor [159]. Most of the tunicate species have a biphasic life cycle, with a swimming larva, with the chordate synapomorphies, and either a benthic (like the group of ascidians) or a planktonic (in the order thaliaceans) post-metamorphic phase. During this phase, the chordate features are lost and the larva turns into a filter feeder sac-like body structure with two tubular openings, known as inhalant and exhalant siphons (Figure 4). While solitary tunicates reproduce strictly sexually and have limited regenerative capabilities, colonial species can also reproduce asexually and regenerate an entire body via diverse modes of budding, also referred to as non-embryonic development. The result is often a colony of connected, genetically identical zooids [160,161]. Budding can be part of the life cycle, which accounts for colony growth, replication and reproduction, or regeneration, i.e., passive forms triggered by injury [162].

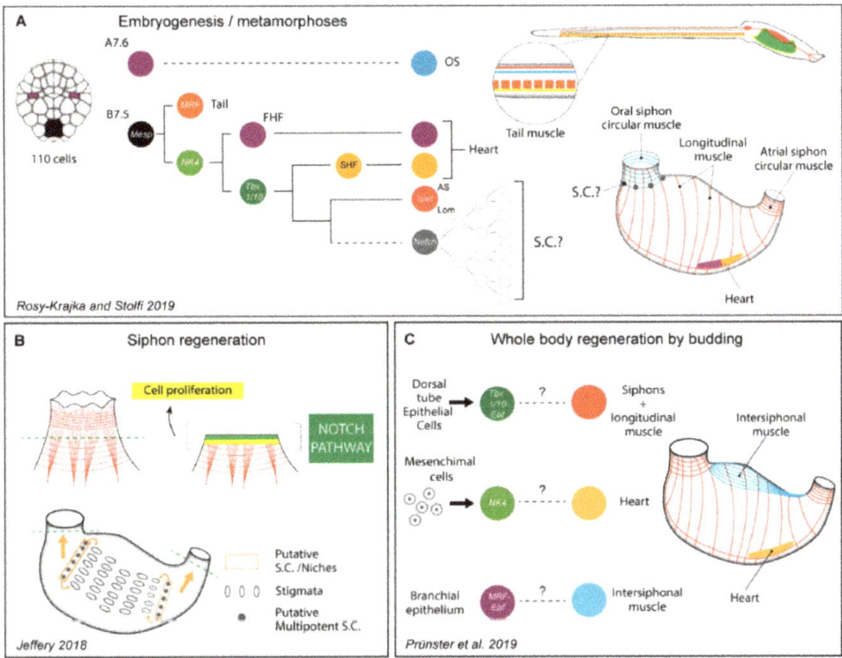

Figure 4. Ciona embryogenesis and regeneration. (**A**) Simplified scheme of muscle development during ciona embryogenesis and metamorphoses. (modified from [163]). (**B**) Scheme of oral siphon regeneration in ciona (modified from [164]). (**C**) Simplified scheme of myogenesis during budding in *Botryllus schlosseri*. FHF: first heart field; SHF: second heart field; OS: oral siphon; AS: atrial siphon; LoM: longitudinal muscles; SC: stem cells.

10.1. Myogenesis during Embryogenesis and Metamorphoses

The typical tunicate larva carries in the tail two bands of mononucleated muscle cells, one on each side of the notochord. The myofibrils of each cell are connected through intercellular junctions, similarly to the vertebrate cardiomyocytes. Each row of cells behaves as a syncytium, allowing the swimming movement. The myocytes are arranged in a striated manner and express myosin heavy chain. After metamorphosis, the sarcomere organized musculature gets reabsorbed along with the tail, and the musculature of the formed zooid generally consists of longitudinal and circular multinucleated fibers that run throughout the mantle along the body and around the two siphons (Figure 4). While ultrastructural studies show unstriated morphology, the adult body musculature seems to have intermediate characteristics between the vertebrate smooth and striated muscles [165]. For instance, the post-metamorphic musculature uses a troponin-tropomyosin complex similar to vertebrate striated muscles, their myocytes are specified via MRFs that generally controls vertebrate skeletal muscle development, and lacks on smooth muscle specification [165,166]. Post metamorphic zooids also develop a tubular heart, which consists of a single chamber formed from two epithelial monolayers. The pericardium is a non-contractile epithelium, whereas the myocardium is a single layer epithelium of mononucleated striated cardiomyocytes.

Most of the information on the molecular bases of muscle cell specification and differentiation comes from a copious amount of studies focused on few solitary model species namely *Ciona intestinalis*, *Halocynthia roretzi*, and Molgula, and have been recently summarized in a comprehensive review by Razy-Krajka and Stolfi (2019) [163]. Briefly, during embryogenesis, the maternal deposition of a zinc finger family member (Zic-r.ca) and the activation of a T-box transcription factor is necessary to induce the development of tail muscles. At the stage of 110-cells, a couple of blastomeres, the B7.5, acquire a cardiopharyngeal mesodermal fate and give rise to both heart and part of the adult body musculature, specifically the exhalant siphon and the longitudinal muscles (Figure 4). The gene regulatory network that governs this cell lineage specification is highly conserved between tunicates and vertebrates cardiopharyngeal myogenesis [167]. Another couple of blastomeres (the A7.6) follow a different fate but are partially regulated by the same transcription factors involved in the cardiopharyngeal specification [107] and give rise to the muscles of the inhalant siphon (Figure 4A).

10.2. Myogenesis during Budding and Regeneration

Many solitary ascidians can repair and regenerate efficiently both the exhalant and inhalant siphons, including the associated muscle fibers [168]. Interestingly, during its progressive differentiation, the B7.5 cell lineage gives rise also to a population of cells that do not express the MRF, but seems to maintain an undifferentiated state and settle around the exhalant siphon [169,170]. These muscle precursors maintain a stem cell-like state via a Notch-mediated lateral inhibition, a mechanism that has been also reported in drosophila and vertebrates to control muscle differentiation [170]. In addition, the A7.6 is multi-lineage, but it is not clear if such multipotency is retained in the fully developed adult. So far, the link between the B7.5 and A7.6 myogenic lineage and the siphon muscle regeneration has not yet been explored. Recently, Jeffery (2018) suggested that, in *Ciona intestinalis*, the siphons repair and regeneration are triggered by the mobilization of multipotent progenitors that migrates from niches located in the branchial sac rather than around the very same siphon [164]. The very same stem cells are also responsible for the regeneration of the central nervous system. In addition, the nature and the dynamics of these stem cells have not been yet described (Figure 4B). While there are no recent studies on *bona fide* heart regeneration in tunicates, growth regions have been reported in the ciona myocardium [167,171]. In these area, clusters of proliferating undifferentiated cells start to accumulate myofilaments and eventually mature into cardiomyocytes. The nature of these precursors, i.e., transdifferentiating cells or cardiac stem cells, remains to be studied.

Although way less studied, the embryogenesis of most colonial tunicates seems to occur in the same way than the solitary ones [172,173] and, at least in the model *Botryllus schlosseri* the myogenic regulative modules and mechanisms appears to be conserved [174]. The blastozooid, the adult produced by non-embryonic development, generally has a bauplan and a muscle architecture that is comparable to the oozooids, i.e., the individuals formed by embryonic development [165,174]. However, budding bypass fertilization, embryogenesis, larval stage, and metamorphosis [24,160,171]. Contrary to their embryonic development, which displays a remarkable level of conservation among almost all the tunicate orders, non-embryonic development encompasses a clade-specific assortment of cells, tissues, and ontogenesis, all displaying different degrees of interaction between epithelial and mesenchymal cells [161].

In *Botryllus schlosseri*, the blastozooid musculature is formed de novo during morphogenesis by partially co-opting and re-wiring the embryonic cardiopharyngeal regulatory network [174]. The body muscle fate seems to be regulated by a kernel of genes expressed in progenitor cells located in a transitory structure, the dorsal tube, which has also neurogenic potential [173]. Instead, the hierarchy of the expression of specific cardiomyogenic transcription factors suggests that the heart is specified by different mesenchymal precursors, located in another domain of the developing bud. Therefore, the reshuffling of the embryonic cardiopharyngeal regulatory modules is also linked with uncoupling of the body muscle and heart muscle precursors [174]. It does remain unclear if these populations of precursors are renewed every budding cycle or persist and pass over asexual generations, or if the same precursor is responsible for the myogenesis during other forms of partial or total regeneration [175] (Figure 4C).

In the other two ascidian species, *Botrylloides leachii* and *Perophora viridis*, a population of adult pluripotent stem cells circulating in the hemolymph seems to be responsible of the regeneration of the whole body, including the entire musculature [176,177]. As for solitary species, the study of these cell populations is still in its infancy.

11. Vertebrates: The Zebrafish

The zebrafish (*Danio rerio*) is one of the most widely used vertebrate model for regeneration studies. Zebrafish are capable of regenerating many of their organs and tissues, including heart, central nervous system, retina, lateral line hair cells, caudal fin, kidney, pancreas, liver, and skeletal muscle (reviewed in [178–180]; Figure 5).

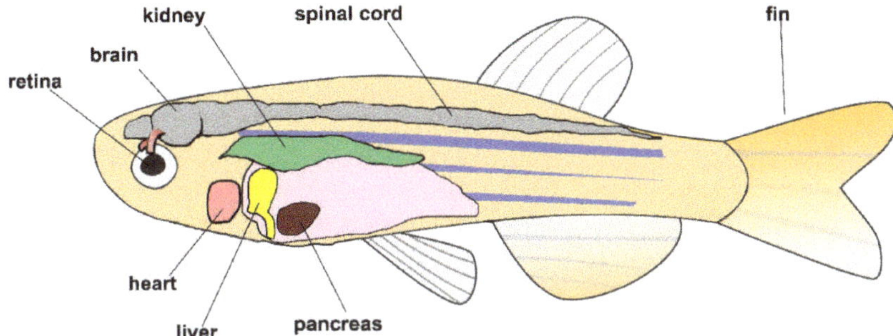

Figure 5. Regeneration in zebrafish. Schematic drawing of a zebrafish adult showing organs used for regeneration studies.

11.1. Zebrafish Skeletal Muscle Regeneration

Zebrafish trunk is composed of spatially separated slow and fast muscle fibers, and slow myofibers are embryonically mononucleated (reviewed in [181,182]). The trunk muscles are arranged in repeated chevron-shaped segments, along the head-to-tail axis. The thin partition between each pair of adjacent somites is named 'vertical myoseptum' and the one between the dorsal and ventral halves is named 'horizontal myoseptum'. The myosepta are anchoring structures for muscle fibers, enabling force transmission [183]. Two main Pax7-positive muscle stem cell populations were characterized in zebrafish. The first population is formed after somitogenesis at the external surface of the myotome, the external cell layer (ECL), expresses Pax3 and Pax7, and contributes to muscle growth throughout the zebrafish lifespan by secondary myogenesis [184]. The second population is functionally equivalent to the amniote satellite-cell population, scattered between myofibers throughout most of the myotome and serves as a source of new muscle fibers during adult zebrafish muscle repair and regeneration. These satellite-cells are mainly enriched in slow muscle near the myosepta, have dense heterochromatin and express Pax7, Pax3, and Met. In response to muscle injury, they divide asymmetrically to form two distinct cell pools: proliferative cells that fuse to form de novo myofibers or repair damaged muscle fibers, and proliferative cells that self-renew to ensure the preservation of a satellite stem cell pool [180,185]. Zebrafish skeletal muscle is a heterogeneous tissue, composed of slow, fast, and intermediate myofibers. However, at variance with mammals' intermixed muscle bundles, zebrafish trunk is composed of spatially separated slow and fast muscle fibers, and slow myofibers are embryonically mononucleated (reviewed in [181,182]). The trunk muscles are arranged in repeated chevron-shaped segments, along the head-to-tail axis. The thin partition between each pair of adjacent somites is named 'vertical myoseptum' and the one between the dorsal and ventral halves is named 'horizontal myoseptum'. The myosepta are anchoring structures for muscle fibers, enabling force transmission [183].

Myogenesis events are fundamentally common to all vertebrates. The myogenic regulatory factors (MRFs) that direct myogenic lineage development and muscle differentiation (i.e., Myf5, MyoD, Myogenin, Mrf4) are highly conserved in fish and mammals (reviewed in [178–180,186]). Two main

Pax7-positive muscle stem cell populations were characterized in zebrafish. The first population, the external cell layer (ECL), is formed after somitogenesis at the external surface of the myotome. It expresses Pax3 and Pax7 and contributes to hyperplastic muscle growth throughout the zebrafish lifespan by secondary myogenesis [184]. The second population is functionally equivalent to the amniote satellite-cell population, scattered between myofibers throughout most of the myotome and serves as a source of new muscle fibers during adult zebrafish muscle repair and regeneration. These satellite-cells are mainly enriched in slow muscle near the myosepta, have dense heterochromatin and express Pax7, Pax3, and Met. In response to muscle injury, they divide asymmetrically to form two distinct cell pools: proliferative cells that fuse to form de novo myofibers or repair damaged muscle fibers, and proliferative cells that self-renew to ensure the preservation of a satellite stem cell pool [180,185].

11.2. Zebrafish Heart Regeneration

The zebrafish heart is simpler than the mammalian heart and is composed of a single atrium and a single ventricle. Blood exits the heart through the *bulbous arteriosus*, an elastic, non-contractile chamber composed of smooth muscle. The wall of the zebrafish ventricle is lined by the epicardium, an outer mesothelial lining, and an inner endothelial lining, the endocardium. The wall is composed mainly by muscle cells, it is vascularized and innervated, and contain also fibroblast and several other type of cells [187].

Poss et al., in 2002, described for the first time that zebrafish is able to regenerate up to ~20% of its heart ventricle after amputation, thus showing the most robust cardiac regenerative response in a vertebrate [188]. The injury leads to a blood clot formation that is subsequently replaced by fibrin and collagen to preserve the ventricular wall and seal the wound. From 7–9 days post-injury (dpi) to the next following weeks, this fibrotic tissue is replaced by new cardiomyocytes (CM). After 60 dpi the size and shape of the ventricle, as well as the heart beating contractile capability, gets back to normal [189]. The use of genetic fate-mapping approaches allowed establishing that the source of the new muscle cells is from pre-existing muscle cells, stimulated by injury to divide [190,191] but, it is not completely clear what molecular signals are involved in this process. Few insights came from a study of Sande-Melon et al. that identified a subset of sox10-positive cardiomyocytes within the uninjured heart with a high capacity to contribute to the new myocardium [192]. Ablation of these cardiomyocytes confirmed that they play an essential role during the heart regeneration.

The induction of CM proliferation is triggered and controlled by various cells and factors. The first responders to heart injury are inflammatory cells like neutrophils, macrophages, and T-cells [193–195]. The cryoinjury procedure revealed that early macrophage invasion, rapid appearance of angiogenic sprouts into the wound, and transient fibrosis are required for robust cardiomyocyte proliferation [182, 196–199].

Several works assessed that the epicardium is involved in multiple aspects of cardiac repair after injury with the ability to regulate heart regeneration through secretion of soluble growth factors. Indeed, when activated, cells from the epicardium are able to proliferate and migrate to the injury area where they can secrete extracellular matrix components and molecules able to regulate cell proliferation and heart regeneration [189,200–202]. Furthermore, genetic ablation of the epicardium and its derivatives, disrupts CM proliferation and muscle regeneration, but the process renews following epicardium recovery [202].

Information available on the signaling pathways underlying cardiac regeneration in zebrafish is not very extensive but some of the identified factors, e.g., the Hippo pathway and its downstream effectors, the transcriptional co-activators Yes-associated protein (YAP) and transcriptional co-activator with PDZ-binding motif (TAZ), seems to be important in enhancing cardiac regeneration. Hereafter we summarize the main signaling pathways know and their specific functional involvement in cardiac regeneration.

The FGF family is fundamental in regeneration as they initiate a downstream signaling cascade through Ras/MAPK, Akt, and Stat signaling. Impaired heart regeneration was observed in blocked FGF-signaling transgenic fish [203]. Additionally, FGFs stimulate neovascularization and epicardial cell activation during the zebrafish heart regeneration [203,204]. However, the exact role of FGF ligands directly on zebrafish cardiomyocyte proliferation remains to be determined.

The Nrg1 is an extracellular ligand that also activates Ras/MAPK signaling. It is secreted from perivascular and epicardial cells at 7 dpi and has fundamental roles in regulating cardiomyocyte proliferation in both zebrafish and mice. Overexpression of Nrg1 increased cardiomyocytes proliferation after injury and inhibition of Nrg1 receptor caused reduction in cardiomyocyte proliferation after injury [179].

Growth factors are important modulators of zebrafish heart regeneration. The insulin growth factor (IGF) binds to receptors Igf1r and activates downstream signaling pathways that contribute to cell growth, differentiation, and anti-apoptotic pathways. Studies have shown that Igf signaling is a critical stimulator of cardiomyocyte proliferation [205]. Another growth factor, the platelet-derived growth factor (PDGF) is also important in cardiac regeneration as it activates downstream pathways involved in wound healing and proliferation. Studies have shown that PDGF ligands and receptors play an important role in heart regeneration and they are required in coronary vasculature formation during heart regeneration [206]. Apparently, their direct and main function in vivo is to support angiogenesis of the regenerating heart.

Transforming growth Factor-β (TGF-β) family ligands, such as TGF-β, BMPs, and activins seem to be key regulators in cardiomyocyte proliferation and scar formation. These factors operate through two different classes of receptors to phosphorylate distinct Smad transcription factors, which then complex with each other, and additional co-factors that regulate gene expression. Blocking TGF-β receptor activin Receptor-like Kinase 5/4 (Alk5/4) resulted with a significant decrease in pSmad3 and BrdU+ cardiomyocytes near the infarct suggesting that TGF-β is required for cardiomyocyte proliferation [207,208]. Further genetic loss-of-function mutations in activin A (*inhbaa*) showed a significant decrease in cardiomyocyte proliferation after cryoinjury [209].

BMP plays also an important role as chemical inhibition or overexpression of BMP-inhibitor *noggin3* delayed muscle repair, limited CM de-differentiation and cell cycle entry, while Induced global overexpression of *bmp2b* decreased the wound size [210].

Notch signaling pathway is also involved in heart development and cardiomyocyte maturation and proliferation. In zebrafish, following amputation of ventricular apex, Notch receptor expression becomes activated specifically in the endocardium and epicardium. Using a dominant-negative approach, Long Zhao et al. show the exquisite sensitivity of regenerative cardiomyocyte proliferation to perturbations in Notch signaling. They discovered that suppression of Notch signaling profoundly impairs cardiac regeneration and induces scar formation at the amputation site. Unexpectedly, hyperactivation of Notch signaling also suppressed cardiomyocyte proliferation and heart regeneration [211].

Additional regeneration effectors, such as miRNAs, are suggested being able to affect cardiomyocytes proliferation by inhibiting or activating the cell cycle [212]. Epigenetic regulation through chromatin remodeling or histone modification has also been shown to be involved in zebrafish heart regeneration [213] and seem to be important for cardiomyocyte regeneration.

11.3. Zebrafish as a Model of Human Regeneration

Zebrafish is a useful model for studying molecular mechanisms of regeneration [179] due to the availability of genetic data, fully sequenced genome, readiness for genetic manipulations and biosensor and reporter zebrafish lines [214,215]. In addition, the muscle structure and muscle-related gene expression are highly conserved between human and zebrafish and over 70% of human genes have a true ortholog in the zebrafish genome [216]. From a methodological viewpoint, many study on zebrafish regeneration employ larval stages whose transparent body allows easy tracking of structural changes during development and regeneration of the skeletal and heart muscle tissue.

Injured human heart does not regenerate and results in irreversible loss of myocardial cells. The damaged myocardium is replaced by fibrotic scar tissue that undermines pump function and leads to congestive heart failure and arrhythmia. Although several works suggest cardiomyocytes proliferation ability in the human heart [217], this process is not significant in response to injury [218] thus not providing a complete restoration of its functions. The zebrafish is probably one of the most important vertebrate model for studying heart developmental and regenerative properties relevant also to mammalian heart for its impressive regeneration ability following different forms of injury [187] that it is not based on stem cells or transdifferentiation of other cells but on the proliferation of preexisting cardiomyocytes [190,191]. Hence, studying the zebrafish model could expand the knowledge on cardiac regenerative processes and may contribute to the identification of specific molecules able to regulate the proliferation of these cells. This may provide insights for the design of future therapies for cardiac repair after myocardial infarction (MI) and other cardiac injuries in humans.

Concerning skeletal muscles, zebrafish and human share a high similarity at cellular, molecular, histological, and ultrastructural levels. In addition to the genetic tools for the expression of pathological phenotypes such as muscular dystrophy [219], both adult and larval zebrafish muscle have been shown to be a valuable models to study regeneration events by exploiting the possibility of performing selective injury of muscles while imaging morphogenetic processes using for example fluorescent reporter lines [178,220,221]. Notably, these studies also shed light on possible factors limiting mammalians regeneration abilities. Indeed, it has been shown that in order to allow a proper regeneration, and differently from mammals, zebrafish heart and skeletal muscles maintain the ability to activate specific gene regulatory networks (GNRs) in response to injury and perform epigenetic modifications necessary to trigger regeneration. In addition, a fundamental role is also played by the immune system whose harnessing has been shown to promote cardiac regeneration (for a review see [178]).

12. Mammals: Cell Therapy for Skeletal Muscle Regeneration

Mammalian skeletal muscle possesses a certain potential to regenerate, but this process can be compromised in several pathological conditions (e.g., neuromuscular disease, cancer-associated cachexia, or age-dependent sarcopenia) and following trauma the extended loss of muscle fibers cannot be fully recovered. These events lead to a weak regeneration and formation of fibrotic scar tissue, and result in loss of functional muscle mass. Consequently, the ability to perform intense muscular efforts and even easy, daily-life tasks may be impaired [222–224].

In the last decades, the scientific community devoted increasing efforts to develop therapies for the regeneration of damaged tissues in humans. This section aims at providing an overview of the cellular strategies that have been developed for improving skeletal muscle regeneration in mammals, particularly humans. As for the methods employed to study skeletal muscle regeneration, readers can refer to more comprehensive reviews [7,223,225–227].

Regenerative medicine aims at promoting the formation of new functional tissue by delivering precursor cells or bio-engineered tissue patches into the injured area. Indeed, therapeutic cells are isolated from the donor subject, expanded in culture (if needed), integrated into an acellular synthetic scaffold (in case of bio-engineered tissue patches), and transplanted into the recipient tissue. Although this general procedure looks simple, the choice of the specific therapeutic strategy is very complex due to the large number of different cell sources and implantation technologies. As for the cell source, the immunological compatibility between donor and recipient should be taken into account. Therefore, in the clinical setting, autologous transplantation is often preferred over the heterologous or xenologous one [223].

Focusing on skeletal muscle regeneration, the following types of cells have been employed: satellite cells, muscle-derived stem cells, myoblasts, mesoangioblasts, hTERT/Bmi-1- or hTERT/CD4-immortalized muscle precursor cells, pericytes, CD133+ cells, hematopoietic stem cells, mesenchymal stem cells, myoendothelial cells, side-population interstitial cells, myogenic precursor cells, dental pulp pluripotent-like stem cells, and eventually induced pluripotent stem cells (iPSCs) [228,229]. In mammalian animal models, these cells demonstrated a certain ability to proliferate both in vitro and in vivo, and to generate functional, integrated skeletal muscle tissue [228,230]. However, the ideal source of myogenic cells is still debated, due to a number of limitations such as the availability of bioptic biological material, the tumorigenic risk of immortalized cells, the capacity of cells to proliferate and graft into the host tissue, etc.

The possibility to artificially reprogram fully differentiated mammalian cells into iPSCs provides a virtually unlimited source of pluripotent stem cells for almost every individual. Nowadays, iPSCs represent a very useful tool for developing patient-specific regenerative therapies, thanks to the effective and reliable protocols for the expansion and differentiation of these cells both in vitro and in vivo [231–233]. Moreover, the opportunity to generate myoblasts, as well as different types of cells from iPSCs, such as neurons, endothelial cells, pericytes, and to edit their genome through CRISPR/CAS9 and TALEN technologies, further enhances their possible use for therapeutic applications [234,235]. IPSCs are stem-like cells generated by the reprogramming of fully differentiated somatic cells. Reprogramming strategies involve either the delivery of genetic material encoding reprogramming factors (e.g., Oct4, Sox2, Nanog, c-Myc, Klf4, and Lin28) or the administration of specific miRNAs or cocktail of proteins and small molecules into the somatic cells. In the latter case, reprogramming is triggered by direct activation of the endogenous stem-cell factors. The integration of exogenous DNA into the host genome relies on retroviruses, lentiviruses, and piggy Bac transposons; the non-integrating DNA and RNA procedures use adenoviruses, Sendai virus, plasmids, episomal vectors, and mRNA. Once generated, iPSCs can proliferate indefinitely in culture and differentiate into any kind of adult somatic cell by administering the appropriate growth or differentiation factors. In particular, the induction of myogenesis can be achieved either by expressing exogenous muscle-specific transcription factors, such as MYOD, PAX7, and PAX3, into the cells, or by activating endogenous pro-myogenic differentiation pathways (e.g., Wnt and BMPs signaling) by supplying, in the culture media, specific molecules, such as GSK3b inhibitors, bFGF, FGF-2, epidermal growth factor (EGF), DAPT, forskolin, BMP inhibitors, hepatocyte growth factor (HGF), and IGF-1 [236].

From a clinical perspective, the use of genetic manipulation, although more effective, is not safe, due to the risk of genetic recombination. Methods based on the supplementation in the culture media of chemical compounds prompting cell reprogramming and differentiation toward a skeletal muscle phenotype, are preferred. It should also be considered that in vitro, the differentiation process has not a 100% efficiency. Therefore, besides myogenic cells, other cell types, such as neural cells and fibroblasts, originate within the culture. In addition, myogenic cell at different stages of differentiation coexist in the entire cell population and a cell sorting strategy (by means of a fluorescence-activated cell sorter) is required to obtain a pure pool of myogenic precursor cells [237–239]. The selected population of progenitor cells can be transplanted into the host via different methods, such as intramuscular injection, systemic cell delivery, and microsurgical implants. Each of these methods has pros and cons and a unique optimal strategy has not been defined yet. Indeed, systemic delivery results in a highly variable success rate, intramuscular injection requires multiple local treatments, and implants involve more invasive procedures.

The efficacy of these regenerative therapies strictly depends on the capacity of the transplanted cells to engraft the injured site, survive, proliferate, differentiate, and integrate within the native skeletal muscles. To promote the success of these therapies, specific strategies have been developed: artificial co-administration of small molecules and growth factors such as TGF-b and myostatin inhibitors, IGF-I, fibrin, keratin, collagen; tissue engineering and bioprinting for the generation of synthetic scaffolds embedding myogenic cells [228]. The composition and the three-dimensional architecture of the synthetic scaffold provide structural and functional support to the cells and promote the formation of new, functional skeletal muscle tissue, as well as the establishment of a pool of self-renewing stem cells [240–244].

Although regenerative medicine applied to muscle disorders has greatly advanced, a standard for cell-based therapeutic interventions is still lacking. At the moment, we are far away from effective treatments promoting skeletal muscle regeneration in mammals. However, cell therapies, benefiting from the great potential of iPSCs, genome editing, tissue engineering, and cell biology methods, hold great promise for successful skeletal muscle regeneration.

13. Conclusions and Future Perspectives

Model organisms have always been playing a fundamental role to uncover general biological mechanisms common also to humans and historically they have a key role in translational medicine. In the last decades, deeper mechanistic studies of animal diversity have been made possible by the availability of a broader and affordable toolbox of technical resources such as genomics, transcriptomic, connectomics, and many other molecular biology techniques [245]. One emblematic example is the introduction of CRISPR/Cas9 genome editing, which made functional approaches possible in a wider range of so-called non-canonical model organisms. These techniques are allowing scientists to address a broader spectrum of biological problems exploiting the diversity of animal biology [246]. For instance, "simple" model organisms like Planarians and Echinoderms greatly benefited from these advancements and are nowadays considered *bona fide* models in translational research for the possibility of performing cell tracking and expression profiling of their tissue.

Nonetheless, the number and variety of animal species currently used in biomedical research are still rather limited. The reason is certainly not the suitability of particular species to a specific scientific question, but rather to the laboratory amenability (due for example to the flexibility and cost of breeding, the animal availability, the length of life cycles, the optical transparency, the possibility to perform genetic manipulations, etc.) and the familiarity of a critical mass of researchers with established animal models. The result is a scenario where only a few species retain a legitimate biomedical interest, while many others are left aside.

Nowadays, only few animal models (mainly rodents, chicken, and to some extent zebrafish) have been suitably exploited in muscle regeneration studies and various injury protocols (e.g., surgery, chemically induced muscle damage, genetic ablation, denervation-devascularization, intensive exercise, etc.) have been employed [178,225,226,247–249]. Except for zebrafish, none of the animal species presented in this review have been systematically tested for the efficacy and utility of the diverse injury models; indeed, the main methods employed to stimulate regeneration are based on physically- (by surgery or irradiation) and chemically-induced muscle damage, and genetic ablation.

It is important to point out that the application of a specific injury protocol in different animal species, given their diversity in body morphology, physiology and regeneration mechanisms, may not lead to directly comparable results. Thus, a comparative experimental approach taking into account animal diversity is fundamental to explain the variation in regenerative capacity through phylogeny, ontogeny, and even aging [250]. This may provide hallmarks of molecular homology between regenerative events in the metazoans [251] as well as an explanation of the loss or restriction of regeneration abilities occurring in some animals like *D. melanogaster* and *C. elegans*. Notably, even with their limited regeneration abilities, these animals may supply important insights into the negative regulation of this supposedly advantageous attribute.

In this review, we wished to provide the researchers interested in muscle regeneration with a lookout of muscle diversity across animals, and a prospect on their regenerating potentials (see Table 1). This work was not meant to provide 'guidelines' into animal selection when addressing specific regeneration issue but to offer 'insights' on open questions and new standpoints. We thus gave an overview of how cell precursors and regeneration strategies are adopted to partially or completely restore muscular components in various clades or even in different species within a single clade.

Table 1. Overview of muscle cell types and regenerative potentials of the animals treated in this review. * Muscle type discussed in this review in the context of regeneration, ** Induced pluripotent stem cells, can be derived from various species, *** Mostly repair.

Animal Species	Muscle Type *	REG	Cell Precursor	Known Signaling Pathways and Molecular Players
IPSC **	Any	n/a	n/a	MyoD, Pax7/3, Wnt and BMPs
Vertebrates	Striated; Cardiac	yes	Satellite stem cell; Pre-existing muscle cells	Notch, BMP, TGF-β, IGF, FGF family, Ngr1; Pax3/7, Met
Tunicates	Striated, smooth-like, cardiac	yes	Stem cells-like precursors (?)	Notch, Nk4, Tbx1/10, MRF
Cephalocordates	Striated, mononucleated	yes	De-differentiation, Multipotent cells	Pax3/7, Wnt/β-catenin and BMP
Echinoderms	Smooth-like, striated, mononucleated	yes	De-differentiation	BMP/TGFB, HOX, Ependymin, etc.
Arthropods	Striated	no ***	Adult muscle precursor (AMP)	Notch-Delta, Transcription factor Zhf1
Nematodes	Striated, mononucleated	no	—	—
Mollusks	Striated, mononucleated	yes	Sarcoblasts (?)	AChE, Growth factors (EGF, FGFs and VEGF)??
Platyhelmintes	Combine features of both vertebrate skeletal and smooth muscle cells	yes	Neoblasts: Adult pluripotent stem cells	Many known signaling pathways such as PCGs, Wnt/β-catenin, FGF family, insulin/IGF-1, Pax3/7, TGF-β, Hox genes, etc. (see text for references)
Cnidarians	Epitheliomuscular, smooth	yes	Hydrozoan: i-cells in Hydractinia, epithelial stem cells in Hydra Anthozoan: yet to be determined	The myogenic gene repertoire is present in cnidarians, but no experimental evidence relate them to the myogenic trajectory
Porifera	Myocytes	yes	Adult stem cells (ASC) Totypotent, pluripotent cells	ADPRC (ADP-ribosyl cyclase), TGF-β

We showed that the presence and nature of cell precursors giving rise to new muscles have been addressed in most of the clades and seem to be rather heterogeneous although a proper cell fate mapping has in many cases not yet been disclosed. In Tunicates, the sister-group of vertebrates, we saw the involvement of putative pluri- or multipotent stem cells during regenerative processes, and the partial co-option of embryonic myogenesis (the latter highly conserved in vertebrates) during muscle regeneration.

We saw that cell migration and tissue re-organization are crucial for regeneration and, given their rather 'simplicity', animals like Porifera, Cnidarians, and Planarians may represent valuable model to investigate these aspects. Moreover, we saw that animals use different regeneration strategies, e.g., epimorphosis and morphallaxis, that are equally successful and that may also coexist within the same organism, as it happens in amphioxus, to meet tissue specific regeneration requirements. From a translational perspective, these examples reinforce the idea that regenerative medicine should not seek out a 'regeneration blueprint' but rather a set of 'context-dependent' regeneration strategies.

We highlighted how the gene regulation aspect has been studied to various extents in many clades and is particularly well assessed for Echinoderms. On the other hand, the range of epigenetic controls has been investigated only in few species, such as zebrafish. Indeed, epigenetic studies are still at their dawn in non-vertebrates.

Comparing the architectures of a "regeneration permissive" vs. a "non-permissive" gene regulatory network (GRN) between closely related species, or finding signatures of epigenetic control of regeneration can fuel the expanding field of synthetic biology or allow for an ample testing of new classes of drug, targeting molecular and cellular mechanisms conserved in human but more functionally testable in other animal models. In this sense, regenerative therapies might greatly benefit from these comparative studies.

We also showed that, besides genetics and epigenetics, another key factor for a successful regeneration in metazoans is the 'environmental qualification', i.e., the extracellular matrix (ECM) structure, remodeling, composition, collagen content, cytoarchitecture, and secreted factors influencing the heterogeneous population of cells present within the regenerating environment [2,223]. Environmental qualification is currently considered a fundamental aspect of cell engraftment during transplantation and the lack of knowledge on this topic represent one of the current bottlenecks in regenerative medicine. Indeed, to improve regeneration of muscle tissues, transplantation of cells can be done through scaffolds ideally mimicking native tissues. Scaffolds can be made by natural polymers, synthetic polymers, or even decellularized ECM that can be filled after implantation by stem cells to restore muscle morphology. They are used to provide chemical and physical cues to transplanted cells and to create a microenvironment niche that favor survival of the resident cells and engraftment of the transplanted cells [223]. These study are at the forefront of tissue engineering and aim at providing a structural and biochemical framework for regeneration. In this regard, animals such as octopus and zebrafish can be useful models to study the cytological and histological architecture of the regenerating environment thus providing information on how to model and enrich the regenerating niche [109,189,251].

From a more clinical perspective, the information provided by studying and comparing different animal models, without forgetting their phylogenetic framework, can help to address the problem of lack of regeneration in human tissues and might eventually be used to overcome the limits of muscle regenerative therapy.

Author Contributions: L.Z. conceptualized the study; all authors wrote the original draft, reviewed and edited the text. Figures have been assembled by E.R., S.C., S.T. and L.Z. and Tables reviewed and edited by all authors. All authors have read and agreed to the published version of the manuscript.

Funding: S.T. was supported by Agence Nationale de Recherche (ANR14-CE02-0019-01). E.R. was supported by the Seventh Framework Programme (EU—631665). S.C. was supported by the Assemble Plus Project (730984).

Acknowledgments: We would like to thank M. Khamla for his contribution to the artwork (Figure 3). S.C. and M.B. are thankful to H. Escriva and M. Schubert for help obtaining *B. lanceolatum* adults.

Conflicts of Interest: The authors declare no conflict of interest.

References

1. Ciciliot, S.; Schiaffino, S. Regeneration of mammalian skeletal muscle. Basic mechanisms and clinical implications. *Curr. Pharm. Des.* **2010**, *16*, 906–914. [CrossRef] [PubMed]
2. Musaro, A. The basis of muscle regeneration. *Adv. Biol.* **2014**, *2014*, 1–16. [CrossRef]
3. Maden, M. The evolution of regeneration—Where does that leave mammals? *Int. J. Dev. Biol.* **2018**, *62*, 369–372. [CrossRef] [PubMed]
4. Tiozzo, S.; Copley, R. Reconsidering regeneration in metazoans: An evo-devo approach. *Front. Ecol. Evol.* **2015**, *3*, 67. [CrossRef]

5. Steinmetz, P.R.; Kraus, J.E.; Larroux, C.; Hammel, J.U.; Amon-Hassenzahl, A.; Houliston, E.; Worheide, G.; Nickel, M.; Degnan, B.M.; Technau, U. Independent evolution of striated muscles in cnidarians and bilaterians. *Nature* **2012**, *487*, 231–234. [CrossRef]

6. Li, Q.; Yang, H.; Zhong, T. Regeneration across metazoan phylogeny: Lessons from model organisms. *J. Genet. Genom.* **2015**, *42*. [CrossRef]

7. Baghdadi, M.; Tajbakhsh, S. Regulation and phylogeny of skeletal muscle regeneration. *Dev. Biol.* **2017**, *433*. [CrossRef]

8. Hejnol, A. Muscle's dual origins. *Nature* **2012**, *487*, 181–182. [CrossRef]

9. Telford, M.J.; Moroz, L.L.; Halanych, K.M. Evolution: A sisterly dispute. *Nature* **2016**, *529*, 286–287. [CrossRef]

10. Brunet, T.; Fischer, A.H.L.; Steinmetz, P.R.H.; Lauri, A.; Bertucci, P.; Arendt, D. The evolutionary origin of bilaterian smooth and striated myocytes. *eLife* **2016**, *5*, e19607. [CrossRef]

11. Sikes, J.M.; Newmark, P.A. Restoration of anterior regeneration in a planarian with limited regenerative ability. *Nature* **2013**, *500*, 77–80. [CrossRef] [PubMed]

12. Liu, S.Y.; Selck, C.; Friedrich, B.; Lutz, R.; Vila-Farré, M.; Dahl, A.; Brandl, H.; Lakshmanaperumal, N.; Henry, I.; Rink, J.C. Reactivating head regrowth in a regeneration-deficient planarian species. *Nature* **2013**, *500*, 81–84. [CrossRef] [PubMed]

13. Umesono, Y.; Tasaki, J.; Nishimura, Y.; Hrouda, M.; Kawaguchi, E.; Yazawa, S.; Nishimura, O.; Hosoda, K.; Inoue, T.; Agata, K. The molecular logic for planarian regeneration along the anterior-posterior axis. *Nature* **2013**, *500*, 73–76. [CrossRef] [PubMed]

14. Mokalled, M.H.; Poss, K.D. A Regeneration Toolkit. *Dev. Cell* **2018**, *47*, 267–280. [CrossRef] [PubMed]

15. Ruppert, E.E.; Fox, R.S.; Barnes, R.D. *Invertebrate Zoology: A Functional Evolutionary Approach*, 7th ed.; Thomson-Brooks/Cole: Pacific Grove, CA, USA, 2004; pp. 662–664.

16. Dorit, R.L.; Walker, W.F.; Barnes, R.D. *Zoology*; Saunders College Publishers: New York, NY, USA, 1991; pp. 154–196.

17. Nickel, M.; Scheer, C.; Hammel, J.U.; Herzen, J.; Beckmann, F. The contractile sponge epithelium sensu lato–body contraction of the demosponge Tethya wilhelma is mediated by the pinacoderm. *J. Exp. Biol.* **2011**, *214*, 1692–1698. [CrossRef]

18. Borisenko, I.; Adamska, M.; Tokina, D.; Ereskovsky, A. Transdifferentiation is a driving force of regeneration in Halisarca dujardini (Demospongiae, Porifera). *PeerJ* **2015**, *3*, e1211. [CrossRef]

19. Funayama, N. The stem cell system in demosponges: Insights into the origin of somatic stem cells. *Dev. Growth Differ.* **2010**, *52*, 1–14. [CrossRef]

20. Lorenz, B.; Bohnensack, R.; Gamulin, V.; Steffen, R.; Muller, W.E. Regulation of motility of cells from marine sponges by calcium ions. *Cell. Signal.* **1996**, *8*, 517–524. [CrossRef]

21. Katz, A.M. Review of calcium in muscle contraction, cellular, and molecular physiology, by J. C. Rüegg. *J. Mol. Cell. Cardiol.* **1994**, *26*. [CrossRef]

22. Müller, W. *Sponges (Porifera)*; Springer: New York, NY, USA, 2003; Volume 3, pp. 154–196.

23. Padua, A.; Klautau, M. Regeneration in calcareous sponges (Porifera). *J. Mar. Biol. Assoc. UK* **2015**, *96*, 553–558. [CrossRef]

24. Kürn, U.; Rendulic, S.; Tiozzo, S.; Lauzon, R. Asexual propagation and regeneration in colonial ascidians. *Biol. Bull.* **2011**, *221*, 43–61. [CrossRef] [PubMed]

25. Basile, G.; Cerrano, C.; Radjasa, O.K.; Povero, P.; Zocchi, E. ADP-ribosyl cyclase and abscisic acid are involved in the seasonal growth and in post-traumatic tissue regeneration of Mediterranean sponges. *J. Exp. Mar. Biol. Ecol.* **2009**, *381*, 10–17. [CrossRef]

26. Pozzolini, M.; Gallus, L.; Ghignone, S.; Ferrando, S.; Candiani, S.; Bozzo, M.; Bertolino, M.; Costa, G.; Bavestrello, G.; Scarfi, S. Insights into the evolution of metazoan regenerative mechanisms: Roles of TGF superfamily members in tissue regeneration of the marine sponge Chondrosia reniformis. *J. Exp. Biol.* **2019**, *222*, jeb207894. [CrossRef] [PubMed]

27. Zapata, F.; Goetz, F.; Smith, S.; Howison, M.; Siebert, S.; Church, S.; Sanders, S.; Ames, C.L.; McFadden, C.; France, S.; et al. Phylogenomic analyses support traditional relationships within cnidaria. *PLoS ONE* **2015**, *10*, e0139068. [CrossRef] [PubMed]

28. Leclere, L.; Rottinger, E. Diversity of cnidarian muscles: Function, Anatomy, development and regeneration. *Front. Cell Dev. Biol.* **2016**, *4*, 157. [CrossRef]

29. Holstein, T.W.; Hobmayer, E.; Technau, U. Cnidarians: An evolutionarily conserved model system for regeneration? *Dev. Dyn. Off. Publ. Am. Assoc. Anat.* **2003**, *226*, 257–267. [CrossRef]

30. Renfer, E.; Amon-Hassenzahl, A.; Steinmetz, P.R.; Technau, U. A muscle-specific transgenic reporter line of the sea anemone, nematostella vectensis. *Proc. Natl. Acad. Sci. USA* **2010**, *107*, 104–108. [CrossRef]

31. Jahnel, S.M.; Walzl, M.; Technau, U. Development and epithelial organisation of muscle cells in the sea anemone Nematostella vectensis. *Front. Zool.* **2014**, *11*, 44. [CrossRef]

32. Ryan, J.F.; Mazza, M.E.; Pang, K.; Matus, D.Q.; Baxevanis, A.D.; Martindale, M.Q.; Finnerty, J.R. Pre-bilaterian origins of the Hox cluster and the Hox code: Evidence from the sea anemone, nematostella vectensis. *PLoS ONE* **2007**, *2*, e153. [CrossRef]

33. Matus, D.Q.; Pang, K.; Daly, M.; Martindale, M.Q. Expression of Pax gene family members in the anthozoan cnidarian, Nematostella vectensis. *Evol. Dev.* **2007**, *9*, 25–38. [CrossRef]

34. Putnam, N.H.; Srivastava, M.; Hellsten, U.; Dirks, B.; Chapman, J.; Salamov, A.; Terry, A.; Shapiro, H.; Lindquist, E.; Kapitonov, V.V.; et al. Sea anemone genome reveals ancestral eumetazoan gene repertoire and genomic organization. *Science* **2007**, *317*, 86–94. [CrossRef] [PubMed]

35. Saina, M.; Genikhovich, G.; Renfer, E.; Technau, U. BMPs and chordin regulate patterning of the directive axis in a sea anemone. *Proc. Natl. Acad. Sci. USA* **2009**, *106*, 18592–18597. [CrossRef] [PubMed]

36. Magie, C.R.; Pang, K.; Martindale, M.Q. Genomic inventory and expression of Sox and Fox genes in the cnidarian Nematostella vectensis. *Dev. Genes Evol.* **2005**, *215*, 618–630. [CrossRef] [PubMed]

37. Steinmetz, P.R.H.; Aman, A.; Kraus, J.E.M.; Technau, U. Gut-like ectodermal tissue in a sea anemone challenges germ layer homology. *Nat. Ecol. Evol.* **2017**, *1*, 1535–1542. [CrossRef]

38. Genikhovich, G.; Technau, U. Complex functions of Mef2 splice variants in the differentiation of endoderm and of a neuronal cell type in a sea anemone. *Development* **2011**, *138*, 4911–4919. [CrossRef]

39. Martindale, M.Q.; Pang, K.; Finnerty, J.R. Investigating the origins of triploblasty: 'mesodermal' gene expression in a diploblastic animal, the sea anemone Nematostella vectensis (phylum, Cnidaria; class, Anthozoa). *Development* **2004**, *131*, 2463–2474. [CrossRef]

40. Ryan, J.F.; Burton, P.M.; Mazza, M.E.; Kwong, G.K.; Mullikin, J.C.; Finnerty, J.R. The cnidarian-bilaterian ancestor possessed at least 56 homeoboxes: Evidence from the starlet sea anemone, Nematostella vectensis. *Genome Biol.* **2006**, *7*, R64. [CrossRef]

41. Layden, M.J.; Rentzsch, F.; Rottinger, E. The rise of the starlet sea anemone nematostella vectensis as a model system to investigate development and regeneration. *Wiley Interdiscip. Rev. Dev. Biol.* **2016**, *5*, 408–428. [CrossRef]

42. Hand, C.; Uhlinger, K.R. The culture, sexual and asexual reproduction, and growth of the sea anemone nematostella vectensis. *Biol. Bull.* **1992**, *182*, 169–176. [CrossRef]

43. Warner, J.F.; Guerlais, V.; Amiel, A.R.; Johnston, H.; Nedoncelle, K.; Rottinger, E. NvERTx: A gene expression database to compare embryogenesis and regeneration in the sea anemone nematostella vectensis. *Development* **2018**, *145*. [CrossRef]

44. Sebe-Pedros, A.; Saudemont, B.; Chomsky, E.; Plessier, F.; Mailhe, M.P.; Renno, J.; Loe-Mie, Y.; Lifshitz, A.; Mukamel, Z.; Schmutz, S.; et al. Cnidarian Cell type diversity and regulation revealed by whole-organism single-cell RNA-Seq. *Cell* **2018**, *173*, 1520–1534. [CrossRef] [PubMed]

45. Rottinger, E.; Dahlin, P.; Martindale, M.Q. A framework for the establishment of a cnidarian gene regulatory network for "endomesoderm" specification: The inputs of ss-catenin/TCF signaling. *PLoS Genet.* **2012**, *8*, e1003164. [CrossRef] [PubMed]

46. Layden, M.J.; Rottinger, E.; Wolenski, F.S.; Gilmore, T.D.; Martindale, M.Q. Microinjection of mRNA or morpholinos for reverse genetic analysis in the starlet sea anemone, nematostella vectensis. *Nat. Protoc.* **2013**, *8*, 924–934. [CrossRef] [PubMed]

47. Ikmi, A.; McKinney, S.A.; Delventhal, K.M.; Gibson, M.C. TALEN and CRISPR/Cas9-mediated genome editing in the early-branching metazoan Nematostella vectensis. *Nat. Commun.* **2014**, *5*, 5486. [CrossRef]

48. Amiel, A.R.; Foucher, K.; Ferreira, S. Synergic coordination of stem cells is required to induce a regenerative response in anthozoan cnidarians. *bioRxiv* **2019**. [CrossRef]

49. Lin, Y.C.J.; Spencer, A. Localisation of intracellular calcium stores in the striated muscles of the jellyfish Polyorchis penicillatus: Possible involvement in excitation-contraction coupling. *J. Exp. Biol.* **2001**, *204*, 3727–3736.

50. Hyman, L. The invertebrates. (Scientific books: The invertebrates: Protozoa through ctenophora). *Science* **1940**, *92*, 219–220.

51. Boero, F.; Gravili, C.; Pagliara, P.; Piraino, S.; Bouillon, J.; Schmid, V. The cnidarian premises of metazoan evolution: From triploblasty, to coelom formation, to metamery. *Ital. J. Zool.* **1998**, *65*, 5–9. [CrossRef]

52. Satterlie, R.; Thomas, K.; Gray, C. Muscle organization of the cubozoan jellyfish tripedalia cystophora conant 1897. *Biol. Bull.* **2005**, *209*, 154–163. [CrossRef]

53. Helm, R.; Tiozzo, S.; Lilley, M.; Fabien, L.; Dunn, C. Comparative muscle development of scyphozoan jellyfish with simple and complex life cycles. *EvoDevo* **2015**, *6*, 11. [CrossRef]

54. Davis, R.L.; Weintraub, H.; Lassar, A.B. Expression of a single transfected cDNA converts fibroblasts to myoblasts. *Cell* **1987**, *51*, 987–1000. [CrossRef]

55. Andrikou, C.; Pai, C.Y.; Su, Y.H.; Arnone, M.I. Logics and properties of a genetic regulatory program that drives embryonic muscle development in an echinoderm. *eLife* **2015**, *4*. [CrossRef] [PubMed]

56. Hand, C.; Uhlinger, K.R. Asexual reproduction by transverse fission and some anomalies in the sea anemone nematostella vectensis. *Invertebr. Biol.* **1995**, *114*, 9–18. [CrossRef]

57. Reitzel, A.; Burton, P.; Krone, C.; Finnerty, J. Comparison of developmental trajectories in the starlet sea anemone Nematostella vectensis: Embryogenesis, regeneration, and two forms of asexual fission. *Invertebr. Biol.* **2007**, *126*, 99–112. [CrossRef]

58. Passamaneck, Y.; Martindale, M. Cell proliferation is necessary for the regeneration of oral structures in the anthozoan cnidarian nematostella vectensis. *Bmc Dev. Biol.* **2012**, *12*, 34. [CrossRef] [PubMed]

59. Amiel, A.; Johnston, H.; Nedoncelle, K.; Warner, J.; Ferreira, S.; Röttinger, E. Characterization of morphological and cellular events underlying oral regeneration in the sea anemone, nematostella vectensis. *Int. J. Mol. Sci.* **2015**, *2015*, 28449–28471. [CrossRef]

60. Warner, J.; Amiel, A.; Johnston, H.; Röttinger, E. Regeneration is a partial Redeployment of the embryonic Gene Network. *bioRxiv* **2019**. [CrossRef]

61. Morgan, T.H. Regeneration and liability to injury. *Science* **1901**, *14*, 235–248. [CrossRef]

62. Bossert, P.E.; Dunn, M.P.; Thomsen, G.H. A staging system for the regeneration of a polyp from the aboral physa of the anthozoan Cnidarian Nematostella vectensis. *Dev. Dyn. Off. Publ. Am. Assoc. Anat.* **2013**, *242*, 1320–1331. [CrossRef]

63. Cebria, F. Planarian body-wall muscle: Regeneration and function beyond a simple skeletal support. *Front. Cell Dev. Biol.* **2016**, *4*, 8. [CrossRef]

64. Moumen, M.; Chiche, A.; Cagnet, S.; Petit, V.; Raymond, K.; Faraldo, M.M.; Deugnier, M.A.; Glukhova, M.A. The mammary myoepithelial cell. *Int. J. Dev. Biol.* **2011**, *55*, 763–771. [CrossRef]

65. Newmark, P.; Sánchez Alvarado, A. Not your father's planarian: A classic model enters the era of functional genomics. *Nat. Rev. Genet.* **2002**, *3*, 210–219. [CrossRef]

66. Giacomotto, J.; Ségalat, L. High-throughput screening and small animal models, where are we? *Br. J. Pharmacol.* **2010**, *160*, 204–216. [CrossRef] [PubMed]

67. Chan, J.D.; Marchant, J.S. Pharmacological and functional genetic assays to manipulate regeneration of the planarian Dugesia japonica. *J. Vis. Exp.* **2011**. [CrossRef] [PubMed]

68. King, R.S.; Newmark, P.A. Whole-mount in situ hybridization of planarians. *Methods Mol. Biol.* **2018**, *1774*, 379–392. [CrossRef] [PubMed]

69. Ivankovic, M.; Haneckova, R.; Thommen, A.; Grohme, M.A.; Vila-Farré, M.; Werner, S.; Rink, J.C. Model systems for regeneration: Planarians. *Development* **2019**, *146*. [CrossRef] [PubMed]

70. Scimone, M.L.; Cote, L.E.; Reddien, P.W. Orthogonal muscle fibres have different instructive roles in planarian regeneration. *Nature* **2017**, *551*, 623–628. [CrossRef]

71. Witchley, J.N.; Mayer, M.; Wagner, D.E.; Owen, J.H.; Reddien, P.W. Muscle cells provide instructions for planarian regeneration. *Cell Rep.* **2013**, *4*, 633–641. [CrossRef]

72. Cote, L.E.; Simental, E.; Reddien, P.W. Muscle functions as a connective tissue and source of extracellular matrix in planarians. *Nat. Commun.* **2019**, *10*, 1592. [CrossRef]

73. Baguñà, J. The planarian neoblast: The rambling history of its origin and some current black boxes. *Int. J. Dev. Biol.* **2012**, *56*, 19–37. [CrossRef]

74. Kreshchenko, N.D. Pharynx regeneration in planarians. *Ontogenez* **2009**, *40*, 3–18. [CrossRef] [PubMed]

75. Adler, C.E.; Seidel, C.W.; McKinney, S.A.; Sanchez Alvarado, A. Selective amputation of the pharynx identifies a FoxA-dependent regeneration program in planaria. *eLife* **2014**, *3*, e02238. [CrossRef] [PubMed]

76. Guedelhoefer, O.C., IV; Alvarado, A.S. Planarian immobilization, partial irradiation, and tissue transplantation. *JoVE* **2012**. [CrossRef] [PubMed]

77. Zeng, A.; Li, H.; Guo, L.; Gao, X.; McKinney, S.; Wang, Y.; Yu, Z.; Park, J.; Semerad, C.; Ross, E.; et al. Prospectively isolated tetraspanin. *Cell* **2018**, *173*, 1593–1608. [CrossRef] [PubMed]

78. Dattani, A.; Sridhar, D.; Aziz Aboobaker, A. Planarian flatworms as a new model system for understanding the epigenetic regulation of stem cell pluripotency and differentiation. *Semin. Cell Dev. Biol.* **2019**, *87*, 79–94. [CrossRef]

79. Felix, D.; Gutiérrez-Gutiérrez, Ó.; Espada, L.; Thems, A.; González-Estévez, C. It is not all about regeneration: Planarians striking power to stand starvation. *Semin. Cell Dev. Biol.* **2018**, *87*. [CrossRef]

80. Pellettieri, J.; Sánchez Alvarado, A. Cell turnover and adult tissue homeostasis: From humans to planarians. *Annu. Rev. Genet.* **2007**, *41*, 83–105. [CrossRef]

81. Karami, A.; Tebyanian, H.; Goodarzi, V.; Shiri, S. Planarians: An in vivo model for regenerative medicine. *Int. J. Stem. Cells* **2015**, *8*, 128–133. [CrossRef]

82. Guzmán, L.; Alejo-Plata, C. Arms regeneration in lolliguncula panamensis (Mollusca: Cephalopoda). *Lat. Am. J. Aquat. Res.* **2019**, *47*, 356–360. [CrossRef]

83. Imperadore, P.; Fiorito, G. Cephalopod tissue regeneration: Consolidating over a century of knowledge. *Front. Physiol.* **2018**, *9*. [CrossRef]

84. Tressler, J.; Maddox, F.; Goodwin, E.; Zhang, Z.; Tublitz, N. Arm regeneration in two species of cuttlefish Sepia officinalis and Sepia pharaonis. *Invertebr. Neurosci.* **2013**, *14*. [CrossRef] [PubMed]

85. Zullo, L.; Imperadore, P. Regeneration and healing. In *Handbook of Pathogens and Diseases in Cephalopods*; Gestal, C., Pascual, S., Guerra, Á., Fiorito, G., Vieites, J.M., Eds.; Springer International Publishing: Cham, Switzerland, 2019; pp. 193–199. [CrossRef]

86. Fossati, S.; Benfenati, F.; Zullo, L. Morphological characterization of the Octopus vulgaris arm. *Front. Cell Dev. Biol.* **2011**, *61*, 191–195.

87. Zullo, L.; Fossati, S.M.; Benfenati, F. Transmission of sensory responses in the peripheral nervous system of the arm of Octopus vulgaris. *Front. Cell Dev. Biol.* **2011**, *61*, 197–201.

88. Kier, W.; Stella, M. The arrangement and function of octopus arm musculature and connective tissue. *J. Morphol.* **2007**, *268*, 831–843. [CrossRef]

89. Zullo, L.; Eichenstein, H.; Maiole, F.; Hochner, B. Motor control pathways in the nervous system of Octopus vulgaris arm. *Front. Cell Dev. Biol.* **2019**. [CrossRef]

90. Nödl, M.T.; Fossati, S.M.; Domingues, P.; Sánchez, F.J.; Zullo, L. The making of an octopus arm. *Front. Cell Dev. Biol.* **2015**, *6*. [CrossRef]

91. Ochiai, Y. Structural and phylogenetic profiles of muscle actins from cephalopods. *J. Basic Appl. Sci.* **2013**. [CrossRef]

92. Motoyama, K.; Ishizaki, S.; Nagashima, Y.; Shiomi, K. Cephalopod tropomyosins: Identification as major allergens and molecular cloning. *Food Chem. Toxicol.* **2007**, *44*, 1997–2002. [CrossRef]

93. Zullo, L.; Fossati, S.M.; Imperadore, P.; Nödl, M.-T. Molecular determinants of cephalopod muscles and their implication in muscle regeneration. *Front. Cell Dev. Biol.* **2017**, *5*, 53. [CrossRef]

94. Kier, W.M. The musculature of coleoid cephalopod arms and tentacles. *Front. Cell Dev. Biol.* **2016**, *4*, 10. [CrossRef]

95. Matzner, H.; Gutfreund, Y.; Hochner, B. Neuromuscular system of the flexible arm of the octopus: Physiological characterization. *J. Neurophysiol.* **2000**, *83*, 1315–1328. [CrossRef] [PubMed]

96. Rokni, D.; Hochner, B. Ionic currents underlying fast action potentials in the obliquely striated muscle cells of the octopus arm. *J. Neurophysiol.* **2003**, *88*, 3386–3397. [CrossRef] [PubMed]

97. Nesher, N.; Maiole, F.; Shomrat, T.; Hochner, B.; Zullo, L. From synaptic input to muscle contraction: Arm muscle cells of Octopus vulgaris show unique neuromuscular junction and excitation-contraction coupling properties. *Proc. R. Soc. B* **2019**, *286*. [CrossRef]

98. Taylor, D.; Sampaio, L.; Gobin, A. Building new hearts: A review of trends in cardiac tissue engineering. *Am. J. Transplant.* **2014**, *14*. [CrossRef] [PubMed]

99. Sommese, L.; Zullo, A.; Schiano, C.; Mancini, F.; Napoli, C. Possible muscle repair in the human cardiovascular system. *Stem. Cell Rev. Rep.* **2017**, *13*. [CrossRef] [PubMed]

100. Gilly, W.; Renken, C.; Rosenthal, J.; Kier, W. Specialization for rapid excitation in fast squid tentacle muscle involves action potentials absent in slow arm muscle. *J. Exp. Biol.* **2020**, *223*, jeb.218081. [CrossRef]

101. Kang, R.; Guglielmino, E.; Zullo, L.; Branson, D.T.; Godage, I.; Caldwell, D.G. Embodiment design of soft continuum robots. *Adv. Mech. Eng.* **2016**, *8*, 1–13. [CrossRef]

102. Zullo, L.; Hochner, B. A new perspective on the organization of an invertebrate brain. *Commun. Integr. Biol.* **2011**, *4*, 26–29. [CrossRef]

103. Zullo, L.; Sumbre, G.; Agnisola, C.; Flash, T.; Hochner, B. Nonsomatotopic organization of the higher motor centers in octopus. *Curr. Biol.* **2009**, *19*, 1632–1636. [CrossRef]

104. Grimaldi, A.; Tettamanti, G.; Rinaldi, L.; Brivio, M.; Castellani, D.; Eguileor, M. Muscle differentiation in tentacles of Sepia officinalis (Mollusca) is regulated by muscle regulatory factors (MRF) related proteins. *Dev. Growth Differ.* **2004**, *46*, 83–95. [CrossRef]

105. Albertin, C.B.; Simakov, O.; Mitros, T.; Wang, Z.Y.; Pungor, J.R.; Edsinger-Gonzales, E.; Brenner, S.; Ragsdale, C.W.; Rokhsar, D.S. The octopus genome and the evolution of cephalopod neural and morphological novelties. *Nature* **2015**, *524*, 220–224. [CrossRef] [PubMed]

106. Grimaldi, A.; Tettamanti, G.; Acquati, F.; Bossi, E.; Guidali, M.L.; Banfi, S.; Monti, L.; Valvassori, R.; de Eguileor, M. A hedgehog homolog is involved in muscle formation and organization of Sepia officinalis (mollusca) mantle. *Dev. Dyn.* **2008**, *237*, 659–671. [CrossRef] [PubMed]

107. Lange, M.M. On the regeneration and finer structure of the arms of the cephalopods. *J. Exp. Zool.* **1920**, *31*, 1–57. [CrossRef]

108. Shaw, T.J.; Osborne, M.; Ponte, G.; Fiorito, G.; Andrews, P.L.R. Mechanisms of wound closure following acute arm injury in Octopus vulgaris. *Zool. Lett.* **2016**, *2*, 8. [CrossRef]

109. Fossati, S.M.; Carella, F.; De Vico, G.; Benfenati, F.; Zullo, L. Octopus arm regeneration: Role of acetylcholinesterase during morphological modification. *J. Exp. Mar. Biol. Ecol.* **2013**, *447*, 93–99. [CrossRef]

110. Fossati, S.M.; Candiani, S.; Nödl, M.T.; Maragliano, L.; Pennuto, M.; Domingues, P.; Benfenati, F.; Pestarino, M.; Zullo, L. Identification and expression of acetylcholinesterase in octopus vulgaris arm development and regeneration: A conserved role for ACHE? *Mol. Neurobiol.* **2015**, *52*, 45–56. [CrossRef]

111. Vibert, L.; Daulny, A.; Jarriault, S. Wound healing, cellular regeneration and plasticity: The elegans way. *Int. J. Dev. Biol.* **2018**, *62*, 491–505. [CrossRef]

112. Benian, G.; Epstein, H. Caenorhabditis elegans muscle A Genetic and molecular model for protein interactions in the heart. *Circ. Res.* **2011**, *109*, 1082–1095. [CrossRef]

113. Moerman, D.; Fire, A. *Muscle: Structure, Function, and Development*, 2nd ed.; Springer: New York, NY, USA, 1997; Volume 33, pp. 154–196.

114. Kiontke, K.; Sudhaus, W. Ecology of caenorhabditis species. *WormBook* **2006**. [CrossRef]

115. Bird, A.; Bird, J. *The Structure of Nematodes*, 2nd ed.; Academic Press: Boston, MA, USA, 1991; pp. 10–317.

116. Corsi, A.; Wightman, B.; Chalfie, M. A transparent window into biology: A primer on caenorhabditis elegans. *WormBook* **2015**, *200*, 1–31. [CrossRef]

117. White, J.G.; Southgate, E.; Thomson, J.N.; Brenner, S. The structure of the nervous system of the nematode caenorhabditis elegans. *Philos. Trans. R. Soc.* **1986**, *275*, 327–348.

118. Gieseler, K.; Qadota, H.; Benian, G. Development, structure, and maintenance of *C. elegans* body wall muscle. *WormBook* **2016**, *2017*, 1–59. [CrossRef] [PubMed]

119. Stetina, S.E. *Brenner's Encyclopedia of Genetics*, 2nd ed.; Academic Press: Boston, MA, USA, 2013; pp. 469–476.

120. Brouilly, N.; Lecroisey, C.; Martin, E.; Pierson, L.; Mariol, M.C.; Qadota, H.; Labouesse, M.; Streichenberger, N.; Mounier, N.; Gieseler, K. Ultra-structural time-course study in the C. elegans model for Duchenne muscular dystrophy highlights a crucial role for sarcomere-anchoring structures and sarcolemma integrity in the earliest steps of the muscle degeneration process. *Hum. Mol. Genet.* **2015**, *24*, 6428–6445. [CrossRef] [PubMed]

121. Chamberlain, J.S.; Benian, G.M. Muscular dystrophy: The worm turns to genetic disease. *Curr. Biol. CB* **2000**, *10*, R795–R797. [CrossRef]

122. Marden, J.H. Variability in the size, composition, and function of insect flight muscles. *Annu. Rev. Physiol.* **2000**, *62*, 157–178. [CrossRef] [PubMed]

123. Bhakthan, N.; Nair, K.; Borden, J. Fine structure of degenerating and regenerating flight muscles in a bark beetle, Ips confusus. II. Regeneration. *Can. J. Zool.* **1971**, *49*, 85–89. [CrossRef]

124. Bernstein, S.; O'Donnell, P.; Cripps, R. Molecular genetic analysis of muscle development, structure, and function in drosophila. *Int. Rev. Cytol.* **1993**, *143*, 63–152. [CrossRef]

125. Josephson, R.; Malamud, J.; Stokes, D. Asynchronous muscle: A primer. *J. Exp. Biol.* **2000**, *203*, 2713–2722.
126. Osborne, M.P. Supercontraction in the muscles of the blowfly larva: An ultrastructural study. *J. Insect Physiol.* **1967**, *13*, 1471–1482. [CrossRef]
127. Goldstein, M.; Burdette, W. Striated visceral muscle of Drosophila melanogaster. *J. Morphol.* **1971**, *134*, 315–334. [CrossRef]
128. Goldstein, M.A. An ultrastructural study of supercontraction in the body wall muscles of Drosophila melanogaster larvae. *Anat. Rec.* **1971**, *169*, 326.
129. Hardie, J. The tension/length relationship of an insect (Calliphora erythrocephala) supercontracting muscle. *Cell. Mol. Life Sci. CMLS* **1976**, *32*, 714–716. [CrossRef]
130. Herrel, A.; Meyers, J.; Aerts, P.; Nishikawa, K. Functional implications of supercontracting muscle in the chameleon tongue retractors. *J. Exp. Biol.* **2001**, *204*, 3621–3627. [PubMed]
131. Beramendi, A.; Peron, S.; Megighian, A.; Reggiani, C.; Cantera, R. The IκB ortholog cactus is necessary for normal neuromuscular function in drosophila melanogaster. *Neuroscience* **2005**, *134*, 397–406. [CrossRef]
132. Staehling, K.; Hoffmann, F.; Baylies, M.; Rushton, E.; Bate, M. Dpp induces mesodermal gene expression in drosophila. *Nature* **1994**, *372*, 783–786. [CrossRef]
133. Azpiazu, N.; Lawrence, P.A.; Vincent, J.-P.; Frasch, M. Segmentation and specification of the drosophila mesoderm. *Genes Dev.* **1997**, *10*, 3183–3194. [CrossRef]
134. Baylies, M.; Bate, M. twist: A Myogenic Switch in Drosophila. *Science* **1996**, *272*, 1481–1484. [CrossRef]
135. Chaturvedi, D.; Reichert, H.; Gunage, R.; Vijayraghavan, K. Identification and functional characterization of muscle satellite cells in Drosophila. *eLife* **2017**, *6*. [CrossRef]
136. Postigo, A.A.; Ward, E.; Skeath, J.B.; Dean, D.C. zfh-1, the Drosophila homologue of ZEB, is a transcriptional repressor that regulates somatic myogenesis. *Mol. Cell. Biol.* **1999**, *19*, 7255–7263. [CrossRef]
137. Carnevali, M.D.C. Regeneration in echinoderms: Repair, regrowth, cloning. *Invertebr. Surviv. J.* **2006**, *3*, 64–76.
138. Dupont, S.; Thorndyke, M. Bridging the regeneration gap: Insights from echinoderm models. *Nat. Rev. Genet.* **2007**, *8*, 320. [CrossRef]
139. Ziegler, A.; Schröder, L.; Ogurreck, M.; Faber, C.; Stach, T. Evolution of a Novel Muscle Design in Sea Urchins (Echinodermata: Echinoidea). *PLoS ONE* **2012**, *7*, e37520. [CrossRef] [PubMed]
140. Ortiz-Pineda, P.; Ramirez-Gomez, F.; Pérez-Ortiz, J.; González-Díaz, S.; Jesús, F.; Hernández-Pasos, J.; Avila, C.; Rojas-Cartagena, C.; Suárez-Castillo, E.; Tossas, K.; et al. Gene expression profiling of intestinal regeneration in the sea cucumber. *BMC Genom.* **2009**, *10*, 262. [CrossRef]
141. García-Arrarás, J.E.; Dolmatov, I.Y. Echinoderms: Potential model systems for studies on muscle regeneration. *Curr. Pharm. Des.* **2010**, *16*, 942–955. [CrossRef] [PubMed]
142. Quiñones, J.L.; Rosa, R.; Ruiz, D.L.; García-Arrarás, J.E. Extracellular matrix remodeling and metalloproteinase involvement during intestine regeneration in the sea cucumber holothuria glaberrima. *Dev. Biol.* **2002**, *250*, 181–197. [CrossRef]
143. Dolmatov, I.Y.; Eliseikina, M.G.; Ginanova, T.T.; Lamash, N.E.; Korchagin, V.P.; Bulgakov, A.A. Muscle regeneration in the holothurian *Stichopus japonicus*. *Roux's Arch. Dev. Biol.* **1996**, *205*, 486–493. [CrossRef]
144. Holland, L.; Albalat, R.; Azumi, K.; Benito Gutierrez, E.; Blow, M.; Bronner-Fraser, M.; Brunet, F.; Butts, T.; Candiani, S.; Dishaw, L.; et al. The amphioxus genome illuminates vertebrate origins and cephalochordate biology. *Genome Res.* **2008**, *18*, 1100–1111. [CrossRef]
145. Somorjai, I.; Escrivà, H.; GarciaFernandez, J. Amphioxus makes the cut—Again. *Commun. Integr. Biol.* **2012**, *5*, 499–502. [CrossRef]
146. Somorjai, I.; Somorjai, R.; GarciaFernandez, J.; Escrivà, H. Vertebrate-like regeneration in the invertebrate chordate amphioxus. *Proc. Natl. Acad. Sci. USA* **2011**, *109*, 517–522. [CrossRef]
147. Liang, Y.; Rathnayake, D.; Huang, S.; Pathirana, A.; Xu, Q.; Zhang, S. BMP signaling is required for amphioxus tail regeneration. *Development* **2019**, *146*, dev.166017. [CrossRef]
148. Kaneto, S.; Wada, H. Regeneration of amphioxus oral cirri and its skeletal rods: Implications for the origin of the vertebrate skeleton. *J. Exp. Zool. Part. B Mol. Dev. Evol.* **2011**, *316*, 409–417. [CrossRef] [PubMed]
149. Flood, P.; Guthrie, D.; Banks, J. Paramyosin muscle in the notochord of amphioxus. *Nature* **1969**, *222*, 87–88. [CrossRef] [PubMed]
150. Peachey, L. Structure of the longitudinal body muscles of Amphioxus. *J. Biophys. Biochem. Cytol.* **1961**, *10*, 159–176. [CrossRef] [PubMed]

151. Hagiwara, S.; Henkart, M.; Kidokoro, Y. Excitation-contraction coupling in amphioxus muscle cells. *J. Physiol.* **1972**, *219*, 233–251. [CrossRef] [PubMed]

152. Welsch, U. The fine structure of the pharynx, cyrtopodocytes and digestive caecum of amphioxus (*Branchiostoma lanceolatum*). *Symp. Zool. Soc. Lond.* **1975**, *36*, 17–41.

153. Aldea, D.; Subirana, L.; Keime, C.; Meister, L.; Maeso, I.; Marcellini, S.; Gómez-Skarmeta, J.; Bertrand, S.; Escrivà, H. Genetic regulation of amphioxus somitogenesis informs the evolution of the vertebrate head mesoderm. *Nat. Ecol. Evol.* **2019**, *3*. [CrossRef]

154. Meulemans, D.; Bronner-Fraser, M.; Holland, L.; Holland, N. Differential mesodermal expression of two amphioxus MyoD family members (AmphiMRF1 and AmphiMRF2). *Gene Expr. Patterns GEP* **2003**, *3*, 199–202. [CrossRef]

155. Bertrand, S.; Camasses, A.; Somorjai, I.; Belgacem, M.; Chabrol, O.; Escande, M.-L.; Pontarotti, P.; Escrivà, H. Amphioxus FGF signaling predicts the acquisition of vertebrate morphological traits. *Proc. Natl. Acad. Sci. USA* **2011**, *108*, 9160–9165. [CrossRef]

156. Onai, T.; Aramaki, T.; Inomata, H.; Hirai, T.; Kuratani, S. On the origin of vertebrate somites. *Zool. Lett.* **2015**, *1*. [CrossRef]

157. Shenkar, N.; Swalla, B. Global Diversity of Ascidiacea. *PLoS ONE* **2011**, *6*, e20657. [CrossRef]

158. Delsuc, F.; Brinkmann, H.; Chourrout, D.; Philippe, H. Tunicates and not cephalochordates are the closest living relatives of vertebrates. *Nature* **2006**, *439*, 965–968. [CrossRef] [PubMed]

159. Tiozzo, S.; Brown, F.; de Tomaso, A. Regeneration and stem cells in ascidians. In *Stem Cells*; Springer: New York, NY, USA, 2008; pp. 95–112.

160. Alié, A.; Hiebert, L.; Scelzo, M.; Tiozzo, S. The eventful history of non-embryonic development in tunicates. *J. Exp. Zool. Part. B Mol. Dev. Evol.* **2020**. [CrossRef] [PubMed]

161. Nakauchi, M. Asexual development of ascidians: Its biological significance, diversity, and morphogenesis. *Am. Zool.* **1982**, *22*. [CrossRef]

162. Razy-Krajka, F.; Stolfi, A. Regulation and evolution of muscle development in tunicates. *EvoDevo* **2019**, *10*. [CrossRef]

163. Jeffery, W. Progenitor targeting by adult stem cells in ciona homeostasis, injury, and regeneration. *Dev. Biol.* **2018**, *448*. [CrossRef]

164. Degasperi, V.; Gasparini, F.; Shimeld, S.; Sinigaglia, C.; Burighel, P.; Manni, L. Muscle differentiation in a colonial ascidian: Organisation, gene expression and evolutionary considerations. *BMC Dev. Biol.* **2009**, *9*, 48. [CrossRef]

165. Meedel, T.; Chang, P.; Yasuo, H. Muscle development in Ciona intestinalis requires the b-HLH myogenic regulatory factor gene Ci-MRF. *Dev. Biol.* **2007**, *302*, 333–344. [CrossRef]

166. Anderson, H.E.; Christiaen, L. Ciona as a simple chordate model for heart development and regeneration. *J. Cardiovasc. Dev. Dis.* **2016**, *3*, 25. [CrossRef]

167. Jeffery, W. Closing the wounds: One hundred and twenty five years of regenerative biology in the ascidian ciona intestinalis. *Genesis* **2015**, *53*. [CrossRef]

168. Christiaen, L.; Tolkin, T. Rewiring of an ancestral Tbx1/10-Ebf-Mrf network for pharyngeal muscle specification in distinct embryonic lineages. *BioRxiv* **2016**. [CrossRef]

169. Razy-Krajka, F.; Lam, K.; Wang, W.; Stolfi, A.; Joly, M.; Bonneau, R.; Christiaen, L. Collier/OLF/EBF-dependent transcriptional dynamics control pharyngeal muscle specification from primed cardiopharyngeal progenitors. *Dev. Cell* **2014**, *29*. [CrossRef] [PubMed]

170. Davidson, B. Ciona intestinalis as a model for cardiac development. *Semin. Cell Dev. Biol.* **2007**, *18*, 16–26. [CrossRef] [PubMed]

171. Ricci, L.; Cabrera, F.; Lotito, S.; Tiozzo, S. Re-deployment of germ layers related TFs shows regionalized expression during two non-embryonic developments. *Dev. Biol.* **2016**, *416*. [CrossRef] [PubMed]

172. Prünster, M.M.; Ricci, L.; Brown, F.; Tiozzo, S. De novo neurogenesis in a budding chordate: Co-option of larval anteroposterior patterning genes in a transitory neurogenic organ. *Dev. Biol.* **2018**, *448*. [CrossRef]

173. Prünster, M.M.; Ricci, L.; Brown, F.; Tiozzo, S. Modular co-option of cardiopharyngeal genes during non-embryonic myogenesis. *EvoDevo* **2019**, *10*. [CrossRef]

174. Voskoboynik, A.; Simon-Blecher, N.; Soen, Y.; de Tomaso, A.; Ishizuka, K.; Weissman, I. Striving for normality: Whole body regeneration through a series of abnormal generations. *FASEB J.* **2007**, *21*, 1335–1344. [CrossRef]

175. Freeman, G. The role of blood cells in the process of asexual reproduction in the tunicate Perophora. *J. Exp. Zool.* **1964**, *156*, 157–183. [CrossRef]

176. Kassmer, S.; Langenbacher, A.; de Tomaso, A. *Primordial Blasts, a Population of Blood Borne Stem Cells Responsible for Whole Body Regeneration in a basal Chordate*; Springer: New York, NY, USA, 2019. [CrossRef]

177. Marques, I.J.; Lupi, E.; Mercader, N. Model systems for regeneration: Zebrafish. *Development* **2019**, *146*. [CrossRef]

178. Gemberling, M.; Bailey, T.; Hyde, D.; Poss, K. The zebrafish as a model for complex tissue regeneration. *TIG* **2013**, *29*. [CrossRef]

179. Berberoglu, M.; Gallagher, T.; Morrow, Z.; Talbot, J.; Hromowyk, K.; Tenente, I.; Langenau, D.; Amacher, S. Satellite-like cells contribute to pax7-dependent skeletal muscle repair in adult zebrafish. *Dev. Biol.* **2017**, *424*. [CrossRef]

180. Keenan, S.R.; Currie, P.D. The developmental phases of zebrafish myogenesis. *J. Dev. Biol.* **2019**, *7*, 12. [CrossRef] [PubMed]

181. Lai, S.L.; Marin-Juez, R.; Moura, P.L.; Kuenne, C.; Lai, J.K.H.; Tsedeke, A.T.; Guenther, S.; Looso, M.; Stainier, D.Y. Reciprocal analyses in zebrafish and medaka reveal that harnessing the immune response promotes cardiac regeneration. *eLife* **2017**, *6*. [CrossRef] [PubMed]

182. Charvet, B.; Malbouyres, M.; Pagnon-Minot, A.; Ruggiero, F.; le Guellec, D. Development of the zebrafish myoseptum with emphasis on the myotendinous junction. *Cell Tissue Res.* **2011**, *346*, 439–449. [CrossRef] [PubMed]

183. Hammond, C.L.; Hinits, Y.; Osborn, D.P.; Minchin, J.E.; Tettamanti, G.; Hughes, S.M. Signals and myogenic regulatory factors restrict pax3 and pax7 expression to dermomyotome-like tissue in zebrafish. *Dev. Biol.* **2007**, *302*, 504–521. [CrossRef]

184. Gurevich, D.B.; Nguyen, P.D.; Siegel, A.L.; Ehrlich, O.V.; Sonntag, C.; Phan, J.M.; Berger, S.; Ratnayake, D.; Hersey, L.; Berger, J.; et al. Asymmetric division of clonal muscle stem cells coordinates muscle regeneration in vivo. *Science* **2016**, *353*, aad9969. [CrossRef]

185. Rossi, G.; Messina, G. Comparative myogenesis in teleosts and mammals. *Cell. Mol. Life Sci.* **2014**, *71*, 3081–3099. [CrossRef]

186. González-Rosa, J.M.; Burns, C.; Burns, C. Zebrafish heart regeneration: 15 years of discoveries. *Regeneration* **2017**, *4*. [CrossRef]

187. Poss, K.D.; Wilson, L.G.; Keating, M.T. Heart regeneration in zebrafish. *Science* **2002**, *298*, 2188–2190. [CrossRef]

188. Beffagna, G. Zebrafish as a smart model to understand regeneration after heart injury: How fish could help humans. *Front. Cardiovasc. Med.* **2019**, *6*. [CrossRef]

189. Kikuchi, K.; Holdway, J.; Werdich, A.; Anderson, R.; Fang, Y.; Egnaczyk, G.; Evans, T.; Macrae, C.; Stainier, D.; Poss, K. Primary contribution to zebrafish heart regeneration by Gata4 cardiomyocytes. *Nature* **2010**, *464*, 601–605. [CrossRef]

190. Jopling, C.; Sleep, E.; Raya, M.; Marti, M.; Raya, A.; Izpisua Belmonte, J.C. Zebrafish heart regeneration occurs by cardiomyocyte dedifferentiation and proliferation. *Nature* **2010**, *464*, 606–609. [CrossRef] [PubMed]

191. Sande-Melón, M.; Marques, I.; Galardi-Castilla, M.; Langa Oliva, X.; Pérez-López, M.; Botos, M.-A.; Sánchez-Iranzo, H.; Guzmán-Martínez, G.; Francisco, D.; Pavlinic, D.; et al. Adult sox10+ Cardiomyocytes Contribute to Myocardial Regeneration in the Zebrafish. *Cell Rep.* **2019**, *29*, 1041–1054. [CrossRef] [PubMed]

192. Huang, W.-C.; Yang, C.-C.; Chen, I.H.; Liu, L.Y.-m.; Chang, S.J.; Chuang, Y.-J. Treatment of glucocorticoids inhibited early immune responses and impaired cardiac repair in adult zebrafish. *PLoS ONE* **2013**, *8*, e66613. [CrossRef] [PubMed]

193. de Preux Charles, A.-S.; Bise, T.; Baier, F.; Marro, J.; Jazwinska, A. Distinct effects of inflammation on preconditioning and regeneration of the adult zebrafish heart. *Open Biol.* **2016**, *6*, 160102. [CrossRef] [PubMed]

194. Harrison, M.; Bussmann, J.; Huang, Y.; Zhao, L.; Osorio, A.; Burns, C.; Burns, C.; Sucov, H.; Siekmann, A.; Lien, C.-L. Chemokine-guided angiogenesis directs coronary vasculature formation in zebrafish. *Dev. Cell* **2015**, *33*, 442–454. [CrossRef]

195. González-Rosa, J.M.; Martín, V.; Peralta, M.; Torres, M.; Mercader, N. Extensive scar formation and regression during heart regeneration after cryoinjury in zebrafish. *Development* **2011**, *138*, 1663–1674. [CrossRef]

196. Schnabel, K.; Wu, C.C.; Kurth, T.; Weidinger, G. Regeneration of cryoinjury induced necrotic heart lesions in zebrafish is associated with epicardial activation and cardiomyocyte proliferation. *PLoS ONE* **2011**, *6*, e18503. [CrossRef]

197. Marín-Juez, R.; Marass, M.; Gauvrit, S.; Rossi, A.; Lai, S.-L.; Materna, S.; Black, B.; Stainier, D. Fast revascularization of the injured area is essential to support zebrafish heart regeneration. *Proc. Natl. Acad. Sci. USA* **2016**, *113*. [CrossRef]

198. Sánchez-Iranzo, H.; Galardi-Castilla, M.; Sanz-Morejón, A.; González-Rosa, J.M.; Costa, R.; Ernst, A.; Sainz de Aja, J.; Langa Oliva, X.; Mercader, N. Transient fibrosis resolves via fibroblast inactivation in the regenerating zebrafish heart. *Proc. Natl. Acad. Sci. USA* **2018**, *115*, 201716713. [CrossRef]

199. Cao, J.; Poss, K. The epicardium as a hub for heart regeneration. *Nat. Rev. Cardiol.* **2018**, *15*. [CrossRef]

200. Lavine, K.; Yu, K.; White, A.; Zhang, X.; Smith, C.; Partanen, J.; Ornitz, D. Endocardial and epicardial derived FGF signals regulate myocardial proliferation and differentiation in vivo. *Dev. Cell* **2005**, *8*, 85–95. [CrossRef] [PubMed]

201. Wang, J.; Cao, J.; Dickson, A.; Poss, K. Epicardial regeneration is guided by cardiac outflow tract and Hedgehog signalling. *Nature* **2015**, *522*. [CrossRef] [PubMed]

202. Lepilina, A.; Coon, A.N.; Kikuchi, K.; Holdway, J.E.; Roberts, R.W.; Burns, C.G.; Poss, K.D. A dynamic epicardial injury response supports progenitor cell activity during zebrafish heart regeneration. *Cell* **2006**, *127*, 607–619. [CrossRef] [PubMed]

203. González-Rosa, J.M.; Peralta, M.; Mercader, N. Pan-epicardial lineage tracing reveals that epicardium derived cells give rise to myofibroblasts and perivascular cells during zebrafish heart regeneration. *Dev. Biol.* **2012**, *370*, 173–186. [CrossRef] [PubMed]

204. Huang, Y.; Harrison, M.; Osorio, A.; Kim, J.; Baugh, A.; Duan, C.; Sucov, H.; Lien, C.-L. Igf Signaling is required for cardiomyocyte proliferation during zebrafish heart development and regeneration. *PLoS ONE* **2013**, *8*, e67266. [CrossRef]

205. Kim, J.; Wu, Q.; Zhang, Y.; Wiens, K.; Huang, Y.; Rubin, N.; Shimada, H.; Handin, R.; Chao, M.; Tuan, T.-L.; et al. PDGF signaling is required for epicardial function and blood vessel formation in regenerating zebrafish hearts. *Proc. Natl. Acad. Sci. USA* **2010**, *107*, 17206–17210. [CrossRef]

206. Chablais, F.; Jazwinska, A. The regenerative capacity of the zebrafish heart is dependent on TGFβ signaling. *Development* **2012**, *139*, 1921–1930. [CrossRef]

207. Chablais, F.; Jazwinska, A. Induction of myocardial infarction in adult zebrafish using cryoinjury. *J. Vis. Exp. JoVE* **2012**, *62*. [CrossRef]

208. Dogra, D.; Ahuja, S.; Kim, H.-T.; Rasouli, S.J.; Stainier, D.; Reischauer, S. Opposite effects of Activin type 2 receptor ligands on cardiomyocyte proliferation during development and repair. *Nat. Commun.* **2017**, *8*. [CrossRef]

209. Wu, C.C.; Kruse, F.; Dalvoy, M.; Junker, J.; Zebrowski, D.C.; Fischer, K.; Noel, E.; Grün, D.; Berezikov, E.; Engel, F.; et al. Spatially resolved genome-wide transcriptional profiling identifies BMP signaling as essential regulator of zebrafish cardiomyocyte regeneration. *Dev. Cell* **2015**, *36*. [CrossRef]

210. Zhao, L.; Borikova, A.; Ben-Yair, R.; Guner-Ataman, B.; Macrae, C.; Lee, R.; Burns, C.; Burns, C. Notch signaling regulates cardiomyocyte proliferation during zebrafish heart regeneration. *Proc. Natl. Acad. Sci. USA* **2014**, *111*, 1403–1408. [CrossRef] [PubMed]

211. Yin, V.; Thomson, J.; Thummel, R.; Hyde, D.; Hammond, S.; Poss, K. Fgf-dependent depletion of microRNA-133 promotes appendage regeneration in zebrafish. *Genes Dev.* **2008**, *22*, 728–733. [CrossRef] [PubMed]

212. Xiao, C.-L.; Hou, Y.; Xu, C.; Chang, N.; Wang, F.; Hu, K.; He, A.; Luo, Y.; Wang, J.; Peng, J.; et al. Chromatin-remodelling factor Brg1 regulates myocardial proliferation and regeneration in zebrafish. *Nat. Commun.* **2016**, *7*, 13787. [CrossRef] [PubMed]

213. Keßler, M.; Rottbauer, W.; Just, S. Recent progress in the use of zebrafish for novel cardiac drug discovery. *Expert Opin. Drug Discov.* **2015**, *10*. [CrossRef] [PubMed]

214. Moro, E.; Vettori, A.; Porazzi, P.; Schiavone, M.; Rampazzo, E.; Casari, A.; Ek, O.; Facchinello, N.; Astone, M.; Zancan, I.; et al. Generation and application of signaling pathway reporter lines in zebrafish. *Mol. Genet. Genom.* **2013**, *288*, 231–242. [CrossRef]

215. Howe, K.; Clark, M.D.; Torroja, C.F.; Torrance, J.; Berthelot, C.; Muffato, M.; Collins, J.E.; Humphray, S.; McLaren, K.; Matthews, L.; et al. The zebrafish reference genome sequence and its relationship to the human genome. *Nature* **2013**, *496*, 498–503. [CrossRef]

216. Beltrami, A.P.; Urbanek, K.; Kajstura, J.; Yan, S.-M.; Finato, N.; Bussani, R.; Nadal-Ginard, B.; Silvestri, F.; Leri, A.; Beltrami, C.A.; et al. Evidence that human cardiac myocytes divide after myocardial infarction. *N. Engl. J. Med.* **2001**, *344*, 1750–1757. [CrossRef]

217. Pasumarthi, K.B.; Field, L.J. Cardiomyocyte cell cycle regulation. *Circ. Res.* **2002**, *90*, 1044–1054. [CrossRef]

218. Bassett, D.I.; Currie, P.D. The zebrafish as a model for muscular dystrophy and congenital myopathy. *Hum. Mol. Genet.* **2003**, *12*, R265–R270. [CrossRef]

219. Otten, C.; Abdelilah-Seyfried, S. Laser-inflicted injury of zebrafish embryonic skeletal muscle. *JoVE* **2013**. [CrossRef]

220. Saera-Vila, A.; Kish, P.E.; Kahana, A. Fgf regulates dedifferentiation during skeletal muscle regeneration in adult zebrafish. *Cell. Signal.* **2016**, *28*, 1196–1204. [CrossRef] [PubMed]

221. Naranjo, J.; Dziki, J.; Badylak, S. Regenerative medicine approaches for age-related muscle loss and sarcopenia: A mini-review. *Gerontology* **2017**, *63*. [CrossRef] [PubMed]

222. Liu, J.; Saul, D.; Böker, K.; Ernst, J.; Lehmann, W.; Schilling, A. Current methods for skeletal muscle tissue repair and regeneration. *Biomed. Res. Int.* **2018**, *2018*. [CrossRef] [PubMed]

223. Zullo, A.; Mancini, F.P.; Schleip, R.; Wearing, S.; Yahia, L.H.; Klingler, W. The interplay between fascia, skeletal muscle, nerves, adipose tissue, inflammation and mechanical stress in musculo-fascial regeneration. *J. Gerontol. Geriatr.* **2017**, *65*, 271–283.

224. Forcina, L.; Cosentino, M.; Musaro, A. Mechanisms regulating muscle regeneration: Insights into the interrelated and time-dependent phases of tissue healing. *Cells* **2020**, *9*, 1297. [CrossRef]

225. Sicherer, S.; Grasman, J. Recent trends in injury models to study skeletal muscle regeneration and repair. *Bioengineering* **2020**, *7*, 76. [CrossRef]

226. Khodabukus, A.; Prabhu, N.; Wang, J.; Bursac, N. In vitro tissue-engineered skeletal muscle models for studying muscle physiology and disease. *Adv. Healthc. Mater.* **2018**, *7*, 1701498. [CrossRef]

227. Mueller, A.; Bloch, R. Skeletal muscle cell transplantation: Models and methods. *J. Muscle Res. Cell Motil.* **2019**. [CrossRef]

228. Marg, A.; Escobar, H.; Karaiskos, N.; Grunwald, S.; Metzler, E.; Kieshauer, J.; Sauer, S.; Pasemann, D.; Malfatti, E.; Mompoint, D.; et al. Human muscle-derived CLEC14A-positive cells regenerate muscle independent of PAX7. *Nat. Commun.* **2019**, *10*, 5776. [CrossRef]

229. Lingjun, R.; Qian, Y.; Khodabukus, A.; Ribar, T.; Bursac, N. Engineering human pluripotent stem cells into a functional skeletal muscle tissue. *Nat. Commun.* **2018**, *9*. [CrossRef]

230. Chan, S.S.-K.; Arpke, R.; Filareto, A.; Xie, N.; Pappas, M.; Penaloza, J.; Perlingeiro, R.; Kyba, M. Skeletal muscle stem cells from PSC-derived teratomas have functional regenerative capacity. *Cell Stem Cell* **2018**, *23*, 74–85. [CrossRef]

231. Costela, M.C.O.; López, M.G.; López, V.C.; Gallardo, M.E. iPSCs: A powerful tool for skeletal muscle tissue engineering. *J. Cell. Mol. Med.* **2019**, *23*. [CrossRef]

232. Hall, M.; Hall, J.; Cadwallader, A.; Pawlikowski, B.; Doles, J.; Elston, T.; Olwin, B. Transplantation of skeletal muscle stem cells. *Methods Mol. Biol.* **2017**, *1556*, 237–244. [PubMed]

233. Quattrocelli, M.; Swinnen, M.; Giacomazzi, G.; Camps, J.; Barthélémy, I.; Ceccarelli, G.; Caluwé, E.; Grosemans, H.; Thorrez, L.; Pelizzo, G.; et al. Mesodermal iPSC–derived progenitor cells functionally regenerate cardiac and skeletal muscle. *J. Clin. Investig.* **2015**, *125*. [CrossRef] [PubMed]

234. Maffioletti, S.; Sarcar, S.; Henderson, A.; Mannhardt, I.; Pinton, L.; Moyle, L.; Steele, H.; Cappellari, O.; Wells, K.; Ferrari, G.; et al. Three-dimensional human iPSC-derived artificial skeletal muscles model muscular dystrophies and enable multilineage tissue engineering. *Cell Rep.* **2018**, *23*, 899–908. [CrossRef] [PubMed]

235. Danišovič, Ľ.; Galambosova, M.; Csobonyeiová, M. Induced pluripotent stem cells for duchenne muscular dystrophy modeling and therapy. *Cells* **2018**, *7*, 253. [CrossRef]

236. Choi, I.; Lim, H.; Estrellas, K.; Mula, J.; Cohen, T.; Zhang, Y.; Donnelly, C.; Richard, J.-P.; Kim, Y.J.; Kim, H.; et al. Concordant but Varied phenotypes among duchenne muscular dystrophy patient-specific myoblasts derived using a human iPSC-based model. *Cell Rep.* **2016**, *15*, 1–12. [CrossRef]

237. Webster, M.; Fan, C. c-MET regulates myoblast motility and myocyte fusion during adult skeletal muscle regeneration. *PLoS ONE* **2013**, *8*, e81757. [CrossRef]

238. Wal, E.; Herrero-Hernandez, P.; Wan, R.; Broeders, M.; Groen, S.; Gestel, T.; Van Ijcken, W.; Cheung, T.; Ploeg, A.; Schaaf, G.; et al. Large-scale expansion of human iPSC-derived skeletal muscle cells for disease modeling and cell-based therapeutic strategies. *Stem Cell Rep.* **2018**, *10*. [CrossRef]

239. Ong, C.S.; Yesantharao, P.; Huang, C.-Y.; Mattson, G.; Boktor, J.; Fukunishi, T.; Zhang, H.; Hibino, N. 3D bioprinting using stem cells. *Pediatr. Res.* **2017**, *83*. [CrossRef]

240. Mandrycky, C.; Wang, D.Z.; Kim, K.; Kim, D.-H. 3D Bioprinting for engineering complex tissues. *Biotechnol. Adv.* **2015**, *34*. [CrossRef] [PubMed]

241. Pollot, B.; Rathbone, C.; Wenke, J.; Guda, T. Natural polymeric hydrogel evaluation for skeletal muscle tissue engineering: Skeletal muscle engineering natural hydrogels. *J. Biomed. Mater. Res. Part. B Appl. Biomater.* **2017**, *106*. [CrossRef] [PubMed]

242. Fuoco, C.; Petrilli, L.L.; Cannata, S.; Gargioli, C. Matrix scaffolding for stem cell guidance toward skeletal muscle tissue engineering. *J. Orthop. Surg. Res.* **2016**, *11*. [CrossRef] [PubMed]

243. Jiao, A.; Moerk, C.; Penland, N.; Perla, M.; Kim, J.; Smith, A.; Murry, C.; Kim, D.-H. Regulation of skeletal myotube formation and alignment by nanotopographically controlled cell-secreted extracellular matrix. *J. Biomed. Mater. Res. Part. A* **2018**, *106*. [CrossRef]

244. Matthews, B.; Vosshall, L. How to turn an organism into a model organism in 10 'easy' steps. *J. Exp. Biol.* **2020**, *223*, jeb218198. [CrossRef]

245. Dickinson, M.; Vosshall, L.; Dow, J. Genome editing in non-model organisms opens new horizons for comparative physiology. *J. Exp. Biol.* **2020**, *223*, jeb221119. [CrossRef]

246. Chargé, S.B.; Rudnicki, M.A. Cellular and molecular regulation of muscle regeneration. *Physiol. Rev.* **2004**, *84*, 209–238. [CrossRef]

247. Hardy, D.; Besnard, A.; Latil, M.; Jouvion, G.; Briand, D.; Thépenier, C.; Pascal, Q.; Guguin, A.; Gayraud-Morel, B.; Cavaillon, J.-M.; et al. Comparative study of injury models for studying muscle regeneration in mice. *PLoS ONE* **2016**, *11*, e0147198. [CrossRef]

248. Hardy, D.; Latil, M.; Gayraud-Morel, B.; Briand, D.; Jouvion, G.; Rocheteau, P.; Chrétien, F. Choosing the appropriate model for studying muscle regeneration in mice: A comparative study of classical protocols. *Morphologie* **2015**, *99*, 168. [CrossRef]

249. Yun, M. Changes in Regenerative Capacity through Lifespan. *Int. J. Mol. Sci.* **2015**, *16*, 25392–25432. [CrossRef]

250. Alvarado, A.S. Regeneration in the metazoans: Why does it happen? *BioEssays* **2000**, *22*, 578–590. [CrossRef]

251. Imperadore, P.; Shah, S.B.; Makarenkova, H.P.; Fiorito, G. Nerve degeneration and regeneration in the cephalopod mollusc Octopus vulgaris: The case of the pallial nerve. *Sci. Rep.* **2017**, *7*, 46564. [CrossRef] [PubMed]

Review

Genetic Control of Muscle Diversification and Homeostasis: Insights from *Drosophila*

Preethi Poovathumkadavil * and Krzysztof Jagla

Institute of Genetics Reproduction and Development, iGReD, INSERM U1103, CNRS UMR6293, University of Clermont Auvergne, 28 Place Henri Dunant, 63000 Clermont-Ferrand, France; christophe.jagla@uca.fr
* Correspondence: preethi.poovathumkadavil@uca.fr

Received: 30 May 2020; Accepted: 23 June 2020; Published: 25 June 2020

Abstract: In the fruit fly, *Drosophila melanogaster*, the larval somatic muscles or the adult thoracic flight and leg muscles are the major voluntary locomotory organs. They share several developmental and structural similarities with vertebrate skeletal muscles. To ensure appropriate activity levels for their functions such as hatching in the embryo, crawling in the larva, and jumping and flying in adult flies all muscle components need to be maintained in a functionally stable or homeostatic state despite constant strain. This requires that the muscles develop in a coordinated manner with appropriate connections to other cell types they communicate with. Various signaling pathways as well as extrinsic and intrinsic factors are known to play a role during *Drosophila* muscle development, diversification, and homeostasis. In this review, we discuss genetic control mechanisms of muscle contraction, development, and homeostasis with particular emphasis on the contractile unit of the muscle, the sarcomere.

Keywords: *Drosophila*; muscle; genetic control; muscle diversification; muscle homeostasis

1. Introduction

1.1. General Overview

Drosophila melanogaster, a holometabolic insect with a short lifespan, has served as a simple model to study myogenesis [1,2] and contractile proteins [3] for decades. Myogenesis in *Drosophila* occurs in two waves, one during the embryonic stage that gives rise to the larval body wall or somatic muscles and the second during pupal development that gives rise to adult flight, leg, and abdominal muscles [4]. All these muscles are voluntary, syncytial (multinucleate), and striated making them similar to vertebrate skeletal muscles [5]. Multiple signaling pathways, genes, and processes are conserved from *Drosophila* to vertebrates [6,7]. Muscles provide force to ensure various locomotory behaviors such as crawling, walking, jumping, and flying in *Drosophila*. Thus, they need to carry high levels of a mechanical load and are subject to constant strains, which can potentially disrupt homeostasis. Muscle movements need to be precise and coordinated, where communication with other tissues such as the nervous system provides critical inputs [8]. Muscles are the major reservoir for amino acids in the body that contribute to muscle mass and protein homeostasis [9]. All muscle functionalities require that they are correctly formed in the first place to attain a homeostatic state in which they are physiologically active and stable. Muscle intrinsic signaling as well as signaling from external organs contribute to muscle homeostasis. Muscles display a high degree of plasticity or flexibility at the signaling, metabolic, myonuclear, mitochondrial, and stem cell levels.

This review is divided into three parts. The first part presents an overview of the mechanisms of muscle contraction in *Drosophila*. The second part focuses on the development of the larval and adult muscles. In the third part, we discuss the maintenance of muscle homeostasis in normal conditions and the adverse effects of the loss of this homeostasis in pathological conditions. Throughout the review, the focus is on sarcomeres, which are the basic contractile units of the muscle.

1.2. Major Structural Components of the Drosophila Muscle and Their Vertebrate Counterparts

In *Drosophila*, muscle function is coordinated by sensory, excitatory, and mechanical inputs by its connection to the nervous system via neuromuscular junctions and to the epidermis via myotendinous junctions akin to vertebrate systems though they present differences, some of which are outlined below.

1.2.1. Sarcomeres

Sarcomeres are the basic contractile units of the muscle and provide the force for contraction during movements (Figure 1). They are repetitively arranged in a regular pattern that gives a striated appearance under the microscope to vertebrate skeletal muscles as well as *Drosophila* somatic, flight, and leg muscles [10,11]. Sarcomeric length, functional domains, and many component proteins are conserved between invertebrates and vertebrates, although studies also point to interesting differences among species, which appear to be adaptations to individual muscle function [12–15]. Despite structural differences in *Drosophila* sarcomeric proteins in comparison to vertebrate counterparts, they have similar functional interactions and possess conserved functional domains; for example, the PEVK domain of the *Drosophila* titin, Sallimus (Sls) confers elasticity similar to vertebrates [16]. Thus, the sarcomere provides an example of nature reusing and repurposing components across evolution.

1.2.2. Myotendinous Junctions (MTJs)

In *Drosophila*, the MTJ is an attachment formed between the muscle and specialized groups of tendon-like cells of ectodermal origin called tendon cells, also known as apodemes (Figure 1a). Unlike vertebrates, *Drosophila* does not have an internal skeleton and tendon cells help anchor the muscles firmly to the cuticular exoskeleton instead, which helps transmit the contractile forces to the body to generate motion. This makes them functionally similar to vertebrate tendons despite their distinct embryological origins, mesodermal for vertebrates and ectodermal for *Drosophila* [17,18]. The formation and maintenance of the MTJ is mediated through the ECM by specific integrin heterodimers on the muscle and tendon ends in *Drosophila* similar to vertebrates [19–22].

1.2.3. Neuromuscular Junctions (NMJs)

The NMJ is the point of contact between the motor neurons of the nervous system and the muscle, which enables environmental inputs to be transmitted via synapses to the muscle (Figure 1a). The *Drosophila* larval NMJ is an established model for NMJ formation and function. This NMJ is glutamatergic and responds to the neurotransmitter glutamate unlike vertebrate NMJs that are cholinergic and respond to acetylcholine. However, they are of particular interest owing to their similarity to mammalian brain glutamatergic synapses that express multiple genes orthologous to *Drosophila* genes and the ease with which NMJ assembly can be studied in this model [23–25]. It continues to be an active field of study with focus equally shifting to adult motor neurons formed after metamorphosis [26,27].

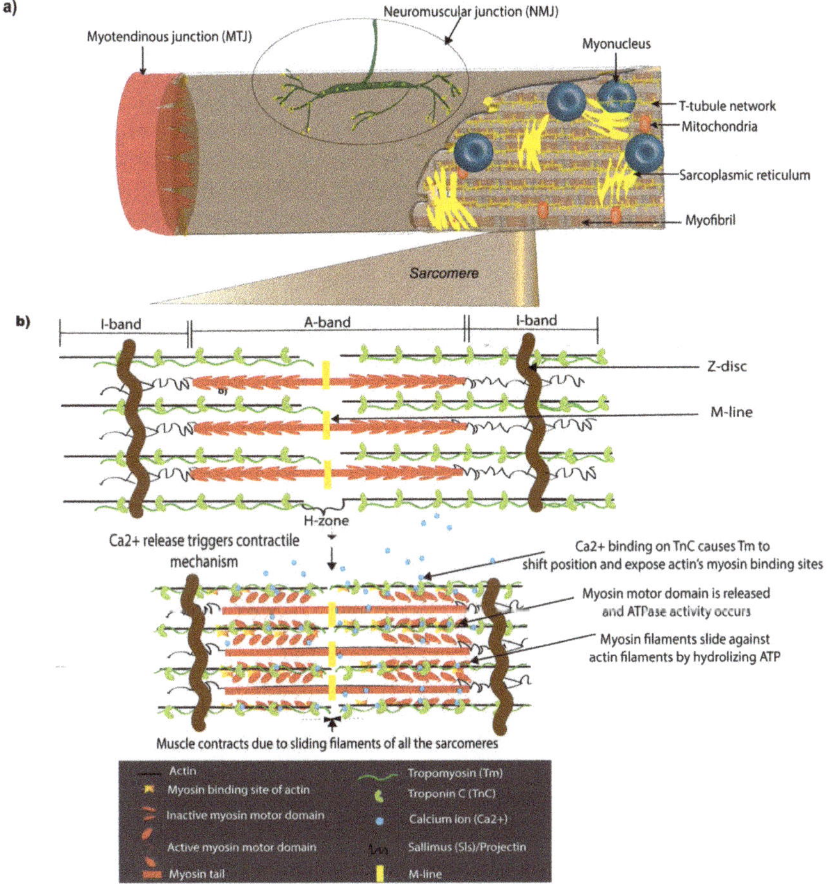

Figure 1. Schematic representation of the larval body wall or somatic muscle structure and the sliding filament theory of muscle contraction. (**a**) Muscle structure with myofibrils and the network of myonuclei, sarcoplasmic reticulum (SR), T-tubules, and mitochondria. The muscle is connected to the nervous system via the neuromuscular junction (NMJ) and to the epidermis via the myotendinous junction (MTJ). Myofibrils are formed of repetitive contractile units, the sarcomeres. (**b**) The structure of a sarcomere and the mechanism of contraction proposed by the sliding filament theory. Ca^{2+} ions released upon neurotransmitter signaling from the NMJ launch a cascade by binding to TroponinC (TnC) on the thin filaments of sarcomeres. This Ca^{2+} binding causes a conformation change in Tropomyosin (Tm) bound to actin, exposing actin's myosin binding sites. This permits the activated myosin motor domain to bind to actin and slide against it by utilizing the energy stored in Adenosine Triphosphate (ATP).

2. The Sarcomere and Molecular Mechanisms of Muscle Contraction

Voluntary muscle contraction is a highly coordinated process that depends on cooperative signaling from sensory neurons via interneurons and motor neurons to the NMJ of the muscle [28–30]. Given that the principal muscle function is to generate movements by contracting, the sarcomeric contractile units are indispensable for muscle function and their maintenance is crucial. The *Drosophila* adult indirect flight muscle (IFM) is established as a model to study sarcomere assembly and the functions of its components [31]. IFMs are built of multiple myofibers and have a stereotypic pattern of sarcomeric proteins forming highly ordered myofibrils similar to human skeletal muscles allowing the study of sarcomere malformations under mutant conditions. The IFM is also a model to study stretch activation (SA) [32]. During SA, there is a high frequency of contraction although the nervous system input frequency is much lower. This is possible due to the delayed increase in tension following muscle stretching. SA is a mechanism found in all muscles though it has particular significance in certain muscle types with rhythmic activity such as human cardiac muscles and the fruit fly flight muscles. In contrast to the multi-fiber IFM muscles of the adult, the somatic muscles in the *Drosophila* embryo and larvae are built of only one muscle fiber per muscle and present a much simpler model to study myofibers.

A sarcomere is a specialized structure adapted for muscle contraction (Figure 1). During myofibrillogenesis, newly formed sarcomeres align in repeating units along the length of a muscle to form a myofibril and multiple myofibrils covered by the plasma membrane form a myofiber. A sarcomere is built of thin-actin and thick-myosin filaments with associated proteins facilitating contraction-relaxation cycles. The thick filaments consist of myosin polymers with each myosin consisting of a myosin tail and two myosin heads, which are capable of attaching to actin during muscle contraction. The two ends of a sarcomere are demarcated by a Z-disc, a huge protein complex that anchors the thin filaments that form I-bands on either side of a sarcomere, while the thick filaments form an A-band in the center (Figure 1). In between the two I-bands is an H-zone lacking myosin heads and in the center of the H-zone is an M-line that corresponds to another large protein complex that anchors the thick filaments [33].

Sarcomere function is intricately linked to other organelles such as the mitochondria [34], myonuclei [35], sarcoplasmic reticulum (SR), and T-tubules [10,36]. The efficient function of sarcomeres is closely coupled with the periodic arrangement of the SR and T-tubules around them [10,36–38]. T-tubules are regular tubular invaginations of the plasma membrane at each sarcomere. The membrane organelle SR is linked to the myonuclei and T-tubules to facilitate the exchange of proteins and ions. The SR is the major intracellular reservoir of calcium (Ca^{2+}) ions in the muscle, which are essential for muscle contraction. The T-tubule and SR form a specialized triad/dyad structure, which is indispensable for correct muscle functioning by excitation-contraction (EC) coupling. This EC coupling enables the transmission of excitation potentials from the NMJ to the SR, which triggers Ca^{2+} release from the SR that in turn initiates sarcomeric sliding movements leading to muscle contraction. Apart from Ca^{2+}, other ions contribute to muscle contraction [39]. The Na^+K^+-ATPase is a Na^+-K^+ pump that can pump Na^+ out of and K^+ into the cells against their normal concentration gradients. In muscles, the concentration of these ions fine-tunes the force of contraction [40]. In *Drosophila*, muscles are one of the major organs that express the Na^+K^+-ATPase α subunit [41]. One form of the Na^+K^+-ATPase β subunit, Nrv1 interacts with Dystroglycan (Dg), which is part of a complex that helps transmit forces into the muscle cell [42].

The mechanism of muscle contraction is explained by the sliding filament theory [43,44], reviewed by Hugh Huxley [33]. This theory proposes that the myosin head domain acts as a motor and slides against the actin filament powered by the energy stored in ATP. This sliding of the central myosin along the thin filaments causes the two I bands on either side to come closer to each other. During contraction, environmental inputs are transmitted by the nervous system to the NMJ leading to Ca^{2+} binding to the Troponin C (TnC) subunit of the Troponin (Tn) complex. This leads to the Troponin T (TnT) subunit that binds to the actin binding protein Tropomyosin (Tm) triggering a conformational change in Tm, thus shifting its position on actin and exposing the myosin binding site of actin [45–47]. Myosin that is turned 'on' by a myosin regulatory light chain (Rlc) phosphorylation [48] liberates the motor domains in the myosin head that were folded onto the myosin tail, thus facilitating its binding to actin. Subsequent ATP hydrolysis and energy release, thanks to its ATPase activity, permits it to move along the thin filament to contract the muscle. For the muscle to relax, the Troponin I (TnI) troponin subunit inhibits the actomyosin interaction [49] so that Tm covers the myosin binding site of actin and the myosin is switched 'off' and folded back onto the myosin tail [50,51]. This coordinated key muscle function highlights the importance of ionic and sarcomeric component homeostasis in muscles, which implies the supply and maintenance of the right quantities of the right ions and sarcomeric components at the right time to ensure muscle functionality.

During contraction, the MTJ helps anchor the myofibrils and transmits forces [19,52]. Tight interactions between sarcomeric components ensure myofibrillar integrity and prevent disintegration due to contractile forces. CapZ binds to the actin barbed end and links it to the Z-disc [13] while Z-disc proteins such as the filamin Cher [53], Zasp, and α-actinin anchor the thin filaments [54]. Similarly, the M-line protein Obscurin that associates with the thick filament [55], Muscle LIM protein at 84B (Mlp84B) that cooperates with Sallimus (Sls) known as the *Drosophila* titin [56], integrins [57], and other proteins ensure muscle integrity. Sarcomeres are subject to constant mechanical strain due to the thin and thick filament friction and need to be consistently replenished to ensure their function over a lifetime. Since these muscles are voluntary, they also need to be able to stop contracting at will and go back to their natural state. Defective sarcomeric formation, maintenance, and homeostasis are associated with muscular diseases [15,58].

3. Muscle Diversification—On the Road to Muscle Homeostasis

Muscle development is a finely orchestrated, synchronized process that occurs in spatial and temporal coordination with the development of other communicating tissues to finally form a homeostatic muscle. There are similarities as well as differences between *Drosophila* and vertebrate myogenesis [59]. During development, each muscle diversifies to attain an identity tailored to its specific functional requirements. The study of muscle diversification during development is of interest in the context of homeostasis for two primary reasons:

(a) Events similar to those occurring during development need to be reinitiated to repair and regenerate an injured muscle and reestablish muscle homeostasis [60]. This is a new field of study in *Drosophila* stemming from the recent discovery of muscle satellite cells in adult flies [61].

(b) The two waves of myogenesis in *Drosophila* result in two homeostatic states, one in the larva and one in the adult. The larval homeostatic states are highly dynamic given the large growth spurt that occurs over the three larval instars. This might provide insights into mechanisms of muscle atrophy and hypertrophy. Forkhead box sub-group O (Foxo), for example, has been shown to inhibit larval muscle growth by repressing *diminutive* (*myc*) [62]. In mice, excess c-Myc has been shown to induce cardiac hypertrophy [63].

3.1. Embryonic Myogenesis of Larval Muscles

Embryonic myogenesis gives rise to monofiber larval somatic muscles whose main function is to aid in hatching and the peristaltic, crawling movements of the larvae. The embryonic and larval somatic musculature consists of a stereotypical pattern of muscles in each segment, with 30 muscles in most abdominal hemisegments (A2–A6) (figure in Table 1). There are fewer muscles in the posterior and first abdominal hemisegment and a slightly different set of muscles in the three thoracic hemisegments (T1–T3). Embryonic muscles arise from the mesoderm germ layer and their development requires intrinsic mesodermal cues and extrinsic cues from the adjacent epidermal and neural cells. Thus, they develop in synchrony with the development of muscle-interactors such as tendon cells and motor neurons and need to 'speak a common language' to communicate for coordinated development and maintenance.

Somatic muscle specification and differentiation have been reviewed extensively in the past [1,2,7,60–62] and this review presents complementary as well as new information that emphasizes the role of developmental factors in future muscle homeostasis.

3.1.1. Muscle Diversification by the Specification of Muscle Founder Cells Expressing Identity Transcription Factors (iTFs)

The embryo undergoes gastrulation by invagination [64], which brings the three germ layers, the ectoderm, the somatic muscle forming mesoderm, and endoderm in juxtaposition with each other. This helps provide extrinsic signals to the developing mesoderm. Following this juxtaposition, the mesoderm is divided into domains by morphogenic signaling [65] giving rise to a somatic muscle domain in which the transcription factor (TF) Twist (Twi) provides a myogenic switch [66]. Subsequently, equivalence or promuscular cell clusters expressing the neurogenic gene *lethal of scute* (*l'sc*) form and one muscle progenitor cell is singled out from each cluster by lateral inhibition involving Notch and Ras/MAPK signaling [67,68]. The remaining Notch activated cells in the equivalence groups become fusion competent myoblasts (FCMs). This process is reminiscent and coincides temporally with the specification of neural lineages from the neurectoderm [69,70], which occurs during embryonic stages 8–11, while muscle cell identity specification occurs during stages 9–11.

The singled-out muscle progenitors divide asymmetrically to give rise to founder cells (FCs), which are believed to carry all the information necessary to give rise to the diversity of muscle types. Asymmetric divisions of progenitors can give rise to two FCs, an FC and a Numb negative adult muscle precursor (AMP) or an FC and a cardiac progenitor, which subsequently migrate away from each other [67,71,72]. Each FC contains the information to establish one muscle's identity since it can form correct attachments and be correctly innervated even in the absence of myoblast fusion with surrounding FCMs [73,74]. It expresses its characteristic code of TFs known as muscle identity transcription factors (iTFs) (Figure 2). The expression of a combinatorial code of iTFs in distinct progenitors is the result of their spatial positioning as well as tissue specific convergence of multiple signaling cascades [75]. For example, Wg signaling from the adjacent developing central nervous system (CNS) is implicated in the specification of Slouch (Slou) positive FCs [76] highlighting the importance of coordinated tissue development.

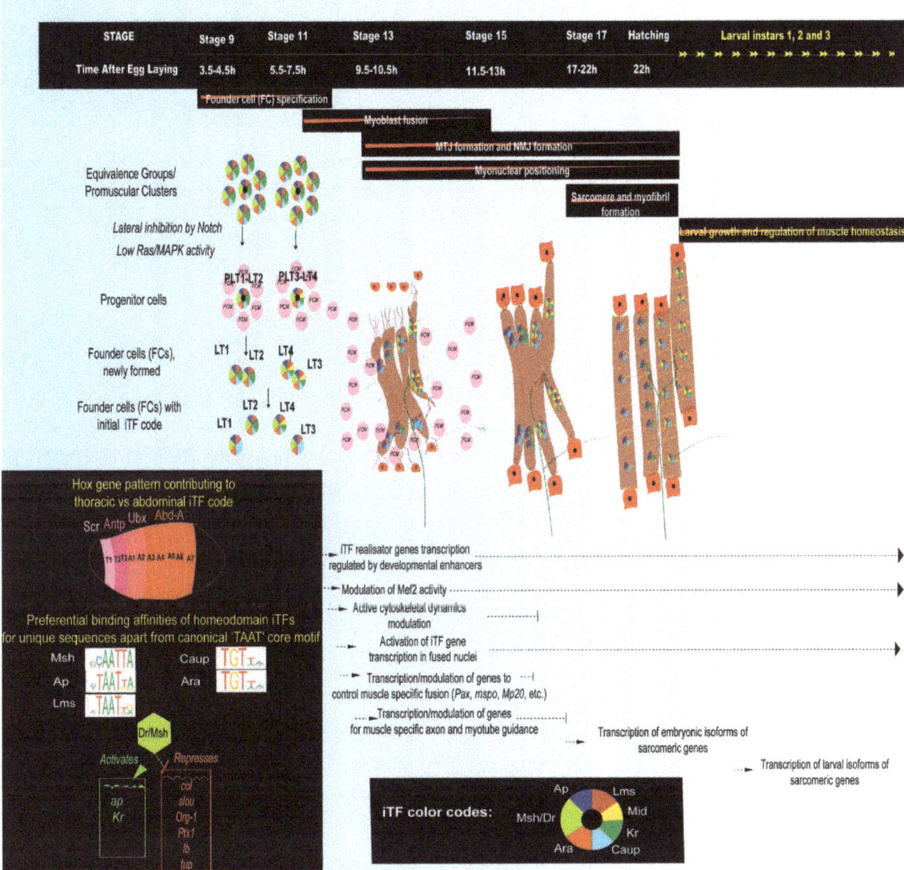

Figure 2. Spatial and temporal expression muscle identity transcription factors (iTFs) of the larval lateral transverse (LT) muscles. Sizes are not up to scale. Following the specification of progenitor cells by a lateral inhibition by Notch and low Ras/MAPK activity, founder cells (FCs) expressing muscle specific iTFs are specified for each LT muscle, LT1, LT2, LT3, and LT4 with a contribution from homeobox (Hox) genes to specify thoracic versus abdominal identities. Each iTF has preferential binding abilities to certain enhancers. The iTF expression is followed by the regulation of transcription and modulation of expression of their realisator genes which establish muscle identity over the course of development. The spatial and temporal expression of iTFs coupled with their modulation of realisator genes, which include generic muscle genes, in collaboration with Mef2 begs the question about their contribution to muscle homeostasis. Abbreviations: FCM: Fusion competent myoblasts; FC: Founder cells; LT: Lateral transverse muscles; iTF: Identity transcription factor.

3.1.2. The Role of iTFs

After the initial discovery of distinct Slou expressing FCs [77], many other TFs expressed in discrete subsets of FCs were subsequently identified and collectively named muscle identity transcription factors or iTFs (Table 1). A loss or gain of iTF function can cause muscle loss [78,79] or transformation of one muscle to another muscle fate [80,81] and impede muscle development [82] thus disrupting muscle patterns. The iTFs such as Ap, Slou, Eve, Kr, Lb, and Lms are also expressed in the CNS [78]. Many identified iTFs such as Dr/Msh, Lms, Ap, Ara, Caup, Lb, Slou, Eve, Ptx1, and Tup are homeodomain TFs that are known to recognize similar canonical TAAT containing binding motifs, but they could have preferential high affinity binding motifs (Figure 2), as has been shown for Slou [83] and Caup [84]. The iTFs from other TF families are Twi, Nau, Kr, Kn/Col, Mid, Six4, Poxm, Org-1, and Vg. Newly identified iTFs for a subset of dorsal muscles are Sine occulis (So), No ocelli (Noc), and the cofactor ETS-domain lacking (Edl) [85], which act sequentially with their cofactors. The iTF Vg also acts with a cofactor, Sd [86].

Table 1. The iTF expression patterns in embryonic somatic muscle founder cells.

iTF	Human Orthologs	FCs Expressing iTF [1]	References	Embryonic Somatic Muscle Pattern
Apterous (Ap)	LHX	LT1, LT2, LT3, LT4, VA2, VA3	[78]	
Araucan (Ara)	IRX	LT1, LT2, LT3, LT4, SBM, DT1-*DO3*	[87]	
Caupolican (Caup)	IRX	LT1, LT2, LT3, LT4, SBM, DT1-*DO3*	[87]	
Collier (Col)/Knot (Kn)	EBF	*DA2*, DA3-*DO5*, DT1-*DO3*, *LL1-DO4*	[82,85,88]	
Drop (Dr)/Muscle segment homeobox (Msh)	MSX	DO1, DO2, LT1-LT2, LT3-LT4, VA2, VA3	[89,90]	
Even-skipped (Eve)	EVX	DA1, *DO2*	[91,92]	
Krüppel (Kr)	KLF	DA1, DO1, *LT1*-LT2, *LT3*-LT4, LL1, *VA1*-VA2, *DO2*, VL3, VO2, VO5	[87,92,93]	External muscles are represented in dark brown, intermediate muscles in a medium shade of brown, and internal muscles in fuchsia.
Ladybird (Lb)	LBX	SBM	[94]	
Lateral muscles scarcer (Lms)	-	LT1-LT2, LT3-LT4	[95]	
Midline (Mid)	TBX20	*LT3*-LT4, LO1, *VA1*-VA2	[96]	
Nautilus (Nau)	MYOD	DO1, DA2, DA3-DO5, DO3, LL1- DO4, LO1, VA1	[79,85,88,97]	
Optomotor-blind-related-1 (Org-1)	TBX1	LO1, VT1, SBM	[98]	
Pox meso (Poxm)	PAX	DT1-*DO3*, VA1-VA2, VA3	[99]	
Ptx1	PITX	Ventral muscles	[100]	
Runt		DO2, VA3, VO4	[92,101]	
Slouch (Slou)/S59	NKX1	DT1-*DO3*, *VA1*-VA2, VA3, VT1, *LO1*	[77,80,87]	
Scalloped (Sd)	TEF-1	*All FCs transiently*, maintained in VL1, VL2, VL3, VL4	[86]	
Vestigial (Vg)	VGLL	DA1-DA2, DA3, LL1, VL1, VL2, VL3, VL4	[86]	
Tailup (Tup)	ISL	DA1, DA2, DO1, DO2	[81]	
Eyes absent (Eya)		Differential temporal expression in multiple FCs	[85,102]	
Six4	SIX	Differential temporal expression in multiple FCs	[102,103]	
Sine occulis (So)	SIX	*DA2*, DA3-DO5, *LL1-DO4*	[85]	
No ocelli (Noc)	ZNF	DA3-DO5	[85]	
ETS-domain lacking (Edl)	-	*DA2*, DA3	[85]	

[1] In the 'FCs Expressing iTF' column, each FC name is shown in the colour corresponding to the muscle it generates as depicted in the figure in the column on the extreme right. FCs known to be generated from an asymmetric division of the same progenitor cell are hyphenated. FCs with transient expression are shown in italics.

The iTF code can be hierarchic and activate other iTFs as has been shown for Org-1 that activates Slou and Lb [98]. In addition to hierarchy, there seems to be isoform specificity in iTF expression [85]. Certain iTFs confer identity by repressing other iTFs. Dr, for example, represses Lb that is normally active only in the SBM muscle and Eve that is normally continually expressed in only the DA1 muscle [104], while Tup represses Col in DA2 [81]. The expression levels of one isoform of the chromatin remodeling factor Sin3A is implicated in modulating the response to iTFs by acting on the *slou* enhancer [105].

The same iTFs can be expressed in different muscles, but with different co-iTFs. The identity code for one specific muscle subset, the lateral transverse or LT muscles, which comprises the four muscles LT1-4, for example, is known to be set up by a combinatorial expression of Dr [89], Ap [78], Kr [93], Lms [95], and the Ara/Caup complex [87] (Figure 2). Dr appears to directly or indirectly activate the transcription of many LT iTFs such as *Kr, ap*, and itself while repressing non-LT iTFs such as *col, slou, Org-1, Ptx1, lb*, and *tup* [83] and its expression is lost by mid embryonic stages. Only Lms is specific to all four LT muscles while others are also expressed in other muscle subsets, although not in combination with the same co-iTFs. This seems to be the way the iTF code is set up, where they are repurposed in different combinations to define the identity of different muscles [81,85,88]. Even amongst the LT muscles, each muscle has a specific combination of these iTFs (Figure 2). Some iTFs such as Lms are persistently expressed while others such as Ap and Kr are transient. A characteristic feature of Kr is that it is transiently expressed and subsequently lost from one of the two sibling FCs that arise from the progenitor that expresses it [87,93].

The thoracic LT muscles have slightly different characteristics such as a different number of myonuclei, which might depend on the iTF code along with individual iTF dynamics in conjunction with other TFs [106]. Homeotic or Hox genes such as *Antennapedia (Antp), Abdominal-A (Abd-A), Abdominal-B (Abd-B)*, and *Ultrabithorax (Ubx)* control the muscle pattern along the anterior-posterior axis and are thus part of the iTF code [106–108] (Figure 2). The mechanism of Hox gene regulation in muscles could be by repressing genes specifying alternative fates by altering the epigenetic landscape in a tissue specific manner [109]. Hox genes could also be involved in the coordination of the proper innervation of muscles [110].

Once an FC initiates its diversification with a specific identity determined by an iTF code, it starts differentiating by activating realisator genes acting downstream of iTFs. Some muscle identity realisator genes have been identified. They include several muscular differentiation genes such as *sallimus (sls), Paxillin (Pax), Muscle protein 20 (Mp20)*, and *M-spondin (mspo)*, which are differentially expressed in muscle subsets to control the acquisition of specific muscle properties such as the number of myoblast fusion events or the specific attachment to tendon cells. [111,112]. Thus, an iTF code and a downstream realisator gene code are both essential to generate a diversity of muscle types with specific functions and to set the foundation for muscle homeostasis.

3.1.3. Mef2, a Key Muscle Differentiation Factor and Its Interactions with iTFs

The *Drosophila* Myocyte enhancer factor 2 (Mef2) acts along with iTFs and their realisator genes to cause the muscle to differentiate. Mef2, similar to its vertebrate ortholog MEF2, is indispensable for muscle differentiation [113–115]. It has an equally important role in fully differentiated muscles and the control of its expression and activity is dynamic. Though it is expressed in the mesoderm during all stages, its loss of function does not prevent initial muscle specification and FC generation, but completely blocks subsequent differentiation so that muscle cells undergo apoptosis in later embryonic stages [116,117]. Mef2 activity levels change over time and appear to be adapted to varying target gene expression requirements during different developmental stages [86,118]. It regulates a vast array of muscle specific genes [119,120], sometimes in cooperation with other TFs such as Cf2 [121,122]. It is itself regulated by various mechanisms including autoregulation [123], signal and TF integration at its specific cis regulatory modules (CRMs) [124], or post transcriptionally by highly conserved miRNAs such as miR-92b [125]. TFs such as Twi and Lameduck (Lmd) [119,126] acting on muscle specific CRMs and Akirin-bearing chromatin remodeling complexes [127] are known regulators of Mef2 transcriptional activity. The RNA modifying enzyme Ten-eleven-translocation family protein (Tet) shows a strong overlap with Mef2 expression in somatic muscles and its depletion in muscle precursors leads to larval locomotion defects [128] though the relationship between the two factors is unclear.

The iTFs Vg and Sd physically interact with each other and with Mef2 either alone or in combination [129]. Each of them has a spatially and temporally controlled expression pattern and altering their expression levels severely affects the development of specific ventral muscles during late stages by affecting the levels of realisator genes. Thus, iTFs could play a key role in the modulation of Mef2 interactions. Given the central role of Mef2 in muscle development, disrupted Mef2 expression can have deleterious consequences at all stages of muscle development and maintenance.

3.1.4. Myoblast Fusion and Myonuclear Positioning

In order to form a differentiated muscle, in the mid-stage embryo, a specific number of neighboring FCMs fuse with the FC to form a syncytium (Figure 2). The formation of syncytial fibers by myoblast fusion is complete by the end of stage 15 [130–132]. As fusion proceeds, the round-shaped FC becomes a myotube that elongates, becomes polarized, and locally sends out filopodia in the presumptive area of MTJ and NMJ formation. Fusion involves complementary cell adhesion molecules (CAMs) such as Dumbfounded (Duf) or its paralogue Roughest (Rst) expressed on FCs [133,134] and Stick and Stones (Sns) or its paralogue Hibris (Hbs) expressed on FCMs [135,136], respectively. They trigger a signaling cascade, thereby modulating cytoskeleton dynamics to form a fusogenic synapse that helps integrate the FCM nucleus into the FC/myotube. In *Drosophila*, the iTF code dictates the number of fusion events by controlling the expression level of fusion genes encoding actin cytoskeleton modulators such as *Muscle Protein 20* (*Mp20*) and *Paxillin* (*Pax*) or the ECM component *m-spondin* (*mspo*) [112]. A recent study provides insights into the FCM-FC transcription dynamics in a syncytial myotube [137]. FCMs appear to be naïve and respond to the local environment that recruits them for fusion. Upon fusion, the FCM adopts an FC transcriptional program triggered by the transcription of certain muscle specific iTFs. However, once fusion is complete, differences in gene transcription among myonuclei within the same muscle are observed. For example, not all myonuclei transcribe the iTFs at a given timepoint, which could help maintain an mRNA-protein balance. Evidence from this study suggests that even after fusion is complete the FC nucleus that seeded the muscle retains a transcriptional program that is distinct from other myonuclei.

As fusion proceeds, at around stage 14, the nuclei of newly fused FCMs start exhibiting characteristic movements until they are positioned peripherally to maximize the internuclear distances. This process, also observed in vertebrate muscles [138,139], has been extensively studied in the LT muscles in the *Drosophila* embryo. In these muscles the new myonuclei initially cluster into two groups, unlike in vertebrates where nuclei cluster in the center of the myotube [140], then disperse and are finally arranged along the periphery of the myotube. Correct myonuclear positioning is dependent on the LINC complex [141] that links the inner nuclear membrane (INM) to the outer nuclear membrane (ONM) and the ONM to the microtubules (MT) and the actin cytoskeleton. Mispositioned myonuclei in *Drosophila* larvae cause locomotion defects, and in humans, are associated with various diseases [142]. This is not surprising considering the close association of myonuclei with muscle structural components such as the NMJ, MTJ, actin cytoskeleton, microtubule, SR, Golgi complex, and T-tubules. During the larval growth spurt following hatching, myonuclei increase in size along with the increasing muscle size by Myc dependent endoreplication to adapt transcription to muscle functionality requirements [62].

3.1.5. Myotendinous Junction (MTJ) Formation

MTJ formation has been previously reviewed [17,143–145]. Once FCs are specified, they migrate towards the ectoderm while tendon precursor cells are specified in the ectoderm in parallel in a muscle independent fashion by the induction of expression of the early growth response factor (Egr)-like zinc finger TF, Stripe (Sr). Interestingly, tendon progenitor cells in mice express the Sr orthologs, the early growth response TFs EGR1 and EGR2 [146]. The StripeB (SrB) isoform is induced during the precursor stage to maintain the tendon cells in a non-differentiated state until later when they differentiate following signals from the approaching muscles. These signals lead to an increase in the expression of the StripeA (SrA) isoform in an integrin dependent manner by promoting *stripe* splicing by the short isoform How(S) of the splice factor interactor How [147]. SrA induces the expression of tendon differentiation markers such as *short stop* (*shot*), *delilah* (*dei*), and *β1-tubulin* (*β1-tub*). At stage 14, tendon cells guide myotubes to their final attachment sites. The targeting of muscles to tendon cells at stage 15 is facilitated by muscle type dependent and generic CAMs as well as signaling molecules. These include Slit-Robo [148] in some ventral muscles, Derailed (Drl) [149] in LT muscles, Kon-tiki (Kon), Glutamate receptor interacting protein (Grip), and Echinoid (Ed), probably involving integrin complexes [149–152]. Once muscles target their tendon cells, integrin complexes assemble on the muscle and tendon cells facilitated by the αPS2-βPS integrin heterodimer on the muscle end and αPS1-βPS on the tendon cell end to stabilize the attachments [153]. Each attachment site is muscle type specific and the iTF code could potentially modulate the expression of genes such as *kon* [137].

MTJ formation is complete by the end of stage 16 and is then further refined to withstand contractile forces. Talin phosphorylation contributes to MTJ refinement [154]. This is followed by myofibril maturation and attachment to the MTJ. Once muscles start contraction, mechanical forces stabilize the MTJ by reducing integrin turnover [155]. The MTJ grows along with the massive larval growth spurt following hatching.

3.1.6. Sarcomere Assembly and Myofibrillogenesis

Sarcomere assembly has been extensively studied in the *Drosophila* indirect flight muscles (IFM) [31] and other invertebrate models as well as in vertebrate models and cultured human cells [156,157]. These studies point to similarities as well as differences in vertebrate and insect muscles. The premyofibril theory that is widely accepted for vertebrate sarcomere assembly proposes the formation of premyofibrils along the cell periphery containing non muscle myosin, which then incorporate muscle myosin to form nascent myofibrils that subsequently form mature myofibrils [158,159]. In the early stages, distinct Mhc positive fibrils and I-Z-I complexes containing thin filaments protruding from α-actinin positive central Z bodies are seen in invertebrates as well as vertebrates [160,161]. In *Drosophila*, it has been proposed that the individual components of the sarcomere are assembled separately as latent complexes and are then assembled into sarcomeres without assembling into premyofibrils [162–164]. Most studies on *Drosophila* sarcomere assembly have been using the IFM as a model and not much attention has been given to embryonic sarcomere development.

In the *Drosophila* embryo, sarcomere assembly is initiated at stage 17. Individual sarcomere constituents are first assembled and then integrated into a mature sarcomere by integrin dependent interdigitation [162]. The precise stage at which each sarcomere component is added is currently not known. Certain sarcomeric proteins such as actin [165] and myosin [166] express sarcomere specific as well as generic cytoplasmic isoforms with roles in other muscle components such as the MTJ. TFs such as Mef2, Chorion factor 2 (Cf2), and E2F transcription factor 1 (E2f1) have been shown to regulate the expression of sarcomeric genes [122,167]. *Drosophila* has six actin isoforms including Act57B and Act87E that are muscle specific and incorporate into larval sarcomeres [165]. Thin filament formation and elongation requires actin binding factors such as the *Drosophila* formin Dishevelled Associated Activator of Morphogenesis (DAAM) [168] and Sarcomere length short (Sals) [169], which localize to the growing thin filament pointed ends. Once the thin actin filament attains its final length, it is capped by a short embryonic isoform of Tmod [169,170]. While non-muscle myosin is a component of premyofibrils in vertebrates, this does not seem to be the case in *Drosophila*, which has only one non-muscle myosin, Zipper (Zip). During stage 16, it colocalizes with PS2 integrin at muscle attachment sites and at stage 17 when sarcomeres form, it also colocalizes to Z-discs and is essential for myofibril formation [162,171]. PS2 integrin follows a similar expression pattern in culture with initial occurrence at contact sites, then at Z-discs [172]. The observation of Zip association with PS2 before sarcomere assembly is significant because myofibrils attach to the MTJ via integrin complexes with Zip acting downstream of PS2 signaling [52]. Therefore, it would appear that the embryo is getting individual components ready for future integration into myofibrils.

By stage 17 of embryogenesis, several sarcomere proteins localize to Z-discs and thin and thick filament organization and myofibril structures are seen. A knockdown of Z-disc proteins Zip, Zasp, and α-actinin at this stage disrupts sarcomerogenesis [162], though Zasp mutant sarcomeres disintegrate after initial correct formation. The myoblast fusion protein Rolling pebbles (Rols7) also colocalizes to the Z-discs during sarcomerogenesis [173], but its function remains to be elucidated. Integrins are distributed along the width of the muscle and align with Z-discs during embryonic sarcomerogenesis. Their loss results in clumping, where I-Z-I body components stay distinct from Mhc containing components. In addition, integrins associate with the ECM and mutant larvae for the ECM type IV collagen Col4a1 present abnormalities in thin-thick filament interdigitation and the degeneration of body wall muscles [162,172,174]. They are also present at epidermal muscle attachment sites along with several Z-disc proteins. Mature sarcomeres align themselves to form myofibrils that attach to the MTJ via the terminal Z-disc to be able to sustain muscle contractions [162,173].

Auld and Folker showed that myonuclear movements are intricately linked to sarcomere and myofibril formation [35]. Their study showed that the Z-disc protein Zasp66, one of the *Drosophila* Zasp family of proteins, localizes as puncta to the cytoplasmic face of the nuclei along with F-actin during initial stages of sarcomerogenesis. At later stages, puncta were observed throughout the muscle. They showed that LINC complex components such as Klarsicht (Klar) and Klaroid (Koi) coordinate initial colocalization of puncta around the nucleus. However, Z-disc-like structures still formed and aligned into myofibrils in LINC component depleted muscles, although they had altered morphology suggesting a specialized role for myonuclei-associated Zasp66 puncta. *sals* mutants display clustered myonuclei at muscle ends as well as myofibrils with numerous shorter sarcomeres suggesting a role for correct myonuclear positioning in myofibril organization [169].

Embryonic myofibrillogenesis within the egg is complete by late stage 17. Asynchronous, episodic contractions occur during the process of myofibril assembly, but coordinated contractions only occur later after mature NMJ formation results in adequate motor inputs [175,176]. Following hatching, during larval stages when the muscles rapidly grow in size, new sarcomeres are generated and organized into myofibrils during an approximately five-day period [62]. In fully mature larval muscles, T-tubules and the SR organize themselves around each sarcomere for excitation-contraction coupling and this organization is Amphiphysin (Amph) dependent [37]. The iTF code could play a role in modulating the muscle specific expression of sarcomeric genes, as has been shown for Vg and Sd that form a complex with Mef2 to modulate Mef2 targets involved in sarcomerogenesis including *Act57B* and *Mhc* [129].

3.1.7. Innervation and Neuromuscular Junction (NMJ) Formation

The development of the NMJ of larval somatic muscles has been previously reviewed [24,25,177–179] and represents another example of intricate communication between two different tissues. After neuroblasts differentiate into motor neurons (MNs) in parallel with FC specification [180–182], their dendrites in the CNS are organized in a 'myotopic map' reflecting the innervation pattern of their target muscles and MNs can reach target locations even in the absence of muscles [182]. Each neuroblast expresses a characteristic code of TFs that defines its identity as is the case for muscle FCs expressing iTFs [26]. By stage 12, MN axons fasciculate in each hemisegment within three peripheral nerves, the intersegmental nerve (ISN), segmental nerve (SN), and transverse nerve (TN) that extend towards specific target muscles from the ventral nerve chord (VNC). The ISN, SN, and TN branch stereotypically as they extend growth cones towards muscles to form MN nerve branches that further defasciculate into axons to innervate muscles. The SN nerve, for example, branches into SNa, SNb, SNc, and SNd with a subset of MNs from the SNa innervating a muscle subset including LT muscles [181].

At around stage 14, each MN extends numerous filopodia from axon growth cones towards muscles to explore their target muscles. Muscles in turn extend myopodia that cluster together on axon growth cone arrival and intermingle with growth cone filopodia. Muscles also form lamellipodia during innervation [183]. Target muscle recognition and contact are facilitated by muscle and MN specific CAMs, Cell Surface and Secreted (CSS) proteins, and other proteins [184]. Certain guidance molecules such as the homophilic Connectin (Con) are expressed in the SNa MN as well as the LT muscles it innervates [185]. Con is also expressed in the DT1 muscle and its expression is potentially modulated by the iTF code [137]. Some MNs and the muscles they innervate express the same iTF, as is the case for the Eve expressing DA1 muscle and its innervating aCC MN in the ISNb [91,181,182,186]. Eve indirectly modulates the MN expression of the Netrin repulsive presynaptic receptor Unc-5 in the ISNb [187] to guide MN axons. Upon MN contact, muscles start to accumulate Glutamate Receptor (GluR) at synaptic zones mediated by Disks large (Dlg) to form primitive synapses in an innervation dependent fashion [188,189]. By the end of stage 17, non-target synapses are pruned and mature synapses form, which exhibit a stereotyped morphology of boutons with active zones for vesicle release on the presynaptic end and novel synthesis and clustering of more GluR on the postsynaptic end [190]. Once NMJ formation is complete, muscles are ready to contract in a coordinated manner.

During larval stages, some MNs are remodeled and this is reflected in the larval CNS myotopic map [191]. Until the third larval instar, the NMJ grows by arborization and addition of boutons, a process that requires the gene *miles to go* (*mtgo*), which is an ortholog of mammalian *FNDC3* genes [192], and integrins [193]. There is also an activity dependent refinement of the synapse mediated by Ca^{2+} [194]. Tenurins, a conserved family of transmembrane proteins enriched in the vertebrate brain that possesses glutamatergic synapses are implicated in *Drosophila* axon guidance as well as synaptic organization and signaling with muscle specific expression [195].

As muscles form, abdominal adult muscle precursors (AMPs) arrange themselves in niches between specific peripheral nerves and muscles. They form an interconnected network connecting to each other and to the peripheral nerves by extending filopodia [196–198]. All embryonic muscle development processes finally lead to the formation of functional larval body wall muscles that closely communicate with the epidermis via the MTJ and with the nervous system via the NMJ to ensure larval locomotion.

3.2. Pupal Myogenesis of Adult Muscles

Adult muscles are generated during a second wave of myogenesis during pupal metamorphosis where most larval muscles are histolyzed. Metamorphosis marks the end of larval muscle homeostatic states. Adult flies have a pair of wings in the thoracic segment T2 and three pairs of legs in thoracic segments T1–T3, which are powered by specialized thoracic flight and appendicular muscles, respectively (figure in Table 2). Adult myogenesis has been reviewed recently in [199,200]. All adult thoracic muscles including the indirect flight muscles (IFMs), direct flight muscles (DFMs), and leg muscles possess a multi-fiber structure similar to vertebrate skeletal muscles. However, unlike heterogenous mammalian skeletal muscles with one muscle composed of slow and fast fiber types, each *Drosophila* muscle appears to have a single fiber type. IFMs are constituted of the dorsoventral muscles (DVMs) and the dorsal longitudinal muscles (DLMs), which facilitate upward and downward wing strokes respectively during flight. The muscle fibers that build the IFM and leg muscles differ in organization and morphology to adapt to different functionalities. The IFMs are fibrillar, asynchronous muscles while the tergal depressor of the trochanter (TDT) or leg jump muscles and DFM are tubular, synchronous muscles [201–203]. Similar to mammals, individual fiber types in the adult fly differ in component constitution such as expressing specific myosin heavy chain isoforms [203,204]. The generation of adult muscles is initiated by a series of coordinated processes again requiring close communication between tissues.

3.2.1. Myoblast Pool Generation by Adult Muscle Precursors (AMPs) during Larval Stages

Embryonic myogenesis sets the foundation for adult muscle development since the asymmetric divisions of embryonic muscle progenitor cells give rise to adult muscle precursors (AMPs) in addition to the embryonic muscle FCs. AMPs are Notch positive, Numb negative cells that remain quiescent with persistent Twist (Twi) expression until initial pupal stages when they get reactivated [205] and contribute to adult muscle development. Abdominal AMPs are closely associated with the larval muscles and with the peripheral nervous system (PNS) enabling crosstalk and providing positional cues to the AMPs that give rise to adult abdominal muscles [196–198]. In the thoracic segments, AMPs associate with wing and leg imaginal discs, which are epidermal cell clusters set aside in the embryo and larva and act as precursors for the future generation of adult wings and legs, respectively [206]. During the first and second instar larval stages, these AMPs undergo symmetric divisions giving rise to an imaginal disc associated monolayer of Twi and Notch positive adepithelial cells. In the abdominal segments, they proliferate while remaining associated with their muscle fibers similar to vertebrate satellite cells [207,208]. During the third larval instar, due to the activation of Wg signaling from the imaginal discs, AMPs undergo asymmetric divisions forming one stem cell and one Numb positive post-mitotic myoblast where Notch signaling is inhibited [209,210]. Thus, a large pool of myoblasts is primed for metamorphosis.

The myoblasts primed to form IFM express high levels of Vestigial (Vg), which represses Notch and promotes IFM differentiation [211], and low levels of the TF Cut (Ct) while DFM myoblasts express high Ct levels, with the levels being governed extrinsically by the ectoderm [212]. DFM myoblasts also express Lms [95]. The myoblasts associated with the leg imaginal disc on the other hand express Ladybird (Lb) similar to vertebrate limb bud myoblasts that express the Lb orthologue LBX1 [213,214], which represses Vg. Mutant *vg*, *ct*, *lms*, and *lb* flies have severely disrupted muscle pattern or function and they thus contribute to the adult muscle iTF code (Table 2). Vg, Lms, and Lb also act as embryonic somatic muscle iTFs expressed in a subset of embryonic muscles [86,94,95] (Table 1). Among embryonic myogenic factors, it was noticed that Apterous (Ap) expression defines all prospective flight muscle epidermal muscle attachment sites in the wing disc [215]. Similar to embryonic stages, Duf positive adult FC specification takes place by the third larval instar, but in contrast to embryos it is driven by Heartless (Htl) mediated Fibroblast growth factor (Fgf) signaling and Hox genes [212,214,216,217].

Table 2. The iTF expression patterns in myoblasts of adult muscles.

Adult iTF	Human Orthologs	Adult Myoblast Expression	Embryonic iTF Function [1]	References	Adult Flight and Leg Muscle Pattern
Vestigial (Vg)	VGLL	IFM	DA1-DA2, DA3, LL1, VL1, VL2, VL3, VL4	[211]	Indirect flight muscles (IFM) are shown in shades of red and the direct flight muscles (DFM) in dark brown. Among the leg muscles, only the tergal depressor of trochanter (TDT) muscles are highlighted in olive green. Other leg muscles are in a light shade of green.
Extradenticle (Exd)	PBX	IFM		[218]	
Homeothorax (Hth)	MEIS	IFM		[218]	
Spalt major (Salm)	SALL	IFM		[219]	
Erect wing (Ewg)	NRF1	IFM		[220]	
Cut (Ct)		DFM		[213,214]	
Lateral muscles scarcer (Lms)	–	DFM	LT1-LT2, LT3-LT4	[95]	
Apterous (Ap)	LHX	DFM	LT1, LT2, LT3, LT4, VA2, VA3	[215]	
Ladybird (Lb)	LBX	Leg muscles	SBM	[214]	

[1] In the 'Adult myoblast expression' column, names are shown in the colour corresponding to the muscles they generate as depicted in the figure in the column on the extreme right. Embryonic FCs known to be generated from an asymmetric division of the same progenitor cell are hyphenated.

3.2.2. Histolysis of Larval Muscles, Adult iTF Code Refinement, and the Contribution of AMPs

During pupal stages, most of the larval muscles are histolyzed in the thoracic as well as abdominal hemisegments [221,222]. Myoblasts generated from AMPs either fuse with non-histolyzed larval muscle scaffolds to which they associate or give rise to adult muscles de novo [221]. In the T2 mesothoracic segment, three larval dorsal oblique muscles, DO1, DO2, and DO3 escape histolysis and serve as templates for the formation of the DLMs while the DVMs and leg muscles are generated de novo. At the end of the third larval instar, the myoblasts start expressing the muscle differentiation factor Mef2 in an ecdysone dependent manner [123]. As with embryonic myogenesis, adult muscle formation is seeded by FCs, with the number of FCs generated corresponding to the number of muscles they will seed [217]. The DLMs are an exception where the three remnant larval muscles serve as FCs and express the marker Duf. Nevertheless, if the larval muscles giving rise to adult DLMs are ablated they still form muscles de novo by an innervation dependent process, although with aberrations [223].

During early pupal stages myoblasts start migrating. MNs play a significant role in initial adult myogenesis by regulating myoblast proliferation during the second larval instar and subsequent myoblast migration during pupal stages. In denervated flies, DVM muscle formation is severely compromised and it leads to the reduction in DLM size when using larval muscles as templates whereas if larval templates are ablated, their de novo formation is abolished [223]. In the abdominal segments, myoblasts migrate and associate with nerves to form adult muscles [224]. In the thoracic segments, the wing and leg discs evaginate and myoblasts migrate along them to reach their destinations where adult muscles are generated. The myoblasts either fuse with FCs or with larval templates using a similar machinery to embryonic myoblast fusion to form fully differentiated adult muscles by 36 h after puparium formation (APF). Muscles extend as they fuse and attach to the MTJ on either end [221,225,226].

Apart from Vg, Ct, and Lb that act as adult muscle iTFs (Table 2) to confer myoblast identity in the imaginal discs during larval stages, the expression of the embryonic iTF Ap is initiated during pupal stages in myoblasts that will give rise to the DFM but not IFM in addition to epidermal attachment sites [215]. Unlike the embryonic FCs, it is expressed in adult FCMs instead of adult FCs, but similar to the embryonic FCs they contribute to the same muscle's iTF code along with Lms. This hints at specific muscle patterning information derived from these iTFs. Ap is necessary for the correct formation of DFMs and continues to be expressed in fully formed DFMs. It is also necessary for IFM attachment by regulating Stripe (Sr) expression which, similar to the embryo, is essential for adult muscle attachment. In *lms* mutants, the wing disc Vg domain is expanded and although muscles seem normal, the adult wings exhibit a held-out phenotype suggesting contraction abnormalities [95]. As fusion begins, the IFM FCs also express the adult iTFs Extradenticle (Exd), Homeothorax (Hth), and Spalt major (Salm), which genetically interact to specify a fibrillar versus tubular fate by regulating fiber specific gene expression and splicing regulated by Arrest (Aret) [201,218,219]. The iTF Erect wing (Ewg) also significantly contributes to IFM identity [220].

3.2.3. MTJ Formation

The wing and leg imaginal discs generate Sr positive tendon-like precursor cell clusters starting from the third larval instar until the beginning of pupation. Sr expression is initiated by Notch signaling [227,228]. Leg muscles attach to the internal tendons on one end and the tendon cells in the exoskeleton on the other end. At about 3 h APF, the leg disc Sr positive tendon precursor cells invaginate into an evaginating leg disc and are closely associated with myoblasts that give rise to leg muscles [18]. Disrupting tendon precursors also disrupts myoblast localization. The epidermal tendon precursor cells' shape changes to form tubular structures during invagination giving rise to internal tendons to which each leg muscle attaches on one end with tendon specificity. DLM muscles that form by the splitting of remnant larval muscles extend filopodia on either end as they grow and split. Still in the process of splitting, their filopodia interdigitate with those of their target tendon cells and initiate MTJ formation that requires Kon, integrins, Tsp, and Talin similar to embryos. DLM filopodia disappear after a mature MTJ forms by 30 h APF and tendon cells elongate due to tension [163]. In the abdomen, MTJ maturation follows a similar process but is complete only by 40 h APF [229].

3.2.4. Sarcomere Assembly

Similar to embryos, premyofibrils are absent in DLM muscles. Mhc positive complexes are observed throughout the muscle by 26 h APF and assemble rapidly and synchronously across the entire muscle into myofibrils at 30 h APF immediately following tension generated by MTJ maturation [163,230]. This myofibril assembly fails in the absence of muscle attachment. The terminal Z-disc attaches to the MTJ mediated by integrins and IAPs [52]. Myofibrils are refined to regular arrays of sarcomeres over the next several hours where more sarcomeres are added. DLM myofibrils are flanked by MT arrays during initial stages of assembly that are dissembled by the end of pupation. The myofibril length then increases without other structural changes to reach its final length shortly after eclosion [231]. In the IFM, distinct transcriptional dynamics are associated with different stages of myofibrillogenesis, with the iTF Salm contributing to the transition after 30 h APF and its expression is maintained to establish IFM fate [230,232]. A similar sequence of myofibrillogenesis occurs in abdominal muscles that form mature MTJ by 50 h APF when myofibril assembly synchronously starts and is refined further to form the transversely aligned sarcomeres seen in abdominal muscles. Thin and thick filament complexes appear separately, then start interdigitating to form immature myofibrils by 46 h when muscles have stably attached to MTJ and exhibit spontaneous contractions. They subsequently assemble into ordered myofibrils by 50 h APF and are refined over the next several hours to begin coordinated contraction [229].

During IFM sarcomere assembly, thin filaments elongate from their pointed ends as is the case during embryonic myogenesis [170]. They initially form a dispersed pattern by the polymerization of actin into nascent thin filaments which become regularly patterned after 30 h APF. At this time, active incorporation of actin at both ends of the thin filament and further refinement and growth occurs by new actin monomer incorporation at the pointed ends of thin filaments and the formation of new thin filaments at the sarcomere periphery. Tmod and Sals that are located to pointed ends are necessary for thin filament length control [170,233]. The nebulin repeat containing protein Lasp regulates thin filament length by regulating its stability [234]. The *Drosophila* formin Fhos mediates thin filament assembly by initially regulating actin monomer incorporation into thin filaments during mid pupal stages and then localizes near Z-discs to facilitate radial growth of thin filament arrays to increase myofibril diameter [233]. In *Drosophila*, IFM thick filaments are associated with many insect-specific proteins such as myofilin [235], arthrin which is a ubiquitinated actin [236], paramyosin [237], minipramyosin [238], and flightin [239,240] not found in vertebrates, which could represent proteins adapted for flight [241]. The insect and IFM specific protein flightin is implicated in regulating the thick filament length by associating with myosin filaments as they grow [242,243]. Z-disc formation fails in the IFM if actin lacks its α-actinin binding domain showing the importance of sarcomere component interdigitation [244]. A downregulation of Sls results in smaller Z-discs around which a normal thick filament assembly occurs with abnormally long thick filaments at the periphery lacking the Z-disc [164,245]. As myofibrils grow, the Z-disc protein Zasp controls the final myofibril diameter by switching to growth restricting isoforms [246]. After complete myofibril growth, coordinated contractions can be initiated after mature NMJ formation.

3.2.5. Innervation and NMJ Formation

Embryonic neuroblast lineages undergo a second larval wave of neurogenesis where embryonic neuroblasts are re-specified to give rise to adult MN lineages whose dendrites are organized in a 'myotopic map' within the CNS that reflects the innervation pattern of their target adult muscles similar to embryonic/larval stages [247–250]. MNs innervate adult muscles in a stereotypical pattern. For DLMs generated from larval templates, the primary larval ISN branch remains while secondary branches are initially retracted, and extensive new branching is generated as the muscles fuse with adult myoblasts and then split. Initial nerve arrival is muscle independent, but subsequent nerve branching occurs only in the presence of the target muscle [251]. Among DVMs, DVM I and DVM II are innervated by new branches arising from the larval ISN while the larval SN innervates DVM III [251,252]. The 14 leg muscles are innervated by around 50 MNs arising from specific neuroblasts in the CNS in a stereotypical pattern [250]. Following initial innervation, the NMJ is formed by extensive branching and synapse formation. The glial cells at the IFM NMJ express the glutamate *Drosophila* Excitatory Amino Acid Transporter 1 (dEAAT1) unlike during other stages for efficient neurotransmission [253]. Muscle iTFs contribute to correct innervation since malformed muscles cause MN branching aberrations as has been shown for Ewg [220].

In the end, a stereotypical muscle pattern along with stereotypical innervation generates fully functional adult muscles.

3.2.6. Programmed Cell Death Following Eclosion of New Adults

Some larval abdominal muscles persist through metamorphosis and are used for the eclosion of new adults. These muscles degenerate after eclosion along with associated nerves [254].

4. The Maintenance of Muscle Homeostasis

4.1. Muscle Homeostasis under Normal Conditions

Functional larval somatic muscles and adult muscles represent two different homeostatic states during the fly lifetime. The embryonic wave of myogenesis takes only one day leading to the formation of functional larval muscles, which undergo continuous growth and refinement during the larval stages spanning five days. Larval muscle homeostasis needs to be coordinated with larval growth during the three larval instars until metamorphosis to ensure functional stability. Following metamorphosis and the pupal wave of myogenesis over a period of five days, adult flies eclose from their pupae and adult muscle homeostasis needs to be maintained during the fly lifespan of several weeks.

The stereotypical muscle pattern is associated with iTFs and their realisator genes that also exhibit tightly controlled spatial and temporal expression patterns in larval and adult muscles. Therefore, some of the iTFs can play a key role in the maintenance of muscle specific homeostasis by regulating the levels of key myogenic factors such as Mef2 as well as the expression of realisator genes [111,112,129,137] (Figure 2). The control of the level of activity of the key differentiation TF Mef2 is quintessential throughout the fly lifetime since this in turn controls the muscle specific levels of its vast array of target genes [118,129]. In the embryo, various genes were shown to require different Mef2 activity, with early expressing genes such as *Act57B* requiring lower levels compared to late expressing genes such as *Mhc* [118]. In the adult, the development and maintenance of the adult DLM muscles have been observed to be sensitive to the levels of Mef2 as well as its antagonist Holes in muscles (Him). Tubular adult muscles such as the TDT and DVM muscles seem to require lower Mef2 activity than the fibrillar DLM muscles since RNAi lines affect these muscles differently [255,256]. TFs such as Cf2 and E2f1 acting along with Mef2 could also contribute to setting the muscle homeostatic state [122,167]. A study identified putative Cf2 and Mef2 binding site clusters for multiple sarcomeric genes including *Mhc, Tm1, Tm2, up, wupA* (or *TnI*), and *paramyosin* (*Prm*) [122]. On Cf2 depletion, the stoichiometry of proteins such as TnT, TnI, and Prm was found to be altered and this imbalance worsened over the course of development. Another study detected E2f binding site enrichment upstream of myogenic genes such as *how, sals, Tm1, Mef2*, etc. This study also showed that E2f1 depletion altered the gene expression levels of *Tm2, Act88F, Mlc2, how,* and *Mef2* [167].

One hallmark of muscle homeostasis in *Drosophila* larval and adult muscles is the expression of fiber specific protein isoforms. Many sarcomeric genes switch between embryonic, larval, and/or adult isoforms during development, with different muscle types also exhibiting isoform specificity. Isoform switching usually occurs by switching to a predominant isoform. Embryonic *Mhc* transcripts contain exon 19, which is spliced out of adult versions and results in a different carboxy terminal [242,257].

Embryonic isoforms lack the functionality for the high ATPase rate and sliding velocity required for adult muscles [258]. The IFM muscles initially express an Mhc isoform containing exon 19 and switch to the adult exon 18 containing isoform during late stages of myofibril assembly [242]. A shorter embryonic/larval isoform of the pointed end capping protein Tmod is associated with actin during pupal sarcomere assembly and there is a switch to a longer Tmod isoform in eclosed adults [170]. Adult *Drosophila* muscles express fiber specific actins, with Act88F being expressed in the IFM and Act79B in the TDT, for example [202]. Two IFM specific Tm1 isoforms are expressed in adult flies [259,260]. Kettin is the predominant *Drosophila* titin isoform in embryos and the IFM muscles switch to the IFM specific predominant long Sls(700) isoform [245]. Zasp52 and other Zasp proteins also switch to adult isoforms [246,261], with Zasp52 expressing an exon 8 containing isoform absent in embryos, but present in the IFM and TDT. Obscurin expresses a single larval isoform and two IFM isoforms [262].

Isoform switches are potentially associated with cis regulatory modules (CRMs) that seem to be arranged in sequential modules mirroring developmental expression and regulation by different TFs and cofactors. Marin et al. identified an upstream regulatory element (URE) and an intronic regulatory element (IRE) in intron 1 of the *wupA* (or *TnI*) gene that acted synergistically and was capable of driving LacZ tagged TnI expression. Mas et al. identified similar elements in the *up* (or *TnT*) gene [263]. They showed that these elements synergistically interact in larval muscles, whereas the contribution of the IRE is higher in adult muscles. In addition, they showed that there was decreasing IRE contribution from the IFM to the jump muscles to the visceral muscles [264]. Garcia-Zaragoza et al. followed up on this study and identified the URE and potential IRE elements of *Tm1*, *Tm2*, and *Mhc*. *Tm1* was previously shown to be coordinately regulated by two intronic enhancers in cooperation with Mef2 and its interactor PAR domain protein 1 (Pdp1) [265–267]. Mature muscles need to ensure the activation and maintenance of the correct protein isoforms [268] since aberrant isoform expression impedes muscle function. For example, transient overexpression of a shorter Tmod isoform during mid-to late IFM assembly leads to normal length thin filaments at the periphery of the myofibrils that are correctly capped by the long Tmod isoform. However, they exhibit shorter core thin filaments within the myofibril caused by the permanent association of the shorter Tmod at their pointed ends, which cannot be dynamically uncapped to permit thin filament elongation. Therefore, this prevents its elongation causing defective sarcomeres that interfere with flight during adult stages [170]. The embryonic Mhc isoform fails to substitute for the IFM isoform due to different physiological properties [258,269].

Post transcriptional mechanisms such as phosphorylation could potentially contribute to muscle homeostasis. Thin and thick filament disruptions, for example, are associated with concomitant flightin phosphorylation deregulations [239]. Tm1 IFM isoforms are phosphorylated only in adult flies, which could have functional implications [260]. Impaired Talin phosphorylation leads to severe muscle detachment at late embryonic stages [154]. This means the right CRM regulatory mechanisms as well as post translational mechanisms such as phosphorylation [48,270] need to be dynamically maintained since specific protein domains are necessary for muscle specific functionality [269,271,272].

The accumulation of insoluble protein aggregates in the muscle is associated with protein aggregate myopathies (PAM) and in *Drosophila*, p38b deficiency leads to the deposition of polyubiquitinated protein aggregates in adult thoracic muscles and to locomotor defects [273]. Loss of components of the proteasome, which mediate protein turnover were shown to cause protein aggregates and progressive muscle atrophy in larval muscles [274]. Ubiquitin protein ligases such as Mind bomb 2 (Mib2) and Ubiquitin protein ligase E3A (Ube3A), which tag proteins for proteasomal degradation, have been associated with muscle defects. The loss of function of *mib2* was shown to trigger embryonic muscle apoptosis [275] and the over or under expression of *Ube3a* alters larval NMJ neurotransmission with associated altered number of active zones [276]. Proteostasis is thus integral to muscle maintenance.

Muscle contraction is associated with multiple biochemical and morphological changes as well as large mechanical strains. This necessitates efficient mechanisms to withstand these forces to prevent muscle disintegration during contraction and to reinstate the stable muscle state (Figure 3). Protein stoichiometry is integral to sarcomere integrity since varying the expression levels of one protein has a cascading effect on the levels of other sarcomeric proteins leading to altered muscle functionality [277,278]. Sarcomeric integrity during contractions is maintained by components such as Mlp84B, Cher, small heat shock proteins (sHsps) such as dCryAB and Hsp67Bc and integrin-mediated adhesions. Mlp84B localizes to the Z-disc and genetically interacts with Sls. Mlp84B-Sls transheterozygotes exacerbate individual mutant phenotypes disrupting myofibrillar integrity [56]. Cher also interacts with Sls in addition to actin stably anchoring them to each other [53]. In addition, Cher interacts physically with dCryAB and a disruption of this interaction affects sarcomeric integrity [279]. The chaperone Hsp67Bc also colocalizes to the Z-disc although its function is unknown [280]. Integrin mediated adhesions maintain sarcomeric integrity and reduced adhesion results in the progressive age-dependent loss of sarcomeric cytoarchitecture [57]. Integrin and IAP stoichiometries at the MTJ are important to respond to different types of forces [166]. The myonuclear LINC complex and associated components such as Msp300 and Spectraplakin, which regulate MT organization, play a role in myonuclear maintenance by providing elasticity to resist contractile forces with the help of the MT network that surrounds it [281–283]. In addition, Msp300 associates with the Z-disc and keeps the mitochondria and SR anchored to the Z-disc during contractions [284]. Its presence around myonuclei near the larval NMJ also regulates glutamate receptor density to control locomotion [285].

NMJ activity perturbations lead to homeostatic synaptic plasticity, which enables compensatory modulations of the NMJ synaptic strength to resist these perturbations and stabilize synaptic activity. Lifelong synaptic plasticity ensures efficient neurotransmission of signals at the NMJ. The NMJ adapts various homeostatic mechanisms to maintain appropriate muscle function levels [286–289]. Mutants for *endophilin* (*endo*) exhibit tremendous synaptic overgrowth, but the overall synaptic strength is stabilized by reducing the active zone number in synaptic buttons, which modulates neurotransmitter release [286]. The NMJ adapts a homeostatic scaling mechanism called presynaptic homeostatic potentiation (PHP), where there is a compensatory increase in neurotransmitter release to maintain muscle excitation in response to abnormally reduced GluR on the postsynaptic end. This compensation appears to be associated with an uncharacteristic multilayer ring of electron dense T-bars in active zones to increase the neurotransmitter release [289]. The PHP maintenance has been shown to require inositol triphosphate (IP$_3$) directed signaling [290]. During the larval growth spurt, NMJ homeostasis needs to be maintained even though the presynaptic end grows slower than the muscle surface that tends to accumulate GluRs. Ziegler et al. showed that the amino acid transporter, Juvenile hormone Inducible-21 (JhI-21) is a gene that coevolved with GluRs, is expressed at presynaptic ends and plays a role in suppressing excess GluR accumulation [288].

The close association of the mesoderm with other germ layers right from the embryonic stage and the continued association with epidermal and nervous tissues over the fly lifetime highlights the importance of coordinated intrinsic and extrinsic signaling for homeostasis.

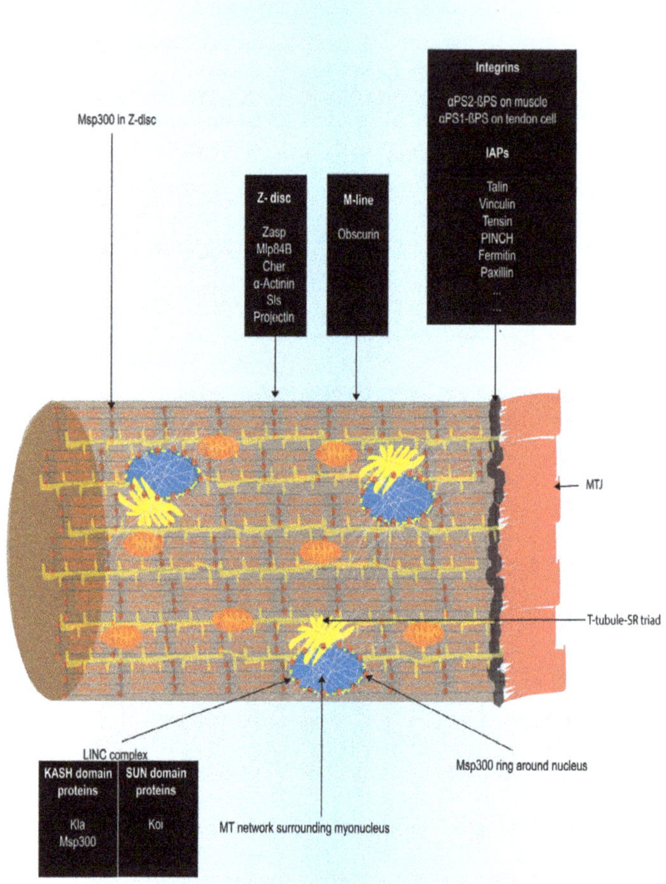

Figure 3. Maintenance of myofibril integrity and homeostasis. The integrin complex links the myofibrils to the MTJ via the extracellular matrix (ECM) and senses the forces transmitted by the MTJ. Integrins and Integrin Associated Proteins (IAPs) constitute the integrin complex. Integrin complex turnover and constitution are adapted to the forces sensed during contraction. The dense microtubule (MT) network anchored to the myonuclei by the Msp300 ring associated with the LINC complex on the nuclear envelope provides myonuclear elasticity during contractions to prevent disintegration of myonuclei and dissociation of the myofibril network. Msp300 in the Z-disc ensure regular spacing of organelles such as mitochondria and the SR for contractions. Z-disc and M-line components provide anchorage and elasticity to ensure sarcomeric integrity.

4.2. Re-Establishment of Muscle Homeostasis Following Muscle Injury

Muscle regeneration has not been described in the larva. However, the larval stem cell-like AMPs that are capable of differentiating and giving rise to adult muscles were noted to have similarities to vertebrate muscle stem cells (MuSCs), also known as satellite cells. Similar to MuSCs, the Notch pathway [209,291] and zinc-finger homeodomain 1 (Zfh1), the *Drosophila* homolog of the vertebrate ZEB1/ZEB2 [292,293], maintain the AMPs in an undifferentiated state and they are capable of self-renewal by asymmetric divisions [209,294]. In addition, they are capable of fusion with existing larval muscle remnants during the formation of DLM muscles, which is reminiscent of muscle repair. It was initially thought that all muscle stem cell-like cells or AMPs are depleted during adult muscle formation and thus adult muscles were believed to lack regenerative capacity. Recently, Chaturvedi et al. identified a population of Zfh1 positive adult stem cells closely apposed to the adult muscle, which appear to possess the ability to proliferate and contribute to muscle regeneration upon injury [61] similar to vertebrate MuSCs [293]. Boukhatmi and Bray subsequently showed that Notch directly regulates Zfh1 to antagonize the differentiation of these cells by expressing a short Zfh1 isoform transcribed from an alternate promoter that is not subject to regulation by the conserved micro RNA, miR-8 [292]. Using the G-TRACE method for cell lineage analysis, their study showed that these cells, which they termed population of progenitors that persist in adults or pMPs, were mitotically active and incorporated into adult muscles even under normal conditions. Thus, they reiterated that these cells contributed to adult muscle homeostasis. Since this is a recent discovery, further studies could provide insights into the extent of repair in *Drosophila* adult muscles and the mechanisms involved in re-establishing and maintaining muscle identity and homeostasis.

4.3. Muscle Homeostasis under Pathological Conditions

Multiple studies in *Drosophila* models have reproduced defects observed in human pathological conditions and could provide important insights into disruptions of muscle homeostasis under pathological conditions. Many myopathies and neuromuscular disorders are associated with or even caused by myonuclear defects and others are associated with sarcomeric defects leading to muscle dysfunction, wasting, and/or degeneration. *Drosophila* models exist for multisystemic disorders such as Myotonic Dystrophy Type 1 (DM1) that is caused by CTG expansions in the *Dystrophia Myotonica Protein Kinase* (*DMPK*) gene leading to the sequestration of RNA binding proteins such as MBNL1 in nuclear foci [295,296]. This causes a disruption of muscle homeostasis as indicated by the progressive muscle degeneration observed in the IFM muscles in a model expressing 480 CTG repeats. The Dystrophin (Dys)-Dystroglycan (Dg) transmembrane complex at the plasma membrane acts as a crucial signaling mediator by relaying information to and from muscles to interacting tissues via the ECM. Mutations in genes constituting this complex or their interactors thus cause a disruption of homeostasis, thereby causing diseases such as Duchenne Muscular Dystrophy (DMD) where there is muscle wasting. In *Drosophila*, Dg was shown to be under miRNA regulation by miR-9a to ensure correct MTJ formation [297]. Large scale genetic and interactome screens in *Drosophila* have identified factors affecting muscle integrity such as stress response components [298] and components of the Hippo signaling pathway [42].

Laminopathies are disorders caused by mutations in the human *LMNA* genes which code for lamins present in the INM providing structural support and regulating gene expression. One *Drosophila* model revealed increased reductive stress due to the nuclear translocation of Nrf2, which is normally sequestered in the cytoplasm and released only during oxidative stress [299]. Chandran et al. observed a loss of muscle proteostasis in a *Drosophila* model of laminopathies and corroborated this by RNA-seq analyses of human muscle biopsy tissues. Interestingly, they were able to rescue the muscular phenotypes by the modulation of the AMPK pathway which could present future therapeutic directions [300]. Apart from laminopathies, other myopathies such as Centronuclear Myopathies (CNM) are associated with myonuclear positioning defects [301]. Muscle development studies in

Drosophila are beginning to unveil mechanisms for myonuclear positioning and factors that disrupt this [141,284,302,303].

Other studies in *Drosophila* are providing insights into pathological features caused by disruptions in sarcomeric components. Muscular phenotypes caused by a mutation in the *Tm2* gene was found to be rescued by a suppressor mutation in the *wupA* gene coding for TnI [304]. *Drosophila* models exist for myosin myopathies such as Inclusion Body Myopathy Type 3 and Laing Distal Myopathy (LDM). A study has shown that the formation of large aggregates in muscles similar to those seen in human patients with *ZASP* mutations is caused by an imbalance in the levels of Zasp isoforms [246]. Dahl-Halvarsson et al. showed that the overexpression of the Thin protein, a homolog of the human TRIM family of proteins that is implicated in maintaining sarcomeric integrity, could alleviate LDM-like phenotypes [305].

5. Discussion

In vertebrates, the loss of skeletal muscle homeostasis is the cause of various muscular disorders. Studies in vertebrate systems are complicated by the presence of large gene families for multiple genes. *Drosophila* is a simple model organism with various conserved pathways and genes to study muscle homeostasis while at the same time mostly having one to a few genes orthologous to large vertebrate gene families that perform functions similar to vertebrate genes. Thus, it appears that genes are reused/repurposed over the course of evolution instead of 'reinventing the wheel'. *Drosophila* muscle development has been studied for decades. The embryonic somatic muscles being uni-fiber muscles present a simple model to study development since all muscles have been well characterized along with their specific attachment sites and innervating MNs [143,180]. The IFM muscles have been equally well characterized [31,221,252]. In addition, a large number of tools are available in *Drosophila* to study in vivo mechanisms [306].

A better understanding of developmental and post developmental processes would help us gain a better understanding of the mechanisms of maintenance and disruption of homeostasis. The short life cycle of the fruit fly facilitates the quick and detailed study of processes making it a valuable model for the study of factors that initiate, maintain, and disrupt muscle homeostasis. The study of muscle regeneration following muscle injury, where developmental processes need to be re-initiated, represents an example of how the *Drosophila* model could help understand the mechanisms of muscle homeostasis. Some potential therapeutic targets have been unveiled by studies in *Drosophila* models of myopathies [300,305]. The recent discovery of stem cell-like cells associated with adult muscles is an exciting new direction of research to study muscle regeneration and homeostasis [61]. In vertebrates, aging is related to a depletion of the MuSC population leading to sarcopenia or age-related gradual loss of muscle mass and function [307] that is also characteristic of pathological conditions such as DMD [308]. The short life span of the *Drosophila* model presents a huge advantage to study homeostatic disruptions during aging.

Large gaps exist in our understanding of pathological mechanisms and simpler models could provide valuable insights and therapeutic directions. In *Drosophila*, although a lot of attention has been given to the major muscle components including the sarcomeres, MTJ and NMJ, muscle organelles that play an equally central role such as the SR, T-tubules, golgi complex, and transport vesicles have received lesser attention, although myonuclei are beginning to be studied in detail. Given the detailed characterization and tools available for this established model system that has already helped advance research [309,310], it would continue to serve as an important backbone for research into various physiological processes including muscle development and homeostasis.

Author Contributions: Conceptualization, P.P.; writing—original draft preparation, P.P.; writing—review and editing, P.P. and K.J.; visualization, P.P.; supervision, K.J. All authors have read and agreed to the published version of the manuscript.

Funding: This work was funded by the AFM-Telethon, grant number 21182 to the MyoNeurAlp Alliance.

Conflicts of Interest: The authors declare no conflict of interest.

References

1. Bate, M. The embryonic development of larval muscles in Drosophila. *Development* **1990**, *110*, 791. [PubMed]
2. Abmayr, S.M.; Erickson, M.S.; Bour, B.A. Embryonic development of the larval body wall musculature of Drosophila melanogaster. *Trends Genet.* **1995**, *11*, 153–159. [CrossRef]
3. Fyrberg, E. Study of contractile and cytoskeletal proteins using Drosophila genetics. *Cell Motil. Cytoskelet.* **1989**, *14*, 118–127. [CrossRef]
4. Dobi, K.C.; Schulman, V.K.; Baylies, M.K. Specification of the somatic musculature in Drosophila. *Wiley Interdiscip. Rev. Dev. Biol.* **2015**, *4*, 357–375. [CrossRef]
5. Mukund, K.; Subramaniam, S. Skeletal muscle: A review of molecular structure and function, in health and disease. *Wiley Interdiscip. Rev. Syst. Biol. Med.* **2020**, *12*, e1462. [CrossRef] [PubMed]
6. Taylor, M. Comparison of muscle development in Drosophila and vertebrates. In *Muscle development in Drosophila*; Sink, H., Ed.; Landes Bioscience/Springer: Georgetown, TX, USA; New York, NY, USA, 2006; pp. 169–203.
7. Piccirillo, R.; Demontis, F.; Perrimon, N.; Goldberg, A.L. Mechanisms of muscle growth and atrophy in mammals and Drosophila. *Dev. Dyn. Off. Publ. Am. Assoc. Anat.* **2014**, *243*, 201–215. [CrossRef]
8. Kohsaka, H.; Guertin, P.A.; Nose, A. Neural Circuits Underlying Fly Larval Locomotion. *Curr. Pharm. Des.* **2017**, *23*, 1722–1733. [CrossRef]
9. Wolfe, R.R. The underappreciated role of muscle in health and disease. *Am. J. Clin. Nutr.* **2006**, *84*, 475–482. [CrossRef]
10. Veratti, E. Investigations on the fine structure of striated muscle fiber read before the Reale Istituto Lombardo, 13 March 1902. *J. Biophys. Biochem. Cytol.* **1961**, *10*, 1–59. [CrossRef]
11. Hanson, J.; Huxley, H.E. Structural Basis of the Cross-Striations in Muscle. *Nature* **1953**, *172*, 530–532. [CrossRef]
12. Royuela, M.; Fraile, B.; Arenas, M.I.; Paniagua, R. Characterization of several invertebrate muscle cell types: A comparison with vertebrate muscles. *Microsc. Res. Tech.* **2000**, *48*, 107–115. [CrossRef]
13. Littlefield, R.S.; Fowler, V.M. Thin filament length regulation in striated muscle sarcomeres: Pointed-end dynamics go beyond a nebulin ruler. *Semin. Cell Dev. Biol.* **2008**, *19*, 511–519. [CrossRef] [PubMed]
14. Lemke, S.B.; Schnorrer, F. Mechanical forces during muscle development. *Mech. Dev.* **2017**, *144*, 92–101. [CrossRef] [PubMed]
15. Wang, L.; Geist, J.; Grogan, A.; Hu, L.-Y.R.; Kontrogianni-Konstantopoulos, A. Thick Filament Protein Network, Functions, and Disease Association. *Compr. Physiol.* **2018**, *8*, 631–709. [CrossRef] [PubMed]
16. Hooper, S.L.; Thuma, J.B. Invertebrate Muscles: Muscle Specific Genes and Proteins. *Physiol. Rev.* **2005**, *85*, 1001–1060. [CrossRef] [PubMed]
17. Schweitzer, R.; Zelzer, E.; Volk, T. Connecting muscles to tendons: Tendons and musculoskeletal development in flies and vertebrates. *Dev. Camb. Engl.* **2010**, *137*, 2807–2817. [CrossRef]
18. Soler, C.; Laddada, L.; Jagla, K. Coordinated Development of Muscles and Tendon-Like Structures: Early Interactions in the Drosophila Leg. *Front. Physiol.* **2016**, *7*, 22. [CrossRef]
19. Lemke, S.B.; Weidemann, T.; Cost, A.-L.; Grashoff, C.; Schnorrer, F. A small proportion of Talin molecules transmit forces at developing muscle attachments in vivo. *PLoS Biol.* **2019**, *17*, e3000057. [CrossRef]
20. Richier, B.; Inoue, Y.; Dobramysl, U.; Friedlander, J.; Brown, N.H.; Gallop, J.L. Integrin signaling downregulates filopodia during muscle-tendon attachment. *J. Cell Sci.* **2018**, *131*, jcs217133. [CrossRef]
21. Nawrotzki, R.; Willem, M.; Miosge, N.; Brinkmeier, H.; Mayer, U. Defective integrin switch and matrix composition at alpha 7-deficient myotendinous junctions precede the onset of muscular dystrophy in mice. *Hum. Mol. Genet.* **2003**, *12*, 483–495. [CrossRef]
22. Marshall, J.L.; Chou, E.; Oh, J.; Kwok, A.; Burkin, D.J.; Crosbie-Watson, R.H. Dystrophin and utrophin expression require sarcospan: Loss of $\alpha 7$ integrin exacerbates a newly discovered muscle phenotype in sarcospan-null mice. *Hum. Mol. Genet.* **2012**, *21*, 4378–4393. [CrossRef] [PubMed]
23. Broadie, K.; Bate, M. The Drosophila NMJ: A genetic model system for synapse formation and function. *Semin. Dev. Biol.* **1995**, *6*, 221–231. [CrossRef]
24. Menon, K.P.; Carrillo, R.A.; Zinn, K. Development and plasticity of the Drosophila larval neuromuscular junction. *WIREs Dev. Biol.* **2013**, *2*, 647–670. [CrossRef] [PubMed]

25. Harris, K.P.; Littleton, J.T. Transmission, Development, and Plasticity of Synapses. *Genetics* **2015**, *201*, 345–375. [CrossRef]

26. Lacin, H.; Truman, J.W. Lineage mapping identifies molecular and architectural similarities between the larval and adult Drosophila central nervous system. *eLife* **2016**, *5*, e13399. [CrossRef]

27. Pérez-Moreno, J.J.; O'Kane, C.J. GAL4 Drivers Specific for Type Ib and Type Is Motor Neurons in Drosophila. *G3 Genes Genomes Genet.* **2019**, *9*, 453. [CrossRef]

28. Dasari, S.; Cooper, R.L. Modulation of sensory–CNS–motor circuits by serotonin, octopamine, and dopamine in semi-intact Drosophila larva. *Neurosci. Res.* **2004**, *48*, 221–227. [CrossRef]

29. Kohsaka, H.; Takasu, E.; Morimoto, T.; Nose, A. A Group of Segmental Premotor Interneurons Regulates the Speed of Axial Locomotion in Drosophila Larvae. *Curr. Biol.* **2014**, *24*, 2632–2642. [CrossRef]

30. Babski, H.; Jovanic, T.; Surel, C.; Yoshikawa, S.; Zwart, M.F.; Valmier, J.; Thomas, J.B.; Enriquez, J.; Carroll, P.; Garcès, A. A GABAergic Maf-expressing interneuron subset regulates the speed of locomotion in Drosophila. *Nat. Commun.* **2019**, *10*, 4796. [CrossRef]

31. Vigoreaux, J.O. Genetics of the Drosophila flight muscle myofibril: A window into the biology of complex systems. *BioEssays* **2001**, *23*, 1047–1063. [CrossRef]

32. Campbell, K.B.; Chandra, M. Functions of stretch activation in heart muscle. *J. Gen. Physiol.* **2006**, *127*, 89–94. [CrossRef] [PubMed]

33. Huxley, H.E. Fifty years of muscle and the sliding filament hypothesis. *Eur. J. Biochem.* **2004**, *271*, 1403–1415. [CrossRef] [PubMed]

34. Wang, Z.-H.; Clark, C.; Geisbrecht, E.R. Analysis of mitochondrial structure and function in the Drosophila larval musculature. *Mitochondrion* **2016**, *26*, 33–42. [CrossRef] [PubMed]

35. Auld, A.L.; Folker, E.S. Nucleus-dependent sarcomere assembly is mediated by the LINC complex. *Mol. Biol. Cell* **2016**, *27*, 2351–2359. [CrossRef] [PubMed]

36. Al-Qusairi, L.; Laporte, J. T-tubule biogenesis and triad formation in skeletal muscle and implication in human diseases. *Skelet. Muscle* **2011**, *1*, 26. [CrossRef]

37. Razzaq, A.; Robinson, I.M.; McMahon, H.T.; Skepper, J.N.; Su, Y.; Zelhof, A.C.; Jackson, A.P.; Gay, N.J.; O'Kane, C.J. Amphiphysin is necessary for organization of the excitation-contraction coupling machinery of muscles, but not for synaptic vesicle endocytosis in Drosophila. *Genes Dev.* **2001**, *15*, 2967–2979. [CrossRef]

38. Ackermann, M.A.; Ziman, A.P.; Strong, J.; Zhang, Y.; Hartford, A.K.; Ward, C.W.; Randall, W.R.; Kontrogianni-Konstantopoulos, A.; Bloch, R.J. Integrity of the network sarcoplasmic reticulum in skeletal muscle requires small ankyrin 1. *J. Cell Sci.* **2011**, *124*, 3619–3630. [CrossRef]

39. Maughan, D.W.; Godt, R.E. Equilibrium distribution of ions in a muscle fiber. *Biophys. J.* **1989**, *56*, 717–722. [CrossRef]

40. Clausen, T. Na^+-K^+ Pump Stimulation Improves Contractility in Damaged Muscle Fibers. *Ann. N. Y. Acad. Sci.* **2006**, *1066*, 286–294. [CrossRef]

41. Lebovitz, R.M.; Takeyasu, K.; Fambrough, D.M. Molecular characterization and expression of the (Na^+ + K^+)-ATPase alpha-subunit in Drosophila melanogaster. *EMBO J.* **1989**, *8*, 193–202. [CrossRef]

42. Yatsenko, A.S.; Kucherenko, M.M.; Xie, Y.; Aweida, D.; Urlaub, H.; Scheibe, R.J.; Cohen, S.; Shcherbata, H.R. Profiling of the muscle-specific dystroglycan interactome reveals the role of Hippo signaling in muscular dystrophy and age-dependent muscle atrophy. *BMC Med.* **2020**, *18*, 8. [CrossRef] [PubMed]

43. Huxley, H.; Hanson, J. Changes in the Cross-Striations of Muscle during Contraction and Stretch and their Structural Interpretation. *Nature* **1954**, *173*, 973–976. [CrossRef]

44. Huxley, A.F.; Niedergerke, R. Structural Changes in Muscle During Contraction: Interference Microscopy of Living Muscle Fibres. *Nature* **1954**, *173*, 971–973. [CrossRef] [PubMed]

45. Potter, J.D.; Sheng, Z.; Pan, B.-S.; Zhao, J. A Direct Regulatory Role for Troponin T and a Dual Role for Troponin C in the Ca^{2+} Regulation of Muscle Contraction. *J. Biol. Chem.* **1995**, *270*, 2557–2562. [CrossRef] [PubMed]

46. Qiu, F.; Lakey, A.; Agianian, B.; Hutchings, A.; Butcher, G.W.; Labeit, S.; Leonard, K.; Bullard, B. Troponin C in different insect muscle types: Identification of two isoforms in Lethocerus, Drosophila and Anopheles that are specific to asynchronous flight muscle in the adult insect. *Biochem. J.* **2003**, *371*, 811–821. [CrossRef] [PubMed]

47. Vibert, P.; Craig, R.; Lehman, W. Steric-model for activation of muscle thin filaments 1 1 Edited by P.E. Wright. *J. Mol. Biol.* **1997**, *266*, 8–14. [CrossRef]

48. Farman, G.P.; Miller, M.S.; Reedy, M.C.; Soto-Adames, F.N.; Vigoreaux, J.O.; Maughan, D.W.; Irving, T.C. Phosphorylation and the N-terminal extension of the regulatory light chain help orient and align the myosin heads in Drosophila flight muscle. *J. Struct. Biol.* **2009**, *168*, 240–249. [CrossRef]

49. Beall, C.J.; Fyrberg, E. Muscle abnormalities in Drosophila melanogaster heldup mutants are caused by missing or aberrant troponin-I isoforms. *J. Cell Biol.* **1991**, *114*, 941–951. [CrossRef]

50. Lee, K.H.; Sulbarán, G.; Yang, S.; Mun, J.Y.; Alamo, L.; Pinto, A.; Sato, O.; Ikebe, M.; Liu, X.; Korn, E.D.; et al. Interacting-heads motif has been conserved as a mechanism of myosin II inhibition since before the origin of animals. *Proc. Natl. Acad. Sci. USA* **2018**, *115*, E1991–E2000. [CrossRef]

51. Jung, H.S.; Komatsu, S.; Ikebe, M.; Craig, R. Head-head and head-tail interaction: A general mechanism for switching off myosin II activity in cells. *Mol. Biol. Cell* **2008**, *19*, 3234–3242. [CrossRef]

52. Green, H.J.; Griffiths, A.G.; Ylänne, J.; Brown, N.H. Novel functions for integrin-associated proteins revealed by analysis of myofibril attachment in Drosophila. *eLife* **2018**, *7*, e35783. [CrossRef] [PubMed]

53. González-Morales, N.; Holenka, T.K.; Schöck, F. Filamin actin-binding and titin-binding fulfill distinct functions in Z-disc cohesion. *PLoS Genet.* **2017**, *13*, e1006880. [CrossRef] [PubMed]

54. Liao, K.A.; González-Morales, N.; Schöck, F. Zasp52, a Core Z-disc Protein in Drosophila Indirect Flight Muscles, Interacts with α-Actinin via an Extended PDZ Domain. *PLoS Genet.* **2016**, *12*, e1006400. [CrossRef] [PubMed]

55. Katzemich, A.; West, R.J.H.; Fukuzawa, A.; Sweeney, S.T.; Gautel, M.; Sparrow, J.; Bullard, B. Binding partners of the kinase domains in Drosophila obscurin and their effect on the structure of the flight muscle. *J. Cell Sci.* **2015**, *128*, 3386–3397. [CrossRef] [PubMed]

56. Clark, K.A.; Bland, J.M.; Beckerle, M.C. The Drosophila muscle LIM protein, Mlp84B, cooperates with D-titin to maintain muscle structural integrity. *J. Cell Sci.* **2007**, *120*, 2066. [CrossRef] [PubMed]

57. Perkins, A.D.; Ellis, S.J.; Asghari, P.; Shamsian, A.; Moore, E.D.W.; Tanentzapf, G. Integrin-mediated adhesion maintains sarcomeric integrity. *Dev. Biol.* **2010**, *338*, 15–27. [CrossRef]

58. Prill, K.; Dawson, J.F. Assembly and Maintenance of Sarcomere Thin Filaments and Associated Diseases. *Int. J. Mol. Sci.* **2020**, *21*, 542. [CrossRef]

59. Ciglar, L.; Furlong, E.E. Conservation and divergence in developmental networks: A view from Drosophila myogenesis. *Cell Differ. Cell Div. Growth Death* **2009**, *21*, 754–760. [CrossRef]

60. Karalaki, M.; Fili, S.; Philippou, A.; Koutsilieris, M. Muscle regeneration: Cellular and molecular events. *Vivo Athens Greece* **2009**, *23*, 779–796.

61. Chaturvedi, D.; Reichert, H.; Gunage, R.D.; VijayRaghavan, K. Identification and functional characterization of muscle satellite cells in Drosophila. *eLife* **2017**, *6*, e30107. [CrossRef]

62. Demontis, F.; Perrimon, N. Integration of Insulin receptor/Foxo signaling and dMyc activity during muscle growth regulates body size in Drosophila. *Dev. Camb. Engl.* **2009**, *136*, 983–993. [CrossRef] [PubMed]

63. Xiao, G.; Mao, S.; Baumgarten, G.; Serrano, J.; Jordan, M.C.; Roos, K.P.; Fishbein, M.C.; MacLellan, W.R. Inducible activation of c-Myc in adult myocardium in vivo provokes cardiac myocyte hypertrophy and reactivation of DNA synthesis. *Circ. Res.* **2001**, *89*, 1122–1129. [CrossRef] [PubMed]

64. Martin, A.C. The Physical Mechanisms of Drosophila Gastrulation: Mesoderm and Endoderm Invagination. *Genetics* **2020**, *214*, 543. [CrossRef] [PubMed]

65. Azpiazu, N.; Lawrence, P.A.; Vincent, J.P.; Frasch, M. Segmentation and specification of the Drosophila mesoderm. *Genes Dev.* **1996**, *10*, 3183–3194. [CrossRef]

66. Baylies, M.K.; Bate, M. twist: A Myogenic Switch in Drosophila. *Science* **1996**, *272*, 1481. [CrossRef] [PubMed]

67. Carmena, A.; Bate, M.; Jiménez, F. Lethal of scute, a proneural gene, participates in the specification of muscle progenitors during Drosophila embryogenesis. *Genes Dev.* **1995**, *9*, 2373–2383. [CrossRef]

68. Carmena, A.; Buff, E.; Halfon, M.S.; Gisselbrecht, S.; Jiménez, F.; Baylies, M.K.; Michelson, A.M. Reciprocal Regulatory Interactions between the Notch and Ras Signaling Pathways in the Drosophila Embryonic Mesoderm. *Dev. Biol.* **2002**, *244*, 226–242. [CrossRef]

69. Doe, C.Q.; Skeath, J.B. Neurogenesis in the insect central nervous system. *Curr. Opin. Neurobiol.* **1996**, *6*, 18–24. [CrossRef]

70. Crews, S.T. Drosophila Embryonic CNS Development: Neurogenesis, Gliogenesis, Cell Fate, and Differentiation. *Genetics* **2019**, *213*, 1111. [CrossRef]

71. Gomez Ruiz, M.; Bate, M. Segregation of myogenic lineages in Drosophila requires numb. *Development* **1997**, *124*, 4857.

72. Liu, J.; Qian, L.; Wessells, R.J.; Bidet, Y.; Jagla, K.; Bodmer, R. Hedgehog and RAS pathways cooperate in the anterior–posterior specification and positioning of cardiac progenitor cells. *Dev. Biol.* **2006**, *290*, 373–385. [CrossRef] [PubMed]

73. Rushton, E.; Drysdale, R.; Abmayr, S.M.; Michelson, A.M.; Bate, M. Mutations in a novel gene, myoblast city, provide evidence in support of the founder cell hypothesis for Drosophila muscle development. *Development* **1995**, *121*, 1979. [PubMed]

74. Prokop, A.; Landgraf, M.; Rushton, E.; Broadie, K.; Bate, M. Presynaptic Development at the Drosophila Neuromuscular Junction: Assembly and Localization of Presynaptic Active Zones. *Neuron* **1996**, *17*, 617–626. [CrossRef]

75. Halfon, M.S.; Carmena, A.; Gisselbrecht, S.; Sackerson, C.M.; Jiménez, F.; Baylies, M.K.; Michelson, A.M. Ras Pathway Specificity Is Determined by the Integration of Multiple Signal-Activated and Tissue-Restricted Transcription Factors. *Cell* **2000**, *103*, 63–74. [CrossRef]

76. Cox, V.T.; Beckett, K.; Baylies, M.K. Delivery of wingless to the ventral mesoderm by the developing central nervous system ensures proper patterning of individual slouch-positive muscle progenitors. *Dev. Biol.* **2005**, *287*, 403–415. [CrossRef] [PubMed]

77. Dohrmann, C.; Azpiazu, N.; Frasch, M. A new Drosophila homeo box gene is expressed in mesodermal precursor cells of distinct muscles during embryogenesis. *Genes Dev.* **1990**, *4*, 2098–2111. [CrossRef]

78. Bourgouin, C.; Lundgren, S.E.; Thomas, J.B. Apterous is a drosophila LIM domain gene required for the development of a subset of embryonic muscles. *Neuron* **1992**, *9*, 549–561. [CrossRef]

79. Keller, C.A.; Grill, M.A.; Abmayr, S.M. A Role for nautilus in the Differentiation of Muscle Precursors. *Dev. Biol.* **1998**, *202*, 157–171. [CrossRef]

80. Knirr, S.; Azpiazu, N.; Frasch, M. The role of the NK-homeobox gene slouch (S59) in somatic muscle patterning. *Development* **1999**, *126*, 4525.

81. Boukhatmi, H.; Frendo, J.L.; Enriquez, J.; Crozatier, M.; Dubois, L.; Vincent, A. Tup/Islet1 integrates time and position to specify muscle identity in Drosophila. *Development* **2012**, *139*, 3572. [CrossRef]

82. Crozatier, M.; Vincent, A. Requirement for the Drosophila COE transcription factor Collier in formation of an embryonic muscle: Transcriptional response to notch signalling. *Development* **1999**, *126*, 1495. [PubMed]

83. Busser, B.W.; Shokri, L.; Jaeger, S.A.; Gisselbrecht, S.S.; Singhania, A.; Berger, M.F.; Zhou, B.; Bulyk, M.L.; Michelson, A.M. Molecular mechanism underlying the regulatory specificity of a Drosophila homeodomain protein that specifies myoblast identity. *Dev. Camb. Engl.* **2012**, *139*, 1164–1174. [CrossRef] [PubMed]

84. Busser, B.W.; Gisselbrecht, S.S.; Shokri, L.; Tansey, T.R.; Gamble, C.E.; Bulyk, M.L.; Michelson, A.M. Contribution of distinct homeodomain DNA binding specificities to Drosophila embryonic mesodermal cell-specific gene expression programs. *PLoS ONE* **2013**, *8*, e69385. [CrossRef] [PubMed]

85. Dubois, L.; Frendo, J.-L.; Chanut-Delalande, H.; Crozatier, M.; Vincent, A. Genetic dissection of the Transcription Factor code controlling serial specification of muscle identities in Drosophila. *eLife* **2016**, *5*, e14979. [CrossRef] [PubMed]

86. Deng, H.; Bell, J.B.; Simmonds, A.J. Vestigial is required during late-stage muscle differentiation in Drosophila melanogaster embryos. *Mol. Biol. Cell* **2010**, *21*, 3304–3316. [CrossRef]

87. Carrasco-Rando, M.; Tutor, A.S.; Prieto-Sánchez, S.; González-Pérez, E.; Barrios, N.; Letizia, A.; Martín, P.; Campuzano, S.; Ruiz-Gómez, M. Drosophila araucan and caupolican integrate intrinsic and signalling inputs for the acquisition by muscle progenitors of the lateral transverse fate. *PLoS Genet.* **2011**, *7*, e1002186. [CrossRef]

88. Enriquez, J.; de Taffin, M.; Crozatier, M.; Vincent, A.; Dubois, L. Combinatorial coding of Drosophila muscle shape by Collier and Nautilus. *Dev. Biol.* **2012**, *363*, 27–39. [CrossRef]

89. Nose, A.; Isshiki, T.; Takeichi, M. Regional specification of muscle progenitors in Drosophila: The role of the msh homeobox gene. *Development* **1998**, *125*, 215.

90. Lord, P.C.W.; Lin, M.-H.; Hales, K.H.; Storti, R.V. Normal Expression and the Effects of Ectopic Expression of the Drosophila muscle segment homeobox (msh) Gene Suggest a Role in Differentiation and Patterning of Embryonic Muscles. *Dev. Biol.* **1995**, *171*, 627–640. [CrossRef]

91. Knirr, S.; Frasch, M. Molecular Integration of Inductive and Mesoderm-Intrinsic Inputs Governs even-skipped Enhancer Activity in a Subset of Pericardial and Dorsal Muscle Progenitors. *Dev. Biol.* **2001**, *238*, 13–26. [CrossRef]

92. Fujioka, M.; Wessells, R.J.; Han, Z.; Liu, J.; Fitzgerald, K.; Yusibova, G.L.; Zamora, M.; Ruiz-Lozano, P.; Bodmer, R.; Jaynes, J.B. Embryonic even skipped-dependent muscle and heart cell fates are required for normal adult activity, heart function, and lifespan. *Circ. Res.* **2005**, *97*, 1108–1114. [CrossRef] [PubMed]

93. Ruiz-Gomez, M.; Romani, S.; Hartmann, C.; Jackle, H.; Bate, M. Specific muscle identities are regulated by Kruppel during Drosophila embryogenesis. *Development* **1997**, *124*, 3407. [PubMed]

94. Jagla, T.; Bellard, F.; Lutz, Y.; Dretzen, G.; Bellard, M.; Jagla, K. ladybird determines cell fate decisions during diversification of Drosophila somatic muscles. *Development* **1998**, *125*, 3699. [PubMed]

95. Müller, D.; Jagla, T.; Bodart, L.M.; Jährling, N.; Dodt, H.-U.; Jagla, K.; Frasch, M. Regulation and functions of the lms homeobox gene during development of embryonic lateral transverse muscles and direct flight muscles in Drosophila. *PLoS ONE* **2010**, *5*, e14323. [CrossRef]

96. Kumar, R.P.; Dobi, K.C.; Baylies, M.K.; Abmayr, S.M. Muscle cell fate choice requires the T-box transcription factor midline in Drosophila. *Genetics* **2015**, *199*, 777–791. [CrossRef]

97. Corbin, V.; Michelson, A.M.; Abmayr, S.M.; Neel, V.; Alcamo, E.; Maniatis, T.; Young, M.W. A role for the Drosophila neurogenic genes in mesoderm differentiation. *Cell* **1991**, *67*, 311–323. [CrossRef]

98. Schaub, C.; Nagaso, H.; Jin, H.; Frasch, M. Org-1, the Drosophila ortholog of Tbx1, is a direct activator of known identity genes during muscle specification. *Development* **2012**, *139*, 1001. [CrossRef]

99. Duan, H.; Zhang, C.; Chen, J.; Sink, H.; Frei, E.; Noll, M. A key role of Pox meso in somatic myogenesis of Drosophila. *Development* **2007**, *134*, 3985. [CrossRef]

100. Vorbrüggen, G.; Constien, R.; Zilian, O.; Wimmer, E.A.; Dowe, G.; Taubert, H.; Noll, M.; Jäckle, H. Embryonic expression and characterization of a Ptx1 homolog in Drosophila. *Mech. Dev.* **1997**, *68*, 139–147. [CrossRef]

101. Drysdale, R.; Rushton, E.; Bate, M. Genes required for embryonic muscle development in Drosophila melanogaster A survey of the X chromosome. *Rouxs Arch. Dev. Biol. Off. Organ EDBO* **1993**, *202*, 276–295. [CrossRef]

102. Liu, Y.-H.; Jakobsen, J.S.; Valentin, G.; Amarantos, I.; Gilmour, D.T.; Furlong, E.E.M. A Systematic Analysis of Tinman Function Reveals Eya and JAK-STAT Signaling as Essential Regulators of Muscle Development. *Dev. Cell* **2009**, *16*, 280–291. [CrossRef] [PubMed]

103. Clark, I.B.N.; Boyd, J.; Hamilton, G.; Finnegan, D.J.; Jarman, A.P. D-six4 plays a key role in patterning cell identities deriving from the Drosophila mesoderm. *Dev. Biol.* **2006**, *294*, 220–231. [CrossRef] [PubMed]

104. Jagla, T.; Bidet, Y.; Da Ponte, J.P.; Dastugue, B.; Jagla, K. Cross-repressive interactions of identity genes are essential for proper specification of cardiac and muscular fates in Drosophila. *Development* **2002**, *129*, 1037. [PubMed]

105. Dobi, K.C.; Halfon, M.S.; Baylies, M.K. Whole-Genome Analysis of Muscle Founder Cells Implicates the Chromatin Regulator Sin3A in Muscle Identity. *Cell Rep.* **2014**, *8*, 858–870. [CrossRef] [PubMed]

106. Capovilla, M.; Kambris, Z.; Botas, J. Direct regulation of the muscle-identity gene apterous by a Hox protein in the somatic mesoderm. *Development* **2001**, *128*, 1221. [PubMed]

107. Michelson, A.M. Muscle pattern diversification in Drosophila is determined by the autonomous function of homeotic genes in the embryonic mesoderm. *Development* **1994**, *120*, 755.

108. Enriquez, J.; Boukhatmi, H.; Dubois, L.; Philippakis, A.A.; Bulyk, M.L.; Michelson, A.M.; Crozatier, M.; Vincent, A. Multi-step control of muscle diversity by Hox proteins in the Drosophila embryo. *Dev. Camb. Engl.* **2010**, *137*, 457–466. [CrossRef]

109. Domsch, K.; Carnesecchi, J.; Disela, V.; Friedrich, J.; Trost, N.; Ermakova, O.; Polychronidou, M.; Lohmann, I. The Hox transcription factor Ubx stabilizes lineage commitment by suppressing cellular plasticity in Drosophila. *eLife* **2019**, *8*, e42675. [CrossRef]

110. Hessinger, C.; Technau, G.M.; Rogulja-Ortmann, A. The Drosophila Hox gene Ultrabithorax acts in both muscles and motoneurons to orchestrate formation of specific neuromuscular connections. *Development* **2017**, *144*, 139. [CrossRef]

111. Junion, G.; Bataillé, L.; Jagla, T.; Da Ponte, J.P.; Tapin, R.; Jagla, K. Genome-wide view of cell fate specification: Ladybird acts at multiple levels during diversification of muscle and heart precursors. *Genes Dev.* **2007**, *21*, 3163–3180. [CrossRef]

112. Bataillé, L.; Delon, I.; Da Ponte, J.P.; Brown, N.H.; Jagla, K. Downstream of Identity Genes: Muscle-Type-Specific Regulation of the Fusion Process. *Dev. Cell* **2010**, *19*, 317–328. [CrossRef] [PubMed]

113. Black, B.L.; Olson, E.N. Transcriptional control of muscle development by myocyte enhancer factor-2 (Mef2) proteins. *Annu. Rev. Cell Dev. Biol.* **1998**, *14*, 167–196. [CrossRef] [PubMed]

114. Pon, J.R.; Marra, M.A. MEF2 transcription factors: Developmental regulators and emerging cancer genes. *Oncotarget* **2016**, *7*, 2297–2312. [CrossRef]

115. Taylor, M.V.; Hughes, S.M. Mef2 and the skeletal muscle differentiation program. *Skelet. Muscle Dev. 30th Anniv. MyoD* **2017**, *72*, 33–44. [CrossRef]

116. Bour, B.A.; O'Brien, M.A.; Lockwood, W.L.; Goldstein, E.S.; Bodmer, R.; Taghert, P.H.; Abmayr, S.M.; Nguyen, H.T. Drosophila MEF2, a transcription factor that is essential for myogenesis. *Genes Dev.* **1995**, *9*, 730–741. [CrossRef] [PubMed]

117. Ranganayakulu, G.; Zhao, B.; Dokidis, A.; Molkentin, J.D.; Olson, E.N.; Schulz, R.A. A Series of Mutations in the D-MEF2 Transcription Factor Reveal Multiple Functions in Larval and Adult Myogenesis in Drosophila. *Dev. Biol.* **1995**, *171*, 169–181. [CrossRef] [PubMed]

118. Elgar, S.J.; Han, J.; Taylor, M.V. mef2 activity levels differentially affect gene expression during Drosophila muscle development. *Proc. Natl. Acad. Sci. USA* **2008**, *105*, 918–923. [CrossRef]

119. Cunha, P.M.F.; Sandmann, T.; Gustafson, E.H.; Ciglar, L.; Eichenlaub, M.P.; Furlong, E.E.M. Combinatorial binding leads to diverse regulatory responses: Lmd is a tissue-specific modulator of Mef2 activity. *PLoS Genet.* **2010**, *6*, e1001014. [CrossRef]

120. Junion, G.; Jagla, T.; Duplant, S.; Tapin, R.; Da Ponte, J.-P.; Jagla, K. Mapping Dmef2-binding regulatory modules by using a ChIP-enriched in silico targets approach. *Proc. Natl. Acad. Sci. USA* **2005**, *102*, 18479. [CrossRef]

121. Tanaka, K.K.K.; Bryantsev, A.L.; Cripps, R.M. Myocyte Enhancer Factor 2 and Chorion Factor 2 Collaborate in Activation of the Myogenic Program in Drosophila. *Mol. Cell. Biol.* **2008**, *28*, 1616. [CrossRef]

122. García-Zaragoza, E.; Mas, J.A.; Vivar, J.; Arredondo, J.J.; Cervera, M. CF2 activity and enhancer integration are required for proper muscle gene expression in Drosophila. *Mech. Dev.* **2008**, *125*, 617–630. [CrossRef] [PubMed]

123. Cripps, R.M.; Lovato, T.L.; Olson, E.N. Positive autoregulation of the Myocyte enhancer factor-2 myogenic control gene during somatic muscle development in Drosophila. *Dev. Biol.* **2004**, *267*, 536–547. [CrossRef] [PubMed]

124. Nguyen, H.T.; Xu, X. Drosophila mef2 Expression during Mesoderm Development Is Controlled by a Complex Array ofcis-Acting Regulatory Modules. *Dev. Biol.* **1998**, *204*, 550–566. [CrossRef] [PubMed]

125. Chen, Z.; Liang, S.; Zhao, Y.; Han, Z. miR-92b regulates Mef2 levels through a negative-feedback circuit during Drosophila muscle development. *Dev. Camb. Engl.* **2012**, *139*, 3543–3552. [CrossRef] [PubMed]

126. Cripps, R.M.; Black, B.L.; Zhao, B.; Lien, C.L.; Schulz, R.A.; Olson, E.N. The myogenic regulatory gene Mef2 is a direct target for transcriptional activation by Twist during Drosophila myogenesis. *Genes Dev.* **1998**, *12*, 422–434. [CrossRef] [PubMed]

127. Nowak, S.J.; Aihara, H.; Gonzalez, K.; Nibu, Y.; Baylies, M.K. Akirin links twist-regulated transcription with the Brahma chromatin remodeling complex during embryogenesis. *PLoS Genet.* **2012**, *8*, e1002547. [CrossRef]

128. Wang, F.; Minakhina, S.; Tran, H.; Changela, N.; Kramer, J.; Steward, R. Tet protein function during Drosophila development. *PLoS ONE* **2018**, *13*, e0190367. [CrossRef]

129. Deng, H.; Hughes, S.C.; Bell, J.B.; Simmonds, A.J. Alternative requirements for Vestigial, Scalloped, and Dmef2 during muscle differentiation in Drosophila melanogaster. *Mol. Biol. Cell* **2009**, *20*, 256–269. [CrossRef]

130. Haralalka, S.; Abmayr, S.M. Myoblast fusion in Drosophila. *Exp. Cell Res.* **2010**, *316*, 3007–3013. [CrossRef]

131. Deng, S.; Azevedo, M.; Baylies, M. Acting on identity: Myoblast fusion and the formation of the syncytial muscle fiber. *Semin. Cell Dev. Biol.* **2017**, *72*, 45–55. [CrossRef]

132. Lee, D.M.; Chen, E.H. Drosophila Myoblast Fusion: Invasion and Resistance for the Ultimate Union. *Annu. Rev. Genet.* **2019**, *53*, 67–91. [CrossRef] [PubMed]

133. Ruiz-Gómez, M.; Coutts, N.; Price, A.; Taylor, M.V.; Bate, M. Drosophila Dumbfounded: A Myoblast Attractant Essential for Fusion. *Cell* **2000**, *102*, 189–198. [CrossRef]

134. Strünkelnberg, M.; Bonengel, B.; Moda, L.M.; Hertenstein, A.; de Couet, H.G.; Ramos, R.G.P.; Fischbach, K.-F. rst and its paralogue kirre act redundantly during embryonic muscle development in Drosophila. *Development* **2001**, *128*, 4229.

135. Bour, B.A.; Chakravarti, M.; West, J.M.; Abmayr, S.M. Drosophila SNS, a member of the immunoglobulin superfamily that is essential for myoblast fusion. *Genes Dev.* **2000**, *14*, 1498–1511. [PubMed]

136. Artero, R.D.; Castanon, I.; Baylies, M.K. The immunoglobulin-like protein Hibris functions as a dose-dependent regulator of myoblast fusion and is differentially controlled by Ras and Notch signaling. *Development* **2001**, *128*, 4251.

137. Bataillé, L.; Boukhatmi, H.; Frendo, J.-L.; Vincent, A. Dynamics of transcriptional (re)-programming of syncytial nuclei in developing muscles. *BMC Biol.* **2017**, *15*. [CrossRef]

138. Azevedo, M.; Baylies, M.K. Getting into Position: Nuclear Movement in Muscle Cells. *Trends Cell Biol.* **2020**, *30*, 303–316. [CrossRef]

139. Roman, W.; Gomes, E.R. Nuclear positioning in skeletal muscle. *SI Nucl. Position.* **2018**, *82*, 51–56. [CrossRef]

140. Cadot, B.; Gache, V.; Gomes, E.R. Moving and positioning the nucleus in skeletal muscle - one step at a time. *Nucl. Austin Tex* **2015**, *6*, 373–381. [CrossRef]

141. Starr, D.A.; Fridolfsson, H.N. Interactions Between Nuclei and the Cytoskeleton Are Mediated by SUN-KASH Nuclear-Envelope Bridges. *Annu. Rev. Cell Dev. Biol.* **2010**, *26*, 421–444. [CrossRef]

142. Folker, E.S.; Baylies, M.K. Nuclear positioning in muscle development and disease. *Front. Physiol.* **2013**, *4*, 363. [CrossRef] [PubMed]

143. Volk, T. Singling out Drosophila tendon cells: A dialogue between two distinct cell types. *Trends Genet.* **1999**, *15*, 448–453. [CrossRef]

144. Schnorrer, F.; Dickson, B.J. Muscle Building: Mechanisms of Myotube Guidance and Attachment Site Selection. *Dev. Cell* **2004**, *7*, 9–20. [CrossRef] [PubMed]

145. Subramanian, A.; Schilling, T.F. Tendon development and musculoskeletal assembly: Emerging roles for the extracellular matrix. *Dev. Camb. Engl.* **2015**, *142*, 4191–4204. [CrossRef]

146. Lejard, V.; Blais, F.; Guerquin, M.-J.; Bonnet, A.; Bonnin, M.-A.; Havis, E.; Malbouyres, M.; Bidaud, C.B.; Maro, G.; Gilardi-Hebenstreit, P.; et al. EGR1 and EGR2 involvement in vertebrate tendon differentiation. *J. Biol. Chem.* **2011**, *286*, 5855–5867. [CrossRef]

147. Volohonsky, G.; Edenfeld, G.; Klämbt, C.; Volk, T. Muscle-dependent maturation of tendon cells is induced by post-transcriptional regulation of stripeA. *Development* **2007**, *134*, 347. [CrossRef]

148. Kramer, S.G.; Kidd, T.; Simpson, J.H.; Goodman, C.S. Switching Repulsion to Attraction: Changing Responses to Slit During Transition in Mesoderm Migration. *Science* **2001**, *292*, 737. [CrossRef]

149. Callahan, C.A.; Bonkovsky, J.L.; Scully, A.L.; Thomas, J.B. derailed is required for muscle attachment site selection in Drosophila. *Development* **1996**, *122*, 2761.

150. Schnorrer, F.; Kalchhauser, I.; Dickson, B.J. The Transmembrane Protein Kon-tiki Couples to Dgrip to Mediate Myotube Targeting in Drosophila. *Dev. Cell* **2007**, *12*, 751–766. [CrossRef]

151. Estrada, B.; Gisselbrecht, S.S.; Michelson, A.M. The transmembrane protein Perdido interacts with Grip and integrins to mediate myotube projection and attachment in the Drosophila embryo. *Development* **2007**, *134*, 4469. [CrossRef]

152. Swan, L.E.; Schmidt, M.; Schwarz, T.; Ponimaskin, E.; Prange, U.; Boeckers, T.; Thomas, U.; Sigrist, S.J. Complex interaction of Drosophila GRIP PDZ domains and Echinoid during muscle morphogenesis. *EMBO J.* **2006**, *25*, 3640–3651. [CrossRef] [PubMed]

153. Chanana, B.; Graf, R.; Koledachkina, T.; Pflanz, R.; Vorbrüggen, G. αPS2 integrin-mediated muscle attachment in Drosophila requires the ECM protein Thrombospondin. *Mech. Dev.* **2007**, *124*, 463–475. [CrossRef] [PubMed]

154. Katzemich, A.; Long, J.Y.; Panneton, V.; Fisher, L.A.B.; Hipfner, D.; Schöck, F. Slik phosphorylation of Talin T152 is crucial for proper Talin recruitment and maintenance of muscle attachment in Drosophila. *Development* **2019**, *146*, dev176339. [CrossRef] [PubMed]

155. Pines, M.; Das, R.; Ellis, S.J.; Morin, A.; Czerniecki, S.; Yuan, L.; Klose, M.; Coombs, D.; Tanentzapf, G. Mechanical force regulates integrin turnover in Drosophila in vivo. *Nat. Cell Biol.* **2012**, *14*, 935–943. [CrossRef]

156. Sparrow, J.C.; Schöck, F. The initial steps of myofibril assembly: Integrins pave the way. *Nat. Rev. Mol. Cell Biol.* **2009**, *10*, 293–298. [CrossRef]

157. Gautel, M.; Djinović-Carugo, K. The sarcomeric cytoskeleton: From molecules to motion. *J. Exp. Biol.* **2016**, *219*, 135. [CrossRef]

158. Rhee, D.; Sanger, J.M.; Sanger, J.W. The premyofibril: Evidence for its role in myofibrillogenesis. *Cell Motil. Cytoskeleton* **1994**, *28*, 1–24. [CrossRef]

159. Jirka, C.; Pak, J.H.; Grosgogeat, C.A.; Marchetii, M.M.; Gupta, V.A. Dysregulation of NRAP degradation by KLHL41 contributes to pathophysiology in nemaline myopathy. *Hum. Mol. Genet.* **2019**, *28*, 2549–2560. [CrossRef]

160. Antin, P.B.; Tokunaka, S.; Nachmias, V.T.; Holtzer, H. Role of stress fiber-like structures in assembling nascent myofibrils in myosheets recovering from exposure to ethyl methanesulfonate. *J. Cell Biol.* **1986**, *102*, 1464–1479. [CrossRef]

161. Epstein, H.; Fischman, D. Molecular analysis of protein assembly in muscle development. *Science* **1991**, *251*, 1039–1044. [CrossRef]

162. Rui, Y.; Bai, J.; Perrimon, N. Sarcomere formation occurs by the assembly of multiple latent protein complexes. *PLoS Genet.* **2010**, *6*, e1001208. [CrossRef] [PubMed]

163. Weitkunat, M.; Kaya-Çopur, A.; Grill, S.W.; Schnorrer, F. Tension and Force-Resistant Attachment Are Essential for Myofibrillogenesis in Drosophila Flight Muscle. *Curr. Biol.* **2014**, *24*, 705–716. [CrossRef] [PubMed]

164. Orfanos, Z.; Leonard, K.; Elliott, C.; Katzemich, A.; Bullard, B.; Sparrow, J. Sallimus and the Dynamics of Sarcomere Assembly in Drosophila Flight Muscles. *J. Mol. Biol.* **2015**, *427*, 2151–2158. [CrossRef] [PubMed]

165. Röper, K.; Mao, Y.; Brown, N.H. Contribution of sequence variation in Drosophila actins to their incorporation into actin-based structures in vivo. *J. Cell Sci.* **2005**, *118*, 3937. [CrossRef] [PubMed]

166. Bulgakova, N.A.; Wellmann, J.; Brown, N.H. Diverse integrin adhesion stoichiometries caused by varied actomyosin activity. *Open Biol.* **2017**, *7*, 160250. [CrossRef] [PubMed]

167. Zappia, M.P.; Frolov, M.V. E2F function in muscle growth is necessary and sufficient for viability in Drosophila. *Nat. Commun.* **2016**, *7*, 10509. [CrossRef] [PubMed]

168. Molnár, I.; Migh, E.; Szikora, S.; Kalmár, T.; Végh, A.G.; Deák, F.; Barkó, S.; Bugyi, B.; Orfanos, Z.; Kovács, J.; et al. DAAM is required for thin filament formation and Sarcomerogenesis during muscle development in Drosophila. *PLoS Genet.* **2014**, *10*, e1004166. [CrossRef]

169. Bai, J.; Hartwig, J.H.; Perrimon, N. SALS, a WH2-Domain-Containing Protein, Promotes Sarcomeric Actin Filament Elongation from Pointed Ends during Drosophila Muscle Growth. *Dev. Cell* **2007**, *13*, 828–842. [CrossRef]

170. Mardahl-Dumesnil, M.; Fowler, V.M. Thin filaments elongate from their pointed ends during myofibril assembly in Drosophila indirect flight muscle. *J. Cell Biol.* **2001**, *155*, 1043–1053. [CrossRef]

171. Bloor, J.W.; Kiehart, D.P. zipper Nonmuscle Myosin-II Functions Downstream of PS2 Integrin in Drosophila Myogenesis and Is Necessary for Myofibril Formation. *Dev. Biol.* **2001**, *239*, 215–228. [CrossRef]

172. Volk, T.; Fessler, L.I.; Fessler, J.H. A role for integrin in the formation of sarcomeric cytoarchitecture. *Cell* **1990**, *63*, 525–536. [CrossRef]

173. Kreiskôther, N.; Reichert, N.; Buttgereit, D.; Hertenstein, A.; Fischbach, K.-F.; Renkawitz-Pohl, R. Drosophila Rolling pebbles colocalises and putatively interacts with alpha-Actinin and the Sls isoform Zormin in the Z-discs of the sarcomere and with Dumbfounded/Kirre, alpha-Actinin and Zormin in the terminal Z-discs. *J. Muscle Res. Cell Motil.* **2006**, *27*, 93. [CrossRef] [PubMed]

174. Kelemen-Valkony, I.; Kiss, M.; Csiha, J.; Kiss, A.; Bircher, U.; Szidonya, J.; Maróy, P.; Juhász, G.; Komonyi, O.; Csiszár, K.; et al. Drosophila basement membrane collagen col4a1 mutations cause severe myopathy. *Matrix Biol.* **2012**, *31*, 29–37. [CrossRef] [PubMed]

175. Katzemich, A.; Liao, K.A.; Czerniecki, S.; Schöck, F. Alp/Enigma family proteins cooperate in Z-disc formation and myofibril assembly. *PLoS Genet.* **2013**, *9*, e1003342. [CrossRef]

176. Crisp, S.J.; Evers, J.F.; Bate, M. Endogenous patterns of activity are required for the maturation of a motor network. *J. Neurosci. Off. J. Soc. Neurosci.* **2011**, *31*, 10445–10450. [CrossRef]

177. Broadie, K.; Bate, M. Development of the embryonic neuromuscular synapse of Drosophila melanogaster. *J. Neurosci.* **1993**, *13*, 144. [CrossRef]

178. Keshishian, H.; Broadie, K.; Chiba, A.; Bate, M. The Drosophila Neuromuscular Junction: A Model System for Studying Synaptic Development and Function. *Annu. Rev. Neurosci.* **1996**, *19*, 545–575. [CrossRef]

179. Ruiz-Cañada, C.; Budnik, V. Introduction on The Use of The Drosophila Embryonic/Larval Neuromuscular Junction as A Model System to Study Synapse Development and Function, and A Brief Summary of Pathfinding and Target Recognition. In *International Review of Neurobiology*; Academic Press: New York, NY, USA, 2006; pp. 1–31, ISBN 0074-7742.

180. Landgraf, M.; Bossing, T.; Technau, G.M.; Bate, M. The origin, location, and projections of the embryonic abdominal motorneurons of Drosophila. *J. Neurosci. Off. J. Soc. Neurosci.* **1997**, *17*, 9642–9655. [CrossRef]

181. Schmid, A.; Chiba, A.; Doe, C.Q. Clonal analysis of Drosophila embryonic neuroblasts: Neural cell types, axon projections and muscle targets. *Development* **1999**, *126*, 4653.

182. Landgraf, M.; Jeffrey, V.; Fujioka, M.; Jaynes, J.B.; Bate, M. Embryonic Origins of a Motor System: Motor Dendrites Form a Myotopic Map in Drosophila. *PLoS Biol.* **2003**, *1*, e41. [CrossRef]

183. Ritzenthaler, S.; Suzuki, E.; Chiba, A. Postsynaptic filopodia in muscle cells interact with innervating motoneuron axons. *Nat. Neurosci.* **2000**, *3*, 1012–1017. [CrossRef] [PubMed]

184. Nose, A. Generation of neuromuscular specificity in Drosophila: Novel mechanisms revealed by new technologies. *Front. Mol. Neurosci.* **2012**, *5*, 62. [CrossRef] [PubMed]

185. Nose, A.; Umeda, T.; Takeichi, M. Neuromuscular target recognition by a homophilic interaction of connectin cell adhesion molecules in Drosophila. *Development* **1997**, *124*, 1433. [PubMed]

186. Patel, N.H.; Ball, E.E.; Goodman, C.S. Changing role of even-skipped during the evolution of insect pattern formation. *Nature* **1992**, *357*, 339–342. [CrossRef] [PubMed]

187. Labrador, J.P.; O'Keefe, D.; Yoshikawa, S.; McKinnon, R.D.; Thomas, J.B.; Bashaw, G.J. The Homeobox Transcription Factor Even-skipped Regulates Netrin-Receptor Expression to Control Dorsal Motor-Axon Projections in Drosophila. *Curr. Biol.* **2005**, *15*, 1413–1419. [CrossRef] [PubMed]

188. Chen, K.; Featherstone, D.E. Discs-large (DLG) is clustered by presynaptic innervation and regulates postsynaptic glutamate receptor subunit composition in Drosophila. *BMC Biol.* **2005**, *3*, 1–13. [CrossRef] [PubMed]

189. Bachmann, A.; Kobler, O.; Kittel, R.J.; Wichmann, C.; Sierralta, J.; Sigrist, S.J.; Gundelfinger, E.D.; Knust, E.; Thomas, U. A perisynaptic ménage à trois between Dlg, DLin-7, and Metro controls proper organization of Drosophila synaptic junctions. *J. Neurosci. Off. J. Soc. Neurosci.* **2010**, *30*, 5811–5824. [CrossRef]

190. Broadie, K.; Bate, M. Innervation directs receptor synthesis and localization in Drosophila embryo synaptogenesis. *Nature* **1993**, *361*, 350–353. [CrossRef]

191. Kim, M.D.; Wen, Y.; Jan, Y.-N. Patterning and organization of motor neuron dendrites in the Drosophila larva. *Dev. Biol.* **2009**, *336*, 213–221. [CrossRef]

192. Syed, A.; Lukacsovich, T.; Pomeroy, M.; Bardwell, A.J.; Decker, G.T.; Waymire, K.G.; Purcell, J.; Huang, W.; Gui, J.; Padilla, E.M.; et al. Miles to go (mtgo) encodes FNDC3 proteins that interact with the chaperonin subunit CCT3 and are required for NMJ branching and growth in Drosophila. *Dev. Biol.* **2019**, *445*, 37–53. [CrossRef]

193. Beumer, K.J.; Rohrbough, J.; Prokop, A.; Broadie, K. A role for PS integrins in morphological growth and synaptic function at the postembryonic neuromuscular junction of Drosophila. *Development* **1999**, *126*, 5833. [PubMed]

194. Vonhoff, F.; Keshishian, H. In Vivo Calcium Signaling during Synaptic Refinement at the Drosophila Neuromuscular Junction. *J. Neurosci. Off. J. Soc. Neurosci.* **2017**, *37*, 5511–5526. [CrossRef]

195. DePew, A.T.; Aimino, M.A.; Mosca, T.J. The Tenets of Teneurin: Conserved Mechanisms Regulate Diverse Developmental Processes in the Drosophila Nervous System. *Front. Neurosci.* **2019**, *13*, 27. [CrossRef] [PubMed]

196. Bate, M.; Rushton, E.; Currie, D.A. Cells with persistent twist expression are the embryonic precursors of adult muscles in Drosophila. *Development* **1991**, *113*, 79. [PubMed]

197. Figeac, N.; Jagla, T.; Aradhya, R.; Da Ponte, J.P.; Jagla, K. Drosophila adult muscle precursors form a network of interconnected cells and are specified by the rhomboid-triggered EGF pathway. *Development* **2010**, *137*, 1965. [CrossRef] [PubMed]

198. Lavergne, G.; Zmojdzian, M.; Da Ponte, J.P.; Junion, G.; Jagla, K. Drosophila adult muscle precursor cells contribute to motor axon pathfinding and proper innervation of embryonic muscles. *Development* **2020**, *147*, dev183004. [CrossRef]

199. Gunage, R.D.; Dhanyasi, N.; Reichert, H.; VijayRaghavan, K. Drosophila adult muscle development and regeneration. *Skelet. Muscle Dev. 30th Anniv. MyoD* **2017**, *72*, 56–66. [CrossRef]

200. Laurichesse, Q.; Soler, C. Muscle development: A view from adult myogenesis in Drosophila. *Semin. Cell Dev. Biol.* **2020**. [CrossRef]

201. Oas, S.T.; Bryantsev, A.L.; Cripps, R.M. Arrest is a regulator of fiber-specific alternative splicing in the indirect flight muscles of Drosophila. *J. Cell Biol.* **2014**, *206*, 895–908. [CrossRef]

202. DeAguero, A.A.; Castillo, L.; Oas, S.T.; Kiani, K.; Bryantsev, A.L.; Cripps, R.M. Regulation of fiber-specific actin expression by the Drosophila SRF ortholog Blistered. *Dev. Camb. Engl.* **2019**, *146*, dev164129. [CrossRef]

203. Schiaffino, S.; Reggiani, C. Fiber Types in Mammalian Skeletal Muscles. *Physiol. Rev.* **2011**, *91*, 1447–1531. [CrossRef] [PubMed]

204. Hastings, G.A.; Emerson, C.P., Jr. Myosin functional domains encoded by alternative exons are expressed in specific thoracic muscles of Drosophila. *J. Cell Biol.* **1991**, *114*, 263–276. [CrossRef] [PubMed]

205. Figeac, N.; Daczewska, M.; Marcelle, C.; Jagla, K. Muscle stem cells and model systems for their investigation. *Dev. Dyn.* **2007**, *236*, 3332–3342. [CrossRef] [PubMed]

206. Beira, J.V.; Paro, R. The legacy of Drosophila imaginal discs. *Chromosoma* **2016**, *125*, 573–592. [CrossRef] [PubMed]

207. Aradhya, R.; Zmojdzian, M.; Da Ponte, J.P.; Jagla, K. Muscle niche-driven Insulin-Notch-Myc cascade reactivates dormant Adult Muscle Precursors in Drosophila. *eLife* **2015**, *4*, e08497. [CrossRef]

208. Tavi, P.; Korhonen, T.; Hänninen, S.L.; Bruton, J.D.; Lööf, S.; Simon, A.; Westerblad, H. Myogenic skeletal muscle satellite cells communicate by tunnelling nanotubes. *J. Cell. Physiol.* **2010**, *223*, 376–383. [CrossRef]

209. Gunage, R.D.; Reichert, H.; VijayRaghavan, K. Identification of a new stem cell population that generates Drosophila flight muscles. *eLife* **2014**, *3*, e03126. [CrossRef]

210. Vishal, K.; Brooks, D.S.; Bawa, S.; Gameros, S.; Stetsiv, M.; Geisbrecht, E.R. Adult Muscle Formation Requires Drosophila Moleskin for Proliferation of Wing Disc-Associated Muscle Precursors. *Genetics* **2017**, *206*, 199. [CrossRef]

211. Bernard, F.; Dutriaux, A.; Silber, J.; Lalouette, A. Notch pathway repression by vestigial is required to promote indirect flight muscle differentiation in Drosophila melanogaster. *Dev. Biol.* **2006**, *295*, 164–177. [CrossRef]

212. Dutta, D.; Umashankar, M.; Lewis, E.B.; Rodrigues, V.; VijayRaghavan, K. Hox Genes Regulate Muscle Founder Cell Pattern Autonomously and Regulate Morphogenesis Through Motor Neurons. *J. Neurogenet.* **2010**, *24*, 95–108. [CrossRef]

213. Sudarsan, V.; Anant, S.; Guptan, P.; VijayRaghavan, K.; Skaer, H. Myoblast Diversification and Ectodermal Signaling in Drosophila. *Dev. Cell* **2001**, *1*, 829–839. [CrossRef]

214. Maqbool, T.; Soler, C.; Jagla, T.; Daczewska, M.; Lodha, N.; Palliyil, S.; VijayRaghavan, K.; Jagla, K. Shaping leg muscles in Drosophila: Role of ladybird, a conserved regulator of appendicular myogenesis. *PLoS ONE* **2006**, *1*, e122. [CrossRef] [PubMed]

215. Ghazi, A.; Anant, S.; Vijay Raghavan, K. Apterous mediates development of direct flight muscles autonomously and indirect flight muscles through epidermal cues. *Development* **2000**, *127*, 5309. [PubMed]

216. Dutta, D.; Shaw, S.; Maqbool, T.; Pandya, H.; VijayRaghavan, K. Drosophila Heartless Acts with Heartbroken/Dof in Muscle Founder Differentiation. *PLoS Biol.* **2005**, *3*, e337. [CrossRef]

217. Dutta, D.; Anant, S.; Ruiz-Gomez, M.; Bate, M.; VijayRaghavan, K. Founder myoblasts and fibre number during adult myogenesis in Drosophila. *Development* **2004**, *131*, 3761. [CrossRef]

218. Bryantsev, A.L.; Duong, S.; Brunetti, T.M.; Chechenova, M.B.; Lovato, T.L.; Nelson, C.; Shaw, E.; Uhl, J.D.; Gebelein, B.; Cripps, R.M. Extradenticle and homothorax control adult muscle fiber identity in Drosophila. *Dev. Cell* **2012**, *23*, 664–673. [CrossRef] [PubMed]

219. Schönbauer, C.; Distler, J.; Jährling, N.; Radolf, M.; Dodt, H.-U.; Frasch, M.; Schnorrer, F. Spalt mediates an evolutionarily conserved switch to fibrillar muscle fate in insects. *Nature* **2011**, *479*, 406–409. [CrossRef]

220. DeSimone, S.; Coelho, C.; Roy, S.; VijayRaghavan, K.; White, K. ERECT WING, the Drosophila member of a family of DNA binding proteins is required in imaginal myoblasts for flight muscle development. *Development* **1996**, *122*, 31.

221. Fernandes, J.; Bate, M.; Vijayraghavan, K. Development of the indirect flight muscles of Drosophila. *Development* **1991**, *113*, 67.

222. Kuleesha, Y.; Puah, W.C.; Wasser, M. Live imaging of muscle histolysis in Drosophila metamorphosis. *BMC Dev. Biol.* **2016**, *16*, 12. [CrossRef]

223. Fernandes, J.J.; Keshishian, H. Motoneurons regulate myoblast proliferation and patterning in Drosophila. *Dev. Biol.* **2005**, *277*, 493–505. [CrossRef]

224. Currie, D.A.; Bate, M. The development of adult abdominal muscles in Drosophila: Myoblasts express twist and are associated with nerves. *Development* **1991**, *113*, 91. [PubMed]

225. Rivlin, P.K.; Schneiderman, A.M.; Booker, R. Imaginal Pioneers Prefigure the Formation of Adult Thoracic Muscles in Drosophila melanogaster. *Dev. Biol.* **2000**, *222*, 450–459. [CrossRef] [PubMed]

226. Mukherjee, P.; Gildor, B.; Shilo, B.-Z.; VijayRaghavan, K.; Schejter, E.D. The actin nucleator WASp is required for myoblast fusion during adult Drosophila myogenesis. *Dev. Camb. Engl.* **2011**, *138*, 2347–2357. [CrossRef] [PubMed]

227. Ghazi, A.; Paul, L.; VijayRaghavan, K. Prepattern genes and signaling molecules regulate stripe expression to specify Drosophila flight muscle attachment sites. *Mech. Dev.* **2003**, *120*, 519–528. [CrossRef]

228. Laddada, L.; Jagla, K.; Soler, C. Odd-skipped and Stripe act downstream of Notch to promote the morphogenesis of long appendicular tendons in Drosophila. *Biol. Open* **2019**, *8*, bio038760. [CrossRef] [PubMed]

229. Weitkunat, M.; Brasse, M.; Bausch, A.R.; Schnorrer, F. Mechanical tension and spontaneous muscle twitching precede the formation of cross-striated muscle in vivo. *Dev. Camb. Engl.* **2017**, *144*, 1261–1272. [CrossRef] [PubMed]

230. Spletter, M.L.; Schnorrer, F. Transcriptional regulation and alternative splicing cooperate in muscle fiber-type specification in flies and mammals. *Dev. Biol.* **2014**, *321*, 90–98. [CrossRef]

231. Reedy, M.C.; Beall, C. Ultrastructure of Developing Flight Muscle in Drosophila. I. Assembly of Myofibrils. *Dev. Biol.* **1993**, *160*, 443–465. [CrossRef]

232. Spletter, M.L.; Barz, C.; Yeroslaviz, A.; Zhang, X.; Lemke, S.B.; Bonnard, A.; Brunner, E.; Cardone, G.; Basler, K.; Habermann, B.H.; et al. A transcriptomics resource reveals a transcriptional transition during ordered sarcomere morphogenesis in flight muscle. *eLife* **2018**, *7*, e34058. [CrossRef]

233. Shwartz, A.; Dhanyasi, N.; Schejter, E.D.; Shilo, B.-Z. The Drosophila formin Fhos is a primary mediator of sarcomeric thin-filament array assembly. *eLife* **2016**, *5*, e16540. [CrossRef] [PubMed]

234. Fernandes, I.; Schöck, F. The nebulin repeat protein Lasp regulates I-band architecture and filament spacing in myofibrils. *J. Cell Biol.* **2014**, *206*, 559–572. [CrossRef]

235. Qiu, F.; Brendel, S.; Cunha, P.M.F.; Astola, N.; Song, B.; Furlong, E.E.M.; Leonard, K.R.; Bullard, B. Myofilin, a protein in the thick filaments of insect muscle. *J. Cell Sci.* **2005**, *118*, 1527. [CrossRef] [PubMed]

236. Ball, E.; Karlik, C.C.; Beall, C.J.; Saville, D.L.; Sparrow, J.C.; Bullard, B.; Fyrberg, E.A. Arthrin, a myofibrillar protein of insect flight muscle, is an actin-ubiquitin conjugate. *Cell* **1987**, *51*, 221–228. [CrossRef]

237. Becker, K.D.; O'Donnell, P.T.; Heitz, J.M.; Vito, M.; Bernstein, S.I. Analysis of Drosophila paramyosin: Identification of a novel isoform which is restricted to a subset of adult muscles. *J. Cell Biol.* **1992**, *116*, 669–681. [CrossRef] [PubMed]

238. Maroto, M.; Arredondo, J.J.; Román, M.S.; Marco, R.; Cervera, M. Analysis of the Paramyosin/Miniparamyosin Gene: Miniparamyosin is an independently transcribed, distinct paramyosin isoform, widely distributed in invertebrates. *J. Biol. Chem.* **1995**, *270*, 4375–4382. [CrossRef]

239. Vigoreaux, J.O. Alterations in flightin phosphorylation inDrosophila flight muscles are associated with myofibrillar defects engendered by actin and myosin heavy-chain mutant alleles. *Biochem. Genet.* **1994**, *32*, 301–314. [CrossRef]

240. Reedy, M.C.; Bullard, B.; Vigoreaux, J.O. Flightin is essential for thick filament assembly and sarcomere stability in Drosophila flight muscles. *J. Cell Biol.* **2000**, *151*, 1483–1500. [CrossRef]

241. Craig, R.; Woodhead, J.L. Structure and function of myosin filaments. *Curr. Opin. Struct. Biol.* **2006**, *16*, 204–212. [CrossRef]

242. Orfanos, Z.; Sparrow, J.C. Myosin isoform switching during assembly of the Drosophila flight muscle thick filament lattice. *J. Cell Sci.* **2013**, *126*, 139. [CrossRef]

243. Contompasis, J.L.; Nyland, L.R.; Maughan, D.W.; Vigoreaux, J.O. Flightin Is Necessary for Length Determination, Structural Integrity, and Large Bending Stiffness of Insect Flight Muscle Thick Filaments. *J. Mol. Biol.* **2010**, *395*, 340–348. [CrossRef]

244. Sparrow, J.; Reedy, M.; Ball, E.; Kyrtatas, V.; Molloy, J.; Durston, J.; Hennessey, E.; White, D. Functional and ultrastructural effects of a missense mutation in the indirect flight muscle-specific actin gene of Drosophila melanogaster. *J. Mol. Biol.* **1991**, *222*, 963–982. [CrossRef]

245. Burkart, C.; Qiu, F.; Brendel, S.; Benes, V.; Hååg, P.; Labeit, S.; Leonard, K.; Bullard, B. Modular Proteins from the Drosophila sallimus (sls) Gene and their Expression in Muscles with Different Extensibility. *J. Mol. Biol.* **2007**, *367*, 953–969. [CrossRef]

246. González-Morales, N.; Xiao, Y.S.; Schilling, M.A.; Marescal, O.; Liao, K.A.; Schöck, F. Myofibril diameter is set by a finely tuned mechanism of protein oligomerization in Drosophila. *eLife* **2019**, *8*, e50496. [CrossRef]

247. Truman, J.W.; Schuppe, H.; Shepherd, D.; Williams, D.W. Developmental architecture of adult-specific lineages in the ventral CNS of Drosophila. *Development* **2004**, *131*, 5167. [CrossRef]

248. Brierley, D.J.; Blanc, E.; Reddy, O.V.; Vijayraghavan, K.; Williams, D.W. Dendritic targeting in the leg neuropil of Drosophila: The role of midline signalling molecules in generating a myotopic map. *PLoS Biol.* **2009**, *7*, e1000199. [CrossRef]

249. Enriquez, J.; Venkatasubramanian, L.; Baek, M.; Peterson, M.; Aghayeva, U.; Mann, R.S. Specification of Individual Adult Motor Neuron Morphologies by Combinatorial Transcription Factor Codes. *Neuron* **2015**, *86*, 955–970. [CrossRef]

250. Baek, M.; Mann, R.S. Lineage and Birth Date Specify Motor Neuron Targeting and Dendritic Architecture in Adult Drosophila. *J. Neurosci.* **2009**, *29*, 6904. [CrossRef]

251. Fernandes, J.J.; Keshishian, H. Nerve-muscle interactions during flight muscle development in Drosophila. *Development* **1998**, *125*, 1769.

252. Fernandes, J.; VijayRaghavan, K. The development of indirect flight muscle innervation in Drosophila melanogaster. *Development* **1993**, *118*, 215.

253. Rival, T.; Soustelle, L.; Cattaert, D.; Strambi, C.; Iché, M.; Birman, S. Physiological requirement for the glutamate transporter dEAAT1 at the adult Drosophila neuromuscular junction. *J. Neurobiol.* **2006**, *66*, 1061–1074. [CrossRef]

254. Kimura, K.I.; Truman, J.W. Postmetamorphic cell death in the nervous and muscular systems of Drosophila melanogaster. *J. Neurosci. Off. J. Soc. Neurosci.* **1990**, *10*, 403–411. [CrossRef]

255. Soler, C.; Taylor, M.V. The Him gene inhibits the development of Drosophila flight muscles during metamorphosis. *Mech. Dev.* **2009**, *126*, 595–603. [CrossRef] [PubMed]

256. Soler, C.; Han, J.; Taylor, M.V. The conserved transcription factor Mef2 has multiple roles in adult Drosophila musculature formation. *Development* **2012**, *139*, 1270. [CrossRef]

257. Zhang, S.; Bernstein, S.I. Spatially and temporally regulated expression of myosin heavy chain alternative exons during Drosophila embryogenesis. *Mech. Dev.* **2001**, *101*, 35–45. [CrossRef]

258. Swank, D.M.; Bartoo, M.L.; Knowles, A.F.; Iliffe, C.; Bernstein, S.I.; Molloy, J.E.; Sparrow, J.C. Alternative exon-encoded regions of Drosophila myosin heavy chain modulate ATPase rates and actin sliding velocity. *J. Biol. Chem.* **2001**, *276*, 15117–15124. [CrossRef]

259. Karlik, C.C.; Fyrberg, E.A. Two Drosophila melanogaster tropomyosin genes: Structural and functional aspects. *Mol. Cell. Biol.* **1986**, *6*, 1965. [CrossRef]

260. Mateos, J.; Herranz, R.; Domingo, A.; Sparrow, J.; Marco, R. The structural role of high molecular weight tropomyosins in dipteran indirect flight muscle and the effect of phosphorylation. *J. Muscle Res. Cell Motil.* **2006**, *27*, 189–201. [CrossRef]

261. Katzemich, A.; Long, J.Y.; Jani, K.; Lee, B.R.; Schöck, F. Muscle type-specific expression of Zasp52 isoforms in Drosophila. *Gene Expr. Patterns* **2011**, *11*, 484–490. [CrossRef]

262. Katzemich, A.; Kreisköther, N.; Alexandrovich, A.; Elliott, C.; Schöck, F.; Leonard, K.; Sparrow, J.; Bullard, B. The function of the M-line protein obscurin in controlling the symmetry of the sarcomere in the flight muscle of Drosophila. *J. Cell Sci.* **2012**, *125*, 3367. [CrossRef]

263. Marín, M.-C.; Rodríguez, J.-R.; Ferrús, A. Transcription of Drosophila Troponin I Gene Is Regulated by Two Conserved, Functionally Identical, Synergistic Elements. *Mol. Biol. Cell* **2004**, *15*, 1185–1196. [CrossRef] [PubMed]

264. Mas, J.-A.; García-Zaragoza, E.; Cervera, M. Two Functionally Identical Modular Enhancers in Drosophila Troponin T Gene Establish the Correct Protein Levels in Different Muscle Types. *Mol. Biol. Cell* **2004**, *15*, 1931–1945. [CrossRef] [PubMed]

265. Gremke, L.; Lord, P.C.W.; Sabacan, L.; Lin, S.-C.; Wohlwill, A.; Storti, R.V. Coordinate Regulation of Drosophila Tropomyosin Gene Expression Is Controlled by Multiple Muscle-Type-Specific Positive and Negative Enhancer Elements. *Dev. Biol.* **1993**, *159*, 513–527. [CrossRef] [PubMed]

266. Lin, M.H.; Nguyen, H.T.; Dybala, C.; Storti, R.V. Myocyte-specific enhancer factor 2 acts cooperatively with a muscle activator region to regulate Drosophila tropomyosin gene muscle expression. *Proc. Natl. Acad. Sci. USA* **1996**, *93*, 4623. [CrossRef] [PubMed]

267. Reddy, K.L.; Wohlwill, A.; Dzitoeva, S.; Lin, M.-H.; Holbrook, S.; Storti, R.V. The Drosophila PAR Domain Protein 1 (Pdp1) Gene Encodes Multiple Differentially Expressed mRNAs and Proteins through the Use of Multiple Enhancers and Promoters. *Dev. Biol.* **2000**, *224*, 401–414. [CrossRef]

268. Vigoreaux, J.O.; Saide, J.D.; Pardue, M.L. Structurally different Drosophila striated muscles utilize distinct variants of Z-band-associated proteins. *J. Muscle Res. Cell Motil.* **1991**, *12*, 340–354. [CrossRef]

269. Zhao, C.; Swank, D.M. The Drosophila indirect flight muscle myosin heavy chain isoform is insufficient to transform the jump muscle into a highly stretch-activated muscle type. *Am. J. Physiol. Cell Physiol.* **2017**, *312*, C111–C118. [CrossRef]

270. Vigoreaux, J.O.; Perry, L.M. Multiple isoelectric variants of flightin in Drosophila stretch-activated muscles are generated by temporally regulated phosphorylations. *J. Muscle Res. Cell Motil.* **1994**, *15*, 607–616. [CrossRef]

271. Daley, J.; Southgate, R.; Ayme-Southgate, A. Structure of the Drosophila projectin protein: Isoforms and implication for projectin filament assembly11Edited by M. F. Moody. *J. Mol. Biol.* **1998**, *279*, 201–210. [CrossRef]

272. Glasheen, B.M.; Ramanath, S.; Patel, M.; Sheppard, D.; Puthawala, J.T.; Riley, L.A.; Swank, D.M. Five Alternative Myosin Converter Domains Influence Muscle Power, Stretch Activation, and Kinetics. *Biophys. J.* **2018**, *114*, 1142–1152. [CrossRef]

273. Belozerov, V.E.; Ratkovic, S.; McNeill, H.; Hilliker, A.J.; McDermott, J.C. In vivo interaction proteomics reveal a novel p38 mitogen-activated protein kinase/Rack1 pathway regulating proteostasis in Drosophila muscle. *Mol. Cell. Biol.* **2014**, *34*, 474–484. [CrossRef] [PubMed]

274. Haas, K.F.; Woodruff, E., 3rd; Broadie, K. Proteasome function is required to maintain muscle cellular architecture. *Biol. Cell* **2007**, *99*, 615–626. [CrossRef] [PubMed]

275. Nguyen, H.T.; Voza, F.; Ezzeddine, N.; Frasch, M. Drosophila mind bomb2 is required for maintaining muscle integrity and survival. *J. Cell Biol.* **2007**, *179*, 219–227. [CrossRef] [PubMed]

276. Valdez, C.; Scroggs, R.; Chassen, R.; Reiter, L.T. Variation in Dube3a expression affects neurotransmission at the Drosophila neuromuscular junction. *Biol. Open* **2015**, *4*, 776–782. [CrossRef] [PubMed]

277. Marco-Ferreres, R.; Arredondo, J.J.; Fraile, B.; Cervera, M. Overexpression of troponin T in Drosophila muscles causes a decrease in the levels of thin-filament proteins. *Biochem. J.* **2005**, *386*, 145–152. [CrossRef] [PubMed]

278. Firdaus, H.; Mohan, J.; Naz, S.; Arathi, P.; Ramesh, S.R.; Nongthomba, U. A cis-regulatory mutation in troponin-I of Drosophila reveals the importance of proper stoichiometry of structural proteins during muscle assembly. *Genetics* **2015**, *200*, 149–165. [CrossRef] [PubMed]

279. Wójtowicz, I.; Jabłońska, J.; Zmojdzian, M.; Taghli-Lamallem, O.; Renaud, Y.; Junion, G.; Daczewska, M.; Huelsmann, S.; Jagla, K.; Jagla, T. Drosophila small heat shock protein CryAB ensures structural integrity of developing muscles, and proper muscle and heart performance. *Development* **2015**, *142*, 994. [CrossRef]

280. Jabłońska, J.; Dubińska-Magiera, M.; Jagla, T.; Jagla, K.; Daczewska, M. Drosophila Hsp67Bc hot-spot variants alter muscle structure and function. *Cell. Mol. Life Sci. CMLS* **2018**, *75*, 4341–4356. [CrossRef]

281. Wang, S.; Reuveny, A.; Volk, T. Nesprin provides elastic properties to muscle nuclei by cooperating with spectraplakin and EB1. *J. Cell Biol.* **2015**, *209*, 529–538. [CrossRef]

282. Wang, S.; Volk, T. Composite biopolymer scaffolds shape muscle nucleus: Insights and perspectives from Drosophila. *Bioarchitecture* **2015**, *5*, 35–43. [CrossRef]

283. Lorber, D.; Rotkopf, R.; Volk, T. In vivo imaging of myonuclei during spontaneous muscle contraction reveals non-uniform nuclear mechanical dynamics in Nesprin/klar mutants. *bioRxiv* **2019**, 643015. [CrossRef]

284. Elhanany-Tamir, H.; Yu, Y.V.; Shnayder, M.; Jain, A.; Welte, M.; Volk, T. Organelle positioning in muscles requires cooperation between two KASH proteins and microtubules. *J. Cell Biol.* **2012**, *198*, 833–846. [CrossRef] [PubMed]

285. Morel, V.; Lepicard, S.N.; Rey, A.; Parmentier, M.-L.; Schaeffer, L. Drosophila Nesprin-1 controls glutamate receptor density at neuromuscular junctions. *Cell. Mol. Life Sci.* **2014**, *71*, 3363–3379. [CrossRef] [PubMed]

286. Goel, P.; Dufour Bergeron, D.; Böhme, M.A.; Nunnelly, L.; Lehmann, M.; Buser, C.; Walter, A.M.; Sigrist, S.J.; Dickman, D. Homeostatic scaling of active zone scaffolds maintains global synaptic strength. *J. Cell Biol.* **2019**, *218*, 1706–1724. [CrossRef] [PubMed]

287. Goel, P.; Dickman, D. Distinct homeostatic modulations stabilize reduced postsynaptic receptivity in response to presynaptic DLK signaling. *Nat. Commun.* **2018**, *9*, 1856. [CrossRef]

288. Ziegler, A.B.; Augustin, H.; Clark, N.L.; Berthelot-Grosjean, M.; Simonnet, M.M.; Steinert, J.R.; Geillon, F.; Manière, G.; Featherstone, D.E.; Grosjean, Y. The Amino Acid Transporter JhI-21 Coevolves with Glutamate Receptors, Impacts NMJ Physiology, and Influences Locomotor Activity in Drosophila Larvae. *Sci. Rep.* **2016**, *6*, 19692. [CrossRef]

289. Hong, H.; Zhao, K.; Huang, S.; Huang, S.; Yao, A.; Jiang, Y.; Sigrist, S.; Zhao, L.; Zhang, Y.Q. Structural Remodeling of Active Zones Is Associated with Synaptic Homeostasis. *J. Neurosci.* **2020**, *40*, 2817. [CrossRef]
290. James, T.D.; Zwiefelhofer, D.J.; Frank, C.A. Maintenance of homeostatic plasticity at the Drosophila neuromuscular synapse requires continuous IP3-directed signaling. *eLife* **2019**, *8*, e39643. [CrossRef]
291. Mourikis, P.; Gopalakrishnan, S.; Sambasivan, R.; Tajbakhsh, S. Cell-autonomous Notch activity maintains the temporal specification potential of skeletal muscle stem cells. *Dev. Camb. Engl.* **2012**, *139*, 4536–4548. [CrossRef]
292. Boukhatmi, H.; Bray, S. A population of adult satellite-like cells in Drosophila is maintained through a switch in RNA-isoforms. *eLife* **2018**, *7*, e35954. [CrossRef]
293. Siles, L.; Ninfali, C.; Cortés, M.; Darling, D.S.; Postigo, A. ZEB1 protects skeletal muscle from damage and is required for its regeneration. *Nat. Commun.* **2019**, *10*, 1364. [CrossRef] [PubMed]
294. Kuang, S.; Kuroda, K.; Le Grand, F.; Rudnicki, M.A. Asymmetric self-renewal and commitment of satellite stem cells in muscle. *Cell* **2007**, *129*, 999–1010. [CrossRef] [PubMed]
295. de Haro, M.; Al-Ramahi, I.; De Gouyon, B.; Ukani, L.; Rosa, A.; Faustino, N.A.; Ashizawa, T.; Cooper, T.A.; Botas, J. MBNL1 and CUGBP1 modify expanded CUG-induced toxicity in a Drosophila model of myotonic dystrophy type 1. *Hum. Mol. Genet.* **2006**, *15*, 2138–2145. [CrossRef] [PubMed]
296. Picchio, L.; Plantie, E.; Renaud, Y.; Poovthumkadavil, P.; Jagla, K. Novel Drosophila model of myotonic dystrophy type 1: Phenotypic characterization and genome-wide view of altered gene expression. *Hum. Mol. Genet.* **2013**, *22*, 2795–2810. [CrossRef]
297. Yatsenko, A.S.; Shcherbata, H.R. Drosophila miR-9a Targets the ECM Receptor Dystroglycan to Canalize Myotendinous Junction Formation. *Dev. Cell* **2014**, *28*, 335–348. [CrossRef]
298. Kucherenko, M.M.; Marrone, A.K.; Rishko, V.M.; Magliarelli, H.d.F.; Shcherbata, H.R. Stress and muscular dystrophy: A genetic screen for Dystroglycan and Dystrophin interactors in Drosophila identifies cellular stress response components. *Dev. Biol.* **2011**, *352*, 228–242. [CrossRef]
299. Dialynas, G.; Shrestha, O.K.; Ponce, J.M.; Zwerger, M.; Thiemann, D.A.; Young, G.H.; Moore, S.A.; Yu, L.; Lammerding, J.; Wallrath, L.L. Myopathic lamin mutations cause reductive stress and activate the nrf2/keap-1 pathway. *PLoS Genet.* **2015**, *11*, e1005231. [CrossRef]
300. Chandran, S.; Suggs, J.A.; Wang, B.J.; Han, A.; Bhide, S.; Cryderman, D.E.; Moore, S.A.; Bernstein, S.I.; Wallrath, L.L.; Melkani, G.C. Suppression of myopathic lamin mutations by muscle-specific activation of AMPK and modulation of downstream signaling. *Hum. Mol. Genet.* **2019**, *28*, 351–371. [CrossRef]
301. Jungbluth, H.; Gautel, M. Pathogenic mechanisms in centronuclear myopathies. *Front. Aging Neurosci.* **2014**, *6*, 339. [CrossRef]
302. Schulman, V.K.; Folker, E.S.; Rosen, J.N.; Baylies, M.K. Syd/JIP3 and JNK signaling are required for myonuclear positioning and muscle function. *PLoS Genet.* **2014**, *10*, e1004880. [CrossRef]
303. Rosen, J.N.; Azevedo, M.; Soffar, D.B.; Boyko, V.P.; Brendel, M.B.; Schulman, V.K.; Baylies, M.K. The Drosophila Ninein homologue Bsg25D cooperates with Ensconsin in myonuclear positioning. *J. Cell Biol.* **2019**, *218*, 524–540. [CrossRef] [PubMed]
304. Naimi, B.; Harrison, A.; Cummins, M.; Nongthomba, U.; Clark, S.; Canal, I.; Ferrus, A.; Sparrow, J.C. A Tropomyosin-2 Mutation Suppresses a Troponin I Myopathy inDrosophila. *Mol. Biol. Cell* **2001**, *12*, 1529–1539. [CrossRef] [PubMed]
305. Dahl-Halvarsson, M.; Olive, M.; Pokrzywa, M.; Ejeskär, K.; Palmer, R.H.; Uv, A.E.; Tajsharghi, H. Drosophila model of myosin myopathy rescued by overexpression of a TRIM-protein family member. *Proc. Natl. Acad. Sci. USA* **2018**, *115*, E6566–E6575. [CrossRef] [PubMed]
306. Ugur, B.; Chen, K.; Bellen, H.J. Drosophila tools and assays for the study of human diseases. *Dis. Model. Mech.* **2016**, *9*, 235–244. [CrossRef] [PubMed]
307. Day, K.; Shefer, G.; Shearer, A.; Yablonka-Reuveni, Z. The depletion of skeletal muscle satellite cells with age is concomitant with reduced capacity of single progenitors to produce reserve progeny. *Dev. Biol.* **2010**, *340*, 330–343. [CrossRef] [PubMed]
308. Jiang, C.; Wen, Y.; Kuroda, K.; Hannon, K.; Rudnicki, M.A.; Kuang, S. Notch signaling deficiency underlies age-dependent depletion of satellite cells in muscular dystrophy. *Dis. Models Mech.* **2014**, *7*, 997. [CrossRef]

309. Letsou, A.; Bohmann, D. Small flies—Big discoveries: Nearly a century of Drosophila genetics and development. *Dev. Dyn.* **2005**, *232*, 526–528. [CrossRef]
310. Bellen, H.J.; Tong, C.; Tsuda, H. 100 years of Drosophila research and its impact on vertebrate neuroscience: A history lesson for the future. *Nat. Rev. Neurosci.* **2010**, *11*, 514–522. [CrossRef]

Review

Genetic Associations with Aging Muscle: A Systematic Review

Jedd Pratt [1,2,*], Colin Boreham [1], Sean Ennis [2,3], Anthony W. Ryan [2] and Giuseppe De Vito [1,4]

1   Institute for Sport and Health, University College Dublin, Dublin, Ireland;
    colin.boreham@ucd.ie (C.B.); giuseppe.devito@ucd.ie (G.D.V.)
2   Genomics Medicine Ireland, Dublin, Ireland; sean.ennis@ucd.ie (S.E.);
    anthony.ryan@genomicsmed.ie (A.W.R.)
3   UCD ACoRD, Academic Centre on Rare Diseases, University College Dublin, Dublin, Ireland
4   Department of Biomedical Sciences, University of Padova, Via F. Marzolo 3, 35131 Padova, Italy
*   Correspondence: jedd.pratt@ucdconnect.ie

Received: 22 November 2019; Accepted: 18 December 2019; Published: 19 December 2019

Abstract: The age-related decline in skeletal muscle mass, strength and function known as 'sarcopenia' is associated with multiple adverse health outcomes, including cardiovascular disease, stroke, functional disability and mortality. While skeletal muscle properties are known to be highly heritable, evidence regarding the specific genes underpinning this heritability is currently inconclusive. This review aimed to identify genetic variants known to be associated with muscle phenotypes relevant to sarcopenia. PubMed, Embase and Web of Science were systematically searched (from January 2004 to March 2019) using pre-defined search terms such as "aging", "sarcopenia", "skeletal muscle", "muscle strength" and "genetic association". Candidate gene association studies and genome wide association studies that examined the genetic association with muscle phenotypes in non-institutionalised adults aged ≥50 years were included. Fifty-four studies were included in the final analysis. Twenty-six genes and 88 DNA polymorphisms were analysed across the 54 studies. The *ACTN3*, *ACE* and *VDR* genes were the most frequently studied, although the *IGF1/IGFBP3*, *TNFα*, *APOE*, *CNTF/R* and *UCP2/3* genes were also shown to be significantly associated with muscle phenotypes in two or more studies. Ten DNA polymorphisms (rs154410, rs2228570, rs1800169, rs3093059, rs1800629, rs1815739, rs1799752, rs7412, rs429358 and 192 bp allele) were significantly associated with muscle phenotypes in two or more studies. Through the identification of key gene variants, this review furthers the elucidation of genetic associations with muscle phenotypes associated with sarcopenia.

Keywords: genotype; genetic variation; muscle phenotypes; sarcopenia; aging

---

1. Introduction

Sarcopenia refers to the progressive deterioration in skeletal muscle mass, strength and physical function with advancing age [1]. The simultaneous presence of low muscle strength, muscle mass and/or physical function forms the diagnostic basis of the recommendations from the European Working Group on Sarcopenia in Older People [2]. These criteria are strong predictors of a multitude of adverse health outcomes, such as cardiovascular disease [3], functional disability [4], fall incidence [5], hospitalisation [6], stroke [7] and mortality [8]. Up to 10% of individuals aged 60–69 years are affected by sarcopenia, with this proportion rising considerably to 40% for adults over 80 years of age [9,10]. The fundamental loss of independence and susceptibility to additional diseases caused by sarcopenia also places a significant burden on public health systems worldwide. This burden is anticipated to grow considerably in coming decades, in line with increases in longevity and the consequent rise in the proportion of elderly [11]. Thus, the consequences of age-related muscle deterioration will become increasingly relevant globally.

While sarcopenia is generally more prevalent among individuals over the age of 60, strong evidence suggests that pronounced changes in muscle tissue begin from around 50 years of age [12]. From this age, muscle mass and strength begin to deteriorate at an annual rate of 1–2% and 1.5–5% respectively [12–14]. Developing an understanding of why and how skeletal muscle deteriorates from this age will be critical to reducing the burden of sarcopenia for patients as well as public health systems.

Currently, it is known that inter-individual variation in muscle phenotypes may be attributed to genetic factors, environmental factors and/or, gene-environment interactions [15,16]. While environmental factors such as physical activity, protein intake [17], sleep quality [18], smoking status [15] and alcohol consumption [19] have been shown to affect muscle phenotypes, heritability studies have highlighted the importance of genetic factors in determining inter-individual variability in skeletal muscle traits [20,21]. These studies have found that genetic factors account for 46–76% and 32–67% of fat-free mass (FFM) and muscle strength variability, respectively [20,21]. Additional longitudinal studies have observed heritability estimates of 64% for change in muscle strength with advancing age [22]. However, while the overall heritability of skeletal muscle phenotypes is well established, the genetic mechanisms underpinning this heritability remain unclear.

Thus, developing a deeper understanding of genetic associations underpinning skeletal muscle phenotypes is of paramount importance in the development of effective treatment interventions to manage age-related changes in muscle structure and function. Furthermore, understanding the genetic mechanisms regulating muscle accrual and loss will help facilitate early screening for susceptibility to sarcopenia, which could allow for preventative measures to be implemented prior to predicted muscle degradation.

Therefore, the purpose of this systematic review was to identify and synthesize the genetic variants associated with muscle phenotypes relevant to sarcopenia in humans.

2. Materials and Methods

Reporting followed the Preferred Reporting Items for Systematic Reviews and Meta-Analyses (PRISMA) statement [23].

2.1. Literature Search and Eligibility Criteria

2.1.1. Inclusion and Exclusion Criteria

To be included in this review, studies had to meet the following criteria:

1. Published between January 2004 and March 2019.
2. Full English text available.
3. Participants must be non-institutionalised human adults, aged 50 years or above.
4. Subjects must have been free from any significant cardiovascular, metabolic or musculoskeletal disorders at the time of the study.
5. Candidate gene association study or genome wide association study (GWAS).

2.1.2. Search Strategy

A systematic literature search of three online databases, PubMed, EMBASE and Web of Science, was conducted on 18 March 2019, for the period between January 2004 and March 2019. This time limit ensured the inclusion of the most pertinent literature. Search terms were selected based off the PEO framework and combined using Boolean operators ("AND", "OR"). Filters were used to limit results to those using human subjects, written in the English language and published within the desired time-frame. The search strategy used was as follows: ("ageing" OR "aged" OR "elderly" OR "older persons" OR "community dwelling") AND ("sarcopenia" OR "skeletal muscle" OR "muscle phenotype" OR "muscle mass" OR "muscle atrophy" OR "muscle strength" OR "grip strength" OR "physical performance" OR "muscle quality" OR "lean mass") AND ("single nucleotide polymorphism" OR "genetic polymorphism" OR "allele" OR "genetic variation" OR "gene variant" OR "mutation" OR "genes" OR "chromosome" OR "genetic predisposition" OR "genetic susceptibility") AND ("genetic association studies" OR "genome-wide association study" OR "GWAS" OR "candidate gene study" OR "genotype" OR "haplotype" OR "heritability"). The scope of the online search was further expanded by assessing bibliographic references of the eligible full text articles for relevant studies.

2.2. Study Selection and Data Extraction

Following the removal of duplicates, titles and abstracts were screened for relevance to the scope of this review. To determine inclusion in this review, the full text of every potentially relevant article was scrutinised for overall content and compliance with the eligibility criteria outlined above. The following data were extracted from each eligible article: authors, year of publication, study design, studied population (number, ethnicity, nationality, sex), gene name, polymorphism, muscle phenotype, main findings of the study.

2.3. Phenotypes

Phenotypic outcomes included in this systematic review were skeletal muscle mass, muscle strength, physical function and sarcopenia prevalence.

2.4. Quality Assessment

The quality and risk of bias of the included studies were assessed using the Quality of Genetic Association Studies (Q-Genie) tool [24]. The Q-Genie tool consists of 11 items that cover the following areas: "rationale for study", "selection and definition of outcome of interest", "selection and comparability of comparison groups", "technical classification of the exposure", "non-technical classification of the exposure", "other source of bias", "sample size and power", "a priori planning of analysis", "statistical methods and control for confounding", "testing of assumptions and inferences for genetic analysis" and "appropriateness of inferences drawn from results". Each area was rated using a 7-point Likert scale ("1 = poor"; "2", "3 = good"; "4", "5 = very good"; "6", "7 = excellent"). The overall quality of the included articles was classified by collating the scores for each theme. Studies with control groups were classified as "poor quality" if the score was ≤ 35, "moderate quality" if the score was > 35 and ≤ 45, and "good quality" if the score was > 45. For studies without control groups, scoring ≤ 32, > 32 and ≤ 40, and > 40 reflected classifications of "poor quality", "moderate quality" and "good quality", respectively.

3. Results

3.1. Search Strategy

The systematic search of the online databases identified 771 papers. Following the addition of filters, removal of duplicates and screening for eligibility, 48 studies remained. Six additional articles were retrieved through the manual search of reference lists, leaving a total of 54 articles to be included in this systematic review. Figure 1 highlights the identification and selection process in accordance with the PRISMA statement.

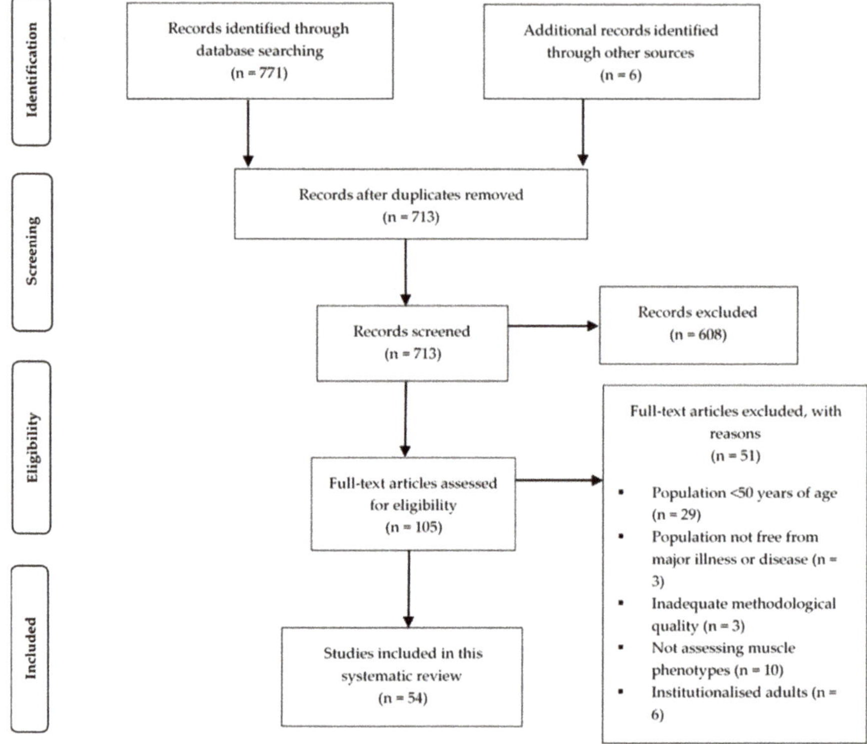

Figure 1. PRISMA flow chart presenting the identification and selection process of articles.

3.2. Quality Assessment

A detailed quality classification for each article is displayed in Table 1. Studies scored between 33 and 50 in the Q-Genie checklist. For studies with control groups ($n = 12$), five were classified as "moderate quality" and seven as "good quality". For non-control group studies ($n = 42$), 17 were classified as "moderate quality" and 25 as "good quality".

Table 1. Q-Genie quality assessment scores for the included studies.

Studies	Items											Total
	1	2	3	4	5	6	7	8	9	10	11	
Arkin, et al., 2006. [25]	4	4	N/A	4	4	3	3	5	4	5	4	40
Bahat, et al., 2010. [26]	4	4	3	4	3	3	3	4	5	4	4	41
Barr, et al., 2010. [27]	5	3	N/A	4	5	2	4	5	5	4	5	42
Bjork, et al., 2019. [28]	4	4	N/A	5	4	3	5	5	5	4	4	43
Buford, et al., 2014. [29]	5	4	3	4	5	3	3	4	6	4	6	47
Bustamante-Ara, et al., 2010. [30]	6	5	N/A	6	5	4	2	4	5	3	5	45
Charbonneau, et al., 2008. [31]	5	5	N/A	6	5	4	2	5	5	3	4	44
Cho, et al., 2017. [32]	4	4	N/A	5	4	3	3	3	4	3	4	37
Crocco, et al., 2011. [33]	5	5	N/A	5	4	3	5	5	5	4	3	44
Da Silva, et al., 2018. [34]	5	4	3	5	4	3	2	4	4	3	5	42
Dato, et al., 2012. [35]	6	5	N/A	5	4	3	4	5	4	4	5	45
De Mars, et al., 2007. [36]	6	5	N/A	5	4	3	3	5	5	4	5	45
Delmonico, et al., 2007. [37]	4	4	N/A	2	3	3	2	4	4	3	4	33
Delmonico, et al., 2008. [38]	5	3	N/A	3	4	3	5	4	5	3	4	39
Garatachea, et al., 2012. [39]	6	5	N/A	7	6	4	2	3	5	3	5	46
Giaccaglia, et al., 2008. [40]	5	5	5	5	4	5	3	4	5	5	4	50

Table 1. *Cont.*

Studies	Items											Total
	1	2	3	4	5	6	7	8	9	10	11	
Gonzalez-Freire, et al., 2010. [41]	5	5	N/A	5	4	3	2	3	4	4	5	40
Gussago, et al., 2016. [42]	5	6	4	5	4	3	3	5	4	4	5	48
Hand, et al., 2007. [43]	5	4	N/A	6	5	3	3	4	6	3	5	44
Heckerman, et al., 2017. [44]	5	4	N/A	5	5	4	2	4	5	5	4	43
Hopkinson, et al., 2008. [45]	4	5	3	4	4	3	3	3	4	4	4	41
Judson, et al., 2010. [46]	5	4	N/A	5	5	2	4	3	4	4	4	40
Keogh, et al., 2015. [47]	5	4	N/A	5	4	3	2	4	3	4	5	39
Klimentidis, et al., 2016. [48]	4	4	N/A	2	4	4	5	5	5	4	4	41
Kikuchi, et al., 2015. [49]	5	4	3	4	4	4	5	5	5	4	5	48
Kostek, et al., 2005. [50]	5	5	N/A	5	5	3	3	4	5	4	5	44
Kostek, et al., 2010. [51]	5	3	3	5	4	2	3	5	5	4	5	44
Kritchevsky, et al., 2005. [52]	5	3	N/A	5	4	4	5	4	5	4	5	44
Li, et al., 2016. [53]	5	6	N/A	4	4	3	5	4	5	3	4	43
Lima, et al., 2011. [54]	5	3	4	5	3	4	3	4	5	4	5	45
Lin, et al., 2014. [55]	5	4	3	5	3	4	4	5	5	4	5	47
Lin, et al., 2014. [56]	5	5	4	5	4	3	3	5	5	5	4	48
Lunardi, et al., 2013. [57]	4	6	N/A	4	3	3	4	5	4	3	5	41
Ma, et al., 2018. [58]	6	5	N/A	5	5	4	5	4	5	3	4	46
McCauley, et al., 2010. [59]	5	5	N/A	5	3	4	3	4	3	3	5	40
Melzer, et al., 2005. [60]	5	5	N/A	4	2	3	4	4	4	4	5	40
Mora, et al., 2011. [61]	4	3	N/A	4	4	4	3	4	5	4	5	40
Onder, et al., 2008. [62]	5	4	N/A	4	4	3	2	5	3	4	4	38
Pereira, et al., 2013. [63]	5	3	N/A	4	4	4	4	3	4	3	5	39
Pereira, et al., 2013. [64]	6	4	N/A	6	4	4	4	5	5	4	5	47
Pereira, et al., 2013. [65]	6	5	N/A	6	5	3	3	5	5	4	5	47
Prakash, et al., 2019. [66]	4	4	3	5	5	3	4	5	4	5	4	46
Roth, et al., 2004. [67]	5	5	N/A	4	3	3	5	4	5	4	5	43
Skoog, et al., 2016. [68]	6	5	N/A	4	3	3	3	4	4	3	5	40
Tiainen, et al., 2012. [69]	6	5	N/A	4	3	4	2	3	3	3	4	37
Urano, et al., 2014. [70]	5	4	N/A	5	3	4	4	4	5	4	5	43
Verghese, et al., 2013. [71]	5	4	N/A	4	3	3	5	4	4	4	5	41
Walsh, et al., 2005. [72]	5	5	N/A	5	4	4	2	4	5	4	4	42
Wu, et al., 2014. [73]	5	5	N/A	4	4	3	4	4	5	3	4	41
Xia, et al., 2019. [74]	5	5	N/A	4	3	4	4	4	5	3	6	43
Yang, et al., 2015. [75]	5	4	N/A	4	3	4	3	4	5	4	5	41
Yoshihara, et al., 2009. [76]	3	4	N/A	3	3	3	4	3	4	3	3	33
Zempo, et al., 2010. [77]	4	4	N/A	3	3	4	3	4	5	4	4	38
Zempo, et al., 2011. [78]	5	4	N/A	3	3	3	4	4	5	4	4	39

Items: 1: Rationale for study, 2: Selection and definition of outcome of interest, 3: Selection and comparability of comparison groups, 4: Technical classification of the exposure, 5: Non-technical classification of the exposure, 6: Other sources of bias, 7: Sample size and power, 8: A priori planning of analysis, 9: Statistical methods and control for confounding, 10: Testing of assumptions and inferences for genetic analyses, 11: Appropriateness of inferences drawn from results. Scoring: 1 to 7, 1 being poor and 7 being excellent. N/A: not applicable.

3.3. Study and Subject Characteristics

Of the 54 studies included in this systematic review, 35 were cross sectional studies while the remaining 19 were longitudinal. A comprehensive description of the characteristics of the cross-sectional studies are presented in Table 2. Of the longitudinal studies, 11 were intervention studies while 8 were observational follow-up studies. The average intervention length was 21.3 weeks (range 10–72 weeks) while the average follow-up was 4.2 years (range 1–10 years). Table 3 presents a detailed description of the characteristics of the longitudinal studies. Out of the 54 studies, 53 were candidate gene association studies and the remaining article was a genome-wide association study.

Table 2. Cross-sectional studies on genetic associations with muscle phenotypes.

Gene	Polymorphism	Population Data	N	Muscle Phenotype	Results	Reference
				Hormone Genes		
VDR	rs2228570 (Fok1) rs1544410 (Bsm1)	Caucasians 46 males and 58 females Mean age 61.8 ± 8.5 years	104	Muscle strength (KE strength)	Individuals homozygous for the F allele of the rs2228570 polymorphism displayed significantly lower KE strength than carriers of ≥ 1 f allele (p = 0.007). KE strength did not differ significantly across rs1544410 genotypes.	Hopkinson, et al., 2008. [45]
VDR	rs2228570 rs1544410	Caucasians (Italians) 87 males and 172 females Aged ≥ 80 years Mean age 85.0 ± 4.5 years	259	Physical function (fall incidence)	Participants homozygous for the b allele of the rs1544410 polymorphism were significantly less likely to fall than carriers of ≥ 1 B allele (p = 0.02). Fall incidence did not differ significantly across rs2228570 genotypes.	Onder, et al., 2008. [62]
VDR	rs2228570 rs1544410	Caucasian females (OPUS cohort) Mean age 66.9 ± 7.0 years	2363	Muscle strength (lower limb power) Physical function (fall incidence, rise from chair)	Individuals with a bb genotype of the rs1544410 polymorphism were significantly less likely to fall than carriers of ≥ 1 B allele (p = 0.025). These individuals also performed significantly better in rise from chair and lower limb power tests (p = 0.03, 0.044 respectively). Fall incidence, muscle power did not differ significantly across rs2228570 genotypes.	Barr, et al., 2010. [27]
VDR	rs2228570 rs1544410 rs731236 (Taq1)	Males living in Turkey Aged 65–93 years Mean age 69 ± 6.9 years	120	Muscle strength (KE and KF peak torque)	KE strength was significantly higher in BB homozygotes compared to carriers of ≥ 1 b allele of the rs1544410 polymorphism (p = 0.038). No significant associations were found for rs2228570 and rs731236 genotypes.	Bahat, et al., 2010. [26]
				Hormone Genes		
VDR	rs2228570 (Fok1) rs1544410 (Bsm1) rs731236 (Taq1) rs7975232 (Apa1) rs7136534	Caucasian male centenarians (Italian) Mean age 102.3 ± 0.3 years	120	Muscle strength (HG strength)	FF homozygotes displayed significantly greater HG than individuals with ≥ 1 f allele of the rs2228570 polymorphism (p = 0.021). HG did not differ significantly between rs1544410, rs7975232 and rs731236 genotypes.	Gussago, et al., 2016. [42]
VDR	rs9729 rs17882106 rs10735810 rs4516035 rs11568820 rs11574024	Males living in Sweden Aged 69–81 years Mean age 75.4 ± 3.2 years	2844	Muscle strength (HG strength) Physical function (fall incidence, 6 m walk test, 20 cm narrow walk test, timed-stand test)	AA homozygotes were significantly less likely to fall compared to carriers of ≥ 1 G allele of rs7136534 (p = 0.002). No other significant associations were found between polymorphism and muscle strength or function tasks.	Bjork, et al., 2019. [28]
VDR	rs2228570 rs1544410	Caucasian males Aged 58–93 years	302	Body composition (FFM, AFFM, SMI) Muscle Strength (KE torque) Sarcopenia (SMI < 7.26 kg/m^2)	Men homozygous for the F allele of the rs2228570 polymorphism had significantly less FFM, AFFM and SMI compared to Ff/ff genotypes (p = 0.002, 0.009, 0.001 respectively). FF homozygotes also had 2.17-fold higher risk of sarcopenia than carriers of ≥ 1 f allele (p = 0.03). No similar associations were found between rs1544410 genotypes.	Roth, et al., 2004. [67]

Table 2. *Cont.*

Gene	Polymorphism	Population Data	N	Muscle Phenotype	Results	Reference
				Hormone Genes		
VDR	rs7975232 (Apa1) rs1544410 (Bsm1) rs2239185 rs3782905	Taiwanese 215 males and 154 females Mean age 74.4 ± 6.3 years (males) and 71.7 ± 4.7 years (females)	369	Muscle strength (HG strength)	Females carrying the AC genotype of rs7975232 polymorphism had significantly lower HG than CC homozygotes ($p < 0.05$). In both men and women, physical inactivity and the minor allele of each polymorphism were jointly associated with increased risk of low HG.	Wu, et al., 2014. [73]
VDR	rs2228570 (Fok1)	Chinese 275 males and 510 females Aged 63.2–72.5 years (males) and 63.1–71.9 years (females)	785	Muscle strength (HG strength) Physical function (4 m gait speed) Body composition (FFM, AFFM, SMI) Sarcopenia (SMI < 7.0 kg/m^2 for men and < 5.4 kg/m^2 for women and either low HG < 26 kg for men and < 18 kg for women or low gait speed < 0.8 m/s for both sexes)	Males who were homozygous for the f allele of the rs2228570 polymorphism had significantly greater HG and SMI when compared to carriers of ≥ 1 F allele ($p = 0.03$, 0.04 respectively). These individuals also had a significantly lower risk of sarcopenia ($p = 0.03$). No similar association was found in the female population.	Xia, et al., 2019. [74]
AR	rs3032358 (CAG repeat)	Caucasian males (STORM cohort) Aged 55-93 years Subjects grouped by repeat number (120 males had < 22 and 174 had $\geq$ 22)	294	Body composition (total FFM and SMI) Muscle strength (KE isometric strength and HG strength)	Men who had ≥ 22 repeats exhibited significantly greater total FFM and SMI than men with < 22 repeats ($p < 0.027$, < 0.019 respectively). A similar association was not found in females. No significant association was observed between repeat number and muscle strength phenotypes.	Walsh, et al., 2005. [72]
TRHR	rs16892496 rs7832552	Brazilian females Aged between 60–82 years Mean age 66.6 ± 5.5 years	241	Body composition (FFM, AFFM and SMI) Muscle strength (KE peak torque) Sarcopenia (SMI < 5.45 kg/m^2)	Subjects who carried the CC variant of rs16892496 had significantly less AFFM and SMI than AA/AC carriers ($p < 0.05$). No significant differences were observed for rs7832552 variants.	Lunardi, et al., 2013. [57]
				Growth Factor and Cytokine Genes		
IGF1	rs35767	Health ABC study cohort Blacks (533 males and 705 females) Whites (925 males and 836 females) Aged 70–79 years	2999	Body composition (FFM) Muscle volume (quadriceps CSA) Muscle strength (KE and HG strength, elbow flexor MVC and 1RM) Physical function (gait speed and single leg chair stands)	Black females with a CC genotype had significantly less FFM and quadriceps CSA compared to TT counterparts (both $p < 0.05$). White males with a CC genotype performed significantly worse in the single leg chair stands compared to CT counterparts ($p < 0.05$).	Kostek, et al., 2010. [51]
IGF1	192 bp allele	Caucasians (Spanish) 144 males and 148 females Mean age 76.7 ± 5.4 years (males) and 77.3 ± 6.4 years (females)	292	Muscle strength (KE isometric strength and HG strength)	No significant associations were observed in either males or females with relation to homozygosity, heterozygosity or non-carrier condition of the 192 bp allele ($p = 0.24$).	Mora, et al., 2011. [61]
IGF1 IGFBP3 IGFBP5	rs6214 rs35767 rs3110697 rs2854744 rs11977526 rs1978346 rs12474719	Taiwanese 251 males and 221 females Aged ≥ 65 years Mean age 74.7 ± 6.4 years (males) and 72.8 ± 5.5 years (females)	472	Body composition (SMI)	Individuals carrying the CC genotype of rs2854744 had a 4.3-fold risk of having low SMI compared with those with the AA genotype ($p < 0.05$). No other significant associations were observed for the other polymorphisms.	Yang, et al., 2015. [75]

Table 2. *Cont.*

Gene	Polymorphism	Population Data	N	Muscle Phenotype	Results	Reference
				Growth Factor and Cytokine Genes		
CNTF	rs948562 rs1800169 rs550942 rs4319530 rs1944055 rs2510559 rs2275993 rs1938596	Caucasian females (North American) Aged 70–79 years	363	Muscle strength (KE, HE and HG strength)	5 polymorphisms (rs948562, rs1800169, rs550942, rs4319530, rs1938596) were associated with HG (p <0.05). Haplotype analysis revealed rs1800169 null allele to fully explain relationship with the haplotype and HG under a recessive model, with homozygotes for the null allele exhibiting 3.80kg lower HG (p<0.01).	Arking, et al., 2006. [25]
CNTF CNTFR	rs1800169 rs3808871 rs2070802 C-174T	Caucasians 99 males and 102 females Aged 60–78 years (males) and 60–80 years (females)	201	Body composition (FFM) Muscle strength (isometric and concentric KE and KF at 60°, 120°, 150°, 180°, 240°)	Females who were G/A heterozygotes for the rs1800169 polymorphism produced significantly lower KE at 150° than both G/G and A/A homozygotes (p = 0.0229). Males who carried the T allele of the rs3808871 polymorphism produced significantly higher KE and KF isometric torque at 120° when compared to CC homozygotes (p < 0.05). Females who carried the T allele of the rs2070802 polymorphism performed better on KF concentric torques at 60°, 180° and 240° than the A/A homozygotes (p = 0.03, 0.04, 0.04 respectively). No significant associations were observed between polymorphisms and FFM.	De Mars, et al. 2007. [36]
				Growth Factor and Cytokine Genes		
CRP IL6 TNFα ICAM1	rs1800947 rs2069829 rs361525 rs5498	Danish twins 200 males and 400 females Aged 73–95 years	600	Physical function (self-reported during a 2-hour interview using a 11-item checklist)	Males who carried ≥1 A allele of the TNFα rs361525 polymorphism had a significantly better physical performance level compared to GG homozygotes (p < 0.001). No other associations were observed between polymorphisms and physical performance.	Tiainen, et al., 2012. [69]
CRP TNFα LTA	rs2794520 rs1205 rs1130864 rs1800947 rs3093059 rs1799964 rs1800629 rs3093662 rs2239704 rs909253 rs1041981	Taiwanese 251 males and 221 females Aged ≥ 65 years Mean age 74.7 ± 6.4 years (males) and 72.8 ± 5.5 years (females)	472	Muscle strength (HG strength)	In females, the main effect of polymorphisms (rs1800947, rs3093059, rs1799964, rs1800629, rs909253, rs1041981) reflected lower HG. In the male population, polymorphisms (rs1130864, rs2239704) produced the same effect.	Li, et al., 2016. [53]
CRP	rs2794520 rs1205 rs1130864 rs1800947 rs3093059	Taiwanese 251 males and 221 females Aged ≥ 65 years Mean age 74.7 ± 6.4 years (males) and 72.8 ± 5.5 years (females)	472	Muscle strength (HG strength)	HG of subjects carrying the CC variant of polymorphisms rs2794520 and rs1205 was lower by 1.24 kg and 1.28 kg, respectively, compared with TT homozygotes. HG was 1.01 kg lower for every additional C allele of rs3093059 polymorphism. Haplotype C-C-C-C-C was significantly associated with lower HG than any other haplotypic formation (p = 0.015).	Lin, et al., 2014. [55]

Table 2. *Cont.*

Gene	Polymorphism	Population Data	N	Muscle Phenotype	Results	Reference
				Growth Factor and Cytokine Genes		
CAV1	rs1997623 rs3807987 rs12672038 rs3757733 rs7804372 rs3807992	Taiwanese 265 males and 237 females Aged $\geq$ 65 years 327 controls, 56 pre-sarcopenic, 63 sarcopenic, 56 severely sarcopenic	502	Body composition (FFM, AFFM, SMI) Muscle strength (HG strength) Muscle function (15 ft walk test) Sarcopenia (SMI < 6.87 kg/m^2 and 5.46 kg/m^2 for males and females, respectively and lowest quintile for muscle strength and function tests)	Subjects carrying $\geq$ 1 A allele of rs3807987 were at a significantly higher risk of sarcopenia than GG homozygotes (p = 0.0235). No other significant associations were observed between the remaining polymorphisms.	Lin, et al., 2014. [56]
MSTN	rs1805065 rs35781413 rs1805086 rs368949692 rs143242500	Caucasian nonagenarians 8 males and 33 females Aged 90–97 years	41	Muscle strength (1RM leg press) Physical function (Tinetti scale measured gait and balance, Barthel index) Body Composition (FFM estimated)	Carriers of the rs1805086 KR genotype were associated with lower FFM compared to KK carriers. The RR homozygote was below the 25th sex specific percentile for FFM and functional capacity.	Gonzalez-Freire, et al., 2010. [41]
ACVR2B	rs2276541	Hispanic (354) and Non-Hispanic (2406) females Mean age 64.1 $\pm$ 7.4 years	2760	Body composition (FFM, AFFM)	Subjects carrying the A allele of rs2276541 had significantly more FFM than G allele carriers (p = 0.006).	Klimentidis, et al., 2016. [49]
				Structural and Metabolic Genes		
ACTN3 ACE	rs1815739 (R577X) rs1799752 (I/D)	Caucasians (Spanish) 8 males and 33 females Aged 90–97 years Mean age 92 $\pm$ 2 years	41	Muscle strength (HG strength and 6–7 RM leg press) Physical function (8 m walk test and 4 step stairs test)	Study phenotypes did not differ significantly between ACE or ACTN3 genotypes (all p > 0.05).	Bustamante, et al., 2010. [30]
				Structural and Metabolic Genes		
ACTN3	rs1815739 (R577X)	Japanese 183 males and 238 females Aged $\geq$ 55 years	421	Muscle strength (HG strength) Physical function (chair stand test, 8 ft walking test)	XX homozygotes performed significantly worse in the chair stand test than RR/RX carriers (p = 0.024, 0.005 respectively). No significant association was found between ACTN3 genotype and 8 ft walk test or HG.	Kikuchi, et al., 2014. [48]
ACTN3	rs1815739	Koreans 62 males and 270 females Aged $\geq$ 65 years Mean age 74.4 $\pm$ 4.6 years (males) and 74.4 $\pm$ 6.6 years (females)	332	Body composition (FFM, AFFM, SMI) Sarcopenia (SMI < 7.0 kg/m^2 and < 5.4 kg/m^2 for men and women respectively)	Sarcopenia prevalence was significantly associated with RX/XX genotypes (p = 0.037, 0.038 respectively). This association remained significant under both a dominant and recessive model (p = 0.043, 0.029 respectively).	Cho, et al., 2017. [32]
ACE	rs1799752 (I/D)	Brazilians 38 males and 53 females Aged 60–95 years Mean age 70.6 $\pm$ 7.2 years	91	Body composition (FFM, AFFM, SMI) Muscle strength (HG strength) Physical function (TUG test) Sarcopenia (based off FFM, muscle strength and physical function)	Sarcopenia prevalence was significantly higher in II genotype carriers compared to individuals with $\geq$ 1 D allele (p = 0.015).	Da Silva, et al., 2018. [34]
ACTN3 ACE	rs1815739 rs1799752	Caucasians (Spanish) 22 males and 59 females Aged 71–93 years Mean age 82.8 $\pm$ 4.8 years	81	Muscle strength (HG strength) Physical function (30s chair stand test, Barthel index) Muscle volume (thigh muscle CSA and muscle quality)	No significant associations were noted between any ACE rs1799752 or ACTN3 rs1815739 genotypes and the tested phenotypes in either males or females (p > 0.05).	Garatachea, et al., 2012. [39]

Table 2. *Cont.*

Gene	Polymorphism	Population Data	N	Muscle Phenotype	Results	Reference
				Structural and Metabolic Genes		
ACTN3	rs1815739 (R577X)	Chinese 686 males and 777 females Aged 70–87 years 2 age groups (70–79 years and 80–87 years)	1463	Muscle strength (HG strength) Physical function (TUG, 5m walk test) Frailty measure (frailty index containing 23 variables)	In the 70–79 age group, male XX homozygotes performed significantly worse than RR carriers in HG, 5 m walk test and TUG (p = 0.012, 0.011 and 0.039 respectively). Females in this age group who carried the XX genotype had a significantly higher frailty index than RR carriers (p = 0.004).	Ma, et al., 2018. [58]
ACTN3 *ACE*	rs1815739 rs1799752 (I/D)	Caucasian males (British) Aged 60–70 years Mean age 65 ± 3 years	100	Body composition (FFM and thigh FFM) Muscle strength (isometric and isokinetic KE strength) Contractile properties (time to peak tension, half-relaxation time, peak rate of force development)	There were no significant associations between either ACE or ACTN3 genotypes and the studied phenotypes.	McCauley, et al., 2010. [59]
ACE	rs1799752	Japanese 228 males and 203 females Aged 76 years	431	Muscle strength (HG strength, isokinetic KE) Physical function (10 s maximal stepping rate, single leg standing time with eyes open, maximum walking speed over 10 m)	Individuals homozygous for the I allele had significantly lower HG than carriers of the D allele (p = 0.004). Although not significant, the ACE rs1799752 polymorphism was also positively associated with 10 m maximum walking speed.	Yoshihara, et al., 2009. [76]
ACTN3	rs1815739	Japanese females Aged 50–78 years Mean age 64.1 ± 6 years	109	Body composition (mid-thigh CSA) Physical function (physical activity was measured using an uniaxial accelerometer)	Thigh muscle CSA was significantly lower in XX homozygotes compared to RX/RR carriers (p = 0.04). Physical activity did not significantly differ between genotypes.	Zempo, et al., 2010. [77]
				Structural and Metabolic Genes		
ACTN3	rs1815739 (R577X)	Japanese females Middle aged group (n = 82) mean age 50.6 ± 0.9 years Older group (n = 80) mean age 66.8 ± 0.5 years	162	Body composition (mid-thigh CSA) Physical activity (physical activity was measured using an uniaxial accelerometer)	In the middle-aged group, no association was observed between ACTN3 genotypes and thigh muscle CSA or physical activity. In the older group, XX homozygotes had significantly lower thigh muscle CSA than RX/RR carriers (p < 0.05).	Zempo, et al., 2010. [78]
UCP3	rs1800849 rs15763	Caucasians (Italians) 221 males and 211 females Aged 65–105 years Mean age 73.37 ± 7.46 years (males) and 73.37 ± 7.69 years (females)	432	Muscle strength (HG strength)	Carriers of the CC genotype of rs1800849 exhibited significantly lower HG than CT/TT genotypes (p = 0.010). No significant association was observed between rs15763 genotypes and HG.	Crocco, et al., 2011. [33]
UCP3	rs11235972 rs1685354 rs3781907 rs647126	Caucasians (Danish 1905 cohort) 265 males and 643 females Aged 93 years	908	Muscle strength (HG strength)	Individuals carrying the AA genotype of rs11235972 showed significantly lower HG than GG homozygotes (p < 0.001). Subjects carrying a GA genotype of rs1685354 displayed significantly greater HG than AA homozygotes (p = 0.016).	Dato, et al., 2012. [35]
PRDM16	rs12409277	Japanese females Mean age 65.1 ± 9.4 years	1081	Body composition (total FFM%)	Individuals who carried CT/CC variants of rs12409277 had a significantly greater FFM% compared to TT homozygotes (p = 0.003).	Urano, et al., 2014. [70]

KE: knee extensor, HE: hip extensor, KF: knee flexor, HG: handgrip, FFM: fat-free mass, AFFM: appendicular fat-free mass, SMI: skeletal muscle index, CSA: cross sectional area, MVC: maximal voluntary contraction, TUG: timed up and go.

Table 3. Longitudinal studies on genetic association with muscle phenotypes.

Gene	Polymorphism	Study Design	Population Data	N	Muscle Phenotype	Results	Reference
			Hormone Genes				
RAMP3	rs3757575 rs2074654 rs1294935 rs11982639 rs12702121	5- and 10-year follow-up	Swedish females (OPRA cohort) Aged 75 years Mean age 75.2 ± 0.1 years	1044	Body composition (total, legs and trunk FFM)	At baseline, C allele carriers of rs2074654 had significantly greater amounts of total and leg FFM ($p = 0.041, 0.038$ respectively) when compared to TT homozygotes. There were no significant associations at follow up.	Prakash, et al., 2019. [66]
			Growth Factor and Cytokine Genes				
IGF1	192 bp allele	10-week intervention of single leg KE RT	Caucasians 32 males and 35 females Mean age 70 ± 6 years (males) and 67 ± 8 years (females)	67	Muscle strength (KE 1RM) Muscle volume (using CT) Muscle quality (1RM/muscle volume)	Carriers of the 192 allele achieved significantly greater KE 1RM improvements than non-carriers ($p = 0.02$). Although not significant, a trend towards greater muscle volume was noted between 192 carriers and non-carriers ($p = 0.08$).	Kostek, et al., 2005. [50]
IGF1	192 bp allele	10-week intervention of single leg KE RT	Blacks (12 males and 21 females) Whites (46 males and 49 females) Aged 50–85 years	128	Muscle strength (KE 1RM) Muscle volume (using CT) Muscle quality (1RM/muscle volume)	Significantly greater KE 1RM improvements were observed in individuals with ≥ 1 192 allele compared to non-carriers ($p < 0.01$). No significant differences in muscle volume or quality were noted.	Hand, et al., 2007. [43]
			Growth Factor and Cytokine Genes				
TNFα *IL6* *IL10*	rs1800629 rs1800795 rs1800896	10-week intervention of either RT or AE	Brazilian females Aged ≥ 65 years 229 RT group and 222 AE group	451	Physical function (TUG and 10 m walking speed test)	Individuals homozygous for the G allele of polymorphism rs1800629 of TNFα achieved significantly greater TUG improvements with exercise compared to AA/AG genotypes ($p < 0.001$). A significant interaction was displayed between the 3 polymorphisms and TUG performance post exercise ($p < 0.001$). No significant interaction was observed between polymorphisms and 10 m walking speed test.	Pereira, et al., 2013. [65]
			Structural and Metabolic Genes				
ACE	rs1799752 (I/D)	10-week intervention of unilateral KE RT	North Americans Whites (65%) and Blacks (35%) 86 males and 139 females Aged 50–85 years (mean age 62 years)	225	Body composition (FFM) Muscle volume (quadriceps) Muscle strength (KE 1RM)	At baseline, carriers of the DD genotype had significantly greater FFM than II homozygotes ($p < 0.05$). DD homozygotes also had greater baseline muscle volume in both the trained and untrained leg than II carriers ($p = 0.02, 0.01$ respectively). No significant associations were observed between genotypes and either 1RM or muscle volume adaptations to RT in either males or females.	Charbonneau, et al., 2008. [31]
ACE	rs1799752	12-month intervention of either PA or health education	Caucasians 97 males and 186 females Aged 70–89 years Mean age 77.2 ± 4.3 years	283	Physical function (400 m gait speed test and SPPB)	A significant difference was observed in gait speed and SPPB post PA in carriers of ≥ 1 D allele ($p = 0.018, 0.015$ respectively), but not in II homozygotes ($p = 0.930, 0.275$ respectively).	Buford, et al., 2014. [29]

Table 3. *Cont.*

Gene	Polymorphism	Study Design	Population Data	N	Muscle Phenotype	Results	Reference
Structural and Metabolic Genes							
ACTN3	rs1815739 (R577X)	10-week intervention of unilateral KE RT	Caucasians 71 males and 86 females Aged 50–85 years Mean age 65 ± 8 years (males) and 64 ± 9 years (females)	157	Body composition (FFM) Muscle volume (quadriceps) Muscle strength (KE 1RM, peak power and velocity)	At baseline, female XX homozygotes had significantly higher absolute and relative KE peak power and peak velocity than carriers of ≥ 1 R allele ($p < 0.05$). In males, change in absolute KE peak power post RT approached significance in RR homozygotes compared to XX carriers ($p = 0.07$). In females, change in relative KE peak power post RT was significantly higher in RR homozygotes compared to XX carriers ($p = 0.02$).	Delmonico, et al., 2007. [37]
ACTN3	rs1815739	5-year follow-up	White North Americans 726 males and 641 females (Health ABC cohort) Aged 70–79 years Loss to follow-up (372)	1367	Muscle volume (thigh muscle CSA) Muscle strength (KE isokinetic torque) Physical function (400 m walk test, SPPB, self-reported functional limitation)	At follow-up, male XX homozygotes had a significantly greater increase in 400 m walk time when compared to RX/RR carriers ($p = 0.03$). Female XX carriers had a 35% greater risk of functional limitation compared to RR homozygotes. No significant associations were noted between genotype and phenotypes at baseline in either males or females ($p > 0.05$).	Delmonico, et al., 2008. [38]
Structural and Metabolic Genes							
ACE	rs1799752 (I/D)	18-month intervention of exercise training (AE and RT)	Caucasians (75%), African-American (22%), Native American, Asian/Pacific Islander, Hispanic (3%)63 males and 150 femalesAged ≥ 65 yearsLoss to follow-up (37)	213	Muscle strength (concentric KE isokinetic strength) Physical function (6 min walk test, self-reported FAST)	Carriers of the DD genotype showed significantly greater improvements in concentric KE strength in response to exercise training than II homozygotes ($p < 0.05$). At baseline, no significant associations were noted between genotypes and measures of muscle strength and physical performance.	Giaccaglia, et al., 2008. [40]
ACTN3	rs1815739 (R577X)	Follow-up (NOSOS 1 year follow up, APOSS 2 year follow up)	Caucasian females (Scottish) NOSOS cohort ($n = 1245$) APOSS cohort ($n = 2918$) Mean age (NOSOS 69.6 ± 5.5 years and APOSS 54.8 ± 2.2 years)	4163	Fall incidences (self-reported for previous year)	In both NOSO and APOSS cohorts, baseline falls were significantly associated with carrying RX/XX genotypes ($p = 0.049, 0.02$ respectively). In a pooled analysis, follow-up fall incidences in the previous year were associated with X allele carriers ($p = 0.01$).	Judson, et al., 2011. [46]

Table 3. *Cont.*

Gene	Polymorphism	Study Design	Population Data	N	Muscle Phenotype	Results	Reference
Structural and Metabolic Genes							
ACE	rs1799752 (I/D)	Follow-up (4.1 year average)	Whites and Blacks 1439 males and 1527 females Aged 70–79 years	2966	Muscle volume (thigh muscle CSA) Muscle strength (maximal and mean isokinetic KE strength) Physical function (physical activity questionnaire, self-reported mobility limitations)	Among individuals with high levels of physical activity II homozygotes developed limitation at a 45% faster rate when compared to ID/DD carriers ($p = 0.01$). ACE genotype did not affect mobility limitation in inactive individuals, nor did it affect any other phenotype in either active or inactive individuals.	Kritchevsky, et al., 2005. [52]
ACTN3 *ACE*	rs1815739 (R577X) rs1799752	24-week intervention of RT	Brazilian females Mean age 66.7 ± 5.5 years	246	Body composition (FFM, relative total FFM, AFFM and SMI) Muscle strength (KE isokinetic peak torque at 60°s)	At baseline, ACE DD homozygotes had significantly greater SMI than I/ID carriers ($p = 0.044$). ACTN3 X allele carriers had significantly more relative total FFM at baseline than RR homozygotes ($p = 0.04$). In response to RT, only ACE II homozygotes significantly increased AFFM ($p < 0.001$).	Lima, et al., 2011. [54]
Structural and Metabolic Genes							
ACTN3 *ACE*	rs1815739 (R577X) rs1799752 (I/D)	12-week intervention of high-speed power training	Caucasian females Mean age 65.5 ± 8.2 years	139	Muscle strength (1RM bench press and leg extension and vertical jump) Physical function (sit-to-stand test)	Post intervention, ACE DD homozygotes showed significantly greater improvements in 1RM bench press and sit-to-stand tests ($p = 0.019, 0.013$ respectively) than II carriers. The same interaction approached significance for vertical jump ($p = 0.052$). ACTN3 RR homozygotes displayed significantly greater improvements across all measures than XX carriers ($p < 0.05$). At baseline, there were no significant differences between ACE or ACTN3 genotype for any phenotype.	Pereira, et al., 2013. [63]
ACTN3 *ACE*	rs1815739 rs1799752	12-week intervention of high-speed power training	Caucasian females Mean age 65.5 ± 8.2 years	139	Muscle function (10 m maximal effort sprints, TUG test)	ACE DD homozygotes displayed significantly greater improvements in 10 m sprint time ($p = 0.012$) than II carriers, but not in GUG performance ($p = 0.331$). Similarly, ACTN3 RR homozygotes improved significantly more than XX carriers in 10m sprint time ($p = 0.044$) but not in TUG performance ($p = 0.477$). At baseline, there were no significant differences between ACE or ACTN3 genotype for any phenotype.	Pereira, et al., 2013 [64]

Table 3. *Cont.*

Gene	Polymorphism	Study Design	Population Data	N	Muscle Phenotype	Results	Reference
Structural and Metabolic Genes							
ACE/UCP2	rs1799752 (I/D) rs659366	12-week intervention of RT, balance and cardiovascular exercises	Caucasians 18 males and 40 females Aged > 60 years Mean age 70.0 ± 5.9 years (males) and 69.7 ± 5.3 years (females)	58	Muscle strength (HG strength) Physical function (30 s sit to stand, 30 s bicep curls, 8 ft TUG, 6 min walk, Purdue pegboard test)	At baseline, ACE II homozygotes performed significantly worse than ID/DD carriers in the 6 min walk and 8 ft TUG tests ($p = 0.008$, $p < 0.001$ respectively). GG carriers of rs659366 performed significantly worse in the 8 ft TUG test compared with AA/GA genotypes ($p = 0.045$). Post intervention, GG carriers of rs659366 had the greatest improvements in 8 ft TUG performance compared to AA/GA carriers ($p = 0.023$), while a trend for greater improvements in bicep strength was noted for ID/DD carriers compared to II carriers ($p = 0.099$).	Keogh, et al., 2015. [47]
APOE	rs7412 rs429358 (e4 status)	6-year follow-up	Caucasians (Dutch) 553 males and 709 females Aged > 65 years Mean age 74.9 ± 5.8 years Loss to follow-up (449)	1262	Physical function (5 chair stand test, 3 m gait speed, self-reported mobility)	At baseline, e4 carriers displayed significantly worse gait speed and chair stand performance ($p = 0.006$, 0.015 respectively) than the e3 group. At follow-up, e4 status was associated with significantly worse chair stand performance ($p = 0.034$) compared to e3 carriers.	Melzer, et al., 2005. [60]
Structural and Metabolic Genes							
APOE	rs7412 rs429358 (e4 status)	4-year follow-up	Swedish 245 males and 364 females Aged 75 years Loss to follow-up (28)	609	Muscle strength (HG strength) Physical function (20 m maximum gait speed, 5 chair stand test, 30 s single leg stand)	Subjects who carried the APOE e4 allele had a significantly larger decline in HG between age 75 and 79 compared to non-carriers ($p = 0.015$). Carriers of the APOE e4 allele had significantly lower HG at age 79 compared to non-carriers ($p = 0.006$). The effect of e4 allele on HG was significantly larger at age 79 than age 75 ($p = 0.033$).	Skoog, et al., 2016. [68]
APOE	rs7412 rs429358 (e4 status)	Follow-up (3-year average)	North Americans (67.8% White and Blacks 27.1%) 235 males and 392 females Mean age 79.4 ± 5.2 years	627	Physical function (15 ft and 20 ft gait speed, disability scale examining ability to perform ADL's)	Males carrying the ε4 allele showed a significantly more rapid decline in gait speed than male non-carriers ($p = 0.04$). This was most significant in white males only ($p = 0.007$). Similarly, males who carried the e4 allele had a significantly greater risk of disability than non-carriers ($p = 0.007$).	Verghese, et al., 2013. [71]

Table 3. *Cont.*

Gene	Polymorphism	Study Design	Population Data	N	Muscle Phenotype	Results	Reference
			Structural and Metabolic Genes				
ZNF295 C2CD2	rs928874 rs1788355	GWAS 2-year follow-up	Italians ilSIRENTE cohort (n = 286) 116 males and 170 females Mean age 86.1 ± 4.9 years Replication cohort inCHIANTI (n = 1055) 440 males and 615 females Mean age 67.8 ± 15.7 years	1341	Body composition (calf circumference, mid-arm muscle circumference) Muscle strength (HG strength) Physical function (4 m walk test, SPPB, ADL)	In the ilSIRENTE cohort, rs928874 and rs1788355 were significantly associated with 4 m gait speed ($p = 5.61 \times 10^{-8}$, 5.73×10^{-8} respectively). This association was not replicated in the inCHIANTI cohort.	Heckerman, et al., 2017. [44]

KE: knee extensor, HG: handgrip, FFM: fat-free mass, AFFM: appendicular fat-free mass, SMI: skeletal muscle index, RT: resistance training, AE: aerobic exercise, CT: computed tomography, CSA: cross sectional area, 1RM: 1 repetition maximum, PA: physical activity, TUG: timed up and go, ADL: activity of daily living, SPPB: short physical performance battery.

A total of 38,112 subjects participated across the 54 studies. Of these, 24,890 (65.3%) were female and 13,222 (34.7%) were male. Thirty-two studies included Caucasians, 13 assessed Asian subjects and the remaining nine studies included Hispanic and African-American participants. As described in the inclusion criteria, all subjects were older than 50 years of age. Thirteen studies included subjects over 50 years of age, 22 studies recruited subjects over 60 years of age and 19 studies included individuals aged 70 years or older.

3.4. Phenotypes and Genotypes

Of the included studies, 26 reported skeletal muscle mass outcomes, 39 studies included muscle strength testing, 27 articles analysed physical function and six examined sarcopenia prevalence. A full description of the phenotypic outcomes in each study are presented in Tables 2 and 3.

In total, 88 DNA polymorphisms in or near to 26 different genes were analysed across the 54 studies included in this review. The Alpha-actinin 3 (*ACTN3*), Angiotensin Converting Enzyme (*ACE*), and Vitamin D Receptor (*VDR*) genes were the most frequently researched, present in 14, 13 and nine articles, respectively. For clarity and ease of interpretation in the present review, genes are categorised into three main groups: hormone genes, growth factor and cytokine genes and structural and metabolic genes.

3.5. Synthesis of Results

3.5.1. Hormone Genes

VDR

Nine studies analysed the association between *VDR* polymorphisms and muscle phenotypes. The first, conducted in 2004 by Roth et al. [67], highlighted a significant association between the rs2228570 (*Fok1*) polymorphism and FFM. Male FF homozygotes had significantly less FFM, appendicular fat-free mass (AFFM) and skeletal muscle index (SMI) compared to f allele carriers ($p = 0.002$, $p = 0.009$, $p = 0.001$ respectively). Furthermore, when classified as sarcopenic, FF carriers were at a two-fold higher risk of being sarcopenic when compared to carriers of the f allele ($p = 0.03$). Hopkinson et al. [45] also found significant interactions between the rs2228570 polymorphism and muscle phenotypes with male FF homozygotes displaying significantly lower knee extensor (KE) strength than f allele carriers ($p = 0.007$). Similarly, Xia et al. [74] found subjects carrying one or more F alleles to have significantly lower handgrip (HG) strength, and FFM ($p = 0.03$, $p = 0.04$ respectively). Furthermore, these individuals had a significantly higher risk of sarcopenia than ff homozygotes ($p = 0.03$). In contrast, a study conducted by Gussago et al. [42] found FF homozygotes to have significantly greater HG strength than f allele carriers ($p = 0.021$).

Significant associations were also identified between the rs1544410 (*Bsm1*) polymorphism and muscle performance phenotypes although, in keeping with the above findings, results were conflicting. In a study conducted by Onder et al. [62], bb homozygotes were significantly less likely to fall than carriers of the B allele ($p = 0.02$). Similarly, in 2010, Barr et al. [27] found females who were homozygous for the b allele to have a significantly lower risk of falling than Bb/BB carriers. These individuals also performed significantly better in the rise from chair and power tests when compared to carriers of B allele ($p = 0.03$, $p = 0.044$ respectively). Contrarily to the above studies, Bahat et al. [26] found KE strength to be significantly higher in BB homozygotes compared to carriers of one or more b alleles ($p = 0.038$).

Additional *VDR* polymorphisms rs7136534 and rs7975232 (*Apa1*) were significantly associated with fall incidence and HG strength respectively ($p = 0.002$, $p < 0.05$) [28,73]. No significant associations were found for the rs731236 (*Taq1*) polymorphism.

Other Genes

Genes encoding the androgen receptor (*AR*), thyrotropin-releasing hormone receptor (*TRHR*) and receptor activity-modifying protein 3 (*RAMP3*) were also shown to associate significantly with skeletal muscle traits (Tables 2 and 3) [57,66,72].

3.5.2. Growth Factor and Cytokine Genes

IGF1 and IGFBP3

The interaction between the Insulin-like Growth Factor 1 (*IGF1*) gene and muscle phenotypes was particularly evident in the intervention studies (Table 3). Both Kostek et al. [50] and Hand et al. [43] demonstrated that carriers of one or more 192 alleles achieved significantly greater KE strength improvements in response to resistance training (RT), compared to non-carriers ($p = 0.02$, $p < 0.01$). However, in a cross-sectional study conducted by Mora et al. [61], no significant differences were observed in muscle strength between carriers and non-carriers of the 192 allele ($p = 0.024$).

Significant associations were also noted for polymorphisms rs35767 of the *IGF1* gene and rs2854744 of the Insulin-like Growth Factor Binding Protein 3 (*IGFBP3*) gene. Kostek et al. [51] observed black females carrying the CC genotype of the rs35767 polymorphism to have significantly less total FFM and muscle cross sectional area (CSA) than TT carriers (both $p < 0.05$). Furthermore, male CC homozygotes performed significantly worse in the single leg chair stand test than carriers of the T allele ($p < 0.05$). In a study conducted by Yang et al. [75], CC carriers of the rs2854744 polymorphism had a 4.3 times higher risk of having low SMI compared to AA carriers ($p < 0.05$).

CNTF and CNTFR

Two studies examined the Ciliary Neurotrophic Factor (*CNTF*) and Ciliary Neurotrophic Factor Receptor (*CNTFR*) genes (Table 2). In 2006, Arking et al. [25] observed five DNA polymorphisms (rs948562, rs1800169, rs550942, rs4319530, rs1938596) of the *CNTF* gene to be significantly associated with HG strength ($p < 0.05$). Further haplotype analysis revealed the null allele (A) of rs1800169 to fully explain this relationship under a recessive model. Individuals homozygous for the A allele had 3.8 kg lower HG strength than G allele carriers ($p < 0.01$). Interestingly, De Mars et al. [36] found only G/A carriers of the rs1800169 polymorphism to have significantly lower KE strength than G/G or A/A carriers ($p = 0.0229$). Additionally, male T allele carriers of the rs3808871 polymorphism produced significantly higher KE and knee flexor (KF) isometric torque at 120° when compared to CC homozygotes ($p < 0.05$). Furthermore, females who carried the T allele of the rs2070802 polymorphism produced greater KF concentric torques than the A/A homozygotes ($p = 0.04$).

TNFα

Three studies were included in this review which investigated the Tumour Necrosis Factor Alpha (*TNFα*) gene, each with significant findings. In 2013, Pereira et al. [65] observed that G allele homozygotes of the rs1800629 polymorphism achieved significantly faster timed up and go (TUG) test results in response to 10 weeks of RT compared to A allele carriers ($p < 0.001$). Additionally, Tiainen et al. [69] found the A allele of the rs361525 polymorphism to be associated with a significantly better physical performance level compared to GG homozygotes ($p < 0.001$). Finally, Li et al. [53] highlighted the interaction between the A allele of the rs1799964 polymorphism with either the G allele of the Tumour Necrosis Factor Beta (*TNF-β*) rs909253 polymorphism or the A allele of the *TNF-β* rs1041981 polymorphism to result in significantly lower handgrip strength among females ($p = 0.005$, $p = 0.006$ respectively).

Other Genes

Polymorphisms rs2276541 of the activin A type IIB receptor (*ACVR2B*) gene, rs3807987 of Caveolin 1 (*CAV1*) gene and rs1805086 of the Myostatin (*MSTN*) gene were all significantly associated with FFM (Table 2) [41,49,56].

3.5.3. Structural and Metabolic Genes

ACTN3 (The Sprint Gene)

In this review, fourteen studies were included which examined the association between the *ACTN3* rs1815739 (*R577X*) polymorphism and skeletal muscle phenotypes. Carrying the X allele was often associated with lower baseline muscle strength and function (Table 2). For example, in a study conducted by Kikuchi et al. [48], homozygosity for the X allele was associated with significantly poorer performance in the chair stand test compared to RR carriers ($p = 0.024$). Ma et al. [58] also found XX homozygotes to perform significantly worse in HG strength ($p = 0.012$), 5 m walk ($p = 0.011$) and TUG ($p = 0.039$) tests and to also have a significantly higher frailty index ($p = 0.004$). Similar results were observed by Judson et al. [46] in a group of 4163 females where RX and XX genotypes were significantly associated with fall incidence ($p = 0.049$, $p = 0.02$ respectively). In contrast, Delmonico et al. [37] found female XX homozygotes to have significantly higher absolute and relative KE peak power and peak velocity than carriers of the R allele ($p < 0.05$).

Individuals carrying the XX genotype were also shown to have significantly lower improvements in one repetition maximum (1RM) bench press and leg extension, vertical jump and sit-to-stand performance in response to speed and power training when compared to RR carriers (all $p < 0.05$) [63]. Pereira et al. [64] also demonstrated XX carriers to have significantly poorer improvements in 10 m sprint times in response to high speed and power training compared to RR homozygotes ($p = 0.044$). Similarly, female XX carriers were observed to have significantly lower improvements in relative KE peak power following RT compared to RR homozygotes ($p = 0.02$) [37]. In the male population, change in absolute KE peak power post RT approached significance when comparing RR and XX genotypes ($p = 0.07$) [37]. In contrast to the above studies, Delmonico et al. [38] found male XX homozygotes had a significantly greater increase in 400 m walk time when compared to RX/RR carriers ($p = 0.03$).

In a study conducted by Zempo et al. [77] XX homozygotes were observed to have significantly lower thigh muscle CSA compared to RR carriers ($p = 0.04$). Interestingly, in a secondary analysis comparing a middle age group with an old age group, XX homozygosity was only associated with low thigh muscle CSA in the old age group ($p < 0.05$), suggesting that the influence of ACTN3 deficiency is heightened with age [78]. Similar results were noted in 2017 by Cho et al. [32], where sarcopenia prevalence was significantly associated with the XX genotype ($p = 0.038$). In contrast, Lima et al. [54] found X allele carriers to have significantly more relative total FFM than RR homozygotes ($p = 0.04$).

Three studies found no significant differences in muscle phenotypes between *ACTN3* rs1815739 genotypes [30,39,59].

ACE

The relationship between the *ACE* rs1799752 (insertion/deletion) polymorphism and skeletal muscle traits has been extensively investigated since the original study of Montgomery et al. in 1998 [79]. Thirteen articles are included in this review. Firstly, Charbonneau et al. [31] found that carriers of the DD genotype had significantly greater total FFM ($p < 0.05$) and lower limb muscle volume ($p = 0.01$) than II homozygotes. Similarly, in a study of 246 Brazilian females, Lima et al. [54] noted DD homozygotes to have a significantly greater SMI than I allele carriers ($p = 0.044$). These findings were further strengthened by Da Silva et al. [34], who demonstrated sarcopenia prevalence to be significantly higher in II genotype carriers compared to D allele carriers ($p = 0.015$) (Table 2). Interestingly, Lima et al. [54] showed that in response to RT, only *ACE* II homozygotes significantly increased AFFM ($p < 0.001$).

The II genotype was also associated with lower muscle strength and functional performance. For example, within a group of 431 Japanese individuals, Yoshihara et al. [76] found II homozygosity to be associated with significantly lower HG strength compared to D allele carriers ($p = 0.004$). Homozygosity for the I allele was also shown to associate with significantly poorer performance in the 6-min walking test and 8 ft TUG test ($p = 0.008$, $p < 0.001$ respectively) when compared to ID/DD genotypes. Furthermore, in response to RT, DD carriers achieved significantly greater improvements in 1RM bench press and sit-to-stand performance ($p = 0.019$, $p = 0.013$ respectively) [63]. Giaccaglia et al. [40] also found that DD genotype carriers achieved significantly greater improvements in concentric KE strength in response to RT compared to II homozygotes ($p < 0.05$). Similarly, Pereira et al. [64] observed that DD homozygotes became significantly quicker performing 10 m sprints ($p = 0.012$) compared to II carriers. Buford et al. [29] also reported that a 12-month exercise intervention evoked significant improvements in 400 m walking speed ($p = 0.018$) and short physical performance battery test (SPPB) scores ($p = 0.015$), but only in D allele carriers. Interestingly, II homozygosity was also significantly associated with developing mobility limitation at a 45% faster rate when compared to ID/DD carriers ($p = 0.01$) [52].

As with the *ACTN3* rs1815739 genotypes, three studies found rs1799752 genotypes to have no significant influence on skeletal muscle traits [30,39,59].

APOE

Three studies demonstrated significant associations between the Apolipoprotein E (*APOE*) gene and muscle phenotypes (Table 3). A 6-year follow-up study conducted by Melzer et al. [60] found that e4 carriers displayed significantly slower gait speed and chair stand performance ($p = 0.006$, $p = 0.015$ respectively) at baseline and significantly slower chair stand performance ($p = 0.034$) at the end of the 6-year follow-up, compared to e3 carriers. The *APOE* e4 allele was also shown to be associated with a significantly larger decline in HG strength between the ages of 75 and 79 over a 4-year period, compared to non-carriers ($p = 0.015$) [68]. Furthermore, carriers of the e4 allele had significantly lower HG strength at age 79 compared to non-carriers ($p = 0.006$). Interestingly, the effect of the e4 allele on HG strength was significantly larger at age 79 than age 75 ($p = 0.033$), suggesting that the e4 allele becomes increasingly influential with age. In a 3-year follow-up study conducted by Verghese et al. [71], males carrying the e4 allele showed a significantly more rapid decline in gait speed and greater risk of disability than male non-carriers ($p = 0.04$, $p = 0.007$ respectively).

UCP2 and *UCP3*

Three studies reported significant interactions between Uncoupling Proteins 2/3 (*UCP2/3*) polymorphisms and skeletal muscle traits. Firstly, in a group of 432 Caucasians, Crocco et al. [33] found carriers of the CC genotype of the *UCP3* rs1800849 polymorphism to exhibit significantly lower HG strength than carriers of the T allele ($p = 0.010$). Dato et al. [35], then showed that individuals carrying the AA genotype of *UCP3* rs11235972 polymorphism have significantly lower HG strength than GG homozygotes ($p < 0.001$). In 2015, Keogh et al. [47] demonstrated that GG carriers of *UCP2* rs659366 polymorphism perform significantly worse in the 8 ft TUG test compared with AA/GA genotypes ($p = 0.045$). However, post RT intervention, GG homozygotes of *UCP2* rs659366 had the greatest improvements in 8 ft TUG performance ($p = 0.023$).

Genome-wide Studies

Other genes that demonstrated significant associations with muscle phenotypes included the PR domain containing 16 (*PRDM16*) gene, Zinc finger protein 295 (*ZNF295*) gene and C2 calcium dependent domain containing 2 (*C2CD2*) gene (Tables 2 and 3) [44,70].

Moreover, a recent GWAS by Hernandez-Cordero et al. [80] evaluated genetic contribution to ALM in the UK Biobank dataset, comparing middle-aged (38–49 years) and elderly (60–74 years) individuals. A total of 182 genome-wide significant regions, many with multiple variants within them, were associated with ALM in middle-aged individuals. Of these, 78% were also associated with ALM in elderly individuals. Variants at three genes, *VCAM*, *ADAMTSL3* and *FTO*, had previously been associated with lean body mass in the UK Biobank [81]. Hernandez Cortez et al. also confirmed, in vitro, a functional role for *CPNE1* and *STC2* in myogenesis. In addition, the study highlighted five genomic regions, containing multiple genes, that are associated with muscle mass in both mice and humans.

4. Discussion

To the best of the authors' knowledge, this is the first systematic review to collate literature on genetic associations with muscle phenotypes relevant to sarcopenia. To date, most research targeting genetic associations with muscle phenotypes has not focused on elderly subjects, and thus, the genetic mechanisms underpinning the age-related changes in skeletal muscle traits are largely uncharted.

Given that the deterioration of skeletal muscle with advancing age can have profound consequences for patients and public health systems, improving our understanding of how genes influence this process is of paramount importance. This review has enhanced our knowledge surrounding the key genes and gene variants that may prove crucial in further developing our understanding of the pathogenesis of sarcopenia and improving prognosis and treatment interventions alike.

4.1. Summary of Findings

The systematic literature search identified 24 genes and 46 DNA polymorphisms whose expression was significantly associated with muscle phenotypes in older adults. Ten of these DNA polymorphisms (rs154410, rs2228570, rs1800169, rs3093059, rs1800629, rs1815739, rs1799752, rs7412, rs429358 and 192 bp allele) were significantly associated with muscle phenotypes in two or more studies. The complex and multifactorial mechanisms underpinning muscle regulation suggest that the accrual and loss of muscle mass and muscle strength is not reducible to one single gene or gene variant. The dynamic interactions between inhibitory and promotory pathways within the human body further highlight the importance of a holistic approach when considering genetic associations with skeletal muscle traits.

Nevertheless, the findings of this systematic review demonstrate that the most compelling current evidence in the field exists for the *ACTN3*, *ACE* and *VDR* genotypes.

4.1.1. ACTN3 (The Sprint Gene)

The *ACTN3* gene is among the most extensively researched genes in relation to muscle phenotypes, and appeared most frequently within this review. The ACTN3 protein encoded by the *ACTN3* gene forms an integral part of the sarcomere Z-line in fast twitch muscle fibres and further aids in coordinating myofiber contractions [82,83]. Up to 20% of humans are deficient in this protein, due to homozygosity for the premature stop codon at the rs1815739 polymorphism [84]. This significant proportion of ACTN3 deficiency among the population suggests that X allele status is a key factor in variability in muscle phenotypes. In this regard, much of the research surrounding the *ACTN3* genotype has focused on athletic performance [85]. Association studies have repeatedly found reduced X allele frequency among elite sprint/power athletes [85–87]. This suggests that the presence of ACTN3 is crucial for the optimal generation of force. Considering that fast twitch muscle fibres are particularly susceptible to age-related atrophy [88], it is plausible that regulation of this protein may also be an important factor in understanding age-related changes in muscle phenotypes. To date, however, limited research has been conducted within elderly populations, with the result that the true impact of the *ACTN3* gene on age-related changes in muscle phenotypes remains inconclusive. Despite this, fourteen of the studies included in this review examining the *ACTN3* genotype reported promising findings. Carriers of the X allele were often found to display lower skeletal muscle mass, strength and functional abilities. This was particularly evident among the Asian population. All five cross-sectional studies that examined Asian participants found significant associations between X allele status and muscle phenotypes [32,48,58,77]. No such association was found in the other three cross-sectional studies that targeted Caucasian individuals [30,39,59], therefore suggesting ethnicity may determine the degree to which *ACTN3* genotypes effect aging muscle. This coincides with existing research whereby X allele frequency and fast twitch fibre composition have been shown to vary across different ethnic groups [89–92]. The Asian population have the highest frequency of the X allele [89], while having the lowest percentage of fast twitch muscle fibres [90–92], two likely contributing factors in the ethnic group having the highest sarcopenia prevalence globally [93]. Unlike above, X allele status was significantly associated with training adaptation within Caucasian, North-American and South-American individuals. Thus, the inconsistencies within this review highlight the need for future research to provide clarification on how ethnicity, *ACTN3* genotypes and muscle phenotypes are associated within the elderly.

4.1.2. ACE

Like the *ACTN3* gene, the *ACE* gene has been widely researched within athletic populations, and knowledge within older populations is limited. There are, however, compelling molecular pathways controlled by the *ACE* gene that suggest its importance in age-related changes in muscle phenotypes. The ACE is expressed by skeletal muscle endothelial cells, and catalyses the production of angiotensin II, known to enhance skeletal muscle hypertrophy [94,95]. To date, research in relation to muscle phenotypes has centred around the *ACE* rs1799752 polymorphism. The D and I alleles have been associated with higher and lower ACE activity respectively [96–98]. The D allele is suggested, therefore, to associate with greater muscle performance. To support this hypothesis, recent studies have focused on the rs1799752 polymorphism in elite athletes, with interesting findings. The I allele has been repeatedly associated with endurance performance, while the D allele associates with strength/power capabilities [99,100]. Findings from this systematic review further strengthen these observations. The D allele was consistently associated with higher baseline muscle strength and functional performance, as well as greater improvements in muscle strength and function in response to RT. Evidence of the association between the *ACE* rs1799752 polymorphism and muscle mass is less definitive. While the D allele was often associated with greater amounts of FFM, contradictory findings were also in evidence, and thus, further research is needed in this area to reach a consensus. Like with *ACTN3* genotypes, frequency of the I and D allele of the *ACE* gene are highly determined by ethnic background. Asians have been shown to have the highest frequency for the undesirable I allele [101], while African-American have the lowest [101], aligning with global sarcopenia prevalence estimates where Asians and African-Americans have the highest and lowest risk respectively [93]. While evidence in this review is insufficient in highlighting a true ethnic impact on the association between *ACE* genotypes and aging muscle phenotypes, the disparity in allele frequency among different ethnicities is promising.

4.1.3. VDR

The true significance of the association between the *VDR* gene and muscle phenotypes is currently unknown. While the *VDR* gene has been extensively researched, findings are often contradictory. Furthermore, due to its crucial role in regulating calcium absorption, much of the existing research has focused on the association between *VDR* genotypes and bone health [102]. However, the *VDR* gene is also known to stimulate changes in muscle protein synthesis through its key regulatory role in the transcription of messenger RNA [103], and thus, the potential of the *VDR* gene as a candidate gene for muscle phenotype associations has been suggested. More specifically, the rs2228570 polymorphism is the only known VDR polymorphism where variation results in structural changes within the VDR protein due to differences in translational initiation sites [104]. The *VDR* f allele results in a full length VDR protein of 427 amino acids [105], while a *VDR* F allele results in a truncated VDR protein with three amino acids less [106]. Interestingly, three of four studies that examined the rs2228570 polymorphism in this review found F allele carriers to perform significantly worse across a range of muscle phenotypes [45,67,74], suggesting the potential importance of the rs2228570 polymorphism.

While compelling evidence exists supporting the importance of the *VDR* gene for muscle phenotypes, many studies have failed to replicate earlier results, and thus, the strength of this association remains to be established [107,108]. Unlike for *ACTN3* and *ACE* polymorphisms, evidence of an ethnic influence on *VDR* polymorphism frequency is conflicting [109,110]. As with most genetic association studies, much of the research surrounding *VDR* polymorphisms and muscle phenotypes has been conducted using Caucasian subjects. Only nine articles examining *VDR* genotypes were included in this review, seven of which focused on Caucasian individuals [26–28,42,45,62,67]. Furthermore, as with the *ACTN3* and *ACE* genes, limited research has been conducted within an elderly population, further limiting the transferability of findings for older adults.

4.1.4. Other Genes of Interest

Other genes with convincing molecular pathways and findings, that warrant future investigation include the *IGF1/IGFBP3*, *TNFα*, *APOE*, *CNTF/R* and *UCP2/3* genes.

4.1.5. IGF1 and IGFBP3

The *IGF* family of genes encode peptides that are crucial in regulating cell proliferation, apoptosis and differentiation [111]. The mitogenic effect of IGF1 is integral to the facilitation of growth in multiple tissues, including skeletal muscle [112]. Considering that advancing age is associated with a decline in circulating IGF1 levels, the *IGF1* gene is a likely candidate to effect muscle phenotypes among the elderly [113]. The current review found significant associations between *IGF1* variants and skeletal muscle mass and strength. Associations were particularly convincing in longitudinal studies, suggesting that the *IGF1* 192 polymorphism may be particularly influential in the strength-training response of skeletal muscle phenotypes as opposed to baseline measurements.

The function of IGF1 is mediated through interactions with binding proteins, mainly, IGFBP3. Research has demonstrated that IGFBP3 is the most prolific potentiator of IGF1, therefore suggesting its importance in explaining inter-individual variation in muscle phenotypes [114]. While only Yang et al. [75] have investigated the impact of the *IGFBP3* gene in an elderly population, the significant findings of that study combined with the relevant gene mechanisms warrants future research.

4.1.6. TNFα

Like the *IGF* family, the *TNFα* gene aids in the regulation of a multitude of biological processes such as cell proliferation, differentiation and apoptosis, and is thus an important candidate gene for aging skeletal muscle [115]. TNFα is also known to be an integral mediator of the inflammatory response to muscle damage [116]. Considering that inflammation is a vital response to RT in facilitating muscle regeneration, the *TNFα* gene is likely to affect the response of skeletal muscle tissue to RT [117]. This is supported by the findings of Pereira et al. [65] who observed that *TNFα* genotypes associate significantly with TUG performance adaptation. While Tiainen et al. [69] also highlighted significant cross-sectional associations, these were based on self-reported measures and should be interpreted with caution. Thus, longitudinal studies focusing on RT response of skeletal muscle may prove most beneficial in understanding the effect of *TNFα* genotypes on the aging muscle.

4.1.7. APOE

APOE protein encoded by the *APOE* gene, is involved in lipid metabolism and is a well-established risk factor for Alzheimer's disease and various other aging disorders such as cardiovascular disease, atherosclerosis, stroke and impaired cognitive function [118]. Considering the associations between muscle phenotypes such as HG strength and these disorders, research has begun to investigate the relationship between the *APOE* gene and skeletal muscle traits. The gene has three common alleles (e2, e3 and e4), with e2 and e4 carriers having the lowest and highest risk of developing such aging disorders respectively [119]. As a result, much of the research in relation to skeletal muscle has centred around the e4 allele. The e4 allele was consistently associated with unfavourable skeletal muscle traits within this review, and therefore, supports the possibility of *APOE* as a candidate gene for explaining variation in muscle phenotypes with advancing age. Interestingly, like for *ACTN3* and *ACE* genotypes, prevalence of the e4 allele is known to be highly varied among different populations [120]. With only three studies were included in this review, the effect of ethnicity on e4 allele frequency and the resulting association with muscle phenotypes is yet to be confirmed.

4.1.8. CNTF and CNTFR

The *CNTF* and *CNTFR* genes are both mediated through a common signal-transducing component, and thus are often examined in parallel [121]. CNTF, located in glial cells, aids in the promotion of motor neuron survival, and is therefore suggested to limit age-related atrophy of skeletal muscle caused by denervation [122]. The CNTFR is largely expressed in skeletal muscle, promoting research to examine the role of the *CNTF* and *CNTFR* genes in the regulation of muscle phenotypes [123]. To date, however, much of this research has been conducted using rats, with limited research being conducted with human populations. Thus, while the current review has highlighted some significant associations with muscle phenotypes, additional research is required to further understand the mechanisms underpinning this association in humans.

4.1.9. UCP2 and UCP3

Uncoupling proteins (UCPs) are mitochondrial transporters, best known for their involvement in thermogenesis and energy utilisation. As a result, UCPs are most commonly researched in relation to obesity-related phenotypes [124,125]. There is, however, evidence that suggests their importance in regulating muscle phenotypes. UCP2 and UCP3 have both been shown to effect skeletal muscle performance through the inhibition of mitochondrial ATP synthesis [126]. Additionally, *UCP2* and *UCP3* genes serve a key purpose in the protection of cells by attenuating mitochondrial reactive oxygen species (ROS) production, known to exert damaging effects on cells [127]. While loss of skeletal muscle mitochondrial content is known to occur with advancing age [128], evidence suggests UCPs are particularly active in the latter stages of life due to an increase in ROS and the associated rise in mitochondrial superoxide [129]. Therefore, *UCP2* and *UCP3* genes may affect how metabolic function of skeletal muscle is retained during the aging process. While the three studies included in this review found significant associations between *UCP2* and *UCP3* variants and muscle phenotypes, other data from human studies are scarce and as a result, the strength of this association remains to be elucidated.

4.2. Strengths and Limitations

This is the first systematic literature review to explore the genetic association with muscle phenotypes among the elderly. Only healthy subjects were included in the review, allowing for any association to be solely attributed to genotype-phenotype interactions rather than disease. All subjects were over the age of 50 years, ensuring relevance towards developing the understanding of the pathogenesis of sarcopenia. While some methodological weaknesses exist, most studies were well designed and conducted.

Findings within this review were at times conflicting. This incongruity may be partly explained by between-study disparities in methodological aspects such as sample size, subject characteristics and false-positive reporting. Furthermore, not all studies utilised the same measure for each muscle phenotype. For example, muscle strength measured through handgrip or leg extension may lead to different results. Evidently, there is a need for genetic association studies to implement more comprehensive and stringent methodology to maximise the potential of identifying genetic variants relevant to aging muscle phenotypes.

Finally, while not necessarily a limitation of this review itself, the overall lack of research currently available regarding the association between genetic variants and muscle phenotypes within the elderly prevents more definitive inferences to be made. As evidenced in this review, most research to date has focused on European populations, thus limiting the transferability of findings to other ethnic groups. Considering the promising ethnic differences in polymorphism frequency previously highlighted, future genetic studies may benefit from including individuals from a variety of ethnic backgrounds. The distinct lack of GWAS targeting aging muscle phenotypes is also contributive towards the uncertainty surrounding this area. A large body of research has utilised a candidate gene approach. Historically, many candidate gene studies have been statistically underpowered, the replication of findings has been problematic and there has been a suspected bias against publication of negative results, which may lead to conflicting findings [130]. Many of these issues have been overcome by GWAS in large, well characterised cohorts [80,131–133]. Therefore, future GWAS may help to further illuminate the genetic basis of aging muscle phenotypes.

5. Conclusions

The ability to maintain skeletal muscle mass, strength and function with advancing age is essential in preventing sarcopenia. Thus, the elucidation of the genetic variants associated with these phenotypes is of paramount importance. Evidently, skeletal muscle mass, strength and function are multifaceted characteristics that vary widely among the elderly. While heritability studies have highlighted that significant proportions of this inter-individual variability are determined by genetic factors, the specific genes involved remain mostly unknown.

The genetic association with muscle phenotypes is relatively under-researched, with only a limited number of candidate genes being explored to date. This review identified and systematically compiled the key genes shown to be significantly associated with muscle phenotypes within an elderly population. While relatively few genes have been identified which significantly contribute towards variation in muscle phenotypes, promising findings pointing to more extensive associations exist. Evidence is particularly supportive of the *ACTN3*, *ACE* and *VDR* genes, while the *IGF1/IGFBP3*, *TNFα*, *APOE*, *CNTF/R* and *UCP2/3* genes have also been shown to be significantly associated with skeletal muscle phenotypes in two or more studies.

To conclude, the findings from this review helped to further illuminate the genetic basis of sarcopenia. While the molecular genetic pathways are often compelling, the limited volume of research within this field is as yet insufficient to demonstrate a clear genetic basis for sarcopenia. Future GWAS could facilitate the identification of novel genetic variants that may have key regulatory roles in aging muscle phenotypes. Further still, a more extensive exploration of the candidate genes highlighted in this review should provide further insight into the pathogenesis of sarcopenia and further aid in the development of effective prognosis, preventive and treatment protocols to combat the profound consequences of sarcopenia for patients and health systems worldwide.

Author Contributions: Conceptualization, J.P.; Literature search and validation, J.P., G.D.V. and C.B.; Analysis, J.P.; Writing—original draft preparation, J.P.; Writing—review and editing, J.P., G.D.V., C.B., S.E. and A.W.R. All authors have read and agreed to the published version of the manuscript.

Funding: This research received no external funding.

Conflicts of Interest: The authors declare no conflict of interest.

References

1. Rosenburg, I.H. Sarcopenia: Origins and clinical relevance. *J. Nutr.* **1997**, *127*, 990S–991S. [CrossRef] [PubMed]
2. Cruz-Jentoft, A.J.; Bahat, G.; Bauer, J.; Boirie, Y.; Bruyère, O.; Cederholm, T.; Cooper, C.; Landi, F.; Rolland, Y.; Sayer, A.A.; et al. Sarcopenia: Revised European consensus on definition and diagnosis. *Age Ageing* **2019**, *48*, 16–31. [CrossRef] [PubMed]
3. Chin, S.O.; Rhee, S.Y.; Chon, S.; Hwang, Y.C.; Jeong, I.K.; Oh, S.; Ahn, K.J.; Chung, H.Y.; Woo, J.T.; Kim, S.W.; et al. Sarcopenia Is independently associated with cardiovascular disease in older Korean adults: The Korea national health and nutrition examination survey (KNHANES) from 2009. *PLoS ONE* **2013**, *8*, e60119. [CrossRef] [PubMed]

4. Janssen, I.; Heymsfield, S.B.; Ross, R. Low relative skeletal muscle mass (sarcopenia) in older persons is associated with functional impairment and physical disability. *J. Am. Geriatr. Soc.* **2002**, *50*, 889–896. [CrossRef] [PubMed]

5. Landi, F.; Liperoti, R.; Russo, A.; Giovannini, S.; Tosato, M.; Capoluongo, E. Sarcopenia as a risk factor for falls in elderly individuals: Results from the ilSIRENTE study. *Clin. Nutr.* **2012**, *31*, 652–658. [CrossRef] [PubMed]

6. Zhang, X.; Zhang, W.; Wang, C.; Tao, W.; Dou, Q.; Yang, Y. Sarcopenia as a predictor of hospitalization among older people: A systematic review and meta-analysis. *BMC Geriatr.* **2018**, *18*, 188. [CrossRef]

7. Kim, T.N.; Choi, K.M. The implications of sarcopenia and sarcopenic obesity on cardiometabolic disease. *J. Cell Biochem.* **2015**, *116*, 1171–1178. [CrossRef]

8. Brown, J.C.; Harhay, M.O.; Harhay, M.N. Sarcopenia and mortality among a population based sample of community-dwelling older adults. *J. Cachexia Sarcopenia Muscle* **2016**, *7*, 290–298. [CrossRef]

9. Shafiee, G.; Keshtkar, A.; Soltani, A.; Ahadi, Z.; Larijani, B.; Heshmat, R. Prevalence of sarcopenia in the world: A systematic review and meta- analysis of general population studies. *J. Diabetes Metab. Disord.* **2017**, *16*, 16–21. [CrossRef]

10. Melton, L.J.; Khosla, S.; Crowson, C.S.; O'Connor, M.K.; O'Fallon, W.M.; Riggs, B.L. Epidemiology of sarcopenia. *J. Am. Geriatr. Soc.* **2000**, *48*, 625–630. [CrossRef]

11. Ethgen, O.; Beaudart, C.; Buckinx, F.; Bruyère, O.; Reginster, J.Y. The future prevalence of sarcopenia in Europe: A claim for public health action. *Calcif. Tissue Int.* **2017**, *100*, 229–234. [CrossRef] [PubMed]

12. Keller, K.; Engelhardt, M. Strength and muscle mass loss with aging process. Age and strength loss. *Muscles Ligaments Tendons J.* **2014**, *3*, 346–350. [CrossRef] [PubMed]

13. Deschenes, M.R. Effects of aging on muscle fibre type and size. *Sports Med.* **2004**, *34*, 809–824. [CrossRef] [PubMed]

14. Doherty, T.J. The influence of aging and sex on skeletal muscle mass and strength. *Curr. Opin. Clin. Nutr. Metab. Care* **2001**, *4*, 503–508. [CrossRef]

15. Prior, S.J.; Roth, S.M.; Wang, X.; Kammerer, C.; Miljkovic-Gacic, I.; Bunker, C.H.; Wheeler, V.W.; Patrick, A.L.; Zmuda, J.M. Genetic and environmental influences on skeletal muscle phenotypes as a function of age and sex in large, multigenerational families of African heritage. *J. Appl. Physiol.* **2007**, *103*, 1121–1127. [CrossRef]

16. Kemp, G.J.; Birrell, F.; Clegg, P.D.; Cuthbertson, D.J.; De Vito, G.; Van Dieën, J.H.; Del Din, S.; Eastell, R.; Garnero, P.; Goljanek–Whysall, K.; et al. Developing a toolkit for the assessment and monitoring of musculoskeletal ageing. *Age Ageing* **2018**, *47*, 1–19. [CrossRef]

17. Franzke, B.; Neubauer, O.; Cameron-Smith, D.; Wagner, K.H. Dietary protein, muscle and physical function in the very old. *Nutrients* **2018**, *10*, 935. [CrossRef]

18. Buchmann, N.; Spira, D.; Norman, K.; Demuth, I.; Eckardt, R.; Steinhagen-Thiessen, E. Sleep, muscle mass and muscle function in older people. *Dtsch. Arzteblatt Int.* **2016**, *113*, 253–260. [CrossRef]

19. Yoo, J.I.; Ha, Y.C.; Lee, Y.K.; Hana-Choi; Yoo, M.J.; Koo, K.H. High prevalence of sarcopenia among binge drinking elderly women: A nationwide population-based study. *BMC Geriatr.* **2017**, *17*, 114. [CrossRef]

20. Abney, M.; McPeek, M.S.; Ober, C. Broad and narrow heritabilities of quantitative traits in a founder population. *Am. J. Hum. Genet.* **2001**, *68*, 1302–1307. [CrossRef]

21. Zempo, H.; Miyamoto-Mikami, E.; Kikuchi, N.; Fuku, N.; Miyachi, M.; Murakami, H. Heritability estimates of muscle strength-related phenotypes: A systematic review and meta-analysis. *Scand. J. Med. Sci. Sports* **2017**, *27*, 1537–1546. [CrossRef] [PubMed]

22. Zhai, G.; Ding, C.; Stankovich, J.; Cicuttini, F.; Jones, G. The genetic contribution to longitudinal changes in knee structure and muscle strength: A sibpair study. *Arthritis Rheum.* **2005**, *52*, 2830–2834. [CrossRef] [PubMed]

23. Liberati, A.; Altman, D.; Tetzlaff, J.; Mulrow, C.; Gøtzsche, P.; Ioannidis, J.; Clarke, M.; Devereaux, P.; Kleijnen, J.; Moher, D. The PRISMA statement for reporting systematic reviews and meta-analyses of studies that evaluate health care interventions: Explanation and elaboration. *BMJ* **2009**, *339*, b2700. [CrossRef] [PubMed]

24. Sohani, Z.N.; Meyre, D.; de Souza, R.J.; Joseph, P.G.; Gandhi, M.; Dennis, B.B.; Norman, G.; Anand, S.S. Assessing the quality of published genetic association studies in meta-analyses: The quality of genetic studies (Q-Genie) tool. *BMC Genet.* **2015**, *16*, 50. [CrossRef] [PubMed]

25. Arking, D.E.; Fallin, D.M.; Fried, L.P.; Li, T.; Beamer, B.A.; Xue, Q.L.; Chakravarti, A.; Walston, J. Variation in the ciliary neurotrophic factor gene and muscle strength in older Caucasian women. *J. Am. Geriatr. Soc.* **2006**, *54*, 823–826. [CrossRef]

26. Bahat, G.; Saka, B.; Erten, N.; Ozbek, U.; Coskunpinar, E.; Yildiz, S.; Sahinkaya, T.; Karan, M.A. BsmI polymorphism in the vitamin D receptor gene is associated with leg extensor muscle strength in elderly men. *Aging Clin. Exp. Res.* **2010**, *22*, 198–205. [CrossRef]

27. Barr, R.; Macdonald, H.; Stewart, A.; McGuigan, F.; Rogers, A.; Eastell, R.; Felsenberg, D.; Glüer, C.; Roux, C.; Reid, D.M. Association between vitamin D receptor gene polymorphisms, falls, balance and muscle power: Results from two independent studies (APOSS and OPUS). *Osteoporos. Int.* **2010**, *21*, 457–466. [CrossRef]

28. Björk, A.; Ribom, E.; Johansson, G.; Scragg, R.; Mellström, D.; Grundberg, E.; Ohlsson, C.; Karlsson, M.; Ljunggren, Ö.; Kindmark, A. Variations in the vitamin D receptor gene are not associated with measures of muscle strength, physical performance, or falls in elderly men. Data from MrOS Sweden. *J. Steroid Biochem. Mol. Biol.* **2019**, *187*, 160–165. [CrossRef]

29. Buford, T.W.; Hsu, F.C.; Brinkley, T.E.; Carter, C.S.; Church, T.S.; Dodson, J.A.; Goodpaster, B.H.; McDermott, M.M.; Nicklas, B.J.; Yank, V.; et al. Genetic influence on exercise-induced changes in physical function among mobility-limited older adults. *Physiol. Genomics* **2014**, *46*, 149–158. [CrossRef]

30. Bustamante-Ara, N.; Santiago, C.; Verde, Z.; Yvert, T.; Gómez-Gallego, F.; Rodríguez-Romo, G.; González-Gil, P.; Serra-Rexach, J.A.; Ruiz, J.R.; Lucia, A. ACE and ACTN3 genes and muscle phenotypes in nonagenarians. *Int. J. Sports Med.* **2010**, *31*, 221–224. [CrossRef]

31. Charbonneau, D.E.; Hanson, E.D.; Ludlow, A.T.; Delmonico, M.J.; Hurley, B.F.; Roth, S.M. ACE genotype and the muscle hypertrophic and strength responses to strength training. *Med. Sci. Sports Exerc.* **2008**, *40*, 677–683. [CrossRef] [PubMed]

32. Cho, J.; Lee, I.; Kang, H. ACTN3 Gene and susceptibility to sarcopenia and osteoporotic status in older Korean adults. *Biomed. Res. Int.* **2017**, *2017*, 4239648. [CrossRef] [PubMed]

33. Crocco, P.; Montesanto, A.; Passarino, G.; Rose, G. A common polymorphism in the UCP3 promoter influences hand grip strength in elderly people. *Biogerontology* **2011**, *12*, 265–271. [CrossRef] [PubMed]

34. Da Silva, J.R.D.; Freire, I.V.; Ribeiro, Í.J.; dos Santos, C.S.; Casotti, C.A.; dos Santos, D.B.; Barbosa, A.A.L.; Pereira, R. Improving the comprehension of sarcopenic state determinants: An multivariate approach involving hormonal, nutritional, lifestyle and genetic variables. *Mech. Ageing Dev.* **2018**, *173*, 21–28. [CrossRef] [PubMed]

35. Dato, S.; Soerensen, M.; Montesanto, A.; Lagani, V.; Passarino, G.; Christensen, K.; Christiansen, L. UCP3 polymorphisms, hand grip performance and survival at old age: Association analysis in two Danish middle aged and elderly cohorts. *Mech. Ageing Dev.* **2012**, *133*, 530–537. [CrossRef]

36. De Mars, G.; Windelinckx, A.; Beunen, G.; Delecluse, C.; Lefevre, J.; Thomis, M.A. Polymorphisms in the CNTF and CNTF receptor genes are associated with muscle strength in men and women. *J. Appl. Physiol.* **2007**, *102*, 1824–1831. [CrossRef]

37. Delmonico, M.J.; Kostek, M.C.; Doldo, N.A.; Hand, B.D.; Walsh, S.; Conway, J.M.; Carignan, C.R.; Roth, S.M.; Hurley, B.F. Alpha-actinin-3 (ACTN3) R577X polymorphism influences knee extensor peak power response to strength training in older men and women. *J. Gerontol. A Biol. Sci. Med. Sci.* **2007**, *62*, 206–212. [CrossRef]

38. Delmonico, M.J.; Zmuda, J.M.; Taylor, B.C.; Cauley, J.A.; Harris, T.B.; Manini, T.M.; Schwartz, A.; Li, R.; Roth, S.M.; Hurley, B.F.; et al. Association of the ACTN3 genotype and physical functioning with age in older adults. *J. Gerontol. A Biol. Sci. Med. Sci.* **2008**, *63*, 1227–1234. [CrossRef]

39. Garatachea, N.; Fiuza-Luces, C.; Torres-Luque, G.; Yvert, T.; Santiago, C.; Gómez-Gallego, F.; Ruiz, J.R.; Lucia, A. Single and combined influence of ACE and ACTN3 genotypes on muscle phenotypes in octogenarians. *Eur. J. Appl. Physiol.* **2012**, *112*, 2409–2420. [CrossRef]

40. Giaccaglia, V.; Nicklas, B.; Kritchevsky, S.; Mychalecky, J.; Messier, S.; Bleecker, E.; Pahor, M. Interaction between angiotensin converting enzyme insertion/deletion genotype and exercise training on knee extensor strength in older individuals. *Int. J. Sports Med.* **2008**, *29*, 40–44. [CrossRef]

41. González-Freire, M.; Rodríguez-Romo, G.; Santiago, C.; Bustamante-Ara, N.; Yvert, T.; Gómez-Gallego, F.; Rexach, J.A.S.; Ruiz, J.R.; Lucia, A. The K153R variant in the myostatin gene and sarcopenia at the end of the human lifespan. *Age* **2010**, *32*, 405–409. [CrossRef] [PubMed]

42. Gussago, C.; Arosio, B.; Guerini, F.R.; Ferri, E.; Costa, A.S.; Casati, M.; Bollini, E.M.; Ronchetti, F.; Colombo, E.; Bernardelli, G.; et al. Impact of vitamin D receptor polymorphisms in centenarians. *Endocrine* **2016**, *53*, 558–564. [CrossRef] [PubMed]

43. Hand, B.D.; Kostek, M.C.; Ferrell, R.E.; Delmonico, M.J.; Douglass, L.W.; Roth, S.M.; Hagberg, J.M.; Hurley, B.F. Influence of promoter region variants of insulin-like growth factor pathway genes on the strength-training response of muscle phenotypes in older adults. *J. Appl. Physiol.* **2007**, *103*, 1678–1687. [CrossRef] [PubMed]

44. Heckerman, D.; Traynor, B.J.; Picca, A.; Calvani, R.; Marzetti, E.; Hernandez, D.; Nalls, M.; Arepali, S.; Ferrucci, L.; Landi, F. Genetic variants associated with physical performance and anthropometry in old age: A genome-wide association study in the ilSIRENTE cohort. *Sci. Rep.* **2017**, *7*, 15879. [CrossRef] [PubMed]

45. Hopkinson, N.S.; Li, K.W.; Kehoe, A.; Humphries, S.E.; Roughton, M.; Moxham, J.; Montgomery, H.; Polkey, M.I. Vitamin D receptor genotypes influence quadriceps strength in chronic obstructive pulmonary disease. *Am. J. Clin. Nutr.* **2008**, *87*, 385–390. [CrossRef]

46. Judson, R.N.; Wackerhage, H.; Hughes, A.; Mavroeidi, A.; Barr, R.J.; Macdonald, H.M.; Ratkevicius, A.; Reid, D.M.; Hocking, L.J. The functional ACTN3 577X variant increases the risk of falling in older females: Results from two large independent cohort studies. *J. Gerontol. A Biol. Sci. Med. Sci.* **2011**, *66*, 130–135. [CrossRef]

47. Keogh, J.W.L.; Palmer, B.R.; Taylor, D.; Kilding, A.E. ACE and UCP2 gene polymorphisms and their association with baseline and exercise-related changes in the functional performance of older adults. *PeerJ* **2015**, *3*, e980. [CrossRef]

48. Kikuchi, N.; Yoshida, S.; Min, S.K.; Lee, K.; Sakamaki-Sunaga, M.; Okamoto, T.; Nakazato, K. The ACTN3 R577X genotype is associated with muscle function in a Japanese population. *Appl. Physiol. Nutr. Metab.* **2015**, *40*, 316–322. [CrossRef]

49. Klimentidis, Y.C.; Bea, J.W.; Thompson, P.; Klimecki, W.T.; Hu, C.; Wu, G.; Nicholas, S.; Ryckman, K.K.; Chen, Z. Genetic variant in ACVR2B is associated with lean mass. *Med. Sci. Sports Exerc.* **2016**, *48*, 1270–1275. [CrossRef]

50. Kostek, M.C.; Delmonico, M.J.; Reichel, J.B.; Roth, S.M.; Douglass, L.; Ferrell, R.E.; Hurley, B.F. Muscle strength response to strength training is influenced by insulin-like growth factor 1 genotype in older adults. *J. Appl. Physiol.* **2005**, *98*, 2147–2154. [CrossRef]

51. Kostek, M.C.; Devaney, J.M.; Gordish-Dressman, H.; Harris, T.B.; Thompson, P.D.; Clarkson, P.M.; Angelopoulos, T.J.; Gordon, P.M.; Moyna, N.M.; Pescatello, L.S.; et al. A polymorphism near IGF1 is associated with body composition and muscle function in women from the Health, Aging, and Body Composition Study. *Eur. J. Appl. Physiol.* **2010**, *110*, 315–324. [CrossRef] [PubMed]

52. Kritchevsky, S.B.; Nicklas, B.J.; Visser, M.; Simonsick, E.M.; Newman, A.B.; Harris, T.B.; Lange, E.M.; Penninx, B.W.; Goodpaster, B.H.; Satterfield, S.; et al. Angiotensin-converting enzyme insertion/deletion genotype, exercise, and physical decline. *Jama* **2005**, *294*, 691–698. [CrossRef] [PubMed]

53. Li, C.I.; Li, T.C.; Liao, L.N.; Liu, C.S.; Yang, C.W.; Lin, C.H.; Hsiao, J.H.; Meng, N.H.; Lin, W.Y.; Wu, F.Y.; et al. Joint effect of gene-physical activity and the interactions among CRP, TNF-alpha, and LTA polymorphisms on serum CRP, TNF-alpha levels, and handgrip strength in community-dwelling elders in Taiwan-TCHS-E. *Age* **2016**, *38*, 46. [CrossRef] [PubMed]

54. Lima, R.M.; Leite, T.K.; Pereira, R.W.; Rabelo, H.T.; Roth, S.M.; Oliveira, R.J. ACE and ACTN3 genotypes in older women: Muscular phenotypes. *Int. J. Sports Med.* **2011**, *32*, 66–72. [CrossRef] [PubMed]

55. Lin, C.C.; Wu, F.Y.; Liao, L.N.; Li, C.I.; Lin, C.H.; Yang, C.W.; Meng, N.H.; Chang, C.K.; Lin, W.Y.; Liu, C.S.; et al. Association of CRP gene polymorphisms with serum CRP level and handgrip strength in community-dwelling elders in Taiwan: Taichung Community Health Study for Elders (TCHS-E). *Exp. Gerontol.* **2014**, *57*, 141–148. [CrossRef] [PubMed]

56. Lin, C.H.; Lin, C.C.; Tsai, C.W.; Chang, W.S.; Yang, M.D.; Bau, D.T. A novel caveolin-1 biomarker for clinical outcome of sarcopenia. *In Vivo* **2014**, *28*, 383–389.

57. Lunardi, C.C.; Lima, R.M.; Pereira, R.W.; Leite, T.K.; Siqueira, A.B.; Oliveira, R.J. Association between polymorphisms in the TRHR gene, fat-free mass, and muscle strength in older women. *Age* **2013**, *35*, 2477–2483. [CrossRef]

58. Ma, T.; Lu, D.; Zhu, Y.S.; Chu, X.F.; Wang, Y.; Shi, G.P.; Wang, Z.D.; Yu, L.; Jiang, X.Y.; Wang, X.F. ACTN3 genotype and physical function and frailty in an elderly Chinese population: The rugao longevity and ageing study. *Age Ageing* **2018**, *47*, 416–422. [CrossRef]

59. McCauley, T.; Mastana, S.S.; Folland, J.P. ACE I/D and ACTN3 R/X polymorphisms and muscle function and muscularity of older Caucasian men. *Eur. J. Appl. Physiol.* **2010**, *109*, 269–277. [CrossRef]

60. Melzer, D.; Dik, M.G.; van Kamp, G.J.; Jonker, C.; Deeg, D.J. The apolipoprotein E e4 polymorphism is strongly associated with poor mobility performance test results but not self-reported limitation in older people. *J. Gerontol. A Biol. Sci. Med. Sci.* **2005**, *60*, 1319–1323. [CrossRef]

61. Mora, M.; Perales, M.J.; Serra-Prat, M.; Palomera, E.; Buquet, X.; Oriola, J.; Puig-Domingo, M.; Mataró Ageing Study Group. Aging phenotype and its relationship with IGF-I gene promoter polymorphisms in elderly people living in Catalonia. *Growth Horm. IGF Res.* **2011**, *21*, 174–180. [CrossRef] [PubMed]

62. Onder, G.; Capoluongo, E.; Danese, P.; Settanni, S.; Russo, A.; Concolino, P.; Bernabei, R.; Landi, F. Vitamin D receptor polymorphisms and falls among older adults living in the community: Results from the ilSIRENTE study. *J. Bone Miner. Res.* **2008**, *23*, 1031–1036. [CrossRef] [PubMed]

63. Pereira, A.; Costa, A.M.; Izquierdo, M.; Silva, A.J.; Bastos, E.; Marques, M.C. ACE I/D and ACTN3 R/X polymorphisms as potential factors in modulating exercise-related phenotypes in older women in response to a muscle power training stimuli. *Age* **2013**, *35*, 1949–1959. [CrossRef] [PubMed]

64. Pereira, A.; Costa, A.M.; Leitão, J.C.; Monteiro, A.M.; Izquierdo, M.; Silva, A.J.; Bastos, E.; Marques, M.C. The influence of ACE ID and ACTN3 R577X polymorphisms on lower-extremity function in older women in response to high-speed power training. *BMC Geriatr.* **2013**, *13*, 131. [CrossRef]

65. Pereira, D.S.; Mateo, E.C.C.; de Queiroz, B.Z.; Assumpção, A.M.; Miranda, A.S.; Felício, D.C.; Rocha, N.P.; dos Anjos, D.M.D.C.; Pereira, D.A.G.; Teixeira, A.L.; et al. TNF-alpha, IL6, and IL10 polymorphisms and the effect of physical exercise on inflammatory parameters and physical performance in elderly women. *Age* **2013**, *35*, 2455–2463. [CrossRef]

66. Prakash, J.; Herlin, M.; Kumar, J.; Garg, G.; Akesson, K.E.; Grabowski, P.S.; Skerry, T.M.; Richards, G.O.; McGuigan, F.E. Analysis of RAMP3 gene polymorphism with body composition and bone density in young and elderly women. *Gene X* **2019**, *2*, 100009. [CrossRef]

67. Roth, S.M.; Zmuda, J.M.; Cauley, J.A.; Shea, P.R.; Ferrell, R.E. Vitamin D receptor genotype is associated with fat-free mass and sarcopenia in elderly men. *J. Gerontol. A Biol. Sci. Med. Sci.* **2004**, *59*, 10–15. [CrossRef]

68. Skoog, I.; Hörder, H.; Frändin, K.; Johansson, L.; Östling, S.; Blennow, K.; Zetterberg, H.; Zettergren, A. Association between APOE genotype and change in physical function in a population-based swedish cohort of older individuals followed over four years. *Front. Aging Neurosci.* **2016**, *8*, 225. [CrossRef]

69. Tiainen, K.; Thinggaard, M.; Jylha, M.; Bladbjerg, E.; Christensen, K.; Christiansen, L. Associations between inflammatory markers, candidate polymorphisms and physical performance in older Danish twins. *Exp. Gerontol.* **2012**, *47*, 109–115. [CrossRef]

70. Urano, T.; Shiraki, M.; Sasaki, N.; Ouchi, Y.; Inoue, S. Large-scale analysis reveals a functional single-nucleotide polymorphism in the 5′-flanking region of PRDM16 gene associated with lean body mass. *Aging Cell* **2014**, *13*, 739–743. [CrossRef]

71. Verghese, J.; Holtzer, R.; Wang, C.; Katz, M.J.; Barzilai, N.; Lipton, R.B. Role of APOE genotype in gait decline and disability in aging. *J. Gerontol. A Biol. Sci. Med. Sci.* **2013**, *68*, 1395–1401. [CrossRef] [PubMed]

72. Walsh, S.; Zmuda, J.M.; Cauley, J.A.; Shea, P.R.; Metter, E.J.; Hurley, B.F.; Ferrell, R.E.; Roth, S.M. Androgen receptor CAG repeat polymorphism is associated with fat-free mass in men. *J. Appl. Physiol.* **2005**, *98*, 132–137. [CrossRef] [PubMed]

73. Wu, F.Y.; Liu, C.S.; Liao, L.N.; Li, C.I.; Lin, C.H.; Yang, C.W.; Meng, N.H.; Lin, W.Y.; Chang, C.K.; Hsiao, J.H.; et al. Vitamin D receptor variability and physical activity are jointly associated with low handgrip strength and osteoporosis in community-dwelling elderly people in Taiwan: The Taichung Community Health Study for Elders (TCHS-E). *Osteoporos. Int.* **2014**, *25*, 1917–1929. [CrossRef] [PubMed]

74. Xia, Z.; Man, Q.; Li, L.; Song, P.; Jia, S.; Song, S.; Meng, L.; Zhang, J. Vitamin D receptor gene polymorphisms modify the association of serum 25-hydroxyvitamin D levels with handgrip strength in the elderly in Northern China. *Nutrition* **2019**, *57*, 202–207. [CrossRef]

75. Yang, C.W.; Li, T.C.; Li, C.I.; Liu, C.S.; Lin, C.H.; Lin, W.Y.; Lin, C.C. Insulin like growth factor-1 and its binding protein-3 polymorphisms predict circulating IGF-1 level and appendicular skeletal muscle mass in Chinese elderly. *J. Am. Med. Dir. Assoc.* **2015**, *16*, 365–370. [CrossRef]

76. Yoshihara, A.; Tobina, T.; Yamaga, T.; Ayabe, M.; Yoshitake, Y.; Kimura, Y.; Shimada, M.; Nishimuta, M.; Nakagawa, N.; Ohashi, M.; et al. Physical function is weakly associated with angiotensin-converting enzyme gene I/D polymorphism in elderly Japanese subjects. *Gerontology* **2009**, *55*, 387–392. [CrossRef]

77. Zempo, H.; Tanabe, K.; Murakami, H.; Iemitsu, M.; Maeda, S.; Kuno, S. ACTN3 polymorphism affects thigh muscle area. *Int. J. Sports Med.* **2010**, *31*, 138–142. [CrossRef]

78. Zempo, H.; Tanabe, K.; Murakami, H.; Iemitsu, M.; Maeda, S.; Kuno, S. Age differences in the relation between ACTN3 R577X polymorphism and thigh-muscle cross-sectional area in women. *Genet. Test. Mol. Biomark.* **2011**, *15*, 639–643. [CrossRef]

79. Montgomery, H.E.; Marshall, R.; Hemingway, H.; Myerson, S.; Clarkson, P.; Dollery, C.; Hayward, M.; Holliman, D.E.; Jubb, M.; World, M.; et al. Human gene for physical performance. *Nature* **1998**, *393*, 221–222. [CrossRef]

80. Cordero, A.I.H.; Gonzales, N.M.; Parker, C.C.; Sokoloff, G.; Vandenbergh, D.J.; Cheng, R.; Abney, M.; Skol, A.; Douglas, A.; Palmer, A.A.; et al. Genome-wide associations reveal human-mouse genetic convergence and modifiers of myogenesis, CPNE1 and STC2. *Am. J. Hum. Genet.* **2019**, *105*, 1222–1236. [CrossRef]

81. Zillikens, M.C.; Demissie, S.; Hsu, Y.H.; Yerges-Armstrong, L.M.; Chou, W.C.; Stolk, L.; Livshits, G.; Broer, L.; Johnson, T.; Koller, D.L.; et al. Large meta-analysis of genome-wide association studies identifies five loci for lean body mass. *Nat. Commun.* **2017**, *8*, 80. [CrossRef] [PubMed]

82. Houweling, P.J.; North, K.N. Sarcomeric α-actinins and their role in human muscle disease. *Future Neurol.* **2009**, *4*, 731–741. [CrossRef]

83. Blanchard, A.; Ohanian, V.; Critchley, D. The structure and function of α-actinin. *J. Muscle Res. Cell Motil.* **1989**, *10*, 280–289. [CrossRef] [PubMed]

84. North, K.N.; Yang, N.; Wattanasirichaigoon, D.; Mills, M.; Easteal, S.; Beggs, A.H. A common nonsense mutation results in alpha-actinin-3 deficiency in the general population. *Nat. Genet.* **1999**, *21*, 353–354. [CrossRef]

85. Yang, N.; MacArthur, D.G.; Gulbin, J.P.; Hahn, A.G.; Beggs, A.H.; Easteal, S.; North, K. ACTN3 genotype is associated with human elite athletic performance. *Am. J. Hum. Genet.* **2003**, *73*, 627–631. [CrossRef]

86. Druzhevskaya, A.M.; Ahmetov, I.I.; Astratenkova, I.V.; Rogozkin, V.A. Association of the ACTN3 R577X polymorphism with power athlete status in Russians. *Eur. J. Appl. Physiol.* **2008**, *103*, 631–634. [CrossRef]

87. Papadimitriou, I.D.; Lucia, A.; Pitsiladis, Y.P.; Pushkarev, V.P.; Dyatlov, D.A.; Orekhov, E.F.; Artioli, G.G.; Guilherme, J.P.L.; Lancha, A.H.; Ginevičienė, V.; et al. ACTN3 R577X and ACE I/D gene variants influence performance in elite sprinters: A multi-cohort study. *BMC Genomics* **2016**, *17*, 285. [CrossRef]

88. Tieland, M.; Trouwborst, I.; Clark, B.C. Skeletal muscle performance and ageing. *J. Cachexia Sarcopenia Muscle* **2018**, *9*, 3–19. [CrossRef]

89. Pickering, C.; Kiely, J. ACTN3: More than just a gene for speed. *Front. Physiol.* **2017**, *8*, 1080. [CrossRef]

90. Kumagai, H.; Tobina, T.; Ichinoseki-Sekine, N.; Kakigi, R.; Tsuzuki, T.; Zempo, H.; Shiose, K.; Yoshimura, E.; Kumahara, H.; Ayabe, M.; et al. Role of selected polymorphisms in determining muscle fiber composition in Japanese men and women. *J. Appl. Physiol.* **2018**, *124*, 1377–1384. [CrossRef]

91. Nielsen, J.; Christensen, D.L. Glucose intolerance in the West African Diaspora: A skeletal muscle fibre type distribution hypothesis. *Acta Physiol.* **2011**, *202*, 605–616. [CrossRef] [PubMed]

92. Ama, P.F.; Simoneau, J.A.; Boulay, M.R.; Serresse, O.; Thériault, G.; Bouchard, C. Skeletal muscle characteristics in sedentary black and Caucasian males. *J. Appl. Physiol.* **1986**, *61*, 1758–1761. [CrossRef] [PubMed]

93. Jeng, C.; Zhao, L.; Wu, K.; Zhou, Y.; Chen, T.; Deng, H.W. Race and socioeconomic effect on sarcopenia and sarcopenic obesity in the Louisiana Osteoporosis Study (LOS). *JCSM Clin. Rep.* **2018**, *3*, e00027. [CrossRef] [PubMed]

94. Rigat, B.; Hubert, C.; Alhenc-Gelas, F.; Cambien, F.; Corvol, P.; Soubrier, F. An insertion/deletion polymorphism in the angiotensin I-converting enzyme gene accounting for half the variance of serum enzyme levels. *J. Clin. Invest.* **1990**, *86*, 1343–1346. [CrossRef]

95. Gordon, S.E.; Davis, B.S.; Carlson, C.J.; Booth, F.W. ANG II is required for optimal overload-induced skeletal muscle hypertrophy. *Am. J. Physiol. Endocrinol. Metab.* **2001**, *280*, E150–E159. [CrossRef]

96. Danser, A.J.; Schalekamp, M.A.; Bax, W.A.; van den Brink, A.M.; Saxena, P.R.; Riegger, G.A.; Schunkert, H. Angiotensin-converting enzyme in the human heart. Effect of the deletion/insertion polymorphism. *Circulation* **1995**, *92*, 1387–1388. [CrossRef]

97. Tiret, L.; Rigat, B.; Visvikis, S.; Breda, C.; Corvol, P.; Cambien, F.; Soubrier, F. Evidence, from combined segregation and linkage analysis, that a variant of the angiotensin I-converting enzyme (ACE) gene controls plasma ACE levels. *Am. J. Hum. Genet.* **1992**, *51*, 197–205.

98. Williams, A.G.; Day, S.H.; Folland, J.P.; Gohlke, P.; Dhamrait, S.; Montgomery, H.E. Circulating angiotensin converting enzyme activity is correlated with muscle strength. *Med. Sci. Sports Exerc.* **2005**, *37*, 944–948. [CrossRef]

99. Myerson, S.; Hemingway, H.; Budget, R.; Martin, J.; Humphries, S.; Montgomery, H. Human angiotensin I-converting enzyme gene and endurance performance. *J. Appl. Physiol.* **1999**, *87*, 1313–1316. [CrossRef]

100. Puthucheary, Z.; Skipworth, J.R.; Rawal, J.; Loosemore, M.; Van Someren, K.; Montgomery, H.E. The ACE gene and human performance: 12 Years on. *Sports Med.* **2011**, *41*, 433–448. [CrossRef]

101. Al-Hinai, A.T.; Hassan, M.O.; Simsek, M.; Al-Barwani, H.; Bayoumi, R. Genotypes and allele frequencies of angiotensin converting enzyme (ACE) insertion/deletion polymorphism among Omanis. *J. Sci. Res. Med. Sci.* **2002**, *4*, 25–27. [PubMed]

102. Wishart, J.M.; Horowitz, M.; Need, A.G.; Scopacasa, F.; Morris, H.A.; Clifton, P.M. Relations between calcium intake, calcitriol, polymorphisms of the vitamin D receptor gene, and calcium absorption in premenopausal women. *Am. J. Clin. Nutr.* **1997**, *65*, 798–802. [CrossRef] [PubMed]

103. Pfeifer, M.; Begerow, B.; Minne, H.W. Vitamin D and muscle function. *Osteoporos. Int.* **2002**, *13*, 187–194. [CrossRef] [PubMed]

104. Whitfield, G.K.; Remus, L.S.; Jurutka, P.W.; Zitzer, H.; Oza, A.K.; Dang, H.T.; Haussler, C.A.; Galligan, M.A.; Thatcher, M.L.; Dominguez, C.E.; et al. Functionally relevant polymorphisms in the human nuclear vitamin D receptor gene. *Mol. Cell Endocrinol.* **2001**, *177*, 145–159. [CrossRef]

105. Baker, A.R.; McDonnell, D.P.; Hughes, M.; Crisp, T.M.; Mangelsdorf, D.J.; Haussler, M.R.; Pike, J.W.; Shine, J.; O'Malley, B.W. Cloning and expression of full-length cDNA encoding human vitamin D receptor. *Proc. Natl. Acad. Sci. USA* **1988**, *85*, 3294–3298. [CrossRef]

106. Arai, H.; Miyamoto, K.I.; Taketani, Y.; Yamamoto, H.; Iemori, Y.; Morita, K.; Tonai, T.; Nishisho, T.; Mori, S.; Takeda, E. A vitamin D receptor gene polymorphism in the translation initiation codon: Effect on protein activity and relation to bone mineral density in Japanese women. *J. Bone Miner. Res.* **1997**, *12*, 915–921. [CrossRef]

107. Moreno Lima, R.; De Abreu, B.S.; Gentil, P.; de Lima Lins, T.C.; Grattapaglia, D.; Pereira, R.W.; De Oliveira, R.J. Lack of association between vitamin D receptor genotypes and haplotypes with fat-free mass in postmenopausal Brazilian women. *J. Gerontol. A Biol. Sci. Med. Sci.* **2007**, *62*, 966–972. [CrossRef]

108. Iki, M.; Saito, Y.; Dohi, Y.; Kajita, E.; Nishino, H.; Yonemasu, K.; Kusaka, Y. Greater trunk muscle torque reduces postmenopausal bone loss at the spine independently of age, body size, and vitamin D receptor genotype in Japanese women. *Calcif. Tissue Int.* **2002**, *71*, 300–307. [CrossRef]

109. Nelson, D.A.; Vande-Vord, P.J.; Wooley, P.H. Polymorphism in the vitamin D receptor gene and bone mass in African-American and white mothers and children: A preliminary report. *Ann. Rheum. Dis.* **2000**, *59*, 626–630. [CrossRef]

110. Fleet, J.C.; Harris, S.S.; Wood, R.J.; Dawson-Hughes, B. The BsmI vitamin D receptor restriction fragment length polymorphism (BB) predicts low bone density in premenopausal black and white women. *J. Bone Miner. Res.* **1995**, *10*, 985–990. [CrossRef]

111. O'Dell, S.D.; Day, I.N. Insulin-like growth factor II (IGF-II). *Int. J. Biochem. Cell Biol.* **1998**, *30*, 767–771. [CrossRef]

112. Stewart, C.E.; Rotwein, P. Growth, differentiation, and survival: Multiple physiological functions for insulin-like growth factors. *Physiol. Rev.* **1996**, *76*, 1005–1026. [CrossRef] [PubMed]

113. Junnila, R.K.; List, E.O.; Berryman, D.E.; Murrey, J.W.; Kopchick, J.J. The GH/IGF-1 axis in ageing and longevity. *Nat. Rev. Endocrinol.* **2013**, *9*, 366–376. [CrossRef] [PubMed]

114. Jones, J.I.; Clemmons, D.R. Insulin-like growth factors and their binding proteins: Biological actions. *Endocr. Rev.* **1995**, *16*, 3–34. [CrossRef]

115. Baxter, G.T.; Kuo, R.C.; Jupp, O.J.; Vandenabeele, P.; MacEwan, D.J. Tumor necrosis factor-alpha mediates both apoptotic cell death and cell proliferation in a human hematopoietic cell line dependent on mitotic activity and receptor subtype expression. *J. Biol. Chem.* **1999**, *274*, 9539–9547. [CrossRef]

116. Bradley, J.R. TNF-mediated inflammatory disease. *J. Pathol.* **2008**, *214*, 149–160. [CrossRef]

117. Costamagna, D.; Costelli, P.; Sampaolesi, M.; Penna, F. Role of inflammation in muscle homeostasis and myogenesis. *Mediat. Inflamm.* **2015**, *2015*, 805172. [CrossRef]

118. Smith, J.D. Apolipoproteins and aging: Emerging mechanisms. *Ageing Res. Rev.* **2002**, *1*, 345–365. [CrossRef]

119. Bertram, L.; McQueen, M.B.; Mullin, K.; Blacker, D.; Tanzi, R.E. Systematic meta-analyses of Alzheimer disease genetic association studies: The AlzGene database. *Nat. Genet.* **2007**, *39*, 17–23. [CrossRef]

120. Fullerton, S.M.; Clark, A.G.; Weiss, K.M.; Nickerson, D.A.; Taylor, S.L.; Stengård, J.H.; Salomaa, V.; Vartiainen, E.; Perola, M.; Boerwinkle, E.; et al. Apolipoprotein E variation at the sequence haplotype level: Implications for the origin and maintenance of a major human polymorphism. *Am. J. Hum. Genet.* **2000**, *67*, 881–900. [CrossRef]

121. Kami, K.; Morikawa, Y.; Sekimoto, M.; Senba, E. Gene expression of receptors for IL-6, LIF, and CNTF in regenerating skeletal muscles. *J. Histochem. Cytochem.* **2000**, *48*, 1203–1213. [CrossRef] [PubMed]

122. Sendtner, M.; Kreutzberg, G.W.; Thoenen, H. Ciliary neurotrophic factor prevents the degeneration of motor neurons after axotomy. *Nature* **1990**, *345*, 440–441. [CrossRef] [PubMed]

123. Sleeman, M.W.; Anderson, K.D.; Lambert, P.D.; Yancopoulos, G.D.; Wiegand, S.J. The ciliary neurotrophic factor and its receptor, CNTFR alpha. *Pharm. Acta Helv.* **2000**, *74*, 265–272. [CrossRef]

124. Schrauwen, P.; Hesselink, M. UCP2 and UCP3 in muscle controlling body metabolism. *J. Exp. Biol.* **2002**, *205*, 2275–2285.

125. Schrauwen, P.; Hoeks, J.; Hesselink, M.K. Putative function and physiological relevance of the mitochondrial uncoupling protein-3: Involvement in fatty acid metabolism? *Prog. Lipid Res.* **2006**, *45*, 17–41. [CrossRef]

126. Dhamrait, S.S.; Williams, A.G.; Day, S.H.; Skipworth, J.; Payne, J.R.; World, M.; Humphries, S.E.; Montgomery, H.E. Variation in the uncoupling protein 2 and 3 genes and human performance. *J. Appl. Physiol.* **2012**, *112*, 1122–1127. [CrossRef]

127. Brand, M.D.; Pamplona, R.; Portero-Otín, M.; Requena, J.R.; Roebuck, S.J.; Buckingham, J.A.; Clapham, J.C.; Cadenas, S. Oxidative damage and phospholipid fatty acyl composition in skeletal muscle mitochondria from mice underexpressing or overexpressing uncoupling protein 3. *Biochem. J.* **2002**, *368*, 597–603. [CrossRef]

128. Seo, D.Y.; Lee, S.R.; Kim, N.; Ko, K.S.; Rhee, B.D.; Han, J. Age-related changes in skeletal muscle mitochondria: The role of exercise. *Integr. Med. Res.* **2016**, *5*, 182–186. [CrossRef]

129. Brand, M.D.; Affourtit, C.; Esteves, T.C.; Green, K.; Lambert, A.J.; Miwa, S.; Pakay, J.L.; Parker, N. Mitochondrial superoxide: Production, biological effects, and activation of uncoupling proteins. *Free Radic. Biol. Med.* **2004**, *37*, 755–767. [CrossRef]

130. Colhoun, H.M.; McKeigue, P.M.; Smith, D.G. Problems of reporting genetic associations with complex outcomes. *Lancet* **2003**, *361*, 865–872. [CrossRef]

131. Duncan, L.E.; Ostacher, M.; Ballon, J. How genome-wide association studies (GWAS) made traditional candidate gene studies obsolete. *Neuropsychopharmacology* **2019**, *44*, 1518–1523. [CrossRef] [PubMed]

132. Visscher, P.M.; Wray, N.R.; Zhang, Q.; Sklar, P.; McCarthy, M.I.; Brown, M.A.; Yang, J. 10 Years of GWAS discovery: Biology, function, and translation. *Am. J. Hum. Genet.* **2017**, *101*, 5–22. [CrossRef] [PubMed]

133. Sawcer, S. The complex genetics of multiple sclerosis: Pitfalls and prospects. *Brain* **2008**, *131*, 3118–3131. [CrossRef] [PubMed]

Review

The Survey of Cells Responsible for Heterotopic Ossification Development in Skeletal Muscles—Human and Mouse Models

Łukasz Pulik [1,†], Bartosz Mierzejewski [2,†], Maria A. Ciemerych [2], Edyta Brzóska [2,*] and Paweł Łęgosz [1,*]

[1]   Department of Orthopaedics and Traumatology, Medical University of Warsaw, Lindley 4 St, 02-005 Warsaw, Poland; lukasz.pulik@wum.edu.pl
[2]   Department of Cytology, Faculty of Biology, University of Warsaw, Miecznikowa 1 St, 02-096 Warsaw, Poland; bmierzejewski@biol.uw.edu.pl (B.M.); ciemerych@biol.uw.edu.pl (M.A.C.)
*   Correspondence: edbrzoska@biol.uw.edu.pl (E.B.); pawel.legosz@wum.edu.pl (P.Ł.); Tel.: +48-22-5542-203 (E.B.); +48-22-5021-514 (P.Ł.)
†   These authors contribute equaly to this work.

Received: 17 April 2020; Accepted: 21 May 2020; Published: 26 May 2020

Abstract: Heterotopic ossification (HO) manifests as bone development in the skeletal muscles and surrounding soft tissues. It can be caused by injury, surgery, or may have a genetic background. In each case, its development might differ, and depending on the age, sex, and patient's conditions, it could lead to a more or a less severe outcome. In the case of the injury or surgery provoked ossification development, it could be, to some extent, prevented by treatments. As far as genetic disorders are concerned, such prevention approaches are highly limited. Many lines of evidence point to the inflammatory process and abnormalities in the bone morphogenetic factor signaling pathway as the molecular and cellular backgrounds for HO development. However, the clear targets allowing the design of treatments preventing or lowering HO have not been identified yet. In this review, we summarize current knowledge on HO types, its symptoms, and possible ways of prevention and treatment. We also describe the molecules and cells in which abnormal function could lead to HO development. We emphasize the studies involving animal models of HO as being of great importance for understanding and future designing of the tools to counteract this pathology.

Keywords: muscles; heterotopic ossification; skeletal muscle stem and progenitor cells; HO precursors

---

1. Introduction

Heterotopic ossification (HO) is a disregulation of skeletal muscle homeostasis and regeneration, which results in mature bone formation in atypical locations. HO could develop in the skeletal muscles, and also in surrounding tissues such as fascia, tendons, skin, and subcutis [1]. HO can be acquired or have genetic origin. The most prevalent is acquired HO which can occur in response to a direct trauma, burn, or amputations. Similarly, iatrogenic trauma, caused by orthopedic surgery such as hip replacement, often triggers HO development [1,2]. Another acquired form of the disease is neurogenic HO (NHO) which is a frequent complication of central nervous system injury [3]. The knowledge about the molecular mechanisms that leads to HO formation and cell precursors engaged in this process is still limited. HO requires the presence of stem or progenitor cells which are able to follow the osteogenic program, although the identity of these cells remains unclear. Many different populations of progenitor cells could be possible precursors in the HO development. The animal studies suggest that progenitor cells can vary depending on the HO subtype. The studies using mouse HO models show that endothelial cells, mesenchymal cells, pericytes present in the skeletal muscles, tendons and connective tissue cells, or even circulating stem/precursor cells could be a source of HO precursors [1,4,5]. It is also known, that trauma or micro-trauma, which leads to a local inflammatory response, delivers the signals to develop HO. Recent studies showed the role of immune response cells, especially monocytes/macrophages, at the early stages of trauma-induced HO development [6]. They confirm the importance of macrophages in the induction of neurogenic and genetic forms of HO [7]. Activated macrophages express osteoinductive signaling factors in the course of HO pathogenesis. Thus, the presence of the cells reflects increased secretion of HO promoting cytokines/chemokines such as interleukin 6 (IL6), IL10, transforming beta-1 growth factor (TGFβ1), and neurotrophin 3 (NT3). A significant dysregulation of macrophage immune checkpoints was proven in HO animal models [8–10]. Finally, both individual predisposition and risk factors also attribute to HO development [11].

Histologically HO formation is similar to the physiological bone fracture healing. During HO development initially soft tissue is infiltrated with the whole spectrum of inflammatory cells. Such infiltration is followed by enhanced fibroblast proliferation, neovascularization, differentiation of chondrocytes, and results in mature bone formation [12]. HO is formed mainly by endochondral ossification. However, an intramembranous mechanism can also be involved. Typically HO formation is characterized by a zonal bone development model called "eggshell calcification" [13]. HO consists mainly of mechanically weak woven bone with an irregular osteoblasts distribution, but mature lamellar bone with Haversian-like canals can be often found. The bone tissue gradually matures with the outer appearance of the cortical bone [14]. Primarily, HO occurs in soft tissues and has no connection with the skeletal bone, but when it grows in the volume, it can attach to the periosteum.

2. Heterotopic Ossification as a Clinical Issue

HO is a diverse pathologic process and its spectrum can range from mild, clinically irrelevant to severe cases. In most of the patients it is minor and symptomless. Unfortunately, in some patients, extensive HO located around joints can cause restrictions in the range of motion (ROM), resulting even in the total ankylosis of the joint. In this group of patients HO can be associated with a significant limitation of daily activities and disability [1].

2.1. Traumatic HO

HO lesions can occur in response to direct trauma, such as connective tissue injury, bone fractures, burns, amputations, and combat-related blast injuries [2]. Approximately 30% of all fractures and dislocations which were subjected to operative treatment can trigger HO formation. It was a recognized clinical problem, in the acetabular and proximal femur fractures and fractures or dislocations of the elbow [1,15]. HO is also a common complication of traumatic limb amputation, both in civilian (22.8%) and in a military setting (62.9%) [16]. After the isolated burn injury HO incidence is relatively low (1%–4%), but it can be underestimated due to the lack of routine x-ray screening of such patients [17,18]. The typical locations of burn-induced HO are the elbow (50.0%), glenohumeral joint (20.3%), and the hip joint (17.6%) [18].

Recognized risk factors for trauma-induced HO are young age, male sex, severe concomitant injuries, compound fractures, extensive surgical approaches, and postponed surgeries [19,20]. The presence of local wound infection is also a well-established factor associated with HO [21]. In the military setting, the high incidence of HO is associated with concomitant brain injury, multiple wounds, and the severity of the injury [2]. The extent of burns, local wound infection, and the duration of intensive therapy are the risk factors for the formation of burn-induced HO. The role of immobilization and iatrogenic paralysis is also under investigation [18,22].

As far as the symptoms are concerned it can be associated with reduced joint mobility, pain, and decreased limb function. In the upper extremity, HO can limit everyday activities such as eating, dressing, and personal grooming, while in the lower limb it can affect gait and cause limp and difficulties in sitting [15]. Patients with amputation associated HO may experience difficulties with prosthesis fitting. Other local complications can occur, such as ulcers, skin graft necrosis, and neurovascular impairment [16]. The contractures and reduction of joint ROM in burn injuries are often caused by soft tissue scarring. However, HO should always be taken into consideration in differential diagnosis [18].

2.2. Surgery-Induced HO

HO is a well-described complication of orthopedic surgical procedures, typically joint replacements. The main indication for this kind of treatment is symptomatic, end-stage joint osteoarthritis. This type of HO usually involves the tissues where the surgical approach is performed, as unavoidable trauma is done to the muscle and fascia. It is most commonly described after a total hip replacement (THR) and cervical total disc arthroplasty (CTDA). It was also reported after the ankle, knee, and shoulder arthroplasty [23–25]. HO is radiologically present in every second patient after total hip replacement (40%–56%), and cervical total disc arthroplasty (44.6%–58.2%). However, high-grade HO occurs only in 2%–7% of total hip replacement and 11%–16% of cervical total disc arthroplasty patients [24–26].

The risk factors for the development of HO in patients undergoing total hip replacement are young age and male sex. The other predisposing conditions are bone and joint diseases such as ankylosing spondylitis, hypertrophic arthritis, and Paget's disease. The impact of the surgical approach, especially micro-invasive surgery (MIS) techniques, is being intensively discussed, but the results are still inconclusive [24,27,28]. Similarly, in patients after the cervical total disc arthroplasty, the male sex is an independent risk factor for HO development. Another important aspect is an artificial disc device type [29].

In the majority of patients suffering from surgery induced HO, small islands of the bone of no clinical significance are observed. However, extensive lesions can affect the biomechanical function of an endoprosthesis and block the movement in the affected joint. In total hip replacement patients, high-grade HO can significantly impact ROM, especially flexion, abduction, and external rotation, and affect the overall function of the hip [30]. In extreme cases, HO can require surgical intervention and excision (3.3%) [27]. In contrast to total hip replacement, in cervical total disc arthroplasty patients, severe HO does not affect patient-related pain, quality of life, or function [31].

2.3. Neurogenic HO

Neurogenic heterotopic ossification (NHO) can occur after the spinal cord injury (SCI) or traumatic brain injury (TBI). Other clinical conditions, such as cerebral stroke, anoxia, and non-traumatic myelopathies, can also attribute to NHO [32–34]. The incidence of NHO in spinal cord injury (40%–50%) was reported to be higher than in traumatic brain injury patients (8%–23%). However, symptomatic NHO is more frequent in traumatic brain injury than in spinal cord injury patients (11% vs. 4%). The incidence of this type of HO in cerebral stroke patients is relatively low (0.5%–1.2%) [35,36]. In contrast to traumatic lesions, NHO lesions typically occur in locations distant from the site of injury. The NHO lesions are usually located around the hip joints in both spinal cord injury (63%) and traumatic brain injury (40%) [36]. Other possible locations of NHO are the shoulder, knee, and elbow joints [37,38].

The demographic risk factors predisposing to NHO after spinal cord injury are male sex and young age. The complete spinal cord injury, high level of rupture, and spasticity can also be associated with an elevated incidence of NHO. Moreover, urinary tract infections and pneumonia significantly increase the risk of NHO [1,38]. NHO may also be associated with human leukocyte antigen B27 (HLA-B27) presence in spinal cord injury patients [39]. In traumatic brain injury patients, the NHO associated conditions are lower walking abilities, spasticity, pressure ulcers, neurogenic bladder, and systemic infections [40].

NHO usually develops two months after a spinal cord injury or traumatic brain injury [36,41,42]. Initially, it is characterized by inflammation-like prodromal symptoms such as swelling, redness of the joint, and low-grade fever. If there is no sensory impairment, the pain can also be present. Usually, two years after the neurological event, the lesions are fully developed [3,36]. Most of the patients do not suffer from NHO associated symptoms. However, when the lesions are extensive, it can affect joint ROM creating problems with moving from sitting to lying position and nursing. Additionally, the risk of bedsores significantly rises [3,37]. Moreover, patients with severe NHO obtain less satisfactory functional results and require prolonged rehabilitation [43].

2.4. Genetic HO

There are also rare, inherited forms of HO, such as fibrodysplasia ossificans progressiva (FOP) and progressive osseous heteroplasia (POH) [44,45]. FOP is autosomal-dominant disorder caused by up to 14 different mutations localised in the type I bone morphogenic protein (BMP) receptor, i.e., activin type 1 receptor (ACVR1; also called activin-like kinase 2, ALK2) gene [44,46]. However, a single mutation, i.e., arginine to histidine at position 206; R206H, is present in the majority of FOP patients [44,46]. The ossification of skeletal muscles in FOP occurs mostly in early childhood and is characterized by inflammation-like symptoms and episodical flareups [44]. POH is the other autosomal dominant inherited form of HO, which is caused by the mutation in gene encoding guanine nucleotide-binding protein, alpha stimulating (GNAS) [47]. The exact incidence of FOP is estimated to be approximately 1 person per 2 million [47]. The epidemiological data of 299 FOP patients from fifty-four countries participating in the International FOP Association (IFOPA) Global Registry will be published soon [48]. In most cases of FOP, it is caused by de novo mutations, but there is also a risk of parental transmission [49]. The incidence of POH is unknown. Similarly to FOP, in POH family transmissions have been documented, but the majority of the patients have spontaneous mutations [47]. There are no identified predisposing factors for inherited HO, including ethnic, racial, or geographic factors [47,50].

Children suffering from FOP are born with characteristic deformities of the toe and then, usually between 5 and 10 years of age, start to present soft tissue swelling which could be spontaneous or caused by minor injuries. With age, mature bone appears at the site of edema in the muscles and surrounding tissues. HO can appear in any location except for the viscera and thoracic diaphragm. The first affected areas are neck and upper back. Progression of HO over time leads to mobility restriction, respiratory problems, and heart failure associated with intercostal and spinal muscle ossification and chest deformation [50]. Recent studies revealed dysmorphology of the hip, spine, and tibiofibular joint, which can predispose to the high incidence of arthropathy in FOP patients [51]. Other aspects of FOP are malnutrition due to temporomandibular joint ossification and hearing problems due to middle ear HO. Most patients use a wheelchair at the end of their second decade of life [47,49]. In contrast to FOP, the HO lesions in POH usually appear early, i.e., during the first year of life. POH starts from the skin and subcutis and later on affects the deeper-lying striated muscles and fascia. POH is characterized by changeable expressivity and somatic mosaicism, including asymptomatic carriers. In some cases with high expression, it can result in early severe disability with joint ankylosis. The sings of POH can also include growth retardation, osteoporosis, and low body weight. The diagnostic criteria for POH have been proposed [45].

2.5. Diagnostic Imaging

Radiography is a first-line diagnostic tool in routine HO detection. The most commonly used HO classification systems, such as the Brooker classification for the hip and Hastings and Graham classification of the elbow, are based on the X-ray assessment [3]. Computed tomography (CT) can provide a more accurate assessment of the relation of HO lesion to the joint and other vascular and neural structures. CT and X-ray examinations remain the gold standard in the imaging diagnosis due to a low cost, simplicity, and high effectiveness in detecting fully developed HO lesions [52]. Nuclear medicine modalities can also be useful and provide metabolic and functional information on developing HO. The scintigraphy, including planar bone scan and single-photon emission computed tomography (SPECT), is proven to be a highly sensitive method in HO detection [53,54]. Similarly, the positron emission tomography (PET) can be useful in the HO diagnosis and successfully identify early HO and chronic lesions [55]. The other diagnostic imaging techniques include magnetic resonance imaging (MRI) that can identify vascularization and increased density in the early phases of HO as early as two days after the onset of clinical symptoms [38]. Recently, the ultrasonography is gaining popularity in HO detection and monitoring due to its safety profile, low cost, and the possibility of bedside-application [56]. In diagnostic imaging, it is critical to distinguish HO from neoplastic processes such as osteosarcoma, deep venous thrombosis (DVT). HO can also mimic gout, avulsion fracture, or local tissue calcifications like dystrophic and tumoral calcification or calcific tendonitis [11,52].

2.6. Biomarkers

The serum alkaline phosphatase enzyme (ALP) was extensively investigated as a potential biomarker of HO in traumatic HO, NHO after spinal cord injury, and total hip replacement induced HO. The elevated ALP level reflects enhanced bone turnover and increases with osteoclast activity. Detection of ALP could serve as a relatively inexpensive and widely available test for HO. Serum ALP concentration increases about two weeks after the operation reaching the peak concentration at week 10–12 and returns to the base level at week 18. However, ALP levels can be normal in the presence of HO development (55.2%), and the usefulness of ALP in HO screening is being discussed. Similarly, a bone-specific isoform of alkaline phosphatase (BAP) can be elevated in HO patients, but BAP levels are normal in most cases (67.8%). The other tested HO biomarkers are urinary excretion of type I collagen cross-linked C-telopeptide (CTX-1) and prostaglandin E2 (PGE2). The classical inflammation marker C-reactive protein, CRP is elevated in 77.0% of HO patients, but it is not specific [39,57,58]. Additionally, cytokine levels are investigated as biomarkers of HO onset. In the mouse model of FOP, the level of monocyte chemoattractant protein 1 (MCP1) (serum, saliva), IL1β (saliva), and tumor necrosis factor α (TNFα) (serum) were significantly increased compared to control group. In the mouse model of trauma-induced HO, the levels of TNFα, IL1β, IL6, and MCP1 were increased in serum samples [59]. In human studies, HO was associated with the level of serum (IL3, IL12) and wound effluent cytokines (IL3, IL13) in combat-injured patients [60].

Recently proteomic biomarkers were analyzed with mass spectrometry in non-genetic HO patients. Significant differences were found in the levels of certain peptides in patients with HO compared to the non-HO group. The researchers point out the protein fragments of osteocalcin (OC), collagen alpha 1 (COL1), osteomodulin (OMD) as potential clinical biomarkers for HO [61]. Another investigated class of HO biomarkers are small non-coding RNA molecules (miRNA). The disregulation of miRNA homeostasis may play a vital role in HO development. For instance, the decreased expression of miRNA-630, which is responsible for endothelial cells transition towards mesenchymal cells, was observed in HO patients [62]. The decreased level of miRNA-421 in humeral fracture patients is associated with BMP2 overexpression and a higher rate of the HO occurrence [63]. The miRNA-203 downregulation leads to an increase in expression of runt-related transcription factor 2 (Runx2), which is a crucial osteoblast differentiation regulator [64]. The miRNA particles are not only possible HO indicators, but they can also be future therapeutic targets.

2.7. Prophylaxis

The standard HO prophylaxis is pharmacological treatment with nonsteroidal anti-inflammatory drugs (NSAIDs) or local external beam radiotherapy (RT). The NSAIDs or radiotherapy prophylaxis is a well-proven and effective method, but it is not specific. Currently, the more targeted pharmacological strategies are being tested and developed for inhibiting specific pathways and molecules responsible for HO. Once the mature lesion is developed, it is not possible to reverse the changes, and the only remaining treatment option is surgical resection [65]. Despite that NSAIDs are effective prophylaxis of HO, they do not present efficacy when HO is fully developed. There is no difference between non-selective NSAIDs and selective NSAIDs in HO treatment [66]. The selective cyclooxygenase-2 (COX-2) inhibitors can significantly decrease discontinuation of treatment due to gastro-intestinal (GI) side effects [67,68]. However, non-selective NSAIDs are the most commonly used in clinical practice (87%) and remain "golden standard" in HO prevention. Indomethacin non-selective COX inhibitor is the most commonly prescribed NSAID for HO prophylaxis (57%) with a daily dose of 100–150 mg with a mean of 30 days of administration [69,70]. In addition to high efficiency, NSAIDs are approximately 45 times more cost-effective compared to RT [71,72].

RT recommended before the surgery or early, up to 72 h post-surgery, is an equally successful method for prophylaxis of HO development as NSAIDs. The multiple fractions RT is more effective in the reduction of HO. It is dose-dependent, but a modification of a biologically effective radiation dose over the >2500 cGy did not result in better effectiveness [73]. In total hip replacement patients, the combination of NSAIDs and RT may also be beneficial [74]. The RT seems to be a safe method of HO prevention in total hip replacement patients regarding local neoplastic processes and aseptic loosening of the implant [75].

The other prophylaxis modalities were also proposed. Taking into account bacterial contamination of wound in traumatic HO, locally administered vancomycin prophylaxis suppressed HO in trauma-induced rats infected with methicillin—resistant *Staphylococcus aureus* (MRSA) [76]. Bisphosphonates that are mainly used as anti-osteoporosis drugs and act by inhibiting calcification, and bone resorption dependent on osteoclasts, have no significantly higher efficacy than NSAIDs [77]. The aspirin, which has both effects of NSAID and the anti-platelet agents, is often used for venous thromboembolism (VTE) prophylaxis in total hip replacement patients and was also shown to be effective in the HO rate reduction [78].

2.8. Treatment

Surgical removal of lesions is currently the only effective method when HO is already formed and gives clinical symptoms. However, the operation itself may induce the formation of new ossifications. Among the indications of HO are pain and reduction of ROM. In most cases, the treatment also includes NSAIDs or radiotherapy as the prevention of relapse. A common strategy is to change the type of prophylaxis or the application of another type of NSAIDs class if the previously used prophylaxis has failed. The standard procedure is simple excision of HO, but it is unclear whether it should be removed completely or only partially [79]. Some authors recommend HO surgery only when the mature bone tissue is formed. However, early intervention minimizes the development of intra-articular changes and HO recurrence, so ossifications should be removed as soon as the mature bone is formed, without unnecessary delay [80–83]. As a result of surgery, the pain level is reduced and ROM increases, which significantly improves the function and often reduces the level of pain [70,84–88]. The total hip replacement is a promising solution for NHO in the area of the hip joint in patients after traumatic brain injury. The standard procedure is the Girdlestone procedure, but total hip replacement seems to give better results than a simple excision. When using THA, ossification has less tendency to relapse and the patient achieves more satisfactory functional results. [89,90].

3. Heterotopic Ossification Precursor Cells

3.1. Stem and Progenitor Cells in Skeletal Muscles

HO development is a complex process engaging many different cell types. Several lines of evidence suggest that the development of HO in skeletal muscle could be a result of pathological differentiation of stem and progenitor cells present in skeletal muscle. The most important cells responsible for postnatal skeletal muscle growth and regeneration are satellite cells (SCs), i.e., unipotent stem cells located between muscle fibers plasmalemma and basal lamina (Figure 1). These cells are activated in response to skeletal muscle injury which results in the cell cycle re-entry [91]. The signals activating satellite cells are provided by damaged muscle fibers, inflammatory cells, and endothelium [92]. Activated SCs start to proliferate, differentiate into myoblasts, i.e., muscle progenitor cells, and then myocytes. The myocytes fuse with existing myofibers or with each other to form myotubes and then, after innervation, myofibers. Many studies showed that SC presence is essential for skeletal muscle regeneration [93]. This multi-step process is accompanied by changes in expression of pair box transcription factors 7 (Pax7) and myogenic regulatory factors (MRFs), such as MYOD, MRF5, myogenin, MRF4, as well as skeletal muscle structural proteins [94]. Importantly, SCs are able to follow two different fates—they could maintain PAX7 and down-regulate MYOD expression to self-renew their population or down-regulate PAX7 and maintain MYOD expression to upregulate MYOGENIN and initiate differentiation [94]. SCs proliferation is regulated by MYOD and MYF5 which control the activity of the genes involved in DNA replication and cell cycle progression, such as cell division cycle 6 protein (CDC6) and minichromosome maintenance complex component 2 (MCM2). MYOD contribution in the myogenic differentiation also involves the induction of miR206 and miR486 which downregulate PAX7 [95]. Moreover, long non-coding RNA linc-RAM promotes the formation of MYOD complex with chromatin modifier BAF60c which enables MYOD binding to promoters of target genes and marks the chromatin for recruitment of chromatin-remodeling complex, i.e., BRG1-based SWItch/Sucrose NonFermentable (SWI/SNF). This MYOD-BAF60c-BRG1 complex remodels the chromatin and activates transcription of MYOD-target genes [96]. Furthermore, MYOD, as stated above, promotes expression of MYOGENIN and MRF4, i.e., transcription factors responsible for myoblast cell cycle exit and their differentiation into myocytes and myotubes. These differentiation steps are accompanied with expression of myosin heavy chains (MHC), enolase 3 (ENO3), and muscle creatine kinase (MCK) [91].

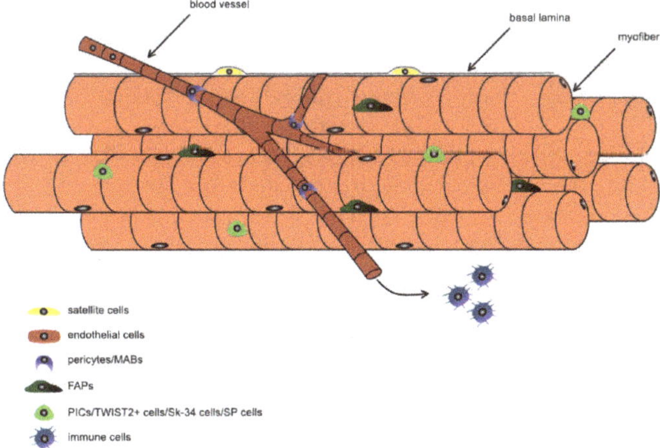

Figure 1. The stem and progenitor cells responsible for skeletal muscle homeostasis. The multinucleated skeletal muscle myofibers are accompanied by several types of stem and progenitor cells, such as satellite cells, endothelial cells, pericytes, mesoangioblasts (MABs), and fibro-adipogenic progenitors (FAPs), which could participate in regeneration. Other populations of muscle interstitial cells, such as, PW1+/PAX7 interstitial cells (PICs), Sk-34 cells, TWIST2+ cells, side population (SP) cells was also shown to be able to follow myogenic program. Moreover, the skeletal muscle reconstruction is accompanied by infiltration by immune cells.

Importantly, the fate of SCs is determined by their interactions with the niche. The quiescent SC niche is formed by myofibers and the extracellular matrix (ECM), i.e., the basal lamina. Such a niche is modified after the skeletal muscle injury and during regeneration. The factors secreted by damaged myofibers, inflammatory cells, endothelial cells, fibroblasts, and fibro-adipogenic progenitors (FAP), present in skeletal muscle, regulate the fate of SCs and myoblasts. Since the inflammation is among the initial responses to muscle injury, resident immune cells, such as mast cells and neutrophils, are activated by factors released by degenerated fibers [97]. The immune cells start to produce pro-inflammatory molecules, such as histamine, TNFα, interferon γ (IFNγ), IL1β, which leads to increased vascular permeability and myeloid cells recruitment. Both neutrophils and macrophages participate in damaged myofibers removal. Simultaneously, factors secreted by neutrophils and macrophages play an important role in the SC activation and myoblast proliferation and differentiation. Thus, the depletion of macrophages reduces the level of hepatocyte growth factor (HGF) and insulin-like growth factor 1 (IGF1), causing impairment of skeletal muscle regeneration [98]. HGF binds with c-met and plays a role in SC activation [99]. IGF-1 promotes myoblasts proliferation and differentiation [100,101]. Macrophages also secrete TNFα and IL6, i.e., factors which promote myoblasts proliferation and differentiation [102]. Other cells that play crucial role in skeletal muscle reconstruction are endothelial cells. They participate in the restoration of vasculature in damaged muscle and secrete pro-angiogenic and pro-myogenic factors, such as apelin, oncostatin, and periostin [103–105]. ECM remodeling that is an important step during muscle regeneration involves fibroblasts and FAPs (also named "mesenchymal progenitors"). These cells also produce pro-myogenic factors, such as: IGF1, IL6, and follistatin [106–108]. FAPs are interstitial non-myogenic progenitors expressing platelet derived growth factor receptor α (PDGFRα) [109–111]. In intact muscle FAPs are quiescent but after an injury they start to proliferate and synthesize ECM proteins, as well as abovementioned factors [112]. In aged muscles and during chronic diseases the FAPs accumulation and differentiation into fibroblasts and adipocytes is observed. Thus, these cells could be engaged in the formation of fibrosis or adipose tissue accumulation [112].

Except for abovementioned cell populations, skeletal muscle interstitium is the source of stem and progenitor cells different form SCs [113]. Their role in skeletal muscle homeostasis is extensively studied using mouse models. However, many studies also focus on human cells [113]. In mouse as well as in human muscles pericytes and mesangioblast are localized peripherally to microvessel endothelium. They are described as PDGFRβ, NG2, CD146 expressing cells [114–119]. Such cells were shown to be able to fuse with myofibers and occupy satellite cell niche in regenerating muscle [114–118]. Moreover, mouse and human pericytes secrete IFG-1 and angiopoetin that are known factors supporting myoblasts differentiation [120]. Other populations detected in mouse muscles are PW1+/PAX7 interstitial cells, i.e., PICs expressing PW1, SCA1, and CD34 [121]. These cells transplanted to injured mouse muscles participated in the regeneration and restoration of SC population [121]. Next, the TWIST2+ progenitor cells expressing transcription factor TWIST2, myoendothelial cells expressing CD34, i.e., Sk34 cells, and side population (SP) cells isolated on the basis of Hoechst day exclusion were identified in mouse muscle interstitium [122–124]. They showed myogenic potential in vitro and formed new myofibers after transplantation into injured muscles [122–124], similarly to CD133+ cells presented in human muscles [125].

3.2. The Osteogenic Potential of Stem and Progenitor Cells Residing in Skeletal Muscle—In Vitro Studies

Few populations of stem and progenitor cells residing in skeletal muscle and described above could follow osteogenic differentiation in vitro. Among them are mouse and human SCs. BMP4 and BMP7 treatment of mouse SCs induced their osteogenic differentiation, which was shown by increased expression of ALP (and also its activity), osteopontin, and osteocalcin, i.e., the markers of osteogenic differentiation. Moreover, SCs were able to undergo spontaneous osteogenic differentiation when cultured in Matrigel [126]. Osteogenic properties were also documented for human SCs after their in vitro culture in osteogenic differentiation medium (OB-1, ZenBio). After 14 days of treatment, cells increased expression of osteogenic differentiation genes, such as, RUNX2 and BGLAP. Moreover, Alizarin Red staining revealed the accumulation of calcium deposits [127]. Further, such staining of mouse skeletal muscle-derived TBX18+ pericytes, cultured in medium supplemented with dexamethasone, L-ascorbic acid-phosphate, β-glycerophosphate, and BMP2, revealed the deposition of mineralized matrix also indicating differentiation in osteogenic lineages [128]. Similarly, human ALP+ pericytes were able to differentiate in osteoblasts after BMP2 treatment in vitro [117]. On the other hand, CD146+/ALP+ progenitors isolated from human skeletal muscles were not able to follow osteogenic program in vivo after transplantation with hydroxyapatite/tricalcium phosphate scaffold [129]. Finally, mouse FAPs characterized by the presence of markers such as TIE2, PDGFRα or SCA1 differentiated into osteoblasts formation after BMP7, BMP2, treatment or when cells were cultured in osteogenic differentiation medium containing dexamethasone, β-glycerophosphate, and ascorbic-acid [109,110,130]. So far, osteogenic differentiation has not been analyzed or documented for other cell populations, such as PIC, TWIST2+ cell, Sk34 cells, as well as human circulating CD133+ cells [121,123,125,131–136].

3.3. The Cells Directly Participating in Heterotropic Ossification Formation In Vivo

Different animal models, which could be divided into two groups, were used to follow the cells responsible for HO development [1,5,21]. The first one consists of genetically modified animals, i.e., mouse engineered to express, in controlled manner, constitutively active ACVR1, which mimic FOP. The second group includes animal models in that trauma was caused by muscle blunt-force or forced ROM damage, muscle dissection, hip surgery or skin burn with Achilles tenotomy. The third group includes animals in which HO develops after BMPs injection or overexpression. The fourth model bases at the spinal cord injury in conjunction with cardiotoxin induced muscle damage. Moreover, a lot of information about the cell types responsible for HO formation was obtained thanks to lineage tracing [1,5,21].

Using the abovementioned models, a few cell populations were designated to be responsible for HO formation. As mentioned, several skeletal muscle cell types, such as human pericytes and mouse SCs, and FAPs present osteogenic potential in vitro. Importantly, in vitro results cannot be directly translated to in vivo situation. Notably, in vivo studies using mouse models proposed that HO precursors could originate from skeletal muscle endothelial, "mesenchymal" or pericyte populations or tendon and connective tissue cells or even circulating stem/precursor cells [1,5,21,137]. Some initial studies suggested that endothelial cells, characterized by the presence of TIE2, which is the tyrosine kinase receptor for angiopoetin, are engaged in HO formation [138,139]. Tracing these cells on the basis of Tie2 expression proved that they participate in HO development after BMP2 intramuscular injection or cardiotoxin induced skeletal muscle injury in transgenic mice that overexpressed BMP4 at neuromuscular junction [138,139]. Importantly, neither SCs (expressing *MyoD*) nor vascular smooth muscle cells (expressing smooth muscle myosin heavy chain) contributed to HO [138]. Moreover, also other lineage tracing and transplantation experiments clearly showed that SCs did not participate in HO development [130,138,140]. The presence of TIE2 expressing cells was observed in human fragments of tissue from FOP patients [139]. However, the studies in which the cells expressing *Cdh5* (VE-cadherin), i.e., endothelial progenitor cells, were traced, showed that in HO lesion such cells were located only peripherally [141]. Thus, the endothelial progenitors did not participate in HO formation [141]. Moreover, it was showed that TIE2 is not unique marker of endothelial cells. It is also expressed by mouse muscle interstitial cells that are able to follow osteogenic program. Next, TIE2+ cells express PDGFRα and SCA1 and do not express CD31 and CD45 [130]. Thus, these cells correspond to the population of FAPs described in human and mouse skeletal muscles [109,142]. The mesodermal origin of HO precursors was also proven by tracing of PRX1+ cells after tenotomy resulting in the formation of HO [143,144]. During embryogenesis *Prx1* gene is expressed in tissues of mesodermal origin and is crucial for cartilage and bone development. *Dermo1* gene expression, on the other hand, is restricted to the perichondrium. Tracing of DERMO+1 cells showed their engagement in HO development [143]. The localization in skeletal muscle interstitium was also demonstrated for MX1 expressing cells that form HO in response to muscle injury in mice expressing constitutively active form of ACVR1 [145]. MX1 is interferon induced GTP binding protein and is expressed in skeletal muscle interstitial cells, bone marrow osteoprogenitors and endothelial cells [145]. Lineage tracing method allowed further characterization of HO precursor cells. Thus, it was shown that GLAST or GLI1 expressing cells form HO [146,147]. GLAST, i.e., glial high affinity glutamate transporter, is expressed in different tissues and among them are interstitial cells of connective tissue and pericytes [147]. GLI1, i.e., glioma-associated oncogene 1, is a transcription factor engaged in HEDGEHOG signaling [146]. In the skeletal muscle interstitium *Glast* or *Gli1* expressing cells were localized close to vasculature, co-expressed fibroblast-specific protein 1 (FSP1), STRO1, and PDGFRα [146,147]. On the other hand, it was also well documented that NG2+ pericytes, similarly to endothelial or hematopoietic cells, did not participate in HO development [145].

The other source of HO precursors is tendon and connective tissue within the skeletal muscle [145,148]. The cells expressing transcription factor scleraxis (*Scx*) were localized in the tendon and ligaments [145]. However, the presence of *Scx* expressing cells was also noticed in the connective tissue within the skeletal muscle [148]. The SCX+ cells also express PDGFRα, SCA1, and S100A4 [148]. *Scx* expressing cells are able to develop HO localized in the tendon and joints spontaneously in mice expressing constitutively active form of ACVR1 and after tendon injury or intramuscular loading of BMP containing-scaffold [145]. In such mice, the HO developed only in injured muscles [148].

Summarizing, the question about the HO precursor cell identity is still open. Evidence presented above allowed us to conclude that potential HO precursors are of mesodermal origin and are located in the skeletal muscle interstitium. In mouse cells able to form HO could be identified on the basis of TIE2, PDGFRα, SCA1, GLAST, FSP1, STRO1, GLI1, and MX1 expression. Moreover, such cells should show many similar features to skeletal muscle FAPs and pericytes. Tendon and connective tissue present within skeletal muscle could be considered as the source of cells responsible for HO formation. In human, however, such cells are not precisely described yet. Moreover, it is also suggested that different types of cells could be responsible for HO development dependently of HO type [1,5,21,137].

4. Possible Signaling Mechanisms of Ectopic Osteogenesis in Skeletal Muscles

The knowledge on molecular mechanisms of HO formation is limited. It is well established that ectopic osteogenesis occurs as a result of traumatic injury, severe burns, and is commonly observed after invasive surgeries, which indicates that it is related to inflammation. However, the precise immune and signaling regulation is poorly understood. One of the best-known regulators of bone development and postnatal bone maintenance are bone morphogenic proteins (BMPs) [149]. BMPs are members of TGFβ superfamily, which also consists of TGFβ, activins or inhibins. Canonical TGFβ/BMP signaling is a linear cascade which involves TGFβ/BMP ligands, two types of receptors (type I and II), and signal transducers—SMADs. Receptor binding to BMP leads to SMADs—SMAD1/5/8, to TGFβ leads to SMAD2/3 phosphorylation. Activated SMADs bind to SMAD4, then the complex is accumulated in nucleus where regulates target gene expression [150]. One of the downstream targets of these pathways is for example gene encoding RUNX2, well-known master regulator of osteogenesis which is also aberrantly expressed in the ossified soft tissues [151–154]. TGFβ dependent activation of SMAD2/3 promotes osteoprogenitors migration and early stages of differentiation, while negatively regulates further steps of osteogenesis. SMAD2/3 phosphorylation inhibits RUNX2 expression and activated SMAD3 recruits class II histone deacetylases (HDACs) 4 and 5 which inhibit RUNX2 function. Although TGFβ-SMAD3 negatively regulates osteoblastogenesis, it also inhibits osteoblast apoptosis and differentiation into osteocytes [155]. On the other hand, there is TGFβ dependent non-SMAD pathway which also contributes to bone formation. TGFβ binding to its receptors can result in activation of MAPK p38 or MAPK ERK1/2 pathways through TAB1-TAK1 complex which leads to positive regulation of RUNX2 activity and favors osteoclast differentiation [156]. It indicates that TGFβ molecule is coupling bone formation, through RUNX2 phosphorylation, osteoprogenitors enrichment or osteoblast proliferation promotion, and inhibition of apoptosis with bone resorption through inhibition of RUNX2 expression and function and osteoclast maturation [157]. BMPs binding to receptor leads to SMAD1/5/8 phosphorylation (except BMP3 which action leads to SMAD2/3 phosphorylation). Activated SMAD1/5/8 bind with SMAD4 and promote expression of many osteogenesis promoting factors like RUNX2, OSX or DLX5. Similarly, to TGFβ, BMPs also can activate SMAD-independent pathway through phosphorylation of TAK1-TAB1 complex and activation of MAPK p38 or MAPK ERK1/2 pathways. In conclusion, most BMP ligands are strong osteogenic agents, acting through both SMAD-dependent and SMAD-independent signaling pathway, which synergize osteogenic transcriptional factors like RUNX2 or OSX [157,158].

In vitro and in vivo exogenous stimulation of TGFβ/BMP signaling (BMP2, BMP4, BMP9 or TGFβ) is commonly used for induction of ossification. Those proteins, especially BMP2 and BMP9, are also highly expressed in human HO [159]. One of the most intensively studied disease which manifests itself in severe HO is FOP. It is still unclear, however, what the cellular and molecular mechanisms are that cause pathological effects. Analyzes of human mesenchymal stromal cells (MSC; expressing CD44, CD73 and CD105), derived from induced pluripotent stem cells (iPSC) obtained from FOP which patients, showed that these cells were characterized by higher activity ofof SMAD1/3/5,

SMAD2/3 and MAPK ERK1/2 when compared to genetically corrected resFOP-iPSCs-MSCs [160]. Most studies suggest that mutation in ACVR1 present in FOP patients cells causes hypersensitivity to BMPs, which results in constitutive phosphorylation of its receptor, and continuous signal transduction via phosphorylation of SMAD1/5/8. As a result downstream targets of BMP signaling, like ID-1, OSX or RUNX2 are expressed [130,161,162]. However, recent report, based on study of murine FAPs, demonstrated that R206H substitution in ACVR1 may be neomorphic and altering signaling specificity to activins. Normally, activins binding to ACVR1 receptor lead to SMAD2/3 phosphorylation. Obtained results suggest that Activin A binding to mutated ACVR1 (R206H) receptor leads to SMAD1/5/8 instead of SMAD2/3 phosphorylation which results in ectopic bone formation [163]. It is well established that TGFβ/BMP signaling crosstalks with other pathways during embryonic and postnatal development, similarly as with MAPK described above. For example, crosstalk between canonical WNT pathway, TLR pathway or mTOR pathway was described [164]. TLR signaling intermediate evolutionary conserved signaling intermediate in toll pathway (ECSIT) is necessary for BMP signaling in formation of mesoderm during mouse embryogenesis [165]. Additionally, β-catenin was shown to be necessary for bone development and osteoblast formation in mouse embryos [151]. Other studies showed that β-catenin complexed with T cell factor 1 (TCF1) directly stimulates Runx2 expression [166]. Another study suggested that β-catenin, together with other proteins like SMAD1, DLX5, Sp7 or SOX6, form an enhanceosome which binds to enhancer of Runx2 gene and promotes its expression [167].

Hypoxia and inflammation are also among the factors implicated in the episodic induction of ectopic bone formation. Notably, mTOR modulates hypoxic and inflammatory signaling during the early stages of HO. At the later stages of HO, however, mTOR signaling is critical for chondrogenesis and osteogenesis [164]. Increase in mTOR signaling was shown using mouse model of FOP, i.e., animals which express constitutively active activin receptor, i.e., ACVR1 [168] and its inhibitor, rapamycin, has been shown to suppress HO formation [168]. Hypoxic environment stabilizes hypoxia-inducible factor 1α (HIF1α) which regulates expression of many proteins, such as VEGF or BMPs, which are involved in HO formation [169]. Analysis of three different mouse FOP/HO models have demonstrated hypoxia and increased HIF1α signaling [144]. Expression of HIF1α was also increased in adipose samples derived from severely burned patients, i.e., those ones being at risk for trauma-induced HO development [144,170]. Interestingly, analysis of human HO tissues, human preosteoblasts (hFOB1.19) and tissues of mice serving as a model of HO, revealed that miRNAs have essential role in osteoblast differentiation and HO development. miRNA-203 have been shown to be negatively correlated with HO and to participate in inhibition of osteoblast differentiation by directly binding to RUNX2 [64]. Nevertheless, the mechanisms underlying the development of HO in patients that not carry any mutations are still obscure. Moreover, even in FOP patients bone formation is not always observed in their soft tissues. Bone formation seems to be rather a result of injuries and inflammation, which strongly suggests a link between immune response and HO. In vivo studies using rabbits showed that bacterial transplantation into tibia bone increased inflammation-driven bone formation. In the same study, lipoteichoic acid (LTA)—the bacterial cell-wall derived toll-like receptor 2 (TLR2) activator—was identified as an osteo-stimulatory factor [171]. Other studies involving FOP patients-derived connective tissue progenitor cells revealed that such cells present much higher expression of TLRs in comparison to cells that are expressing normal ACVR1 receptor. That effect was even more significant after TNFα treatment of examined cells. The same study also revealed that TLR signaling can induce SMAD1/5/8 phosphorylation. Additionally, ECSIT, complex including TAK1 and TRAF6, which plays pivotal role in TLR-mediated NK-kB and SMAD1/5/8 signaling, was identified as a link between TLR-pathway and BMP pathway in human FOP-connective tissue progenitor cells [172,173]. As described above, normal cells, not carrying any mutations in *ACVR1*, can undergo osteogenesis after BMP stimulation. Thus, development of heterotopic bone provides the signaling environment in which BMP level is sufficient enough to stimulate the cells to form the bone [163]. Results of these reports together suggest that main mechanism of HO formation is connected with TGFβ/BMP signaling, especially SMAD1/5/8 action, which leads to expression of osteogenic transcription factors. Factors present in damaged tissue lead also to activation of mTOR, WNT or TLR pathways which may cross-talk with TGFβ/BMP or independently promote osteogenic factors expression and induce HO formation. Even small ossification within tissue provide a BMP rich environment, which further stimulates neighboring cells to follow osteogenic differentiation and support newly creating bone growth. However, it still remains unclear why spontaneous HO is observed or how that process is induced and regulated (Figure 2).

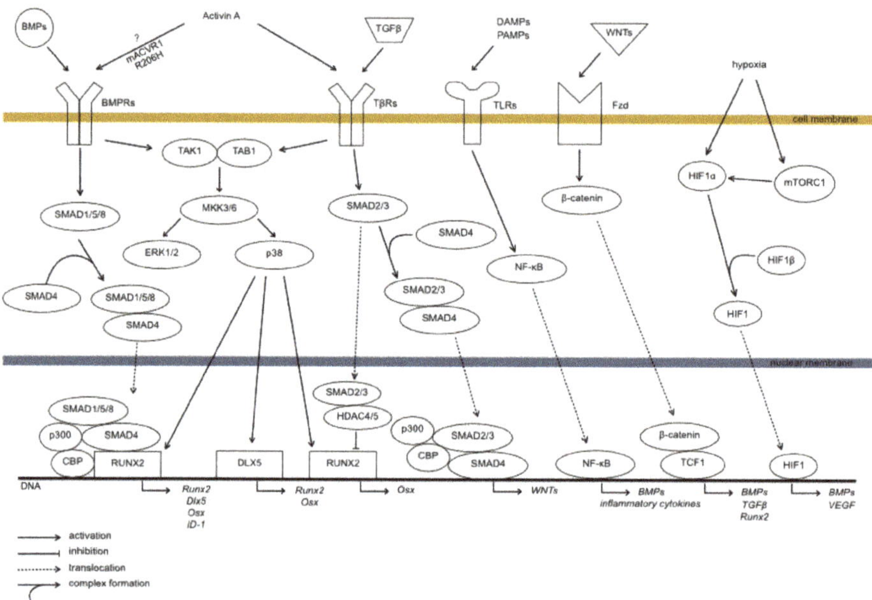

Figure 2. Possible signaling mechanisms of ectopic osteogenesis in skeletal muscles. BMPs bind to homomeric type II receptors which phosphorylate homomeric type I receptor and induce SMAD-dependent and SMAD-independent signaling. In the SMAD-dependent signaling SMADs 1, 5 or 8 complex with SMAD4 and translocate to the nucleus where recruit RUNX2 and other co-factors to regulate osteogenic gene expression. TGFβ binds to complex of two TGFβ types I receptors (TβRI) and two type II receptors (TβRII), which phosphorylate each other and induce SMAD-dependent and SMAD-independent signaling. In the SMAD-dependent signaling activated SMAD2/3 form complex with SMAD4. Complex translocates to the nucleus where recruits co-factors and regulates target gene expression. Activated SMAD3 recruits HDACs which inhbit RUN2 activity. In the SMAD-independent pathway, regardless of the ligand bind to the receptors, TAK1 recruits TAB1 to initiate p38 MAPK or ERK1/2 MAPK signaling cascade. MAPK phosphorylates and activates RUNX2, DLX5, and OSX transcription factors. Activation of TLR singaling pathways by PAMPs and DAMPs lead to activation of nuclear factor-kappaB (NF-κB), which controls the expression of an array of inflammatory cytokine genes and BMPs. WNTs bind to Frizzled (Fzd) receptors and activate the canonical WNT pathway which leads to accumulation of β-catenin in the cytoplasm. β-catenin is translocated to the nucleus where forms complex with TCF1 which acts as transcriptional activator of *Runx2*. Low level of oxygen (hypoxia) induces the mTOR pathway. HIF1α, a downstream intermediate in mTOR signaling, is a key transcriptional regulator of the cellular response to hypoxia. It forms complex with HIF1β and as HIF1 enters to the nuclei where regulates target gene expression.

5. Future Therapeutic Options

Currently, a clinical trial phase 3 of highly specific retinoic acid receptor γ (RARγ) agonist, R667 (palovarotene), carried out by Clementia Pharmaceuticals involves 90 FOP patients (NCT03312634).

RAR is a strong inhibitor of chondrogenesis. Stimulation of its γ subtype reduces in BMP signaling by lowering SMAD1/5/8 phosphorylation and as a result decreases HO formation. The safety profile of this drug is being carefully assessed due to the teratogenic potential of RAR agonists and other side effects, including cheilitis, xerosis, dryness of mucous membranes, inhibition of growth plates in children, hearing and vision impairment [70,174]. Another investigated strategy is blocking mutant ACVR1. The ACRV1 stimulates BMP through SMAD1/5/8 signaling and promote HO. Such approaches involve anti-Activin A antibody (REGN2477) and which is currently at phase 2 of randomized control trial for 44 FOP patients (NCT03188666, Regeneron Pharmaceuticals) [175]. Other ACVR1 direct inhibitors (AZD0530 and PD 161570) are being investigated. The AZD0530 difumarate inhibits both BMP and TGFβ signaling. Phase 2 study involving AZD0530 (Saracatinib) to prevent FOP is currently carried out by VU University Medical Center (NCT04307953) [176]. The other treatment strategies include a local application of apyrase, which influences the BMP-SMAD pathway by the reduction of SMAD1/5/8 phosphorylation [177]. The BMP receptor antagonists, such as noggin, also inhibit HO in animal models [178]. Researchers also suggest that pharmacological inhibition of HIF1α using PX-478, rapamycin, apigenin, or imatinib can reduce pathologic extraskeletal bone formation [144,179]. Recently, gene therapy opportunities raised for HO. Non-virus-mediated transfer of small interfering RNA (siRNA) particles against mRNA encoding *Runx2* and *Smad4* inhibited HO in rats after Achilles tenotomy [180]. The siRNA could also possibly directly block mutant ACVR1 as a therapeutic option in future studies [179]. Additionally, the immune system may be a potential target for HO prevention. Neutralizing antibodies against immune checkpoint proteins (ICs) block limit the extent of HO in animal studies [9].

6. Conclusions

In this review we summarize current knowledge on the development of different forms of HO. Numerous projects involving analysis of patients' tissues and also animal models allowed great advancement in the understanding of this pathology. However, we are still not certain about the precise sources of the osteogenic progenitors involved in this pathology. Additionally, the knowledge on the signaling pathways deregulated despite being enormous is still not sufficient to design the properly targeted treatment. Nevertheless, what we already know allowed us to propose several hope-giving therapeutic approaches which are currently tested. Thus, more work needs to be done, but it seems that we are on the proper path.

Author Contributions: Conceptualization, P.Ł. and E.B.; Writing—Original Draft Preparation, Ł.P., E.B., and B.M.; Writing—Review and Editing, E.B., P.Ł., and M.A.C.; Visualization, B.M.; Supervision, P.Ł. and E.B. All authors have read and agreed to the published version of the manuscript.

Funding: This research received funding from The University of Warsaw (E.B.) and The Medical University of Warsaw (Ł.P.) as part of the mutual Microgrant Program (WUM-UW, 1WE/NUW1/20).

Conflicts of Interest: The authors declare no conflict of interest.

Abbreviations

ACVR1/ALK2	Activin A receptor type 1/activin-like kinase 2
ALP	alkaline phosphatase
BAF60c	60 KDa BRG-1/Brm-Associated Factor Subunit C
BAP	bone-specific isoform of alkaline phosphatase
BMP	bone morphogenetic protein
CD	cluster of differentiation
CDC6	cell division cycle 6 protein
COL-1	collagen alpha 1
COX-2	cyclooxygenase-2
CRP	C-reactive protein
CTDA	cervical total disc arthroplasty
CTX-1	type I collagen cross-linked C-telopeptide

Dlx5	Distal-Less Homeobox 5
DVT	deep venous thrombosis
ECM	extracellular matrix
ENO3	enolase 3
ERK	extracellular signal-regulated kinase
FAP	fibro-adipogenic progenitors
FOP	fibrodysplasia ossificans progressive
FSP1	fibroblast-specific protein 1
GLAST	glial high affinity glutamate transporter
GLI1	glioma-associated oncogene 1
GNAS	guanine nucleotide-binding protein, subunit alpha
HDAC	histone deacetylase
HGF	hepatocyte growth factor
HIF1α	hypoxia-inducible factor 1α
HLA	human leukocyte antigen
HO	heterotopic ossification
ID-1	DNA-Binding Protein Inhibitor ID-1
IFN γ	interferon γ
IFOPA	international FOP Association
IGF1	insulin-like growth factor 1
IL	Interleukin
iPSC	induced pluripotent stem cell
LTA	lipoteichoic acid
MAPK	mitogen-activated protein kinases
MCK	muscle creatine kinase
MCM2	minichromosome maintenance complex component 2
MCP-1	monocyte chemoattractant protein-1
MHC	myosin heavy chains
miRNA	microRNA
MIS	micro-invasive surgery
MRF	myogenic regulatory factor
MRI	magnetic resonance imaging
MRSA	methicillin-resistant Staphylococcus aureus
MSC	mesenchymal stromal cell
mTOR	mammalian target of rapamycin
NF-κB	Nuclear factor-κB
NG2	neural-glial antigen 2
NHO	neurogenic heterotopic ossification
NSAID	nonsteroidal anti-inflammatory drug
NT-3	neurotrophin-3
OC/BGLAP	osteocalcin/Bone Gamma-Carboxyglutamate Protein
OMD	Osteomodulin
OSX/Sp7	Osterix
Pax7	paired box transcription factor 7
PDGFRα	platelet derived growth factor receptor α
PGE2	prostaglandin E2
PIC	PW1 interstitial cell
POH	progressive osseous heteroplasia
Prx1	peroxiredoxin Prx1
ROM	range of motion
RT	Radiotherapy
RUNX2	runt-related transcription factor 2
SC	satellite cell
SCA1	stem cell antigen 1
SCI	spinal cord injury

Scx	Scleraxis
Sox6	SRY-Box Transcription Factor 6
SP cell	side population cell
SPECT	single-photon emission computed tomography
SWI/SNF	SWItch/Sucrose NonFermentable
TAB1	TAK1 binding protein
TBI	traumatic brain injury
TBX18	T-box transcription factor 18
TCF1	T cell factor 1
TGFβ	transforming growth factor β
THR	total hip replacement
TIE2	angiopoietin receptor
TLR	toll-like receptor
TNF-α	tumor necrosis factor-α
TRAF6	TNF Receptor Associated Factor 6
TWIST2	Twist Basic Helix-Loop-Helix Transcription Factor 2/DERMO1
VE-cadherin	vascular endothelial cadherin (Cdh15)
VEGF	vascular endothelial growth factor
VTE	venous thromboembolism
WNT	wingless/integrated

References

1. Meyers, C.; Lisiecki, J.; Miller, S.; Levin, A.; Fayad, L.; Ding, C.; Sono, T.; McCarthy, E.; Levi, B.; James, A.W. Heterotopic Ossification: A Comprehensive Review. *JBMR Plus* **2019**, *3*, e10172. [CrossRef] [PubMed]
2. Nauth, A.; Giles, E.; Potter, B.K.; Nesti, L.J.; O'brien, F.P.; Bosse, M.J.; Anglen, J.O.; Mehta, S.; Ahn, J.; Miclau, T.; et al. Heterotopic ossification in orthopaedic trauma. *J. Orthop. Trauma* **2012**, *26*, 684–688. [CrossRef] [PubMed]
3. van Kuijk, A.A.; Geurts, A.C.H.; van Kuppevelt, H.J.M. Neurogenic heterotopic ossification in spinal cord injury. *Spinal Cord.* **2002**, *40*, 313–326. [CrossRef] [PubMed]
4. Łęgosz, P.; Drela, K.; Pulik, Ł.; Sarzyńska, S.; Małdyk, P. Challenges of heterotopic ossification-Molecular background and current treatment strategies. *Clin. Exp. Pharm. Physiol.* **2018**, *45*, 1229–1235. [CrossRef] [PubMed]
5. Lees-Shepard, J.B.; Goldhamer, D.J. Stem cells and heterotopic ossification: Lessons from animal models. *Bone* **2018**, *109*, 178–186. [CrossRef] [PubMed]
6. Convente, M.R.; Wang, H.; Pignolo, R.J.; Kaplan, F.S.; Shore, E.M. The immunological contribution to heterotopic ossification disorders. *Curr. Osteoporos. Rep.* **2015**, *13*, 116–124. [CrossRef]
7. Levesque, J.P.; Sims, N.A.; Pettit, A.R.; Alexander, K.A.; Tseng, H.W.; Torossian, F.; Genet, F.; Lataillade, J.J.; Le Bousse-Kerdiles, M.C. Macrophages Driving Heterotopic Ossification: Convergence of Genetically-Driven and Trauma-Driven Mechanisms. *J. Bone Miner. Res.* **2018**, *33*, 365–366. [CrossRef]
8. Sorkin, M.; Huber, A.K.; Hwang, C.; Carson, W.F.; Menon, R.; Li, J.; Vasquez, K.; Pagani, C.; Patel, N.; Li, S.; et al. Regulation of heterotopic ossification by monocytes in a mouse model of aberrant wound healing. *Nat. Commun.* **2020**, *11*, 722. [CrossRef]
9. Kan, C.; Yang, J.; Na, D.; Xu, Y.; Yang, B.; Zhao, H.; Lu, H.; Li, Y.; Zhang, K.; McGuire, T.L.; et al. Inhibition of immune checkpoints prevents injury-induced heterotopic ossification. *Bone Res.* **2019**, *7*, 33. [CrossRef]
10. Zhang, J.; Wang, L.; Chu, J.; Ao, X.; Jiang, T.; Bin, Y.; Huang, M.; Zhang, Z. Macrophage-derived neurotrophin-3 promotes heterotopic ossification in rats. *Lab. Investig.* **2020**. [CrossRef]
11. Bossche, L.V.; Vanderstraeten, G. Heterotopic ossification: A review. *J. Rehabil. Med.* **2005**, *37*, 129–136. [CrossRef] [PubMed]
12. Foley, K.L.; Hebela, N.; Keenan, M.A.; Pignolo, R.J. Histopathology of periarticular non-hereditary heterotopic ossification. *Bone* **2018**, *109*, 65–70. [CrossRef] [PubMed]
13. Fleckenstein, J.L.; Crues, J.V.; Reimers, C.D. *Muscle Imaging in Health and Disease*; Springer: New York, NY, USA, 1996; ISBN 978-1-4612-2314-6.

14. Ohlmeier, M.; Krenn, V.; Thiesen, D.M.; Sandiford, N.A.; Gehrke, T.; Citak, M. Heterotopic Ossification in Orthopaedic and Trauma surgery: A Histopathological Ossification Score. *Sci. Rep.* **2019**, *9*, 18401. [CrossRef]

15. Barfield, W.R.; Holmes, R.E.; Hartsock, L.A. Heterotopic Ossification in Trauma. *Orthop. Clin. N. Am.* **2017**, *48*, 35–46. [CrossRef] [PubMed]

16. Eisenstein, N.; Stapley, S.; Grover, L. Post-Traumatic Heterotopic Ossification: An Old Problem in Need of New Solutions. *J. Orthop. Res.* **2018**, *36*, 1061–1068. [CrossRef]

17. Chen, H.-C.; Yang, J.-Y.; Chuang, S.-S.; Huang, C.-Y.; Yang, S.-Y. Heterotopic ossification in burns: Our experience and literature reviews. *Burns* **2009**, *35*, 857–862. [CrossRef]

18. Thefenne, L.; de Brier, G.; Leclerc, T.; Jourdan, C.; Nicolas, C.; Truffaut, S.; Lapeyre, E.; Genet, F. Two new risk factors for heterotopic ossification development after severe burns. *PLoS ONE* **2017**, *12*, e0182303. [CrossRef]

19. Hong, C.C.; Nashi, N.; Hey, H.W.; Chee, Y.H.; Murphy, D. Clinically relevant heterotopic ossification after elbow fracture surgery: A risk factors study. *Orthop. Traumatol. Surg. Res.* **2015**, *101*, 209–213. [CrossRef]

20. Firoozabadi, R.; O'Mara, T.J.; Swenson, A.; Agel, J.; Beck, J.D.; Routt, M. Risk Factors for the Development of Heterotopic Ossification After Acetabular Fracture Fixation. *Clin. Orthop. Relat. Res.* **2014**, *472*, 3383–3388. [CrossRef]

21. Dey, D.; Wheatley, B.M.; Cholok, D.; Agarwal, S.; Yu, P.B.; Levi, B.; Davis, T.A. The traumatic bone: Trauma-induced heterotopic ossification. *Transl. Res.* **2017**, *186*, 95–111. [CrossRef]

22. Orchard, G.R.; Paratz, J.D.; Blot, S.; Roberts, J.A. Risk Factors in Hospitalized Patients With Burn Injuries for Developing Heterotopic Ossification—A Retrospective Analysis. *J. Burn Care Res.* **2015**, *36*, 465–470. [CrossRef] [PubMed]

23. Bemenderfer, T.B.; Davis, W.H.; Anderson, R.B.; Wing, K.; Escudero, M.I.; Waly, F.; Penner, M. Heterotopic Ossification in Total Ankle Arthroplasty: Case Series and Systematic Review. *J. Foot Ankle Surg.* **2020**. [CrossRef] [PubMed]

24. Zhu, Y.; Zhang, F.; Chen, W.; Zhang, Q.; Liu, S.; Zhang, Y. Incidence and risk factors for heterotopic ossification after total hip arthroplasty: A meta-analysis. *Arch. Orthop. Trauma Surg.* **2015**, *135*, 1307–1314. [CrossRef] [PubMed]

25. Chen, J.; Wang, X.; Bai, W.; Shen, X.; Yuan, W. Prevalence of heterotopic ossification after cervical total disc arthroplasty: A meta-analysis. *Eur. Spine J.* **2012**, *21*, 674–680. [CrossRef] [PubMed]

26. Lee, K.-B.; Cho, Y.-J.; Park, J.-K.; Song, E.-K.; Yoon, T.-R.; Seon, J.-K. Heterotopic Ossification After Primary Total Ankle Arthroplasty. *JBJS* **2011**, *93*, 751–758. [CrossRef]

27. Łęgosz, P.; Sarzyńska, S.; Pulik, Ł.; Stępiński, P.; Niewczas, P.; Kotela, A.; Małdyk, P. Heterotopic ossification and clinical results after total hip arthroplasty using the anterior minimally invasive and anterolateral approaches. *Arch. Med. Sci.* **2018**. [CrossRef]

28. Anthonissen, J.; Ossendorf, C.; Hock, J.L.; Steffen, C.T.; Goetz, H.; Hofmann, A.; Rommens, P.M. The role of muscular trauma in the development of heterotopic ossification after hip surgery: An animal-model study in rats. *Injury* **2016**, *47*, 613–616. [CrossRef]

29. Yi, S.; Shin, D.A.; Kim, K.N.; Choi, G.; Shin, H.C.; Kim, K.S.; Yoon, D.H. The predisposing factors for the heterotopic ossification after cervical artificial disc replacement. *Spine J.* **2013**, *13*, 1048–1054. [CrossRef]

30. Kocic, M.; Lazovic, M.; Mitkovic, M.; Djokic, B. Clinical significance of the heterotopic ossification after total hip arthroplasty. *Orthopedics* **2010**, *33*, 16. [CrossRef]

31. Sundseth, J.; Jacobsen, E.A.; Kolstad, F.; Sletteberg, R.O.; Nygaard, O.P.; Johnsen, L.G.; Pripp, A.H.; Andresen, H.; Fredriksli, O.A.; Myrseth, E.; et al. Heterotopic ossification and clinical outcome in nonconstrained cervical arthroplasty 2 years after surgery: The Norwegian Cervical Arthroplasty Trial (NORCAT). *Eur. Spine J.* **2016**, *25*, 2271–2278. [CrossRef]

32. McCarthy, E.F.; Sundaram, M. Heterotopic ossification: A review. *Skelet. Radiol.* **2005**, *34*, 609–619. [CrossRef] [PubMed]

33. Nalbantoglu, M.; Tuncer, O.G.; Acik, M.E.; Matur, Z.; Altunrende, B.; Ozgonenel, E.; Ozgonenel, L. Neurogenic heterotopic ossification in Guillain-Barre syndrome: A rare case report. *J. Musculoskelet. Neuronal Interact.* **2020**, *20*, 160–164. [PubMed]

34. Zhang, Y.; Zhan, Y.; Kou, Y.; Yin, X.; Wang, Y.; Zhang, D. Identification of biological pathways and genes associated with neurogenic heterotopic ossification by text mining. *PeerJ* **2020**, *8*, e8276. [CrossRef] [PubMed]

35. Pek, C.H.; Lim, M.C.; Yong, R.; Wong, H.P. Neurogenic heterotopic ossification after a stroke: Diagnostic and radiological challenges. *Singap. Med. J.* **2014**, *55*, e119–e122. [CrossRef]

36. Reznik, J.; Biros, E.; Marshall, R.; Jelbart, M.; Milanese, S.; Gordon, S.; Galea, M. Prevalence and risk-factors of neurogenic heterotopic ossification in traumatic spinal cord and traumatic brain injured patients admitted to specialised units in Australia. *J. Musculoskelet. Neuronal Interact.* **2014**, *14*, 19–28.

37. Citak, M.; Suero, E.M.; Backhaus, M.; Aach, M.; Godry, H.; Meindl, R.; Schildhauer, T.A. Risk Factors for Heterotopic Ossification in Patients With Spinal Cord Injury: A Case-Control Study of 264 Patients. *Spine* **2012**, *37*, 1953–1957. [CrossRef]

38. Sullivan, M.P.; Torres, S.J.; Mehta, S.; Ahn, J. Heterotopic ossification after central nervous system trauma: A current review. *Bone Jt. Res.* **2013**, *2*, 51–57. [CrossRef]

39. Citak, M.; Grasmücke, D.; Suero, E.M.; Cruciger, O.; Meindl, R.; Schildhauer, T.A.; Aach, M. The roles of serum alkaline and bone alkaline phosphatase levels in predicting heterotopic ossification following spinal cord injury. *Spinal Cord.* **2016**, *54*, 368–370. [CrossRef]

40. Dizdar, D.; Tiftik, T.; Kara, M.; Tunç, H.; Ersöz, M.; Akkuş, S. Risk factors for developing heterotopic ossification in patients with traumatic brain injury. *Brain Inj.* **2013**, *27*, 807–811. [CrossRef]

41. Ohlmeier, M.; Suero, E.M.; Aach, M.; Meindl, R.; Schildhauer, T.A.; Citak, M. Muscle localization of heterotopic ossification following spinal cord injury. *Spine J.* **2017**, *17*, 1519–1522. [CrossRef]

42. Suero, E.M.; Meindl, R.; Schildhauer, T.A.; Citak, M. Clinical Prediction Rule for Heterotopic Ossification of the Hip in Patients with Spinal Cord Injury. *Spine* **2018**, *43*, 1572–1578. [CrossRef] [PubMed]

43. Johns, J.S.; Cifu, D.X.; Keyser-Marcus, L.; Jolles, P.R.; Fratkin, M.J. Impact of Clinically Significant Heterotopic Ossification on Functional Outcome after Traumatic Brain Injury. *J. Head Trauma Rehabil.* **1999**, *14*, 269–276. [CrossRef] [PubMed]

44. Shore, E.M.; Xu, M.; Feldman, G.J.; Fenstermacher, D.A.; Cho, T.-J.; Choi, I.H.; Connor, J.M.; Delai, P.; Glaser, D.L.; LeMerrer, M.; et al. A recurrent mutation in the BMP type I receptor ACVR1 causes inherited and sporadic fibrodysplasia ossificans progressiva. *Nat. Genet.* **2006**, *38*, 525–527. [CrossRef] [PubMed]

45. Adegbite, N.S.; Xu, M.; Kaplan, F.S.; Shore, E.M.; Pignolo, R.J. Diagnostic and mutational spectrum of progressive osseous heteroplasia (POH) and other forms of GNAS-based heterotopic ossification. *Am. J. Med. Genet. Part A* **2008**, *146*, 1788–1796. [CrossRef]

46. Valer, J.A.; Sanchez-de-Diego, C.; Pimenta-Lopes, C.; Rosa, J.L.; Ventura, F. ACVR1 Function in Health and Disease. *Cells* **2019**, *8*, 1366. [CrossRef]

47. Pignolo, R.J.; Ramaswamy, G.; Fong, J.T.; Shore, E.M.; Kaplan, F.S. Progressive osseous heteroplasia: Diagnosis, treatment, and prognosis. *Appl. Clin. Genet.* **2015**, *8*, 37–48. [CrossRef]

48. Pignolo, R.J.; Cheung, K.; Kile, S.; Fitzpatrick, M.A.; De Cunto, C.; Al Mukaddam, M.; Hsiao, E.C.; Baujat, G.; Delai, P.; Eekhoff, E.M.W.; et al. Self-reported baseline phenotypes from the International Fibrodysplasia Ossificans Progressiva (FOP) Association Global Registry. *Bone* **2020**. [CrossRef]

49. Pignolo, R.J.; Shore, E.M.; Kaplan, F.S. Fibrodysplasia Ossificans Progressiva: Clinical and Genetic Aspects. *Orphanet J. Rare Dis.* **2011**, *6*, 80. [CrossRef] [PubMed]

50. Kaplan, F.S.; Le Merrer, M.; Glaser, D.L.; Pignolo, R.J.; Goldsby, R.E.; Kitterman, J.A.; Groppe, J.; Shore, E.M. Fibrodysplasia ossificans progressiva. *Best Pr. Res. Clin. Rheumatol.* **2008**, *22*, 191–205. [CrossRef]

51. Towler, O.W.; Shore, E.M.; Kaplan, F.S. Skeletal malformations and developmental arthropathy in individuals who have fibrodysplasia ossificans progressiva. *Bone* **2020**, *130*, 115116. [CrossRef]

52. Mujtaba, B.; Taher, A.; Fiala, M.J.; Nassar, S.; Madewell, J.E.; Hanafy, A.K.; Aslam, R. Heterotopic ossification: Radiological and pathological review. *Radiol. Oncol.* **2019**, *53*, 275–284. [CrossRef] [PubMed]

53. Ghanem, M.A.; Dannoon, S.; Elgazzar, A.H. The added value of SPECT-CT in the detection of heterotopic ossification on bone scintigraphy. *Skelet. Radiol.* **2020**, *49*, 291–298. [CrossRef] [PubMed]

54. Lima, M.C.; Passarelli, M.C.; Dario, V.; Lebani, B.R.; Monteiro, P.H.S.; Ramos, C.D. The use of spect/ct in the evaluation of heterotopic ossification in para/tetraplegics. *Acta Ortopédica Bras.* **2014**, *22*, 12–16. [CrossRef] [PubMed]

55. Botman, E.; Raijmakers, P.G.H.M.; Yaqub, M.; Teunissen, B.; Netelenbos, C.; Lubbers, W.; Schwarte, L.A.; Micha, D.; Bravenboer, N.; Schoenmaker, T.; et al. Evolution of heterotopic bone in fibrodysplasia ossificans progressiva: An [18F]NaF PET/CT study. *Bone* **2019**, *124*, 1–6. [CrossRef]

56. Rosteius, T.; Suero, E.; Grasmücke, D.; Aach, M.; Gisevius, A.; Ohlmeier, M.; Meindl, R.; Schildhauer, T.; Citak, M. The sensitivity of ultrasound screening examination in detecting heterotopic ossification following spinal cord injury. *Spinal Cord.* **2017**, *55*, 71–73. [CrossRef]

57. Shehab, D.; Elgazzar, A.H.; Collier, B.D. Heterotopic ossification. *J. Nucl. Med.* **2002**, *43*, 346–353.

58. Łęgosz, P.; Pulik, Ł.; Stępiński, P.; Janowicz, J.; Wirkowska, A.; Kotela, A.; Sarzyńska, S.; Małdyk, P. The Use of Type I Collagen Cross-Linked C-Telopeptide (CTX-1) as a Biomarker Associated with the Formation of Periprosthetic Ossifications Following Total Hip JoInt. Arthroplasty. *Ann. Clin. Lab. Sci.* **2018**, *48*, 183–190.

59. Sung Hsieh, H.H.; Chung, M.T.; Allen, R.M.; Ranganathan, K.; Habbouche, J.; Cholok, D.; Butts, J.; Kaura, A.; Tiruvannamalai-Annamalai, R.; Breuler, C.; et al. Evaluation of Salivary Cytokines for Diagnosis of both Trauma-Induced and Genetic Heterotopic Ossification. *Front. Endocrinol.* **2017**, *8*. [CrossRef]

60. Forsberg, J.A.; Potter, B.K.; Polfer, E.M.; Safford, S.D.; Elster, E.A. Do Inflammatory Markers Portend Heterotopic Ossification and Wound Failure in Combat Wounds? *Clin. Orthop. Relat. Res.* **2014**, *472*, 2845–2854. [CrossRef]

61. Edsberg, L.E.; Crowgey, E.L.; Osborn, P.M.; Wyffels, J.T. A survey of proteomic biomarkers for heterotopic ossification in blood serum. *J. Orthop. Surg. Res.* **2017**, *12*, 69. [CrossRef]

62. Sun, Y.; Cai, J.; Yu, S.; Chen, S.; Li, F.; Fan, C. MiR-630 Inhibits Endothelial-Mesenchymal Transition by Targeting Slug in Traumatic Heterotopic Ossification. *Sci. Rep.* **2016**, *6*, 22729. [CrossRef] [PubMed]

63. Ju, C.; Lv, Z.; Zhang, C.; Jiao, Y. Regulatory effect of miR-421 on humeral fracture and heterotopic ossification in elderly patients. *Exp. Ther. Med.* **2019**, *17*, 1903–1911. [CrossRef] [PubMed]

64. Tu, B.; Liu, S.; Yu, B.; Zhu, J.; Ruan, H.; Tang, T.; Fan, C. miR-203 inhibits the traumatic heterotopic ossification by targeting Runx2. *Cell Death Dis.* **2016**, *7*, e2436. [CrossRef] [PubMed]

65. Li, F.; Mao, D.; Pan, X.; Zhang, X.; Mi, J.; Rui, Y. Celecoxib cannot inhibit the progression of initiated traumatic heterotopic ossification. *J. Shoulder Elb. Surg.* **2019**, *28*, 2379–2385. [CrossRef] [PubMed]

66. Joice, M.; Vasileiadis, G.I.; Amanatullah, D.F. Non-steroidal anti-inflammatory drugs for heterotopic ossification prophylaxis after total hip arthroplasty. *Bone Jt. J.* **2018**, *100*, 915–922. [CrossRef] [PubMed]

67. Xue, D.; Zheng, Q.; Li, H.; Qian, S.; Zhang, B.; Pan, Z. Selective COX-2 inhibitor versus nonselective COX-1 and COX-2 inhibitor in the prevention of heterotopic ossification after total hip arthroplasty: A meta-analysis of randomised trials. *Int. Orthop.* **2011**, *35*, 3–8. [CrossRef]

68. Kan, S.L.; Yang, B.; Ning, G.Z.; Chen, L.X.; Li, Y.L.; Gao, S.J.; Chen, X.Y.; Sun, J.C.; Feng, S.Q. Nonsteroidal Anti-inflammatory Drugs as Prophylaxis for Heterotopic Ossification after Total Hip Arthroplasty: A Systematic Review and Meta-Analysis. *Medicine* **2015**, *94*, e828. [CrossRef]

69. Winkler, S.; Wagner, F.; Weber, M.; Matussek, J.; Craiovan, B.; Heers, G.; Springorum, H.R.; Grifka, J.; Renkawitz, T. Current therapeutic strategies of heterotopic ossification—A survey amongst orthopaedic and trauma departments in Germany. *BMC Musculoskelet. Disord.* **2015**, *16*, 313. [CrossRef]

70. Neal, B.C.; Rodgers, A.; Clark, T.; Gray, H.; Reid, I.R.; Dunn, L.; MacMahon, S.W. A systematic survey of 13 randomized trials of non-steroidal anti-inflammatory drugs for the prevention of heterotopic bone formation after major hip surgery. *Acta Orthop. Scand.* **2000**, *71*, 122–128. [CrossRef]

71. Strauss, J.B.; Chen, S.S.; Shah, A.P.; Coon, A.B.; Dickler, A. Cost of radiotherapy versus NSAID administration for prevention of heterotopic ossification after total hip arthroplasty. *Int. J. Radiat. Oncol. Biol. Phys.* **2008**, *71*, 1460–1464. [CrossRef]

72. Vavken, P.; Dorotka, R. Economic evaluation of NSAID and radiation to prevent heterotopic ossification after hip surgery. *Arch. Orthop. Trauma Surg.* **2011**, *131*, 1309–1315. [CrossRef] [PubMed]

73. Milakovic, M.; Popovic, M.; Raman, S.; Tsao, M.; Lam, H.; Chow, E. Radiotherapy for the prophylaxis of heterotopic ossification: A systematic review and meta-analysis of randomized controlled trials. *Radiother. Oncol.* **2015**, *116*, 4–9. [CrossRef] [PubMed]

74. Wu, F.; Gao, H.; Huang, S.; Wang, G.; Jiang, X.; Li, J.; Shou, Z. NSAIDs combined with radiotherapy to prevent heterotopic ossification after total hip arthroplasty. *Zhongguo Gu Shang China J. Orthop. Traumatol.* **2018**, *31*, 538–542.

75. Pakos, E.E.; Papadopoulos, D.V.; Gelalis, I.D.; Tsantes, A.G.; Gkiatas, I.; Kosmas, D.; Tsekeris, P.G.; Xenakis, T.A. Is prophylaxis for heterotopic ossification with radiation therapy after THR associated with early loosening or carcinogenesis? *Hip. Int.* **2019**, 1120700019842724. [CrossRef] [PubMed]

76. Seavey, J.G.; Wheatley, B.M.; Pavey, G.J.; Tomasino, A.M.; Hanson, M.A.; Sanders, E.M.; Dey, D.; Moss, K.L.; Potter, B.K.; Forsberg, J.A. Early local delivery of vancomycin suppresses ectopic bone formation in a rat model of trauma-induced heterotopic ossification. *J. Orthop. Res.* **2017**, *35*, 2397–2406. [CrossRef]

77. Vasileiadis, G.I.; Sakellariou, V.I.; Kelekis, A.; Galanos, A.; Soucacos, P.N.; Papagelopoulos, P.J.; Babis, G.C. Prevention of heterotopic ossification in cases of hypertrophic osteoarthritis submitted to total hip arthroplasty. Etidronate or Indomethacin? *J. Musculoskelet. Neuronal Interact.* **2010**, *10*, 159–165.

78. Haykal, T.; Kheiri, B.; Zayed, Y.; Barbarawi, M.; Miran, M.S.; Chahine, A.; Katato, K.; Bachuwa, G. Aspirin for venous thromboembolism prophylaxis after hip or knee arthroplasty: An updated meta-analysis of randomized controlled trials. *J. Orthop.* **2019**, *16*, 294–302. [CrossRef]

79. Brouwer, K.M.; Lindenhovius, A.L.; de Witte, P.B.; Jupiter, J.B.; Ring, D. Resection of heterotopic ossification of the elbow: A comparison of ankylosis and partial restriction. *J. Hand Surg.* **2010**, *35*, 1115–1119. [CrossRef]

80. Almangour, W.; Schnitzler, A.; Salga, M.; Debaud, C.; Denormandie, P.; Genet, F. Recurrence of heterotopic ossification after removal in patients with traumatic brain injury: A systematic review. *Ann. Phys. Rehabil. Med.* **2016**, *59*, 263–269. [CrossRef]

81. Genet, F.; Marmorat, J.L.; Lautridou, C.; Schnitzler, A.; Mailhan, L.; Denormandie, P. Impact of late surgical intervention on heterotopic ossification of the hip after traumatic neurological injury. *J. Bone Jt. Surg. Br. Vol.* **2009**, *91*, 1493–1498. [CrossRef]

82. Koh, K.H.; Lim, T.K.; Lee, H.I.; Park, M.J. Surgical treatment of elbow stiffness caused by post-traumatic heterotopic ossification. *J. Shoulder Elb. Surg.* **2013**, *22*, 1128–1134. [CrossRef] [PubMed]

83. Jayasundara, J.A.; Punchihewa, G.L.; de Alwis, D.S.; Renuka, M.D. Short-term outcome after resection of neurogenic heterotopic ossification around the hips and elbow following encephalitis. *Singap. Med. J.* **2012**, *53*, e97–e100.

84. Pontell, M.E.; Sparber, L.S.; Chamberlain, R.S. Corrective and reconstructive surgery in patients with postburn heterotopic ossification and bony ankylosis: An evidence-based approach. *J. Burn Care Res.* **2015**, *36*, 57–69. [CrossRef]

85. Rubayi, S.; Gabbay, J.; Kruger, E.; Ruhge, K. The Modified Girdlestone Procedure With Muscle Flap for Management of Pressure Ulcers and Heterotopic Ossification of the Hip Region in Spinal Injury Patients: A 15-Year Review With Long-term Follow-up. *Ann. Plast. Surg.* **2016**, *77*, 645–652. [CrossRef] [PubMed]

86. Pansard, E.; Schnitzler, A.; Lautridou, C.; Judet, T.; Denormandie, P.; Genet, F. Heterotopic ossification of the shoulder after central nervous system lesion: Indications for surgery and results. *J. Shoulder Elb. Surg.* **2013**, *22*, 767–774. [CrossRef] [PubMed]

87. Mitsionis, G.I.; Lykissas, M.G.; Kalos, N.; Paschos, N.; Beris, A.E.; Georgoulis, A.D.; Xenakis, T.A. Functional outcome after excision of heterotopic ossification about the knee in ICU patients. *Int. Orthop.* **2009**, *33*, 1619–1625. [CrossRef] [PubMed]

88. Akman, S.; Sonmez, M.M.; Erturer, R.E.; Seckin, M.F.; Kara, A.; Ozturk, I. The results of surgical treatment for posttraumatic heterotopic ossification and ankylosis of the elbow. *Acta Orthop. Traumatol. Turc.* **2010**, *44*, 206–211. [CrossRef]

89. Denormandie, P.; de l'Escalopier, N.; Gatin, L.; Grelier, A.; Genêt, F. Resection of neurogenic heterotopic ossification (NHO) of the hip. *Orthop. Traumatol. Surg. Res.* **2018**, *104*, S121–S127. [CrossRef]

90. Łęgosz, P.; Stępiński, P.; Pulik, Ł.; Kotela, A.; Małdyk, P. Total hip replacement vs femoral neck osteotomy in the treatment of heterotopic ossifications, neurogenic in the IV degree by scale Brooker–comparison of treatment results. *Chir. Narz. Ruchu Ortop. Pol.* **2017**, *82*, 28–40.

91. Forcina, L.; Miano, C.; Pelosi, L.; Musaro, A. An Overview about the Biology of Skeletal Muscle Satellite Cells. *Curr. Genom.* **2019**, *20*, 24–37. [CrossRef]

92. Brzoska, E.; Ciemerych, M.A.; Przewozniak, M.; Zimowska, M. Regulation of muscle stem cells activation: The role of growth factors and extracellular matrix. *Vitam. Horm.* **2011**, *87*, 239–276. [CrossRef] [PubMed]

93. Relaix, F.; Zammit, P.S. Satellite cells are essential for skeletal muscle regeneration: The cell on the edge returns centre stage. *Development* **2012**, *139*, 2845–2856. [CrossRef] [PubMed]

94. Zammit, P.S. Function of the myogenic regulatory factors Myf5, MyoD, Myogenin and MRF4 in skeletal muscle, satellite cells and regenerative myogenesis. *Semin. Cell Dev. Biol.* **2017**, *72*, 19–32. [CrossRef] [PubMed]

95. Dey, B.K.; Gagan, J.; Dutta, A. miR-206 and -486 induce myoblast differentiation by downregulating Pax7. *Mol. Cell Biol.* **2011**, *31*, 203–214. [CrossRef] [PubMed]

96. Forcales, S.V.; Albini, S.; Giordani, L.; Malecova, B.; Cignolo, L.; Chernov, A.; Coutinho, P.; Saccone, V.; Consalvi, S.; Williams, R.; et al. Signal-dependent incorporation of MyoD-BAF60c into Brg1-based SWI/SNF chromatin-remodelling complex. *Embo J.* **2012**, *31*, 301–316. [CrossRef]

97. Chazaud, B. Inflammation during skeletal muscle regeneration and tissue remodeling: Application to exercise-induced muscle damage management. *Immunol. Cell Biol.* **2016**, *94*, 140–145. [CrossRef]

98. Liu, X.; Liu, Y.; Zhao, L.; Zeng, Z.; Xiao, W.; Chen, P. Macrophage depletion impairs skeletal muscle regeneration: The roles of regulatory factors for muscle regeneration. *Cell Biol. Int.* **2017**, *41*, 228–238. [CrossRef]

99. Tatsumi, R.; Anderson, J.E.; Nevoret, C.J.; Halevy, O.; Allen, R.E. HGF/SF is present in normal adult skeletal muscle and is capable of activating satellite cells. *Dev. Biol.* **1998**, *194*, 114–128. [CrossRef]

100. Galvin, C.D.; Hardiman, O.; Nolan, C.M. IGF-1 receptor mediates differentiation of primary cultures of mouse skeletal myoblasts. *Mol. Cell Endocrinol.* **2003**, *200*, 19–29. [CrossRef]

101. Tidball, J.G.; Welc, S.S. Macrophage-Derived IGF-1 Is a Potent Coordinator of Myogenesis and Inflammation in Regenerating Muscle. *Mol. Ther. J. Am. Soc. Gene Ther.* **2015**, *23*, 1134–1135. [CrossRef]

102. Chen, S.E.; Jin, B.; Li, Y.P. TNF-alpha regulates myogenesis and muscle regeneration by activating p38 MAPK. *Am. J. Physiol. Cell Physiol.* **2007**, *292*, C1660–C1671. [CrossRef] [PubMed]

103. Latroche, C.; Weiss-Gayet, M.; Chazaud, B. Investigating the Vascular Niche: Three-Dimensional Co-culture of Human Skeletal Muscle Stem Cells and Endothelial Cells. *Methods Mol. Biol.* **2019**, *2002*, 121–128. [CrossRef] [PubMed]

104. Abou-Khalil, R.; Mounier, R.; Chazaud, B. Regulation of myogenic stem cell behavior by vessel cells: The "menage a trois" of satellite cells, periendothelial cells and endothelial cells. *Cell Cycle* **2010**, *9*, 892–896. [CrossRef] [PubMed]

105. Christov, C.; Chretien, F.; Abou-Khalil, R.; Bassez, G.; Vallet, G.; Authier, F.J.; Bassaglia, Y.; Shinin, V.; Tajbakhsh, S.; Chazaud, B.; et al. Muscle satellite cells and endothelial cells: Close neighbors and privileged partners. *Mol. Biol. Cell* **2007**, *18*, 1397–1409. [CrossRef] [PubMed]

106. Fiore, D.; Judson, R.N.; Low, M.; Lee, S.; Zhang, E.; Hopkins, C.; Xu, P.; Lenzi, A.; Rossi, F.M.; Lemos, D.R. Pharmacological blockage of fibro/adipogenic progenitor expansion and suppression of regenerative fibrogenesis is associated with impaired skeletal muscle regeneration. *Stem Cell Res.* **2016**, *17*, 161–169. [CrossRef]

107. Mathew, S.J.; Hansen, J.M.; Merrell, A.J.; Murphy, M.M.; Lawson, J.A.; Hutcheson, D.A.; Hansen, M.S.; Angus-Hill, M.; Kardon, G. Connective tissue fibroblasts and Tcf4 regulate myogenesis. *Development* **2011**, *138*, 371–384. [CrossRef]

108. Murphy, M.M.; Lawson, J.A.; Mathew, S.J.; Hutcheson, D.A.; Kardon, G. Satellite cells, connective tissue fibroblasts and their interactions are crucial for muscle regeneration. *Development* **2011**, *138*, 3625–3637. [CrossRef] [PubMed]

109. Uezumi, A.; Fukada, S.; Yamamoto, N.; Takeda, S.; Tsuchida, K. Mesenchymal progenitors distinct from satellite cells contribute to ectopic fat cell formation in skeletal muscle. *Nat. Cell Biol.* **2010**, *12*, 143–152. [CrossRef] [PubMed]

110. Joe, A.W.; Yi, L.; Natarajan, A.; Le Grand, F.; So, L.; Wang, J.; Rudnicki, M.A.; Rossi, F.M. Muscle injury activates resident fibro/adipogenic progenitors that facilitate myogenesis. *Nat. Cell Biol.* **2010**, *12*, 153–163. [CrossRef] [PubMed]

111. Wosczyna, M.N.; Konishi, C.T.; Perez Carbajal, E.E.; Wang, T.T.; Walsh, R.A.; Gan, Q.; Wagner, M.W.; Rando, T.A. Mesenchymal Stromal Cells Are Required for Regeneration and Homeostatic Maintenance of Skeletal Muscle. *Cell Rep.* **2019**, *27*, 2029–2035.e5. [CrossRef]

112. Forcina, L.; Miano, C.; Scicchitano, B.M.; Musaro, A. Signals from the Niche: Insights into the Role of IGF-1 and IL-6 in Modulating Skeletal Muscle Fibrosis. *Cells* **2019**, *8*, 232. [CrossRef] [PubMed]

113. Mierzejewski, B.; Archacka, K.; Grabowska, I.; Florkowska, A.; Ciemerych, M.A.; Brzoska, E. Human and mouse skeletal muscle stem and progenitor cells in health and disease. *Semin. Cell Dev. Biol.* **2020**. [CrossRef] [PubMed]

114. Birbrair, A.; Delbono, O. Pericytes are Essential for Skeletal Muscle Formation. *Stem Cell Rev. Rep.* **2015**, *11*, 547–548. [CrossRef] [PubMed]

115. Birbrair, A.; Zhang, T.; Wang, Z.M.; Messi, M.L.; Mintz, A.; Delbono, O. Type-1 pericytes participate in fibrous tissue deposition in aged skeletal muscle. *Am. J. Physiol. Cell Physiol.* **2013**, *305*, C1098–C1113. [CrossRef] [PubMed]

116. Dellavalle, A.; Maroli, G.; Covarello, D.; Azzoni, E.; Innocenzi, A.; Perani, L.; Antonini, S.; Sambasivan, R.; Brunelli, S.; Tajbakhsh, S.; et al. Pericytes resident in postnatal skeletal muscle differentiate into muscle fibres and generate satellite cells. *Nat. Commun.* **2011**, *2*, 499. [CrossRef] [PubMed]

117. Dellavalle, A.; Sampaolesi, M.; Tonlorenzi, R.; Tagliafico, E.; Sacchetti, B.; Perani, L.; Innocenzi, A.; Galvez, B.G.; Messina, G.; Morosetti, R.; et al. Pericytes of human skeletal muscle are myogenic precursors distinct from satellite cells. *Nat. Cell Biol.* **2007**, *9*, 255–267. [CrossRef] [PubMed]

118. Gautam, J.; Yao, Y. Pericytes in Skeletal Muscle. *Adv. Exp. Med. Biol.* **2019**, *1122*, 59–72. [CrossRef]

119. Persichini, T.; Funari, A.; Colasanti, M.; Sacchetti, B. Clonogenic, myogenic progenitors expressing MCAM/CD146 are incorporated as adventitial reticular cells in the microvascular compartment of human post-natal skeletal muscle. *PLoS ONE* **2017**, *12*, e0188844. [CrossRef]

120. Kostallari, E.; Baba-Amer, Y.; Alonso-Martin, S.; Ngoh, P.; Relaix, F.; Lafuste, P.; Gherardi, R.K. Pericytes in the myovascular niche promote post-natal myofiber growth and satellite cell quiescence. *Development* **2015**, *142*, 1242–1253. [CrossRef]

121. Mitchell, K.J.; Pannerec, A.; Cadot, B.; Parlakian, A.; Besson, V.; Gomes, E.R.; Marazzi, G.; Sassoon, D.A. Identification and characterization of a non-satellite cell muscle resident progenitor during postnatal development. *Nat. Cell Biol.* **2010**, *12*, 257–266. [CrossRef] [PubMed]

122. Tamaki, T.; Akatsuka, A.; Ando, K.; Nakamura, Y.; Matsuzawa, H.; Hotta, T.; Roy, R.R.; Edgerton, V.R. Identification of myogenic-endothelial progenitor cells in the interstitial spaces of skeletal muscle. *J. Cell Biol.* **2002**, *157*, 571–577. [CrossRef] [PubMed]

123. Liu, N.; Garry, G.A.; Li, S.; Bezprozvannaya, S.; Sanchez-Ortiz, E.; Chen, B.; Shelton, J.M.; Jaichander, P.; Bassel-Duby, R.; Olson, E.N. A Twist2-dependent progenitor cell contributes to adult skeletal muscle. *Nat. Cell Biol.* **2017**, *19*, 202–213. [CrossRef] [PubMed]

124. Gussoni, E.; Soneoka, Y.; Strickland, C.D.; Buzney, E.A.; Khan, M.K.; Flint, A.F.; Kunkel, L.M.; Mulligan, R.C. Dystrophin expression in the mdx mouse restored by stem cell transplantation. *Nature* **1999**, *401*, 390–394. [CrossRef] [PubMed]

125. Torrente, Y.; Belicchi, M.; Sampaolesi, M.; Pisati, F.; Meregalli, M.; D'Antona, G.; Tonlorenzi, R.; Porretti, L.; Gavina, M.; Mamchaoui, K.; et al. Human circulating AC133(+) stem cells restore dystrophin expression and ameliorate function in dystrophic skeletal muscle. *J. Clin. Investig.* **2004**, *114*, 182–195. [CrossRef]

126. Asakura, A.; Komaki, M.; Rudnicki, M. Muscle satellite cells are multipotential stem cells that exhibit myogenic, osteogenic, and adipogenic differentiation. *Differentiation* **2001**, *68*, 245–253. [CrossRef]

127. Castiglioni, A.; Hettmer, S.; Lynes, M.D.; Rao, T.N.; Tchessalova, D.; Sinha, I.; Lee, B.T.; Tseng, Y.H.; Wagers, A.J. Isolation of progenitors that exhibit myogenic/osteogenic bipotency in vitro by fluorescence-activated cell sorting from human fetal muscle. *Stem Cell Rep.* **2014**, *2*, 92–106. [CrossRef]

128. Guimaraes-Camboa, N.; Cattaneo, P.; Sun, Y.; Moore-Morris, T.; Gu, Y.; Dalton, N.D.; Rockenstein, E.; Masliah, E.; Peterson, K.L.; Stallcup, W.B.; et al. Pericytes of Multiple Organs Do Not Behave as Mesenchymal Stem Cells In Vivo. *Cell Stem Cell* **2017**, *20*, 345–359.e5. [CrossRef]

129. Sacchetti, B.; Funari, A.; Remoli, C.; Giannicola, G.; Kogler, G.; Liedtke, S.; Cossu, G.; Serafini, M.; Sampaolesi, M.; Tagliafico, E.; et al. No Identical "Mesenchymal Stem Cells" at Different Times and Sites: Human Committed Progenitors of Distinct Origin and Differentiation Potential Are Incorporated as Adventitial Cells in Microvessels. *Stem Cell Rep.* **2016**, *6*, 897–913. [CrossRef]

130. Wosczyna, M.N.; Biswas, A.A.; Cogswell, C.A.; Goldhamer, D.J. Multipotent progenitors resident in the skeletal muscle interstitium exhibit robust BMP-dependent osteogenic activity and mediate heterotopic ossification. *J. Bone Miner. Res.* **2012**, *27*, 1004–1017. [CrossRef]

131. Pannerec, A.; Formicola, L.; Besson, V.; Marazzi, G.; Sassoon, D.A. Defining skeletal muscle resident progenitors and their cell fate potentials. *Development* **2013**, *140*, 2879–2891. [CrossRef]

132. Tamaki, T.; Akatsuka, A.; Yoshimura, S.; Roy, R.R.; Edgerton, V.R. New fiber formation in the interstitial spaces of rat skeletal muscle during postnatal growth. *J. Histochem. Cytochem.* **2002**, *50*, 1097–1111. [CrossRef] [PubMed]

133. Tamaki, T.; Uchiyama, Y.; Okada, Y.; Ishikawa, T.; Sato, M.; Akatsuka, A.; Asahara, T. Functional recovery of damaged skeletal muscle through synchronized vasculogenesis, myogenesis, and neurogenesis by muscle-derived stem cells. *Circulation* **2005**, *112*, 2857–2866. [CrossRef] [PubMed]

134. Negroni, E.; Riederer, I.; Chaouch, S.; Belicchi, M.; Razini, P.; Di Santo, J.; Torrente, Y.; Butler-Browne, G.S.; Mouly, V. In vivo myogenic potential of human CD133+ muscle-derived stem cells: A quantitative study. *Moleculus* **2009**, *17*, 1771–1778. [CrossRef] [PubMed]

135. Torrente, Y.; Belicchi, M.; Marchesi, C.; D'Antona, G.; Cogiamanian, F.; Pisati, F.; Gavina, M.; Giordano, R.; Tonlorenzi, R.; Fagiolari, G.; et al. Autologous transplantation of muscle-derived CD133+ stem cells in Duchenne muscle patients. *Cell Transpl.* **2007**, *16*, 563–577. [CrossRef] [PubMed]

136. Benchaouir, R.; Meregalli, M.; Farini, A.; D'Antona, G.; Belicchi, M.; Goyenvalle, A.; Battistelli, M.; Bresolin, N.; Bottinelli, R.; Garcia, L.; et al. Restoration of human dystrophin following transplantation of exon-skipping-engineered DMD patient stem cells into dystrophic mice. *Cell Stem Cell* **2007**, *1*, 646–657. [CrossRef]

137. Kaji, D.A.; Tan, Z.; Johnson, G.L.; Huang, W.; Vasquez, K.; Lehoczky, J.A.; Levi, B.; Cheah, K.S.E.; Huang, A.H. Cellular Plasticity in Musculoskeletal Development, Regeneration, and Disease. *J. Orthop. Res.* **2020**, *38*, 708–718. [CrossRef]

138. Lounev, V.Y.; Ramachandran, R.; Wosczyna, M.N.; Yamamoto, M.; Maidment, A.D.; Shore, E.M.; Glaser, D.L.; Goldhamer, D.J.; Kaplan, F.S. Identification of progenitor cells that contribute to heterotopic skeletogenesis. *J. Bone Jt. Surg. Am.* **2009**, *91*, 652–663. [CrossRef]

139. Medici, D.; Shore, E.M.; Lounev, V.Y.; Kaplan, F.S.; Kalluri, R.; Olsen, B.R. Conversion of vascular endothelial cells into multipotent stem-like cells. *Nat. Med.* **2010**, *16*, 1400–1406. [CrossRef]

140. Kan, L.; Liu, Y.; McGuire, T.L.; Berger, D.M.; Awatramani, R.B.; Dymecki, S.M.; Kessler, J.A. Dysregulation of local stem/progenitor cells as a common cellular mechanism for heterotopic ossification. *Stem Cells* **2009**, *27*, 150–156. [CrossRef]

141. Hwang, C.; Marini, S.; Huber, A.K.; Stepien, D.M.; Sorkin, M.; Loder, S.; Pagani, C.A.; Li, J.; Visser, N.D.; Vasquez, K.; et al. Mesenchymal VEGFA induces aberrant differentiation in heterotopic ossification. *Bone Res.* **2019**, *7*, 36. [CrossRef]

142. Uezumi, A.; Fukada, S.; Yamamoto, N.; Ikemoto-Uezumi, M.; Nakatani, M.; Morita, M.; Yamaguchi, A.; Yamada, H.; Nishino, I.; Hamada, Y.; et al. Identification and characterization of PDGFRalpha+ mesenchymal progenitors in human skeletal muscle. *Cell Death Dis.* **2014**, *5*, e1186. [CrossRef] [PubMed]

143. Regard, J.B.; Malhotra, D.; Gvozdenovic-Jeremic, J.; Josey, M.; Chen, M.; Weinstein, L.S.; Lu, J.; Shore, E.M.; Kaplan, F.S.; Yang, Y. Activation of Hedgehog signaling by loss of GNAS causes heterotopic ossification. *Nat. Med.* **2013**, *19*, 1505–1512. [CrossRef] [PubMed]

144. Agarwal, S.; Loder, S.; Brownley, C.; Cholok, D.; Mangiavini, L.; Li, J.; Breuler, C.; Sung, H.H.; Li, S.; Ranganathan, K.; et al. Inhibition of Hif1alpha prevents both trauma-induced and genetic heterotopic ossification. *Proc. Natl. Acad. Sci. USA* **2016**, *113*, E338–E347. [CrossRef] [PubMed]

145. Dey, D.; Bagarova, J.; Hatsell, S.J.; Armstrong, K.A.; Huang, L.; Ermann, J.; Vonner, A.J.; Shen, Y.; Mohedas, A.H.; Lee, A.; et al. Two tissue-resident progenitor lineages drive distinct phenotypes of heterotopic ossification. *Sci. Transl. Med.* **2016**, *8*, 366ra163. [CrossRef]

146. Kan, C.; Chen, L.; Hu, Y.; Ding, N.; Li, Y.; McGuire, T.L.; Lu, H.; Kessler, J.A.; Kan, L. Gli1-labeled adult mesenchymal stem/progenitor cells and hedgehog signaling contribute to endochondral heterotopic ossification. *Bone* **2018**, *109*, 71–79. [CrossRef]

147. Kan, L.; Peng, C.Y.; McGuire, T.L.; Kessler, J.A. Glast-expressing progenitor cells contribute to heterotopic ossification. *Bone* **2013**, *53*, 194–203. [CrossRef]

148. Agarwal, S.; Loder, S.J.; Cholok, D.; Peterson, J.; Li, J.; Breuler, C.; Cameron Brownley, R.; Hsin Sung, H.; Chung, M.T.; Kamiya, N.; et al. Scleraxis-Lineage Cells Contribute to Ectopic Bone Formation in Muscle and Tendon. *Stem Cells* **2017**, *35*, 705–710. [CrossRef]

149. Chen, D.; Zhao, M.; Mundy, G.R. Bone morphogenetic proteins. *Growth Factors* **2004**, *22*, 233–241. [CrossRef]

150. Guo, X.; Wang, X.F. Signaling cross-talk between TGF-beta/BMP and other pathways. *Cell Res.* **2009**, *19*, 71–88. [CrossRef]

151. Komori, T. Runx2, an inducer of osteoblast and chondrocyte differentiation. *Histochem. Cell Biol.* **2018**, *149*, 313–323. [CrossRef]

152. Montecino, M.; Stein, G.; Stein, J.; Zaidi, K.; Aguilar, R. Multiple levels of epigenetic control for bone biology and pathology. *Bone* **2015**, *81*, 733–738. [CrossRef]

153. Uchida, K.; Yayama, T.; Cai, H.X.; Nakajima, H.; Sugita, D.; Guerrero, A.R.; Kobayashi, S.; Yoshida, A.; Chen, K.B.; Baba, H. Ossification process involving the human thoracic ligamentum flavum: Role of transcription factors. *Arthritis Res. Ther.* **2011**, *13*, R144. [CrossRef] [PubMed]

154. Lin, L.; Shen, Q.; Xue, T.; Yu, C. Heterotopic ossification induced by Achilles tenotomy via endochondral bone formation: Expression of bone and cartilage related genes. *Bone* **2010**, *46*, 425–431. [CrossRef] [PubMed]

155. Kang, J.S.; Alliston, T.; Delston, R.; Derynck, R. Repression of Runx2 function by TGF-beta through recruitment of class II histone deacetylases by Smad3. *Embo J.* **2005**, *24*, 2543–2555. [CrossRef] [PubMed]

156. Lee, K.S.; Hong, S.H.; Bae, S.C. Both the Smad and p38 MAPK pathways play a crucial role in Runx2 expression following induction by transforming growth factor-beta and bone morphogenetic protein. *Oncogene* **2002**, *21*, 7156–7163. [CrossRef]

157. Wu, M.; Chen, G.; Li, Y.P. TGF-beta and BMP signaling in osteoblast, skeletal development, and bone formation, homeostasis and disease. *Bone Res.* **2016**, *4*, 16009. [CrossRef]

158. Rahman, M.S.; Akhtar, N.; Jamil, H.M.; Banik, R.S.; Asaduzzaman, S.M. TGF-beta/BMP signaling and other molecular events: Regulation of osteoblastogenesis and bone formation. *Bone Res.* **2015**, *3*, 15005. [CrossRef]

159. Grenier, G.; Leblanc, E.; Faucheux, N.; Lauzier, D.; Kloen, P.; Hamdy, R.C. BMP-9 expression in human traumatic heterotopic ossification: A case report. *Skelet. Muscle* **2013**, *3*, 29. [CrossRef]

160. Matsumoto, Y.; Ikeya, M.; Hino, K.; Horigome, K.; Fukuta, M.; Watanabe, M.; Nagata, S.; Yamamoto, T.; Otsuka, T.; Toguchida, J. New Protocol to Optimize iPS Cells for Genome Analysis of Fibrodysplasia Ossificans Progressiva. *Stem Cells* **2015**, *33*, 1730–1742. [CrossRef]

161. Medici, D.; Olsen, B.R. The role of endothelial-mesenchymal transition in heterotopic ossification. *J. Bone Miner. Res.* **2012**, *27*, 1619–1622. [CrossRef]

162. Agarwal, S.; Loder, S.; Li, J.; Brownley, C.; Peterson, J.R.; Oluwatobi, E.; Drake, J.; Cholok, D.; Ranganathan, K.; Sung, H.H.; et al. Diminished Chondrogenesis and Enhanced Osteoclastogenesis in Leptin-Deficient Diabetic Mice (ob/ob) Impair Pathologic, Trauma-Induced Heterotopic Ossification. *Stem Cells Dev.* **2015**, *24*, 2864–2872. [CrossRef] [PubMed]

163. Lees-Shepard, J.B.; Yamamoto, M.; Biswas, A.A.; Stoessel, S.J.; Nicholas, S.E.; Cogswell, C.A.; Devarakonda, P.M.; Schneider, M.J., Jr.; Cummins, S.M.; Legendre, N.P.; et al. Activin-dependent signaling in fibro/adipogenic progenitors causes fibrodysplasia ossificans progressiva. *Nat. Commun.* **2018**, *9*, 471. [CrossRef] [PubMed]

164. Wu, J.; Ren, B.; Shi, F.; Hua, P.; Lin, H. BMP and mTOR signaling in heterotopic ossification: Does their crosstalk provide therapeutic opportunities? *J. Cell Biochem.* **2019**, *120*, 12108–12122. [CrossRef] [PubMed]

165. Moustakas, A.; Heldin, C.H. Ecsit-ement on the crossroads of Toll and BMP signal transduction. *Genes Dev.* **2003**, *17*, 2855–2859. [CrossRef]

166. Gaur, T.; Lengner, C.J.; Hovhannisyan, H.; Bhat, R.A.; Bodine, P.V.; Komm, B.S.; Javed, A.; van Wijnen, A.J.; Stein, J.L.; Stein, G.S.; et al. Canonical WNT signaling promotes osteogenesis by directly stimulating Runx2 gene expression. *J. Biol. Chem* **2005**, *280*, 33132–33140. [CrossRef] [PubMed]

167. Kawane, T.; Komori, H.; Liu, W.; Moriishi, T.; Miyazaki, T.; Mori, M.; Matsuo, Y.; Takada, Y.; Izumi, S.; Jiang, Q.; et al. Dlx5 and mef2 regulate a novel runx2 enhancer for osteoblast-specific expression. *J. Bone Miner. Res.* **2014**, *29*, 1960–1969. [CrossRef]

168. Hino, K.; Horigome, K.; Nishio, M.; Komura, S.; Nagata, S.; Zhao, C.; Jin, Y.; Kawakami, K.; Yamada, Y.; Ohta, A.; et al. Activin-A enhances mTOR signaling to promote aberrant chondrogenesis in fibrodysplasia ossificans progressiva. *J. Clin. Investig.* **2017**, *127*, 3339–3352. [CrossRef]

169. Huang, Y.; Wang, X.; Lin, H. The hypoxic microenvironment: A driving force for heterotopic ossification progression. *Cell Commun. Signal.* **2020**, *18*, 20. [CrossRef]

170. Peterson, J.R.; De La Rosa, S.; Sun, H.; Eboda, O.; Cilwa, K.E.; Donneys, A.; Morris, M.; Buchman, S.R.; Cederna, P.S.; Krebsbach, P.H.; et al. Burn injury enhances bone formation in heterotopic ossification model. *Ann. Surg.* **2014**, *259*, 993–998. [CrossRef]

171. Croes, M.; Kruyt, M.C.; Boot, W.; Pouran, B.; Braham, M.V.; Pakpahan, S.A.; Weinans, H.; Vogely, H.C.; Fluit, A.C.; Dhert, W.J.; et al. The role of bacterial stimuli in inflammation-driven bone formation. *Eur. Cells Mater.* **2019**, *37*, 402–419. [CrossRef]

172. Wang, H.; Behrens, E.M.; Pignolo, R.J.; Kaplan, F.S. ECSIT links TLR and BMP signaling in FOP connective tissue progenitor cells. *Bone* **2018**, *109*, 201–209. [CrossRef] [PubMed]

173. Wi, S.M.; Moon, G.; Kim, J.; Kim, S.T.; Shim, J.H.; Chun, E.; Lee, K.Y. TAK1-ECSIT-TRAF6 complex plays a key role in the TLR4 signal to activate NF-kappaB. *J. Biol. Chem* **2014**, *289*, 35205–35214. [CrossRef] [PubMed]

174. Shimono, K.; Tung, W.E.; Macolino, C.; Chi, A.H.; Didizian, J.H.; Mundy, C.; Chandraratna, R.A.; Mishina, Y.; Enomoto-Iwamoto, M.; Pacifici, M.; et al. Potent inhibition of heterotopic ossification by nuclear retinoic acid receptor-gamma agonists. *Nat. Med.* **2011**, *17*, 454–460. [CrossRef] [PubMed]

175. Chakkalakal, S.A.; Uchibe, K.; Convente, M.R.; Zhang, D.; Economides, A.N.; Kaplan, F.S.; Pacifici, M.; Iwamoto, M.; Shore, E.M. Palovarotene Inhibits Heterotopic Ossification and Maintains Limb Mobility and Growth in Mice With the Human ACVR1(R206H) Fibrodysplasia Ossificans Progressiva (FOP) Mutation. *J. Bone Miner. Res.* **2016**, *31*, 1666–1675. [CrossRef]

176. Hino, K.; Zhao, C.; Horigome, K.; Nishio, M.; Okanishi, Y.; Nagata, S.; Komura, S.; Yamada, Y.; Toguchida, J.; Ohta, A. An mTOR signaling modulator suppressed heterotopic ossification of Fibrodysplasia Ossificans Progressiva. *Stem Cell Rep.* **2018**, *11*, 1106–1119. [CrossRef] [PubMed]

177. Peterson, J.R.; De La Rosa, S.; Eboda, O.; Cilwa, K.E.; Agarwal, S.; Buchman, S.R.; Cederna, P.S.; Xi, C.; Morris, M.D.; Herndon, D.N.; et al. Treatment of heterotopic ossification through remote ATP hydrolysis. *Sci. Transl. Med.* **2014**, *6*, 255ra132. [CrossRef]

178. Hannallah, D.; Peng, H.; Young, B.; Usas, A.; Gearhart, B.; Huard, J. Retroviral delivery of Noggin inhibits the formation of heterotopic ossification induced by BMP-4, demineralized bone matrix, and trauma in an animal model. *J. Bone Jt. Surg. Am. Vol.* **2004**, *86*, 80–91. [CrossRef]

179. Kaplan, F.S.; Pignolo, R.J.; Al Mukaddam, M.M.; Shore, E.M. Hard targets for a second skeleton: Therapeutic horizons for fibrodysplasia ossificans progressiva (FOP). *Expert Opin. Orphan Drugs* **2017**, *5*, 291–294. [CrossRef]

180. Xue, T.; Mao, Z.; Lin, L.; Hou, Y.; Wei, X.; Fu, X.; Zhang, J.; Yu, C. Non-virus-mediated transfer of siRNAs against Runx2 and Smad4 inhibit heterotopic ossification in rats. *Gene Ther.* **2010**, *17*, 370–379. [CrossRef]

Article

High-Dimensional Single-Cell Quantitative Profiling of Skeletal Muscle Cell Population Dynamics during Regeneration

Lucia Lisa Petrilli [1,2,†], Filomena Spada [1,†], Alessandro Palma [1], Alessio Reggio [1], Marco Rosina [1], Cesare Gargioli [1], Luisa Castagnoli [1], Claudia Fuoco [1,*] and Gianni Cesareni [1,3]

[1] Department of Biology, University of Rome "Tor Vergata", 00133 Rome, Italy; lucialisa.petrilli@opbg.net (L.L.P.); filomena.spada86@gmail.com (F.S.); alessandro.palma@live.it (A.P.); alessio.reggio@uniroma2.it (A.R.); marco.rosina90@gmail.com (M.R.); Cesare.Gargioli@uniroma2.it (C.G.); castagnoli@uniroma2.it (L.C.); cesareni@uniroma2.it (G.C.)

[2] Department of Onco-hematology, Gene and Cell Therapy—Bambino Gesù Children's Hospital—IRCCS, 00146 Rome, Italy

[3] Fondazione Santa Lucia Istituto di Ricovero e Cura a Carattere Scientifico (IRCCS), 00143 Rome, Italy

*   Correspondence: claudia.fuoco@uniroma2.it; Tel.: +39-0672594814

†   These authors contributed equally to the study.

Received: 17 April 2020; Accepted: 14 July 2020; Published: 18 July 2020

Abstract: The interstitial space surrounding the skeletal muscle fibers is populated by a variety of mononuclear cell types. Upon acute or chronic insult, these cell populations become activated and initiate finely-orchestrated crosstalk that promotes myofiber repair and regeneration. Mass cytometry is a powerful and highly multiplexed technique for profiling single-cells. Herein, it was used to dissect the dynamics of cell populations in the skeletal muscle in physiological and pathological conditions. Here, we characterized an antibody panel that could be used to identify most of the cell populations in the muscle interstitial space. By exploiting the mass cytometry resolution, we provided a comprehensive picture of the dynamics of the major cell populations that sensed and responded to acute damage in wild type mice and in a mouse model of Duchenne muscular dystrophy. In addition, we revealed the intrinsic heterogeneity of many of these cell populations.

Keywords: single-cell; mass cytometry; skeletal muscle regeneration; skeletal muscle homeostasis; fibro/adipogenic progenitors; myogenic progenitors; muscle populations

---

1. Introduction

In physiological conditions, the adult skeletal muscle has a relatively low cell turnover [1]. However, physical activity, trauma, or muscle pathologies, undermining tissue integrity, trigger a tightly controlled regeneration process. Although satellite cells (SCs) are the main actors of myofiber regeneration after damage [2–6], successful muscle healing requires the participation of additional cell types that directly or indirectly contribute to this process. In this context, immune cells and fibro/adipogenic progenitors (FAPs) play a prominent role in supporting the clearance of the damaged tissue, while assisting SCs in their regenerative role [7–11]. However, the orchestrated crosstalk of the regeneration machinery gradually fails in patients affected by muscle-related disorders, such as dystrophies [12]. Here, the accumulation of intrinsic cell defects and the changes in the stem cell niche lead to infiltrations of fat and fibrotic deposition, compromising muscle functions [9,13–16].

Over the past decades, the complex cell crosstalk occurring during muscle regeneration has been studied in detail [17–19]. However, to date, most studies have mainly relied on the analysis of bulk cell populations identified by the expression of a few specific markers and sorted for ex vivo analysis.

As a consequence, due to the lack of technologies suitable to address this issue, little is known about muscle cell population heterogeneity. Only recently, the development of technologies to determine the transcriptome of single-cells or their exposed antigen repertoires has permitted to reveal the extent of this heterogeneity and its possible implication in muscle physiology and pathology [20–24]. Moreover, the dynamic changes and the relative abundance of muscle cell populations, upon acute or chronic damage, still remain largely uncharacterized. To fill this gap, we explored, via single-cell mass cytometry, the changes in the multidimensional antigen repertoires of the main players, colonizing the muscle stem cell niche after injury. Here, we described, at single-cell resolution, the time-dependent changes of muscle population dynamics upon myotoxin-induced damage in wild type (wt) and in a mouse model of Duchenne muscular dystrophy (the mdx model).

2. Materials and Methods

2.1. Mouse Strains and Animal Procedures

C57BL/6J (RRID:IMSR_JAX:000664) and C57BL/10ScSn-Dmd^{mdx}/J mice (RRID:IMSR_JAX:001801), hereafter referred to as wt and mdx mice, respectively, were purchased from the Jackson Laboratory.

Mice were bred respecting the standard animal facility procedures, and all the procedures were conducted in accordance with rules of good animal experimentation I.A.C.U.C. n°432 of 12 March 2006 and under ethical approval released on 23/October/2017 from the Italian Ministry of Health, protocol #820/2017-PR.

For muscle injury, 45-day-old wt and mdx mice were anesthetized with an intramuscular injection of saline solution containing ketamine (5 mg/mL) and xylazine (1 mg/mL) prior to the intramuscular administration of 20 μL of 10 μM cardiotoxin solution, isolated from *Naja Pallida* (Latoxan L8102, Portes les valence, France), into *tibialis anterior, quadriceps,* and *gastrocnemius* muscles.

2.2. Histological Analysis

Tibialis anterior (TA) muscles were collected, embedded in optimal cutting temperature compound (Killik—O.C.T., Bio Optica, Milan, Italy), and snap-frozen in liquid nitrogen for 10 s. Embedded muscles were stored at −80 °C for transverse cryo-sectioning with a Leica cryostat. Cryosections (10 μm thickness) were collected on Superfrost glass slides (Thermo Fisher Scientific, Monza, Italy), and tissue slides were stained with hematoxylin and eosin (H&E).

For the H&E, cryosections were fixed with 4% paraformaldehyde (PFA, Santa Cruz Biotechnology, D.B.A. Italia S.r.l., Segrate Milan, Italy) for 15 min at room temperature (RT). After washing in 1X PBS, tissue slides were incubated in the hematoxylin solution for 15 min and rinsed for 5 min in tap water. Cryosections were then counterstained with an alcoholic solution of eosin for 30 min. Following the eosin staining, cryosections were dehydrated in increasing concentrations of alcohol, clarified with the histo-clear solution (Agar Scientific Ltd, Stansted, UK), and finally mounted on coverslips, using the resinous Eukitt mounting medium (Electron Microscopy Sciences, Hatfield Township, PA, USA).

H&E images were captured using the Zeiss Lab A1 AX10 microscope at the 20× magnification in the bright field.

2.3. Skeletal Muscle Mononuclear Cell Purification

Isolation of mononuclear cell populations was performed as in Spada et al. [25]. Mice were sacrificed by cervical dislocation, and the hind limbs were washed with 70% ethanol. Mice hind limbs were then dissected and finely minced in Hank's balanced salt solution (HBSS) with calcium and magnesium (Gibco- Thermo Fisher Scientific, Monza, Italy) supplemented with 0.2% bovine serum albumin (BSA) (AppliChem, Cinisello Balsamo, Milan, Italy) and 1% penicillin-streptomycin (P/S) (Life Technologies, Monza, Italy, 10,000 U/mL) (HBSS$^+$) under a sterile hood. The homogenized tissue preparation was centrifuged at 70× *g* for 10 min at 4 °C to separate fat and subjected to enzymatic digestion for 1 h at 37 °C, with gentle mixing in a solution containing 2 μg/μL collagenase A

(Roche- Merck KGaA, Darmstadt, Germany), 2.4 U/mL dispase II (Roche- Merck KGaA, Darmstadt, Germany), and 10 μg/mL DNase I (Roche- Merck KGaA, Darmstadt, Germany) diluted in Dulbecco's phosphate-buffered saline (D-PBS) with calcium and magnesium (Gibco-Thermo Fisher Scientific, Monza, Italy). The reaction was inactivated with HBSS$^+$, and the cell suspension was subjected to three sequential filtrations through 100 μm, 70 μm, and 40 μm cell strainers (BD Falcon, BD Italia, Milan, Italy) and centrifugations at 700× g for 5 min. The lysis of red blood cells was performed by incubating with RBC Lysis Buffer (Santa Cruz Biotechnology, D.B.A. Italia S.r.l., Segrate, Milan, Italy) for 150 s on ice, prior to the 40 μm filtration step.

2.4. Single-Cell Mass Cytometry

For single-cell mass cytometry experiments, 3×10^6 cells were used for each condition. Each time point was analyzed in triplicate, starting from mononuclear cells purified from three different mice. Cells were centrifuged at 600× g for 5 min and washed in D-PBS w/o calcium and magnesium (BioWest-VWR INTERNATIONAL PBI S.r.l., Milan, Italy). To minimize the inter-sample antibody staining variation, we applied a mass-tag barcoding protocol on fixed cells. Cells were fixed with 1 mL of Fix I Buffer (Fluidigm, South San Francisco, CA, USA) and then incubated for 10 min at RT. The fixation was quenched with Barcode Perm Buffer (Fluidigm, South San Francisco, CA, USA). The different samples were barcoded by individually incubating the cell suspensions with the appropriate combination of palladium isotopes from the Cell-ID^{TM} 20-Plex Pd Barcoding Kit (Fluidigm, South San Francisco, CA, USA) in Barcode Perm Buffer for 30 min at RT. The staining was quenched with MaxPar Cell Staining Buffer (Fluidigm, South San Francisco, CA, USA).

The antibody staining with metal-tagged antibodies that target surface and intracellular antigens was performed on the samples pooled after mass-tag barcoding. Samples were collected in a single tube, and the surface antibody staining protocol was performed according to manufacturers' instructions for 30 min at RT. Surface-stained cells were then washed twice with MaxPar Cell Staining Buffer (Fluidigm South San Francisco, CA, USA) and permeabilized with ice-cold methanol for 10 min on ice. Membrane-permeabilized cells were washed twice with MaxPar Cell Staining Buffer (Fluidigm, South San Francisco, CA, USA) and incubated with antibodies against intracellular antigens for 30 min at RT according to manufacturers' instructions. The full list of antibodies is detailed in Table 1. All the antibodies listed were purchased from Fluidigm (South San Francisco, CA, USA). After intracellular antibody staining, cells were washed twice with MaxPar Cell Staining Buffer and stained for 1 h at RT with the intercalation solution, composed of Cell-ID Intercalator-Ir (191Ir and 193Ir, Fluidigm South San Francisco, CA, USA) in MaxPar Fix and Perm Buffer (Fluidigm, South San Francisco, CA, USA) at a final concentration of 125 nM. Cells were washed twice with MaxPar Cell Staining Buffer and MaxPar Water.

For mass cytometry analysis, cells were resuspended at the final concentration of 2.5×10^5 cells/mL in MaxPar Water containing 10% of EQ^{TM} Four Element Calibration Beads (Fluidigm, South San Francisco, CA, USA) and filtered through a 30-μm filter-cap FACS tube. Samples were kept on ice prior to the acquisition by using the mass cytometry platform CyTOF2 System (Fluidigm, South San Francisco, CA, USA).

Table 1. List of the metal-tagged antibodies used in the mass cytometry experiments.

Antibody	Metal
Anti-mouse CD45	147Sm
Anti-mouse Ly-6A/E (SCA1)	164Dy
Anti-mouse CD90.2 (Thy-1.2)	156Gd
Anti-mouse CD146	141Pr
Anti-mouse F4/80	146Nd
Anti-mouse CD140α	148Nd
Anti-mouse CD140β	151Eu
Anti-mouse α7-integrin	161Dy
Anti-mouse CD206	169Tm
Anti-mouse CD34	144Nd
Anti-mouse CXCR4	159Tb
Anti-mouse CD4	172Yb
Anti-mouse CD25 (IL-2R)	150Nd
Anti-vimentin	154Sm
Anti-CD31 (PECAM-1)	165Ho
Anti-pan-actin	175Lu
Anti-mouse interleukin-6 (IL-6)	167Er
Anti-phospho-Akt (S473)	152Sm
Anti-phospho-Stat1 (Y701)	153Eu
Anti-phospho-Erk1/2 (T202/Y204)	171Yb
Anti-phospho-Stat3 (Y705)	158Gd
Anti-cleaved caspase3	142Nd
Anti-phospho-Creb	176Yb

CD: Cluster Differentiation; SCA1: Stem Cell Antigen1; CXCR4: C-X-C Motif Chemokine Receptor 4; PECAM-1: Platelet Endothelial Cell adhesion-1; Akt: RAC alpha serine/threonine-protein kinase; Stat1: Signal Transducer and activator of transcription 1; Erk1/2: Extracellular signal-regulated kinases 1, 2; Stat3: Signal transducer and activator of transcription 3; cleaved caspase3: cysteine-aspartic proteases; Creb: cAMP response element-binding protein.

2.5. CyTOF Data Analysis

Following data acquisition, channel intensity was normalized using calibration beads [26], and the normalized *fcs* file was de-barcoded by using the Debarcoder software (Fluidigm, South San Francisco, CA, USA). Data have been pre-processed using the Cytobank software platform [27]. Cells were manually gated from debris on the basis of DNA content monitored by the incorporation of the iridium (Ir) intercalator. Doublets were then excluded according to the event length parameter, and single live cells were finally manually gated by using the cisplatin (Pt) intercalator signal. Manually gated singlet ($191Ir^+$ $193Ir^+$), viable ($195Pt^-$) events were imported into Cytofkit for further analysis [28]. Cytofkit [28] parameters were set as follows: 14 biomarkers were included for clustering all the detected live cells per sample ("all" merge method); transformation method: cytofAsinh; FlowSOM was used as clustering algorithm with $k = 15$, tSNE perplexity set to 30; 2000 iterations and seed: 42.

Data were analyzed with Cytofkit shiny app [28] https://github.com/JinmiaoChenLab/cytofkit) and R scripts. Gating for marker-positive cells was performed by setting the mean +/- standard deviation as a threshold, depending on the expression value distribution of the specific marker.

2.6. Statistical Analysis

The experiments were performed at least in biological triplicates, that is, from at least 3 independent mononuclear cell preparations for each experiment. Only for the time point at day 5 in the wild type time series, we had just two biological repeats. Results were presented as mean ± SEM unless otherwise mentioned. Statistical evaluation was done by using One-way or Two-way ANOVA. Comparisons were considered statistically significant at * $p < 0.05$; ** $p < 0.01$; *** $p < 0.001$; **** $p < 0.0001$. All statistical analysis was performed using Prism 6 (GraphPad, San Diego, CA, USA).

3. Results

3.1. Histological Profiling of Skeletal Muscle Tissue from wt and mdx Muscles Following Acute Damage

To gain insights into the skeletal muscle repair process, we aimed at describing the dynamics of muscle cell populations following acute damage. To induce muscle injury, we used a well-established protocol based on the injection of the snake (*Naja pallida*) myotoxin (cardiotoxin, CTX) into the hind limb muscles of wt and mdx dystrophic mice (Figure 1A) [29,30]. Cardiotoxin, by inhibiting protein kinase C (PKC), induced the increase of cytosolic calcium, causing myofiber myolysis that, in turn, triggers regeneration [31–33]. First, we monitored, by hematoxylin and eosin staining, the changes in the skeletal muscle architecture at five different time points that were chosen to monitor the key events of the muscle healing process after damage: necrosis, inflammation, regeneration, and remodeling (Figure 1B,C).

Sections of uninjured wt muscles were characterized by polygonal fibers of uniform size containing peripheral nuclei (Figure 1B) [34]. Upon CTX injection, the skeletal muscle underwent degeneration, setting in motion the regeneration process. The degeneration of the muscle architecture was clearly observable at day 1 after injury, while interstitial cells, either resident cell populations [8] or infiltrating immune cells [35], became conspicuous at day 3 after damage. The regeneration process was completed after 20 days, as highlighted by the presence of multinucleated regenerated myofibers.

The mdx skeletal muscle, on the other hand, even in the absence of acute insult, was characterized by infiltrating inflammatory interstitial cells and centrally nucleated myofibers of different sizes, hallmarks of the dystrophic pathology (Figure 1C) [12,36,37]. Following CTX injection, the injured tissue underwent a regeneration process, without being apparently impacted further by extensive necrosis. If anything, the mdx muscle seemed to be more resilient to the myotoxin-induced damage and did not undergo the massive structural damage that was observed early after cardiotoxin injection in the wt muscle. Altogether, the histological analysis showed that the muscles from the two genetic backgrounds responded differently to CTX-induced injury.

3.2. Single-Cell Quantitative Profiling of Skeletal Muscle Populations Following Myotoxin-Induced Injury

Next, we sought to characterize the different mononuclear cell populations in the two mouse models and monitor their abundance changes during the regeneration process. To this end, we resorted to using mass cytometry [8].

In wt muscles, the number of isolated mononuclear cells increased after damage and peaked at day 3 to return to almost baseline levels at day 10 (Figure 1D). In contrast, consistent with the resilience of the mdx muscle observed in the histological analysis, this response was not detected in the injured muscles of mdx mice, where the mononuclear cell number remained constant over the whole regeneration process (Figure 1E).

The mononuclear cell samples from both animal models were separately barcoded and labeled with a panel of 23 metal-tagged antibodies (Table 1) targeting antigens expressed by muscle resident cells and/or by cell populations from the hematopoietic compartment. Mononuclear cells were purified from the uninjured and injured muscles at five time-points and, after barcoding and labeling, analyzed with a CyTOF2 mass cytometer in a single-run experiment [38]. Signals were debarcoded, and live cells were identified using the cisplatin (Pt) intercalator signal. Live/dead cell analysis highlighted that dead wt cells significantly increased at day 3 (Figure S1A), while for mdx muscles, the live/dead cell ratio remained constant (Figure S1B).

Single-cell data were analyzed by applying, as a dimensionality reduction method, the t-distributed stochastic neighbor embedding (t-SNE) algorithm implemented in Cytofkit [28,39].

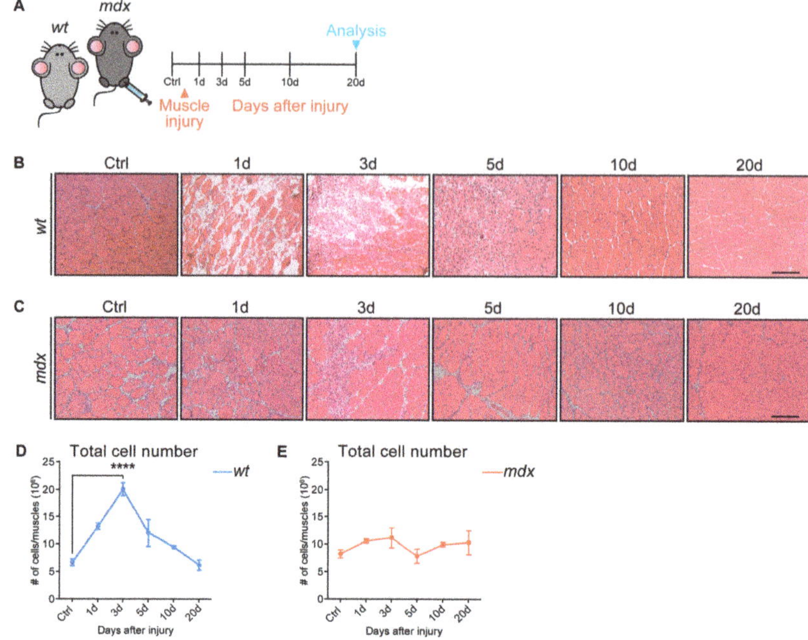

Figure 1. Cardiotoxin-induced injury on wild type (wt) and a mouse model of Duchenne muscular dystrophy (mdx) skeletal muscle tissue. (**A**) Experimental procedure. 45-day-old wt and mdx mice were injected intramuscularly with cardiotoxin (CTX) (10 μM), and the skeletal muscles were analyzed 1, 3, 5, 10, and 20 days (d) after injury. (**B**) Representative hematoxylin and eosin staining of uninjured wt *tibialis anterior* (TA) muscles and regenerating wt TA muscles at 1, 3, 5, 10, and 20 days after intramuscular CTX injection. Regenerating muscles were characterized by centrally located nuclei at day 5, but reconstituted multinucleated myofibers by day 10. (**C**) Representative hematoxylin and eosin staining on histological sections of uninjured and regenerating mdx TA muscles at 1, 3, 5, 10, and 20 days after intramuscular CTX injection. All along the considered time points, mdx muscles were characterized by infiltrating inflammatory interstitial cells and centrally nucleated myofibers of different sizes. (**D,E**) The number of cells (in millions) extracted from uninjured and CTX-injured wt (**D**) and mdx (**E**) mice (n=3; for 5d wt time point, n=2). All data were represented as mean ± SEM, and the statistical significance was estimated by two-way ANOVA (**** $p < 0.0001$). (**B,C**) 20× magnification; scale bar: 100 μm.

As the readouts of phospho-antibodies were barely above the background signal, albeit cell-specific (Figure S2), they were not considered in this analysis. The readouts of the 14 antigens in Figure 2A were used as input for the t-SNE algorithm. This approach yielded a two-dimensional map of the antigenic expression profiles of mononuclear cell populations in the wt skeletal muscle (Figure 2A) and led to the identification of 15 different cell clusters (Figure 2B).

The 15 identified clusters (Figure 2B) were further grouped into eight cell types by matching their expression profile to that of cell types already described in the literature [20,40]. More specifically, we were able to identify populations expressing antigens typical of immune cells (CD45+) (clusters 7, 8, 12, 13, 14, and 15), macrophages (CD45+ and F4/80+) (clusters 7, 13, and 14), myogenic progenitors (MPs) (α7-integrin+) (cluster 1), fibro/adipogenic progenitors (FAPs) (SCA1+, CD34+, CD140α+, CD90.2+, and vimentin+) (cluster 10), endothelial progenitor cells (CD31+) (clusters 3 and 4), pericyte-like cells (CD146+ and CD140β+) (cluster 11), and mesenchymal-like cells (CD90.2+) (cluster 5). The expression profiles of three remaining non-abundant clusters (clusters 2, 6, and 9) could not be matched to any of the already described muscle cell types and were collectively dubbed "others" (Figure 2B,C).

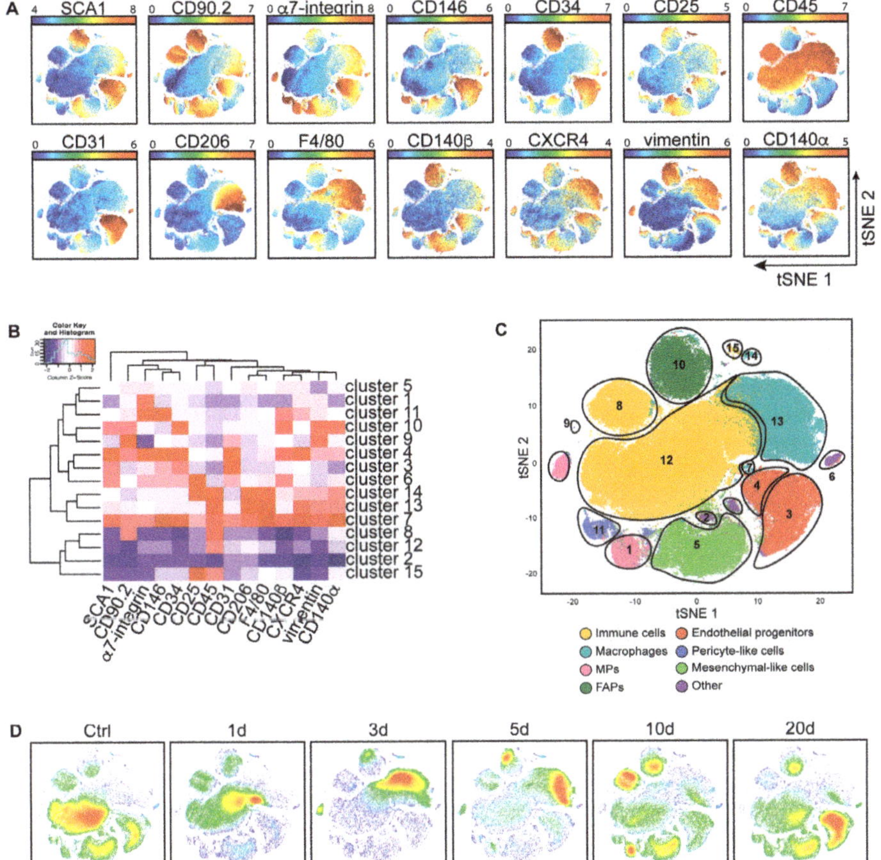

Figure 2. Dynamic changes of mononuclear cell subpopulations in wt muscles during regeneration. (**A**) The different mononuclear cell samples from uninjured and CTX-injured wt hind limb muscles were merged to create a single t-distributed stochastic neighbor embedding (t-SNE) map; colored according to expression levels of SCA1, CD90.2, α7-integrin, CD146, CD34, CD25, CD45, CD31, CD206, F4/80, CD140β, CXCR4, vimentin, and CD140α (blue: low expression; red: high expression of the selected marker). (**B**) FlowSOM heatmap of column normalized (Z-score) marker expression for each of the 15 identified clusters. Colors varied according to the expression level of the considered marker in a blue to red scale, indicating low to high expression, respectively. (**C**) Cell clusters defined by the FlowSOM analysis were assigned to arbitrary colors (yellow: immune cells; light blue: macrophages; pink: myogenic progenitors (MPs); dark green: fibro/adipogenic progenitors (FAPs); red: endothelial progenitors; blue: pericyte-like cells; light green: mesenchymal-like cells; purple: other). (**D**) Time course of the variation in cell subpopulation abundance upon CTX-induced injury. The dynamic changes were illustrated by density plots, colored according to cell density (blue: low density; red: high density).

As we analyzed samples at different time points, we could also monitor the dynamic of cell populations after CTX damage. The picture in Figure 2C, in fact, was not static as the relative abundance of the different cell populations changed along the regeneration process. These could be best appreciated by comparing the bidimensional t-SNE maps at different time points (Figure 2D). Here the colors, relating to cell density, from blue (low density) to red (high density), allowed to appreciate the significant changes in the cell population proportions during regeneration.

To quantitatively describe the dynamics of these changes, we first classified the observed cell populations into two main clusters according to CD45 expression: (i) the immune (CD45$^+$) and the (ii) non-immune (CD45$^-$) subpopulations of mononuclear cells. The data were shown as population relative abundance in the mononuclear cell preparations analyzed in the CyTOF. However, this data representation could be easily transformed into changes in the absolute numbers of each cell population in the mouse muscle as the total number of mononuclear cells in each condition was known (Figures S3 and S4). The curve trends in the two representations were similar.

We first focused on CD45$^+$ hematopoietic cells as they play a critical role in the regeneration process by sending regulatory signals by removing the damaged-fiber debris and by stimulating proliferation and differentiation of myogenic progenitors [19,41–45]. In a homeostatic muscle, approximately 60% of the mononuclear cells exposed the CD45 antigen. This cell compartment increased in number, after damage, to peak at days 1 and 3 (Figure 3A).

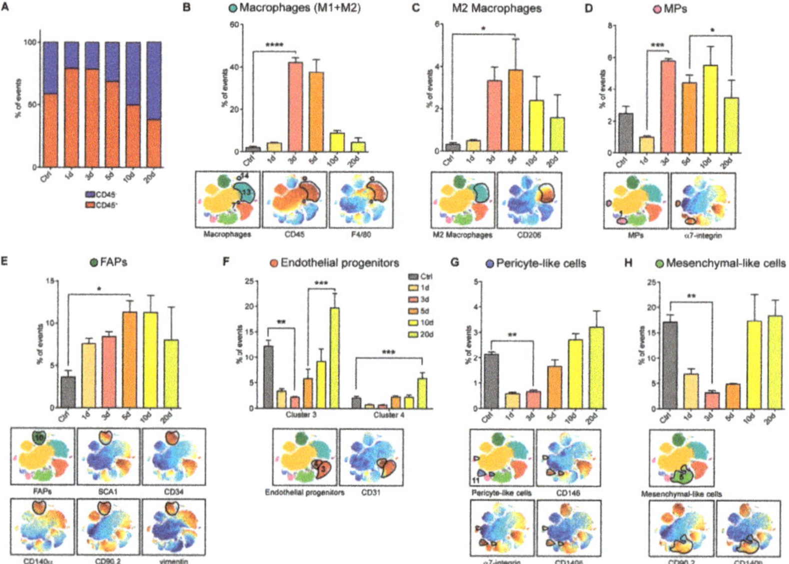

Figure 3. Characterization of the cell-population density rearrangements induced by CTX in the wt skeletal muscle after 1, 3, 5, 10, 20 days. (**A**) Bar plots, showing the CD45$^-$ (blue) / CD45$^+$ (red) ratio in mononuclear cells from wt skeletal muscles at different time points after CTX injury. Data were represented as mean, while the statistical significance was estimated by two-way ANOVA. (**B–H**) Identification in the t-SNE maps of mononuclear cell populations. The different plots were color-coded according to the expression of surface antigens that characterize the relevant cell types. The bar plots quantitated the variation in population abundance in the wt limb muscles at different times during regeneration. Cell percentages were assessed on the total number of cells in each sample (n = 3; for 5d wt time point, n = 2). The statistical significance was estimated by one-way ANOVA. All data were represented as mean ± SEM, and the statistical significance was defined as * $p < 0.05$; ** $p < 0.01$; *** $p < 0.001$; **** $p < 0.0001$.

In physiological conditions, a fraction of the hematopoietic cells, less than 2% of the total recorded events, also expressed the F4/80$^+$ antigen, a pan-macrophage marker (Figure 3B). However, already 3 days after injury, the macrophage population significantly increased, reaching a maximum of approximately 40%. At later times, macrophage abundance gradually decreased, returning close to baseline levels at day 20. The macrophage population is not homogeneous as it contains pro-inflammatory (M1) and anti-inflammatory (M2) macrophages that differ in the expression of the CD206 antigen and play a different role in muscle regeneration [46,47].

Noteworthy, CD206$^+$ M2 macrophages, which are responsible for the resolution of the inflammatory response [48–50], increased significantly on day 5 at the expense of inflammatory macrophages M1 that peaked at day 3 (Figure 3B,C).

We next focused on the cell populations that did not express the CD45 antigen. This group included the two main players of the regeneration process: myogenic progenitors (MPs) and fibro/adipogenic progenitors (FAPs) [5,6,8,51,52]. Our single-cell analysis revealed that both cell types more than doubled in number from day 3 to 10 (Figure 3D,E).

We could not obtain mass cytometry grade antibodies, specifically recognizing the paired box protein PAX7 antigen, which labels satellite cells. Thus, we resorted to using the α7-integrin antigen as a marker of the myogenic progenitor (MP) cluster (Figure 3D), including both satellite cells and myoblasts. Within the population expressing α7-integrin, we could identify two smaller clusters expressing different levels of vimentin (Figure S1C). During regeneration, the MP cell population, after an initial decrease, became more populated (Figure 3D). The relative abundance of the two subpopulations changed in time with the vimentin expressing MP, significantly increasing in number at day 3 (Figure S1C). At day 3 post-injury, the MPs accounted for about 6% of the total recorded cell events.

The FAP compartment was defined in our t-SNE map by a cluster of cells expressing SCA1, CD34, CD140α, CD90.2, and vimentin. Similar to the MP population, albeit with a different trend, the FAP population became more numerous as the regeneration process progressed, reaching the maximum expansion between 5–10 days after CTX injury (about 11.3% of total events), to return to almost control levels at day 20 (Figure 3E).

Our analysis also allowed us to characterize the kinetic of vessel-associated populations during regeneration [53,54]. These were identified, among CD45 negative cells, as they expressed the CD31 antigen. We were able to distinguish two subpopulations of cells expressing additional markers of endothelial populations at different levels (SCA1, CD146, and CD34) (Figure S1D). CD31 cluster 4 was considered a myoendothelial cell subpopulation, containing cells that also express high levels of the α7-integrin antigen [55]. Overall, we observed that the whole endothelial progenitor pool followed a kinetic that was different from the populations examined so far. They decreased significantly in number in the first few days after damage and then increased over the homeostatic level toward the end of the regeneration process (Figure 3F).

We also considered a population of cells whose expression profile was reminiscent of that of pericytes [56,57]. They were characterized by the expression of CD146, CD140β, and α7-integrin (Figure 3G). These pericyte-like cells followed a trend that was very similar to that of endothelial cells, with a sharp decrease early after damage and a rapid increase in the late regeneration phase, when vascularization took place (Figure 3G). We also looked at a cluster of cells that we were not able to match to any of the cell types described so far. We dubbed this population as mesenchymal-like cells, as they were highly positive for CD90.2 and CD140β markers. The abundance of this population followed a kinetic that seemed to be governed by the regeneration process and was also very similar to that of the classical endothelial cells (Figure 3H).

3.3. Single-Cell Profiling of the mdx Muscle

We further aimed at characterizing the response to acute damage and the ensuing regeneration process in the mdx dystrophic muscle. Mononuclear cells were isolated from the uninjured and CTX-injured skeletal muscles of mdx mice, following the same procedure described for the wt in the previous section (Figure 4A–D). Mononuclear cells were separated from muscle fibers, barcoded, labeled with the same antibody panel, and prepared for mass cytometry. The resulting single-cell antigen expression profile was processed, as described for the wt cells, in order to obtain a two-dimensional representation of antigen expression in the different cell types. As the wt and mdx mass cytometry analyses were performed at different times, the resulting t-SNE maps could not be directly compared (Figures 2A and 4A). However, by comparing the antigen expression profiles in the different clusters, we were able to associate each cluster to one of the main muscle mononuclear cell types and match it to those observed in the wt (Figure 4B,C). As observed in injured-recovering wt muscles, the relative numerosity of the different cell populations also changed after damage and during regeneration in the mdx model (Figure 4D).

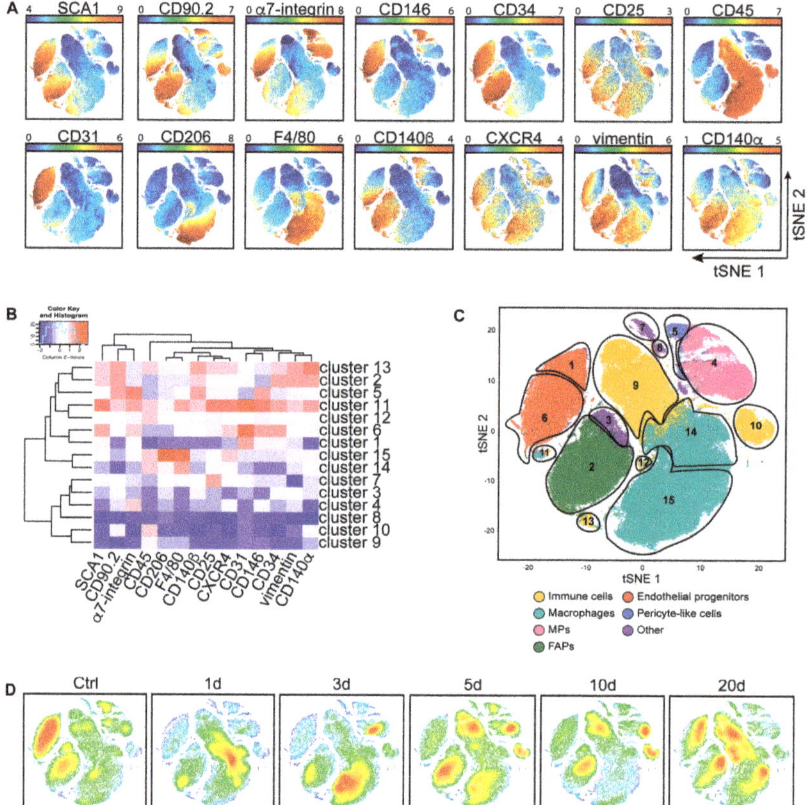

Figure 4. Identification of the main mdx skeletal muscle cell populations upon CTX-induced injury. (**A**) A mononuclear cell suspension was purified from uninjured and CTX-injured mdx hind limb muscles. The different samples were combined, barcoded, and analyzed to create a single t-SNE map colored according to SCA1, CD90.2, α7-integrin, CD146, CD34, CD25, CD45, CD31, CD206, F4/80, CD140β, CXCR4, vimentin, and CD140α expression levels (blue: low expression; high expression of the selected marker). (**B**) FlowSOM heatmap of column normalized (Z-score) marker expression for each of the 15 clusters identified by Cytofkit analysis. Colors varied according to the expression level of each considered marker in a blue to red scale, indicating low and high expression, respectively. (**C**) Cell population clusters defined by overlapping the expression for each of the different analyzed antigens were projected onto t-SNE space and assigned to specific colors (yellow: immune cells; light blue: macrophages; pink: myogenic progenitors (MPs); dark green: fibro/adipogenic progenitors (FAPs); red: endothelial progenitors; blue: pericyte-like cells; purple: other). (**D**) Density plots colored by density (blue: low density; red: high density), showing cell abundance variations at different times during the regeneration process.

The number of cells in the CD45$^+$ compartment (clusters 9, 10, 11, 12, 13, 14, and 15), as a whole, did not change significantly along the regeneration process (Figure 5A). However, the distribution of cells in the different populations in the compartment was found to be highly dynamic. In particular, macrophages (clusters 11, 14, and 15), which were already abundant in the mdx muscle before acute damage, increased significantly in the early days after CTX treatment, as observed in the wt model, to return to mdx baseline levels, approximately 30% of total mononuclear cells, at day 20 (Figure 5B). M2 macrophages, on the other hand, remained rather constant early after an injury to increase only late in the regeneration process (Figure 5C).

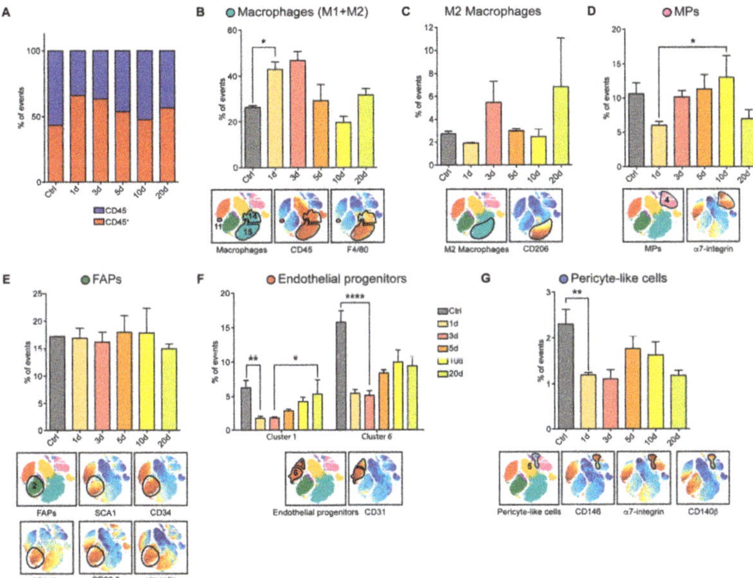

Figure 5. Characterization of the population dynamics induced by CTX injury in the mdx skeletal muscle at day 1, 3, 5, 10, 20 after injury. (**A**) Bar plots, showing the CD45$^-$ (blue) / CD45$^+$ (red) ratio in the mdx mononuclear cells at different time points after CTX injury. Data were represented as mean, and the statistical significance was estimated by two-way ANOVA. (**B**) t-SNE maps, showing the population gated as macrophage, colored for CD45 and F4/80 expression (red: high expression; blue: low expression) together with the macrophage trend observed in uninjured and CTX-injured mdx hind limb muscles. (**C**) M2 macrophages, expressing the CD206 antigen. (**D**) t-SNE maps of the myogenic progenitor (MP) population colored according to the levels of α7-integrin expression (red: high expression) and MP population dynamics during regeneration. (**E**) t-SNE maps, representing fibro/adipogenic progenitors (FAPs) identified among uninjured and CTX-injured mdx hind limb muscles, colored according to the expression of SCA1, CD34, CD140α, CD90.2, and vimentin (red: high expression; blue: low expression). The bar plot illustrated the population dynamics during regeneration. (**F**) Bar plots, showing the abundance of two different subclusters (cluster 1 and cluster 6) of endothelial progenitor cells at different time points. (**G**) t-SNE maps of pericyte-like cells identified in uninjured and CTX-injured mdx hind limb cell populations. Cell clusters were colored according to the expression of CD146, α7-integrin, and CD140β (red: high expression; blue: low expression). The bar plot illustrated the population dynamics during regeneration. Population abundance was assessed by calculating the percentage of cells in any given population over the total number of mononuclear cells in each sample (n = 3). The statistical significance was estimated by one-way ANOVA. All data were represented as mean ± SEM, and the statistical significance was defined as * $p < 0.05$; ** $p < 0.01$; **** $p < 0.001$.

When compared to wt, myogenic progenitors (MPs) (cluster 4) and fibro/adipogenic progenitors (FAPs) (cluster 2) followed a different trend. As the mdx muscle was under chronic stress, we observed that both progenitor cell populations were more populated in the absence of acute damage. However, while MPs decreased in number as early as one day after acute damage to return to unperturbed levels at later times, FAPs remained at a constant high level along the whole regeneration process (Figure 5D,E).

As already observed for the wt, also for the mdx mouse model, the CD31 expressing cells (clusters 1 and 6), defining endothelial progenitors, comprised two sub-populations differing for the expression of α7-integrin and other markers (CD90.2, CD140β, and CXCR4) (Figure S1E). The population expressing higher levels of α7-integrin was twice as abundant as the other. Nevertheless, both sub-populations reacted similarly to acute damage and first dropped in abundance by a factor of approximately three, to recover at the late stages of the regeneration process (Figure 5F). However, consistently, the cluster enriched in α7-integrin expressing cells (cluster 6) had not fully recovered at day 20, suggesting that the vascularization was still ongoing at times when histology seemed to indicate completion of the regeneration process. This differed from what was observed in the wt. Consistent with this consideration, and differently from what observed in the wt, also the pericyte-like population (cluster 5) behaved similarly, dropping in abundance immediately after damage and then slowly recovering without, however, reaching full recovery at day 20 (Figure 5G).

4. Discussion

The skeletal muscle has a remarkable capacity to self-repair if damaged [7–10,49,58]. However, this healing process may fail, owing to excessive damage, aging, or genetic disorders [12,59–62]. As a consequence of this failure in the repair process, as in muscle dystrophies, the tissue undergoes degeneration, leading to progressive muscle wasting and weakness characterized by chronic inflammation and, at later stages, fat and fibrotic tissue infiltrations [9,13,14].

Here, we exploited the resolution power of mass cytometry to characterize the modulation of the profile of the mononuclear cell population following chronic or acute damage [63–65]. To this end, we assembled a panel of 23 metal-tagged antibodies and characterized, at the single-cell level, the dynamics of muscle mononuclear cell populations after acute damage in wild type (wt) and in a mouse model of Duchenne muscular dystrophy (mdx). The regeneration process was monitored by examining samples of muscle mononuclear cells at 1, 3, 5, 10, and 20 days after cardiotoxin injection in wt or mdx mice [29,30].

This approach yielded a reach multiparametric dataset, disclosing the details of how the composition and heterogeneity of mononuclear cell populations changed in time as the muscle healing process proceeded. By applying a dimensionality reduction technique, such as the t-distributed stochastic neighbor embedding (t-SNE) algorithm, we generated bidimensional maps, providing a visual description of the regeneration process. Furthermore, this representation contributed to revealing subtle differences in the expression of specific markers in subpopulations within the major clusters that identified "classical" muscle populations. This population heterogeneity could not be only explained by experimental variability, even if any functional implication in muscle physiology or pathology remains to be established. In this report, we only dwelt upon a few of these t-SNE map features, while the dataset remains as a resource for additional analysis for the community.

A time-dependent variation in the abundance of muscle mononuclear cells was observed in wt mice, starting at day 1, indicating that the system sensed the damage and promptly responded to restore muscle tissue homeostasis. Inflammatory cells were the predominant population 3 days after the injury, accounting for over 40% of the total mononuclear cells. The activation of the inflammatory compartment was limited to the first few days after the injury as macrophages returned to baseline levels after 10 days.

Fibro/adipogenic progenitors (FAPs), here identified as SCA1+, CD34+, CD140α+, CD90.2+, and vimentin+ [66], stimulate satellite cell activation and differentiation, thus playing a positive role in muscle regeneration [8]. FAPs are quiescent in intact muscles, while they proliferate in response to injury [8,67,68]. Consistently, we observed that FAPs, after CTX injection, rapidly expanded, in wt muscles, reaching a peak between days 5 and 10, to return eventually to baseline level, having accomplished their function of providing a transient production of pro-differentiation signals and of depositing extracellular matrix for muscle remodeling.

In response to the secreted inflammatory and fibro/adipogenic stimulating signals, the myogenic compartment also promptly activated at day 3, as confirmed by vimentin expression, which marked activated MPs or myoblasts [69]. After an initial drop in concentration, $\alpha 7$-integrin$^+$ MPs, including satellite cells and myoblasts, expanded on day 3, still remaining high on day 10, when they started to decrease.

Different kinetics was observed for endothelial progenitor cells, here, characterized as CD31, SCA1, CD34, and CD146 expressing cells, and for pericyte-like cells, mainly identified as CD146 and CD140β-positive cells. The endothelial and pericyte-like cell contribution to skeletal muscle recovery only took place at the end of the regeneration process; once the inflammatory cells were removed, FAPs decreased in number, and MPs differentiated.

Due to the incomplete nature of our panel, some cell clusters in our t-SNE maps remained loosely defined. One cluster included cells with an antigen repertoire reminiscent of mesenchymal cells, as they were positive for the mesenchymal markers CD90.2 and CD140β [70]. This population had a clear kinetic, suggesting a late intervention of the mesenchymal population in the muscle regeneration process.

We also investigated the response to acute injury of a muscle environment that was already chronically perturbed as in the mdx mice. Mononuclear cells extracted from the muscle of mdx mice did not experience any significant increase in number following cardiotoxin injury. However, by looking into the details of the population profiles at each time point, we observed a significant modulation of the population distributions as the increase in the number of the cells in one population was counterbalanced by the decrease in another one. For instance, while we observed that the cells in the inflammatory compartment, essentially macrophages, expanded early after damage, the endothelial and pericyte-like clusters dropped in numerosity to recover the initial values only late in the process. On the other hand, differently from wt, the myogenic and fibro/adipogenic compartments, the two main players in muscle regeneration, showed little variation during the regeneration process, probably because they were already chronically activated.

Overall, our multiparametric analysis offered a comprehensive description of both muscle tissue homeostasis and the rearrangements induced in the mononuclear cell population profile by a perturbation of the muscle system, be it a chronic condition, as in the case of mdx mice, or acute stress, as that triggered by cardiotoxin injection.

Supplementary Materials: The following are available online at http://www.mdpi.com/2073-4409/9/7/1723/s1. Figures S1–S4 and "FCS folder" with two subfolders ("MDX_FCS" and "WT_FCS") containing the mass cytometry raw data (.fcs files).

Author Contributions: Conceptualization, L.L.P., F.S., C.G., L.C., C.F., and G.C.; methodology, L.L.P., F.S., A.P., A.R., M.R., C.G., and C.F.; software, L.L.P., F.S., and A.P.; validation, L.L.P., F.S., A.P., A.R., M.R., C.G., and C.F.; formal analysis, L.L.P., F.S., A.P., A.R., and M.R.; investigation, L.L.P., F.S., A.P., A.R., and M.R; resources, C.G., L.C., C.F., and G.C.; data curation, L.L.P., F.S., and A.P.; writing—original draft preparation, L.L.P. and F.S.; writing—review and editing, L.L.P., F.S., C.G., L.C., C.F., and G.C.; visualization, L.L.P., F.S., C.G., L.C., C.F., and G.C.; supervision, C.G, L.C., C.F., and G.C.; project administration, L.C., C.F., and G.C.; funding acquisition, G.C. All authors have read and agreed to the published version of the manuscript.

Funding: This research was funded by a grant of the European Research Council Grant N. 322749 (G.C.).

Acknowledgments: We acknowledge the Umberto Veronesi Foundation for awarding C. Fuoco with post-doctoral fellowship 2019.

Conflicts of Interest: The authors declare no conflict of interest.

References

1. Decary, S.; Mouly, V.; Hamida, C.B.; Sautet, A.; Barbet, J.P.; Butler-Browne, G.S. Replicative potential and telomere length in human skeletal muscle: Implications for satellite cell-mediated gene therapy. *Hum. Gene Ther.* **1997**, *8*, 1429–1438. [CrossRef] [PubMed]
2. Mauro, A. Satellite cell of skeletal muscle fibers. *J. Biophys. Biochem. Cytol.* **1961**, *9*, 493–495. [CrossRef] [PubMed]
3. Yin, H.; Price, F.; Rudnicki, M.A. Satellite cells and the muscle stem cell niche. *Physiol. Rev.* **2013**, *93*, 23–67. [CrossRef]
4. Lepper, C.; Partridge, T.A.; Fan, C.-M. An absolute requirement for Pax7-positive satellite cells in acute injury-induced skeletal muscle regeneration. *Development* **2011**, *138*, 3639–3646. [CrossRef] [PubMed]
5. Sambasivan, R.; Yao, R.; Kissenpfennig, A.; Van Wittenberghe, L.; Paldi, A.; Gayraud-Morel, B.; Guenou, H.; Malissen, B.; Tajbakhsh, S.; Galy, A. Pax7-expressing satellite cells are indispensable for adult skeletal muscle regeneration. *Development* **2011**, *138*, 3647–3656. [CrossRef]
6. Seale, P.; Sabourin, L.A.; Girgis-Gabardo, A.; Mansouri, A.; Gruss, P.; Rudnicki, M.A. Pax7 is required for the specification of myogenic satellite cells. *Cell* **2000**, *102*, 777–786. [CrossRef]
7. Heredia, J.E.; Mukundan, L.; Chen, F.M.; Mueller, A.A.; Deo, R.C.; Locksley, R.M.; Rando, T.A.; Chawla, A. Type 2 innate signals stimulate fibro/adipogenic progenitors to facilitate muscle regeneration. *Cell* **2013**, *153*, 376–388. [CrossRef]
8. Joe, A.W.B.; Yi, L.; Natarajan, A.; Le Grand, F.; So, L.; Wang, J.; Rudnicki, M.A.; Rossi, F.M.V. Muscle injury activates resident fibro/adipogenic progenitors that facilitate myogenesis. *Nat. Cell Biol.* **2010**, *12*, 153–163. [CrossRef]
9. Uezumi, A.; Ito, T.; Morikawa, D.; Shimizu, N.; Yoneda, T.; Segawa, M.; Yamaguchi, M.; Ogawa, R.; Matev, M.M.; Miyagoe-Suzuki, Y.; et al. Fibrosis and adipogenesis originate from a common mesenchymal progenitor in skeletal muscle. *J. Cell Sci.* **2011**, *124*, 3654–3664. [CrossRef]
10. Murphy, M.M.; Lawson, J.A.; Mathew, S.J.; Hutcheson, D.A.; Kardon, G. Satellite cells, connective tissue fibroblasts and their interactions are crucial for muscle regeneration. *Development* **2011**, *138*, 3625–3637. [CrossRef]
11. Malecova, B.; Gatto, S.; Etxaniz, U.; Passafaro, M.; Cortez, A.; Nicoletti, C.; Giordani, L.; Torcinaro, A.; Bardi, M.; Bicciato, S.; et al. Dynamics of cellular states of fibro-adipogenic progenitors during myogenesis and muscular dystrophy. *Nat. Commun.* **2018**, *9*, 1–12. [CrossRef] [PubMed]
12. Emery, A.E.H. The muscular dystrophies. *Lancet* **2002**, *359*, 687–695. [CrossRef]
13. 1Marden, F.A.; Connolly, A.M.; Siegel, M.J.; Rubin, D.A. Compositional analysis of muscle in boys with Duchenne muscular dystrophy using MR imaging. *Skelet. Radiol.* **2004**, *34*, 140–148. [CrossRef]
14. Contreras, O.; Rebolledo, D.L.; Oyarzún, J.E.; Olguin, H.C.; Brandan, E. Connective tissue cells expressing fibro/adipogenic progenitor markers increase under chronic damage: Relevance in fibroblast-myofibroblast differentiation and skeletal muscle fibrosis. *Cell Tissue Res.* **2016**, *364*, 647–660. [CrossRef] [PubMed]
15. Hogarth, M.W.; Defour, A.; Lazarski, C.; Gallardo, E.; Diaz-Manera, J.; Partridge, T.A.; Nagaraju, K.; Jaiswal, J.K. Fibroadipogenic progenitors are responsible for muscle loss in limb girdle muscular dystrophy 2B. *Nat. Commun.* **2019**, *10*, 1–13. [CrossRef]
16. Madaro, L.; Passafaro, M.; Sala, D.; Etxaniz, U.; Lugarini, F.; Proietti, D.; Alfonsi, M.V.; Nicoletti, C.; Gatto, S.; De Bardi, M.; et al. Denervation-activated STAT3-IL-6 signalling in fibro-adipogenic progenitors promotes myofibres atrophy and fibrosis. *Nat. Cell Biol.* **2018**, *20*, 917–927. [CrossRef] [PubMed]
17. Sirabella, D.; De Angelis, L.; Berghella, L. Sources for skeletal muscle repair: From satellite cells to reprogramming. *J. Cachexia Sarcopenia Muscle* **2013**, *4*, 125–136. [CrossRef]
18. Dey, D.; Goldhamer, D.J.; Yu, P.B. Contributions of muscle-resident progenitor cells to homeostasis and disease. *Curr. Mol. Biol. Rep.* **2015**, *1*, 175–188. [CrossRef]
19. Pillon, N.J.; Bilan, P.J.; Fink, L.N.; Klip, A. Cross-talk between skeletal muscle and immune cells: Muscle-derived mediators and metabolic implications. *Am. J. Physiol. Endocrinol. Metab.* **2013**, *304*, E453–E465. [CrossRef]
20. Giordani, L.; He, G.J.; Negroni, E.; Sakai, H.; Law, J.Y.C.; Siu, M.M.; Wan, R.; Corneau, A.; Tajbakhsh, S.; Cheung, T.H.; et al. High-dimensional single-cell cartography reveals novel skeletal muscle-resident cell populations. *Mol. Cell* **2019**, *74*, 609–621.e6. [CrossRef]

21. Gatto, S.; Puri, P.L.; Malecova, B. Single cell gene expression profiling of skeletal muscle-derived cells. *Methods Mol. Biol.* **2017**, *1556*, 191–219. [PubMed]

22. Rubenstein, A.B.; Smith, G.R.; Raue, U.; Begue, G.; Minchev, K.; Ruf-Zamojski, F.; Nair, V.D.; Wang, X.; Zhou, L.; Zaslavsky, E.; et al. Single-cell transcriptional profiles in human skeletal muscle. *Nat. Publ. Group* **2020**, *10*, 229. [CrossRef]

23. Dell'Orso, S.; Juan, A.H.; Ko, K.-D.; Naz, F.; Perovanovic, J.; Gutierrez-Cruz, G.; Feng, X.; Sartorelli, V. Single cell analysis of adult mouse skeletal muscle stem cells in homeostatic and regenerative conditions. *Development* **2019**, *146*, dev174177. [CrossRef] [PubMed]

24. Marinkovic, M.; Fuoco, C.; Sacco, F.; Cerquone Perpetuini, A.; Giuliani, G.; Micarelli, E.; Pavlidou, T.; Petrilli, L.L.; Reggio, A.; Riccio, F.; et al. Fibro-adipogenic progenitors of dystrophic mice are insensitive to NOTCH regulation of adipogenesis. *Life Sci. Alliance* **2019**, *2*, e201900437. [CrossRef] [PubMed]

25. Spada, F.; Fuoco, C.; Pirrò, S.; Paoluzi, S.; Castagnoli, L.; Gargioli, C.; Cesareni, G. Characterization by mass cytometry of different methods for the preparation of muscle mononuclear cells. *New Biotechnol.* **2016**, *33*, 514–523. [CrossRef]

26. Finck, R.; Simonds, E.F.; Jager, A.; Krishnaswamy, S.; Sachs, K.; Fantl, W.; Pe'er, D.; Nolan, G.P.; Bendall, S.C. Normalization of mass cytometry data with bead standards. *Cytom. Part* **2013**, *83*, 483–494. [CrossRef]

27. Kotecha, N.; Krutzik, P.O.; Irish, J.M. Web-based analysis and publication of flow cytometry experiments. *Curr. Protoc. Cytom.* **2010**. [CrossRef]

28. Chen, H.; Lau, M.C.; Wong, M.T.; Newell, E.W.; Chen, J. Cytofkit: A bioconductor package for an integrated mass cytometry data analysis pipeline. *PLoS Comput. Biol.* **2016**, *12*, e1005112. [CrossRef]

29. Ramadasan-Nair, R.; Gayathri, N.; Mishra, S.; Sunitha, B.; Mythri, R.B.; Nalini, A.; Subbannayya, Y.; Harsha, H.C.; Kolthur-Seetharam, U.; Srinivas Bharath, M.M. Mitochondrial alterations and oxidative stress in an acute transient mouse model of muscle degeneration: Implications for muscular dystrophy and related muscle pathologies. *J. Biol. Chem.* **2014**, *289*, 485–509. [CrossRef]

30. Duchen, L.W.; Excell, B.J.; Patel, R.; Smith, B. Changes in motor end-plates resulting from muscle fibre necrosis and regeneration. A light and electron microscopic study of the effects of the depolarizing fraction (cardiotoxin) of Dendroaspis jamesoni venom. *J. Neurol. Sci.* **1974**, *21*, 391–417. [CrossRef]

31. Wang, H.-X.; Lau, S.-Y.; Huang, S.-J.; Kwan, C.-Y.; Wong, T.-M. Cobra venom cardiotoxin induces perturbations of cytosolic calcium homeostasis and hypercontracture in adult rat ventricular myocytes. *J. Mol. Cell. Cardiol.* **1997**, *29*, 2759–2770. [CrossRef] [PubMed]

32. Raynor, R.L.; Zheng, B.; Kuo, J.F. Membrane interactions of amphiphilic polypeptides mastoparan, melittin, polymyxin B, and cardiotoxin. Differential inhibition of protein kinase C, Ca2+/calmodulin-dependent protein kinase II and synaptosomal membrane Na,K-ATPase, and Na+ pump and differentiation of HL60 cells. *J. Biol. Chem.* **1991**, *266*, 2753–2758. [PubMed]

33. Suh, B.C.; Song, S.K.; Kim, Y.K.; Kim, K.T. Induction of cytosolic Ca2+ elevation mediated by Mas-7 occurs through membrane pore formation. *J. Biol. Chem.* **1996**, *271*, 32753–32759. [CrossRef]

34. Bentzinger, C.F.; Wang, Y.X.; Dumont, N.A.; Rudnicki, M.A. Cellular dynamics in the muscle satellite cell niche. *EMBO Rep.* **2013**, *14*, 1062–1072. [CrossRef] [PubMed]

35. St Pierre, B.A.; Tidball, J.G. Differential response of macrophage subpopulations to soleus muscle reloading after rat hindlimb suspension. *J. Appl. Physiol.* **1994**, *77*, 290–297. [CrossRef]

36. Sacco, A.; Mourkioti, F.; Tran, R.; Choi, J.; Llewellyn, M.; Kraft, P.; Shkreli, M.; Delp, S.; Pomerantz, J.H.; Artandi, S.E.; et al. Short telomeres and stem cell exhaustion model Duchenne muscular dystrophy in mdx/mTR mice. *Cell* **2010**, *143*, 1059–1071. [CrossRef]

37. Anderson, J.E.; Ovalle, W.K.; Bressler, B.H. Electron microscopic and autoradiographic characterization of hindlimb muscle regeneration in the mdx mouse. *Anat. Rec.* **1987**, *219*, 243–257. [CrossRef]

38. Zunder, E.R.; Finck, R.; Behbehani, G.K.; Amir, E.-A.D.; Krishnaswamy, S.; Gonzalez, V.D.; Lorang, C.G.; Bjornson, Z.; Spitzer, M.H.; Bodenmiller, B.; et al. Palladium-based mass tag cell barcoding with a doublet-filtering scheme and single-cell deconvolution algorithm. *Nat. Protoc.* **2015**, *10*, 316–333. [CrossRef]

39. Amir, E.-A.D.; Davis, K.L.; Tadmor, M.D.; Simonds, E.F.; Levine, J.H.; Bendall, S.C.; Shenfeld, D.K.; Krishnaswamy, S.; Nolan, G.P.; Pe'er, D. VISNE enables visualization of high dimensional single-cell data and reveals phenotypic heterogeneity of leukemia. *Nat. Biotechnol.* **2013**, *31*, 545–552. [CrossRef]

40. Palma, A.; Cerquone Perpetuini, A.; Ferrentino, F.; Fuoco, C.; Gargioli, C.; Giuliani, G.; Iannuccelli, M.; Licata, L.; Micarelli, E.; Paoluzi, S.; et al. Myo-REG: A portal for signaling interactions in muscle regeneration. *Front. Physiol.* **2019**, *10*, 1216. [CrossRef]

41. Arnold, L.; Henry, A.; Poron, F.; Baba-Amer, Y.; Van Rooijen, N.; Plonquet, A.; Gherardi, R.K.; Chazaud, B. Inflammatory monocytes recruited after skeletal muscle injury switch into antiinflammatory macrophages to support myogenesis. *J. Exp. Med.* **2007**, *204*, 1057–1069. [CrossRef] [PubMed]

42. McLennan, I.S. Degenerating and regenerating skeletal muscles contain several subpopulations of macrophages with distinct spatial and temporal distributions. *J. Anat.* **1996**, *188*, 17–28.

43. Lolmede, K.; Campana, L.; Vezzoli, M.; Bosurgi, L.; Tonlorenzi, R.; Clementi, E.; Bianchi, M.E.; Cossu, G.; Manfredi, A.A.; Brunelli, S.; et al. Inflammatory and alternatively activated human macrophages attract vessel-associated stem cells, relying on separate HMGB1- and MMP-9-dependent pathways. *J. Leukoc. Biol.* **2009**, *85*, 779–787. [CrossRef]

44. Chazaud, B.; Sonnet, C.; Lafuste, P.; Bassez, G.; Rimaniol, A.-C.; Poron, F.; Authier, F.-J.; Dreyfus, P.A.; Gherardi, R.K. Satellite cells attract monocytes and use macrophages as a support to escape apoptosis and enhance muscle growth. *J. Cell Biol.* **2003**, *163*, 1133–1143. [CrossRef] [PubMed]

45. Dumont, N.; Frenette, J. Macrophages protect against muscle atrophy and promote muscle recovery in vivo and in vitro: A mechanism partly dependent on the insulin-like growth factor-1 signaling molecule. *Am. J. Pathol.* **2010**, *176*, 2228–2235. [CrossRef] [PubMed]

46. Martinez, F.O.; Gordon, S. The M1 and M2 paradigm of macrophage activation: Time for reassessment. *F1000Prime Rep.* **2014**, *6*, 13. [CrossRef] [PubMed]

47. Kharraz, Y.; Guerra, J.; Mann, C.J.; Serrano, A.L.; Muñoz-Cánoves, P. Macrophage plasticity and the role of inflammation in skeletal muscle repair. *Mediat. Inflamm.* **2013**. [CrossRef] [PubMed]

48. Mantovani, A.; Biswas, S.K.; Galdiero, M.R.; Sica, A.; Locati, M. Macrophage plasticity and polarization in tissue repair and remodelling. *J. Pathol.* **2013**, *229*, 176–185. [CrossRef] [PubMed]

49. Saclier, M.; Yacoub-Youssef, H.; Mackey, A.L.; Arnold, L.; Ardjoune, H.; Magnan, M.; Sailhan, F.; Chelly, J.; Pavlath, G.K.; Mounier, R.; et al. Differentially activated macrophages orchestrate myogenic precursor cell fate during human skeletal muscle regeneration. *Stem Cells* **2013**, *31*, 384–396. [CrossRef] [PubMed]

50. Stout, R.D.; Jiang, C.; Matta, B.; Tietzel, I.; Watkins, S.K.; Suttles, J. Macrophages sequentially change their functional phenotype in response to changes in microenvironmental influences. *J. Immunol.* **2005**, *175*, 342–349. [CrossRef]

51. Fu, X.; Wang, H.; Hu, P. Stem cell activation in skeletal muscle regeneration. *Cell. Mol. Life Sci.* **2015**, *72*, 1663–1677. [CrossRef]

52. Nguyen, H.X.; Tidball, J.G. Interactions between neutrophils and macrophages promote macrophage killing of rat muscle cells in vitro. *J. Physiol. (Lond.)* **2003**, *547*, 125–132. [CrossRef] [PubMed]

53. Musarò, A. The basis of muscle regeneration. *Adv. Biol.* **2014**, *2014*, 1–16. [CrossRef]

54. Hansen-Smith, F.M.; Hudlicka, O.; Egginton, S. In vivo angiogenesis in adult rat skeletal muscle: Early changes in capillary network architecture and ultrastructure. *Cell Tissue Res.* **1996**, *286*, 123–136. [CrossRef] [PubMed]

55. Zheng, B.; Cao, B.; Crisan, M.; Sun, B.; Li, G.; Logar, A.; Yap, S.; Pollett, J.B.; Drowley, L.; Cassino, T.; et al. Prospective identification of myogenic endothelial cells in human skeletal muscle. *Nat. Biotechnol.* **2007**, *25*, 1025–1034. [CrossRef] [PubMed]

56. Dellavalle, A.; Sampaolesi, M.; Tonlorenzi, R.; Tagliafico, E.; Sacchetti, B.; Perani, L.; Innocenzi, A.; Galvez, B.G.; Messina, G.; Morosetti, R.; et al. Pericytes of human skeletal muscle are myogenic precursors distinct from satellite cells. *Nat. Cell Biol.* **2007**, *9*, 255–267. [CrossRef] [PubMed]

57. Birbrair, A.; Zhang, T.; Wang, Z.-M.; Messi, M.L.; Enikolopov, G.N.; Mintz, A.; Delbono, O. Role of pericytes in skeletal muscle regeneration and fat accumulation. *Stem Cells Dev.* **2013**, *22*, 2298–2314. [CrossRef] [PubMed]

58. Christov, C.; Chrétien, F.; Abou-Khalil, R.; Bassez, G.; Vallet, G.; Authier, F.-J.; Bassaglia, Y.; Shinin, V.; Tajbakhsh, S.; Chazaud, B.; et al. Muscle satellite cells and endothelial cells: Close neighbors and privileged partners. *Mol. Biol. Cell* **2007**, *18*, 1397–1409. [CrossRef] [PubMed]

59. Almada, A.E.; Wagers, A.J. Molecular circuitry of stem cell fate in skeletal muscle regeneration, ageing and disease. *Nat. Rev. Mol. Cell Biol.* **2016**, *17*, 267–279. [CrossRef]

60. Blau, H.M.; Cosgrove, B.D.; Ho, A.T.V. The central role of muscle stem cells in regenerative failure with aging. *Nat. Med.* **2015**, *21*, 854–862. [CrossRef]

61. Kuswanto, W.; Burzyn, D.; Panduro, M.; Wang, K.K.; Jang, Y.C.; Wagers, A.J.; Benoist, C.; Mathis, D. Poor repair of skeletal muscle in aging mice reflects a defect in local, interleukin-33-dependent accumulation of regulatory T cells. *Immunity* **2016**, *44*, 355–367. [CrossRef]
62. Pastoret, C.; Sebille, A. Age-related differences in regeneration of dystrophic (mdx) and normal muscle in the mouse. *Muscle Nerve* **1995**, *18*, 1147–1154. [CrossRef] [PubMed]
63. Ornatsky, O.; Bandura, D.; Baranov, V.; Nitz, M.; Winnik, M.A.; Tanner, S. Highly multiparametric analysis by mass cytometry. *J. Immunol. Methods* **2010**, *361*, 1–20. [CrossRef] [PubMed]
64. Bendall, S.C.; Nolan, G.P.; Roederer, M.; Chattopadhyay, P.K. A deep profiler's guide to cytometry. *Trends Immunol.* **2012**, *33*, 323–332. [CrossRef] [PubMed]
65. Bandura, D.R.; Baranov, V.I.; Ornatsky, O.I.; Antonov, A.; Kinach, R.; Lou, X.; Pavlov, S.; Vorobiev, S.; Dick, J.E.; Tanner, S.D. Mass cytometry: Technique for real time single cell multitarget immunoassay based on inductively coupled plasma time-of-flight mass spectrometry. *Anal. Chem.* **2009**, *81*, 6813–6822. [CrossRef] [PubMed]
66. Reggio, A.; Rosina, M.; Palma, A.; Cerquone Perpetuini, A.; Petrilli, L.L.; Gargioli, C.; Fuoco, C.; Micarelli, E.; Giuliani, G.; Cerretani, M.; et al. Adipogenesis of skeletal muscle fibro/adipogenic progenitors is affected by the WNT5a/GSK3/β-catenin axis. *Cell Death Differ.* **2020**, *108*, 1–21. [CrossRef] [PubMed]
67. Uezumi, A.; Fukada, S.-I.; Yamamoto, N.; Takeda, S.; Tsuchida, K. Mesenchymal progenitors distinct from satellite cells contribute to ectopic fat cell formation in skeletal muscle. *Nat. Cell Biol.* **2010**, *12*, 143–152. [CrossRef]
68. Lemos, D.R.; Babaeijandaghi, F.; Low, M.; Chang, C.-K.; Lee, S.T.; Fiore, D.; Zhang, R.-H.; Natarajan, A.; Nedospasov, S.A.; Rossi, F.M.V. Nilotinib reduces muscle fibrosis in chronic muscle injury by promoting TNF-mediated apoptosis of fibro/adipogenic progenitors. *Nat. Med.* **2015**, *21*, 1–11. [CrossRef]
69. Vater, R.; Cullen, M.J.; Harris, J.B. The expression of vimentin in satellite cells of regenerating skeletal muscle in vivo. *Histochem. J.* **1994**, *26*, 916–928. [CrossRef]
70. Chirieleison, S.M.; Feduska, J.M.; Schugar, R.C.; Askew, Y.; Deasy, B.M. Human muscle-derived cell populations isolated by differential adhesion rates: Phenotype and contribution to skeletal muscle regeneration in Mdx/SCID mice. *Tissue Eng. Part* **2012**, *18*, 232–241. [CrossRef]

Article

Older Adults with Physical Frailty and Sarcopenia Show Increased Levels of Circulating Small Extracellular Vesicles with a Specific Mitochondrial Signature

Anna Picca [1,†], Raffaella Beli [2,†], Riccardo Calvani [1,*], Hélio José Coelho-Júnior [3], Francesco Landi [1,3], Roberto Bernabei [1,3], Cecilia Bucci [2], Flora Guerra [2,*] and Emanuele Marzetti [1,3]

[1] Fondazione Policlinico Universitario "Agostino Gemelli" IRCCS, 00168 Rome, Italy; anna.picca@guest.policlinicogemelli.it (A.P.); francesco.landi@unicatt.it (F.L.); roberto.bernabei@unicatt.it (R.B.); emanuele.marzetti@policlinicogemelli.it (E.M.)

[2] Department of Biological and Environmental Sciences and Technologies, Università del Salento, 73100 Lecce, Italy; raffaella.beli@unisalento.it (R.B.); cecilia.bucci@unisalento.it (C.B.)

[3] Università Cattolica del Sacro Cuore, 00168 Rome, Italy; coelhojunior@hotmail.com.br

[*] Correspondence: riccardo.calvani@guest.policlinicogemelli.it (R.C.); guerraflora@gmail.com (F.G.); Tel.: +39-06-3015-5559 (R.C.); +39-08-3229-8610 (F.G.); Fax: +39-06-3051-911 (R.C.)

[†] These authors contributed equally.

Received: 22 March 2020; Accepted: 13 April 2020; Published: 15 April 2020

Abstract: Mitochondrial dysfunction and systemic inflammation are major factors in the development of sarcopenia, but the molecular determinants linking the two mechanisms are only partially understood. The study of extracellular vesicle (EV) trafficking may provide insights into this relationship. Circulating small EVs (sEVs) from serum of 11 older adults with physical frailty and sarcopenia (PF&S) and 10 controls were purified and characterized. Protein levels of three tetraspanins (CD9, CD63, and CD81) and selected mitochondrial markers, including adenosine triphosphate 5A (ATP5A), mitochondrial cytochrome C oxidase subunit I (MTCOI), nicotinamide adenine dinucleotide reduced form (NADH):ubiquinone oxidoreductase subunit B8 (NDUFB8), NADH:ubiquinone oxidoreductase subunit S3 (NDUFS3), succinate dehydrogenase complex iron sulfur subunit B (SDHB), and ubiquinol-cytochrome C reductase core protein 2 (UQCRC2) were quantified by Western immunoblotting. Participants with PF&S showed higher levels of circulating sEVs relative to controls. Protein levels of CD9 and CD63 were lower in the sEV fraction of PF&S older adults, while CD81 was unvaried between groups. In addition, circulating sEVs from PF&S participants had lower amounts of ATP5A, NDUFS3, and SDHB. No signal was detected for MTCOI, NDUFB8, or UQCRC2 in either participant group. Our findings indicate that, in spite of increased sEV secretion, lower amounts of mitochondrial components are discarded through EV in older adults with PF&S. In-depth analysis of EV trafficking might open new venues for biomarker discovery and treatment development for PF&S.

Keywords: aging; biomarkers; mitophagy; mitochondrial dynamics; mitochondrial quality control; mitochondrial-derived vesicles (MDVs); exosomes; mitochondrial-lysosomal axis

1. Introduction

Advancing age is associated with declining muscle mass, function, and strength, a condition referred to as sarcopenia which increases the risk of incurring negative health-related outcomes (e.g., disability, loss of independence, institutionalization, death) [1]. Hence, sarcopenia and its clinical correlates are major public health priorities. Physical activity, nutritional interventions, and multi-component programs have proven to be valuable strategies for managing sarcopenia [2–4]. Yet, no effective pharmacological treatments are currently available to prevent, delay, or treat sarcopenia, which is mostly due to the incomplete knowledge of the underlying pathophysiology [2]. To further complicate the matter, at the clinical level, sarcopenia shows remarkable overlap with frailty, a "multidimensional syndrome characterized by a decrease in physiological reserve and reduced resistance to stressors", often envisioned as a pre-disability condition [5]. Hence, the two conditions have been merged into a new entity, referred to as physical frailty and sarcopenia (PF&S) [6].

Mitochondrial dysfunction and sterile inflammation are invoked among the pathogenic factors of PF&S [7,8]. Derangements at different levels of the mitochondrial quality control (MQC) machinery have been reported in older adults with PF&S [7]. However, whether and how cell-based alterations may spread at the systemic level and impact muscle homeostasis is presently unknown.

One of the mechanisms by which cells communicate with each other involves a conserved delivery system based on the generation and release of extracellular vesicles (EVs) [9]. These vesicles transfer information between cells through several categories of cargo-enriched biomolecules (i.e., proteins, lipids, nucleic acids, and sugars), each of them selectively influencing different cellular domains [10]. This shuttle system also contributes to degradative pathways responsible for eliminating oxidized cell components, including mitochondria, by establishing inter-organelle contact sites [11]. In particular, in the setting of incomplete mitochondrial depolarization, cells may either delay autophagy to remove mildly damaged organelles or shift from mitophagy to the extrusion of mitochondrial components within EVs [12,13]. As such, the generation and release of mitochondrial-derived vesicles (MDVs) may represent a complement to MQC systems before whole-sale organelle is triggered [13,14].

Cell-free mitochondrial DNA (mtDNA) has been identified among the molecules released within exosomes that may act as damage-associated molecular patterns (DAMPs) [15]. One of the biological roles for these molecules is the activation of innate immunity through binding of their hypomethylated CpG motifs, resembling those of bacterial DNA, to membrane- or cytoplasmic-pattern recognition receptors (PRRs), including Toll-like receptor (TLR), nucleotide-binding oligomerization domain (NOD)-like receptor (NLR) [16], and cytosolic cyclic GMP-AMP synthase (cGAS)-stimulator of interferon genes (STING) DNA sensing system-mediated pathways [17]. However, mtDNA is not the only mitochondrial constituent that may be displaced via MDVs and trigger these responses. Recently, the extrusion of mitochondrial components other than mtDNA has been reported within small EVs (sEVs) purified from the serum of older adults with Parkinson's disease (PD) [14]. However, whether and how this mechanism is in place in the setting of PF&S is unexplored.

In the present study, we purified sEVs from older adults with and without PF&S, quantified their amount, and characterized their content for the presence of mitochondrial components. The identification of specific derangements in sEVs in PF&S may shed light on its pathophysiology as well as suggest new biomarkers and possible biological targets for drug development.

2. Materials and Methods

2.1. Participants

Older adults aged 70+ with and without PF&S were recruited among the participants of the "BIOmarkers associated with Sarcopenia and Physical frailty in EldeRly pErsons" (BIOSPHERE) study [18]. BIOSPHERE was designed to determine and validate a panel of PF&S biomarkers through multivariate statistical modeling of biomolecules pertaining to inflammation, redox homeostasis, amino acid metabolism, neuromuscular junction dysfunction, and muscle remodeling pathways [18–20].

The operational definition used in the "Sarcopenia and Physical fRailty IN older people: multi-componenT Treatment strategies" (SPRINTT) project [21,22] was applied to diagnose PF&S: (a) physical frailty, based on a summary score on the Short Physical Performance Battery (SPPB) [23] between 3 and 9, (b) low appendicular muscle mass (aLM), according to the cut-points proposed by the Foundation for the National Institutes of Health (FNIH) sarcopenia project [24], and (c) absence of mobility disability (i.e., inability to complete the 400-m walk test) [25]. The present investigation involved a convenience sample of 21 participants, 11 older adults with PF&S and 10 non-sarcopenic non-frail (non-PF&S) controls. Participants were randomly chosen from the cohort of the BIOSPHERE study [18], among those from whom serum was available for vesicle purification.

The study was approved by the Ethics Committee of the Università Cattolica del Sacro Cuore (Rome, Italy; protocol number BIOSPHERE: 8498/15) and all participants signed an informed consent prior to inclusion. Study procedures and criteria for participant selection were described thoroughly elsewhere [18].

2.2. Measurement of Appendicular Lean Mass by Dual X-Ray Absorptiometry

Appendicular lean mass was quantified through whole-body Dual X-Ray Absorptiometry (DXA) scans on a Hologic Discovery A densitometer (Hologic, Inc., Bedford, MA, USA) according to the manufacturer's procedures. Criteria for low aLM were as follows: (a) aLM to body mass index (BMI) ratio (aLM_{BMI}) < 0.789 in men and <0.512 in women, or (b) crude aLM < 19.75 kg in men and <15.02 kg in women [24].

2.3. Blood Sampling

Blood samples were collected in the morning by venipuncture of the median cubital vein after overnight fasting, using commercial collection tubes (BD Vacutainer®; Becton, Dickinson and Co., Franklin Lakes, NJ, USA). One blood tube was delivered to the centralized diagnostic laboratory of the Fondazione Policlinico Universitario "Agostino Gemelli" IRCCS (Rome, Italy) for standard blood biochemistry. The remaining tubes were processed for serum collection in the Biogerontology lab of the Università Cattolica del Sacro Cuore (Rome, Italy). Serum separation was obtained after 30 min of clotting at room temperature and subsequent centrifugation at $1000\times g$ for 15 min at 4 °C. The upper clear fraction (serum) was collected in 0.5-mL aliquots and stored at −80 °C until analysis.

2.4. Small Extracellular Vesicles Isolation and Characterization

2.4.1. Purification of Small Extracellular Vesicles by Differential Ultracentrifugation

Small EVs/exosomes were purified through differential centrifugation as previously described [14,26]. Briefly, serum samples were diluted with equal volumes of phosphate-buffered saline (PBS) to reduce fluid viscosity. Diluted samples were centrifuged at $2000\times g$ at 4 °C for 30 min and pellets were discarded to remove cell contaminants. Subsequently, supernatants were centrifuged at $12,000\times g$ at 4 °C for 45 min to remove apoptotic bodies, mitochondrial fragments, cell debris, and large vesicles (mean size > 200 nm). Supernatants were collected and ultracentrifuged at $110,000\times g$ at 4 °C for 2 h. Pellets were recovered and resuspended in PBS, filtered through a 0.22-μm filter, and ultracentrifuged at $110,000\times g$ at 4 °C for 70 min to eliminate contaminant proteins. Pellets enriched in purified sEVs were finally resuspended in 100 μL of PBS. To quantify sEVs, total protein concentration was measured using the Bradford assay [27].

2.4.2. Western Immunoblot Analysis of Small Extracellular Vesicles

Western immunoblot analysis was performed to assess the purity of sEV isolation, to determine the type of sEVs on the basis of the expressed tetraspanins, and to characterize their protein cargo as previously described [14,28]. Briefly, equal amounts (1.25 μg) of sEV proteins were separated by sodium dodecyl sulphate polyacrylamide gel electrophoresis (SDS-PAGE) and subsequently electroblotted onto polyvinylidenefluoride (PVDF) Immobilon-P (Millipore, Burlington, MA, USA). Membranes were probed with primary antibodies against tetraspanins CD63 (1:200), CD9 (1:200), CD81 (1:200), a specific cocktail of antibodies (1:250) targeting mitochondrial markers (Table 1), flotilin (1:200), and heterogeneous nuclear ribonucleoprotein A1 (HNRNPA1; 1:1000). Technical specifications of primary antibodies used for Western immunoblotting are detailed in Supplementary Table S1.

Table 1. Mitochondrial components and related electron transport chain complexes assayed in purified small extracellular vesicles by Western immunoblotting.

Mitochondrial Marker	ETC Complex
ATP5A	V
MTCOI	IV
NDUFB8	I
NDUFS3	I
SDHB	II
UQCRC2	III

Abbreviations: ATP5A, adenosine triphosphate 5A; ETC, electron transport chain; MTCOI, mitochondrial cytochrome C oxidase subunit I; NDUFB8, nicotinamide adenine dinucleotide reduced form (NADH):ubiquinone oxidoreductase subunit B8; NDUFS3, NADH:ubiquinone oxidoreductase subunit S3; SDHB, succinate dehydrogenase complex iron sulfur subunit B; UQCRC2, ubiquinol-cytochrome C reductase core protein 2.

The following day, membranes were incubated for 1 h at room temperature with anti-mouse peroxidase-conjugated secondary antibodies (1:2000) (Bio-Rad Laboratories, Inc., Hercules, CA, USA). Blots were visualized using the Clarity Max ECL Western Blotting Substrate (Bio-Rad Laboratories) and images were acquired by the ChemiDoc MP Imaging System and analyzed by Image Lab ™ software version 6.0.1 (Bio-Rad Laboratories). Values of optical density (OD) units of each protein band immunodetected were normalized for the amount of sEV total proteins, as determined by the Bradford assay, and related to the control group, whose OD was set at 100%.

2.4.3. Analysis of Small Extracellular Vesicles by Scanning Electron Microscopy Imaging

Small EVs were fixed in a solution of 3.7% glutaraldehyde (Sigma–Aldrich, St. Luis, MO, USA) in PBS for 15 min, washed twice with PBS, and dehydrated through a series of ascending grades of ethanol (i.e., 40%, 60%, 80%, 96%–98%). Subsequently, samples were mounted on carbon adhesive stubs (Agar Scientifics, Stansted, UK) and left at room temperature for 24 h to obtain complete ethanol evaporation. Samples were gold-coated with a Balzers SCD 040 sputter coater (BAL–TEC AG, Balzers, Lichtenstein, Germany; thickness of gold layer: 40 nm) and analyzed at 132.21 K× magnification by a ZEISS EVO HD 15 Scanning Electron Microscope (Carl Zeiss Microscopy GmbH, Oberkochen, Germany) operating under high-vacuum at an accelerating voltage of 5 kV.

2.5. Statistical Analysis

Descriptive statistics were run on all data. Differences in demographic, anthropometric, and clinical parameters between PF&S and control participants were assessed via t-test statistics and χ^{-2} or Fisher exact tests, for continuous and categorical variables, respectively. All tests were two-sided, with statistical significance set at $p < 0.05$. Analyses were performed using the GraphPrism 5.03 software (GraphPad Software, Inc., San Diego, CA, USA).

3. Results

3.1. Characteristics of the Study Participants

The subset of participants included in the present study was representative of the whole BIOSPHERE cohort in terms of age, sex distribution, clinical characteristics, and body composition and functional parameters [8]. The main characteristics of study participants are presented in Table 2. Sex distribution, BMI, number of comorbid conditions and medications, total serum protein concentrations, and albumin levels did not differ between older adults with and without PF&S. PF&S participants tended to be older than controls, but the difference did not reach statistical significance. As per the selection criteria, SPPB scores and aLM either crude or adjusted by BMI were lower in older adults with PF&S relative to non-PF&S participants.

Table 2. Participant characteristics according to the presence of physical frailty and sarcopenia.

Characteristic	Non-PF&S (n = 10)	PF&S (n = 11)	*p*-Value
Age (years), mean ± SD	73.9 ± 2.7	77.7 ± 5.4	0.0557
Gender (female), n (%)	5 (50)	8 (73)	0.5344
BMI (kg/m^2), mean ± SD	28.1 ± 2.8	30.3 ± 4.3	0.1891
SPPB summary score, mean ± SD	12.0 ± 1.0	7.0 ± 0.3	<0.0001
aLM (kg), mean ± SD	20.21 ± 4.10	15.84 ± 3.63	0.0390
aLM$_{BMI}$, mean ± SD	0.81 ± 0.32	0.51 ± 0.11	0.0118
Albumin (g/L), mean ± SD	45.4 ± 12.7	39.8 ± 1.2	0.1536
Total serum protein concentration (g/L), mean ± SD	71.8 ± 4.6	75.5 ± 3.1	0.0914
Number of diseases [¥], mean	3.2 ± 1.6	3.1 ± 1.2	0.8647
Number of medications [#], mean ± SD	2.9 ± 1.6	3.2 ± 1.8	0.7061

Abbreviations: aLM, appendicular lean mass; aLM$_{BMI}$, aLM adjusted by body mass index (BMI); non-PF&S, non-physically frail non-sarcopenic; PF&S: physical frailty & sarcopenia; SD: standard deviation; SPPB: short physical performance battery. [¥] includes hypertension, coronary artery disease, prior stroke, peripheral vascular disease, diabetes, chronic obstructive pulmonary disease, and osteoarthritis. [#] includes prescription and over-the-counter drugs

3.2. Characterization of Small Extracellular Vesicles from the Serum of Participants with and without Physical Frailty and Sarcopenia

3.2.1. Verification of the Purity of Serum Small Extracellular Vesicles

The purity of sEVs obtained by serum ultracentrifugation was ascertained according to the guidelines of the International Society of Extracellular Vesicles [29]. In particular, the presence of the cytosolic protein flotilin (positive control) and the absence of the non-sEV component HNRNPA1 (negative control) were verified (Figure 1A). The purified biospecimen was also analyzed by scanning electron microscopy (SEM) to confirm enrichment in sEVs. Small EVs appear in the scanning electron micrographs as objects of spherical shape and less than 100 nm in size (Figure 1B).

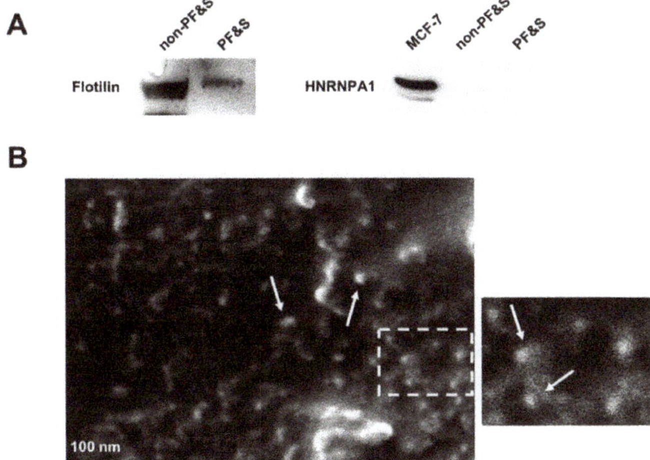

Figure 1. (**A**) Blots of the cytosolic protein flotilin and heterogeneous nuclear ribonucleoprotein A1 (HNRNPA1) as positive and negative markers respectively, in purified small extracellular vesicles (sEVs) obtained by serum ultracentrifugation from participants with physical frailty and sarcopenia (PF&S) and non-physically frail non-sarcopenic (non-PF&S) controls. The Michigan Cancer Foundation-7 (MCF-7) cell extract was used as the positive control for the anti-HNRNPA1 antibody. (**B**) Scanning electron micrographs of purified sEVs. The white-dashed box delimitates the area zoomed on the right. White arrows indicate some of the sEVs found in the observation field. Scale bar: 100 nm.

3.2.2. Quantification of the Amount of Circulating Small Extracellular Vesicles

The total amount of sEVs purified from the serum of PF&S participants was significantly greater than in non-PF&S controls ($p < 0.0001$, Figure 2).

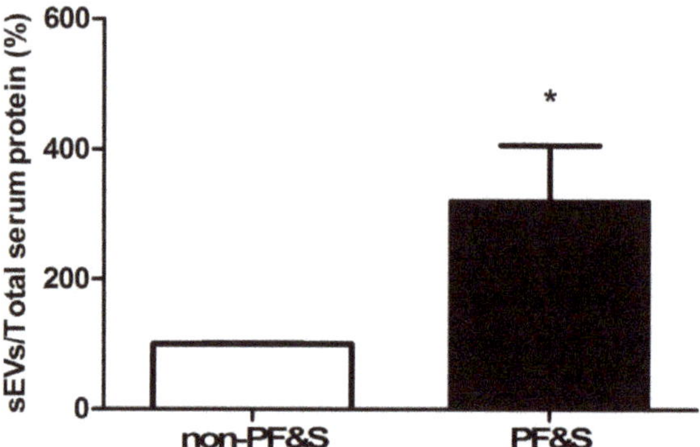

Figure 2. Serum levels of small extracellular vesicles (sEVs) in non-physically frail non-sarcopenic (non-PF&S) controls (n = 10) and participants with physical frailty and sarcopenia (PF&S; n = 11). Data were normalized for the amount of total serum proteins and are shown as percentage of the control group set at 100%. Bars represent mean values (±standard error of the mean). * $p < 0.05$ versus non-PF&S.

3.2.3. Characterization of the Origin and Cargo of Small Extracellular Vesicles

Protein levels of the two tetraspanins, CD9 and CD63, were lower in participants with PF&S than in non-PF&S controls (Figure 3A,B), while CD81 content was unvaried between groups (Figure 3C).

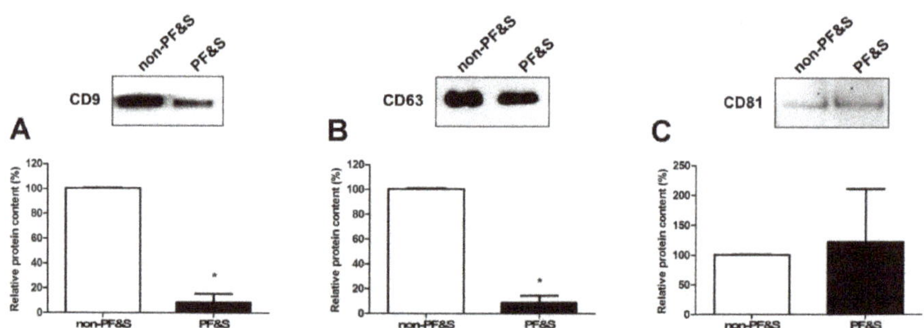

Figure 3. Protein expression of (**A**) CD9, (**B**) CD63, and (**C**) CD81 in purified small extracellular vesicles (sEVs) from non-physically frail non-sarcopenic (non-PF&S) controls (n = 10) and participants with physical frailty and sarcopenia (PF&S; n = 11). Data were normalized for the amount of sEV total proteins and are shown as percentage of the control group set at 100%. Bars represent mean values (±standard error of the mean). * $p < 0.0001$ versus non-PF&S.

As for sEV cargo characterization, protein levels of adenosine triphosphate 5A (ATP5A; complex V), nicotinamide adenine dinucleotide reduced form (NADH):ubiquinone oxidoreductase subunit S3 (NDUFS3; complex I), and succinate dehydrogenase complex iron sulfur subunit B (SDHB; complex II) were lower in participants with PF&S than in non-PF&S controls (Figure 4A–C). No signal was detected for mitochondrial cytochrome C oxidase subunit I (MTCOI, complex IV), NADH:ubiquinone oxidoreductase subunit B8 (NDUFB8; complex I), or ubiquinol-cytochrome C reductase core protein 2 (UQCRC2; complex III) in either participant group.

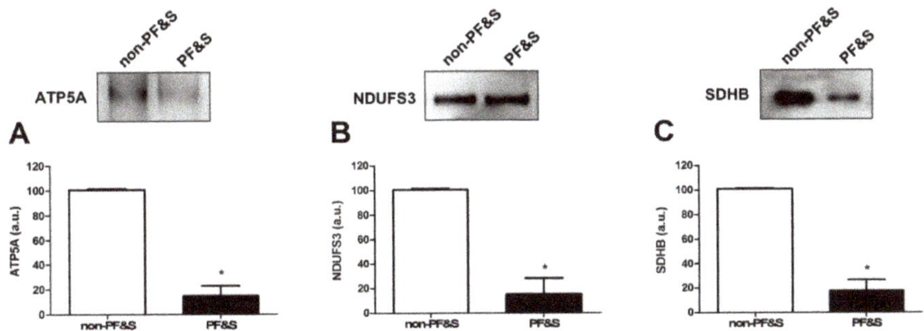

Figure 4. Protein expression of (**A**) adenosine triphosphate 5A (ATP5A), (**B**) nicotinamide adenine dinucleotide reduced form (NADH):ubiquinone oxidoreductase subunit S3 (NDUFS3), and (**C**) succinate dehydrogenase complex iron sulfur subunit (SDHB) in purified small extracellular vesicles (sEVs) from non-physically frail non-sarcopenic (non-PF&S) controls (n = 10) and participants with physical frailty and sarcopenia (PF&S; n = 11). Data were normalized for the amount of sEV total proteins and are shown as percentage of the control group set at 100%. Bars represent mean values (±standard error of the mean). * $p < 0.0001$ versus non-PF&S.

4. Discussion

Among the factors involved in muscle degeneration associated with PF&S, mitochondrial dysfunction and the accrual of abnormal organelles have been indicated as relevant players [30]. However, the exact mechanisms underlying mitochondrial decay are not completely deciphered.

Derangements in MQC processes have been reported in older adults with PF&S [7,31,32]. Nevertheless, alterations in sEV trafficking, which might contribute to MQC dyshomeostasis in muscle [33], have remained largely unexplored. To start filling this gap in knowledge, we purified sEVs from the serum of older adults with and without PF&S and, after ascertaining purity of the preparation, we determined the overall quantity of the mixed sEV population. Our results show a greater amount of sEVs in serum of PF&S participants compared with non-PF&S controls (Figure 2). The verification of the three tetraspanins, CD9, CD63, and CD81, in purified sEVs allowed these vesicles to be identified as a fraction of endosome-derived vesicles, referred to as exosomes, originating from the fusion of multivesicular bodies with the plasma membrane [28]. A lower protein expression of CD9 and CD63 was found in the exosome fraction purified from participants with PF&S (Figure 3), while levels of CD81 were comparable between groups. These observations are in keeping with the heterogenous composition of exosomes themselves, likely reflecting a different vesicle trafficking regulation [34]. Indeed, RAB27A, a guanosine triphosphatase (GTPase) that modulates exosome secretion, has been shown to regulate the secretion of CD63-positive exosomes, but not of those positive for CD9 [35]. Notably, exosomes derived by B-cells are characterized by the tetraspanin markers CD9 and CD81, while CD63 is absent [36]. A previous report by our group showed that RAB7A, a small GTPase and a master regulator of the late endocytic pathway, was able to modulate secretion of CD9- and CD81-positive exosomes [37]. The decreased expression of tetraspanin CD63 found in the present study may therefore be indicative of an altered late endocytic pathway [38], possibly suggesting disarrangements in late endocytic trafficking in PF&S.

The identification of mitochondrial components within the purified material allowed for classification of MDVs among sEVs. In particular, lower levels of the mitochondrial components ATP5A (complex V), NDUFS3 (complex I), and SDHB (complex II) were found in participants with PF&S (Figure 4). With the intent of preserving mitochondrial homeostasis, mitochondrial hyper-fission segregates severely damaged or unnecessary organelles [39,40] that are subsequently disposed via mitophagy [41]. However, mitochondrial-lysosomal crosstalk may dispose mildly oxidized mitochondria via MDV release [42]. Such a mechanism may therefore restore mitochondrial homeostasis before whole-sale organelle degradation is triggered [42]. Though, in the case of defective mitophagy or disruption of the mitochondrial-lysosomal axis, accrual of damaged mitochondria, misfolded proteins, and lipofuscin may occur as a result of inefficient cellular quality control [43]. Therefore, the increased sEV secretion in participants with PF&S (Figure 2) might reflect the cell's attempt to extrude dysfunctional mitochondria. However, the reduced secretion of MDVs in the same participant group (Figure 4) may indicate that the MQC flux is impaired or that the damage to mitochondria is too severe to be disposed via MDVs. This idea is in keeping with previous reports by our group showing derangements in the expression of key proteins of the MQC machinery in old hip-fractured patients with sarcopenia [7,31].

The retrieval of mitochondrial components within sEVs is particularly relevant as it provides novel insights into the mechanisms of sterile inflammation, an age-associated inflammatory response mounted in the absence of infections [44]. This process is framed within the innate immune response and has been included as part of the "danger theory" of inflammation [45]. According to this view, misplaced noxious material from injured cells (i.e., damage-associated molecular patterns (DAMPs)) triggers caspase-1 activation and the secretion of pro-inflammatory cytokines [46]. The release of MDV content (e.g., mitochondrial proteins, mtDNA) can activate inflammatory pathways by interacting with several receptors/systems including TLRs, family pyrin domain-containing 3 (NLRP3) inflammasome, and cGAS-STING DNA sensing system [47].

Recently, we described the existence of a frailty "cytokinome" in older adults with PF&S defined by higher levels of P-selectin, C-reactive protein, and interferon-γ-induced protein 10, and lower levels of myeloperoxidase, interleukin 8, monocyte chemoattractant protein-1, macrophage inflammatory protein 1-α, and platelet-derived growth factor BB [8]. Pro-sarcopenic/pro-disability effects have traditionally been attributed to inflammation [48,49] as much as to dysfunction of anti-inflammatory pathways [49,50]. Furthermore, circulating MDVs have been identified in serum of older adults with PD and associated with a specific inflammatory profile [14]. However, the liaison among failing mitochondrial fidelity pathways, MDV secretion, and systemic inflammation may not be exclusive of neurodegeneration. Indeed, other conditions, such as HIV infection, a model of accelerated and accentuated aging [51], are characterized by pyroptotic bystander cell death and release of DAMPs that may trigger the same pathways as those identified in PD and inflamm-aging [52]. In addition, a massive release of DAMPs is acknowledged as a factor in the development of multiorgan failure in patients with severe injuries or during hemorrhagic shock [53]. Although the pathophysiology of multiple organ failure syndrome, neurodegeneration, and PF&S is heterogeneous, the release of mitochondrial DAMPs might be a converging mechanism shared by all of them. Should this assumption hold true, the scavenging of circulating mitochondrial DAMPs might represent a yet unexplored therapeutic option for the management of age-associated disarrangements, including PF&S. From this perspective, our findings are in line with the geroscience hypothesis, according to which the roots of most chronic diseases may reside in perturbations of a set of basic mechanisms (i.e., hallmarks of aging), including mitochondrial dysfunction [54].

Albeit presenting novel and promising findings, our work has limitations that need to be discussed. First of all, the cross-sectional design of the study precludes establishing cause–effect or temporal relationships between the analyzed pathways and PF&S pathophysiology. Also, although participants were carefully selected and thoroughly characterized, we cannot rule out the possibility that unknown comorbidities may have affected our results. In addition, our study provides an initial characterization of the heterogeneous population of circulating sEVs. Indeed, the analysis of the MDV cargo was limited to selected components/subunits of the mitochondrial electron transport chain. Hence, we cannot exclude that the analysis of other biomolecules, including mtDNA, that may be transported along the same road could provide additional insights into the relationship between sEV trafficking and PF&S. Finally, a deeper characterization of sEVs for their structure and content by means of transmission electron microscopy analysis is needed to confirm and expand our findings as well as to gain further information into the dynamic regulation of vesicle trafficking in PF&S.

Supplementary Materials: The following is available online at http://www.mdpi.com/2073-4409/9/4/973/s1: Table S1: Technical specifications of the primary antibodies used for Western immunoblotting.

Author Contributions: Conceptualization, A.P., C.B., E.M., F.G., and R.C.; Data curation, A.P., F.G., and R.B. (Raffaella Beli); Methodology, A.P., F.G., H.J.C.-J., and R.B. (Raffaella Beli); Writing—original draft preparation, A.P., E.M., and R.C.; Writing—review and editing, C.B., F.G., F.L., and R.B. (Raffaella Beli); Supervision, F.L., and R.B. (Roberto Bernabei); Funding acquisition, C.B. and R.B. (Roberto Bernabei). All authors have read and agreed to the published version of the manuscript.

Funding: This work was supported by Innovative Medicine Initiative-Joint Undertaking (IMI-JU #115621), AIRC (Associazione Italiana per la Ricerca sul Cancro) Investigator grant 2016 #19068 to C.B., Ministero dell'Istruzione, dell'Università e della Ricerca (MIUR) to Consorzio Interuniversitario Biotecnologie (DM 587, 08/08/2018; CIB N. 112/19 to C.B.), 2HE-PONa3_00334 grant for the Zeiss EVO HD 15 SEM, intramural research grants from the Università Cattolica del Sacro Cuore (D3.2 2013 and D3.2 2015), and the nonprofit research foundation "Centro Studi Achille e Linda Lorenzon".

Conflicts of Interest: The authors declare no conflict of interest. The funders had no role in the design of the study; in the collection, analyses, or interpretation of data; in the writing of the manuscript, or in the decision to publish the results.

References

1. Landi, F.; Calvani, R.; Cesari, M.; Tosato, M.; Martone, A.M.; Ortolani, E.; Savera, G.; Salini, S.; Sisto, A.; Picca, A.; et al. Sarcopenia: An overview on current definitions, diagnosis and treatment. *Curr. Protein Pept. Sci.* **2018**, *19*, 633–638. [CrossRef] [PubMed]

2. Calvani, R.; Miccheli, A.; Landi, F.; Bossola, M.; Cesari, M.; Leeuwenburgh, C.; Sieber, C.C.; Bernabei, R.; Marzetti, E. Current nutritional recommendations and novel dietary strategies to manage sarcopenia. *J. Frailty Aging* **2013**, *2*, 38 53. [CrossRef] [PubMed]

3. Chan, D.C.D.; Tsou, H.H.; Chang, C.B.; Yang, R.S.; Tsauo, J.Y.; Chen, C.Y.; Hsiao, C.F.; Hsu, Y.T.; Chen, C.H.; Chang, S.F.; et al. Integrated care for geriatric frailty and sarcopenia: A randomized control trial. *J. Cachexia Sarcopenia Muscle* **2017**, *8*, 78–88. [CrossRef] [PubMed]

4. Bauer, J.M.; Verlaan, S.; Bautmans, I.; Brandt, K.; Donini, L.M.; Maggio, M.; McMurdo, M.E.T.; Mets, T.; Seal, C.; Wijers, S.L.; et al. Effects of a vitamin D and leucine-enriched whey protein nutritional supplement on measures of sarcopenia in older adults, the PROVIDE study: A randomized, double-blind, placebo-controlled trial. *J. Am. Med. Dir. Assoc.* **2015**, *16*, 740–747. [CrossRef] [PubMed]

5. Cesari, M.; Calvani, R.; Marzetti, E. Frailty in older persons. *Clin. Geriatr. Med.* **2017**, *33*, 293–303. [CrossRef] [PubMed]

6. Cesari, M.; Landi, F.; Calvani, R.; Cherubini, A.; Di Bari, M.; Kortebein, P.; Del Signore, S.; Le Lain, R.; Vellas, B.; Pahor, M.; et al. Rationale for a preliminary operational definition of physical frailty and sarcopenia in the SPRINTT trial. *Aging Clin. Exp. Res.* **2017**, *29*, 81–88. [CrossRef]

7. Marzetti, E.; Calvani, R.; Lorenzi, M.; Tanganelli, F.; Picca, A.; Bossola, M.; Menghi, A.; Bernabei, R.; Landi, F. Association between myocyte quality control signaling and sarcopenia in old hip-fractured patients: Results from the Sarcopenia in HIp FracTure (SHIFT) exploratory study. *Exp. Gerontol.* **2016**, *80*, 1–5. [CrossRef]

8. Marzetti, E.; Picca, A.; Marini, F.; Biancolillo, A.; Coelho-Junior, H.J.; Gervasoni, J.; Bossola, M.; Cesari, M.; Onder, G.; Landi, F.; et al. Inflammatory signatures in older persons with physical frailty and sarcopenia: The frailty "cytokinome" at its core. *Exp. Gerontol.* **2019**, *122*, 129–138. [CrossRef] [PubMed]

9. Stahl, P.D.; Raposo, G. Extracellular vesicles: Exosomes and microvesicles, integrators of homeostasis. *Physiology* **2019**, *34*, 169–177. [CrossRef] [PubMed]

10. Maas, S.L.N.; Breakefield, X.O.; Weaver, A.M. Extracellular vesicles: Unique intercellular delivery vehicles. *Trends Cell Biol.* **2017**, *27*, 172–188. [CrossRef] [PubMed]

11. Picca, A.; Calvani, R.; Coelho-Junior, H.J.; Landi, F.; Bernabei, R.; Marzetti, E. Inter-organelle membrane contact sites and mitochondrial quality control during aging: A geroscience view. *Cells* **2020**, *9*, 598. [CrossRef] [PubMed]

12. Bowling, J.L.; Skolfield, M.C.; Riley, W.A.; Nolin, A.P.; Wolf, L.C.; Nelson, D.E. Temporal integration of mitochondrial stress signals by the PINK1:Parkin pathway. *BMC Mol. Cell Biol.* **2019**, *20*, 33. [CrossRef] [PubMed]

13. Soubannier, V.; McLelland, G.-L.; Zunino, R.; Braschi, E.; Rippstein, P.; Fon, E.A.; McBride, H.M. A vesicular transport pathway shuttles cargo from mitochondria to lysosomes. *Curr. Biol.* **2012**, *22*, 135–141. [CrossRef] [PubMed]

14. Picca, A.; Guerra, F.; Calvani, R.; Marini, F.; Biancolillo, A.; Landi, G.; Beli, R.; Landi, F.; Bernabei, R.; Bentivoglio, A.R.; et al. Mitochondrial signatures in circulating extracellular vesicles of older adults with Parkinson's disease: Results from the EXosomes in PArkiNson's Disease (EXPAND) study. *J. Clin. Med.* **2020**, *9*, 504. [CrossRef]

15. Picca, A.; Lezza, A.M.S.; Leeuwenburgh, C.; Pesce, V.; Calvani, R.; Landi, F.; Bernabei, R.; Marzetti, E. Fueling inflamm-aging through mitochondrial dysfunction: Mechanisms and molecular targets. *Int. J. Mol. Sci.* **2017**, *18*, 933. [CrossRef]

16. Collins, L.V.; Hajizadeh, S.; Holme, E.; Jonsson, I.-M.; Tarkowski, A. Endogenously oxidized mitochondrial DNA induces in vivo and in vitro inflammatory responses. *J. Leukoc. Biol.* **2004**, *75*, 995–1000. [CrossRef]

17. Cai, X.; Chiu, Y.H.; Chen, Z.J. The cGAS-cGAMP-STING pathway of cytosolic DNA sensing and signaling. *Mol. Cell* **2014**, *54*, 289–296. [CrossRef]

18. Calvani, R.; Picca, A.; Marini, F.; Biancolillo, A.; Cesari, M.; Pesce, V.; Lezza, A.M.S.; Bossola, M.; Leeuwenburgh, C.; Bernabei, R.; et al. The "BIOmarkers associated with Sarcopenia and PHysical frailty in EldeRly pErsons" (BIOSPHERE) study: Rationale, design and methods. *Eur. J. Intern. Med.* **2018**, *56*, 19–25. [CrossRef]

19. Calvani, R.; Picca, A.; Marini, F.; Biancolillo, A.; Gervasoni, J.; Persichilli, S.; Primiano, A.; Coelho-Junior, H.J.; Bossola, M.; Urbani, A.; et al. A distinct pattern of circulating amino acids characterizes older persons with physical frailty and sarcopenia: Results from the BIOSPHERE study. *Nutrients* **2018**, *10*, 1691. [CrossRef]

20. Picca, A.; Ponziani, F.R.; Calvani, R.; Marini, F.; Biancolillo, A.; Coelho-Junior, H.J.; Gervasoni, J.; Primiano, A.; Putignani, L.; Del Chierico, F.; et al. Gut microbial, inflammatory and metabolic signatures in older people with physical frailty and sarcopenia: Results from the BIOSPHERE study. *Nutrients* **2019**, *12*, 65. [CrossRef]

21. Marzetti, E.; Calvani, R.; Landi, F.; Hoogendijk, E.; Fougère, B.; Vellas, B.; Pahor, M.; Bernabei, R.; Cesari, M. Innovative medicines initiative: The SPRINTT project. *J. Frailty Aging* **2015**, *4*, 207–208. [CrossRef] [PubMed]

22. Marzetti, E.; Cesari, M.; Calvani, R.; Msihid, J.; Tosato, M.; Rodriguez-Mañas, L.; Lattanzio, F.; Cherubini, A.; Bejuit, R.; Di Bari, M.; et al. The "Sarcopenia and Physical fRailty IN older people: Multi-componenT Treatment strategies" (SPRINTT) randomized controlled trial: Case finding, screening and characteristics of eligible participants. *Exp. Gerontol.* **2018**, *113*, 48–57. [CrossRef] [PubMed]

23. Guralnik, J.M.; Simonsick, E.M.; Ferrucci, L.; Glynn, R.J.; Berkman, L.F.; Blazer, D.G.; Scherr, P.A.; Wallace, R.B. A short physical performance battery assessing lower extremity function: Association with self-reported disability and prediction of mortality and nursing home admission. *J. Gerontol.* **1994**, *49*, M85–M94. [CrossRef] [PubMed]

24. Studenski, S.A.; Peters, K.W.; Alley, D.E.; Cawthon, P.M.; McLean, R.R.; Harris, T.B.; Ferrucci, L.; Guralnik, J.M.; Fragala, M.S.; Kenny, A.M.; et al. The FNIH sarcopenia project: Rationale, study description, conference recommendations, and final estimates. *J. Gerontol. A Biol. Sci. Med. Sci.* **2014**, *69*, 547–558. [CrossRef]

25. Newman, A.B.; Simonsick, E.M.; Naydeck, B.L.; Boudreau, R.M.; Kritchevsky, S.B.; Nevitt, M.C.; Pahor, M.; Satterfield, S.; Brach, J.S.; Studenski, S.A.; et al. Association of long-distance corridor walk performance with mortality, cardiovascular disease, mobility limitation, and disability. *J. Am. Med. Assoc.* **2006**, *295*, 2018–2026. [CrossRef]

26. Picca, A.; Guerra, F.; Calvani, R.; Bucci, C.; Lo Monaco, M.R.; Bentivoglio, A.R.; Landi, F.; Bernabei, R.; Marzetti, E. Mitochondrial-derived vesicles as candidate biomarkers in Parkinson's disease: Rationale, design and methods of the exosomes in PArkiNson Disease (EXPAND) Study. *Int. J. Mol. Sci.* **2019**, *20*, 2373. [CrossRef]

27. Théry, C.; Amigorena, S.; Raposo, G.; Clayton, A. Isolation and characterization of exosomes from cell culture supernatants and biological fluids. *Curr. Protoc. Cell Biol.* **2006**, *30*. [CrossRef]

28. Kowal, E.J.K.; Ter-Ovanesyan, D.; Regev, A.; Church, G.M. Extracellular vesicle isolation and analysis by Western blotting. *Methods Mol. Biol.* **2017**, *1660*, 143–152. [CrossRef]

29. Théry, C.; Witwer, K.W.; Aikawa, E.; Alcaraz, M.J.; Anderson, J.D.; Andriantsitohaina, R.; Antoniou, A.; Arab, T.; Archer, F.; Atkin-Smith, G.K.; et al. Minimal information for studies of extracellular vesicles 2018 (MISEV2018): A position statement of the International Society for Extracellular Vesicles and update of the MISEV2014 guidelines. *J. Extracell. Vesicles* **2018**, *7*, 1535750. [CrossRef]

30. Picca, A.; Calvani, R.; Bossola, M.; Allocca, E.; Menghi, A.; Pesce, V.; Lezza, A.M.S.; Bernabei, R.; Landi, F.; Marzetti, E. Update on mitochondria and muscle aging: All wrong roads lead to sarcopenia. *Biol. Chem.* **2018**, *399*, 421–436. [CrossRef]

31. Picca, A.; Calvani, R.; Lorenzi, M.; Menghi, A.; Galli, M.; Vitiello, R.; Randisi, F.; Bernabei, R.; Landi, F.; Marzetti, E. Mitochondrial dynamics signaling is shifted toward fusion in muscles of very old hip-fractured patients: Results from the Sarcopenia in HIp FracTure (SHIFT) exploratory study. *Exp. Gerontol.* **2017**, *96*, 63–67. [CrossRef] [PubMed]

32. Romanello, V.; Sandri, M. Mitochondrial quality control and muscle mass maintenance. *Front. Physiol.* **2016**, *6*, 422. [CrossRef] [PubMed]

33. Picca, A.; Guerra, F.; Calvani, R.; Bucci, C.; Lo Monaco, M.R.; Bentivoglio, A.R.; Coelho-Júnior, H.J.; Landi, F.; Bernabei, R.; Marzetti, E. Mitochondrial dysfunction and aging: Insights from the analysis of extracellular vesicles. *Int. J. Mol. Sci.* **2019**, *20*, 805. [CrossRef]

34. Andreu, Z.; Yáñez-Mó, M. Tetraspanins in extracellular vesicle formation and function. *Front. Immunol.* **2014**, *5*, 442. [CrossRef] [PubMed]

35. Ostrowski, M.; Carmo, N.B.; Krumeich, S.; Fanget, I.; Raposo, G.; Savina, A.; Moita, C.F.; Schauer, K.; Hume, A.N.; Freitas, R.P.; et al. Rab27a and Rab27b control different steps of the exosome secretion pathway. *Nat. Cell Biol.* **2010**, *12*, 19–30. [CrossRef]

36. Saunderson, S.C.; Schuberth, P.C.; Dunn, A.C.; Miller, L.; Hock, B.D.; MacKay, P.A.; Koch, N.; Jack, R.W.; McLellan, A.D. Induction of exosome release in primary B cells stimulated via CD40 and the IL-4 Receptor. *J. Immunol.* **2008**, *180*, 8146–8152. [CrossRef]

37. Guerra, F.; Paiano, A.; Migoni, D.; Girolimetti, G.; Perrone, A.M.; De Iaco, P.; Fanizzi, F.P.; Gasparre, G.; Bucci, C. Modulation of RAB7A protein expression determines resistance to cisplatin through late endocytic pathway impairment and extracellular vesicular secretion. *Cancers* **2019**, *11*, 52. [CrossRef]

38. Guerra, F.; Bucci, C. Role of the RAB7 protein in tumor progression and cisplatin chemoresistance. *Cancers* **2019**, *11*, 1096. [CrossRef]

39. Twig, G.; Hyde, B.; Shirihai, O.S. Mitochondrial fusion, fission and autophagy as a quality control axis: The bioenergetic view. *Biochim. Biophys. Acta* **2008**, *1777*, 1092–1097. [CrossRef]

40. Marzetti, E.; Calvani, R.; Cesari, M.; Buford, T.W.; Lorenzi, M.; Behnke, B.J.; Leeuwenburgh, C. Mitochondrial dysfunction and sarcopenia of aging: From signaling pathways to clinical trials. *Int. J. Biochem. Cell Biol.* **2013**, *45*, 2288–2301. [CrossRef]

41. Youle, R.J.; Narendra, D.P. Mechanisms of mitophagy. *Nat. Rev. Mol. Cell Biol.* **2011**, *12*, 9–14. [CrossRef] [PubMed]

42. Miyamoto, Y.; Kitamura, N.; Nakamura, Y.; Futamura, M.; Miyamoto, T.; Yoshida, M.; Ono, M.; Ichinose, S.; Arakawa, H. Possible existence of lysosome-like organella within mitochondria and its role in mitochondrial quality control. *PLoS ONE* **2011**, *6*, e16054. [CrossRef] [PubMed]

43. Terman, A.; Kurz, T.; Navratil, M.; Arriaga, E.A.; Brunk, U.T. Mitochondrial turnover and aging of long-lived postmitotic cells: The mitochondrial-lysosomal axis theory of aging. *Antioxid. Redox Signal.* **2010**, *12*, 503–535. [CrossRef] [PubMed]

44. Chen, G.Y.; Nuñez, G. Sterile inflammation: Sensing and reacting to damage. *Nat. Rev. Immunol.* **2010**, *10*, 826–837. [CrossRef] [PubMed]

45. Zhang, Q.; Raoof, M.; Chen, Y.; Sumi, Y.; Sursal, T.; Junger, W.; Brohi, K.; Itagaki, K.; Hauser, C.J. Circulating mitochondrial DAMPs cause inflammatory responses to injury. *Nature* **2010**, *464*, 104–107. [CrossRef] [PubMed]

46. Krysko, D.V.; Agostinis, P.; Krysko, O.; Garg, A.D.; Bachert, C.; Lambrecht, B.N.; Vandenabeele, P. Emerging role DAMPs derived from mitochondria in inflammation. *Trends Immunol.* **2011**, *32*, 157–164. [CrossRef] [PubMed]

47. Picca, A.; Lezza, A.M.S.; Leeuwenburgh, C.; Pesce, V.; Calvani, R.; Bossola, M.; Manes-Gravina, E.; Landi, F.; Bernabei, R.; Marzetti, E. Circulating mitochondrial DNA at the crossroads of mitochondrial dysfunction and inflammation during aging and muscle wasting disorders. *Rejuvenation Res.* **2018**, *21*, 350–359. [CrossRef]

48. Franceschi, C.; Garagnani, P.; Parini, P.; Giuliani, C.; Santoro, A. Inflammaging: A new immune–metabolic viewpoint for age-related diseases. *Nat. Rev. Endocrinol.* **2018**, *14*, 576–590. [CrossRef]

49. Furman, D.; Campisi, J.; Verdin, E.; Carrera-Bastos, P.; Targ, S.; Franceschi, C.; Ferrucci, L.; Gilroy, D.W.; Fasano, A.; Miller, G.W.; et al. Chronic inflammation in the etiology of disease across the life span. *Nat. Med.* **2019**, *25*, 1822–1832. [CrossRef]

50. Wilson, D.; Jackson, T.; Sapey, E.; Lord, J.M. Frailty and sarcopenia: The potential role of an aged immune system. *Ageing Res. Rev.* **2017**, *36*, 1–10. [CrossRef]

51. Cesari, M.; Marzetti, E.; Canevelli, M.; Guaraldi, G. Geriatric syndromes: How to treat. *Virulence* **2017**, *8*, 577–585. [CrossRef] [PubMed]

52. Heil, M.; Brockmeyer, N.H. Self-DNA sensing fuels HIV-1-associated inflammation. *Trends Mol. Med.* **2019**, *25*, 941–954. [CrossRef] [PubMed]

53. Aswani, A.; Manson, J.; Itagaki, K.; Chiazza, F.; Collino, M.; Wupeng, W.L.; Chan, T.K.; Wong, W.S.F.; Hauser, C.J.; Thiemermann, C.; et al. Scavenging circulating mitochondrial DNA as a potential therapeutic option for multiple organ dysfunction in trauma hemorrhage. *Front. Immunol.* **2018**, *9*, 891. [CrossRef] [PubMed]

54. López-Otín, C.; Blasco, M.A.; Partridge, L.; Serrano, M.; Kroemer, G. The hallmarks of aging. *Cell* **2013**, *153*, 1194–1217. [CrossRef] [PubMed]

Article

TGF-β Regulates Collagen Type I Expression in Myoblasts and Myotubes via Transient *Ctgf* and *Fgf-2* Expression

Michèle M. G. Hillege, Ricardo A. Galli Caro, Carla Offringa, Gerard M. J. de Wit, Richard T. Jaspers * and Willem M. H. Hoogaars

Laboratory for Myology, Department of Human Movement Sciences, Faculty of Behavioural and Movement Sciences, Vrije Universiteit Amsterdam, Amsterdam Movement Sciences, 1081 BT Amsterdam, The Netherlands; m.m.g.hillege@vu.nl (M.M.G.H.); r.a.galli@amsterdamumc.nl (R.A.G.C.); c.offringa@vu.nl (C.O.); g.m.j.de.wit@vu.nl (G.M.J.d.W.); w.m.h.hoogaars@umcg.nl (W.M.H.H.)
* Correspondence: r.t.jaspers@vu.nl; Tel.: +31-(0)20-5988463

Received: 3 January 2020; Accepted: 3 February 2020; Published: 6 February 2020

Abstract: Transforming Growth Factor β (TGF-β) is involved in fibrosis as well as the regulation of muscle mass, and contributes to the progressive pathology of muscle wasting disorders. However, little is known regarding the time-dependent signalling of TGF-β in myoblasts and myotubes, as well as how TGF-β affects collagen type I expression and the phenotypes of these cells. Here, we assessed effects of TGF-β on gene expression in C2C12 myoblasts and myotubes after 1, 3, 9, 24 and 48 h treatment. In myoblasts, various myogenic genes were repressed after 9, 24 and 48 h, while in myotubes only a reduction in *Myh3* expression was observed. In both myoblasts and myotubes, TGF-β acutely induced the expression of a subset of genes involved in fibrosis, such as *Ctgf* and *Fgf-2*, which was subsequently followed by increased expression of *Col1a1*. Knockdown of *Ctgf* and *Fgf-2* resulted in a lower *Col1a1* expression level. Furthermore, the effects of TGF-β on myogenic and fibrotic gene expression were more pronounced than those of myostatin, and knockdown of TGF-β type I receptor *Tgfbr1*, but not receptor *Acvr1b*, resulted in a reduction in *Ctgf* and *Col1a1* expression. These results indicate that, during muscle regeneration, TGF-β induces fibrosis via *Tgfbr1* by stimulating the autocrine signalling of *Ctgf* and *Fgf-2*.

Keywords: *Acvr1b*; *Tgfbr1*; myostatin; *Col1a1*; skeletal muscle; fibrosis; myogenesis; atrophy

---

1. Introduction

Muscle wasting disorders, such as sarcopenia, cachexia and muscle dystrophies, are characterised by muscle fibre injury or atrophy, which results in the gradual replacement of muscle fibres by adipose and fibrotic tissue [1,2]. This leads to progressive muscle weakness and loss of contractile function. Transforming Growth Factor β (TGF-β) is known for its role in the regulation of skeletal muscle size as well as fibrosis and contributes to the progressive pathology of muscle wasting disorders such as Duchenne Muscular Dystrophy (DMD) [3,4].

TGF-β functions by regulating expression of target genes via specific binding of type II and type I receptor kinases and subsequent activation of intracellular receptor-regulated SMAD2 and SMAD3 proteins (R-SMADS) [5]. TGF-β is expressed by multiple cell types, such as macrophages, monocytes, neutrophils, fibroblasts and bone cells [6–9]. While TGF-β is transiently expressed during skeletal muscle regeneration following injury [10], prolonged elevated TGF-β protein levels are associated with pathologies such as DMD [3], limb girdle muscular dystrophy and amyotrophic lateral sclerosis (ALS), as well as sarcopenia [11–13]. TGF-β may affect skeletal muscle size by the inhibition of muscle stem cell (MuSC) differentiation and the induction of the atrophy of muscle fibres. In vitro studies have shown that TGF-β inhibits myoblast differentiation through the repression of myogenic gene expression, whereas differentiated myotubes seem to be insensitive to TGF-β-induced myogenic inhibition [14–16]. Muscle-specific overexpression of TGF-β in mice stimulates the expression of E3 ligase (i.e., atrogin-1) and concomitant muscle atrophy [17,18]. However, whether the induction of atrogin-1 and muscle atrophy is a direct effect of TGF-β expression or an indirect effect via the stimulation of other paracrine factors remains to be assessed.

TGF-β is also known to be involved in fibrosis. Overexpression of TGF-β in mouse skeletal muscle results in excessive collagen deposition [17]. In addition, antibody treatment to neutralise TGF-β in murine X-linked muscular dystrophy (mdx) mice reduces connective tissue deposition compared to that of untreated mdx mice [19]. Moreover, C2C12 myoblasts overexpressing TGF-β transdifferentiate into fibrotic cells after transplantation into skeletal muscle, which indicates that muscle cells may contribute to fibrosis [20].

Another TGF-β family member, muscle specific cytokine myostatin, has been shown to inhibit myoblast differentiation via a similar mechanism as via TGF-β [21]. Furthermore, myostatin is a well-known regulator of muscle mass and has been suggested to be involved in muscle fibrosis [22]. Myostatin signals via distinct type II and type I receptors than TGF-β does, but also through phosphorylation of SMAD2/3 [23,24]. TGF-β signals mainly via the type I receptor TGF-β receptor type-I (TGFR-1) [24]. While in muscle cells myostatin signals mainly via type I receptor Activin receptor type-1B (ACTR-1B), in fibroblasts myostatin signals mainly via TGFR-1 [23,25]. Both proteins have been indicated as possible therapeutic targets for muscle wasting disorders.

While transient TGF-β expression may contribute to muscle regeneration after injury, the chronic elevated expression of TGF-β in skeletal muscle may be detrimental [cf.10]. Although the role of TGF-β in muscle mass regulation and skeletal muscle fibrosis has been studied extensively, the effects on myoblasts and differentiated muscle cells and underlying mechanisms are not well understood. The aim of this study was to assess the time-dependent effects of TGF-β signalling and downstream signalling on the expression of myogenic, atrophic and fibrotic genes in both myoblasts and myotubes. Furthermore, taking into account the functional and mechanistic similarities between TGF-β and myostatin, as well as the fact that both ligands have been implied as possible therapeutic targets for muscle wasting disorders, the effects of TGF-β and myostatin signalling in myoblasts were compared. Our data indicate that TGF-β inhibits myogenic gene expression in both myoblasts and myotubes but does not affect myotube size. Most importantly, our results show that TGF-β stimulates collagen type I, alpha 1 (*Col1a1*) mRNA expression in both myoblasts and myotubes, which is largely induced via autocrine expression of connective tissue growth factor (*Ctgf*) and fibroblast growth factor-2 (*Fgf-2*). Lastly, the effects of TGF-β on myogenic and fibrotic signalling are more pronounced than those of myostatin, and only TGF-β receptor type-I (*Tgfbr1*) mRNA knockdown, but not Activin receptor type-1B (*Acvr1b*) mRNA knockdown, decreased *Ctgf* and *Col1a1* expression levels, suggesting that myoblasts are more sensitive to TGF-β than to myostatin.

2. Materials and Methods

2.1. C2C12 Cell Culture

The C2C12 mouse muscle myoblast cell line (ATCC CRL-1772) was obtained from ATCC (Wesel, Germany). Cells were cultured in growth medium (DMEM, 4.5% glucose (Gibco, 11995, Waltham, MA, USA), containing 10% fetal bovine serum (Biowest, S181B, Nuaillé, France), 1% penicillin/streptomycin (Gibco, 15140, Waltham, MA, USA), and 0.5% amphotericin B (Gibco, 15290-026, Waltham, MA, USA)) at 37 °C, 5% CO_2. The cells were used for experiments between passage 4–14. All experiments with C2C12 cells were performed on collagen-coated plates (collagen I rat protein, tail (Gibco, A10483-01, Waltham, MA, USA) diluted in 0.02N acetic acid). C2C12 myoblasts were cultured in differentiation medium (DMEM, 4.5% glucose, 2% horse serum (HyClone, 10407223, Marlborough, MA, USA), 1% penicillin/streptomycin, 0.5% Amphotericin B) for 16 h or allowed to differentiate for 3 days before treatment. Cells were treated with 10 ng/mL TGF-β1 (Peprotech, 100-21C, London, UK) or 300 ng/mL myostatin (Peprotech, 120-00, London, UK) for 0, 1, 3, 9, 24 or 48 h, unless indicated differently. The cells were treated with 10μM Ly364947 (dissolved in dimethyl sulfoxide (DMSO), 1mM). As a control, cells were treated with 0.1% DMSO.

2.2. Isolation of the Extensor Digitorum Longus (EDL) Muscle and Primary Myoblast Culture

EDL muscles were obtained from 6-week to 4-month old mice of a C57BL/6 background. The muscles were incubated in collagenase type I (Sigma-Aldrich, C0130, Saint Louis, MO, USA) at 37 °C, 5% CO_2 for 2 h. The muscles were washed in DMEM, 4.5% glucose (Gibco, 11995, Waltham, MA, USA), containing 1% penicillin/streptomycin (Gibco, 15140, Waltham, MA, USA) and incubated in 5% Bovine serum albumin (BSA)-coated dishes containing DMEM (4.5% glucose, 1% penicillin/streptomycin) for 30 min at 37 °C, 5% CO_2 to inactivate collagenase. Single muscle fibres were separated by gently blowing with a blunt ended sterilized Pasteur pipette. Subsequently, muscle fibres were seeded in a thin layer matrigel (VWR, 734-0269, Radnor, PA, USA)-coated 6-well plate containing growth medium (DMEM, 4.5% glucose (Gibco, 11995, Waltham, MA, USA), 1% penicillin/streptomycin (Gibco, 15140, Waltham, MA, USA), 10% horse serum (HyClone, 10407223, Marlborough, MA, USA), 30% fetal bovine serum (Biowest, S181B, Nuaillé, France), 2.5ng/mL recombinant human fibroblast growth factor (rhFGF) (Promega, G5071, Madison, WI, USA), and 1% chicken embryonic extract (Seralab, CE-650-J, Huissen, The Netherlands)). Primary myoblasts were allowed to proliferate and migrate off the muscle fibres for 3–4 days at 37 °C, 5% CO_2. After gentle removal of the muscle fibres, myoblasts were cultured in matrigel-coated flasks until passage 5. Cells were pre-plated in an uncoated flask for 15 min with each passage to reduce the number of fibroblasts in culture. Cell population was 99% Pax7 I. All experiments with primary myoblasts were performed on matrigel-coated plates. Primary myoblasts were cultured in differentiation medium for 6 h or allowed to differentiate for 2 days before treatment with 10 ng/mL TGF-β1 (Peprotech, 100-21C, London, UK) or 300 ng/mL myostatin (Peprotech, 120-00, London, UK).

2.3. Tgfbr1 and Acvr1b siRNA Assay

C2C12 cells were seeded at a density of 7900 cells/cm^2 in a 12-well plate (Greiner Bio-One, 665180, Alphen aan den Rijn, The Netherlands) in antibiotic-free growth medium (DMEM, 1% glucose (Gibco, 31885, Waltham, MA, USA), 10% fetal bovine serum (Biowest, S181B, Nuaillé, France)) at 37 °C, 5% CO_2 and allowed to adhere overnight. SiRNA with a final concentration of 25 nM was prepared according to manufacturer's protocol. Then, 50 nM siControl, 25 nM siAcvr1b + 25 nM siControl, 25 nM siTgfbr1 + 25 nM or 25 nM siAcvr1b + 25 nM siTgfbr1 was added to the medium of the cells. We used 2 μL DharmaFECT1 per well. The cells were treated with siRNA for 24 h in antibiotic-free growth medium. Subsequently, cells were treated with siRNA for 48 h in antibiotic-free differentiation medium (DMEM, 1% glucose (Gibco, 31885, Waltham, MA, USA), 2% horse serum (HyClone, 10407223, Marlborough, MA, USA)). The following reagents for transfection were obtained from Dharmacon (Lafayette, Colorado): ON-TARGET plus Non-targeting Pool (D-001810-10), DharmaFECT1 (T-2001), 5X siRNA Buffer (B-002000-UB-100), mouse ON-TARGET plus *Tgfbr1* siRNA (J-040617-05), and mouse ON-TARGET plus *Acvr1b* siRNA(J-043507-08)

2.4. Ctgf and Fgf-2 siRNA Assay

C2C12 myoblast cells were seeded at a density of 4200 cells/cm^2 and cultured in antibiotic-free growth medium (DMEM, 4.5% glucose (Gibco, 11995, Waltham, MA, USA), 10% fetal bovine serum (Biowest, S181B, Nuaillé, France)) at 37°C, 5% CO_2. The cells were transfected with siRNA targeting *Ctgf* or *Fgf-2* (Ambion® Silencer® Select Pre-Design siRNA, Ctgf siRNA ID: s66077, Fgf-2 siRNA ID: s201344, Carlsbad, CA, USA) or a siRNA-negative control (Silencer® Select Negative Control #1 siRNA, Invitrogen 4390843, Carlsbad, CA, USA). SiRNA was re-suspended to a final concentration of 10 μM and lipofectamine transfection reagent (Lipofectamine® RNAiMAX Reagent, Invitrogen 13778100, Carlsbad, CA, USA) was used to prepare the siRNA–lipid complex according to manufacturer's protocol for a 24-well plate set-up. The cells were cultured for 24 h in antibiotic-free growth medium and transfected with *Ctgf* or *Fgf-2* siRNA–lipid complex for another 24 h. Cells were transfected a second time in antibiotic-free differentiation medium (DMEM, 4.5% glucose, 2% horse serum (HyClone, 10407223, Marlborough, MA, USA). After 16 h, the cells were treated with TGF-β1 (10ng/mL) for 0 h, 3 h and 48 h.

2.5. RNA Isolation and Reverse Transcription

Cells were lysed in TRI reagent (Invitrogen, 11312940, Carlsbad, CA, USA). After this, 10% bromochloropropane (Sigma-Aldrich, B9673, Saint Louis, MO, USA) was added. Lysates were inverted and incubated at room temperature for 5 min and centrifuged (4 °C, 12,000 g, 10 min). The RNA containing supernatant was transferred to a new centrifuge tube and washed with 100% ethanol 2:1. RNA was further isolated using the RiboPure RNA purification kit (Thermo Fisher Scientific, AM1924, Waltham, MA, USA). Then, 500 ng RNA and 4 µL SuperScript VILO Mastermix (Invitrogen, 12023679, Carlsbad, CA, USA) were diluted to 20 µL in RNAse free water and reverse transcription was performed in a 2720 thermal cycler (Applied Biosystems, Foster City, CA, USA), using the following program: 10 min 25 °C, 60 min 42 °C, 5 min 85 °C. The cDNA was diluted 10x in RNAse free water.

2.6. Quantitative Real Time PCR

We added 7.5 µL Fast SYBR Green master mix (Fischer Scientific, 10556555, Pittsburgh, PA, USA), 2.5 µL primer mix and 5 µL cDNA in duplo in a 48-well plate. The program ran on the StepOne real time PCR (Applied Biosystems, Foster City, CA, USA) was 20 s at 95 °C holding stage, 40 times 3 s 95 °C step 1 and 30 s 60 °C step 2 cycle stage, 15 s 95 °C, 1 min 60 °C and 15 s 95 °C. *Gapdh* was used as a housekeeping gene to correct for cDNA input. The efficiency of all used primers (Table 1) was tested.

Table 1. Primers for qPCR.

Primer	Sequence
mGapdh-forward	TCCATGACAACTTTGGCATTG
mGapdh-reverse	TCACGCCACAGCTTTCCA
Myod1-forward	AGCACTACAGTGGCGACTCA
Myod1-reverse	GCTCCACTATGCTGGACAGG
Myog-forward	CCCAACCCAGGAGATCATTT
Myog-reverse	GTCTGGGAAGGCAACAGACA
Myh3-forward	CGCAGAATCGCAAGTCAATA
Myh3-reverse	CAGGAGGTCTTGCTCACTCC
Ctgf-forward	CCACCCGAGTTACCAATGAC
Ctgf-reverse	GCTTGGCGATTTTAGGTGTC
Fgf-2-forward	AAGCGGCTCTACTGCAAGAA
Fgf-2-reverse	GTAACACACTTAGAAGCCAGCAG
Col1a1-forward	ATGTTCAGCTTTGTGGACCT
Col1a1-reverse	CAGCTGACTTCAGGGATGT
Id1-forward	ACCCTGAACGGCGAGATCA
Id1-reverse	TCGTCGGCTGGAACACAT
Nox4-forward	CTTTTCATTGGGCGTCCTC
Nox4-reverse	GGGTCCACAGCAGAAAACTC

2.7. Western Blotting

Cells were lysed in RIPA buffer (Sigma-Aldrich, R0278, Saint Louis, MO, USA) containing 1 tablet of protease inhibitor (Sigma-Aldrich, 11836153001, Saint Louis, MO, USA) and 1 tablet of phosStop (Sigma-Aldrich, 04906837001, Saint Louis, MO, USA) per 10 mL. The total protein concentration in the lysates was determined using a Pierce BCA Protein Assay kit (Thermo Scientific, 23225, Waltham, MA, USA). The absorbance was measured using a microplate spectrophotometer (Epoch Biotek, Winooski, VT, USA) and the protein concentration was calculated using Gen5 software (BioTek, Winooski, VT,

USA). An 8% polyacrylamide gel was made. Then, 15 µL sample mix, containing 9 µg total protein and 5 µL sample buffer (5.7 mL water, 1.6 mL glycerol, 1.1 mL 10% SDS, 1.3 mL 0.5 M Tris (pH6.8), 25 mg dithiotreitol (DTT), 300 µL bromophenol blue) was heated to 90 °C for 5 min, cooled on ice and loaded onto the gel. The gel was run in electrophoresis buffer (25 mM Tris base, 190 mM glycine, 0.1% SDS) at 70 V until the samples reached the separating gel and then run at 150 V until the samples reached the bottom of the gel. Next, the proteins were transferred onto a polyvinylidene fluoride (PVDF) membrane (GE Healthcare, 15269894, Chicago, IL, USA) for 1 h at 80V on ice in cold blot buffer (25 mM Tris base, 190 mM glycine, 20% ethanol). The membrane was rinsed in water and washed 2x in Tris-buffered saline and Tween-20 (TBS-T) (20 mM Tris/HCl, 137 mM NaCl, 0.1% Tween-20). The membrane was incubated for 1 h in 2% enhanced chemiluminescence (ECL) prime blocking agent (GE Healthcare, RPN418, Chicago, IL, USA) in TBS-T at 4 °C while shaking. Subsequently, the membrane was incubated overnight in 2% blocking agent in TBS-T with primary antibody (Table 2) at 4 °C while shaking. The membrane was washed 3 × 5 min in TBS-T and incubated in 2% blocking agent in TBS-T with secondary antibody (Table 2) for 1 h at room temperature. ECL solution A and B (GE Healthcare, RPN2235, Chicago, IL, USA) were mixed 1:1 at room temperature and the membrane was incubated for 5 min. Images were taken by the ImageQuant LAS500 (GE healthcare, life sciences, Chicago, IL, USA) and relative intensity of protein bands was quantified using ImageJ [26]. Pan actin was used as a loading control.

Table 2. Antibodies for Western Blotting and immunofluorescence.

AB	Dilution	Experiment	Company
Phospho-SMAD2 (Ser465/467) Rabbit mAb	1:1000	WB	cell signaling/3108, Leiden. The Netherlands
SMAD2 Rabbit mAb	1:1000, 1:200	WB, IF	cell signaling/5339
Phospho-SMAD3 (S423/425) Rabbit mAb	1:1000	WB	cell signaling/9520
SMAD3 Rabbit mAb	1:1000	WB	cell signaling/9523
Phospho-Akt (Ser473) Rabbit mAb	1:2000	WB	cell signaling/4060
Akt (pan) Rabbit mAb	1:1000	WB	cell signaling/4691
Phospho-ERK1/2 Rabbit mAb	1:2000	WB	cell signaling/4370
ERK1/2 Rabbit mAb	1:4000	WB	cell signaling/4695
Pan actin Rabbit mAb	1:1000	WB	cell signaling/8456
Myosin, sarcomere (MHC)	2.5 µg/mL	IF	DSHB/MF20-s, Iowa City, IA, USA
Anti-Rabbit IgG-POD (LumiLightPLUS Western Blotting Kit)	1:2000	WB	Roche/12015218001, Basel, Switzerland
Goat anti-Rabbit IgG (H + L), Alexa Fluor® 555 conjugate	1:500	IF	ThermoFisher Scientific/A21428, Waltham, MA, USA
Goat anti-Mouse IgG (H + L), Alexa Fluor® 488 conjugate	1:250	IF	ThermoFisher Scientific/A11001

2.8. Immunofluorescence

Cells were washed 2x with cold phosphate-buffered saline (PBS) (Gibco, 14190250, Waltham, MA, USA) and fixated for 10 min in 4% paraformaldehyde (PFA) (Fisher Scientific, Pittsburgh, PA, USA) at room temperature. Cells were washed 3x in PBS and permeabilised in 0.1% Triton X-100 in PBS for 10 min. After this, the cells were washed 3x in PBS with 0.05% Tween20 (PBS-T) and incubated for 1 h in 5% normal goat serum (ThermoFisher Scientific, 50062Z, Waltham, MA, USA) in PBS at room temperature. The cells were incubated overnight with primary antibody (Table 2) in 5% normal goat serum in PBS at 4 °C. Then, the cells were washed 3 × 5 min in PBS-T and incubated with secondary antibody (Table 2) in PBS-T for 1 h at room temperature. Cells were washed again 3 × 5 min in PBS-T and incubated in PBS with 4′,6-diamidino-2-phenylindole (DAPI) (100 ng/mL). After this, the cells were rinsed with PBS and stored in PBS at 4 °C. Images were taken with a fluorescent microscope (Zeiss Axiovert 200M, Hyland Scientific, Stanwood, WA, USA) using the program Slidebook 5.0 (Intelligent Imaging Innovations, Göttingen, Germany). The images were analysed using ImageJ [26].

2.9. Statistical Analysis

Graphs were made in Prism version 8 (GraphPad software, San Diego, CA, USA). All data were presented as mean + standard error of the mean (SEM). The data were normalised by the values of a control group or of control cells at 0 h. In graphs of time-dependent relative mRNA expression, values of control cells at 0 h were not presented. Statistical analysis was performed in SPSS version 25 (IBM,

Amsterdam, The Netherlands). Significance in the difference between two groups was determined by independent t-test. Statistical significance for multiple comparisons was determined by one-way analysis of variance (ANOVA) or two-way ANOVA with post-hoc Bonferroni corrections. Significance was set at * $p < 0.05$.

3. Results

3.1. TGF-β Inhibits Expression of Myogenic Genes in both C2C12 Myoblasts and Myotubes

TGF-β reduced both the fusion index (number of myotubes with two or more nuclei per total number of nuclei) and differentiation index (number of nuclei within the myotubes per total number of nuclei) of C2C12 cells (Figure 1a–d). After 1 h of TGF-β treatment, in both myoblasts and myotubes SMAD2 and SMAD3 were phosphorylated (Figure 1e–i), indicating that both myoblasts and myotubes are sensitive to TGF-β. SMAD phosphorylation was inhibited by TGF-β receptor type I inhibitor Ly364947.

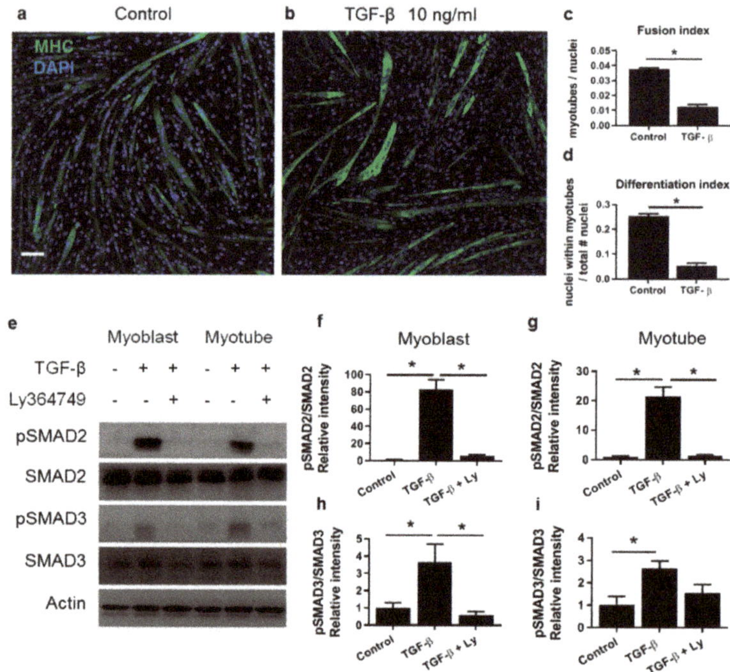

Figure 1. Transforming Growth Factor β (TGF-β) inhibits C2C12 differentiation. (**a,b**) C2C12 cells were induced to differentiate in control medium (**a**) or medium supplemented with TGF-β (**b**). Myotubes stained for myosin heavy chain (MHC) (green). Nuclei were stained using DAPI (blue). Scale indicates 100 μm. (**c,d**) Fusion index, defined as number of myotubes ≥ 2 nuclei/total number of nuclei and differentiation index defined as number of nuclei within MHC+ myotubes/total number of nuclei were reduced after TGF-β. (**e,f,g,h,i**) In both myoblasts and myotubes, phosphorylation levels of SMAD2 (**f,g**) and SMAD3 (**h,i**) were increased upon 1 h of TGF-β treatment. Pan actin served as loading control. Phosphorylation levels are displayed as relative intensity of pSMAD/total SMAD. Data were normalized to values of control condition. Error bars indicate standard error of the mean; * indicates significant difference at $p < 0.05$; $n = 4$ experiments per condition.

Subsequently, to assess acute and delayed effects in both myoblasts and myotubes, the time-dependent effects of TGF-β on myogenic gene expression were examined after 1, 3, 9, 24 and 48 h of treatment. After 9, 24 and 48 h of TGF-β treatment, *Myod* mRNA expression levels in myoblasts were reduced compared to those in untreated cells, although, after 48 h, *Myod* mRNA expression levels were increased compared to those at earlier time points (Figure 2a). After 24 and 48 h, myogenin (*Myog*) and embryonic myosin heavy chain (*Myh3*) mRNA expression levels in myoblasts were reduced compared to those in untreated cells, although expression levels did gradually increase compared to earlier time points (Figure 2b,d). These results show that TGF-β does not acutely reduce the expression levels of *Myod*, *Myog* and *Myh3* in myoblasts, but rather reduces or attenuates differentiation-related increases in mRNA expression levels of *Myod*, *Myog* and *Myh3* at later time points. In myotubes, *Myog* mRNA expression levels were not significantly affected by TGF-β (Figure 2c). However, after 24 and 48 h of TGF-β treatment, *Myh3* expression levels were reduced compared to those in untreated myotubes (Figure 2e). Thus, TGF-β represses *Myh3* mRNA expression, even in differentiated myotubes.

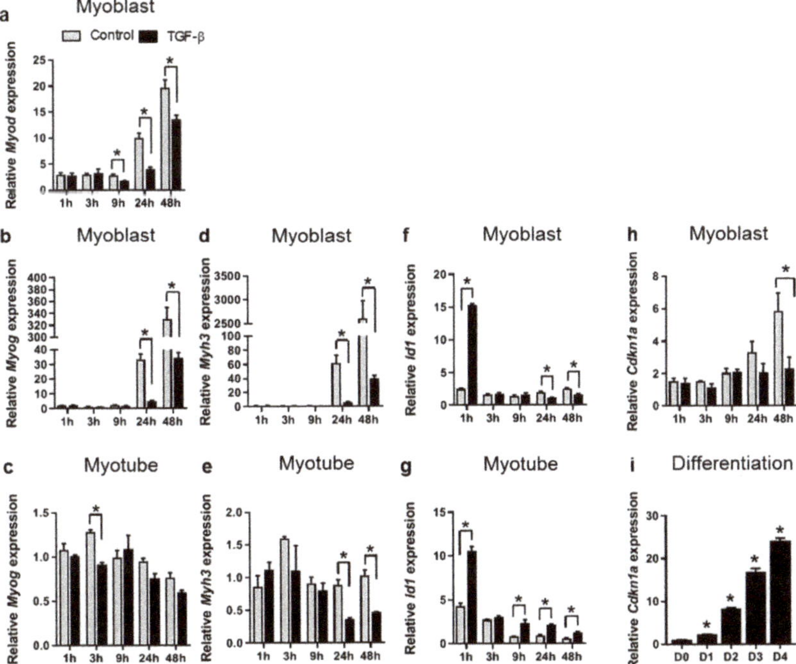

Figure 2. TGF-β reduces myogenic gene expression. (**a,b,d,f**) In myoblasts, expression levels of *Myod* (**a**) were reduced by TGF-β after 9 h compared to those of untreated cells, while expression levels of *Myog* (**b**) and *Myh3* (**d**) were reduced after 24 h. *Id1* expression levels (**f**) were induced after 1 h and repressed after 24 and 48 h. (**c,e,g**) In myotubes, the expression levels of *Myog* (**c**) were unaffected by TGF-β, while expression levels of *Myh3* (**e**) were reduced compared to those of untreated cells after 24 and 48 h. Expression levels of *Id1* (**g**) were induced after 1 h and remained slightly elevated at later time points. (**h**) In untreated myoblasts, the expression levels of *Cdkn1a* significantly increased, while this increase was inhibited in TGF-β treated cells. (**i**) *Cdkn1a* mRNA expression increased during myoblast differentiation, where D0 is the start of differentiation and D1, 2, 3 and 4 are days 1 to 4 of differentiation. *Gapdh* served as housekeeping gene. Data were normalized to values of control cells at 0 h; * indicates significant difference at $p < 0.05$; $n = 3$ experiments per condition.

Regarding the mechanisms underlying effects on differentiation, inhibitor of differentiation 1 (*Id1*) overexpression has been suggested to inhibit differentiation [27]. Since TGF-β induces *Id1* expression in various cell types via SMAD1/5 [28], we quantified *Id1* expression levels. In both C2C12 myoblast and myotubes, TGF-β transiently upregulated *Id1* expression after 1 h (Figure 2f,g), which corresponded with observed SMAD1/5 phosphorylation (Figure S1). In myoblasts, after 24 and 48 h of TGF-β treatment *Id1* mRNA expression levels were slightly reduced compared to those in untreated cells, whereas in myotubes *Id1* mRNA expression levels remained elevated. Based on these results, together with the known function of *Id1*, it is conceivable that *Id1* is involved in TGF-β mediated inhibition of differentiation.

In addition, effects of TGF-β on cell cycle inhibitor cyclin-dependent kinase inhibitor 1A (*Cdk1na*) mRNA expression was examined, because myostatin has been suggested to inhibit myoblast differentiation through inhibition of cyclin-dependent kinase inhibitor 1A [21]. In untreated C2C12 myoblasts, after 48 h *Cdk1na* mRNA expression levels were increased, while this increase was inhibited by TGF-β treatment (Figure 2h). However, *Cdk1na* expression increased during myoblast differentiation (Figure 2i) and no significant effects were observed at earlier time points. This indicates that effects on *Cdkn1a* mRNA expression were likely related to inhibited differentiation, rather than a direct effect of TGF-β on *Cdkn1a* mRNA expression. This suggests that TGF-β does not inhibit differentiation via the regulation of *Cdkn1a* expression.

3.2. TGF-β Does Not Affect Myotube Size In Vitro

TGF-β does not only negatively regulate muscle mass via the inhibition of myoblast differentiation, but TGF-β overexpression in adult mouse muscle has also been shown to result in increased expression of E3 ligase atrogin-1, as well as a reduction in muscle fibre cross sectional area [18]. Furthermore, myostatin is well known to stimulate the expression of E3 ligases, both in adult muscle as well as in myotubes in vitro [29]. E3 ligases are involved in protein degradation via Akt/FOXO signalling and play a role in muscle atrophy [30]. These studies indicate that TGF-β may induce protein degradation and subsequent muscle fibre atrophy via a similar mechanism as myostatin does. In addition, TGF-β-induced protein degradation in differentiating myoblasts may attenuate further myoblast differentiation. Since it remains to be assessed whether TGF-β induces muscle atrophy directly via upregulation of E3 ligase expression, time-dependent effects of TGF-β treatment on expression of muscle specific E3 ligases were determined. In myoblasts, after 3, 9, 24 and 48 h of TGF-β treatment mRNA expression levels of muscle RING-finger 1 (*Murf-1*) were reduced, while after 24 h *Atrogin-1* mRNA expression was transiently repressed (Figure 3a,b). In myotubes, the expression levels of *Atrogin-1* were not affected by TGF-β, whereas after 24 and 48 h mRNA expression levels of *Murf-1* were reduced compared to those in untreated myotubes (Figure 3c,d). These results suggest that TGF-β may protect myotubes against E3 ligase-induced protein degradation. However, our results also show that the endogenous expression of *Murf-1* and *Atrogin-1* increased during differentiation, which suggests that the observed effects of TGF-β on *Murf-1* and *Atrogin-1* expression levels were likely related to its inhibitory effect on differentiation (Figure 3e,f). In both myoblasts and myotubes, TGF-β transiently increased the expression levels of the ligase *Musa1* (Figure 3g,h). In myoblasts, *Musa1* expression levels were significantly increased after 9 h. In myotubes, expression levels were increased after 3 and 9 h.

Subsequently, myotube thickness was measured in C2C12 myoblasts that were differentiated in the presence or absence of TGF-β for three days (cells shown in Figure 1a). There was no significant difference in diameter between myotubes treated with TGF-β and controls (Figure 3n). Furthermore, while SMAD2 and SMAD3 were phosphorylated after 1 h of TGF-β treatment, no significant effects on Akt or ERK1/2 phosphorylation were observed (Figure 3j–m). Together, these results indicate that in vitro TGF-β alone does not affect myotube size.

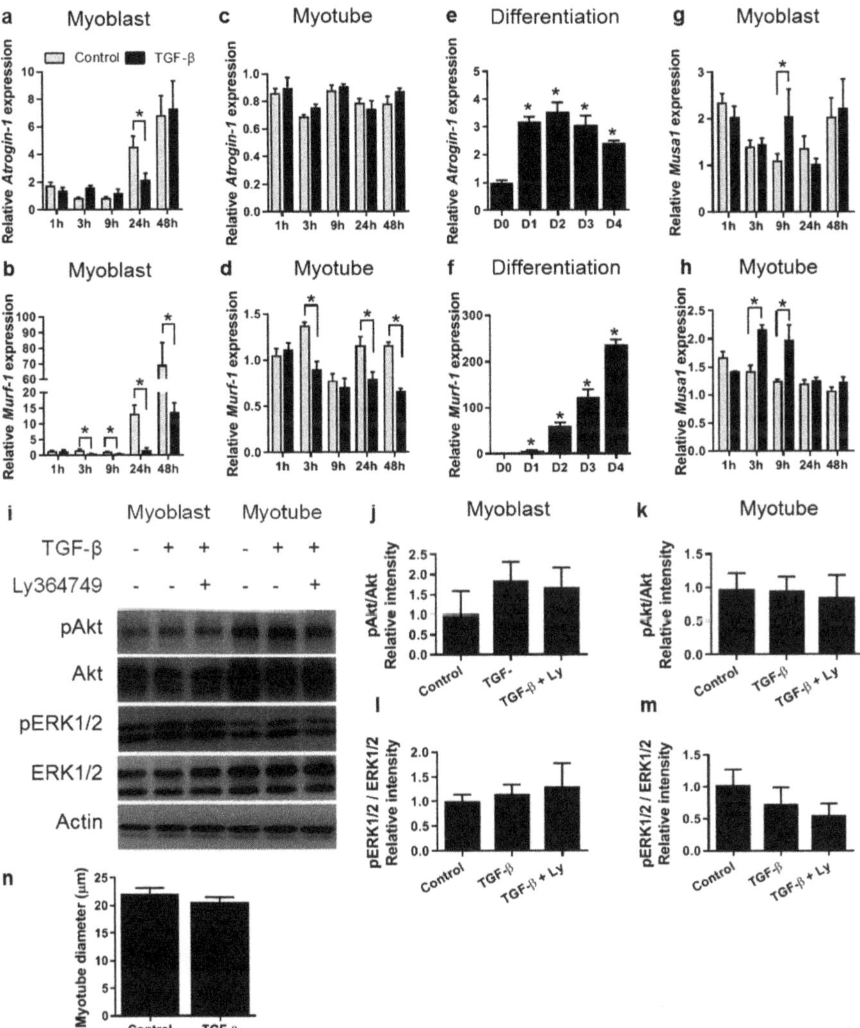

Figure 3. TGF-β does not affect myotube size. (**a,b**) In myoblasts, relative expression levels of *Atrogin-1* (**a**) and *Murf-1* (**b**) were inhibited by TGF-β. (**c,d**) In myotubes, *Atrogin-1* (**c**) expression was unaffected by TGF-β, while *Murf-1* (**d**) expression levels were inhibited after 3, 24 and 48 h. (**e,f**) mRNA expression of *Atrogin-1* (**e**) and *Murf-1* (**f**) increased during differentiation, where D0 is the start of differentiation and D1, 2, 3 and 4 are days 1 to 4 of differentiation. (**g,h**) In both myoblasts (**g**) and myotubes (**h**), TGF-β transiently increased *Musa1* expression levels. *Gapdh* served as housekeeping gene. Data were normalized to values of control cells at 0 h; * indicates significant difference at $p < 0.05$; $n = 3$ experiments per condition (**a–h**). (**i,j**) Western blot quantification of Akt phosphorylation in myoblasts, (**k**) Akt phosphorylation in myotubes, (**l**) ERK1/2 phosphorylation in myoblasts, (**m**) ERK1/2 phosphorylation in myotubes. Pan actin served as loading control. Phosphorylation levels are displayed as relative intensity of phospho/total protein. Data were normalized to values of the control condition. Error bars indicate standard error of the mean; $n = 4$ experiments per condition. **n** After 3 days of TGF-β treatment, the myotube diameters displayed in Figure 1a were not significantly different from those of control condition. $n = 80$ per experimental condition.

3.3. TGF-β Affects Fibrotic Gene Expression in a Time-Dependent Manner in Both Myoblasts and Myotubes

Time-dependent effects of TGF-β on fibrotic gene expression in C2C12 myoblasts and myotubes were studied. In both myoblasts and myotubes, TGF-β acutely and transiently induced the expression of *Ctgf* and *Fgf-2* (Figure 4a–d). Expression levels peaked between 3 and 9 h of treatment and remained significantly increased for at least 48 h compared to levels in untreated cells. In myoblasts, after 3 h TGF-β treatment, *Col1a1* expression levels were 1.9-fold higher compared to those in untreated cells. This effect gradually increased, and after 48 h, *Col1a1* expression levels were 5.6-fold higher in comparison to levels in untreated cells (Figure 4e). In myotubes, *Col1a1* mRNA expression levels were 10-fold higher compared to those in myoblasts (Figure 4i). After 9 and 48 h of TGF-β treatment, *Col1a1* expression levels were 1.5-fold higher compared to those in untreated cells (Figure 4f). NAPDH oxidase 4 (*Nox4*) is a TGF-β target gene that is required for TGF-β-induced expression of components of extracellular matrix (ECM) [31]. Our results show that in both myoblasts and myotubes, *Nox4* mRNA expression levels were significantly higher compared to those in untreated cells, after 9 or 3 h of TGF-β treatment, respectively. The effect of TGF-β treatment gradually increased and after 48 h, in myoblasts, *Nox4* expression levels were 7.9-fold higher and, in myotubes, 3.1-fold higher compared to those of untreated cells (Figure 4g,h). These results suggest that TGF-β stimulates fibrosis by increasing collagen type I expression in both myoblasts and myotubes.

3.4. TGF-β Induces Col1a1 Expression via Ctgf and Fgf-2 in Myoblasts

To investigate whether TGF-β induces *Col1a1* expression in C2C12 myoblasts directly or via the autocrine expression of *Ctgf* or *Fgf-2*, the effects of TGF-β treatment on *Col1a1* expression were studied in the presence of siRNA targeting *Ctgf* or *Fgf-2*. At all time points, treatment with siRNA reduced *Ctgf* or *Fgf-2* mRNA expression levels compared to levels of control siRNA treatment by >90% and >80%, respectively (Figure 4j,m). At 48 h of TGF-β treatment the induction of *Col1a1* mRNA expression was substantially lower (approximately 50%) in the presence of siRNA targeting either *Ctgf* or *Fgf-2* compared to controls (Figure 4k,n), suggesting that *Col1a1* mRNA expression is at least in part regulated by TGF-β dependent *Ctgf* and *Fgf-2* expression. In addition, after 3 h of TGF-β treatment, *Ctgf* knockdown did not affect *Fgf-2* mRNA expression, although after 48 h *Fgf-2* mRNA expression was significantly lower (approximately 70%) in the presence of siRNA targeting *Ctgf*, compared to controls (Figure 4l). *Ctgf* mRNA expression was significantly lower (>55%) in the presence of siRNA against *Fgf-2* compared to controls at all time points (Figure 4o).

3.5. TGF-β Has a Larger Effect on Muscle Differentiation and Fibrosis than Myostatin

Due to the functional and mechanistic similarities between TGF-β and myostatin, the effects of myostatin and TGF-β on C2C12 and primary myoblasts were studied. C2C12 and primary myoblasts, as well as myotubes, were treated with different doses of myostatin or TGF-β. Although a higher concentration of myostatin was needed compared to that of TGF-β in C2C12 and primary myoblasts, as well as myotubes, both proteins induced the translocation of SMAD2 to the nucleus. Figure 5a shows that in primary myotubes and undifferentiated myoblasts, 1 h of 10 ng/mL TGF-β or 300 ng/mL myostatin treatment resulted in the nuclear translocation of SMAD2. Little effect was observed for 0.01 ng/mL TGF-β or 10 ng/mL myostatin. Both of these ligands have a molecular weight of 25 kDa. In primary myoblasts, the comparison of effects of 3 and 48 h myostatin or TGF-β treatment on myogenic and fibrotic gene expression levels showed that after 48 h TGF-β reduced *Myh3* expression by approximately twofold compared to controls, while myostatin did not affect *Myh3* expression (Figure 5b,c). Furthermore, although in primary myoblasts after 3 h of treatment both myostatin and TGF-β significantly enhanced *Ctgf* mRNA expression levels, TGF-β increased *Ctgf* expression levels by 2.2-fold, while myostatin increased *Ctgf* expression levels only by 1.6-fold (Figure 5d,e). These results indicate that TGF-β has a stronger effect on fibrotic and myogenic gene expression levels than myostatin.

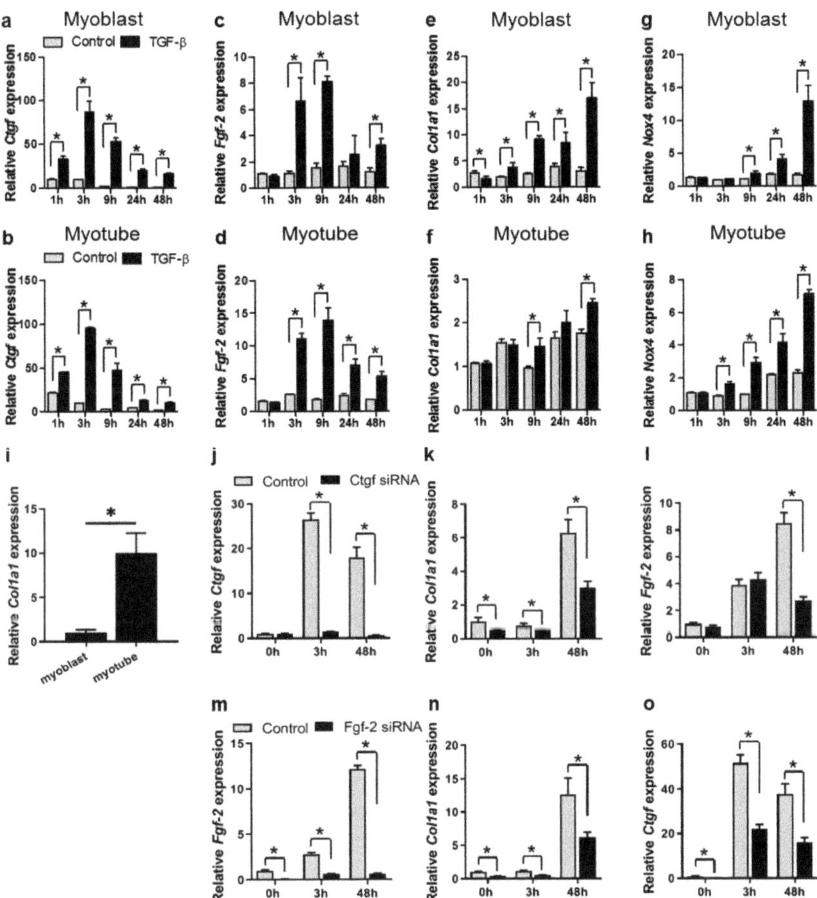

Figure 4. TGF-β affects fibrotic gene expression levels in myoblasts and myotubes in a time-dependent matter. (**a–h**) mRNA expression levels of (**a**) *Ctgf* in myoblasts, (**b**) *Ctgf* in myotubes, (**c**) *Fgf-2* in myoblasts, (**d**) *Fgf-2* in myotubes, (**e**) *Col1a1* in myoblasts, (**f**) *Col1a1* in myotubes, (**g**) *Nox4* in myoblasts, (**h**) *Nox4* in myotubes. (**a–d**) In myoblasts and myotubes, expression levels of *Ctgf* and *Fgf-2* were acutely induced by TGF-β. (**e–h**) *Col1a1* and *Nox4* expression levels were gradually induced by TGF-β. *Gapdh* served as housekeeping gene; data were normalized to values of control cells at 0 h; * indicates significant difference at $p < 0.05$; $n = 3$ experiments per condition (**a–h**). (**i**) *Col1a1* mRNA expression levels in myotubes were approximately 10-fold higher compared to those in myoblasts. *Gapdh* served as housekeeping gene; data were normalized to expression values in myoblasts. * p indicates significant difference at <0.05; $n = 6$ experiments per condition. (**j**) At all time points after siRNA treatment, *Ctgf* expression was knocked down by >90%. (**k**) *Col1a1* expression was reduced in the presence of siRNA targeting *Ctgf*. (**l**) After 3 h of TGF-β treatment, *Fgf-2* expression increased independent of *Ctgf*. After 48 h of TGF-β treatment, in the presence of siRNA targeting *Ctgf*, *Fgf-2* expression was significantly reduced. **m** At all time points after siRNA treatment, *Fgf-2* expression was knocked down by >81%. Both *Col1a1* (**n**) and *Ctgf* (**o**) expression levels were significantly reduced in the presence of siRNA targeting *Fgf-2* compared to those of control siRNA condition. *Gapdh* served as housekeeping gene; data were normalized to values of control cells at 0 h; * indicates significant difference at $p < 0.05$; $n = 6$ experiments per condition (**j–o**).

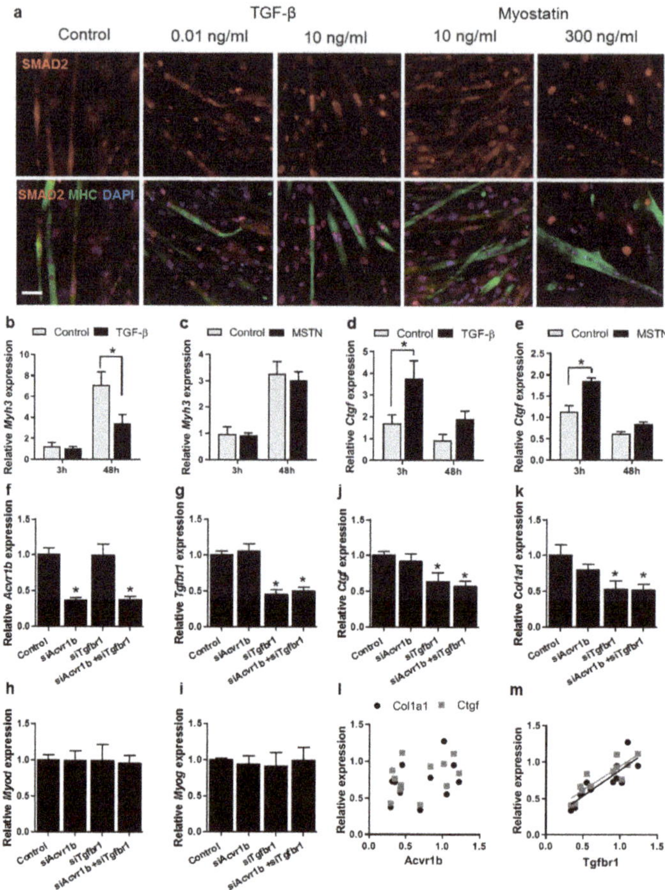

Figure 5. TGF-β has a larger effect on myoblasts compared to myostatin. (**a**) Primary cells were induced to differentiate for 2 days and subsequently treated with TGF-β (0.01 or 10 ng/mL) or myostatin (10 or 300 ng/mL). Myotubes were stained for MHC (green) and nuclei were stained using DAPI (blue). SMAD2 is visible in red. The scale indicates 100 μm. After TGF-β or myostatin treatment, SMAD2 was translocated to the nucleus in both myotubes and undifferentiated myoblasts compared to controls. (**b,c**) In primary mouse myoblasts, expression levels of *Myh3* mRNA were reduced after 48 h TGF-β treatment compared to those in untreated cells (**b**), while no differences were observed for myostatin (MSTN) (**c**). (**d,e**) Expression levels of *Ctgf* were increased after 3 h TGF-β (**d**) or myostatin treatment (**e**) compared to those in untreated cells, although TGF-β had a larger effect. (**f,g**) Treatment with specific siRNAs reduced levels of *Acvr1b* (**f**) or *Tgfbr1* (**g**). (**h,i**) Knockdown of *Acvr1b* or *Tgfbr1* did not affect *Myod* (**h**) or *Myog* (**i**) expression levels. (**j,k**) *Tgfbr1* knockdown slightly reduced *Ctgf* (**j**) and *Col1a1* (**k**) expression levels, while *Acvr1b* knockdown had little effect. The combined knockdown of *Tgfbr1* and *Acvr1b* did significantly reduce *Ctgf* and *Col1a1* expression levels. *Gapdh* served as housekeeping gene; error bars indicate standard error of the mean; * indicates significant difference at $p < 0,05$; $n = 3$ experiments per condition; data were normalized to values of control cells at 0 h. (**l,m**) There is a significant correlation between *Tgfbr1* expression level and *Ctgf* and *Col1a1* expression levels (**l**), while no such correlations were found between *Acvr1b* expression levels and *Ctgf* or *Col1a1* expression levels (**m**).

3.6. Tgfbr1 Levels Correlate with Ctgf and Col1a1 Expression Levels

To further examine the effects of myostatin and TGF-β, type I receptors *Acvr1b* and *Tgfbr1* were individually or simultaneously blocked in myoblasts using specific siRNAs. C2C12 myoblasts were treated with siRNA against *Acvr1b* or *Tgfbr1* for 24 h in growth medium and were additionally treated with siRNA for 48 h in differentiation medium. *Acvr1b* and *Tgfbr1* siRNA reduced receptor mRNA levels by >60% and >50%, respectively, without affecting expression of the other receptor (Figure 5f,g). No significant effects of siRNA treatment on *Myod* or *Myog* expression levels were observed (Figure 5h,i). In line with these results, receptor blocking during differentiation using chemical blocker Ly364947 did not affect fusion or differentiation index nor myotube thickness after 3 days of differentiation (Figure 6a–e). In addition, Ly364947 treatment did not affect *Myh3* expression levels after 48 h of differentiation. However, when C2C12 myoblasts were simultaneously treated with TGF-β and Ly364749, *Myh3* mRNA expression levels were significantly increased compared to those in TGF-β treated cells and similar to those in control cells. In line with observations in primary myoblasts, in C2C12 cells myostatin treatment had no significant effect on *Myh3* expression (Figure 6f). In addition, when the receptors were blocked with Ly364947 during proliferation for 24 h and subsequent differentiation for 2 days, *Myh3* expression was significantly increased (Figure 6g). Knockdown of *Acvr1b* did not significantly affect levels of *Ctgf* and *Col1a1* mRNA, whereas *Tgfbr1* knockdown reduced expression levels of *Ctgf* and *Col1a1* mRNA. Combined knockdown of *Acvr1b* and *Tgfbr1* did not reduce *Ctgf* or *Col1a1* mRNA levels significantly further than *Tgfbr1* knockdown (Figure 5j,k). In addition, *Tgfbr1* mRNA expression levels significantly correlated with both *Ctgf* and *Col1a1* mRNA expression levels (Figure 5l,m). Taken together, these results indicate that TGF-β signalling via *Tgfbr1* has a stronger effect on muscle fibrosis compared to myostatin.

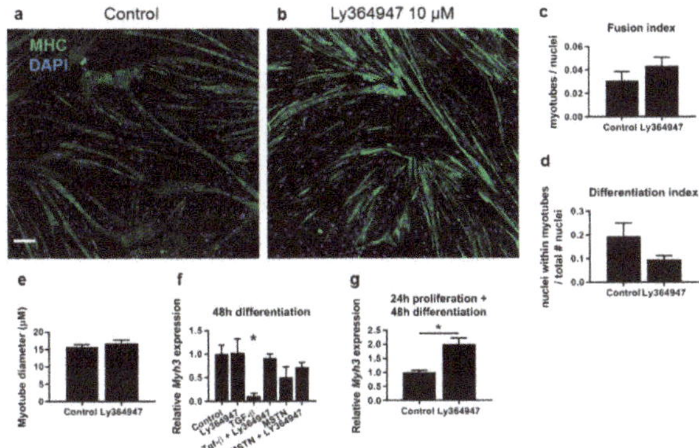

Figure 6. Effects of type I receptor blocking on myoblast differentiation is time-dependent. (**a**,**b**) C2C12 cells were induced to differentiate in control medium (**a**) or in medium supplemented with the TGF-β receptor inhibitor Ly364947 (**b**). Myotubes were stained for MHC (green) and nuclei were stained using DAPI (blue). Scale indicates 100 μm. (**c**) Fusion index, defined as number of myotubes ≥ 2 nuclei/total number of nuclei and (**d**) differentiation index defined as number of nuclei within MHC+ myotubes/total number of nuclei were not significantly affected by Ly364947 treatment. Error bars indicate standard error of the mean; * indicates significant difference at $p < 0.05$; $n = 4$ experiments per condition. (**e**) Myotube thickness was not affected by Ly36447 treatment, compared to control condition. (**f**) 48 h of Ly364947 treatment did not affect *Myh3* expression levels in differentiating C2C12 myoblasts. *Myh3* expression levels were significantly increased in cells simultaneously treated with TGF-β and Ly364947 compared to those of cells treated with TGF-β and similar to those of untreated cells. Myostatin did not significantly affect *Myh3* expression levels. (**g**) *Myh3* expression levels were significantly increased when C2C12 myoblasts were treated with Ly364947 during proliferation for 24 h and subsequent culture in differentiation medium for 48 h.

4. Discussion

The aim of this study was to assess the time-dependent effects of TGF-β signalling on gene expression in myoblasts and myotubes and compare the effects of TGF-β and myostatin signalling in myoblasts. Here we show that in vitro TGF-β treatment inhibits the expression of a subset of myogenic genes in both myoblasts and myotubes, but does not affect myotube thickness. Most importantly, our results show that TGF-β regulates the expression of fibrotic genes in both myoblasts and myotubes in a time-dependent manner. TGF-β regulates *Col1a1* mRNA expression at least in part via *Ctgf* and *Fgf-2* and, in addition, *Ctgf* and *Fgf-2* are also required to induce the expression of each other. Moreover, our results show a more prominent role for TGF-β in SMAD signalling, as well as myogenic and fibrotic gene expression in comparison to myostatin.

4.1. TGF-β Affects Myogenic Gene Expression in Both Myoblasts and Myotubes

TGF-β is known for its inhibitory effect on myoblast differentiation in vitro through inhibition of MyoD [14,32]. As expected, TGF-β inhibited myoblast differentiation and myogenic gene expression. Also, in myotubes, a reduction in *Myh3* expression was observed after 24 and 48 h of TGF-β treatment. Embryonic myosin heavy chain (eMHC), which is encoded by *Myh3*, is normally only expressed during embryonic/fetal and neonatal development, but is transiently re-expressed during muscle regeneration. The loss of eMHC in adult muscle in vivo has been shown to change MHC isoform expression, while in vitro *Myh3* knockdown may result in reduced fusion index and a reduced number of reserve cells, which suggests that loss of *Myh3* results in the early differentiation of MuSCs, depleting the MuSC pool [33]. Together, these results suggest that long-term TGF-β expression in muscle fibres after injury or in chronic disease may impede proper regeneration through repression of *Myh3* expression.

Additionally, TGF-β has been known from previous studies to interfere with MyoD function via two different mechanisms. First, TGF-β-induced SMAD3 can directly interact with MyoD. Second, TGF-β/SMAD3 interferes with the interaction between MyoD and myocyte enhancer factor 2 (MEF2), which is required for the expression of many myogenic genes [14,32]. Here, we show that TGF-β induces *Id1* expression acutely and transiently in both myoblasts and myotubes. *Id1* is known to inhibit myoblast differentiation by interfering with the formation of MyoD/E complexes, which are required for MyoD function [27]. Our data suggest that the upregulation of *Id1* mRNA may be another mechanism through which TGF-β interferes with MyoD function.

4.2. TGF-β Does Not Affect Myotube Size In Vitro

TGF-β overexpression within mouse muscle has been shown to result in the stimulation of atrogin-1 expression and atrophy in vivo [17,18]. To investigate whether this increase in atrogin-1 expression was a direct or indirect effect of TGF-β, time-dependent effects of TGF-β on E3 ligase mRNA expression were studied. In contrast to what has been shown in vivo, C2C12 myotubes did not show evidence for any effect of TGF-β on muscle atrophy. TGF-β treatment resulted in a reduction in *Atrogin-1* and *Murf-1* mRNA expression, rather than an increase. Moreover, an increase in both *Atrogin-1* and *Murf-1* expression was observed during differentiation, which suggests that the observed TGF-β-induced effects on E3 ligase mRNA expression were likely related to inhibition of differentiation. In both myoblasts and myotubes, expression levels of the ligase *Musa1* were transiently increased. Furthermore, TGF-β did not affect Akt or ERK1/2 phosphorylation nor myotube size. Together, these data indicate that in C2C12 myotubes, TGF-β does not directly contribute to atrophy. However, in vivo long term overexpression of TGF-β may lead to a reduction in muscle fibre size [17,18]. Based on our data, this observed in vivo TGF-β overexpression-induced atrophy is possibly mediated via *Musa1* rather than by elevated *Murf-1* or *Atrogin-1* expression levels. Furthermore, we show that TGF-β stimulates *Nox4* and *Id1* mRNA expression. These genes have been implied to play a role in muscle atrophy [34,35]. TGF-β has been shown to induce caspase 3 expression and DNA fragmentation in C2C12 cells [36]. As such, myonuclear apoptosis and loss of muscle stem cells induced by TGF-β may contribute to muscle atrophy as well. The role of TGF-β in the regulation of muscle fibre size requires further investigation.

4.3. TGF-β Contributes to Fibrosis by Stimulation of Fibrotic Gene Expression in Myoblasts and Myotubes

Our data show that both myoblasts and myotubes express various pro-fibrotic genes and TGF-β stimulates the expression of these genes in a time-dependent manner. This suggests that, in addition to its effect on fibroblasts, TGF-β likely also contributes to muscle fibrosis through effects on myoblasts and muscle fibres. The stimulatory effects of TGF-β on *Col1a1* mRNA expression in myotubes were relatively small compared to those in myoblasts. Nevertheless, myotubes may contribute substantially to collagen type I production. Basal expression levels of *Col1a1* in myotubes were approximately 10-fold higher than in myoblasts. Moreover, MuSCs comprise approximately 2%–5% of the myonuclei within mature muscle [37] and the number of fibroblasts is roughly 10-fold lower than the number of myonuclei [38,39]. Therefore, it is conceivable that within mature skeletal muscle, differentiating myoblasts and muscle fibres contribute substantially to the production of collagen type I.

Collagen type I is found in the endo-, peri- and epimysium surrounding muscle fibre [40,41]. Collagen fibres reinforce the ECM surrounding muscle fibres, which is essential in providing a niche for MuSCs, giving structure to the muscle and is even crucial for proper muscle function [42–44]. It is conceivable that during myogenesis and muscle regeneration, muscle fibres will secrete collagen type I to contribute to the deposition of connective tissue that provides a scaffold for the regenerating parts of the muscle fibre. However, chronic high expression of TGF-β in skeletal muscle may contribute to muscle fibrosis via the continuous elevated expression of collagen. In muscular dystrophies and aged muscle, TGF-β expression in damaged areas of the muscle may cause excessive collagen deposition. This may result in locally enhanced stiffness along the muscle fibre, which may cause strain distributions along the length of the muscle fibre. As a consequence, muscle fibres are likely to become susceptible to further injuries. In addition, excessive collagen deposition will result in enhanced stiffness of the muscle stem cell niche and likely alter MuSC mechanosensitivity, which may reduce myoblast differentiation and thus impair muscle regeneration capacity [44–47].

Besides pro-fibrotic growth factors and ECM genes, TGF-β also induced the expression of *Nox4*. *Nox4* expression is induced by TGF-β within various cell types such as endothelial cells or lung mesenchymal cells [31,48]. *Nox4* is part of an enzyme family which catalyses the reduction of oxygen into reactive oxygen species (ROS). In lung fibrosis, *Nox4*-dependent H_2O_2 generation is required for TGF-β mediated myofibroblast differentiation and ECM production [31]. Furthermore, *Nox4* is a known source for oxidative stress in many tissues and in chronic kidney disease both *Nox4* and oxidative damage markers are increased in muscle [49]. Therefore, we suggest that prolonged TGF-β expression in muscle wasting disorders may contribute to oxidative damage via *Nox4* upregulation.

4.4. TGF-β Induces Col1a1 Expression via Autocrine Ctgf and Fgf-2 Signalling

In lung fibrosis, TGF-β is known to induce collagen 1 expression via CTGF [50–52]. This, in combination with the observed expression patterns for *Ctgf*, *Fgf-2* and *Col1a1* in our myoblasts and myotubes, raised the question regarding whether in muscle cells TGF-β directly induced *Col1a1* expression or indirectly via enhancement of expression of these growth factors. *Ctgf* and *Fgf-2* were significantly knocked down using siRNA. After 48 h of TGF-β treatment, *Col1a1* mRNa expression levels were significantly reduced when *Ctgf* or *Fgf-2* was knocked down. This suggests that *Col1a1* expression is at least in part dependent on both *Ctgf* and *Fgf-2* expression in an autocrine manner. In corneal endothelial cells and human vertebral bone marrow stem cells, FGF-2 has been implied to stimulate collagen production [53,54], while in muscle FGF-2 is best known to stimulate MuSC activation and proliferation [55,56]. In this study, we show for the first time that in C2C12 muscle cells *Fgf-2* is required for TGF-β induced *Col1a1* mRNA expression.

Our results show that after 3 h of TGF-β treatment, *Ctgf* knockdown did not significantly affect *Fgf-2* expression; however, *Fgf-2* expression was significantly reduced after 48 h of TGF-β treatment in the presence of siRNA against *Ctgf*. These data suggest that TGF-β acutely induces *Fgf-2* expression independently of changes in *Ctgf* expression, though chronic expression of *Fgf-2* depends on *Ctgf* expression levels. *Ctgf* expression was shown to depend on *Fgf-2* levels both acutely and chronically. To the best of our knowledge, this interaction has not been reported before. See Figure 7 for a schematic of the proposed mechanism for TGF-β induced regulation of *Ctgf*, *Fgf-2* and *Col1a1*. We suggest that TGF-β stimulates *Col1a1* expression largely via the autocrine and paracrine signalling of *Ctgf* and *Fgf-2* and that *Ctgf* and *Fgf-2* may regulate the expression of each other via a positive feedback loop.

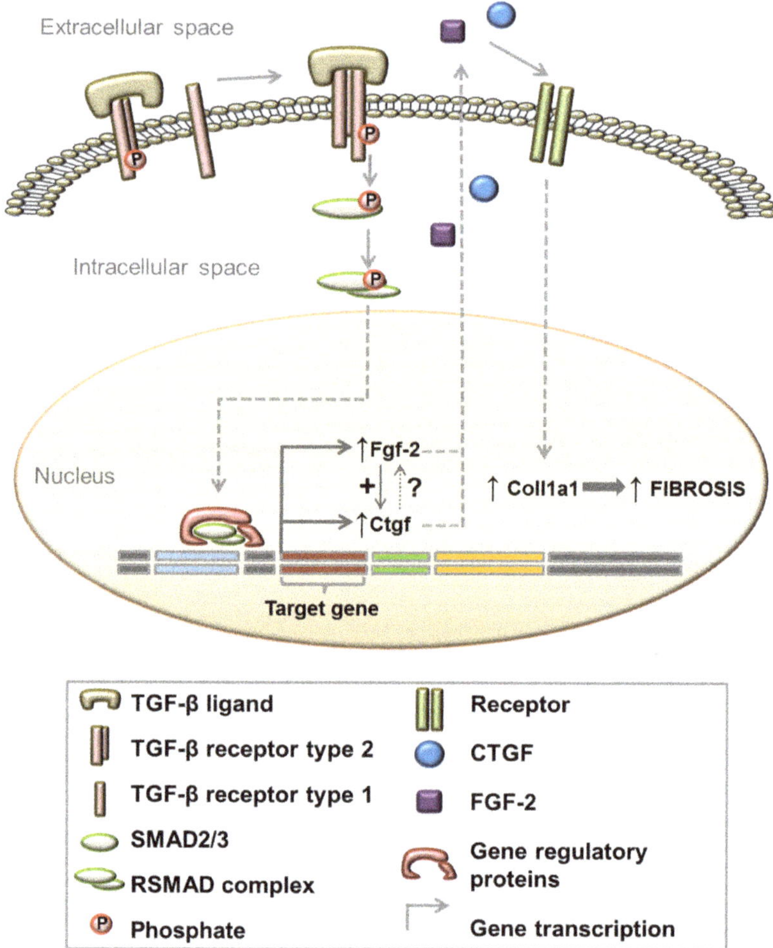

Figure 7. Schematic illustration of a proposed mechanism of how TGF-β regulates *Col1a1* mRNA expression. TGF-β binds to its receptors and activates downstream SMAD2/3 signalling. Subsequently, R-SMAD complexes translocate into the nucleus to regulate mRNA expression of growth factors such as *Ctgf* and *Fgf-2*. CTGF and FGF-2 proteins are secreted by the muscle cell and subsequently induce *Col1a1* expression via autocrine or paracrine signalling. Furthermore, expression levels of *Fgf-2* and *Ctgf* are dependent on each other.

4.5. TGF-β Has a More Pronounced Effect than Myostatin on Myoblast Differentiation and Fibrotic Gene Expression

Because of overlap in functional implications and mechanistic similarities between TGF-β and myostatin signalling in myoblasts, we compared the effects of both growth factors on myogenic and fibrotic gene expression. In order to induce downstream activation of SMAD2 signalling, a higher concentration of myostatin was required compared to TGF-β. Furthermore, in myoblasts, TGF-β had a larger effect on *Myh3* and *Ctgf* expression compared to myostatin. To further compare effects of TGF-β and myostatin signalling on myoblasts, these ligands were inhibited by using siRNA against their type I receptors. TGF-β is best known to signal via the TGF-β type I receptor TGFR-1 [24]. In epithelial cells, it has been shown that myostatin can signal via TGFR-1, as well as via ACTR-1B [23]. In mouse myoblasts, myostatin has been shown to signal mainly via ACTR-1B and not via TGFR-1, while in mouse fibroblasts myostatin signals mainly via ACTR-1B [25]. Together, these studies suggest that the knockdown of *Tgfbr1* mainly inhibits TGF-β signalling, while *Acvr1b* knockdown inhibits myostatin signalling. Here, we show in C2C12 myoblasts that *Ctgf* and *Col1a1* mRNA levels correlate with *Tgfbr1* mRNA expression levels, but not with *Acvr1b* expression levels. Moreover, no synergistic effects on the expression of pro-fibrotic genes were observed for combined receptor knockdown. Together, these data indicate that in muscle cells TGF-β has a more pronounced effect on fibrosis than myostatin and that pro-fibrotic gene expression in muscle is mainly mediated via *Tgfbr1*, and not via *Acvr1b*.

Acvr1b and *Tgfbr1* inhibition using siRNA did not affect the expression of myogenic genes. Furthermore, we showed that receptor blocking during differentiation using chemical blocker Ly364947 did not affect the differentiation or fusion index after 3 days, nor the expression of *Myh3* after 2 days. However, in cells treated with both TGF-β and Ly364947, *Myh3* expression levels were similar to those of control cells, which indicates that Ly364947 cancels out the negative effect of TGF-β on *Myh3* expression levels. In addition, when *Acvr1b* and *Tgfbr1* receptors were blocked by Ly364947 during proliferation for 24 h and subsequent differentiation for 2 days, *Myh3* expression levels were significantly increased compared to those of controls. This indicates that the negative effects of TGF-β on myoblast differentiation were cancelled by *Tgfbr1* blocking. The role of *Acvr1b* in myoblast differentiation cannot be concluded based on these results.

Under differentiation conditions, receptor blocking does not further enhance the expression of myogenic genes, which indicates that effects of TGF-β and possibly myostatin on differentiation are dose-dependent and time-dependent. The serum levels (i.e., growth factors such as TGF-β) are relatively low in the differentiation medium compared to the levels in growth medium. This suggests that low concentrations of TGF-β have a minor effect on myogenic gene expression and myoblast differentiation. Note that there is a difference between the chemical blocker Ly364947 and the siRNAs targeting *Acvr1b* and *Tgfbr1* in interference with type I receptor function. While Ly364947 blocks TGF-β signalling within one hour, as demonstrated in Figure 1, siRNAs interfere with the translation of the target mRNA, which may result in a delayed knockdown of type I receptors (Figure 5). This may explain why the presence of siRNA in growth medium did not affect myogenic gene expression. Based on our results, it seems that myoblast differentiation is less sensitive to myostatin signalling than to TGF-β signalling.

4.6. Implications in Therapeutic Treatments

Altogether, our results demonstrate that TGF-β signalling has an inhibitory effect on myoblast differentiation and contributes substantially to fibrosis. Therefore, the TGF-β pathway proves to be an interesting potential therapeutic target for treatment of muscular dystrophies. The inhibition of the TGF-β pathway may relieve and attenuate progressive muscle pathology characterized by severe fibrosis and loss of muscle mass. However, taking into account that TGF-β affects various cellular processes throughout the body, generic inhibition of the protein may have serious consequences. Our data show that TGF-β inhibits differentiation and induces fibrosis directly via its receptor in myoblasts and differentiated myotubes. This indicates that the inhibition of TGF-β exclusively within muscle tissue may be an effective approach to improve muscle regeneration in muscular dystrophy. Furthermore, our data demonstrate that TGF-β has a larger effect on differentiation and fibrosis than myostatin. Moreover, *Tgfbr1*, but not *Acvr1b* inhibition, significantly reduced *Ctgf* and *Col1a1* mRNA expression levels, while simultaneous receptor knockdown did not reduce expression levels even further. This suggests that solely blocking *Tgfbr1* and concomitant inhibition of TGF-β signalling may be sufficient to reduce fibrosis in muscular dystrophy. However, when in pathological conditions, both the inhibition of fibrosis and improved regeneration are required, thus simultaneous blocking of the *Tgfbr1* and *Acvr1b* receptor may be desirable. It has been shown that both myostatin and activins signal via *Acvr1b* and that these ligands synergistically inhibit regulation of muscle size [57,58]. Thus, simultaneous targeting of *Tgfbr1* and *Acvr1b* in vivo may still have a synergistic effect on overall muscle function improvement.

5. Conclusions

In conclusion, our data show that TGF-β inhibits myogenic gene expression in both myoblasts and myotubes, but does not affect myotube size in vitro. Most importantly, our results show that TGF-β stimulates *Col1a1* mRNA expression largely via autocrine expression of *Ctgf* and *Fgf-2*. Moreover, the effects of TGF-β on myogenic and fibrotic signalling are more pronounced than those of myostatin. Knockdown of *Tgfbr1* was sufficient to decrease *Ctgf* and *Col1a1* expression levels, while knockdown of *Acvr1b* had little effect. These results indicate that during muscle regeneration, TGF-β induces fibrosis via *Tgfbr1* by stimulating autocrine signalling of *Ctgf* and *Fgf-2*.

Supplementary Materials: The following are available online at http://www.mdpi.com/2073-4409/9/2/375/s1, Figure S1: TGF-β supplementation results in SMAD1/5 phosphorylation in C2C12 myoblasts and myotubes.

Author Contributions: Conceptualization, M.M.G.H., R.T.J. and W.M.H.H.; methodology, M.M.G.H. and W.M.H.H.; formal analysis, M.M.G.H.; investigation, M.M.G.H., R.A.G.C., C.O. and G.M.J.d.W.; resources, R.T.J. and W.M.H.H.; writing—original draft preparation, M.M.G.H.; writing—review and editing, M.M.G.H., R.A.G.C., R.T.J. and W.M.H.H.; visualization, M.M.G.H. and R.A.G.C.; supervision, R.T.J. and W.M.H.H.; project administration, W.M.H.H.; funding acquisition, W.M.H.H. and R.T.J. All authors have read and agreed to the published version of the manuscript.

Funding: This research was funded by the Prinses Beatrix Spierfonds, grant number W.OR14-17.

Acknowledgments: We thank students M. Bulut and K. Doetjes for their contribution to this research.

Conflicts of Interest: The authors declare no conflict of interest. The funders had no role in the design of the study; in the collection, analyses, or interpretation of data; in the writing of the manuscript, or in the decision to publish the results.

References

1. Ryall, J.G.; Schertzer, J.D.; Lynch, G.S. Cellular and molecular mechanisms underlying age-related skeletal muscle wasting and weakness. *Biogerontology* **2008**, *9*, 213–228. [CrossRef] [PubMed]

2. Lima, J.; Simoes, E.; de Castro, G.; Morais, M.; de Matos-Neto, E.M.; Alves, M.J.; Pinto, N.I.; Figueredo, R.G.; Zorn, T.M.T.; Felipe-Silva, A.S.; et al. Tumour-derived transforming growth factor-beta signalling contributes to fibrosis in patients with cancer cachexia. *J. Cachexia Sarcopenia Muscle* **2019**, *10*, 1045–1059. [CrossRef]

3. Bernasconi, P.; Torchiana, E.; Confalonieri, P.; Brugnoni, R.; Barresi, R.; Mora, M.; Cornelio, F.; Morandi, L.; Mantegazza, R. Expression of transforming growth factor-beta 1 in dystrophic patient muscles correlates with fibrosis. Pathogenetic role of a fibrogenic cytokine. *J. Clin. Investig.* **1995**, *96*, 1137–1144. [CrossRef]

4. Chen, Y.W.; Nagaraju, K.; Bakay, M.; McIntyre, O.; Rawat, R.; Shi, R.; Hoffman, E.P. Early onset of inflammation and later involvement of tgfbeta in duchenne muscular dystrophy. *Neurology* **2005**, *65*, 826–834. [CrossRef]

5. Shi, Y.; Massague, J. Mechanisms of tgf-beta signaling from cell membrane to the nucleus. *Cell* **2003**, *113*, 685–700. [CrossRef]

6. Robertson, T.A.; Maley, M.A.; Grounds, M.D.; Papadimitriou, J.M. The role of macrophages in skeletal muscle regeneration with particular reference to chemotaxis. *Exp. Cell Res.* **1993**, *207*, 321–331. [CrossRef]

7. Grotendorst, G.R.; Smale, G.; Pencev, D. Production of transforming growth factor beta by human peripheral blood monocytes and neutrophils. *J. Cell. Physiol.* **1989**, *140*, 396–402. [CrossRef]

8. Lawrence, D.A.; Pircher, R.; Kryceve-Martinerie, C.; Jullien, P. Normal embryo fibroblasts release transforming growth factors in a latent form. *J. Cell. Physiol.* **1984**, *121*, 184–188. [CrossRef]

9. Shur, I.; Lokiec, F.; Bleiberg, I.; Benayahu, D. Differential gene expression of cultured human osteoblasts. *J. Cell. Biochem.* **2001**, *83*, 547–553. [CrossRef]

10. Zimowska, M.; Duchesnay, A.; Dragun, P.; Oberbek, A.; Moraczewski, J.; Martelly, I. Immunoneutralization of tgfbeta1 improves skeletal muscle regeneration: Effects on myoblast differentiation and glycosaminoglycan content. *Int. J. Cell Biol.* **2009**, *2009*, 659372. [CrossRef]

11. Pasteuning-Vuhman, S.; Putker, K.; Tanganyika-de Winter, C.L.; Boertje-van der Meulen, J.W.; van Vliet, L.; Overzier, M.; Plomp, J.J.; Aartsma-Rus, A.; van Putten, M. Natural disease history of mouse models for limb girdle muscular dystrophy types 2d and 2f. *PLoS ONE* **2017**, *12*, e0182704. [CrossRef]

12. Gonzalez, D.; Contreras, O.; Rebolledo, D.L.; Espinoza, J.P.; van Zundert, B.; Brandan, E. ALS skeletal muscle shows enhanced tgf-beta signaling, fibrosis and induction of fibro/adipogenic progenitor markers. *PLoS ONE* **2017**, *12*, e0177649. [CrossRef]

13. Carlson, M.E.; Hsu, M.; Conboy, I.M. Imbalance between pSmad3 and Notch induces CDK inhibitors in old muscle stem cells. *Nature* **2008**, *454*, 528–532. [CrossRef]

14. Liu, D.; Black, B.L.; Derynck, R. Tgf-beta inhibits muscle differentiation through functional repression of myogenic transcription factors by smad3. *Genes Dev.* **2001**, *15*, 2950–2966. [CrossRef]

15. Olson, E.N.; Sternberg, E.; Hu, J.S.; Spizz, G.; Wilcox, C. Regulation of myogenic differentiation by type beta transforming growth factor. *J. Cell Biol.* **1986**, *103*, 1799–1805. [CrossRef]

16. Massague, J.; Cheifetz, S.; Endo, T.; Nadal-Ginard, B. Type beta transforming growth factor is an inhibitor of myogenic differentiation. *Proc. Natl. Acad. Sci. USA* **1986**, *83*, 8206–8210. [CrossRef]

17. Narola, J.; Pandey, S.N.; Glick, A.; Chen, Y.W. Conditional expression of tgf-beta1 in skeletal muscles causes endomysial fibrosis and myofibers atrophy. *PLoS ONE* **2013**, *8*, e79356. [CrossRef]

18. Mendias, C.L.; Gumucio, J.P.; Davis, M.E.; Bromley, C.W.; Davis, C.S.; Brooks, S.V. Transforming growth factor-beta induces skeletal muscle atrophy and fibrosis through the induction of atrogin-1 and scleraxis. *Muscle Nerve* **2012**, *45*, 55–59. [CrossRef]

19. Andreetta, F.; Bernasconi, P.; Baggi, F.; Ferro, P.; Oliva, L.; Arnoldi, E.; Cornelio, F.; Mantegazza, R.; Confalonieri, P. Immunomodulation of tgf-beta 1 in mdx mouse inhibits connective tissue proliferation in diaphragm but increases inflammatory response: Implications for antifibrotic therapy. *J. Neuroimmunol.* **2006**, *175*, 77–86. [CrossRef]

20. Li, Y.; Foster, W.; Deasy, B.M.; Chan, Y.; Prisk, V.; Tang, Y.; Cummins, J.; Huard, J. Transforming growth factor-beta1 induces the differentiation of myogenic cells into fibrotic cells in injured skeletal muscle: A key event in muscle fibrogenesis. *Am. J. Pathol.* **2004**, *164*, 1007–1019. [CrossRef]

21. Langley, B.; Thomas, M.; Bishop, A.; Sharma, M.; Gilmour, S.; Kambadur, R. Myostatin inhibits myoblast differentiation by down-regulating myod expression. *J. Biol. Chem.* **2002**, *277*, 49831–49840. [CrossRef]

22. Li, Z.B.; Kollias, H.D.; Wagner, K.R. Myostatin directly regulates skeletal muscle fibrosis. *J. Biol. Chem.* **2008**, *283*, 19371–19378. [CrossRef]

23. Rebbapragada, A.; Benchabane, H.; Wrana, J.L.; Celeste, A.J.; Attisano, L. Myostatin signals through a transforming growth factor beta-like signaling pathway to block adipogenesis. *Mol. Cell. Biol.* **2003**, *23*, 7230–7242. [CrossRef]

24. ten Dijke, P.; Yamashita, H.; Ichijo, H.; Franzen, P.; Laiho, M.; Miyazono, K.; Heldin, C.H. Characterization of type I receptors for transforming growth factor-beta and activin. *Science* **1994**, *264*, 101–104. [CrossRef]

25. Kemaladewi, D.U.; de Gorter, D.J.; Aartsma-Rus, A.; van Ommen, G.J.; ten Dijke, P.; t Hoen, P.A.; Hoogaars, W.M. Cell-type specific regulation of myostatin signaling. *FASEB J.* **2012**, *26*, 1462–1472. [CrossRef]

26. Schneider, C.A.; Rasband, W.S.; Eliceiri, K.W. NIH image to imagej: 25 years of image analysis. *Nat. Methods* **2012**, *9*, 671–675. [CrossRef]

27. Jen, Y.; Weintraub, H.; Benezra, R. Overexpression of Id protein inhibits the muscle differentiation program: In vivo association of Id with E2A proteins. *Genes Dev.* **1992**, *6*, 1466–1479. [CrossRef]

28. Ramachandran, A.; Vizan, P.; Das, D.; Chakravarty, P.; Vogt, J.; Rogers, K.W.; Muller, P.; Hinck, A.P.; Sapkota, G.P.; Hill, C.S. Tgf-beta uses a novel mode of receptor activation to phosphorylate smad1/5 and induce epithelial-to-mesenchymal transition. *Elife* **2018**, *7*, e31756. [CrossRef]

29. McFarlane, C.; Plummer, E.; Thomas, M.; Hennebry, A.; Ashby, M.; Ling, N.; Smith, H.; Sharma, M.; Kambadur, R. Myostatin induces cachexia by activating the ubiquitin proteolytic system through an NF-kappaB-independent, FoxO1-dependent mechanism. *J. Cell. Physiol.* **2006**, *209*, 501–514. [CrossRef]

30. Glass, D.J. Skeletal muscle hypertrophy and atrophy signaling pathways. *Int. J. Biochem. Cell Biol.* **2005**, *37*, 1974–1984. [CrossRef]

31. Hecker, L.; Vittal, R.; Jones, T.; Jagirdar, R.; Luckhardt, T.R.; Horowitz, J.C.; Pennathur, S.; Martinez, F.J.; Thannickal, V.J. NAPDH oxidase-4 mediates myofibroblast activation and fibrogenic responses to lung injury. *Nat. Med.* **2009**, *15*, 1077–1081. [CrossRef] [PubMed]

32. Liu, D.; Kang, J.S.; Derynck, R. TGF-beta-activated Smad3 represses MEF2-dependent transcription in myogenic differentiation. *EMBO J.* **2004**, *23*, 1557–1566. [CrossRef] [PubMed]

33. Sharma, A.; Agarwal, M.; Kumar, A.; Kumar, P.; Saini, M.; Kardon, G.; Mathew, S.J. Myosin heavy chain-embryonic is a crucial regulator of skeletal muscle development and differentiation. *bioRxiv* **2018**.

34. Kadoguchi, T.; Shimada, K.; Koide, H.; Miyazaki, T.; Shiozawa, T.; Takahashi, S.; Aikawa, T.; Ouchi, S.; Kitamura, K.; Sugita, Y.; et al. Possible role of NAPDH oxidase 4 in angiotensin II-induced muscle wasting in mice. *Front. Physiol.* **2018**, *9*, 340. [CrossRef]

35. Gundersen, K.; Merlie, J.P. Id-1 as a possible transcriptional mediator of muscle disuse atrophy. *Proc. Natl. Acad. Sci. USA* **1994**, *91*, 3647–3651. [CrossRef]

36. Cencetti, F.; Bernacchioni, C.; Tonelli, F.; Roberts, E.; Donati, C.; Bruni, P. Tgfbeta1 evokes myoblast apoptotic response via a novel signaling pathway involving S1P4 transactivation upstream of Rho-kinase-2 activation. *FASEB J.* **2013**, *27*, 4532–4546. [CrossRef]

37. Zammit, P.S.; Heslop, L.; Hudon, V.; Rosenblatt, J.D.; Tajbakhsh, S.; Buckingham, M.E.; Beauchamp, J.R.; Partridge, T.A. Kinetics of myoblast proliferation show that resident satellite cells are competent to fully regenerate skeletal muscle fibers. *Exp. Cell Res.* **2002**, *281*, 39–49. [CrossRef]

38. Mackey, A.L.; Magnan, M.; Chazaud, B.; Kjaer, M. Human skeletal muscle fibroblasts stimulate in vitro myogenesis and in vivo muscle regeneration. *J. Physiol.* **2017**, *595*, 5115–5127. [CrossRef]

39. Frese, S.; Ruebner, M.; Suhr, F.; Konou, T.M.; Tappe, K.A.; Toigo, M.; Jung, H.H.; Henke, C.; Steigleder, R.; Strissel, P.L.; et al. Long-term endurance exercise in humans stimulates cell fusion of myoblasts along with fusogenic endogenous retroviral genes in vivo. *PLoS ONE* **2015**, *10*, e0132099. [CrossRef]

40. Listrat, A.; Picard, B.; Geay, Y. Age-related changes and location of type I, III, IV, V and VI collagens during development of four foetal skeletal muscles of double-muscled and normal bovine animals. *Tissue Cell* **1999**, *31*, 17–27. [CrossRef]

41. Light, N.; Champion, A.E. Characterization of muscle epimysium, perimysium and endomysium collagens. *Biochem. J.* **1984**, *219*, 1017–1026. [CrossRef] [PubMed]

42. Purslow, P.P. The structure and functional significance of variations in the connective tissue within muscle. *Comp. Biochem. Physiol. A Mol. Integr. Physiol.* **2002**, *133*, 947–966. [CrossRef]

43. Huijing, P.A.; Jaspers, R.T. Adaptation of muscle size and myofascial force transmission: A review and some new experimental results. *Scand J. Med. Sci. Sports* **2005**, *15*, 349–380. [CrossRef] [PubMed]

44. Thomas, K.; Engler, A.J.; Meyer, G.A. Extracellular matrix regulation in the muscle satellite cell niche. *Connect. Tissue Res.* **2015**, *56*, 1–8. [CrossRef] [PubMed]

45. Gillies, A.R.; Lieber, R.L. Structure and function of the skeletal muscle extracellular matrix. *Muscle Nerve* **2011**, *44*, 318–331. [CrossRef]

46. Boers, H.E.; Haroon, M.; Le Grand, F.; Bakker, A.D.; Klein-Nulend, J.; Jaspers, R.T. Mechanosensitivity of aged muscle stem cells. *J. Orthop. Res.* **2018**, *36*, 632–641.

47. Romanazzo, S.; Forte, G.; Ebara, M.; Uto, K.; Pagliari, S.; Aoyagi, T.; Traversa, E.; Taniguchi, A. Substrate stiffness affects skeletal myoblast differentiation in vitro. *Sci. Technol. Adv. Mater.* **2012**, *13*, 064211. [CrossRef]

48. Yan, F.; Wang, Y.; Wu, X.; Peshavariya, H.M.; Dusting, G.J.; Zhang, M.; Jiang, F. Nox4 and redox signaling mediate tgf-beta-induced endothelial cell apoptosis and phenotypic switch. *Cell Death Dis.* **2014**, *5*, e1010. [CrossRef]

49. Avin, K.G.; Chen, N.X.; Organ, J.M.; Zarse, C.; O'Neill, K.; Conway, R.G.; Konrad, R.J.; Bacallao, R.L.; Allen, M.R.; Moe, S.M. Skeletal muscle regeneration and oxidative stress are altered in chronic kidney disease. *PLoS ONE* **2016**, *11*, e0159411. [CrossRef]

50. Lin, C.H.; Yu, M.C.; Tung, W.H.; Chen, T.T.; Yu, C.C.; Weng, C.M.; Tsai, Y.J.; Bai, K.J.; Hong, C.Y.; Chien, M.H.; et al. Connective tissue growth factor induces collagen I expression in human lung fibroblasts through the Rac1/MLK3/JNK/AP-1 pathway. *Biochim. Biophys. Acta* **2013**, *1833*, 2823–2833. [CrossRef]

51. Yang, Z.; Sun, Z.; Liu, H.; Ren, Y.; Shao, D.; Zhang, W.; Lin, J.; Wolfram, J.; Wang, F.; Nie, S. Connective tissue growth factor stimulates the proliferation, migration and differentiation of lung fibroblasts during paraquat-induced pulmonary fibrosis. *Mol. Med. Rep.* **2015**, *12*, 1091–1097. [CrossRef]

52. Ponticos, M.; Holmes, A.M.; Shi-wen, X.; Leoni, P.; Khan, K.; Rajkumar, V.S.; Hoyles, R.K.; Bou-Gharios, G.; Black, C.M.; Denton, C.P.; et al. Pivotal role of connective tissue growth factor in lung fibrosis: MAPK-dependent transcriptional activation of type I collagen. *Arthritis Rheum.* **2009**, *60*, 2142–2155. [CrossRef]

53. Ko, M.K.; Kay, E.P. Regulatory role of FGF-2 on type I collagen expression during endothelial mesenchymal transformation. *Invest. Ophthalmol. Vis. Sci.* **2005**, *46*, 4495–4503. [CrossRef]

54. Park, D.S.; Park, J.C.; Lee, J.S.; Kim, T.W.; Kim, K.J.; Jung, B.J.; Shim, E.K.; Choi, E.Y.; Park, S.Y.; Cho, K.S.; et al. Effect of FGF-2 on collagen tissue regeneration by human vertebral bone marrow stem cells. *Stem Cells Dev.* **2015**, *24*, 228–243. [CrossRef]

55. Yablonka-Reuveni, Z.; Rivera, A.J. Proliferative dynamics and the role of FGF2 during myogenesis of rat satellite cells on isolated fibers. *Basic Appl. Myol.* **1997**, *7*, 189–202.

56. Liu, Y.; Schneider, M.F. FGF2 activates TRPC and Ca(2+) signaling leading to satellite cell activation. *Front. Physiol.* **2014**, *5*, 38. [CrossRef]

57. Chen, J.L.; Walton, K.L.; Hagg, A.; Colgan, T.D.; Johnson, K.; Qian, H.; Gregorevic, P.; Harrison, C.A. Specific targeting of tgf-beta family ligands demonstrates distinct roles in the regulation of muscle mass in health and disease. *Proc. Natl. Acad. Sci. USA* **2017**, *114*, E5266–E5275.

58. Watt, K.I.; Jaspers, R.T.; Atherton, P.; Smith, K.; Rennie, M.J.; Ratkevicius, A.; Wackerhage, H. Sb431542 treatment promotes the hypertrophy of skeletal muscle fibers but decreases specific force. *Muscle Nerve* **2010**, *41*, 624–629. [CrossRef]

 *cells*

Article

The Transcription Factor Nfix Requires RhoA-ROCK1 Dependent Phagocytosis to Mediate Macrophage Skewing during Skeletal Muscle Regeneration

Marielle Saclier, Michela Lapi, Chiara Bonfanti, Giuliana Rossi [†], Stefania Antonini and Graziella Messina *

Department of Biosciences, University of Milan, via Celoria 26, 20133 Milan, Italy;
marielle.saclier@unimi.it (M.S.); michela.lapi@unimi.it (M.L.); chiara.bonfanti@unimi.it (C.B.);
giuliana.rossi@epfl.ch (G.R.); stefania.antonini@unimi.it (S.A.)
* Correspondence: graziella.messina@unimi.it; Tel.: +3902 503 14800
† Current affiliation: Laboratory of Stem Cell Bioengineering, Institute of Bioengineering, School of Life Sciences
and School of Engineering, École Polytechnique Fédérale de Lausanne (EPFL),
1015 Lausanne, Vaud, Switzerland.

Received: 19 February 2020; Accepted: 12 March 2020; Published: 13 March 2020

Abstract: Macrophages (MPs) are immune cells which are crucial for tissue repair. In skeletal muscle regeneration, pro-inflammatory cells first infiltrate to promote myogenic cell proliferation, then they switch into an anti-inflammatory phenotype to sustain myogenic cells differentiation and myofiber formation. This phenotypical switch is induced by dead cell phagocytosis. We previously demonstrated that the transcription factor Nfix, a member of the nuclear factor I (Nfi) family, plays a pivotal role during muscle development, regeneration and in the progression of muscular dystrophies. Here, we show that Nfix is mainly expressed by anti-inflammatory macrophages. Upon acute injury, mice deleted for Nfix in myeloid line displayed a significant defect in the process of muscle regeneration. Indeed, Nfix is involved in the macrophage phenotypical switch and macrophages lacking Nfix failed to adopt an anti-inflammatory phenotype and interact with myogenic cells. Moreover, we demonstrated that phagocytosis induced by the inhibition of the RhoA-ROCK1 pathway leads to Nfix expression and, consequently, to acquisition of the anti-inflammatory phenotype. Our study identified Nfix as a link between RhoA-ROCK1-dependent phagocytosis and the MP phenotypical switch, thus establishing a new role for Nfix in macrophage biology for the resolution of inflammation and tissue repair.

Keywords: macrophages; Nfix; skeletal muscle; phagocytosis; RhoA-ROCK1

---

1. Introduction

During their lifetime, tissues encounter physiological and non-physiological damages and an effective regeneration is necessary to make these tissues able to continuously sustain their biological functions. Macrophage-mediated inflammation is a fundamental step for tissue recovery. Macrophages (MPs) are immune cells required for tissue regeneration, as their depletion prevents regeneration of different tissues/organs, such as liver [1], spinal cord [2] and skeletal muscle [3]. In several regenerative processes, two populations of MPs have been described. The first population reaching the damaged tissue is the pro-inflammatory population, also called M1 MPs. Pro-inflammatory MPs secrete pro-inflammatory molecules, being the main actors of dead cell clearance. The second population is composed of the anti-inflammatory MPs, named M2 MPs, that come from pro-inflammatory MPs and are involved in the resolution of the inflammation, wound healing and tissue regeneration or repair [4–7]. Several studies have shown that an impaired or a precocious phenotypical switch from M1 to M2 MPs results in defective tissue regeneration [1,8,9]. Interestingly, it has been observed that phagocytosis is at the basis of the pro to anti-inflammatory phenotypical switch in MPs [3,8,10–13]. Although the process of the induction of phagocytosis is well-known ("find-me", "eat-me" and "don't eat-me" signals), the molecular and transcriptional pathways between phagocytosis and the phenotypical switch are still unexplored [10,11,14,15].

Interestingly, the interplay between MPs and tissue regeneration has been widely documented in skeletal muscle [7,16–20]. In vertebrates, muscle progenitors originate from pre-somitic and cranial mesoderm. In pre-natal period, two myogenic waves are necessary for muscle establishment: the first forms the basic muscle pattern and is called primary or "embryonic" myogenesis, while the second or "fetal" myogenesis is characterized by muscle maturation and growth [21]. Adult skeletal muscle is able to regenerate thanks to resident stem cells called satellite cells (SCs), located under the basal lamina of myofibers [22]. Upon injury, SCs exit from quiescence, proliferate, differentiate in myoblasts and fuse to reform myofibers [23]. Nuclear factor I X (Nfix) is a transcription factor belonging to the highly conserved DNA-binding nuclear factor one family (Nfi) together with Nfia, Nfib and Nfic [24]. Nfix has a key role in prenatal myogenesis by driving the transcriptional switch from embryonic to fetal myogenesis [25,26]. Nfix is also required for adult myogenesis upon injury, since its absence leads to defect of SC differentiation [27]. Finally, we recently demonstrated that the deletion of Nfix in two mouse models of muscular dystrophy induces a significant morphological and functional amelioration of the pathology by slowing-down muscle regeneration and promoting a switch towards a more oxidative musculature [28].

During muscle regeneration, myogenic cells and MPs closely interact [7]. Soon after injury, activated SCs attract blood monocytes that infiltrate damaged muscle and differentiate in pro-inflammatory MPs that stimulate the proliferation of myoblasts. Then, by removing dead cells, MPs switch to an anti-inflammatory phenotype that sustains myogenic differentiation [3,29]. While MPs are required for muscle regeneration, preventing MPs infiltration in dystrophic disease decreases muscle damage [30]. Thus, depending on a context of acute or chronic injury, MPs adopt a complete opposite function toward muscle cells and environment [31,32].

In this study, we address the role of Nfix in MPs during skeletal muscle regeneration, by using a mouse model in which Nfix is deleted specifically in MPs. We report that mice lacking Nfix in myeloid lineages exhibit a delay of muscle regeneration upon acute injury. We demonstrated that the RhoA-ROCK1-dependent phagocytosis induces Nfix, whose expression is necessary for the acquisition of anti-inflammatory phenotype and thus pro-regenerative properties through myogenic cells. Indeed, during the process of muscle regeneration, in the absence of Nfix, MPs are able to phagocyte, but failed to adopt an anti-inflammatory phenotype necessary for the resolution of inflammation and muscle regeneration.

2. Materials and Methods

2.1. Animal Models and In Vivo Experimentations

WT, Nfix$^{fl/fl}$ and LysM^{CRE}:Nfix$^{fl/fl}$ mice were used in this study. LysM^{CRE}:Nfix$^{fl/fl}$ mice were generated, crossing Nfix$^{fl/fl}$ mice obtained from Prof. Richard M. Gronostajski [33] and LysM^{CRE} mice obtained from Dr. Rémi Mounier [8]. All LysM^{CRE}:Nfix$^{fl/fl}$ mice analyzed were heterozygous for the LysM^{CRE}. Muscle regeneration was realized by the injection of 20 uL of 100 uM cardiotoxin (CTX, Latoxan, L8102) in the *Tibialis anterior* (TA) of 2-month-old mice. For the in vivo analysis of satellite cells and myoblasts proliferation, EdU (5-ethynyl-2'-deoxyuridine) was injected in Nfix$^{fl/fl}$ and LysM^{CRE}:Nfix$^{fl/fl}$ mice in intraperitoneal, 12 h before the sacrifice of the mice (100 µL of 6mg/mL EdU solution for 20 g of mouse weight) (Click-iT EdU Imaging Kits Alexa Fluor 594, Thermo Fisher A10044, Paisley, UK). Mice were kept in pathogen-free conditions and all procedures conformed to Italian law (D. Lgs n° 2014/26, implementation of the 2010/63/UE) and approved by the University of Milan Animal Welfare Body and by the Italian Minister of Health.

2.2. Isolation of MPs from Skeletal Muscle

Fascia of the TA muscles was removed. Muscles were dissociated and digested in RPMI medium containing 0.2% of collagenase B (Roche Diagnostics GmbH 11088815001) at 37 °C for 1 h and passed through a 70 μm and a 30 μm cell strainer. $CD45^+$ cells were isolated using magnetic beads (Miltenyi Biotec 130-052-301) and incubated with FcR blocking reagent (Miltenyi Biotec 130-059-901) for 20 min at 4 °C in PBS 2% FBS. Cells were then stained with Ly6C-PE (eBioscience 12-5932) and CD64-APC (BD Pharmingen 558539) antibodies for 30 min at 4 °C. MPs were analyzed or sorted using a FACS Aria III cell sorter (BD Biosciences) (gating strategy is shown Figure S1). In some experiments, $Ly6C^+$ and $Ly6C^-$ MPs were cytospined on starfrost (Knitterglaser, Bielefeld, Germany) slides and immunostained.

2.3. Histology and Immunofluorescence Analyses

The fascia of TA muscles was removed and the muscles were frozen in liquid nitrogen-cooled isopentane (VWR) and placed at −80 °C until cut. Then, 8 μm-thick cryosections were stained for hematoxylin-eosin (H&E) and immunofluorescence. H&E (Sigma-Aldrich, Saint-Louis, MO 63103, USA) staining was processed according to standard protocols. For immunofluorescence analysis, sections or cells were fixed for 15 min with 4% paraformaldehyde (except for F4/80 and eMyHC staining). Then, samples were permeabilized with 0.5% Triton X-100 (Sigma-Aldrich) in PBS for 10 min and blocked with 4% BSA (Sigma-Aldrich) in PBS at RT for 1 h. Primary antibodies were incubated O/N at 4 °C in PBS. After three washes of 5 min with PBS, samples were incubated with secondary antibodies (1:500, Jackson Laboratory. Fluorochromes used: 488, 594, 546 and 647) and Hoechst (1:500, Sigma-Aldrich) in PBS for 45 min at RT, then washed four times for 5 min with PBS and mounted with Fluorescence Mounting Medium (Dako). For Nfix-F4/80 double immunolabeling, cryosections were labelled with antibodies against F4/80 (1:400, Novus Biologicals NB300-605) overnight at 4 °C and Nfix labelling using (1:200, Novus Biologicals NBP2-15039) the antibody was performed for 2 h at 37 °C. For EdU-Pax7-laminin immunolabelling, after fixation and permeabilization of muscle sections, we followed the manufacturer's instructions of the Click-iT EdU Imaging Kits Alexa Fluor 594 (Thermo Fisher A10044) to reveal the DNA integrated EdU. Then, for Pax7 immunostaining, antigen retrieval was performed by incubating muscle sections in boiling 10 mM citrate buffer pH6 for 20 min. Muscle sections were then incubated O/N with Pax7 (1:2, Hybridoma, DSHB, Iowa City, IA 52242, USA) and laminin (1:200, Sigma L9393). The other antibodies used were eMyHC (1:2, Hybridome), MyoD (1:50, Santacruz Biotechnology sc-377460), TNFα (1:50, Abcam ab34839), CCL3 (1:500, Abcam ab32609, Cambridge, UK), iNOS (1:25, Novus Biologicals NB300-605, Centennial, CO 80112, USA), CD163 (1:50, Santacruz Biotechnology sc-33560), CD206 (1:50, Bio-Rad MCA2235GA), TGFβ (1:100, Abcam ab64715), Arginase I (1:100, Santacruz Biotechnology, Cambridge, UK).

2.4. Bone Marrow Derived MPs (BMDM) Culture

Total mouse bone marrow was obtained by flushing femur and tibiae with DMEM. Cells were cultured in DMEM containing 20% Fetal Bovine Serum (FBS) and 30% of L929 cell line-derived conditioned medium (enriched in CSF-1) for 6 to 7 days. MPs were polarized using 50 ng/mL IFNγ (for M1 polarization) (Peprotech #315-05), 10 ng/mL IL10 (for M2c polarization) (Peprotech #210-10), in DMEM (10% FBS) for 3 days. After washing three times, DMEM serum-free medium was added for 24 h, and supernatants were recovered and centrifuged to obtain macrophage-conditioned medium. For some experiments, cells were directly used for various analyses. In some experiments, DMSO or 10 μM of ROCK inhibitor Y27632 (Santacruz sc-3536) was added on MPs.

2.5. Myogenic Progenitor Cells (mpc) Culture

Murine WT myoblast progenitor cells (mpcs) were obtained from TA muscle and cultured in DMEM/F12 (Gibco, Paisley, UK), containing 20% FBS and 2.5 ng/mL of human FGF-basic (Peptrotech, 100-18B). For the proliferation assay, mpcs were seeded at 10 000 cell/cm^2 on Matrigel

(1/10) and incubated for 1 day with macrophage-conditioned medium + 2.5% FBS. Then, cells were incubated with the anti-Ki-67 antibody (1/50, BD Biosciences 550609). For differentiation assay, mpcs were seeded at 30 000 cell/cm^2 on Matrigel (diluted 1/10 in DMEM/F12) and incubated for 3 days with macrophage-conditioned medium containing 2% horse serum. Then, cells were incubated with a pan-myosin antibody (1:2, Hybridoma).

2.6. Phagocytosis Assay

Mpcs were labelled using the CellVue Claret Far Red kit (Sigma-Alrich MinClaret) by following the manufacturer's instructions (Sigma-Aldrich) and treated with staurosporin at 5 µM for 4 h, in order to induce apoptosis. M1 and M2c polarized MPs were incubated with apoptotic mpcs at a 1:3 ratio for 30 min at 4 °C or 6 h or 16 h at 37 °C. After three PBS washings, MPs were detached using trypsin and a cell scraper and cells were labelled with a CD64-APC (BD Pharmingen 558539) and analyzed by flow cytometry using a FACS Aria III cell sorter (BD Biosciences). The double-positive cells (CD64$^+$/Far Red$^+$ cells) were phagocytic MPs, whereas the CD64$^+$/Far Red$^-$ cells were nonphagocytic MPs. To exclude MPs that have bound, but not ingested, apoptotic cells, we subtracted the percentage of double-positive cells observed at 4 °C from the value observed at 37 °C. In some experiments, MPs were treated with 1 µg/mL of cytochalasin D (Sigma-Aldrich C8273), 45 min before adding apoptotic mpcs, and with the added mpcs.

2.7. Lentiviral Transduction

BMDM from WT mice were transduced with a lentivirus carrying a scrambled sequence or a shNfix [27]. Transduction was performed in suspension (in DMEM 20% FBS), at a MOI of 10 and in the presence of Polybrene (8 µg/mL, Sigma-Aldrich). After O/N incubation, the medium was changed and cells were treated with puromycin (2 µg/mL, Sigma-Aldrich).

2.8. RNA Extraction and qRT-PCR

RNA was isolated from the sorted apoptotic mpcs, non-phagocyted and phagocyted MPs, by using TRIzol Reagent (Invitrogen 15596026, Bleiswijk, Netherlands), according to the manufacturer's instructions. RNA was quantified using a NanoPhotoneter (Implen). For retro-transcription, 500 ng of RNA was used with the iScript Reverse Transcription Supermix for RT-quantitative qPCR (Bio-Rad 1708840). For qRT PCR, cDNA was diluted 1:10, and 5 µL of the diluted cDNA was loaded in a total volume of 20 µL (SYBR Green Supermix (Bio-Rad 172-5124) and run on the Bio-RAD CFX Connect Real-Time System. The relative quantification of gene expression was determined by the comparative CT method, and normalized to Cyclophiline A. Primers used were: Nfix for CTGGCTTACTTTGTCCACACTC; Nfix rev CCAGCTCTGTCACATTCCAGAC; Myogenin for CTGGGGACCCCTGAGCATTG; Myogenin rev ATCGCGCTCCTCCTGGTTGA; Cyclo A for GTGACTTTACACGCCATAATG; Cyclo A rev ACAAGATGCCAGGACCTGTAT.

2.9. Protein Extraction and Western Blot

Protein extracts were obtained from cultured MPs lysed using RIPA buffer (10 mM Tris-HCl pH 8.0, 1 mM EDTA, 1% Triton-X, 0.1% sodium deoxycholate, 0.1% sodium dodecylsulphate (SDS), 150 mM NaCl, in deionised water), plus protease and phosphatase inhibitors for 30 min on ice. Then, samples were centrifuged at 11.000× g for 10 min at 4 °C, and the supernatants collected for protein quantification (DC Protein Assays Bio-Rad 5000111). 40 µg protein of each sample were denatured at 95 °C for 5 min using SDS PAGE sample-loading buffer (100 mM Tris pH 6.8, 4% SDS, 0.2% bromophenol blue, 20% glycerol, 10 mM dithiothreitol) and loaded into 8% SDS acrylamide gels. After electrophoresis, the protein was blotted into nitrocellulose membranes (Protran nitrocellulose transfer membrane; Whatman) for 2 h at 70 V at 4 °C. Membranes were then blocked for 1 h with 5% milk in Tris-buffered saline, plus 0.02% Tween20 (Sigma-Aldrich). Membranes were incubated with the primary antibodies O/N at 4 °C, using the following antibodies: rabbit anti-Nfix (1:1000,

Novus Biologicals NBP2-15039), mouse anti-vinculin (1:2500, Sigma-Aldrich V9131), rabbit anti-MYPT1 phosphorylated in Thr696 (1:500, SantaCruz Biotechnology sc-17556-R), and rabbit anti-Tot MYPT1 (1:500; SantaCruz Biotechnology, H-130). After incubation with the primary antibodies, the membranes were washed 3 times for 5 min and incubated with the secondary antibodies (1:10,000, IgG-HRP, Bio-Rad) for 45 min at RT, and then washed again 5 times for 5 min. Bands were revealed using ECL detection reagent (ThermoFisher), with images acquired using the ChemiDoc MP system (Bio-Rad). The Image Lab software was used to measure and quantify the bands of at least three independent western blot experiments. The obtained absolute quantity was compared with the reference band (Vinculin) and expressed in the graphs as normalized volume (Norm. Vol. Int.).

2.10. Image Acquisition and Quantification

Images were acquired with an inverted microscope (Leica-DMI6000B) equipped with Leica DFC365FX and DFC400 cameras and 20× and 40× magnification objectives. Necrotic myofibers were defined as pink pale patchy fibers, and phagocyted myofibers were defined as pink pale fibers invaded by basophilic single cells (MPs). For the quantification of CSA, analyses were done on damaged TA, which presented at least 75% of injured muscle. At least 8 pictures in different fields were taken and at least 500 myofibers were analyzed. For each condition of each experiment, at least 8 fields chosen randomly were counted. The number of labelled MPs or mpcs was calculated using the cell tracker in ImageJ software and expressed as a percentage of total MPs or mpcs. Fusion index was the number of nuclei within myotubes divided by the total number of nuclei.

2.11. Statistical Analysis

All data shown in the graph are expressed as mean ± SEM. All experiments were performed using at least three different cultures or animals in independent experiments. A statistical analysis was performed using two-tailed unpaired Student's t-Test, one-way ANOVA or two-way ANOVA. * $p < 0.05$; ** $p < 0.01$; *** $p < 0.001$; confidence intervals 95%, alpha level 0.05.

3. Results

3.1. Nfix is Expressed by Anti-Inflammatory MPs

To understand if the transcription factor Nfix could be involved in MP function, we first analyzed Nfix expression in MPs during normal muscle regeneration in WT mice. We induced muscle injury by cardiotoxin (CTX) injection in the *Tibialis Anterior* (TA) and looked at the number of MPs (F4/80$^+$ cells) positive for Nfix (Figure 1a). While the number of Nfix-positive MPs was identical between day two (D2) and day four (D4) after injury, we observed an increase of MPs expressing Nfix at D7 after CTX injection (Figure 1a). During muscle regeneration, two populations of MPs are present in the damaged tissue. First, the Ly6C$^+$ pro-inflammatory MPs appear and then, they switch into Ly6C$^-$ anti-inflammatory population [3,9,29,34]. Thus, we asked if Nfix could be expressed by one subset of MPs. We sorted MPs after CTX injury at different time points and looked at Nfix expression by immunolabelling. We firstly enriched for CD45$^+$ cells (mainly composed of MPs and neutrophils) using magnetic beads and then we used the known markers CD64 and Ly6C to separate the pro-inflammatory (CD64$^+$/Ly6C$^+$ cells) to the anti-inflammatory MPs (CD64$^+$/Ly6C$^-$ cells) (Figure S1). We observed that the percentage of pro-inflammatory MPs expressing Nfix does not change over the time of the regeneration (Figure 1b). On the contrary, the percentage of CD64$^+$/Ly6C$^-$ cells positive for Nfix always increased over time (Figure 1b). We also isolated BMDMs (bone marrow derived MPs) from WT mice and polarized them in pro- or anti-inflammatory phenotype. We observed that anti-inflammatory MPs express more Nfix compared to pro-inflammatory MPs (Figure 1c). Therefore, we can conclude that in both in vitro and in vivo analyses, Nfix is more expressed by anti-inflammatory MPs.

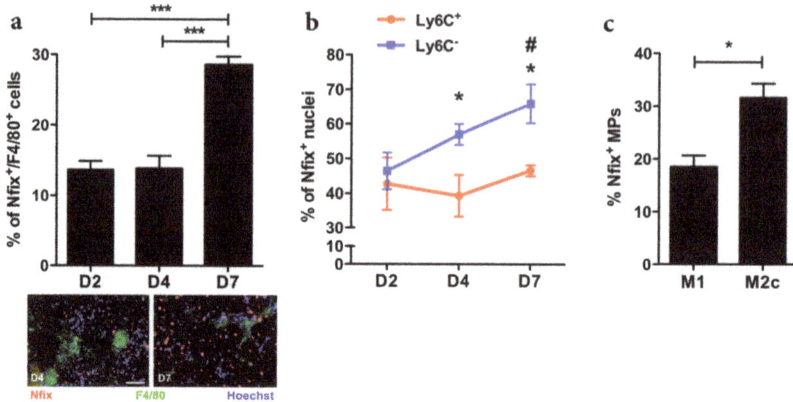

Figure 1. Nfix is mainly expressed by anti-inflammatory MPs. (**a**) Percentage of F4/80$^+$ MPs positive for Nfix in *Tibialis Anterior* muscles (TA) of WT mice injected by CTX at D2, D4 and D7, post-injury. Immunostaining for F4/80 (green), Nfix (red) and DAPI (blue) at D4 and D7 after CTX injection; (**b**) Percentage of Ly6C$^+$ and Ly6C$^-$ sorted MPs positive for Nfix in TA muscles of WT mice injected by CTX at D2, D4 and D7 post-injury; (**c**) Percentage of Nfix$^+$ MPs after M1 and M2c polarization (with IFNγ and IL10, respectively). * $p < 0.05$; *** $p < 0.001$; for (b) * $p < 0.05$ Ly6C+ vs. Ly6C$^+$ at D4 and D7; # $p < 0.05$ Ly6C$^-$ D7 vs. D2. Results are means ± SEM of at least three independent experiments. Scale bar = 50 µm.

3.2. Nfix Expression in MPs is Essential for Muscle Regeneration

We previously demonstrated that Nfix is necessary for the correct differentiation of SCs and, as a consequence, muscle regeneration [27]. In order to understand whether Nfix expression by anti-inflammatory MPs is required for this process, we generated the LysM^{CRE}:Nfix$^{fl/fl}$ mice to obtain an animal model deleted for Nfix only in the myeloid line. Once the proper deletion of Nfix in BMDM and CD45$^+$ infiltrated cells two days after muscle injury was verified (Figure S2a), we evaluated the overall phenotype of this new animal model, with particular interest in skeletal muscle morphology at one and two months of life (Figure S2b). No differences were observed between the Nfix$^{fl/fl}$ control and the LysM^{CRE}:Nfix$^{fl/fl}$ mice model in terms of general mouse growth, TA/mouse weight and myofiber size (CSA: Cross Sectional Area) (Figure S2b,c). Additionally, we did not observe significant differences in the number of resident MPs expressing Nfix between the Nfix$^{fl/fl}$ and the LysM^{CRE}: Nfix$^{fl/fl}$ mice (Figure S2d,e). Therefore, the specific deletion of Nfix in MPs in the LysM^{CRE}: Nfix$^{fl/fl}$ mice does not influence the general development of the mice. Notably, in the LysM^{CRE}: Nfix$^{fl/fl}$ animals resident MPs expressed Nfix similarly to control mice, meaning that the expression of LysM is not required for the establishment of resident MPs.

We then induced muscle injury in control and LysM^{CRE}: Nfix$^{fl/fl}$ mice by CTX injection in TA and we quantified the number of necrotic, phagocyted and regenerating myofibers (Figure 2a,b). Two days after injury, all the myofibers of Nfix$^{fl/fl}$ and LysM^{CRE}: Nfix$^{fl/fl}$ mice were in necrosis or phagocyted (Figure 2b). At D4 in the control mice, some necrotic and phagocyted myofibers were present, but already 60% of the fibers were centronucleated (Figure 2b). On the contrary, in the LysM^{CRE}:Nfix$^{fl/fl}$ mice, we observed a significant decrease in the percentage of centronucleated myofibers (−31%) (Figure 2b). While at D7, almost all myofibers were in regeneration in the control mice, the LysM^{CRE}:Nfix$^{fl/fl}$ mice still exhibited an increase of the percentage of necrotic and phagocyted myofibers and a decrease of centronucleated myofibers (+282%, +150% and −30% respectively), suggesting a delay in the process of muscle regeneration in the absence of Nfix (Figure 2b). We also quantified the CSA of myofibers at D14 and D28 after CTX injury and, in both cases, we observed a decrease of the caliber of myofibers in the LysM^{CRE}:Nfix$^{fl/fl}$ compared to the Nfix$^{fl/fl}$ mice, due to a decrease of the number of big myofibers and an increase of small myofibers (Figure S3a and Figure 2c). These results demonstrated that the expression of Nfix by MPs is necessary for the proper process of muscle regeneration upon acute injury.

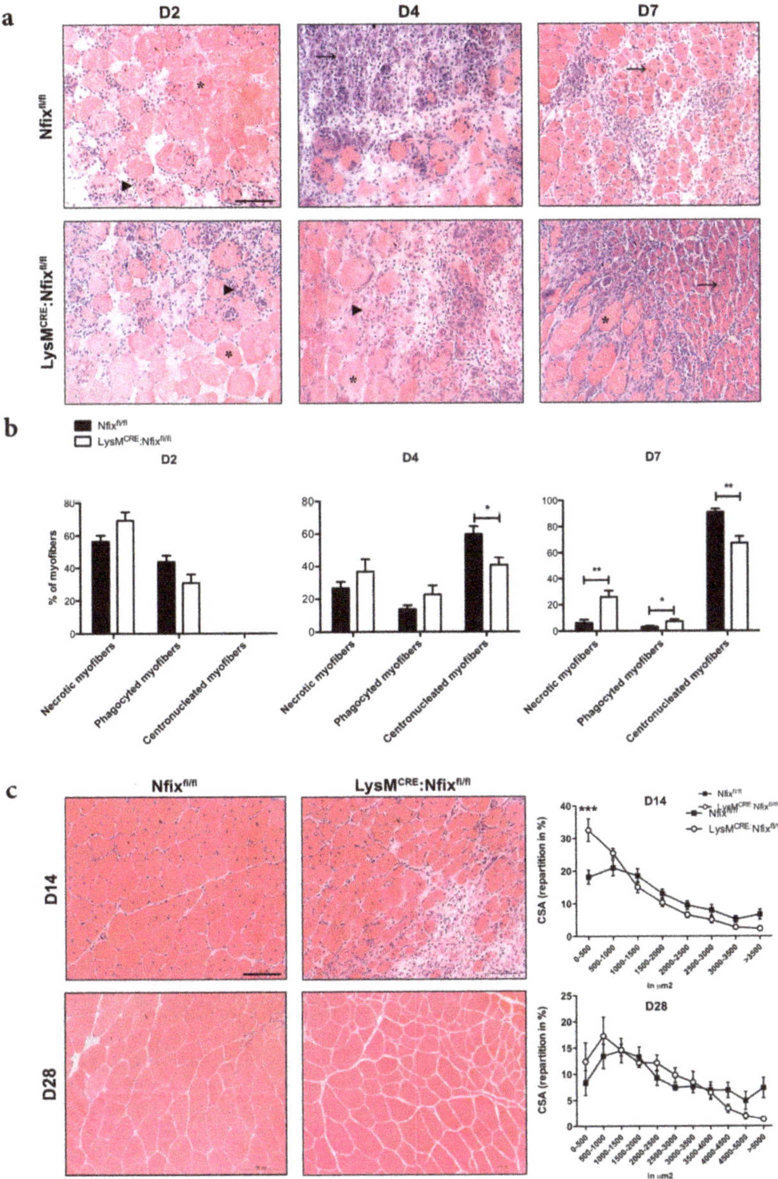

Figure 2. Lack of Nfix in MPs induces a delay of skeletal muscle regeneration. (**a**) Hematoxylin-eosin staining of Nfix[fl/fl] and LysM[CRE]:Nfix[fl/fl] TA muscles injected by CTX at D2, D4, D7 postinjury; (**b**) Quantification of necrotic (asterisk), phagocyted (arrowhead) and centrally-nucleated (arrow) myofibers, expressed as percentage out of total myofibers; (**c**) Hematoxylin-eosin staining of Nfix[fl/fl]

and LysM^{CRE}:Nfix$^{fl/fl}$ TA muscles injected by CTX at D14 and D28 postinjury and repartition in percentage of the cross-sectional area (CSA). * $p < 0.05$, ** $p < 0.01$, *** $p < 0.001$. Results are means ± SEM of at least three independent experiments. Scale bar = 100 μm.

3.3. Nfix is Required for MP Phenotypical Switch In Vivo and In Vitro

Defects of muscle regeneration due to MP dysfunction are usually linked to a defect of phenotype acquisition [8,9,35,36]. Thus, we looked at the switch from pro- to anti-inflammatory phenotype after CTX injury at different time points by FACS (Figure 3a). First, we did not observe any differences in neutrophils and MPs infiltration between the two mouse models at all time points analyzed (Figure S3b). At two days after injury, a majority of Ly6C$^+$ pro-inflammatory MPs was observed in the control Nfix$^{fl/fl}$ mice (Figure 3b). Then, at D4 and D7, the ratio between Ly6C$^+$/Ly6C$^-$ MPs decreased due to the switch from pro- to anti-inflammatory phenotype (Figure 3b). Interestingly, the ratio between Ly6C$^+$/Ly6C$^-$ in the LysM^{CRE}:Nfix$^{fl/fl}$ mice was always higher compared to the Nfix$^{fl/fl}$ control, meaning that Nfix is necessary for the switch from the pro- to anti-inflammatory phenotype (Figure 3b).

We also silenced Nfix in WT BMDM (Bone Marrow Derived MPs) by using a lentiviral vector carrying a small hairpin RNA targeting Nfix (shNfix), or a scrambled sequence as a control (shScramble) [27]. The decrease of Nfix expression in shNfix infected MPs was confirmed by qRT-PCR (Figure S3c). We polarized transduced MPs in pro-inflammatory (M1) and anti-inflammatory (M2c) phenotype (with IFN-γ and IL-10, respectively), and we looked at the expression of several pro- and anti-inflammatory markers by immunofluorescence. As expected, M1 shScramble MPs expressed significantly more TNFα, iNOS and CCl3 pro-inflammatory markers than M2c shScramble MPs (Figure 3c). Conversely, M2c shScramble MPs expressed more ArgI, TGFβ, CD163 and CD206 anti-inflammatory markers than M1 shScramble MPs (Figure 3c). Interestingly, in the absence of Nfix we observed an increase of pro-inflammatory markers (except for CCl3) in MPs polarized to M2c phenotype (Figure 3c). We also observed a decrease of MPs positive for anti-inflammatory markers in the polarized M2c MPs lacking Nfix (Figure 3b). These results clearly show that Nfix is necessary for the proper adoption of an anti-inflammatory phenotype and that, without Nfix, MPs remain in a pro-inflammatory status.

3.4. Nfix is Required for Macrophage Function on Mpcs In Vivo and In Vitro

Since depending on their phenotype MPs act differentially on WT myogenic progenitor cells (mpcs), we set experiments of conditioned medium (CM) coming from pro- or anti-inflammatory MPs [37]. We added CM coming from BMDM derived from Nfix$^{fl/fl}$ and LysM^{CRE}:Nfix$^{fl/fl}$ mice on proliferating or differentiating mpcs. We looked at mpc proliferation by mean of Ki67 staining and at their differentiation by the quantification of the fusion index. As expected, CM coming from M1 Nfix$^{fl/fl}$ MPs stimulated the proliferation of mpcs, while M2c CM had no effect (Figure 4a and Figure S4a). Similarly, CM coming from M1 LysM^{CRE}:Nfix$^{fl/fl}$ MPs stimulated at the same extent the proliferation of mpcs (Figure 4a and Figure S4a). Interestingly, CM from M2c LysM^{CRE}:Nfix$^{fl/fl}$ MPs stimulated the proliferation of mpcs as M1 CM does (Figure 4a and Figure S4b), whereas CM coming from both Nfix$^{fl/fl}$ and LysM^{CRE}:Nfix$^{fl/fl}$ M1 MPs had no effect on the fusion index of mpcs (Figure 4b and Figure S4b). While M2c Nfix$^{fl/fl}$ MPs CM increased the fusion of mpcs compared to M1 Nfix$^{fl/fl}$ CM, the CM from M2c LysM^{CRE}:Nfix$^{fl/fl}$ MPs lost its pro-fusion effect (Figure 4b and Figure S4b). We also investigated if the lack of Nfix in MPs affects myogenic cells in vivo. To quantify mpcs proliferation, we injured, using CTX, the TA of both Nfix$^{fl/fl}$ and LysM^{CRE}:Nfix$^{fl/fl}$ mice and we injected EdU 12 h before sacrifice. The proliferation of SCs (EdU$^+$/Pax7$^+$ cells) was identical between the two models at all the time points analyzed (Figure S4c). On the contrary, the percentage of EdU$^+$/MyoD$^+$ cells at D2 and D4 was higher in the LysM^{CRE}:Nfix$^{fl/fl}$ mice compared to Nfix$^{fl/fl}$ mice (Figure 4c). To quantify the number of newly formed myofibers, we performed an immunofluorescence for eMyHC (embryonic Myosin Heavy Chain) on both injured mouse models (Figure 4d). At D4, the percentage of myofibers positive for eMyHC was around 80%, meaning that almost all myofibers were formed de novo (Figure 4d). At D7, 20% of the Nfix$^{fl/fl}$ myofibers were positive for the eMyHC and only 6.5% at D14 (Figure 4d). On the contrary, at D7 and D14 we observed an increase in the number of eMyHC$^+$ myofibers in the LysM^{CRE}:Nfix$^{fl/fl}$ mice (34.7% and 15.8% respectively) (Figure 4d). To conclude, Nfix is necessary to MPs to adopt an anti-inflammatory phenotype and, consequently, function. The defect of the phenotypical switch due to the absence of Nfix results in a persistence of pro-inflammatory MPs in the injured muscle, leading to a continuous proliferation of MyoD$^+$ cells. The absence of anti-inflammatory MPs induces a delay in the differentiation of new myofibers and, therefore, in the proper muscle regeneration.

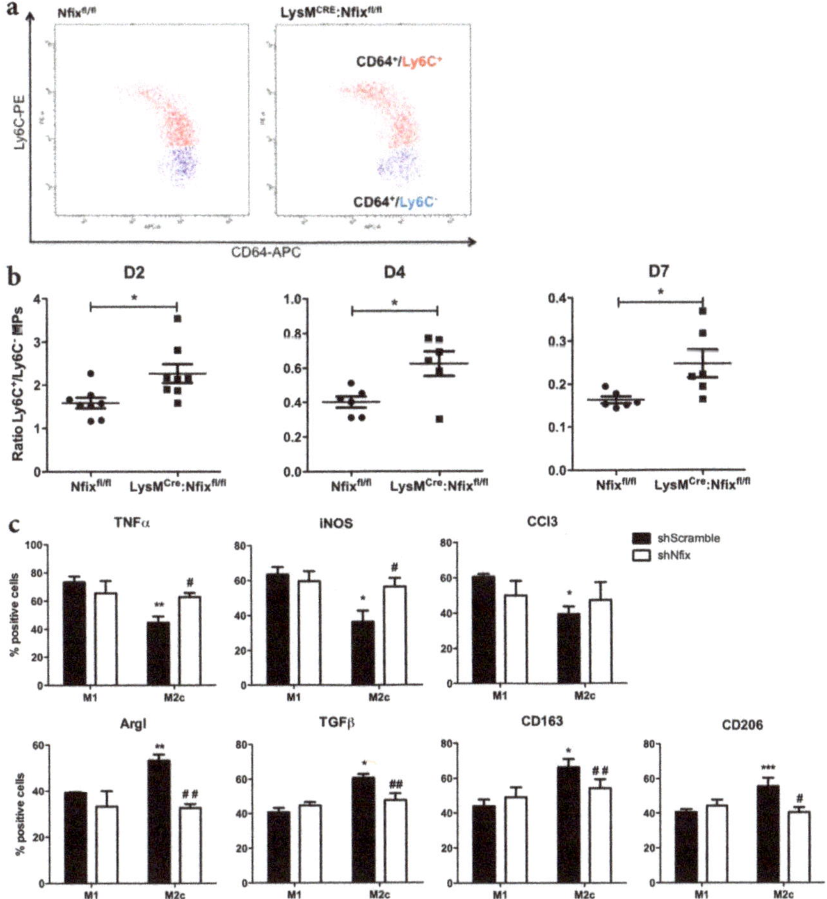

Figure 3. MPs lacking Nfix are unable to adopt an anti-inflammatory phenotype in vivo and in vitro. (**a**) Representative FACS (Fluorescence-Activated Cell Sorting) gate of pro- and anti-inflammatory CD64$^+$ MP populations in TA of Nfix$^{fl/fl}$ and LysM^{CRE}:Nfix$^{fl/fl}$ mice at D2 after CTX injection. (CD64$^+$/Ly6C$^+$ and CD64$^+$/Ly6C$^-$ respectively); (**b**) Ratio of Ly6C$^+$/Ly6C$^-$ MPs sorted from TA of Nfix$^{fl/fl}$ and LysM^{CRE}:Nfix$^{fl/fl}$ mice at D2, D4 and D7, after CTX injection; (**c**) WT BMDM (Bone Marrow Derived Macrophages) were transduced by shScramble and shNfix lentiviral vectors and then polarized into M1 and M2c MPs with IFNγ and IL10 treatment, respectively. MPs were immunolabeled for pro-inflammatory markers (TNFα, iNOS and CCl3) and anti-inflammatory markers (ArgI, TGFβ, CD163 and CD206). The number of positive cells is expressed as percentage out of total cells. * $p < 0.05$;

** $p < 0.01$; *** $p < 0.001$ vs. shScramble M1 MPs. # $p < 0.05$, ## $p < 0.01$, ### $p < 0.001$ vs. shScramble M2c MPs. Results are means ± SEM of at least three independent experiments.

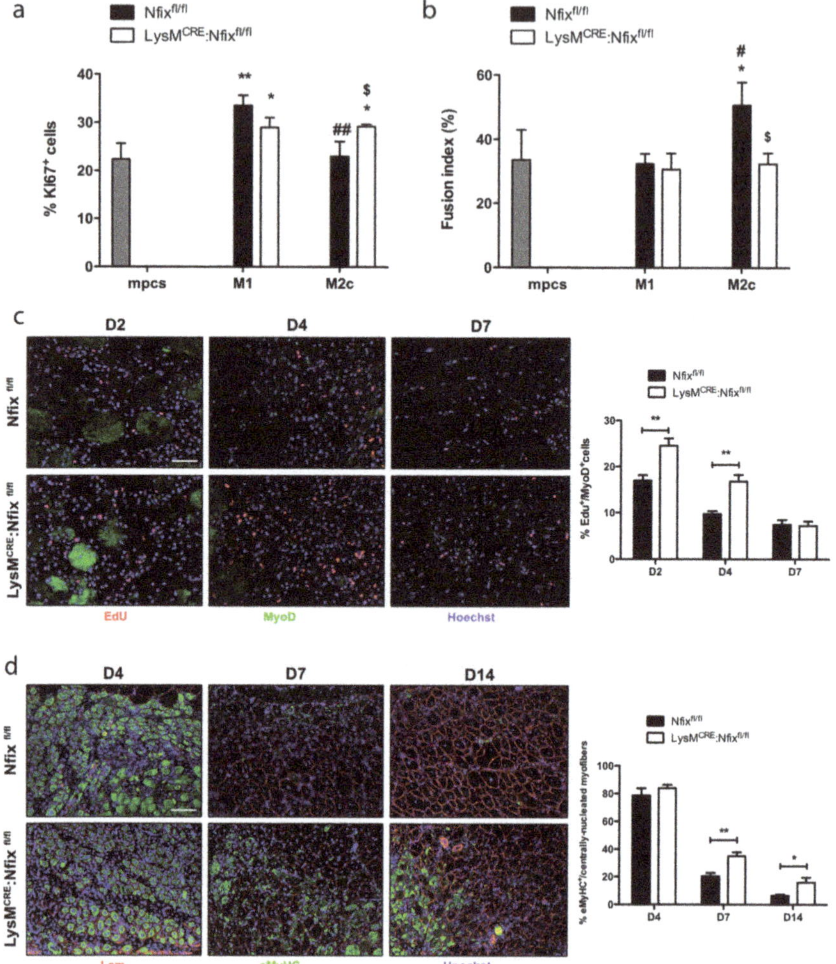

Figure 4. M2 MPs lacking Nfix display M1 MP features on myoblasts in vitro and in vivo. (**a**) Conditioned medium of M1 or M2c polarized Nfix^fl/fl and LysM^CRE:Nfix^fl/fl BMDM was added on mpcs, and after 24 h, mpc proliferation was measured as a percentage of Ki67+ cells; (**b**) Conditioned medium of M1 or M2c polarized Nfix^fl/fl and LysM^CRE:Nfix^fl/fl BMDM was added on mpcs and after 72 h, mpcs fusion index was calculated after sarcomeric MyHC staining (% of MyHC+ nuclei into myotubes out of the total nuclei). * $p < 0.05$, ** $p < 0.01$ vs. mpcs. # $p < 0.05$, ## $p < 0.01$ versus M1 Nfix^fl/fl. $ $p < 0.05$ versus same Nfix^fl/fl polarization; (**c**) Immunostaining for EdU (red), MyoD (green) and Hoechst (blue) of Nfix^fl/fl and LysM^CRE:Nfix^fl/fl TA injected by CTX, at D2, D4 and D7 post-injury and quantification of EdU+/MyoD+ cells. EdU was injected in Nfix^fl/fl and LysM^CRE:Nfix^fl/fl mice 8 h before sacrifice; (**d**) Immunostaining for Lam (red), eMyHC (green) and Hoechst (blue) of Nfix^fl/fl and LysM^CRE:Nfix^fl/fl TA injected by CTX, at D4, D7 and D14 post-injury and quantification of eMyHC+/centrally-nucleated myofibers. * $p < 0.05$, ** $p < 0.01$ Results are means ± SEM of at least three independent experiments. Scale bar = 50 μm.

3.5. Phagocytosis Induces the Expression of Nfix

It has been shown in literature that the phagocytosis of apoptotic cells is the process driving the switch from pro- to anti-inflammatory phenotype, and several studies have demonstrated that MPs presenting a switch defect have a decrease of phagocytic capacity [8,9,35,38]. So, we decided to investigate whether the phagocytosis is altered in LysM^{CRE}:Nfix$^{fl/fl}$ MPs compared to Nfix$^{fl/fl}$ MPs. Primary mpcs previously labelled with CellVue-647 were induced to apoptosis and added on Nfix$^{fl/fl}$ or LysM^{CRE}:Nfix$^{fl/fl}$ MPs. After 6 h in culture, we used a CD64 antibody to discriminate MPs from mpcs: apoptotic mpcs are CellVue-647$^+$/CD64$^-$, non-phagocytic MPs are CellVue-647$^-$/CD64$^+$ and phagocytic MPs are CellVue-647$^+$/CD64$^+$ (Figure S5a). Surprisingly, we did not observe any difference in the phagocytic capacity of Nfix$^{fl/fl}$ and LysM^{CRE}:Nfix$^{fl/fl}$ MPs (Figure 5a). Interestingly, while Nfix$^{fl/fl}$ MPs in contact with apoptotic mpcs adopted an anti-inflammatory phenotype (Figure S5b), LysM^{CRE}:Nfix$^{fl/fl}$ MPs failed to switch from a pro- to anti-inflammatory phenotype (Figure S5c). Thus, we hypothesized that phagocytosis could induce Nfix expression. To answer to this question, we did the same experiment of phagocytosis using WT MPs and we sorted MPs according to their phagocytic capability (phagocytic and non-phagocytic WT MPs, respectively) (Figure 5b). To verify that no mpcs were sorted with MPs, we first analyzed the expression of myogenin in apoptotic mpcs and in both non-phagocytic and phagocytic MPs. Apoptotic mpcs highly expressed myogenin compared to non-phagocytic and phagocytic MPs and no differences in myogenin expression was observed between the two populations of MPs (Figure S5d). Interestingly, we observed an increase of Nfix expression and MPs positive for Nfix in phagocytic MPs compared to the non-phagocytic ones (Figure 5b). On the contrary, treatment of MPs with cytochalasin D (an inhibitor of phagocytosis) prevents the increase of Nfix positive MPs (Figure 5c and Figure S5e).

We recently demonstrated that the inhibition of the RhoA-ROCK1 pathway induces Nfix expression in fetal myoblasts and numerous studies have shown that the inhibition of RhoA-ROCK1 increases the phagocytosis, while its stimulation prevents phagocytosis [39–43]. Thus, we treated WT MPs with Y27632, an inhibitor of ROCK1, and after 1 h of treatment, the phosphorylation of ROCK1-target Mypt decreased, meaning that the inhibition of RhoA-ROCK1 pathway was effective (Figure S5e). After 16 h of treatment, Y27632-treated WT MPs exhibited an increase of Nfix protein (Figure 5d); most importantly, this increase led to a reduction of pro-inflammatory markers and an increase of anti-inflammatory markers (Figure 5e). On the contrary, this switch through an anti-inflammation phenotype did not occur in LysM^{CRE}:Nfix$^{fl/fl}$ MPs treated with Y27632 (Figure 5f). These results show that RhoA-ROCK-dependent phagocytosis induces the expression of Nfix, which in turn is necessary to promote the phenotypical switch of MPs from pro- to anti-inflammatory.

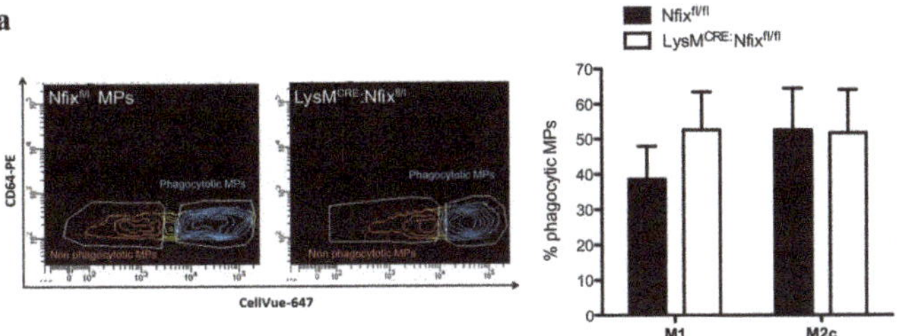

Figure 5. *Cont.*

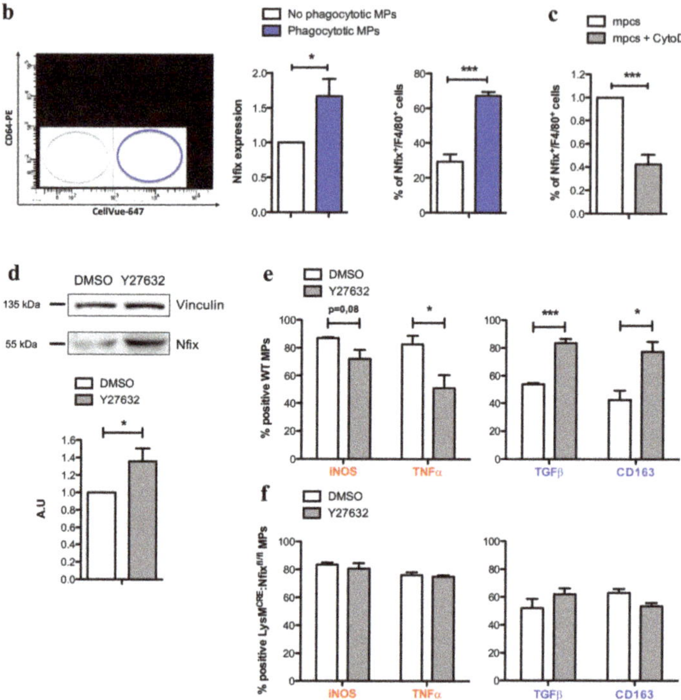

Figure 5. Nfix is expressed after phagocytosis and drive MP phenotypical switch. (**a**) Phagocytosis assay of M1 and M2c Nfix[fl/fl] and LysM[CRE]:Nfix[fl/fl] MPs cocultured 8h with apoptotic mpc. Representative FACS gate of phagocytotic M2c Nfix[fl/fl] and LysM[CRE]:Nfix[fl/fl] MPs (CD64+CellVue+) and percentage of phagocytotic M1 and M2c MPs coming from Nfix[fl/fl] and LysM[CRE]:Nfix[fl/fl] BMDM; (**b**) WT MPs were cocultured 16h with apoptotic mpcs. Representative FACS gate of non-phagocytotic (CD64+CellVue−) and phagocytotic (CD64+CellVue+) WT MPs. Quantification of Nfix expression realized by RT-qPCR on sorted non-phagocytotic and phagocytotic WT MPs and quantification of MPs positive for Nfix (Nfix+/F4/80+) realized by IF on non-phagocytotic and phagocytotic WT MPs; (**c**) WT MPs were cocultured for 16 h with apoptotic mpcs, with or without addition of Cytochalasin D. Quantification of F4/80+ MPs were positive for Nfix on a total of F4/80+ MPs; (**d**) Western blot of Nfix expression in WT MPs treated with DMSO (Dimethyl sulfoxide) or Y27632 for 16 h and quantification. Vinculin was used to normalize; (**e**) WT MPs were treated with DMSO or Y27632 for 16 h and were immunolabeled for pro-inflammatory markers (iNOS and TNFα) and anti-inflammatory markers (TGFβ and CD163). The number of positive cells is expressed as percentage out of total cells; (**f**) LysM[CRE]:Nfix[fl/fl] MPs were treated with DMSO or Y27632 for 16 h and were immunolabeled for pro-inflammatory markers (iNOS and TNFα) and anti-inflammatory markers (TGFβ and CD163). The number of positive cells is expressed as percentage out of total cells. * $p < 0.05$, *** $p < 0.001$. Results are means ± SEM of at least three independent experiments.

4. Discussion

Skeletal muscle regeneration requires specific temporal steps for the efficacious tissue reconstruction and MPs are the immune cells that are necessary to this process [3,29]. Previous work from our group demonstrated that *Nfix* null mice exhibit a delay of muscle regeneration, due to a defect of SC differentiation [27]. In this study, we show that the transcription factor Nfix is also expressed by MPs and that mice lacking Nfix in the myeloid lineage have defects in muscle regeneration upon acute injury. We observed that Nfix is preferentially expressed by anti-Ly6C− MPs and that its expression increases in time with the progression of the regenerative process. Using an shNfix strategy, we observed that M2c MPs silenced for Nfix express higher levels of pro-inflammatory markers (TNFα and Cox2), while they express lower levels of anti-inflammatory markers (CD163, CD206, ArgI and TGFβ) than polarized M2c control MPs. Importantly, we observed in vitro that LysM^{CRE}:Nfixf$^{l/fl}$ M2c MPs act as M1 MPs on myogenic cells: they stimulate myogenic proliferation and are unable to sustain myogenic differentiation. These two features also occur in vivo since without Nfix, MPs exhibit a defect of phenotypical switch from pro-Ly6C$^+$ to anti-Ly6C$^-$ MPs, and within the injured muscle, there is a persistence of myoblast (MyoD$^+$ cells) proliferation and a delay of newly formed myofibers (eMyHC$^+$ myofibers). Previous studies showed that the temporal window of the phenotype skewing is a critical step of an effective regeneration. The switch defect [8,35,36] or early appearance of anti-inflammatory MPs impairs muscle regeneration [9]. In line with this evidence, the impairment in the acquisition of an anti-inflammatory phenotype in MPs lacking Nfix leads to a muscle regenerative delay.

So far, the function of Nfix was mainly analyzed in myogenic and neural cells during both development and adult life [25,27,44–46]. Recently, Nfix was also shown to play a positive role in the survival of hematopoietic stem and progenitor cells (HSPC) [47], but also to be involved in the fate decision between early B lymphopoiesis and myelopoiesis from blood HSPC [48]. During development, yolk sac gives rise to tissue resident MPs and fetal liver to HSPC, from which blood monocytes and damaged-infiltrating MPs are derived [49]. In our experiments, no differences in the number of infiltrating MPs between control and LysM^{CRE}:Nfix$^{fl/fl}$ mice were observed, meaning that the delay observed is due to a defect of macrophage features within the damaged muscle, but not in terms of failed HSPC development.

Little is known about Nfix up-stream regulation, but recently our laboratory identified ERK and RhoA-ROCK1 pathways as, respectively, positive and negative regulators of Nfix expression in pre-natal muscle development. The inhibition of RhoA-ROCK1 induces Nfix expression, promoting myoblasts fusion which is a reflect of myogenesis progression [39]. In MPs, the inhibition of the RhoA-ROCK1 pathway increases the clearance of dead cell phagocytosis, while the constitutive activation of RhoA reduces their phagocytic capacity [40,41]. Importantly, phagocytosis is the process responsible for the induction of the pro- to anti-phenotypical switch in MPs. While numerous studies investigated the mechanisms involved in the progression or inhibition of phagocytosis, how apoptotic cells attract MPs and how MPs recognize them is still unknown [10,14,15,50,51]. In our study, phagocytosis of LysM^{CRE}:Nfix$^{fl/fl}$ MPs was not impaired compared to control cells. We observed that upon phagocytosis, MPs exhibit an increase in Nfix expression and, conversely, the inhibition of phagocytosis, by using the inhibitor of actin polymerization cytochalasin D, prevents Nfix expression. The stimulation of phagocytosis using the ROCK1 inhibitor Y27632 increases Nfix protein, therefore decreasing pro-inflammatory markers and increasing anti-inflammatory markers in WT MPs. On the contrary, either after phagocytosis or after ROCK1 inhibitor treatment, we observed that MPs lacking Nfix do not have a decrease of pro-inflammatory markers and an increase of anti-inflammatory markers. Thus, the inhibition of the RhoA-ROCK1 pathway induces phagocytosis, leading to Nfix expression that, in turn, drives the MP phenotypical switch.

This study is particularly relevant in light of the recent role for Nfix in muscular dystrophies (MDs)[28]. We indeed demonstrated that the lack of Nfix in two different dystrophic animal models improves both morphological and functional parameters associated to the disease, by promoting a more oxidative musculature and by slowing down muscle regeneration [28]. Different studies have shown that the improvement of dystrophies correlates with a decrease of MPs infiltration [30,32]. While MPs are necessary for muscle regeneration upon acute injury, they are deleterious in the case of chronic injury. Indeed, in muscle myopathies, as in several chronic injured pathologies, MPs are at the origin of fibrosis [4,32,52,53]. In the context of acute injury regeneration, pro-inflammatory MPs secrete TNFα that stimulates myoblast proliferation and fibroblast apoptosis, whereas anti-inflammatory MPs secrete TGFβ that promotes myoblast fusion, but also fibroblast proliferation [29,54]. In muscular dystrophies, numerous studies demonstrated that the fibrosis establishment is linked to an over-activation of the TGFβ pathway that stimulates collagen expression by fibroblasts, and in a dystrophic context, more than 75% of MPs express TGFβ [54–62]. Thus, in muscle tissue, MPs closely interact with fibroblasts, promoting normal matrix reformation upon acute injury and fibrosis in chronic injury. With this study, we identified Nfix as a new actor of MPs, demonstrating that Nfix is the link between phagocytosis and the phenotypical switch, a necessary step for the resolution of inflammation and tissue repair. Increasing knowledge about signals and factors controlling MP phenotype and, consequently, functions, will help us to understand and control their function in fibrotic pathologies.

Supplementary Materials: The following are available online at http://www.mdpi.com/2073-4409/9/3/708/s1, **Figure S1**: Gating strategy to isolate MPs from CTX-injured muscles. Figure S2: Characterization of the LysM^{CRE}:Nfix$^{fl/fl}$ mice. Figure S3: CSA quantification, NT and MPs infiltration in Nfix$^{fl/fl}$ and LysM^{CRE}:Nfix$^{fl/fl}$ mice after CTX injury and Nfix silencing in WT BMDM. Figure S4: In vitro proliferation and differentiation assay. Proliferation of Pax7$^+$ cells after CTX injury. Figure S5: Phagocytosis strategy, inhibition and stimulation.

Author Contributions: Conceptualization, M.S. and G.M.; methodology, M.S., M.L., C.B and S.A.; validation, M.S., M.L., C.B. and S.A.; formal analysis, M.S.; investigation, M.S., M.L., C.B, G.R. and S.A.; resources, G.M.; data curation, M.S.; writing—original draft preparation, M.S.; writing—review and editing, G.M.; visualization, M.S.; supervision, G.M.; project administration, G.M.; funding acquisition, G.M. All authors have read and agreed to the published version of the manuscript.

Funding: This research was funded by the European Community, ERCStG2011 (RegeneratioNfix 280611) and the Association Française contre les Myopathies AFM-Telethon (Grant number 20002).

Acknowledgments: We thank Richard Gronostajski for the kind exchange of information and animal models. We are also grateful to Bénédicte Chazaud and Rémi Mounier for helpful discussions and the exchange of animal models.

Conflicts of Interest: The authors declare no conflict of interest.

References

1. Duffield, J.S.; Forbes, S.J.; Constandinou, C.M.; Clay, S.; Partolina, M.; Vuthoori, S.; Wu, S.; Lang, R.; Iredale, J.P. Selective depletion of macrophages reveals distinct, opposing roles during liver injury and repair. *J. Clin. Invest.* **2005**, *115*, 56–65. [CrossRef] [PubMed]

2. Shechter, R.; Miller, O.; Yovel, G.; Rosenzweig, N.; London, A.; Ruckh, J.; Kim, K.W.; Klein, E.; Kalchenko, V.; Bendel, P.; et al. Recruitment of Beneficial M2 Macrophages to Injured Spinal Cord Is Orchestrated by Remote Brain Choroid Plexus. *Immunity* **2013**, *38*, 555–569. [CrossRef] [PubMed]

3. Arnold, L.; Henry, A.; Poron, F.; Baba-Amer, Y.; Van Rooijen, N.; Plonquet, A.; Gherardi, R.K.; Chazaud, B. Inflammatory monocytes recruited after skeletal muscle injury switch into antiinflammatory macrophages to support myogenesis. *J. Exp. Med.* **2007**, *204*, 1057–1069. [CrossRef] [PubMed]

4. Wynn, T.A.; Vannella, K.M. Macrophages in Tissue Repair, Regeneration, and Fibrosis. *Immunity* **2016**, *44*, 450–462. [CrossRef]

5. Vannella, K.M.; Wynn, T.A. Mechanisms of Organ Injury and Repair by Macrophages. *Annu. Rev. Physiol.* **2017**, *79*, 593–617. [CrossRef]

6. Chazaud, B. Macrophages: Supportive cells for tissue repair and regeneration. *Immunobiology* **2014**, *219* 172–178. [CrossRef]

7. Saclier, M.; Cuvellier, S.; Magnan, M.; Mounier, R.; Chazaud, B. Monocyte/macrophage interactions with myogenic precursor cells during skeletal muscle regeneration. *FEBS J.* **2013**, *280*, 4118–4130. [CrossRef]

8. Mounier, R.; Théret, M.; Arnold, L.; Cuvellier, S.; Bultot, L.; Göransson, O.; Sanz, N.; Ferry, A.; Sakamoto, K.; Foretz, M.; et al. AMPKα1 regulates macrophage skewing at the time of resolution of inflammation during skeletal muscle regeneration. *Cell Metab.* **2013**, *18*, 251–264. [CrossRef]

9. Perdiguero, E.; Sousa-Victor, P.; Ruiz-Bonilla, V.; Jardí, M.; Caelles, C.; Serrano, A.L.; Muñoz-Cánoves, P. p38/MKP-1-regulated AKT coordinates macrophage transitions and resolution of inflammation during tissue repair. *J. Cell Biol.* **2011**, *195*, 307–322. [CrossRef]

10. Lemke, G. How macrophages deal with death. *Nat. Rev. Immunol.* **2019**, *19*, 539–549. [CrossRef]

11. Elliott, M.R.; Ravichandran, K.S. The Dynamics of Apoptotic Cell Clearance. *Dev. Cell* **2016**, *38*, 147–160. [CrossRef] [PubMed]

12. Xiao, Y.Q.; Freire-de-Lima, C.G.; Schiemann, W.P.; Bratton, D.L.; Vandivier, R.W.; Henson, P.M. Transcriptional and Translational Regulation of TGF-β Production in Response to Apoptotic Cells. *J. Immunol.* **2008**, *181*, 3575–3585. [CrossRef] [PubMed]

13. Johann, A.M.; Barra, V.; Kuhn, A.M.; Weigert, A.; Von Knethen, A.; Brüne, B. Apoptotic cells induce arginase II in macrophages, thereby attenuating NO production. *FASEB J.* **2007**, *21*, 2704–2712. [CrossRef] [PubMed]

14. Hochreiter-Hufford, A.; Ravichandran, K.S. Clearing the dead: Apoptotic cell sensing, recognition, engulfment, and digestion. *Cold Spring Harb. Perspect. Biol.* **2013**, *5*. [CrossRef]

15. Freeman, S.A.; Grinstein, S. Phagocytosis: Receptors, signal integration, and the cytoskeleton. *Immunol. Rev.* **2014**, *262*, 193–215. [CrossRef]

16. Tidball, J.G.; Wehling-Henricks, M. Damage and inflammation in muscular dystrophy: Potential implications and relationships with autoimmune myositis. *Curr. Opin. Rheumatol.* **2005**, *17*, 707–713. [CrossRef]

17. Dort, J.; Fabre, P.; Molina, T.; Dumont, N.A. Macrophages Are Key Regulators of Stem Cells during Skeletal Muscle Regeneration and Diseases. *Stem Cells Int.* **2019**, *2019*. [CrossRef]

18. Farup, J.; Madaro, L.; Puri, P.L.; Mikkelsen, U.R. Interactions between muscle stem cells, mesenchymal-derived cells and immune cells in muscle homeostasis, regeneration and disease. *Cell Death Dis.* **2015**, *6*, e1830. [CrossRef]

19. Chazaud, B.; Brigitte, M.; Yacoub-Youssef, H.; Arnold, L.; Gherardi, R.; Sonnet, C.; Lafuste, P.; Chretien, F. Dual and beneficial roles of macrophages during skeletal muscle regeneration. *Exerc. Sport Sci. Rev.* **2009**, *37*, 18–22. [CrossRef]

20. Rigamonti, E.; Zordan, P.; Sciorati, C.; Rovere-Querini, P.; Brunelli, S. Macrophage plasticity in skeletal muscle repair. *Biomed Res. Int.* **2014**, *2014*. [CrossRef]

21. Biressi, S.; Molinaro, M.; Cossu, G. Cellular heterogeneity during vertebrate skeletal muscle development. *Dev. Biol.* **2007**, *308*, 281–293. [CrossRef] [PubMed]

22. Mauro, A. Satellite Cell of Skeletal Muscle Fibers. *J. Biophys Biochem Cytol* **1961**, *9*, 493–498. [CrossRef] [PubMed]

23. Dumont, N.A.; Bentzinger, C.F.; Sincennes, M.C.; Rudnicki, M.A. Satellite cells and skeletal muscle regeneration. *Compr. Physiol.* **2015**, *5*, 1027–1059. [PubMed]

24. Gronostajski, R.M. Roles of the NFI/CTF gene family in transcription and development. *Gene* **2000**, *249*, 31–45. [CrossRef]

25. Messina, G.; Biressi, S.; Monteverde, S.; Magli, A.; Cassano, M.; Perani, L.; Roncaglia, E.; Tagliafico, E.; Starnes, L.; Campbell, C.E.; et al. Nfix Regulates Fetal-Specific Transcription in Developing Skeletal Muscle. *Cell* **2010**, *140*, 554–566. [CrossRef] [PubMed]

26. Pistocchi, A.; Gaudenzi, G.; Foglia, E.; Monteverde, S.; Moreno-Fortuny, A.; Pianca, A.; Cossu, G.; Cotelli, F.; Messina, G. Conserved and divergent functions of Nfix in skeletal muscle development during vertebrate evolution. *Development* **2013**, *140*, 2443. [CrossRef]

27. Rossi, G.; Antonini, S.; Bonfanti, C.; Monteverde, S.; Vezzali, C.; Tajbakhsh, S.; Cossu, G.; Messina, G. Nfix Regulates Temporal Progression of Muscle Regeneration through Modulation of Myostatin Expression. *Cell Rep.* **2016**, *14*, 2238–2249. [CrossRef]

28. Rossi, G.; Bonfanti, C.; Antonini, S.; Bastoni, M.; Monteverde, S.; Innocenzi, A.; Saclier, M.; Taglietti, V.; Messina, G. Silencing Nfix rescues muscular dystrophy by delaying muscle regeneration. *Nat. Commun.* **2017**, *8*. [CrossRef]

29. Saclier, M.; Yacoub-Youssef, H.; Mackey, A.L.; Arnold, L.; Ardjoune, H.; Magnan, M.; Sailhan, F.; Chelly, J.; Pavlath, G.K.; Mounier, R.; et al. Differentially activated macrophages orchestrate myogenic precursor cell fate during human skeletal muscle regeneration. *Stem Cells* **2013**, *31*, 384–396. [CrossRef]

30. Wehling, M.; Spencer, M.J.; Tidball, J.G. A nitric oxide synthase transgene ameliorates muscular dystrophy in mdx mice. *J. Cell Biol.* **2001**, *155*, 123–131. [CrossRef]

31. Kharraz, Y.; Guerra, J.; Mann, C.J.; Serrano, A.L.; Muñoz-Cánoves, P. Macrophage plasticity and the role of inflammation in skeletal muscle repair. *Mediators Inflamm.* **2013**, *2013*. [CrossRef] [PubMed]

32. Muñoz-Cánoves, P.; Serrano, A.L. Macrophages decide between regeneration and fibrosis in muscle. *Trends Endocrinol. Metab.* **2015**, *26*, 449–450. [CrossRef] [PubMed]

33. Campbell, C.E.; Piper, M.; Plachez, C.; Yeh, Y.-T.; Baizer, J.S.; Osinski, J.M.; Litwack, E.D.; Richards, L.J.; Gronostajski, R.M. The transcription factor Nfix is essential for normal brain development. *BMC Dev. Biol.* **2008**, *8*, 52. [CrossRef] [PubMed]

34. Varga, T.; Mounier, R.; Horvath, A.; Cuvellier, S.; Dumont, F.; Poliska, S.; Ardjoune, H.; Juban, G.; Nagy, L.; Chazaud, B. Highly Dynamic Transcriptional Signature of Distinct Macrophage Subsets during Sterile Inflammation, Resolution, and Tissue Repair. *J. Immunol.* **2016**, *196*, 4771–4782. [CrossRef] [PubMed]

35. Ruffell, D.; Mourkioti, F.; Gambardella, A.; Kirstetter, P.; Lopez, R.G.; Rosenthal, N.; Nerlov, C. A CREB-C/EBPbeta cascade induces M2 macrophage-specific gene expression and promotes muscle injury repair. *Proc. Natl. Acad. Sci. USA* **2009**, *106*, 17475–17480. [CrossRef] [PubMed]

36. Nie, M.; Liu, J.; Yang, Q.; Seok, H.Y.; Hu, X.; Deng, Z.-L.; Wang, D.-Z. MicroRNA-155 facilitates skeletal muscle regeneration by balancing pro- and anti-inflammatory macrophages. *Cell Death Dis.* **2016**, *7*, e2261. [CrossRef]

37. Saclier, M.; Theret, M.; Mounier, R.; Chazaud, B. Effects of macrophage conditioned-medium on murine and human muscle cells: Analysis of proliferation, differentiation, and fusion. *Proc. Natl. Acad. Sci. USA* **2009**, *106*, 17475–17480.

38. Arnold, L.; Perrin, H.; de Chanville, C.B.; Saclier, M.; Hermand, P.; Poupel, L.; Guyon, E.; Licata, F.; Carpentier, W.; Vilar, J.; et al. CX3CR1 deficiency promotes muscle repair and regeneration by enhancing macrophage ApoE production. *Nat. Commun.* **2015**, *6*, 8972. [CrossRef]

39. Taglietti, V.; Angelini, G.; Mura, G.; Bonfanti, C.; Caruso, E.; Monteverde, S.; Le Carrou, G.; Tajbakhsh, S.; Relaix, F.; Messina, G. RhoA and ERK signalling regulate the expression of the transcription factor Nfix in myogenic cells. *Development* **2018**, *145*. [CrossRef]

40. Bros, M.; Haas, K.; Moll, L. Grabbe RhoA as a Key Regulator of Innate and Adaptive Immunity. *Cells* **2019**, *8*, 733. [CrossRef]

41. Nakaya, M.; Tanaka, M.; Okabe, Y.; Hanayama, R.; Nagata, S. Opposite effects of Rho family GTPases on engulfment of apoptotic cells by macrophages. *J. Biol. Chem.* **2006**, *281*, 8836–8842. [CrossRef] [PubMed]

42. Königs, V.; Jennings, R.; Vogl, T.; Horsthemke, M.; Bachg, A.C.; Xu, Y.; Grobe, K.; Brakebusch, C.; Schwab, A.; Bähler, M.; et al. Mouse Macrophages completely lacking Rho subfamily GTPases (RhoA, RhoB, and RhoC) have severe lamellipodial retraction defects, but robust chemotactic navigation and altered motility. *J. Biol. Chem.* **2014**, *289*, 30772–30784. [CrossRef] [PubMed]

43. Kim, S.Y.; Kim, S.; Bae, D.J.; Park, S.Y.; Lee, G.Y.; Park, G.M.; Kim, I.S. Coordinated balance of Rac1 and RhoA plays key roles in determining phagocytic appetite. *PLoS ONE* **2017**, *12*. [CrossRef] [PubMed]

44. Harris, L.; Zalucki, O.; Gobius, I.; McDonald, H.; Osinki, J.; Harvey, T.J.; Essebier, A.; Vidovic, D.; Gladwyn-Ng, I.; Burne, T.H.; et al. Transcriptional regulation of intermediate progenitor cell generation during hippocampal development. *Development* **2016**, *143*, 4620–4630. [CrossRef]

45. Harris, L.; Dixon, C.; Cato, K.; Heng, Y.H.E.; Kurniawan, N.D.; Ullmann, J.F.P.; Janke, A.L.; Gronostajski, R.M.; Richards, L.J.; Burne, T.H.J.; et al. Heterozygosity for Nuclear Factor One X Affects Hippocampal-Dependent Behaviour in Mice. *PLoS ONE* **2013**, *8*. [CrossRef]

46. Fraser, J.; Essebier, A.; Gronostajski, R.M.; Boden, M.; Wainwright, B.J.; Harvey, T.J.; Piper, M. Cell-type-specific expression of NFIX in the developing and adult cerebellum. *Brain Struct. Funct.* **2017**, *222*, 2251–2270. [CrossRef]

47. Holmfeldt, P.; Pardieck, J.; Saulsberry, A.C.; Nandakumar, S.K.; Finkelstein, D.; Gray, J.T.; Persons, D.A.; Mckinney-freeman, S. Nfix is a novel regulator of murine hematopoietic stem and progenitor cell survival. *Blood* **2015**, *122*, 2987–2997. [CrossRef]

48. O'Connor, C.; Campos, J.; Osinski, J.M.; Gronostajski, R.M.; Michie, A.M.; Keeshan, K. Nfix expression critically modulates early B lymphopoiesis and myelopoiesis. *PLoS ONE* **2015**, *10*, 1–15. [CrossRef]

49. Gomez Perdiguero, E.; Klapproth, K.; Schulz, C.; Busch, K.; de Bruijn, M.; Rodewald, H.R.; Geissmann, F. The Origin of Tissue-Resident Macrophages: When an Erythro-myeloid Progenitor Is an Erythro-myeloid Progenitor. *Immunity* **2015**, *43*, 1023–1024. [CrossRef]

50. Park, S.Y.; Kim, I.S. Engulfment signals and the phagocytic machinery for apoptotic cell clearance. *Exp. Mol. Med.* **2017**, *49*. [CrossRef]

51. Haney, M.S.; Bohlen, C.J.; Morgens, D.W.; Ousey, J.A.; Barkal, A.A.; Tsui, C.K.; Ego, B.K.; Levin, R.; Kamber, R.A.; Collins, H.; et al. Identification of phagocytosis regulators using magnetic genome-wide CRISPR screens. *Nat. Genet.* **2018**, *50*, 1716–1727. [CrossRef] [PubMed]

52. Ngambenjawong, C.; Gustafson, H.H.; Pun, S.H. Progress in tumor-associated macrophage (TAM)-targeted therapeutics. *Adv. Drug Deliv. Rev.* **2017**, *114*, 206–221. [CrossRef]

53. Tang, P.M.K.; Nikolic-Paterson, D.J.; Lan, H.Y. Macrophages: Versatile players in renal inflammation and fibrosis. *Nat. Rev. Nephrol.* **2019**, *15*, 144–158. [CrossRef] [PubMed]

54. Lemos, D.R.; Babaeijandaghi, F.; Low, M.; Chang, C.-K.; Lee, S.T.; Fiore, D.; Zhang, R.-H.; Natarajan, A.; Nedospasov, S.A.; Rossi, F.M. V Nilotinib reduces muscle fibrosis in chronic muscle injury by promoting TNF-mediated apoptosis of fibro/adipogenic progenitors. *Nat. Med.* **2015**, *21*, 786–794. [CrossRef]

55. Ueha, S.; Shand, F.H.W.; Matsushima, K. Cellular and molecular mechanisms of chronic inflammation-associated organ fibrosis. *Front. Immunol.* **2012**, *3*, 1–6. [CrossRef]

56. Tan, R.J.; Liu, Y. Macrophage-derived TGF-beta in renal fibrosis: Not a macro- impact after all. *Am. J. Physiol. Renal Physiol.* **2013**, 1–7.

57. Mann, C.J.; Perdiguero, E.; Kharraz, Y.; Aguilar, S.; Pessina, P.; Serrano, A.L.; Muñoz-Cánoves, P. Aberrant repair and fibrosis development in skeletal muscle. *Skelet. Muscle* **2011**, *1*, 21. [CrossRef]

58. Vidal, B.; Serrano, A.L.; Tjwa, M.; Suelves, M.; Ardite, E.; De Mori, R.; Baeza-Raja, B.; De Lagrán, M.M.; Lafuste, P.; Ruiz-Bonilla, V.; et al. Fibrinogen drives dystrophic muscle fibrosis via a TGFβ/alternative macrophage activation pathway. *Genes Dev.* **2008**, *22*, 1747–1752. [CrossRef] [PubMed]

59. Tidball, J.G.; Wehling-Henricks, M. Shifts in macrophage cytokine production drive muscle fibrosis. *Nat. Med.* **2015**, *21*, 665–666. [CrossRef]

60. Pakshir, P.; Hinz, B. The big five in fibrosis: Macrophages, myofibroblasts, matrix, mechanics, and miscommunication. *Matrix Biol.* **2018**, *68–69*, 81–93. [CrossRef]

61. Smith, L.R.; Barton, E.R. Regulation of fibrosis in muscular dystrophy. *Matrix Biol.* **2018**, *68–69*, 602–615. [CrossRef] [PubMed]

62. Biernacka, A.; Dobaczewski, M.; Frangogiannis, N.G. TGF-β signaling in fibrosis. *Growth Factors* **2011**, *29*, 196–202. [CrossRef] [PubMed]

Review

"The Social Network" and Muscular Dystrophies: The Lesson Learnt about the Niche Environment as a Target for Therapeutic Strategies

Ornella Cappellari [†], Paola Mantuano [†] and Annamaria De Luca [*]

Section of Pharmacology, Department of Pharmacy-Drug Sciences, University of Bari "Aldo Moro", via Orabona 4—Campus, 70125 Bari, Italy; ornella.cappellari@uniba.it (O.C.); paola.mantuano@uniba.it (P.M.)
* Correspondence: annamaria.deluca@uniba.it; Tel.: +39-080-544-22-45
† These authors contributed equally to this work.

Received: 31 May 2020; Accepted: 2 July 2020; Published: 9 July 2020

Abstract: The muscle stem cells niche is essential in neuromuscular disorders. Muscle injury and myofiber death are the main triggers of muscle regeneration via satellite cell activation. However, in degenerative diseases such as muscular dystrophy, regeneration still keep elusive. In these pathologies, stem cell loss occurs over time, and missing signals limiting damaged tissue from activating the regenerative process can be envisaged. It is unclear what comes first: the lack of regeneration due to satellite cell defects, their pool exhaustion for degeneration/regeneration cycles, or the inhibitory mechanisms caused by muscle damage and fibrosis mediators. Herein, Duchenne muscular dystrophy has been taken as a paradigm, as several drugs have been tested at the preclinical and clinical levels, targeting secondary events in the complex pathogenesis derived from lack of dystrophin. We focused on the crucial roles that pro-inflammatory and pro-fibrotic cytokines play in triggering muscle necrosis after damage and stimulating satellite cell activation and self-renewal, along with growth and mechanical factors. These processes contribute to regeneration and niche maintenance. We review the main effects of drugs on regeneration biomarkers to assess whether targeting pathogenic events can help to protect niche homeostasis and enhance regeneration efficiency other than protecting newly formed fibers from further damage.

Keywords: muscle regeneration; muscle stem cells; stem cells niche; muscle homeostasis; neuromuscular disorders; Duchenne muscular dystrophy; pharmacological approach

---

1. The Muscle Tissue: Development and Insight

Skeletal muscle is a complex and heterogeneous tissue with a high regeneration potential and plasticity. Muscle regeneration recapitulates skeletal muscle ontogenesis for many aspects. Myogenesis can be divided into different phases, which comprehend embryonic (from E10.5 to E12.5 of mouse development) and fetal (from E14.5 to E17.5) phases [1]. First, muscle fibers are generated during embryonic myogenesis in the somites, transient mesodermal units, to which other fibers are subsequentially added for following differentiation into ventral sclerotome and a dorsal dermomyotome [2]. Myogenic progenitors appear at the end of the somitogenesis and respond to signals from the neural tube, such as Wnts (wingless-type MMTV integration site family) and Sonic hedgehog (Shh), which activate the basic helix–loop–helix transcription factors, such as myogenic factor 5 (Myf5) and myoblast determination protein 1 (MyoD) which commit cells to myogenesis [3]. The embryology of skeletal muscle is out of the scope of this review and excellent reviews on the topic are available [3].

Importantly, as previously stated, skeletal muscle is formed in successive and distinct, though overlapping waves, involving different types of myoblasts (embryonic, fetal myoblasts, and satellite cells). The progressive growth of muscles occurring during late embryonic (E10.5–12.5), fetal (E14.5–17.5), and postnatal life was recently attributed to a population of muscle progenitors that can be found already at embryonic stage [4–7]. These might derive from a paired box gene (*Pax*) *3/7* positive population of myogenic progenitors, residing in the central part of the dermomyotome. Around E11.5 of mouse development, embryonic myoblasts enter the myotome and fuse into myotubes. More or less at the same stage, during a phase referred to as primary myogenesis, myogenic progenitors (migrated from the dermomyotome to the limb), start to differentiate into multinucleated muscle fibers, commonly known as primary fibers. A second wave of myogenesis (from E14.5 and E17.5 in mouse) known as secondary myogenesis, is characterized by fetal myoblasts fusing with each other [8–10]. At the end of this phase, satellite cells can be morphologically identified as mononucleated cells located between the basal lamina and the sarcolemma. During perinatal and also postnatal development, satellite cells start dividing at a slow pace. Most of the progeny fuse with the adjacent fibers, with new nuclei contributing to growing muscle fibers (whose nuclei are not able to divide). Because of this process, it is possible to think that the majority of the nuclei of a mature muscle are probably derived from satellite cells. Then, when postnatal growth is finished, satellite cells enter a phase of quiescence, but they can be activated when the muscle tissue is damaged or in response to further growth demands. In these cases, satellite cells exit the quiescent state, and undergo a number of cells divisions, thereby producing fusion competent cells that are able either to fuse with damaged fibers or to form new ones. Moreover, part of the cells return instead to quiescence, thereby maintaining the progenitor pool. This ability has led to the suggestion that they represent a type of stem cells [11]. Many factors influence satellite cells' population during myogenesis, such as obesity, diabetes, and other metabolism-related problems. A very important one, for example, is represented by nutrient administration in the maternal stage, which seems to have a direct role in perinatal muscle growth, as extensively explained in Fiorotto and Davis [12].

2. Muscle Stem Cell Niche: Role in Tissue Homeostasis and Muscle Regeneration

Satellite cells occupy an exclusive niche within the muscle tissue, with both stem-like properties and demonstrated myogenic activities. As previously stated, satellite cells are able to remain quiescent or they can be activated in response either to growth/regenerative signal/injuries [13]. After this activation, they re-enter the cell cycle and undergo an asymmetric division to maintain self-renewal. Self-renewal is perpetuated via symmetric cell expansion (generating two identical daughter stem cells) or through an asymmetric cell division (generating both a stem cell and a committed progenitor daughter cell) [14].

Of the two formed daughter cells, one goes back replenishing the niche, then becoming quiescent again; meanwhile, the other participates in the muscle regeneration/growth/homeostasis process. This mechanism is finely regulated. In fact, satellite cell fate is tuned by mechanisms involving both cell-autonomous and external stimuli, in concert with the programmed expression and action of various transcription factors [15,16]. The complex processes governing satellite cell activation and myogenesis have attracted much interest over the years and have been beautifully revised [16,17]. Notably, the decision to undergo symmetric or asymmetric self-renewal is a critical step in satellite cell fate determination, and a deregulation of this process could potentially have detrimental consequences on the execution of a muscle regeneration program. Satellite cells are located beneath the basal lamina in a quiescent state, in which they express Pax7 and Myf5 [18]. When they are activated and differentiate into myoblasts, they express MyoD and myogenin (Myog). If a Pax7+ cell population is deleted, skeletal muscle regeneration is impaired, thereby reinforcing the importance of these cells in this process [19]. After muscle injury, there is a time-dependent and well-organized inflammatory response that happens together with satellite cell activation and through their differentiation process. The recruitment of immune cells to the site of injury is pivotal to obtaining complete skeletal muscle regeneration [20]. The acute inflammatory response following muscle injury usually begins with neutrophils infiltration [21]. This is usually followed by an infiltration of macrophages carrying an

M1 phenotype, which produces mostly inflammatory cytokines, such as tumor necrosis factor-alpha (TNF-α), interleukin 1 beta (IL-1β), and interferon-gamma (IFN-γ) [22].

The addition of these cytokines in primary myoblasts' culture remarkably increases cell proliferation, supporting the assumption that the early increase in M1 macrophage population and the first phase of inflammation actively participate in satellite cell activation [23]. Afterward, an important expansion of M2 macrophages does occur, which is associated with tissue repair and satellite cell differentiation [24]. In fact, M2 macrophages produce different cytokines, such as interleukins IL-4 and IL-10, which improve myoblast differentiation in vitro and increase Myog expression levels, the transcription factor that is essential for satellite cell terminal differentiation [25]. Therefore, M1 and M2 macrophages' kinetics are critical for the early steps of muscle regeneration.

Epigenetic regulation mechanisms can also play a role in satellite cell pool maintenance by modulating proliferation and differentiation. A complete epigenetic profiling of quiescent satellite cells, obtained by Liu and colleagues via chip-seq analysis, has shown that during quiescence, chromatin is maintained in a transcriptionally permissive state, thereby allowing various epigenetic modifications, leading to increased expression of genes involved in satellite cells' proliferation. In particular, DNA methylation of some genes promoters (e.g., *Notch*, Notch homolog 1, translocation-associated), has been found to cause changes in satellite cell renewal, maintenance, and homeostasis [26–28].

In addition, mitochondrial functions, particularly fission and fusion, have been recently reported to play a role in maintaining and dictating satellite cells' fates. In fact, mitochondria are strongly connected to metabolic programming during quiescence, activation, self-renewal, proliferation, and differentiation. Interestingly, mitochondrial adaptation might take place to modify satellite cells' fates and function in the presence of different environmental cues, and under different metabolic states [29,30]. Therefore, satellite cells' functional outcomes are strongly associated with mitochondrial energy output [31,32]. Mitochondrial functions are so broad that some of them, including regeneration, could be interesting targets for pharmacological therapy.

However, while muscle repair after damage is an efficient process in healthy muscle, its probability of success appears low in many muscle disorders. The maintenance of an efficient regeneration process is guaranteed by both satellite cells' niche environment and satellite cells' pool. By disrupting either one of the two or both, the impairment in muscle regeneration suddenly happens, as likely occurs in many muscular dystrophies. In Duchenne muscular dystrophy (DMD), the most frequent and most studied one, it is still unclear which phenomenon comes first, also in relation to the different roles that cytokines play on adult myofiber and satellite cells and their complex crosstalk. However, the plethora of data collected in DMD over the last decades, also with the extensive use and characterization of the *mdx* mouse, its main animal model [33]. allow a deeper insight into the possibility to improve regeneration efficiency as a consequence of therapeutic approaches, from classical drugs to cell therapy.

Pharmacological approaches, even if unable to restore the primary defect, can target disease pathophysiology and progression, by acting at different stages of the inflammatory cascade and therefore slowing down the necrotic process. Yet, the outcomes of such strategies on regeneration efficiency are still unclear. Similarly, other approaches aim at enhancing regeneration with a direct drug action on satellite cell activation and differentiation, although such an effect without a parallel mechanism to minimize the damage of mature myofiber appears to be weak. Herein, we critically revised some of the main approaches used preclinically and clinically in DMD in the attempt to assess their potential outcomes in maintaining or enhancing the regeneration potential, also in relation to the mechanism of action. The overview of these effects may help the community to go back to basic scientific research with a better understanding of the imbalance in the social network governing muscle repair and stem cell niche in relation to disease mechanisms to better address therapeutic intervention for tissue repair.

3. Duchenne Muscular Dystrophy: Is There a Role for Satellite Cells and Their Niche?

3.1. DMD General Picture

DMD is a lethal progressive pediatric muscle disorder. It is genetically inherited as an X-linked disease caused by mutations in the dystrophin gene. The *DMD* gene (which encodes the protein dystrophin) is affected by point mutations, duplications, and deletions of parts of the gene, causing alterations in the reading frame and consequent truncation of the dystrophin protein, which is then rapidly degraded. Dystrophin protein is mainly expressed in skeletal and cardiac muscle and to a lesser extent in smooth muscle and the brain. Dystrophin represents an essential component of the large oligomeric dystrophin-glycoprotein complex (DGC) [34,35]. The DGC acts as a connector between the actin cytoskeleton of the myofiber and the surrounding extracellular matrix (ECM) through the sarcolemma. In the absence of dystrophin, DGC assembly is impaired which weakens the muscle fibers, rendering them highly susceptible to injury. At this point, muscle contraction-induced stress results in constant cycles of degeneration and regeneration [36]. Eventual accumulation of inflammation and fibrosis lead to progressive muscle weakening and loss of muscle mass and function [37]. The efficiency of regeneration appears to be low [38]. This progressive muscle wasting condition leads to severe disability and follows with premature death in affected individuals due to respiratory and/or cardiac failure, typically by or before the age of 30.

3.2. DMD and Stem Cell Polarity

DMD has also been considered a stem cell disease, as a failing regeneration is a typical feature. In fact, there is still a debate about the main determinant of the regeneration failure; it is not clear whether the lack of dystrophin impairs satellite cells' ability to repair the muscle, or the disruption of the stem cells niche environment, or the two altogether. The role of the niche for stem cells in muscle repair is, in general, crucial. In the case of DMD, the niche environment is believed to be compromised by the cascade of events due to constant inflammation and muscle degeneration [39]. However, the absence of dystrophin can play a key role, by affecting asymmetric division. In general, cell cortex polarization and specification of the mitotic spindle orientation are critical steps for the asymmetric localization of cell fate determinants [40].

Importantly, dystrophin interacts with the cell polarity-regulating serine/threonine-protein kinase MARK2 at the muscle fiber membrane [41]. Moreover, it has been demonstrated that activated satellite cells also express dystrophin [42]. Interestingly, dystrophin protein expression has been found to be polarized in satellite cells, and apparently, it is expressed at a very high level when cells are about to undergo cell division. Therefore, dystrophin has a pivotal role in regulating polarity in asymmetric satellite cell division. In support of this view, Chang et al. described a significant reduction in asymmetric division, with a consequent progressive loss of myogenic progenitor, in myoblast-derived satellite cells isolated from *mdx* mouse [43]. This phenomenon seems due to the *mdx*-derived loss of polarity of *mdx* satellite cells, which then resulted in defective mitotic spindle orientation, causing an impairment in asymmetric cell division. This observation supports the hypothesis that the absence of dystrophin, even at the satellite cell level and during asymmetric division, is a significant contributing factor in the failing repair efficiency manifestation of DMD phenotype [43].

However, the niche surrounding environment can also play a key role in the effectiveness of regeneration. This will be addressed in the following paragraph. An important question that remains unaddressed is to determine what the fate is of dystrophic satellite cells that are unable to undergo proper cell division. In accordance with the observation that *mdx* satellite cells exhibit a reduced ability to commit to the myogenic program, another recent study found that satellite cells from *mdx* mice have reduced myogenic potential and initiate a fibrogenic program. It is conceivable that satellite cell dysfunction in DMD can also account for enhanced fibrosis.

3.3. Inflammation and Regeneration Efficiency in DMD

In DMD, muscle tissue goes through continuous cycles of degeneration/regeneration. In the latest stages of the pathology, muscle tissue is substituted by fibrotic and adipose tissue mostly due to the inability of satellite cells to repair muscle damage. Chronic inflammation is a typical hallmark of DMD and may contribute to impaired skeletal muscle regeneration. Although many cells are involved in chronic inflammation, one of the most important roles is played by macrophages, since these cells are associated not only to satellite cells' activation but also to the survival of fibro/adipogenic progenitors (FAPs), which outcompete the satellite cell population during inflammation [44]. By competing with satellite cells, FAPs can increment the fibrotic process; an imbalance between the two populations ultimately leads to the accumulation of FAPs in skeletal muscles with consequent aberrant production of pro-fibrotic factors (e.g., ECM components). Thus, a pharmacological reduction of FAP accumulation could potentially help in preserving satellite cells and their ability to repair injured muscles.

Adding to the role of macrophages in modulating satellite cell proliferation and differentiation, a recent study demonstrated that the cytokines, produced by both M1 and M2, infiltrated macrophages in the injured skeletal muscle, are able to modulate ECM production through FAPs [45]. In particular, it has been shown that in physiological condition, ECM components' production by FAPs was regulated by TNF-α or transforming growth factor-beta 1 (TGF-β1), which are both secreted by M1 and M2 macrophages. On top of that, the M1 and M2 macrophage kinetics after muscle damage supported the initial accumulation followed by FAP apoptosis, avoiding the aberrant deposition of ECM in skeletal muscle. In this regard, an increase in both cytokines can be responsible for excessive ECM accumulation, thereby leading to poor or non-effective skeletal muscle regeneration. Based on those previous studies, FAPs and macrophages have been characterized as some of cells associated with generation and maintenance of the microenvironment responsible for satellite cells' activation and differentiation, i.e., the satellite cell niche, pivotal during skeletal muscle regeneration process. Even though the acute inflammatory response is associated with proper skeletal muscle regeneration, chronic and non-resolute inflammation, which is observed in the skeletal muscle of idiopathic inflammatory myopathies, dystrophies, and aging, is strongly associated with the impaired functions of satellite cells, immune cells, and FAPs, leading then to fibrosis accumulation and poor skeletal muscle regeneration [45,46].

Regarding M1 macrophage over-activation, some in vitro studies show that higher levels of the cytokines produced by these cells (e.g., IL-1β, TNF-α, IFN-γ) are able to mitigate or abrogate myoblast proliferation [23,47]. Moreover, the continuous stimulation of myogenic progenitor cells by IFN-γ leads to the suppression of genes responsible for terminal differentiation. This suppression accounts for the activity of the histone methyltransferase EZH2, which is mediated by the class II major histocompatibility complex transactivator [48]. A chronic increase in IFN-γ and class II major histocompatibility complex transactivator levels repress the expression of genes related to the late stages of satellite cell differentiation by enhancing the promoter region of these genes [48]. Although the level and chronicity of IFN-γ required to start these epigenetics effects in vivo are unknown, these findings suggest that a persistent increase in M1 macrophages expressing IFN-γ for a long time can definitely impair skeletal muscle regeneration. Table 1 summarizes the main cytokines involved in the early and late damaging signals, and in the pro-fibrotic pathways.

Table 1. Cytokines involved in satellite cell regeneration. List of cytokines involved in the inflammatory pathway for muscle regeneration and maintenance. The listed cytokines take part either in proliferation or differentiation of the satellite cell population aiming to repair muscle tissue after injury. Their role in relation to the phase of regeneration is indicated.

Cytokines	Effects on Satellite Cells (Early Phase)	Effects on Myoblasts (Later Phases)	References
IL-1β	Pro-inflammatory; increases SCs proliferation and coordinates interactions between SCs and microenvironment	Reduces myogenic differentiation	[23,49]
IL-4	Improves myoblast differentiation in vitro and increases Myog expression	Plays a role in SCs fusion and growth	[50]
IL-6	Pro-inflammatory; induces SCs proliferation	Stimulates hypertrophy and promotes myoblast differentiation	[51]
IL-7	None reported	Possible involvement in inhibiting differentiation (limited data available)	[52]
IL-10	Anti-inflammatory, counteracts IL-6; no effects on proliferation	Stimulates differentiation	[53]
IL-13	Pro-inflammatory; increases SCs proliferation	Fusion-promoting activity	[48,54]
IFN-γ	Pro-inflammatory; increases SCs proliferation	Impairs differentiation via inhibition of Myog expression	[48]
TGF-β1	Pro-fibrotic; maintains and induces SCs quiescence	Inhibits differentiation	[55]
TNF-α	Pro-inflammatory; increases SCs proliferation, activates SCs to enter the cell cycle via p38 MAPK activation	Inhibits differentiation and fusion	[23]

Abbreviations: SCs, satellite cells; IL, interleukin; IFN-γ, interferon γ; Myog, myogenin; TGF-β1, transforming growth factor β1; TNF-α, tumor necrosis factor α.

4. Pharmacological Approaches Targeting Niche Homeostasis: What We Learned from DMD

For many years, scientists have been putting a lot of effort into finding an effective and definitive treatment for DMD patients. Although pharmacological and technological progress has been made, there is still no absolute cure for this severe disease. Several promising gene and molecular therapies are currently under investigation, aimed at targeting the primary defect. These include gene replacement, exon skipping, and suppression of stop codons [56–58]. More recently, the promising gene-editing tool CRISPR/Cas9 has been offering exciting perspectives for restoring dystrophin expression in patients with DMD [59–61].

In parallel, various therapeutic strategies have been explored with drugs able to target the complex secondary mechanisms responsible for DMD pathogenesis. The aim is to find drugs safer than the current standard of care represented by corticosteroids. Indeed, glucocorticoids are beneficial to prolonging ambulation in DMD boys and are initiated early before other symptomatic therapies. The main efficacy of steroids is believed to be related to the control of inflammation [62–65].

Along this line, a large number of drugs have been investigated in DMD, many of them aimed at reducing inflammation and fibrosis, and they are then able to shut down the pathological loop leading to progressive damage [58]. Among this surge of new experimental pharmacotherapies, in this review, we will revise available data to assess whether drugs can help to maintain a proper equilibrium in satellite cell self-renewal via direct action, or mainly by regulating the inflammatory response and controlling fibrosis. In fact, to date, it is still unclear whether there is a pharmacological treatment that can help in maintaining better muscle homeostasis and improving satellite cell efficiency.

4.1. Biomarkers of Regeneration in DMD: Advantages and Limitations

From the perspective of evaluating the potential ability of old and new pharmacological strategies to modulate skeletal muscle regeneration in DMD, it is crucial to rely on a robust panel of translational biomarkers to obtain a more complete view on and assessment of the regenerative process in preclinical research. Recently, besides several valuable indices commonly used to quantify regeneration, degeneration, and repair efficiency, the advances in technology offered many possibilities to implement the number of regenerative biomarkers to be assessed. All the identified biomarkers described in this paragraph are summarized in Table 2.

The histological evaluation of dystrophic muscles in DMD patients and animal models is the most traditional way to quantify and characterize damage and regeneration. The classical and standardized hematoxylin and eosin (H&E) staining protocol (TREAT-NMD Standard Operating Procedures for the *mdx* mouse model; DMD_M.1.2.007) enables evaluating histopathology. In this context, the proportion of centronucleated fibers (CNF) represents a common index of regeneration, in parallel to a morphological change in size of nascent muscle cells (DMD_M.1.2.001) [66]. However, this structural assessment of the regenerative process is characterized by intrinsic limitations and variables which can complicate data interpretation; e.g., concerning how the level of centronucleation is associated with a still efficient/non-efficient repair system, the identification of activated satellite cells and the amount of fibrosis depending on pathology stage. Part of these uncertainties can be solved with immunohistochemistry (IHC) and immunofluorescence (IF) techniques that allow us to appreciate the indices of satellite cell activation and regeneration, and the presence of specific markers of regeneration in fused myotubes. In this frame of knowledge, a robust regenerative biomarker in different muscles of *mdx* mice at different ages is the presence of developmental myosin heavy chains (embryonic and neonatal MyHCs), usually assessed by IF imaging [67].

High levels of MyHCs are also considered valuable indicators of disease severity which correlate well with functional impairment in DMD boys. However, this index is also subjected to misinterpretation, since MyHCs may be occasionally present in non-regenerating fibers and are differentially expressed throughout the regeneration process. IHC and IF can allow detecting the presence and cellular localization of any cytokine and transcription factor potentially involved in regeneration, and then help to quickly characterize the efficiency and extent of the process in natural disease history and as effects of therapeutics. qRT-PCR and gene array platforms, together with immunoblotting, ELISA, and proteomic arrays, are also widely used to detect regeneration biomarkers while researchers are intending to gain better insight into the mechanisms behind the regenerative process in DMD, and to assess whether drugs can modulate the expression of these indicators of regeneration in DMD. The transcription factor Pax7 is frequently assessed with various imaging and quantitative approaches, since its expression is directly related to the maintenance of the satellite cell pool, in parallel with the relative expression of other myogenic regulatory factors Myf5, MyoD, and Myog, in relation to the stage of the regenerative process [68,69].

A regeneration-associated biomarker of growing interest is represented by utrophin, an autosomal analogue of dystrophin (80% similarity between the two proteins), physiologically abundant in early developing muscles, and progressively replaced by dystrophin at the sarcolemma level towards birth. In DMD and *mdx* muscles, utrophin is upregulated because of the repairing process, but not to the extent to efficiently compensate for dystrophin absence [68]. As pointed out recently, utrophin sarcolemmal localization and the homogeneity of its signal across the whole muscle in correlation with ongoing regeneration, are critical to assess potential protection resulting from direct or indirect stimulation of its upregulation. Thus, the importance of combining imaging techniques to identify utrophin located at myofibers sarcolemma with the traditional assessment of regenerating fibers and their size is increasing [67,68,70,71].

Furthermore, since inflammation is crucial in modulating the muscle regeneration microenvironment, a more detailed view of the ongoing inflammatory process can be obtained by checking cytokine expression and their intracellular signaling, and by the parallel assessment and relative proportion of M1 and M2 macrophage phenotypes in relation to a drug treatment [68,72]. In parallel, considering the existing cross-talk between myogenesis and angiogenesis during muscle regeneration, orchestrated by restorative macrophages in vivo, the levels of growth factors and particularly of vascular endothelial growth factor (VEGF) along with its receptors, represent another set of biomarkers of interest to monitoring the progression of the microvasculature [73].

The multifunctional cell-surface protein neural cell adhesion molecule (NCAM) is expressed in activated satellite cells and during myogenic differentiation, representing a useful tool to evaluate active muscle regeneration following spontaneous and/or induced degeneration, and the proportion of adult myogenic cells already committed to differentiation [74,75].

Blood-circulating biomarkers are also becoming increasingly attractive for monitoring DMD disease progression and the efficacy of experimental therapeutic options. Among these, an emerging candidate for evaluating muscle regeneration in dystrophic animal models and DMD patients is serum osteopontin (OPN), an inflammatory cytokine and myogenic factor which has been recently found to be highly elevated in the early disease phase of CXMD$_J$ (canine X-linked muscular dystrophy) dogs in Japan. Importantly, high serum OPN levels correlate well with phenotypic severity in CXMD$_J$ dogs [76,77]. Similarly, in DMD patients, a single-nucleotide polymorphism (SNP, rs28357094T>G referred to as the G allele) in the promoter of OPN gene *SPP1* has been identified as a genetic modifier of disease severity by modifying OPN activity [78,79]. It has been recently reported that, in the *mdx* mouse, OPN exacerbates the dystrophic phenotype by skewing macrophage polarization and promoting TGF-β1 activation via matrix metalloproteinase-9 (MMP-9) [80]. This extracellular protease and its inhibitor TIMP-1 are also strongly suggested as DMD progression plasma biomarkers. In fact, high serum levels of both MMP-9 and TIMP-1 are associated with dystrophic pathology; however, only MMP-9 has been shown to increase age-dependently, thereby becoming a marker of late-stage disease in older, non-ambulant patients [81]. Importantly, although the precise mechanisms by which MMP-9 regulates dystrophic muscle regeneration are still unclear, the knock-out of MMP-9 in *mdx*$^{Mmp9-/-}$ mice has been found to augment satellite cell proliferation and transplanted myoblast engraftment in muscles, accompanied by a significant reduction of M1 macrophages with a concomitant increase in the number of pro-myogenic M2 macrophages [82].

Several microRNAs (miRNAs) are specifically expressed in healthy skeletal muscle fibers, playing a crucial role in muscle development; DMD patients and *mdx* mice share a common altered signature of muscle-specific miRNAs [83]. Consequently, miRNAs are attracting increasing interest in recent years. In particular, bloodstream levels of specific regeneration and degeneration miRNAs have been proposed as bona fide markers for DMD diagnosis and therapeutic outcome [84]. In particular, miR-1 and miR-133, normally expressed in mature muscle fibers, are 2-fold reduced in the DMD muscle signature, whereas the levels of regeneration miRNAs, including miR-206, are doubled in satellite cells and proliferating myoblasts of dystrophic muscles. In parallel, the high serum levels of all these miRNAs in DMD patients and animal models compared to controls derive from the intensive degeneration occurring in DMD muscles. Interestingly, high levels of circulating miRNAs correspond to poor ambulant activity in patients [84].

In addition, other less canonical biomarkers of regeneration can come from functional studies at the cellular level. For instance, the expression and function of specific ion channels in myofibers may be useful indicators of the repairing process, and of activation of myogenic process and myofiber differentiation. These include various voltage-gated ion channels, such as Nav1.4, Kir, Cav1.1 [85]. One of such biomarkers we had the chance to better characterize in the frame of degeneration/regeneration events in *mdx* muscles is the macroscopic conductance to ClC-1 muscle chloride channel (gCl). ClC-1 channel is a skeletal muscle-specific channel of key importance for its role in setting sarcolemmal excitability. Its expression and function are strictly developmentally, phenotypically, and nerve regulated. The gCl is directly sensitive to inflammation, as shown by its decrease in response to inflammatory cytokines, chronic exercise, and angiotensin II (ANGII) in wt and *mdx* animals. In parallel, gCl is increased during active regeneration phases and by regeneration-promoting factors, such as IGF-1, as also shown in response to drugs with anti-inflammatory properties, such as gold standard α-methylprednisolone (PDN) [86–88]. The main disadvantage of functional biophysical biomarkers resides in the complex and time-consuming methodology required, which limits validation.

Finally, the evaluation of the expression of main regulators of stem cells division and polarization (e.g., partitioning-defective Par1b and Pard3) to monitor the ability of satellite cells to enter the myogenic program, maintain cell polarity, and ensure a proper asymmetric division is of the utmost importance to assess the effects of pharmacological approaches on dystrophic muscle stem cell niche balance [42,88].

Since none of these biomarkers may unambiguously identify the regenerative state in dystrophic muscles, it is essential to complementarily use these indices and to implement research to identify new ones, with the final purpose of obtaining an exhaustive view of these highly-orchestrated mechanisms, their alteration in the pathology, and the effects of drugs. In light of these observations, the following paragraphs will review the most relevant results obtained by standard therapy (i.e., glucocorticoids) and novel pharmacological approaches in DMD, with a particular focus on preclinical findings highlighting the ability of these drugs to enhance regeneration efficiency in dystrophic muscles, via the analysis of predictive biomarkers. Considering the great amount of preclinical data available on the *mdx* mouse model and the plethora of new compounds proposed and repurposed as potentially effective treatments in DMD, specific attention has been devoted to describing the impacts of drugs targeting muscle stem cell niche homeostasis in the regenerative process, particularly of those directed against pro-inflammatory and pro-fibrotic mediators, and of drugs directly aimed at directly modulating satellite cell self-renewal.

Table 2. Biomarkers of regeneration in DMD. List of tissue and circulating biomarkers identified for the assessment of regeneration in dystrophic animal models and also in DMD patients. The main techniques to perform their assessment at the structural and molecular level and the meaning of each biomarker in the regenerative process in relation to disease stages are also indicated.

Regenerative Biomarkers in DMD					
Biomarker	**Sample Type**	**Detection Method**	**Disease Phase**	**Role-Meaning**	**References**
Centronucleation and variation in fiber size	Skeletal muscle	Histology (H&E)	Early stage	Index of degeneration/ regeneration cycles	*TREAT-NMD SOPs DMD_M.1.2.007, DMD_M.1.2.001;* [66,68]
Embryonic and neonatal MyHCs	Skeletal muscle	IHC, IF imaging	Differential expression depending on muscle and age	Indicator of muscle damage; correlates with functional impairment	[67]
Macrophage phenotypes (M1, M2)	Skeletal muscle	IHC, IF imaging	Early stage	Immune response during degeneration/ regeneration	[68,72]
Pax7, Myf5, MyoD, Myog	Skeletal muscle	IHC, IF imaging; qRT-PCR, gene arrays; WB, ELISA, protein arrays	Differential expression depending on myogenesis stage	Myogenic regulatory factors	[68,69]
Par1b, Pard3	Skeletal muscle	IHC, IF imaging; qRT-PCR, gene arrays; WB, ELISA, protein arrays	Early stage	Regulators of stem cells asymmetric division and polarization	[42,68]
Utrophin	Skeletal muscle	IHC, IF imaging (for sarcolemmal localization); qRT-PCR; WB	Early stage	Abundant in early developing muscles and during repair	[68]
NCAM	Skeletal muscle	IHC, IF imaging	Early stage	Marks adult myogenic cells committed to differentiation	[74,75]

Table 2. *Cont.*

Regenerative Biomarkers in DMD					
Biomarker	**Sample Type**	**Detection Method**	**Disease Phase**	**Role-Meaning**	**References**
VEGF	Skeletal muscle	IHC, IF imaging	Early stage	Indicator of microvasculature progression	[73]
Osteopontin	Serum, Skeletal muscle	ELISA, IF imaging	Early stage	Secreted by myoblasts and macrophages after injury; correlates with disease severity	[76–78]
MMP-9, TIMP-1	Serum	ELISA	Late stage (age-dependent increase of MMP-9)	Remodeling of ECM; activation of latent TGF-β1; inhibition of MMP-9 increases SCs proliferation	[81,82,89]
MicroRNAs signature (miR-1, miR-133, and miR-206)	Serum, Skeletal muscle	qRT-PCR	Differential expression in plasma/muscle depending on regeneration level	Specifically expressed in muscle and released in the bloodstream as a consequence of fibers degeneration	[83,84]
Ion channel biophysics, i.e., macroscopic conductance to ClC-1 chloride channel (gCl)	Skeletal muscle	Intracellular recordings with glass microelectrodes	Early and late stages	Biophysical index directly sensitive to inflammation; increased by regeneration and anti-inflammatory drugs	[86,87]

Abbreviations: ELISA, enzyme-linked immunosorbent assay; H&E, hematoxylin and eosin; IF, immunofluorescence; IHC, immunohistochemistry; MMP-9, matrix metalloproteinase-9; Myf5, myogenic factor 5; MyHC, myosin heavy chain; MyoD, myoblast determination protein 1; NCAM, neural cell adhesion molecule; Par1b, partitioning-defective 1b; Pard3, partitioning-defective 3 homolog; Pax7, paired box protein 7; TIMP-1, tissue inhibitor of metalloproteinases 1; VEGF, vascular endothelial growth factor; WB, Western blot.

4.2. Glucocorticoids: Disease-Related Effects on Degeneration/Regeneration Efficiency for an Old Class

Glucocorticoids are currently the only established supportive therapy used in DMD boys at early pathology stages, although their severe side effects negatively impact on patients' quality of life. The beneficial effects of gold standard medications (i.e., prednisone, prednisolone, deflazacort) are well documented: they control inflammation, delay pathology progression, and increase loss of ambulation up to 2 years in young DMD patients [62,64,65]. Despite this, the precise molecular mechanisms behind their ability to alleviate DMD symptoms remain largely unknown. In this context, several preclinical and clinical studies suggest that glucocorticoids may exert their primary effects by controlling muscle inflammation and fibrosis, and regeneration. We and others collected extensive evidence about the effects of PDN administration to *mdx* mice and its impact on biomarkers of regeneration (see Table 3).

In multiple studies, we observed that treating *mdx* mice from 4–5 weeks of age with PDN (1 mg/kg; for 4 or 8 weeks), resulted in a potent anti-inflammatory action, as shown by the reduced levels of activated p65 nuclear factor-κB (NF-κB) by IHC, and in a marked reduction of reactive oxygen species (ROS), measured by dihydroethidium IF staining in dystrophic muscles. This was accompanied by a notable increase in extensor digitorum longus (EDL) myofibers gCl [86,87,90]. NF-κB modulation by PDN was also confirmed at transcript levels by qRT-PCR in other studies [91]. Importantly, PDN was also able to increase utrophin expression, measured by IF in *mdx* gastrocnemius (GC) muscle, as a direct index of improved regeneration; in parallel, α and β-dystroglycan were found increased by Western blot, as indices of improved membrane stability [71].

Table 3. Glucocorticoid therapy and degeneration-regeneration efficiency in DMD. Preclinical and clinical observations collected about the impact of glucocorticoid supportive treatments on biomarkers of regeneration in DMD. The observed direct and indirect effects on the regenerative process are briefly listed, and the techniques used for the detection.

Standards of Care for DMD and Regeneration		
Glucocorticoid Drugs	**Direct/Indirect Effects on Regenerative Biomarkers**	**References**
α-methyl-prednisolone (PDN)	4—8 weeks of treatment in *mdx* mice (from 4 weeks of age) • Reduced NF-κB expression and activation (qRT-PCR, IHC) • Increased utrophin expression (IF) • Increased EDL myofibers gCl (electrophysiology) • Increased α- and β-dystroglycan (WB) • Reduced macrophage infiltration (H&E)	[71,86,90]
prednisone	• Weekly treatment in 6-month-old *mdx* mice (for 4 weeks) increased expression of Annexins *A1*, *A6* (gene arrays) • 6-month treatment in **DMD patients** increased muscle satellite cells, reducing fibroblasts and dendritic cells (H&E)	[92,94]
deflazacort	• Weekly treatment in 6-month-old *mdx* mice (for 4 weeks) increased expression of Annexins *A1*, *A6* (gene arrays) • 3-month treatment in **DMD patients** increased the gene and protein expression of Pax7, Myf5, MyoD, C-MET and reduced neonatal MyHC, TNF-α and macrophage-related CD68 (qRT-PCR, IHC)	[92,93]

Abbreviations: c-MET, tyrosine-protein kinase Met; NF-κB, nuclear factor kappa-light-chain-enhancer of activated B cells.

Furthermore, in a recent study by McNally and colleagues, the weekly administration of prednisone or deflazacort to *mdx* mice was associated with more consistent expression of muscle repair markers Annexins *A1* and *A6* compared to daily treatment, in parallel to a reduction of inflammatory macrophage infiltration and fibrosis, suggesting that dystrophic muscle remodeling may be also regimen-specific; therefore an appropriate dose frequency could further enhance muscle recovery and proper regeneration, possibly reducing side effects [92].

In the clinical setting, muscle biopsies from DMD patients treated with deflazacort for 3 months, gene and protein expression analyses of selected regenerative, and regulatory biomarkers showed that the drug increased the most important mediators of myogenesis and myofiber regeneration (Pax7, Myf5, MyoD, C-MET) and reduced neonatal MyHC, indicating an improved maturation process. In parallel, deflazacort strongly decreased TNF-α and macrophage-related *CD68* [93]. Accordingly, a 6-month treatment with prednisone in DMD boys was associated with an increased number of satellite cells, paralleled by a decrease in fibroblasts and dendritic cells [94].

Interestingly, the SNP of *SPP1* identified as a determinant of DMD disease severity has been recently associated with an alteration in response to deflazacort treatment in patients, with an increase in serum OPN levels [79,95], further supporting the existence of a cross-talk between regenerative pathways and glucocorticoids pharmacological actions in dystrophic muscles.

All these findings suggest that the effect of glucocorticoid therapy in dystrophic muscles is at least partially mediated by an improved regeneration and that this effect is "paradoxical" compared to that observed in individuals with functional dystrophin, where glucocorticoids are known to trigger muscle atrophy.

4.3. Pharmacological Approaches Targeting Inflammatory Mediators and Pathways

Among the new therapeutic avenues explored to pharmacologically modulate secondary events in DMD pathogenesis, several attempts have been focused on drugs potentially more effective and safer than standard glucocorticoids, targeted against pro-inflammatory molecules (see Table 4). As already stated in previous paragraphs, some of these mediators are key players of regeneration efficiency due to their direct role in muscle wasting, and then, in the modulation of functional and structural indices. For most of them, a clear impact on regeneration efficiency is lacking, and in some cases, an expected, reduction of centronucleated myofibers has been observed, as a clear consequence of necrosis.

In this general frame, a key target is the transcription factor NF-κB, a key regulator of pro-inflammatory responses in skeletal muscle. Its active form, p65 NF-κB, is highly expressed in dystrophic muscles before symptoms onset. Vamorolone (VBP15)—a Δ9,11 glucocorticoid analogue now under evaluation in Phase II clinical trials on DMD boys (NCT02760264, NCT03038399, NCT02760277) acts as a dissociative steroid, a retaining membrane, and has anti-inflammatory properties of classical steroids but loses the transactivation sub-activity associated with their side effects. First identified by Kanneboyina Nagaraju and colleagues, VBP15 strongly reduced NF-κB and TNF-α expression and percentage of inflammatory foci in 8-week-old *mdx* mice muscles, with a parallel amelioration of functional readouts [91]. A specific search of regenerative biomarkers would be certainly useful to gain more insight into the clinical efficacy of this promising therapeutic agent. Other anti-inflammatory compounds of increasing interest for DMD are edasalonexent (formerly CAT-1004, now being tested in a Phase II trial), NCT02439216 [96], and CAT-1041, two inhibitors of the IκB kinase (IKK)/NF-κB complex. In 4-week-old *mdx* mice, a 20-week treatment with each drug reduced p65 NF-κB, interleukin-6 (IL-6) and OPN protein levels, without modifying utrophin expression.

Histopathology was ameliorated, with a reduction of the total area of damage and of inflammatory macrophage infiltrates, in parallel to in vivo and ex vivo functional indices [97]. In *mdx* mice, IKK conditional deletion clarified that NF-κB functions in activated macrophages to promote inflammation and muscle necrosis, reducing regeneration via inhibition of muscle progenitor cells [98]. In 5-week-old *mdx* mice muscles, a 4-week treatment with the IKK inhibitor NEMO-binding-domain (NBD) peptide, induced strong decreases in macrophage infiltration and p65 NF-κB, measured by electrophoretic mobility shift assay (EMSA). Interestingly, increments in embryonic MyHC positive myofibers and CNF proportion, measured by IF and H&E respectively, were here taken as positive indices of increased regenerative potential, since they were accompanied by a notable inhibition of damaging pathways [99,100].

Approved drugs targeting TNF-α have been evaluated for possibly repurposing DMD, with interesting findings concerning regeneration. A 4-week treatment with etanercept (Enbrel®), a chimaera compound bearing the TNF-α soluble receptor, improved EDL myofibers gCl in adult *mdx* mice, while the histological profile was only modestly ameliorated. Histopathology was also analyzed in GC muscles from *mdx* mice treated from two weeks of age, when the first spontaneous degeneration cycle occurred, showing a significant reduction in the proportion of degenerating fibers; however, no clear index of regeneration was observed [101].

Several studies have been performed to assess the role of non-steroidal anti-inflammatory drugs (NSAIDs), inhibitors of cyclooxygenase (COX) enzymes, in dystrophic muscles, considering the canonical role of prostanoids in sustaining early and chronic inflammation. However, the effects of these drugs in DMD are controversial, likely in relation to the differential and tissue-specific roles of COX-related prostanoids. In fact, the various prostaglandins (PGs) have differential and opposite effects on regeneration and myogenesis, which complicate the outcomes of drugs inhibiting either or both COX-1 and COX-2 isoforms. In *mdx* mice, NSAIDs and COXIBs (e.g., ibuprofen, flurbiprofen, parecoxib) contributed to reducing macrophage infiltration to a different extent, without affecting functional indices or the percentage of regenerating myofibers. Our group could not confirm these effects for meloxicam, a COX-2 selective inhibitor, in *mdx* mice, likely in relation to the inhibition of PGE_2, which exerts a key pro-myogenesis action [101,102]. PGD_2, unlike other prostaglandins, inhibits myogenesis and its metabolites are increased in DMD patients; therefore, increased muscle fiber regeneration can be achieved by specific PGD_2 inhibition. Recently, HQL-79, a potent selective inhibitor of hematopoietic PGD synthase (HPGDS), was found to suppress muscle necrosis in *mdx* mice, without interfering with cytoprotective PGE_2 and other pro-myogenic PGs [103]. A highly-selective HPGDS inhibitor, TAS-205, was found to reduce necrosis and improve locomotor activity in *mdx* mice [104]; although no more specific results are available on its regenerative potential, a recent Phase I trial provided early evidence of a modest if any, potential therapeutic activity in DMD population (NCT02246478).

An interesting drug target is IL-6, a myokine known to induce a harmful inflammatory milieu in human and murine dystrophic muscles, by promoting the transition from acute neutrophil infiltration to chronic mononuclear cell infiltration; however, IL-6 has also been reported to participate in muscle regeneration by promoting myoblast differentiation [105]. In 4-week-old *mdx* mice, the IL-6 pharmacological blockade via neutralizing antibody, ameliorated functional performance, modulated inflammation via NF-κB inhibition, and improved homeostatic maintenance of dystrophic muscles, as evidenced by the significant gene upregulation of *Pax7*, *MyoD1*, *Myog*, *IL-4*, and *Wnt7a*, a secreted factor which drives the "planar cell polarity pathway" to promote satellite stem cell expansion via symmetric division. In another study, IL-6 blockade increased inflammation with no functional improvement, suggesting that attention should be paid to its dual role in dystrophic muscles, concerning any possible drug intervention [106–108].

Additionally, compounds with anti-inflammatory properties related to multiple intracellular actions may exert clear effects on regeneration in DMD. Among these, flavocoxid, a mixed flavonoid with antioxidant and NF-κB inhibiting properties, was described to exert an early, remarkable morphological benefit evidenced by H&E staining in 5-week-old *mdx* mice muscles, by reducing necrosis and mononuclear cell infiltrate, with an increased regenerating area defined as an increase in the number of Myog-positive nuclei by IHC, while CNF were comparable to vehicle [109].

Another wide-acting drug is pentoxifylline (PTX), a non-selective phosphodiesterase inhibitor exerting anti-inflammatory, anti-cytokine, and anti-fibrotic actions linked to a specific ability to inhibit abnormal calcium entry in dystrophic myofibers [74]. We found that a 4-week treatment with PTX restored calcium homeostasis and reduced markers of oxidative stress in *mdx* mice. Histopathology and fibrosis were improved in a muscle-specific manner. Interestingly, although no significant variation in CNF percentage was observed, PTX significantly increased the NCAM-positive area in diaphragm (DIA) and GC. Then, a wide action of PTX can be envisaged: A reduction of muscle necrosis by controlling inflammation-related oxidative stress and calcium homeostasis, while stimulating regeneration via reducing pro-fibrotic signaling and activating satellite cells. In vitro experiments with PTX in C_2C_{12} cell cultures further supported the potential ability of PTX to directly activate satellite cells and promote their growth, likely resulting from cAMP increase in satellite cells [74]. Whitehead et al. provided evidence that the anti-oxidant compound *N*-acetylcysteine (NAC) ameliorates skeletal muscle pathophysiology in GC muscles from 8-week-old *mdx* mice, by reducing ROS production, NF-κB activation, and CNF, and importantly, increasing utrophin and β-dystroglycan levels at the sarcolemma [70].

4.4. Pharmacological Approaches Targeting Pro-Fibrotic Mediators

TGF-β1 is a multifunctional cytokine playing a role as a master regulator of both ECM remodeling and myogenesis. In healthy muscles, a timely activation of TGF-β1 and satellite cells is thought to be critical for muscle recovery and development. In DMD patients and *mdx* mice, high levels of TGF-β1 correlate with disease severity and induce excessive collagen deposition, contributing to fibrosis progression (see Table 4). In parallel, the TGF-β1-SMAD (small mother against decapentaplegic) 2/3 pathway can also inhibit the activation of myogenic regulatory factors, thereby inhibiting proliferation and differentiation of satellite cells [13,110]. Therefore, agents able to prevent fibrosis by reducing TGF-β1 pathway, either directly or via modulation of upstream activating signals or epigenetic mechanisms, may be potentially beneficial to improving regeneration in DMD.

A main anti-fibrotic action has been attempted with halofunginone (HT-100), an anti-coccidial drug that inhibits TGF-β1 downstream signaling. This drug was described to promote satellite cell activation and survival in vitro in cultured myofibers from 6-week-old *mdx* mice, as shown by increased MyoD expression, with parallel reduction of pro-apoptotic Bax and Bcl2 [111].

As previously discussed, MMP-9 is aberrantly regulated in both DMD patients and *mdx* mice and likely involved in the cross-talk between inflammation (NF-κB activation augments MMP-9 expression) and fibrosis (MMP-9 cleaves latent TGF-β1). Early 5-week treatment with the MMPs inhibitor batimastat (BB-94) was able to significantly reduce the mRNA expression of a variety of MMPs, including *MMP-9*, *NF-kB*, *TNF-α*, and *TGF-β1* in *mdx* mice; at the histological level, batimastat-treated GC muscles showed significantly reduced fibrosis; and accumulation of macrophages, CNF, and embryonic MyHC-stained myofibers, paralleled by an increase in utrophin protein expression, measured by Western blot [89,112]. TGF-β1-mediated fibrosis and prevention of proper muscle tissue regeneration is also sustained by the increased expression of connective tissue growth factor (CTGF/CCN2) in DMD patients and *mdx* mice, where CTGF is mainly detected in regenerating fibers and in the interstitium between damaged fibers [110]. FG-3019 is an anti-CTGF antibody, found to control muscle damage and improve function in *mdx* mice after a 2-month treatment. However, the ability of FG-3019 to reduce necrosis (less uptake of Evans Blue dye) and fibrosis, implied also a concomitant reduction of regenerating activity, as shown by decreased levels of embryonic MyHC and Myog [113]. FG-3019 is now being tested in Phase II clinical trials on DMD boys (NCT02606136); again, the search of biomarkers of regeneration in patients would be useful.

The pharmacological inhibition of TGF-β superfamily member myostatin is considered as another attractive therapeutic option for DMD patients, for both increasing muscle mass and helping regeneration via reduction of the non-permissive pro-fibrotic environment [114]. Very recently, Wagner et al. demonstrated that the delivery of a myostatin inhibitor (RK35) in tibialis anterior (TA) muscle of dystrophic *mdx-5*^{Cv} mice, mediated by a biological hydrogel scaffold, was able to modulate muscle's immune microenvironment, by promoting a pro-regenerative macrophage polarization, facilitating the M1 to M2 transition, and facilitating the consequent production of anti-inflammatory cytokine IL-10 [115].

In this general frame, also the epigenetic modulation of myostatin/follistatin axis via histone deacetylase inhibitors (HDACi) deserves attention. In particular, the HDACi givinostat was found to induce a functional improvement in vivo in *mdx* mice, and importantly, a reduction of neutrophil granulocytes evaluated by IF for myeloperoxidase in TA muscle used to quantitate the magnitude of inflammation associated with muscle degeneration [116]. In the clinical setting, a 12-month treatment with givinostat in a Phase II study on DMD boys, significantly decreased total fibrosis, necrosis, and adipose tissue replacement, in parallel to increasing myofibers size, although no direct regenerative biomarker was assessed. Now, a safety and efficacy Phase III multicentre study is ongoing in ambulant patients (NCT02851797; [117]). By the way, the ability of HDACi to enhance regeneration also via nitric oxide (NO) pathways [118] is a main working hypothesis that would deserve to be specifically verified at both preclinical and clinical levels.

Other important approaches to control dystrophic muscle fibrosis are those acting through a multifaceted mechanism or on prime signals in fibrotic pathways, such as angiotensin II-related ones [87]. In *mdx* mice, a 6-month treatment with the antihypertensive losartan, an angiotensin-II type 1 receptor blocker, was found to decrease angiotensin II-mediated TGF-β1 signaling, with marked in vivo functional improvement. At the histological level, it attenuated disease progression and improved regeneration, measured as an increase in neonatal MyHC [119]. Interestingly, we reported that early treatment in *mdx* mice with the ACE inhibitor enalapril exerted mainly an anti-oxidant and anti-inflammatory action (via NF-κB inhibition). CNF percentage was reduced in GC muscle of treated mice, while the increase in gCl of EDL muscle could be related to a reduction of the direct action of ANGII on ClC-1 channel, more than to an enhanced regeneration [86]. This underlines how the disease phase is relevant in determining a different drug response and to observe an effect on regeneration efficiency, which can be more likely to be appreciated after a long-lasting control of the niche environment.

This simple hypothesis is not fully supported by data with other drugs which may exert an anti-fibrotic action in dystrophic muscles. Metformin (MET), a well-known anti-diabetic drug, has been repurposed in combination with NO-donors in clinical trials on DMD patients (NCT01995032). Recent reports described the ability of MET to directly inhibit TGF-β1-SMAD 2/3 mediated fibrosis [120]. Accordingly, we disclosed that a 20-week treatment with MET in *mdx* mice exerted a potent, metabolism-independent anti-fibrotic action evidenced by a significant functional and structural improvement, accompanied by decreased TGF-β1 levels in GC muscle. However, no significant changes were observed on histological biomarkers of regeneration, i.e., CNF proportion. Then, it is feasible that the molecular mechanism underlying the anti-fibrotic action (still under clarification for MET) or other parallel mechanisms can define the outcome on regeneration efficiency [121]. Interestingly, Pavlidou et al. recently reported that, in C57BL/6 mice, MET delayed satellite cell activation and maintained quiescence (as shown by reduced Pax7 protein expression) [44].

In our laboratory, the effects on fibrosis and regeneration biomarkers in *mdx* mice were also investigated after a treatment of 4 or 12 weeks GLPG0492, a non-steroidal selective androgen receptor modulator (SARM), proposed as a potential anabolic therapy for DMD patients. GLPG0492 reduced fibrosis and TGF-β1 levels in DIA muscle; however, neither histological signs of regeneration nor an increase in the expression of regeneration-related genes (*Myog, follistatin*, or *IGF-1*) was found [122], in spite of the fact that androgen receptor modulation is supposed to enhance myogenesis [123]. In parallel, the anticancer drug tamoxifen, a selective oestrogen receptor modulator (SERM), was shown to act as a ROS scavenger and inhibitor of fibroblast proliferation. Dorchies et al. tested the effects of long-term treatment with tamoxifen in the *mdx*-5^{Cv} strain, which was found to induce a slower dystrophic phenotype, by reducing DIA muscle fibrosis and increasing CNF proportion [124]. Tamoxifen has been granted the designation of orphan drug by European Medicines Agency in 2017, and is currently under evaluation in a Phase III multicentre trial in DMD patients (NCT03354039). The apparent controversial results can be due to the different pathways modulated by the drugs that need in turn to be interpreted in the frame of the pathology-related events. Interestingly, estrogen receptor EERα in skeletal muscle is known to be an auxiliary co-activator of PGC-1α in enhancing endogenous anti-oxidant response and mitochondrial oxidative metabolism [125–127]. A role of the latter in satellite cells' activation and stem cell niche has been proposed [127].

4.5. Pharmacological Approaches to Enhance Satellite Cell Myogenic Capacity

Besides cell-based therapies and gene replacement strategies aimed at satellite cell reprogramming in DMD, new pharmacological interventions have been recently explored to target muscle stem cell microenvironment and stimulating intrinsic repair (see Table 4). Asymmetric cell division plays a pivotal role in the maintenance of the satellite cell pool. The granulocyte colony-stimulating factor receptor (G-CSFR) is asymmetrically segregated during muscle stem cell division and the G-CSF/G-CSFR axis supports long-term muscle regeneration in mice. Filgrastim, a G-CSF analogue currently being tested for efficacy and safety in a Phase I study on DMD patients (NCT02814110), increased myocytes and improved regeneration in *mdx* mice [128].

A treatment with the secreted factor Wnt7a, which drives the symmetric expansion of satellite cells [129,130], resulted in increased specific muscle force and reduced contractile damage in *mdx* mice. In parallel, it induced hypertrophy and a shift toward slow-twitch in human primary myotubes [19].

β1-integrin is another essential niche molecule that maintains satellite cell homeostasis, sustaining the expansion and self-renewal of the stem cell pool during regeneration. The exogenous administration of β1-integrin enhanced regeneration in vitro and also muscle function in vivo in *mdx* mice [131].

Interesting preclinical results were also obtained via pharmacological inhibition of p38MAPK, which is aberrantly regulated in regenerating dystrophic muscles, although the exact mechanism and the link with regeneration and myogenesis remains to be better determined. Treatment with the p38MAPK-inhibitor SB731445 in the $Sgcd^{-/-}$ mouse model was able to ameliorate the dystrophic phenotype and to improve the self-renewal of satellite cells [132].

Finally, unacylated ghrelin (UnAG) is a circulating hormone that protects muscle from atrophy, promotes myoblast differentiation, and enhances ischemia-induced muscle regeneration. UnAG was found to reduce muscle degeneration, improve muscle function, and increase dystrophin-null SC self-renewal in *mdx* mice, maintaining the satellite cell pool [133]. These first preclinical observations support the use of drugs aiming to directly restore polarity and proper mitotic division of satellite cells, as part of DMD therapy in the future, although a larger body of evidence regarding the mechanisms underlying their effects in dystrophic muscles will be needed to improve data translatability.

Table 4. Pharmacological approaches targeting niche homeostasis in DMD. Synthetic overview of drugs, new or repurposed, targeting the muscle stem cell niche microenvironment in dystrophic muscles by acting on inflammation, fibrosis, or self-renewal, and of their effects on regenerative biomarkers in *mdx* mice muscles. For drugs translated into clinical settings, the stage of development in DMD patients is indicated, and for repurposed drugs, the approval for other pathological conditions. * ClinicalTrials.Gov identifiers.

Some Novel Pharmacological Strategies Potentially Targeting the Niche Microenvironment in DMD				
Drug	**Molecular Target**	**Direct/Indirect Effects on Regeneration (*mdx* Mouse Model)**	**Clinical Status**	**References**
Inhibition of inflammation				
vamorolone (VBP15)	NF-κB	• Reduced NF-κB and TNF-α expression (qRT-PCR, IF) • Reduction of inflammatory foci (H&E)	Phase II NCT02760264, NCT03038399, NCT02760277 *	[91]
CAT-1004 (edasalonexent) CAT-1041	IκB kinase/NF-κB complex	• Reduced activated p65 NF-κB, IL-6 and osteopontin protein expression (WB)	Phase II (edasalonexent) NCT02439216	[96,97]
NEMO-binding-domain peptide	IκB kinase	• Reduced activated p65 NF-κB (EMSA) • Reduced macrophage infiltrates (H&E)	-	[99,100]
etanercept (Enbrel®)	TNF-α	• Increased EDL myofibers gCl (electrophysiology) • No direct regeneration index observed	FDA-approved for rheumatoid arthritis and psoriasis, no trials for DMD	[101]

Table 4. *Cont.*

Some Novel Pharmacological Strategies Potentially Targeting the Niche Microenvironment in DMD				
Drug	**Molecular Target**	**Direct/Indirect Effects on Regeneration (*mdx* Mouse Model)**	**Clinical Status**	**References**
NSAIDs and COXIBs	COX1 and/or COX2	• Reduced macrophage infiltrates (H&E) • No confirmed effects on regeneration • Meloxicam: possible interference with cytoprotective prostaglandin PGE_2	Anti-inflammatory agents	[101,102]
HQL-79, TAS-205	hematopoietic prostaglandin D synthase	• Suppressed muscle necrosis (H&E) • No interference with myogenic PGs • No specific results available on their regenerative potential	Phase I (TAS-205) NCT02246478	[103,104]
IL-6 neutralizing antibody	IL-6	• Improved the homeostatic maintenance (upregulation of *Pax7*, *MyoD1*, *Myog*, *IL-4*, and *Wnt7a* gene expression) • Increased inflammation with no functional improvement also reported	-	[106,108]
IL-1Ra anakinra (Kineret®)	IL-1β pathway	• No significant modification of disease-related regenerative parameters	FDA-approved for arthritis	[134]
flavocoxid	COX1, COX2, 5-lipoxygenase	• Reduced necrosis and macrophage infiltrates; no variation in CNF percentage (H&E) • Increased number of Myog-positive nuclei (IHC)	-	[109]
pentoxifylline	phosphodiesterase enzymes	• Improved histopathology with no variation in CNF percentage (H&E) • Increased NCAM-positive area (IHC) • Increased cAMP in satellite cells in vitro	Antithrombotic agent	[74]
N-acetylcysteine	wide anti-oxidant action	• Reduced NF-κB activation and ROS • Reduced CNF percentage (H&E) • Increased utrophin and β-dystroglycan levels at sarcolemma	Mainstay therapy for acetaminophen toxicity	[70]

Table 4. *Cont.*

Some Novel Pharmacological Strategies Potentially Targeting the Niche Microenvironment in DMD				
Drug	**Molecular Target**	**Direct/Indirect Effects on Regeneration (*mdx* Mouse Model)**	**Clinical Status**	**References**
Inhibition of Fibrosis				
halofunginone (HT-100)	TGF-β1 signalling	• Promoted satellite cell activation (increased MyoD protein expression) and survival (reduced Bax, Bcl2 protein expression) in vitro (IF, WB)	Anti-coccidial agent	[111]
batimastat (BB-94)	MMP-9	• Reduced mRNA expression of *MMP-9, NF-kB, TNF-α* and *TGF-β1* (qRT-PCR) • Reduced MMP-9 enzymatic activity • Reduced fibrosis, macrophage infiltrates and CNF (H&E, Sirius Red) • Reduced embryonic MyHC and increased utrophin expression (WB)	Anticancer agent	[89,112]
FG-3019 antibody	CTGF	• Reduced muscle necrosis (Evans Blue) • Decreased regeneration (lower levels of embryonic MyHC and Myog)	Phase II NCT02606136	[113]
RK35	myostatin	• Biological scaffold–mediated delivery • Promoted M1 to M2 macrophage transition and increased IL-10 release (IHC, IF, qRT-PCR, in vivo/in vitro assays)	-	[115]
givinostat	histone deacetylase (HDAC)	• Reduction of neutrophil granulocytes (IF for myeloperoxidase) • In DMD boys: successfully completed Phase II study; no direct biomarker of regeneration assessed	Phase III NCT02851797	[116–118]
losartan	ANG II type 1 receptor blocker	• Decreased ANG II-mediated TGF-β1 signalling pathway • Increased neonatal MyHC (H&E, IF)	Antihypertensive agent	[119]
enalapril	angiotensin-converting enzyme	• Early treatment reduced CNF (H&E) • Increased EDL myofibers gCl	Antihypertensive agent	[86]

Table 4. *Cont.*

Some Novel Pharmacological Strategies Potentially Targeting the Niche Microenvironment in DMD				
Drug	Molecular Target	Direct/Indirect Effects on Regeneration (*mdx* Mouse Model)	Clinical Status	References
metformin	AMPK	• Decreased muscular TGF-β1 (ELISA) • No changes in structural regenerative biomarkers (e.g., CNF proportion) • Maintained quiescence and reduced Pax 7 in healthy mice (IF, WB)	Phase III NCT01995032	[44,121]
GLPG0492	androgen receptor	• Reduced muscular TGF-β1(ELISA) • No increase of *Myog*, *follistatin* or *IGF-1* (qRT-PCR)	-	[122]
tamoxifen	oestrogen receptor	• Reduced muscle fibrosis and increased CNF proportion (H&E)	EMA Orphan Drug Designation (2017) Phase III NCT03354039	[124]
Promotion of Self-renewal				
filgrastim (G-CSF analogue)	G-CSFR	• Increased satellite cells and Pax 7 (IF)	Phase I NCT02814110	[128]
Wnt7a	activation of "planar cell polarity pathway"	• Hypertrophy and slow-twitch fiber shift (in human myoblasts cultures)	-	[129]
β1-integrin	MAPK Erk, AKT	• Enhanced regeneration in vitro • Maintained the responsiveness of the niche to Fgf2	-	[131]
SB731445	p38MAPK	• Treatment in the *Sgcd*$^{-/-}$ dystrophic mouse model improved satellite cells self-renewal	-	[132]
unacylated ghrelin	GHS-R; pleiotropic, tissue-specific hormonal activity	• Reduced muscle degeneration • Preserved the satellite cell pool at later stage of pathology	-	[133]

Abbreviations: ANG II, angiotensin II; AMPK, AMP-dependent protein kinase; Bax, Bcl-2-associated X protein; Bcl2, B-cell lymphoma 2; CNF, centronucleated fibers; COX1, COX2, cyclooxygenase 1 and 2; COXIBs, cyclooxygenase 2 inhibitors; CTGF, connective tissue growth factor; EMSA, electrophoretic mobility shift assay; Fgf2, fibroblast growth factor 2; G-CSFR, granulocyte colony stimulating factor; GHS-R, growth hormone secretagogue receptor; IL-1Ra, interleukin 1 receptor antagonist; NSAIDs, nonsteroidal anti-inflammatory drugs; PG, prostaglandin; ROS, reactive oxygen species; Wnt7a, wingless-type MMTV integration site family, member 7A.

5. Discussion

Satellite cells are muscle-committed stem cells resident in skeletal muscle, importantly contributing to muscle growth and differentiation, and allowing an efficient repair process after damage. Multiple intrinsic and extrinsic factors are involved in the orchestration of this complex process, thereby fascinating scientists for potential applications in the field of regenerative medicine, and for developing drugs able to counter progressive muscle wasting disorders, by enhancing an efficient repairing process in inherited or acquired conditions. DMD is a prototype of the efforts in this field, due to the intense research aimed at identifying potential therapies. DMD in fact has no cure; the progressive muscle degeneration is directly related to the events following the primary defect, which are related in a complex cross-talk, to the inefficient regeneration process. Accordingly, satellite cells are believed to be pivotal in determining disease outcome, since the exhaustion of the satellite cell pool causes the absence of a turning point of this muscle disorder. Intrinsic defects of satellite cells, due to the absence of dystrophin, have been described, and the role of the niche in the exhaustion of the satellite cell pool is still debated. As a matter of fact, the niche environment seems to be paramount in the ability of satellite cells to repair the continuous cycles of degeneration/regeneration happening in DMD [15,17,20]. Satellite cells' niche is disrupted over time by the inflammation and fibrosis occurring as the disease progress. Therefore, while the effort to treat the primary defect is still the main target of the scientific community, drugs able to act on muscle environment deserve to be taken into consideration. This approach concerns both novel drugs and repurposed ones, the latter having the additional advantage of a more rapid translational potential from bench to clinic. Drugs with the best clinical potential would ideally target main pathogenic events, reducing damage in parallel, making repair efficient. Knowing all that, we took advantage of the large amount of preclinical data obtained in our and other laboratories with the main aim of summarizing the available evidence of a potential drug action on regenerative efficiency via a direct or indirect action on stem cell niche and satellite cells.

As reviewed here, there are many promising compounds able to improve muscle regeneration, and even glucocorticoids have a positive outcome in improving myofiber regeneration and enhancing the maturation process, mostly in relation to the regimen approach, which may in part account for their clinical efficacy. Importantly, the complex and not fully clear action of glucocorticoids in dystrophic muscle, and mostly in the satellite cell niche, deserves to be better investigated, as these drugs are able to counter degeneration while sustaining the regeneration process.

In particular, drugs acting on different ILs and other inflammatory cytokines are promising [90, 100,104] and should be investigated in more depth. These drugs can take advantage of studies performed in DMD patients, and possibly gain better insight by looking at biomarkers of regeneration. Importantly, great progress has been made to identify reliable non-invasive biomarkers of both pathology progression and regeneration and drug efficacy. These will greatly help the translational assessment of therapy efficacy in the stem cell niche [42,66–83]. In a general situation, even if it is not clear yet whether is the exhaustion of the satellite cell pool or niche disruption drives the pathological progression, we know that drugs acting either on the amelioration of the niche environment or satellite cells' asymmetric division could be good candidates to slow down DMD.

At the same time, what we have learned from a complex disease such as DMD is that the outcome of drug action on regeneration efficiency is not always straightforward in spite of robust hypotheses and a clear mechanism of action. Interestingly, among novel potential therapies, our analysis of literature data enlightened that drugs purely directed against inflammatory molecules (e.g., TNF-α, NF-κB) are mostly able to reduce muscle damage but not improve regeneration, whilst several therapeutic interventions inhibiting or modulating molecules with a pleiotropic action seem to positively impact regenerative biomarkers, in parallel with controlling damage. These molecules, i.e., myokine IL-6, pro-fibrotic TGF-β1, c-AMP dependent pathways and estrogen receptors, and all self-renewal mediators (e.g., G-CSF, Wnt7a, β1-integrin, p38MAPK, ghrelin derivatives), represent potential therapeutic targets of new/repurposed drugs for DMD to specifically support regeneration efficiency, via a direct action on satellite cells or by improving niche environment. In parallel, the effects on regeneration via the pharmacological modulation of other promising targets, e.g., HDACs and ANGII pathways, need to be further explored. In particular, this topic will grant further insight into the specific roles in the regeneration process of different HDAC isoenzymes, and roles in alternative pathways of the RAS system, such as those mediated by ANG 1-7 via MAS receptor [86,87,118].

Then, we still need to understand key aspects of this multi-actor process that in turn have to be approached at different levels. In fact, there are major key players in disease progression and homeostasis maintenance, and it is reductive to assess that the disease progression is caused by the exhaustion of the satellite cells' niche. Then, combined strategies may work in synergy and such synergy may also occur with innovative molecular or cell therapies able to restore dystrophin expression. In fact, drugs able to address regeneration efficiency, although they will not cure, create ideal circumstances to sustain therapies aimed at correcting primary DMD defects, by creating a healthier environment. This has not been done yet, and the possible outcomes are not predictable, although they would ideally be of high clinical relevance.

In addition, the possibility to enhance our understanding of drug targetable events in myogenesis may also help to improve the in vitro approach of tissue engineering for building up both experimental platforms for drug discovery and simulation of reparative medicine.

Funding: This research was funded by Fondazione CON IL SUD "Brains to South" project n. 2018-PDR-00351 granted to O.C. entitled "Optogenetic engineered artificial muscle", by PRIN-MIUR (Research Projects of National Interest Ministry of Education, University and Research) project n. 2017FJSM9S_005 entitled "New pharmacological strategies modulating PGC1alpha signalling and mitochondrial biogenesis to restore skeletal and cardiac muscle functionality in Duchenne Muscular Dystrophy", granted to A.D.L., and by The Dutch Duchenne Parent Project NL (DPP NL) Small Project entitled "Optogenetic engineered artificial muscle to enhance preclinical studies in DMD", granted to O.C.

Conflicts of Interest: The authors declare no conflict of interest.

References

1. Tajbakhsh, S. Skeletal muscle stem cells in developmental versus regenerative myogenesis. *J. Intern. Med.* **2009**, *266*, 372–389. [CrossRef] [PubMed]
2. Sambasivan, R.; Tajbakhsh, S. Skeletal muscle stem cell birth and properties. *Semin. Cell Dev. Biol.* **2007**, *18*, 870–882. [CrossRef] [PubMed]
3. Biressi, S.; Molinaro, M.; Cossu, G. Cellular heterogeneity during vertebrate skeletal muscle development. *Dev. Biol.* **2007**, *308*, 281–293. [CrossRef]
4. Gros, J.; Manceau, M.; Thomé, V.; Marcelle, C. A common somitic origin for embryonic muscle progenitors and satellite cells. *Nature* **2005**, *435*, 954–958. [CrossRef] [PubMed]
5. Kassar-Duchossoy, L.; Giacone, E.; Gayraud-Morel, B.; Jory, A.; Gomès, D.; Tajbakhsh, S. Pax3/Pax7 mark a novel population of primitive myogenic cells during development. *Genes Dev.* **2005**, *19*, 1426–1431. [CrossRef] [PubMed]
6. Relaix, F.; Rocancourt, D.; Mansouri, A.; Buckingham, M. A Pax3/Pax7-dependent population of skeletal muscle progenitor cells. *Nature* **2005**, *435*, 948–953. [CrossRef]
7. Schienda, J.; Engleka, K.A.; Jun, S.; Hansen, M.S.; Epstein, J.A.; Tabin, C.J.; Kunkel, L.M.; Kardon, G. Somitic origin of limb muscle satellite and side population cells. *Proc. Natl. Acad. Sci. USA* **2006**, *103*, 945–950. [CrossRef] [PubMed]
8. Duxson, M.J.; Usson, Y.; Harris, A.J. The origin of secondary myotubes in mammalian skeletal muscles: Ultrastructural studies. *Dev. Camb. Engl.* **1989**, *107*, 743–750.
9. Dunglison, G.F.; Scotting, P.J.; Wigmore, P.M. Rat embryonic myoblasts are restricted to forming primary fibres while later myogenic populations are pluripotent. *Mech. Dev.* **1999**, *87*, 11–19. [CrossRef]
10. Evans, D.; Baillie, H.; Caswell, A.; Wigmore, P. During fetal muscle development, clones of cells contribute to both primary and secondary fibers. *Dev. Biol.* **1994**, *162*, 348–353. [CrossRef] [PubMed]
11. Collins, C.A.; Partridge, T.A. Self-renewal of the adult skeletal muscle satellite cell. *Cell Cycle* **2005**, *4*, 1338–1341. [CrossRef] [PubMed]
12. Fiorotto, M.L.; Davis, T.A. Critical Windows for the Programming Effects of Early-Life Nutrition on Skeletal Muscle Mass. *Nestle Nutr. Inst. Workshop Ser.* **2018**, *89*, 25–35. [CrossRef]

13. Forcina, L.; Miano, C.; Musarò, A. The physiopathologic interplay between stem cells and tissue niche in muscle regeneration and the role of IL-6 on muscle homeostasis and diseases. *Cytokine Growth Factor Rev.* **2018**, *41*, 1–9. [CrossRef] [PubMed]

14. Yin, H.; Price, F.; Rudnicki, M.A. Satellite cells and the muscle stem cell niche. *Physiol. Rev.* **2013**, *93*, 23–67. [CrossRef] [PubMed]

15. Chargé, S.B.P.; Rudnicki, M.A. Cellular and molecular regulation of muscle regeneration. *Physiol. Rev.* **2004**, *84*, 209–238. [CrossRef] [PubMed]

16. Mashinchian, O.; Pisconti, A.; Le Moal, E.; Bentzinger, C.F. The Muscle Stem Cell Niche in Health and Disease. *Curr. Top. Dev. Biol.* **2018**, *126*, 23–65. [CrossRef] [PubMed]

17. Hardy, D.; Besnard, A.; Latil, M.; Jouvion, G.; Briand, D.; Thépenier, C.; Pascal, Q.; Guguin, A.; Gayraud-Morel, B.; Cavaillon, J.-M.; et al. Comparative Study of Injury Models for Studying Muscle Regeneration in Mice. *PLoS ONE* **2016**, *11*, e0147198. [CrossRef]

18. Gayraud-Morel, B.; Chrétien, F.; Jory, A.; Sambasivan, R.; Negroni, E.; Flamant, P.; Soubigou, G.; Coppée, J.-Y.; Di Santo, J.; Cumano, A.; et al. Myf5 haploinsufficiency reveals distinct cell fate potentials for adult skeletal muscle stem cells. *J. Cell Sci.* **2012**, *125*, 1738–1749. [CrossRef]

19. Von Maltzahn, J.; Jones, A.E.; Parks, R.J.; Rudnicki, M.A. Pax7 is critical for the normal function of satellite cells in adult skeletal muscle. *Proc. Natl. Acad. Sci. USA* **2013**, *110*, 16474–16479. [CrossRef]

20. Tidball, J.G. Regulation of muscle growth and regeneration by the immune system. *Nat. Rev. Immunol.* **2017**, *17*, 165–178. [CrossRef]

21. Schneider, B.S.P.; Tiidus, P.M. Neutrophil infiltration in exercise-injured skeletal muscle: How do we resolve the controversy? *Sports Med.* **2007**, *37*, 837–856. [CrossRef]

22. Mosser, D.M.; Edwards, J.P. Exploring the full spectrum of macrophage activation. *Nat. Rev. Immunol.* **2008**, *8*, 958–969. [CrossRef]

23. Otis, J.S.; Niccoli, S.; Hawdon, N.; Sarvas, J.L.; Frye, M.A.; Chicco, A.J.; Lees, S.J. Pro-inflammatory mediation of myoblast proliferation. *PLoS ONE* **2014**, *9*, e92363. [CrossRef] [PubMed]

24. Chazaud, B.; Sonnet, C.; Lafuste, P.; Bassez, G.; Rimaniol, A.-C.; Poron, F.; Authier, F.-J.; Dreyfus, P.A.; Gherardi, R.K. Satellite cells attract monocytes and use macrophages as a support to escape apoptosis and enhance muscle growth. *J. Cell Biol.* **2003**, *163*, 1133–1143. [CrossRef] [PubMed]

25. Meadows, E.; Cho, J.-H.; Flynn, J.M.; Klein, W.H. Myogenin regulates a distinct genetic program in adult muscle stem cells. *Dev. Biol.* **2008**, *322*, 406–414. [CrossRef] [PubMed]

26. Liu, L.; Cheung, T.H.; Charville, G.W.; Hurgo, B.M.C.; Leavitt, T.; Shih, J.; Brunet, A.; Rando, T.A. Chromatin modifications as determinants of muscle stem cell quiescence and chronological aging. *Cell Rep.* **2013**, *4*, 189–204. [CrossRef]

27. Segalés, J.; Perdiguero, E.; Muñoz-Cánoves, P. Epigenetic control of adult skeletal muscle stem cell functions. *FEBS J.* **2015**, *282*, 1571–1588. [CrossRef]

28. Terragni, J.; Zhang, G.; Sun, Z.; Pradhan, S.; Song, L.; Crawford, G.E.; Lacey, M.; Ehrlich, M. Notch signaling genes: Myogenic DNA hypomethylation and 5-hydroxymethylcytosine. *Epigenetics* **2014**, *9*, 842–850. [CrossRef]

29. Katajisto, P.; Döhla, J.; Chaffer, C.L.; Pentinmikko, N.; Marjanovic, N.; Iqbal, S.; Zoncu, R.; Chen, W.; Weinberg, R.A.; Sabatini, D.M. Stem cells. Asymmetric apportioning of aged mitochondria between daughter cells is required for stemness. *Science* **2015**, *348*, 340–343. [CrossRef]

30. Bhattacharya, D.; Scimè, A. Mitochondrial Function in Muscle Stem Cell Fates. *Front. Cell Dev. Biol.* **2020**, *8*, 480. [CrossRef]

31. Lyons, C.N.; Leary, S.C.; Moyes, C.D. Bioenergetic remodeling during cellular differentiation: Changes in cytochrome c oxidase regulation do not affect the metabolic phenotype. *Biochem. Cell Biol.* **2004**, *82*, 391–399. [CrossRef]

32. Folmes, C.D.L.; Dzeja, P.P.; Nelson, T.J.; Terzic, A. Mitochondria in control of cell fate. *Circ. Res.* **2012**, *110*, 526–529. [CrossRef]

33. Grounds, M.D.; Radley, H.G.; Lynch, G.S.; Nagaraju, K.; De Luca, A. Towards developing standard operating procedures for pre-clinical testing in the mdx mouse model of Duchenne muscular dystrophy. *Neurobiol. Dis.* **2008**, *31*, 1–19. [CrossRef] [PubMed]

34. Campbell, K.P.; Kahl, S.D. Association of dystrophin and an integral membrane glycoprotein. *Nature* **1989**, *338*, 259–262. [CrossRef] [PubMed]

35. Ervasti, J.M.; Ohlendieck, K.; Kahl, S.D.; Gaver, M.G.; Campbell, K.P. Deficiency of a glycoprotein component of the dystrophin complex in dystrophic muscle. *Nature* **1990**, *345*, 315–319. [CrossRef]

36. Petrof, B.J.; Shrager, J.B.; Stedman, H.H.; Kelly, A.M.; Sweeney, H.L. Dystrophin protects the sarcolemma from stresses developed during muscle contraction. *Proc. Natl. Acad. Sci. USA* **1993**, *90*, 3710–3714. [CrossRef] [PubMed]

37. Kharraz, Y.; Guerra, J.; Pessina, P.; Serrano, A.L.; Muñoz-Cánoves, P. Understanding the Process of Fibrosis in Duchenne Muscular Dystrophy. *BioMed Res. Int.* **2014**, *2014*, 965631. [CrossRef]

38. Pascual Morena, C.; Martinez-Vizcaino, V.; Álvarez-Bueno, C.; Fernández Rodríguez, R.; Jiménez López, E.; Torres-Costoso, A.I.; Cavero-Redondo, I. Effectiveness of pharmacological treatments in Duchenne muscular dystrophy: A protocol for a systematic review and meta-analysis. *BMJ Open* **2019**, *9*, e029341. [CrossRef]

39. Boldrin, L.; Zammit, P.S.; Morgan, J.E. Satellite cells from dystrophic muscle retain regenerative capacity. *Stem Cell Res.* **2015**, *14*, 20–29. [CrossRef]

40. Dewey, E.B.; Taylor, D.T.; Johnston, C.A. Cell Fate Decision Making through Oriented Cell Division. *J. Dev. Biol.* **2015**, *3*, 129–157. [CrossRef]

41. Yamashita, K.; Suzuki, A.; Satoh, Y.; Ide, M.; Amano, Y.; Masuda-Hirata, M.; Hayashi, Y.K.; Hamada, K.; Ogata, K.; Ohno, S. The 8th and 9th tandem spectrin-like repeats of utrophin cooperatively form a functional unit to interact with polarity-regulating kinase PAR-1b. *Biochem. Biophys. Res. Commun.* **2010**, *391*, 812–817. [CrossRef] [PubMed]

42. Dumont, N.A.; Wang, Y.X.; von Maltzahn, J.; Pasut, A.; Bentzinger, C.F.; Brun, C.E.; Rudnicki, M.A. Dystrophin expression in muscle stem cells regulates their polarity and asymmetric division. *Nat. Med.* **2015**, *21*, 1455–1463. [CrossRef] [PubMed]

43. Chang, N.C.; Chevalier, F.P.; Rudnicki, M.A. Satellite Cells in Muscular Dystrophy—Lost in Polarity. *Trends Mol. Med.* **2016**, *22*, 479–496. [CrossRef] [PubMed]

44. Pavlidou, T.; Marinkovic, M.; Rosina, M.; Fuoco, C.; Vumbaca, S.; Gargioli, C.; Castagnoli, L.; Cesareni, G. Metformin Delays Satellite Cell Activation and Maintains Quiescence. *Stem Cells Int.* **2019**, *2019*, 5980465. [CrossRef]

45. Lemos, D.R.; Babaeijandaghi, F.; Low, M.; Chang, C.-K.; Lee, S.T.; Fiore, D.; Zhang, R.-H.; Natarajan, A.; Nedospasov, S.A.; Rossi, F.M.V. Nilotinib reduces muscle fibrosis in chronic muscle injury by promoting TNF-mediated apoptosis of fibro/adipogenic progenitors. *Nat. Med.* **2015**, *21*, 786–794. [CrossRef]

46. Perandini, L.A.; Chimin, P.; da Lutkemeyer, D.S.; Câmara, N.O.S. Chronic inflammation in skeletal muscle impairs satellite cells function during regeneration: Can physical exercise restore the satellite cell niche? *FEBS J.* **2018**, *285*, 1973–1984. [CrossRef]

47. Li, Y.-P. TNF-alpha is a mitogen in skeletal muscle. *Am. J. Physiol. Cell Physiol.* **2003**, *285*, C370–C376. [CrossRef]

48. Londhe, P.; Davie, J.K. Interferon-γ resets muscle cell fate by stimulating the sequential recruitment of JARID2 and PRC2 to promoters to repress myogenesis. *Sci. Signal.* **2013**, *6*, ra107. [CrossRef]

49. Chaweewannakorn, C.; Tsuchiya, M.; Koide, M.; Hatakeyama, H.; Tanaka, Y.; Yoshida, S.; Sugawara, S.; Hagiwara, Y.; Sasaki, K.; Kanzaki, M. Roles of IL-1α/β in regeneration of cardiotoxin-injured muscle and satellite cell function. *Am. J. Physiol. Regul. Integr. Comp. Physiol.* **2018**, *315*, R90–R103. [CrossRef]

50. Van de Vyver, M.; Myburgh, K.H. Cytokine and satellite cell responses to muscle damage: Interpretation and possible confounding factors in human studies. *J. Muscle Res. Cell Motil.* **2012**, *33*, 177–185. [CrossRef]

51. Kurosaka, M.; Machida, S. Interleukin-6-induced satellite cell proliferation is regulated by induction of the JAK2/STAT3 signalling pathway through cyclin D1 targeting. *Cell Prolif.* **2013**, *46*, 365–373. [CrossRef] [PubMed]

52. Haugen, F.; Norheim, F.; Lian, H.; Wensaas, A.J.; Dueland, S.; Berg, O.; Funderud, A.; Skålhegg, B.S.; Raastad, T.; Drevon, C.A. IL-7 is expressed and secreted by human skeletal muscle cells. *Am. J. Physiol. Cell Physiol.* **2010**, *298*, C807–C816. [CrossRef] [PubMed]

53. Deng, B.; Wehling-Henricks, M.; Villalta, S.A.; Wang, Y.; Tidball, J.G. IL-10 triggers changes in macrophage phenotype that promote muscle growth and regeneration. *J. Immunol.* **2012**, *189*, 3669–3680. [CrossRef] [PubMed]

54. Fu, X.; Xiao, J.; Wei, Y.; Li, S.; Liu, Y.; Yin, J.; Sun, K.; Sun, H.; Wang, H.; Zhang, Z.; et al. Combination of inflammation-related cytokines promotes long-term muscle stem cell expansion. *Cell Res.* **2015**, *25*, 655–673. [CrossRef] [PubMed]

55. Rathbone, C.R.; Yamanouchi, K.; Chen, X.K.; Nevoret-Bell, C.J.; Rhoads, R.P.; Allen, R.E. Effects of transforming growth factor-beta (TGF-β1) on satellite cell activation and survival during oxidative stress. *J. Muscle Res. Cell Motil.* **2011**, *32*, 99–109. [CrossRef]

56. Aartsma-Rus, A.; Ginjaar, I.B.; Bushby, K. The importance of genetic diagnosis for Duchenne muscular dystrophy. *J. Med. Genet.* **2016**, *53*, 145–151. [CrossRef]

57. Duan, D. Systemic AAV Micro-dystrophin Gene Therapy for Duchenne Muscular Dystrophy. *Mol. Ther. J. Am. Soc. Gene Ther.* **2018**, *26*, 2337–2356. [CrossRef]

58. Waldrop, M.A.; Flanigan, K.M. Update in Duchenne and Becker muscular dystrophy. *Curr. Opin. Neurol.* **2019**, *32*, 722–727. [CrossRef] [PubMed]

59. Salmaninejad, A.; Valilou, S.F.; Bayat, H.; Ebadi, N.; Daraei, A.; Yousefi, M.; Nesaei, A.; Mojarrad, M. Duchenne muscular dystrophy: An updated review of common available therapies. *Int. J. Neurosci.* **2018**, *128*, 854–864. [CrossRef]

60. Amoasii, L.; Hildyard, J.C.W.; Li, H.; Sanchez-Ortiz, E.; Mireault, A.; Caballero, D.; Harron, R.; Stathopoulou, T.-R.; Massey, C.; Shelton, J.M.; et al. Gene editing restores dystrophin expression in a canine model of Duchenne muscular dystrophy. *Science* **2018**, *362*, 86–91. [CrossRef]

61. Koo, T.; Lu-Nguyen, N.B.; Malerba, A.; Kim, E.; Kim, D.; Cappellari, O.; Cho, H.-Y.; Dickson, G.; Popplewell, L.; Kim, J.-S. Functional Rescue of Dystrophin Deficiency in Mice Caused by Frameshift Mutations Using Campylobacter jejuni Cas9. *Mol. Ther.* **2018**, *26*, 1529–1538. [CrossRef] [PubMed]

62. Birnkrant, D.J.; Bushby, K.; Bann, C.M.; Apkon, S.D.; Blackwell, A.; Brumbaugh, D.; Case, L.E.; Clemens, P.R.; Hadjiyannakis, S.; Pandya, S.; et al. Diagnosis and management of Duchenne muscular dystrophy, part 1: Diagnosis, and neuromuscular, rehabilitation, endocrine, and gastrointestinal and nutritional management. *Lancet Neurol.* **2018**, *17*, 251–267. [CrossRef]

63. McDonald, C.M.; Henricson, E.K.; Abresch, R.T.; Duong, T.; Joyce, N.C.; Hu, F.; Clemens, P.R.; Hoffman, E.P.; Cnaan, A.; Gordish-Dressman, H.; et al. Long-term effects of glucocorticoids on function, quality of life, and survival in patients with Duchenne muscular dystrophy: A prospective cohort study. *Lancet Lond. Engl.* **2018**, *391*, 451–461. [CrossRef]

64. Bushby, K.; Finkel, R.; Birnkrant, D.J.; Case, L.E.; Clemens, P.R.; Cripe, L.; Kaul, A.; Kinnett, K.; McDonald, C.; Pandya, S.; et al. Diagnosis and management of Duchenne muscular dystrophy, part 2: Implementation of multidisciplinary care. *Lancet Neurol.* **2010**, *9*, 177–189. [CrossRef]

65. Birnkrant, D.J.; Bushby, K.; Bann, C.M.; Apkon, S.D.; Blackwell, A.; Colvin, M.K.; Cripe, L.; Herron, A.R.; Kennedy, A.; Kinnett, K.; et al. Diagnosis and management of Duchenne muscular dystrophy, part 3: Primary care, emergency management, psychosocial care, and transitions of care across the lifespan. *Lancet Neurol.* **2018**, *17*, 445–455. [CrossRef]

66. Willmann, R.; Luca, A.D.; Nagaraju, K.; Rüegg, M.A. Best Practices and Standard Protocols as a Tool to Enhance Translation for Neuromuscular Disorders. *J. Neuromuscul. Dis.* **2015**, *2*, 113–117. [CrossRef]

67. Guiraud, S.; Edwards, B.; Squire, S.E.; Moir, L.; Berg, A.; Babbs, A.; Ramadan, N.; Wood, M.J.; Davies, K.E. Embryonic myosin is a regeneration marker to monitor utrophin-based therapies for DMD. *Hum. Mol. Genet.* **2019**, *28*, 307–319. [CrossRef]

68. Guiraud, S.; Davies, K.E. Regenerative biomarkers for Duchenne muscular dystrophy. *Neural Regen. Res.* **2019**, *14*, 1317–1320. [CrossRef]

69. Ribeiro, A.F.; Souza, L.S.; Almeida, C.F.; Ishiba, R.; Fernandes, S.A.; Guerrieri, D.A.; Santos, A.L.F.; Onofre-Oliveira, P.C.G.; Vainzof, M. Muscle satellite cells and impaired late stage regeneration in different murine models for muscular dystrophies. *Sci. Rep.* **2019**, *9*, 11842. [CrossRef]

70. Whitehead, N.P.; Pham, C.; Gervasio, O.L.; Allen, D.G. N-Acetylcysteine ameliorates skeletal muscle pathophysiology in mdx mice. *J. Physiol.* **2008**, *586*, 2003–2014. [CrossRef]

71. Tamma, R.; Annese, T.; Capogrosso, R.F.; Cozzoli, A.; Benagiano, V.; Sblendorio, V.; Ruggieri, S.; Crivellato, E.; Specchia, G.; Ribatti, D.; et al. Effects of prednisolone on the dystrophin-associated proteins in the blood-brain barrier and skeletal muscle of dystrophic mdx mice. *Lab. Investig.* **2013**, *93*, 592–610. [CrossRef] [PubMed]

72. Villalta, S.A.; Nguyen, H.X.; Deng, B.; Gotoh, T.; Tidball, J.G. Shifts in macrophage phenotypes and macrophage competition for arginine metabolism affect the severity of muscle pathology in muscular dystrophy. *Hum. Mol. Genet.* **2009**, *18*, 482–496. [CrossRef] [PubMed]

73. Latroche, C.; Weiss-Gayet, M.; Muller, L.; Gitiaux, C.; Leblanc, P.; Liot, S.; Ben-Larbi, S.; Abou-Khalil, R.; Verger, N.; Bardot, P.; et al. Coupling between Myogenesis and Angiogenesis during Skeletal Muscle Regeneration Is Stimulated by Restorative Macrophages. *Stem Cell Rep.* **2017**, *9*, 2018–2033. [CrossRef]

74. Burdi, R.; Rolland, J.-F.; Fraysse, B.; Litvinova, K.; Cozzoli, A.; Giannuzzi, V.; Liantonio, A.; Camerino, G.M.; Sblendorio, V.; Capogrosso, R.F.; et al. Multiple pathological events in exercised dystrophic mdx mice are targeted by pentoxifylline: Outcome of a large array of in vivo and ex vivo tests. *J. Appl. Physiol. (1985)* **2009**, *106*, 1311–1324. [CrossRef] [PubMed]

75. Capkovic, K.L.; Stevenson, S.; Johnson, M.C.; Thelen, J.J.; Cornelison, D.D.W. Neural cell adhesion molecule (NCAM) marks adult myogenic cells committed to differentiation. *Exp. Cell Res.* **2008**, *314*, 1553–1565. [CrossRef] [PubMed]

76. Kuraoka, M.; Kimura, E.; Nagata, T.; Okada, T.; Aoki, Y.; Tachimori, H.; Yonemoto, N.; Imamura, M.; Takeda, S. Serum Osteopontin as a Novel Biomarker for Muscle Regeneration in Duchenne Muscular Dystrophy. *Am. J. Pathol.* **2016**, *186*, 1302–1312. [CrossRef] [PubMed]

77. Hathout, Y.; Liang, C.; Ogundele, M.; Xu, G.; Tawalbeh, S.M.; Dang, U.J.; Hoffman, E.P.; Gordish-Dressman, H.; Conklin, L.S.; van den Anker, J.N.; et al. Disease-specific and glucocorticoid-responsive serum biomarkers for Duchenne Muscular Dystrophy. *Sci. Rep.* **2019**, *9*, 12167. [CrossRef]

78. Pegoraro, E.; Hoffman, E.P.; Piva, L.; Gavassini, B.F.; Cagnin, S.; Ermani, M.; Bello, L.; Soraru, G.; Pacchioni, B.; Bonifati, M.D.; et al. SPP1 genotype is a determinant of disease severity in Duchenne muscular dystrophy. *Neurology* **2011**, *76*, 219–226. [CrossRef]

79. Kyriakides, T.; Pegoraro, E.; Hoffman, E.P.; Piva, L.; Cagnin, S.; Lanfranchi, G.; Griggs, R.C.; Nelson, S.F. SPP1 genotype is a determinant of disease severity in Duchenne muscular dystrophy: Predicting the severity of Duchenne muscular dystrophy: Implications for treatment. *Neurology* **2011**, *77*, 1858. [CrossRef]

80. Kramerova, I.; Kumagai-Cresse, C.; Ermolova, N.; Mokhonova, E.; Marinov, M.; Capote, J.; Becerra, D.; Quattrocelli, M.; Crosbie, R.H.; Welch, E.; et al. Spp1 (osteopontin) promotes TGFβ processing in fibroblasts of dystrophin deficient muscles through matrix metalloproteinases. *Hum. Mol. Genet.* **2019**. [CrossRef]

81. Nadarajah, V.D.; van Putten, M.; Chaouch, A.; Garrood, P.; Straub, V.; Lochmüller, H.; Ginjaar, H.B.; Aartsma-Rus, A.M.; van Ommen, G.J.B.; den Dunnen, J.T.; et al. Serum matrix metalloproteinase-9 (MMP-9) as a biomarker for monitoring disease progression in Duchenne muscular dystrophy (DMD). *Neuromuscul. Disord. NMD* **2011**, *21*, 569–578. [CrossRef]

82. Hindi, S.M.; Shin, J.; Ogura, Y.; Li, H.; Kumar, A. Matrix metalloproteinase-9 inhibition improves proliferation and engraftment of myogenic cells in dystrophic muscle of mdx mice. *PLoS ONE* **2013**, *8*, e72121. [CrossRef] [PubMed]

83. Greco, S.; De Simone, M.; Colussi, C.; Zaccagnini, G.; Fasanaro, P.; Pescatori, M.; Cardani, R.; Perbellini, R.; Isaia, E.; Sale, P.; et al. Common micro-RNA signature in skeletal muscle damage and regeneration induced by Duchenne muscular dystrophy and acute ischemia. *FASEB J.* **2009**, *23*, 3335–3346. [CrossRef] [PubMed]

84. Cacchiarelli, D.; Legnini, I.; Martone, J.; Cazzella, V.; D'Amico, A.; Bertini, E.; Bozzoni, I. miRNAs as serum biomarkers for Duchenne muscular dystrophy. *EMBO Mol. Med.* **2011**, *3*, 258–265. [CrossRef] [PubMed]

85. Hoffman, E.P. Voltage-gated ion channelopathies: Inherited disorders caused by abnormal sodium, chloride, and calcium regulation in skeletal muscle. *Annu. Rev. Med.* **1995**, *46*, 431–441. [CrossRef]

86. Cozzoli, A.; Nico, B.; Sblendorio, V.T.; Capogrosso, R.F.; Dinardo, M.M.; Longo, V.; Gagliardi, S.; Montagnani, M.; De Luca, A. Enalapril treatment discloses an early role of angiotensin II in inflammation- and oxidative stress-related muscle damage in dystrophic mdx mice. *Pharmacol. Res.* **2011**, *64*, 482–492. [CrossRef]

87. Cozzoli, A.; Liantonio, A.; Conte, E.; Cannone, M.; Massari, A.M.; Giustino, A.; Scaramuzzi, A.; Pierno, S.; Mantuano, P.; Capogrosso, R.F.; et al. Angiotensin II modulates mouse skeletal muscle resting conductance to chloride and potassium ions and calcium homeostasis via the AT1 receptor and NADPH oxidase. *Am. J. Physiol. Cell Physiol.* **2014**, *307*, C634–C647. [CrossRef]

88. Pierno, S.; Camerino, G.M.; Cannone, M.; Liantonio, A.; De Bellis, M.; Digennaro, C.; Gramegna, G.; De Luca, A.; Germinario, E.; Danieli-Betto, D.; et al. Paracrine effects of IGF-1 overexpression on the functional decline due to skeletal muscle disuse: Molecular and functional evaluation in hindlimb unloaded MLC/mIgf-1 transgenic mice. *PLoS ONE* **2014**, *8*, e65167. [CrossRef]

89. Ogura, Y.; Tajrishi, M.M.; Sato, S.; Hindi, S.M.; Kumar, A. Therapeutic potential of matrix metalloproteinases in Duchenne muscular dystrophy. *Front. Cell Dev. Biol.* **2014**, *2*, 11. [CrossRef]

90. Capogrosso, R.F.; Cozzoli, A.; Mantuano, P.; Camerino, G.M.; Massari, A.M.; Sblendorio, V.T.; De Bellis, M.; Tamma, R.; Giustino, A.; Nico, B.; et al. Assessment of resveratrol, apocynin and taurine on mechanical-metabolic uncoupling and oxidative stress in a mouse model of duchenne muscular dystrophy: A comparison with the gold standard, α-methyl prednisolone. *Pharmacol. Res.* **2016**, *106*, 101–113. [CrossRef]

91. Heier, C.R.; Damsker, J.M.; Yu, Q.; Dillingham, B.C.; Huynh, T.; Van der Meulen, J.H.; Sali, A.; Miller, B.K.; Phadke, A.; Scheffer, L.; et al. VBP15, a novel anti-inflammatory and membrane-stabilizer, improves muscular dystrophy without side effects. *EMBO Mol. Med.* **2013**, *5*, 1569–1585. [CrossRef] [PubMed]

92. Quattrocelli, M.; Barefield, D.Y.; Warner, J.L.; Vo, A.H.; Hadhazy, M.; Earley, J.U.; Demonbreun, A.R.; McNally, E.M. Intermittent glucocorticoid steroid dosing enhances muscle repair without eliciting muscle atrophy. *J. Clin. Investig.* **2017**, *127*, 2418–2432. [CrossRef] [PubMed]

93. Jensen, L.; Petersson, S.J.; Illum, N.O.; Laugaard-Jacobsen, H.C.; Thelle, T.; Jørgensen, L.H.; Schrøder, H.D. Muscular response to the first three months of deflazacort treatment in boys with Duchenne muscular dystrophy. *J. Musculoskelet. Neuronal Interact.* **2017**, *17*, 8–18. [PubMed]

94. Hussein, M.R.A.; Abu-Dief, E.E.; Kamel, N.F.; Mostafa, M.G. Steroid therapy is associated with decreased numbers of dendritic cells and fibroblasts, and increased numbers of satellite cells, in the dystrophic skeletal muscle. *J. Clin. Pathol.* **2010**, *63*, 805–813. [CrossRef] [PubMed]

95. Vianello, S.; Pantic, B.; Fusto, A.; Bello, L.; Galletta, E.; Borgia, D.; Gavassini, B.F.; Semplicini, C.; Sorarù, G.; Vitiello, L.; et al. SPP1 genotype and glucocorticoid treatment modify osteopontin expression in Duchenne muscular dystrophy cells. *Hum. Mol. Genet.* **2017**, *26*, 3342–3351. [CrossRef] [PubMed]

96. Donovan, J.M.; Zimmer, M.; Offman, E.; Grant, T.; Jirousek, M. A Novel NF-κB Inhibitor, Edasalonexent (CAT-1004), in Development as a Disease-Modifying Treatment for Patients With Duchenne Muscular Dystrophy: Phase 1 Safety, Pharmacokinetics, and Pharmacodynamics in Adult Subjects. *J. Clin. Pharmacol.* **2017**, *57*, 627–639. [CrossRef] [PubMed]

97. Hammers, D.W.; Sleeper, M.M.; Forbes, S.C.; Coker, C.C.; Jirousek, M.R.; Zimmer, M.; Walter, G.A.; Sweeney, H.L. Disease-modifying effects of orally bioavailable NF-κB inhibitors in dystrophin-deficient muscle. *JCI Insight* **2016**, *1*, e90341. [CrossRef]

98. Acharyya, S.; Villalta, S.A.; Bakkar, N.; Bupha-Intr, T.; Janssen, P.M.L.; Carathers, M.; Li, Z.-W.; Beg, A.A.; Ghosh, S.; Sahenk, Z.; et al. Interplay of IKK/NF-kappaB signaling in macrophages and myofibers promotes muscle degeneration in Duchenne muscular dystrophy. *J. Clin. Investig.* **2007**, *117*, 889–901. [CrossRef]

99. Reay, D.P.; Yang, M.; Watchko, J.F.; Daood, M.; O'Day, T.L.; Rehman, K.K.; Guttridge, D.C.; Robbins, P.D.; Clemens, P.R. Systemic delivery of NEMO binding domain/IKKγ inhibitory peptide to young mdx mice improves dystrophic skeletal muscle histopathology. *Neurobiol. Dis.* **2011**, *43*, 598–608. [CrossRef] [PubMed]

100. Peterson, J.M.; Kline, W.; Canan, B.D.; Ricca, D.J.; Kaspar, B.; Delfín, D.A.; DiRienzo, K.; Clemens, P.R.; Robbins, P.D.; Baldwin, A.S.; et al. Peptide-Based Inhibition of NF-κB Rescues Diaphragm Muscle Contractile Dysfunction in a Murine Model of Duchenne Muscular Dystrophy. *Mol. Med.* **2011**, *17*, 508–515. [CrossRef] [PubMed]

101. Pierno, S.; Nico, B.; Burdi, R.; Liantonio, A.; Didonna, M.P.; Cippone, V.; Fraysse, B.; Rolland, J.-F.; Mangieri, D.; Andreetta, F.; et al. Role of tumour necrosis factor alpha, but not of cyclo-oxygenase-2-derived eicosanoids, on functional and morphological indices of dystrophic progression in mdx mice: A pharmacological approach. *Neuropathol. Appl. Neurobiol.* **2007**, *33*, 344–359. [CrossRef] [PubMed]

102. Serra, F.; Quarta, M.; Canato, M.; Toniolo, L.; De Arcangelis, V.; Trotta, A.; Spath, L.; Monaco, L.; Reggiani, C.; Naro, F. Inflammation in muscular dystrophy and the beneficial effects of non-steroidal anti-inflammatory drugs. *Muscle Nerve* **2012**, *46*, 773–784. [CrossRef]

103. Mohri, I.; Aritake, K.; Taniguchi, H.; Sato, Y.; Kamauchi, S.; Nagata, N.; Maruyama, T.; Taniike, M.; Urade, Y. Inhibition of Prostaglandin D Synthase Suppresses Muscular Necrosis. *Am. J. Pathol.* **2009**, *174*, 1735–1744. [CrossRef]

104. Hoxha, M. Duchenne muscular dystrophy: Focus on arachidonic acid metabolites. *Biomed. Pharmacother. Biomedecine Pharmacother.* **2019**, *110*, 796–802. [CrossRef] [PubMed]

105. Muñoz-Cánoves, P.; Scheele, C.; Pedersen, B.K.; Serrano, A.L. Interleukin-6 myokine signaling in skeletal muscle: A double-edged sword? *Febs J.* **2013**, *280*, 4131–4148. [CrossRef] [PubMed]

106. Pelosi, L.; Berardinelli, M.G.; De Pasquale, L.; Nicoletti, C.; D'Amico, A.; Carvello, F.; Moneta, G.M.; Catizone, A.; Bertini, E.; De Benedetti, F.; et al. Functional and Morphological Improvement of Dystrophic Muscle by Interleukin 6 Receptor Blockade. *EBioMedicine* **2015**, *2*, 285–293. [CrossRef] [PubMed]

107. Guadagnin, E.; Mázala, D.; Chen, Y.-W. STAT3 in Skeletal Muscle Function and Disorders. *Int. J. Mol. Sci.* **2018**, *19*, 2265. [CrossRef] [PubMed]

108. Kostek, M.C.; Nagaraju, K.; Pistilli, E.; Sali, A.; Lai, S.-H.; Gordon, B.; Chen, Y.-W. IL-6 signaling blockade increases inflammation but does not affect muscle function in the mdx mouse. *BMC Musculoskelet. Disord.* **2012**, *13*, 106. [CrossRef] [PubMed]

109. Messina, S.; Bitto, A.; Aguennouz, M.; Mazzeo, A.; Migliorato, A.; Polito, F.; Irrera, N.; Altavilla, D.; Vita, G.L.; Russo, M.; et al. Flavocoxid counteracts muscle necrosis and improves functional properties in mdx mice: A comparison study with methylprednisolone. *Exp. Neurol.* **2009**, *220*, 349–358. [CrossRef]

110. Ismaeel, A.; Kim, J.-S.; Kirk, J.S.; Smith, R.S.; Bohannon, W.T.; Koutakis, P. Role of Transforming Growth Factor-β in Skeletal Muscle Fibrosis: A Review. *Int. J. Mol. Sci.* **2019**, *20*, 2446. [CrossRef]

111. Barzilai-Tutsch, H.; Bodanovsky, A.; Maimon, H.; Pines, M.; Halevy, O. Halofuginone promotes satellite cell activation and survival in muscular dystrophies. *Biochim. Biophys. Acta* **2016**, *1862*, 1–11. [CrossRef]

112. Kumar, A.; Bhatnagar, S.; Kumar, A. Matrix metalloproteinase inhibitor batimastat alleviates pathology and improves skeletal muscle function in dystrophin-deficient mdx mice. *Am. J. Pathol.* **2010**, *177*, 248–260. [CrossRef] [PubMed]

113. Morales, M.G.; Gutierrez, J.; Cabello-Verrugio, C.; Cabrera, D.; Lipson, K.E.; Goldschmeding, R.; Brandan, E. Reducing CTGF/CCN2 slows down mdx muscle dystrophy and improves cell therapy. *Hum. Mol. Genet.* **2013**, *22*, 4938–4951. [CrossRef]

114. Harish, P.; Malerba, A.; Lu-Nguyen, N.; Forrest, L.; Cappellari, O.; Roth, F.; Trollet, C.; Popplewell, L.; Dickson, G. Inhibition of myostatin improves muscle atrophy in oculopharyngeal muscular dystrophy (OPMD). *J. Cachexia Sarcopenia Muscle* **2019**, *10*, 1016–1026. [CrossRef] [PubMed]

115. Estrellas, K.M.; Chung, L.; Cheu, L.A.; Sadtler, K.; Majumdar, S.; Mula, J.; Wolf, M.T.; Elisseeff, J.H.; Wagner, K.R. Biological scaffold–mediated delivery of myostatin inhibitor promotes a regenerative immune response in an animal model of Duchenne muscular dystrophy. *J. Biol. Chem.* **2018**, *293*, 15594–15605. [CrossRef] [PubMed]

116. Consalvi, S.; Mozzetta, C.; Bettica, P.; Germani, M.; Fiorentini, F.; Del Bene, F.; Rocchetti, M.; Leoni, F.; Monzani, V.; Mascagni, P.; et al. Preclinical studies in the mdx mouse model of duchenne muscular dystrophy with the histone deacetylase inhibitor givinostat. *Mol. Med. Camb. Mass* **2013**, *19*, 79–87. [CrossRef] [PubMed]

117. Bettica, P.; Petrini, S.; D'Oria, V.; D'Amico, A.; Catteruccia, M.; Pane, M.; Sivo, S.; Magri, F.; Brajkovic, S.; Messina, S.; et al. Histological effects of givinostat in boys with Duchenne muscular dystrophy. *Neuromuscul. Disord. NMD* **2016**, *26*, 643–649. [CrossRef]

118. Colussi, C.; Mozzetta, C.; Gurtner, A.; Illi, B.; Rosati, J.; Straino, S.; Ragone, G.; Pescatori, M.; Zaccagnini, G.; Antonini, A.; et al. HDAC2 blockade by nitric oxide and histone deacetylase inhibitors reveals a common target in Duchenne muscular dystrophy treatment. *Proc. Natl. Acad. Sci. USA* **2008**, *105*, 19183–19187. [CrossRef]

119. Cohn, R.D.; van Erp, C.; Habashi, J.P.; Soleimani, A.A.; Klein, E.C.; Lisi, M.T.; Gamradt, M.; ap Rhys, C.M.; Holm, T.M.; Loeys, B.L.; et al. Angiotensin II type 1 receptor blockade attenuates TGF-beta-induced failure of muscle regeneration in multiple myopathic states. *Nat. Med.* **2007**, *13*, 204–210. [CrossRef]

120. Juban, G.; Saclier, M.; Yacoub-Youssef, H.; Kernou, A.; Arnold, L.; Boisson, C.; Ben Larbi, S.; Magnan, M.; Cuvellier, S.; Théret, M.; et al. AMPK Activation Regulates LTBP4-Dependent TGF-β1 Secretion by Pro-inflammatory Macrophages and Controls Fibrosis in Duchenne Muscular Dystrophy. *Cell Rep.* **2018**, *25*, 2163.e6–2176.e6. [CrossRef]

121. Mantuano, P.; Sanarica, F.; Conte, E.; Morgese, M.G.; Capogrosso, R.F.; Cozzoli, A.; Fonzino, A.; Quaranta, A.; Rolland, J.-F.; De Bellis, M.; et al. Effect of a long-term treatment with metformin in dystrophic mdx mice: A reconsideration of its potential clinical interest in Duchenne muscular dystrophy. *Biochem. Pharmacol.* **2018**, *154*, 89–103. [CrossRef] [PubMed]

122. Cozzoli, A.; Capogrosso, R.F.; Sblendorio, V.T.; Dinardo, M.M.; Jagerschmidt, C.; Namour, F.; Camerino, G.M.; De Luca, A. GLPG0492, a novel selective androgen receptor modulator, improves muscle performance in the exercised-mdx mouse model of muscular dystrophy. *Pharmacol. Res.* **2013**, *72*, 9–24. [CrossRef] [PubMed]

123. Singh, R.; Artaza, J.N.; Taylor, W.E.; Braga, M.; Yuan, X.; Gonzalez-Cadavid, N.F.; Bhasin, S. Testosterone Inhibits Adipogenic Differentiation in 3T3-L1 Cells: Nuclear Translocation of Androgen Receptor Complex with β-Catenin and T-Cell Factor 4 May Bypass Canonical Wnt Signaling to Down-Regulate Adipogenic Transcription Factors. *Endocrinology* **2006**, *147*, 141–154. [CrossRef] [PubMed]

124. Dorchies, O.M.; Reutenauer-Patte, J.; Dahmane, E.; Ismail, H.M.; Petermann, O.; Patthey- Vuadens, O.; Comyn, S.A.; Gayi, E.; Piacenza, T.; Handa, R.J.; et al. The anticancer drug tamoxifen counteracts the pathology in a mouse model of duchenne muscular dystrophy. *Am. J. Pathol.* **2013**, *182*, 485–504. [CrossRef] [PubMed]

125. Jung, S.; Kim, K. Exercise-induced PGC-1α transcriptional factors in skeletal muscle. *Integr. Med. Res.* **2014**, *3*, 155–160. [CrossRef]

126. Dinulovic, I.; Furrer, R.; Beer, M.; Ferry, A.; Cardel, B.; Handschin, C. Muscle PGC-1α modulates satellite cell number and proliferation by remodeling the stem cell niche. *Skelet. Muscle* **2016**, *6*, 39. [CrossRef]

127. Furrer, R.; Handschin, C. Optimized Engagement of Macrophages and Satellite Cells in the Repair and Regeneration of Exercised Muscle. In *Hormones, Metabolism and the Benefits of Exercise*; Spiegelman, B., Ed.; Springer: Cham, Switzerland, 2017; ISBN 978-3-319-72789-9.

128. Hayashiji, N.; Yuasa, S.; Miyagoe-Suzuki, Y.; Hara, M.; Ito, N.; Hashimoto, H.; Kusumoto, D.; Seki, T.; Tohyama, S.; Kodaira, M.; et al. G-CSF supports long-term muscle regeneration in mouse models of muscular dystrophy. *Nat. Commun.* **2015**, *6*, 6745. [CrossRef]

129. Le Grand, F.; Jones, A.E.; Seale, V.; Scimè, A.; Rudnicki, M.A. Wnt7a Activates the Planar Cell Polarity Pathway to Drive the Symmetric Expansion of Satellite Stem Cells. *Cell Stem Cell* **2009**, *4*, 535–547. [CrossRef]

130. Brack, A.S.; Rando, T.A. Tissue-specific stem cells: Lessons from the skeletal muscle satellite cell. *Cell Stem Cell* **2012**, *10*, 504–514. [CrossRef]

131. Rozo, M.; Li, L.; Fan, C.-M. Targeting β1-integrin signaling enhances regeneration in aged and dystrophic muscle in mice. *Nat. Med.* **2016**, *22*, 889–896. [CrossRef]

132. Wissing, E.R.; Boyer, J.G.; Kwong, J.Q.; Sargent, M.A.; Karch, J.; McNally, E.M.; Otsu, K.; Molkentin, J.D. P38α MAPK underlies muscular dystrophy and myofiber death through a Bax-dependent mechanism. *Hum. Mol. Genet.* **2014**, *23*, 5452–5463. [CrossRef] [PubMed]

133. Reano, S.; Angelino, E.; Ferrara, M.; Malacarne, V.; Sustova, H.; Sabry, O.; Agosti, E.; Clerici, S.; Ruozi, G.; Zentilin, L.; et al. Unacylated Ghrelin Enhances Satellite Cell Function and Relieves the Dystrophic Phenotype in Duchenne Muscular Dystrophy mdx Model. *Stem Cells Dayt. Ohio* **2017**, *35*, 1733–1746. [CrossRef] [PubMed]

134. Klimek, M.E.; Sali, A.; Rayavarapu, S.; Van der Meulen, J.H.; Nagaraju, K. Effect of the IL-1 Receptor Antagonist Kineret® on Disease Phenotype in mdx Mice. *PLoS ONE* **2016**, *11*, e0155944. [CrossRef]

Article

Platelet-Rich Plasma Modulates Gap Junction Functionality and Connexin 43 and 26 Expression During TGF-β1–Induced Fibroblast to Myofibroblast Transition: Clues for Counteracting Fibrosis

Roberta Squecco [1,†], Flaminia Chellini [2,†], Eglantina Idrizaj [1], Alessia Tani [2,†], Rachele Garella [1], Sofia Pancani [2], Paola Pavan [3], Franco Bambi [3], Sandra Zecchi-Orlandini [2] and Chiara Sassoli [2,*,†]

[1] Department of Experimental and Clinical Medicine, Section of Physiological Sciences, University of Florence, 50134 Florence, Italy; roberta.squecco@unifi.it (R.S.); eglantina.idrizaj@unifi.it (E.I.); rachele.garella@unifi.it (R.G.)

[2] Department of Experimental and Clinical Medicine, Section of Anatomy and Histology, University of Florence, 50134 Florence, Italy; flaminia.chellini@unifi.it (F.C.); alessia.tani@unifi.it (A.T.); sofia.pancani@stud.unifi.it (S.P.); sandra.zecchi@unifi.it (S.Z.-O.)

[3] Transfusion Medicine and Cell Therapy Unit, "A. Meyer" University Children's Hospital, 50134 Florence, Italy; paola.pavan@meyer.it (P.P.); franco.bambi@meyer.it (F.B.)

[*] Correspondence: chiara.sassoli@unifi.it; Tel.: +39-0552-7580-63

[†] Interuniversity Institute of Myology (IIM) (https://www.coram-iim.it).

Received: 7 April 2020; Accepted: 8 May 2020; Published: 12 May 2020

Abstract: Skeletal muscle repair/regeneration may benefit by Platelet-Rich Plasma (PRP) treatment owing to PRP pro-myogenic and anti-fibrotic effects. However, PRP anti-fibrotic action remains controversial. Here, we extended our previous researches on the inhibitory effects of PRP on in vitro transforming growth factor (TGF)-β1-induced differentiation of fibroblasts into myofibroblasts, the effector cells of fibrosis, focusing on gap junction (GJ) intercellular communication. The myofibroblastic phenotype was evaluated by cell shape analysis, confocal fluorescence microscopy and Western blotting analyses of α-smooth muscle actin and type-1 collagen expression, and electrophysiological recordings of resting membrane potential, resistance, and capacitance. PRP negatively regulated myofibroblast differentiation by modifying all the assessed parameters. Notably, myofibroblast pairs showed an increase of voltage-dependent GJ functionality paralleled by connexin (Cx) 43 expression increase. TGF-β1-treated cells, when exposed to a GJ blocker, or silenced for Cx43 expression, failed to differentiate towards myofibroblasts. Although a minority, myofibroblast pairs also showed not-voltage-dependent GJ currents and coherently Cx26 expression. PRP abolished the TGF-β1-induced voltage-dependent GJ current appearance while preventing Cx43 increase and promoting Cx26 expression. This study adds insights into molecular and functional mechanisms regulating fibroblast-myofibroblast transition and supports the anti-fibrotic potential of PRP, demonstrating the ability of this product to hamper myofibroblast generation targeting GJs.

Keywords: α-smooth muscle actin; confocal microscopy; connexin 43; connexin 26; fibrosis; gap junctions; myofibroblasts; Platelet-Rich Plasma; skeletal muscle; transforming growth factor (TGF)-β1

---

1. Introduction

Adult skeletal muscle can efficiently repair/regenerate after focal damages [1]. Several studies showed that many different cell types endowed with inducible myogenic potential, residing within the muscle tissue or recruited via the blood, might contribute to the formation of nascent contractile myofibers [2–6]. Nevertheless, muscle resident satellite cells are widely recognized as the main players in the repair/regenerative processes [7,8]. After focal injuries, satellite cells undergo activation to essentially recapitulate the steps of embryonic and fetal myogenesis forming new myofibers or fusing with injured myofibers to repair the damage [1,9]. To accomplish their task, satellite cells (but even the myogenic non-satellite cells), require the establishment of a suitable and conductive surrounding microenvironment. This essentially includes pro-myogenic factors, biochemical and physical pro-myogenic signals, juxtacrine, and paracrine interaction with different interstitial nursing cells and a spatially and temporally limited reparative fibrotic response [1,10–15].

1.1. Fibrotic Response in Skeletal Muscle

The activation of fibrogenic pathways represents an adaptive physiological response of tissues, including skeletal muscle, to damage. A crucial process in such a fibrotic response is represented by the differentiation of fibroblasts resident in the extracellular matrix (ECM) towards myofibroblasts [16]. This is essentially promoted by the combined action of pro-fibrogenic agents, such as transforming growth factor (TGF)-β1, mainly released by infiltrating inflammatory cells (particularly macrophages) at the site of the injury and by the fibroblasts/myofibroblasts itself, and mechanical stimuli coming from the damaged microenvironment [16,17]. Myofibroblasts are characterized by a prominent rough endoplasmic reticulum, typical of collagen-synthetically active fibroblasts, by the de novo expression of α-smooth muscle actin (α-sma) within well-assembled stress fibers that confers contractile properties to the cells. Stress fibers are anchored to fibronexus, a specialized focal adhesion complex on the myofibroblast surface, to link intracellular actin filaments with extracellular fibronectin fibrils. Through fibronexus, the force generated by stress fibers can be transmitted to the surrounding ECM, and vice-versa the ECM mechanical signals can be transduced via this mechano-transduction system into intracellular signals [18]. Moreover, even if myofibroblasts are not regarded as electrically excitable cells, they show peculiar biophysical properties and trans-membrane ion currents typical of smooth muscle cells. In this regard, it has been reported that human atrial myofibroblasts can express a Na$^+$ current (I$_{Na}$) and biophysical properties that could give rise to regenerative action potentials [19,20]. In addition, myofibroblasts typically show the inward-rectifier K$^+$ current (I$_{kir}$), which especially increases under TGF-β1-treatment [21–24]. In physiological conditions, the permanence and function of myofibroblasts after muscle damage are temporally and spatially limited. Indeed, they are responsible for the deposition of ECM components to form a transient contractile scar essentially required to rapidly restore tissue integrity and preserve muscle function, to support activated SCs and the nascent myofibers mechanically. Once the tissue regeneration has taken place, the scar will be degraded thanks to the balanced and finely tuned activity of proteolytic enzymes selectively digesting individual components of ECM, namely matrix metalloproteinases (MMPs), and of their specific tissue inhibitors (TIMPs), that are mainly secreted by different cells including, among others, fibroblasts and inflammatory cells [25]. Myofibroblasts progressively disappear, undergoing apoptosis and/or senescence or reverting to a quiescent state [26,27]. By contrast, the persistence of myofibroblasts in an activated state has been associated with an aberrant maladaptive reparative response to chronic or extended damage leading to the formation of a permanent scar replacing the normal functional tissue and hampering the endogenous cell mediated-mechanism of muscle regeneration [10,16,17,28,29]. Therefore, therapies aimed to limit myofibroblast generation and functionality may result strategical and effective for preventing tissue fibrosis development and thus promoting the regeneration of damaged muscles.

1.2. PRP as an Anti-fibrotic Agent

In this regard, Platelet-Rich Plasma (PRP)—defined as a plasma fraction with a concentration of platelets above baseline levels and representing a source of numerous biologically active molecules—may offer promising perspectives [30]. Indeed, many in vitro and in vivo studies have demonstrated the anti-fibrotic potential of this blood product in different tissues [31–38], including skeletal muscle [30,39–46], and have indicated the fibroblast-myofibroblast transition as the cell process target of its action [31–33,36,38,44,47,48]. Furthermore, the positive contribution of PRP to skeletal muscle regeneration has been demonstrated either in vivo or in vitro, thanks to its capability to promote the myogenic program [30]. Nevertheless, the anti-fibrotic potential of PRP needs to be investigated more in-depth, and the molecular targets of the action of this plasma product need to be clearly identified.

1.3. Gap Junction Intercellular Communication (GJIC)

The gap junction (GJ) channels are dynamic membrane domains built of two docking hemichannels called connexons assembled in the plasma membranes of two adjacent cells. Each connexon is a hexameric structure consisting of six transmembrane proteins named connexins (Cxs) that may have different molecular weights [49] and form an aqueous pore. The opening of these channels allows the flow of ions and small molecules (less than 1000 MW molecular weight size fractions) such as sugars, amino acids, oxygen, as well as second messengers such as cAMP, inositol phosphates, and calcium directly from one cell to another. The type of molecules (second messengers) passing through GJs can be influenced by the Cx isoform composition of the GJ channels. When the Cx isoforms are of the same type within a hemichannel, the resulting structure is called homomeric, whereas it is called heteromeric if more than one Cx isoform is present. The GJ channels composed of two identical hemichannels are named homotypic, and those consisting of two different hemichannels are named heterotypic. These two types of GJ channels exhibit peculiar and different gating properties influencing their voltage sensitivity [50]. GJIC is proposed to play a role in regulating the fibroblasts transition towards myofibroblasts as well as to be involved in the functional coupling of myofibroblasts to coordinate their activity [18,51–54]; however, GJ channel functionality and Cx composition during the phenotypic progression of fibroblasts into the myofibroblasts have not been fully elucidated yet and deserve more attention.

In the present in vitro study, by combining morphological, biomolecular, biochemical, and electrophysiological analyses, we extended our previous researches further exploring the potential molecular targets of the inhibitory action of PRP on myofibroblast generation. In particular, we focused the attention on the GJIC. The experimental model to evaluate fibroblast to myofibroblast transition has been previously validated [23,24,48,55–57] and consists in the culture of the cells in low serum conditions in the presence of TGF-β1. The treatment with PRP was also conducted as previously reported [48,57,58].

Here, while confirming the anti-fibrotic potential of PRP we provide the first experimental evidence that voltage-dependent GJ functionality and the expression of Cx43, a typical Cx forming voltage-dependent connexons, are important mechanisms by which TGF-β1 endorses fibroblasts differentiation towards myofibroblasts, and that PRP treatment hampers this effect. Moreover, we also demonstrated the involvement of not-voltage dependent GJs and Cx26, a typical Cx type forming not (or at least, scarcely) voltage-dependent connexons.

2. Materials and Methods

2.1. Platelet-Rich Plasma (PRP) Preparation

For the present experiments, thawed ready-to-use activated PRP aliquots classified as not suitable for transfusion-infusion purposes previously prepared and stored at −80 °C were used [48]. Briefly, PRP was collected from the blood of healthy adult donors subjected to plasma-platelet apheresis (Haemonetics MCS®, Haemonetics, Milan, Italy) as previously reported in detail [57]. The final platelet concentration in each PRP (without leukocytes) aliquot was 2×10^6 platelets/µL. Platelet activation was induced by the addition of a calcium digluconate solution (10%). The donors gave their written informed consent to allow the use of PRP for in vitro experimentations for which the Ethical Committee's approval is not required. PRP treatment was performed as previously reported [48,57].

2.2. Cell Culture and Treatments

Murine NIH/3T3 fibroblastic cells were obtained from American Type Culture Collection (ATCC, Manassas, VA, USA). The cells were grown in proliferation medium (PM), consisting of Dulbecco's Modified Eagle's Medium (DMEM; Sigma, Milan, Italy) containing 4.5 g/L glucose supplemented with 10% fetal bovine serum (FBS) and 1% penicillin/streptomycin (Sigma), at 37 °C in a humidified atmosphere of 5% CO_2. Fibroblastic cells were induced to differentiate into myofibroblasts by shifting them in differentiation medium (DM) consisting of DMEM supplemented with 2% FBS and 2 ng/mL TGF-β1 (PeproTech, Inc., Rocky Hill, NJ, USA) for 48 h and 72 h, as previously reported [48]. In parallel experiments, to estimate the influence of PRP on fibroblast-myofibroblast transition, PRP was added to DM (1:50) [48,57]. In some experiments, the cells were cultured in PM, DM, or DM + PRP in the presence of 1 mM heptanol (Sigma), a specific GJ blocker, to evaluate the involvement of GJs in myofibroblastic differentiation.

2.3. Electrophysiological Records

Cell pairs were analyzed by the dual whole-cell patch-clamp technique, as previously reported [59–61]. Both passive membrane properties and GJ functionality were investigated. To this aim, cells were plated on glass coverslips (50,000 cells on each glass coverslips) to be located in the recording chamber and continuously superfused at a rate of 1.8 ml/min by a Pump 33 (Harvard Apparatus) with a physiological bath solution containing (mM) 140 NaCl, 5.4 KCl, 1.8 $CaCl_2$, 1.2 $MgCl_2$, 10 D-glucose, and 5 HEPES (pH set at 7.4 with NaOH). The patch electrodes, pulled from borosilicate glass (GC 150–15; Clark, Reading, UK), were filled with the following solution (mM): 130 KCl, 10 NaH_2PO_4, 0.2 $CaCl_2$, 1 EGTA, 5 MgATP, and 10 HEPES (pH was set to 7.2 with KOH). When filled, the pipette resistance ranged between 1.5 to 3.0 MΩ. Experiments were achieved at room temperature (22 °C). The set up for electrophysiological measurements was as previously reported [61] and consisted of the Axopatch 200 B amplifier (Axon Instruments, Union City, CA), an analog-to-digital/digital-to-analog interface (Digidata 1200; Axon Instruments), and pClamp 6 software (Axon Instruments). Currents were low-pass filtered at 1 kHz with a Bessel filter. The passive membrane properties, membrane resistance (R_m), and membrane linear capacitance (C_m) were consistently estimated in voltage-clamp starting from a holding potential (HP) of -70 mV and applying a 10-mV positive and negative step pulse. In brief, R_m was calculated using the relation: $R_m = (\Delta V - I_m R_a)/I_m$, where ΔV is the command voltage step amplitude, I_m is the steady-state membrane current, and R_a the access resistance [23,62]. C_m was calculated from $C_m = \Delta Q(R_m + R_a)/R_m \Delta V$, corrected according to a previous report [63]. To properly compare the currents recorded from different cells, their values were normalized to C_m, assuming that the specific C_m was constant at 1 μF/cm^2. The ratio I/C_m was intended as current density (in pA/pF). The junction potential of the electrode was estimated before making the patch (about −10 mV) and then was subtracted from the recorded membrane potential. The resting membrane potential (RMP) was recorded in current-clamp mode with a stimulus waveform: I = 0 pA. The protocol of stimulation used to record the currents flowing through GJs in voltage clamp and the recording procedure have been previously reported [61,64,65]. In brief, cell 1 of the pair was stepped from a holding potential (HP) of 0 mV, using a bipolar 5 s pulse protocol starting at trans-junctional voltage Vj = ±10 mV and ongoing at 20 mV increments up to ± 150 mV. The transjunctional current flowing through GJs is indicated as I_j. Precisely, the amplitude of I_j determined at the peak was named $I_{j,inst}$ (instantaneous transjunctional current), whereas that measured at the end of each pulse is indicated as $I_{j,ss}$ (steady-state transjunctional current). These values were used to calculate the related gap junctional conductances, $G_{j,inst}$ and $G_{j,ss}$, by the ratios: $G_{j,inst} = I_{j,inst}/V_j$ and $G_{j,ss} = I_{j,ss}/V_j$, respectively. The mean values of $G_{j,ss}$ were normalized to those of $G_{j,inst}$, plotted as a function of V_j and fitted, when possible, with a Boltzmann function using the equation: $G_j = (G_{max} - G_{min})/(1 + \exp(-A(V_j - V_0))) + G_{min}$. In a set of experiments, heptanol (1 mM) was acutely applied to the bath solution to block gap junctional currents.

2.4. Silencing of Cx43 Expression by Short Interfering RNA

To inhibit the expression of Cx43, the cells were cultured either in a 6-wells/plate or on sterile glass coverslips put on the bottom of a 6-wells/plate in PM till a confluence of 80% and then transfected with a mix of short interfering RNA duplexes (siRNA; Santa Cruz Biotechnology, Santa Cruz, CA) corresponding to 3 distinct regions of the DNA sequence of mouse Cx43 gene (NM_010288): 5′CCCAACUGAACCUUAAGAA3′, 5′CCUCACCAAAUGAUUUCUA3′, and 5′CCUACCAGUUUCUUCAAGU3′ and/or with a non-specific scrambled (SCR)-siRNA (Santa Cruz Biotechnology) used as control. The siRNA transfections were performed according to manufacturer's instructions (Santa Cruz Biotechnology) and as previously reported [59]. Briefly, the cells were transfected with Cx43-siRNA duplexes or SCR-siRNA (20 nM) for 24 h and then shifted in fresh PM for additional 5 h. Thereafter, the transfected cells were cultured in DM with the addition or not of PRP for 48 h before being processed for Western blotting or immunofluorescence analysis of Cx43 and α-sma.

2.5. Reverse Transcription - Polymerase Chain Reaction (RT-PCR)

Cellular expression levels of Cx43 were evaluated by RT-PCR, as previously reported [48]. Briefly, according to manufacturer's instructions, total RNA was extracted from the cells cultured in the different experimental conditions on the wells of 6-wells/plates, by using TRIzol Reagent (Invitrogen, Life Technologies, Grand Island, NY, USA). One μg of total extracted RNA was reverse transcribed and amplified by using SuperScript One-Step RT-PCR System (Invitrogen, Life Technologies). cDNA synthesis was performed at 55 °C for 30 min; the samples were pre-denatured at 94 °C for 2 min and then subjected to 40 cycles of PCR performed at 94 °C for 15 s, alternating with 55 °C for 30 s and 72 °C for 1 min; the final extension step was performed at 72 °C for 5 min. The mouse gene-specific primers used were as follow: Cx43 (X61576.1), forward 5′-AACAGTCTGCCTTTCGCTGT-3′ and reverse 5′-ATCTTCACCTTGCCGTGTTC-3′; β-actin (NM_007393), forward 5′-ACTGGGACGACATGGAGAAG-3′ and reverse 5′-ACCAGAGGCATACAGGGACA-3′. β-actin mRNA was used as an internal standard. Blank controls, consisting of no template (water), were performed in each run. The amplified samples were electrophoresed on 1.8% agarose gel containing ethidium bromide staining, and the intensity of the related bands was quantified by densitometric analysis by using ImageJ 1.49v software (NIH, https://imagej.nih.gov/ij/). Each band intensity was normalized to the relative β-actin.

2.6. Confocal Laser Scanning Microscopy

Cells grown on sterile glass coverslips in the different experimental conditions were fixed with paraformaldehyde (PFA) 0.5% diluted in PBS for 10 min at room temperature. Fixed cells were washed and permeabilized with cold acetone for 3 min, incubated with a blocking solution containing 0.5% bovine serum albumin (BSA, Sigma) and 3% glycerol in PBS for 20 min and thereafter incubated overnight at 4 °C with the following antibodies: mouse monoclonal anti-α-sma (1:100; Abcam, Cambridge, UK), rabbit polyclonal anti-Cx43 (1:250; Chemicon, Temecula, CA, USA), rabbit polyclonal anti-type-1 collagen (1:50; Santa Cruz Biotechnology), or mouse monoclonal anti-Cx26 (1:50; Sigma). The immunoreactions were revealed by specific anti-mouse Alexa Fluor 488- or 568 conjugated IgG or anti-rabbit Alexa Fluor 488- conjugated IgG (1:200; Molecular Probes, Eugene, OR, USA). In some experiments the fixed cells were incubated with Alexa Fluor 488-conjugated wheat germ agglutinin (WGA, 1:100; Molecular Probes) for 10 min at room temperature, which binds glycoconjugates present on cell membranes, or counterstained with propidium iodide (PI, 1:30 for 30 s; Molecular Probes), to detect nuclei. Negative controls were carried out by replacing the primary antibodies with non-immune serum, while cross-reactivity of the secondary antibodies was evaluated in control experiments in which primary antibodies were omitted. The immunolabeled samples were washed and mounted with an antifade mounting medium (Biomeda Gel mount, Electron Microscopy Sciences,

Foster City, CA, USA) to allow the observation under a confocal Leica TCS SP5 microscope equipped with a HeNe/Ar laser source for fluorescence measurements and differential interference contrast (DIC) optics (Leica Microsystems, Mannheim, Germany). Observations were performed by means of a Leica Plan Apo 63×/1.43NA oil immersion objective. A series of optical sections (1024 × 1024 pixels each; pixel size 204.3 nm) 0.4 μm in thickness were taken throughout the depth of the cells preparations at intervals of 0.4 μm, and the images were projected onto a single 'extended focus' image. Densitometric analyses of the intensity of α-sma, type-1 collagen, Cx43 and Cx26 fluorescence signals were performed on digitized images using ImageJ 1.49v software (NIH, https://imagej.nih.gov/ij/) in 20 regions of interest (ROI) of 100 μm^2 for each confocal stack (at least 10).

2.7. Western Blotting

Total proteins extracted from the cells in the different experimental conditions were quantified, as reported previously [48]. Forty μg of total proteins were subjected to electrophoresis on NuPAGE®4%–12% Bis-Tris Gel (Invitrogen, Life Technologies; 200 V, 40 min) and blotted onto polyvinylidene difluoride (PVDF) membranes (Invitrogen, Life Technologies; 30 V, 1 h). The membranes were incubated with mouse monoclonal anti-α-sma (1:1000; Abcam), rabbit polyclonal anti-Cx43 (1:2500; Chemicon), and mouse monoclonal anti-Cx26 (1:500; Sigma) overnight at 4 °C. Immunodetection was performed according to the Western Breeze®Chromogenic Western Blot Immunodetection Kit protocol (Invitrogen, Life Technologies). The same membranes were subjected to the immunodetection of the expression of α-tubulin (rabbit polyclonal anti α-tubulin, 1:1000; Merck, Milan, Italy), assumed as control invariant protein. Densitometric analysis of the bands was performed using ImageJ 1.49v software (NIH, https://imagej.nih.gov/ij/), and the values normalized to control.

2.8. Statistical Analysis

Data were expressed as means ± standard error of the mean (S.E.M.) as a result of at least 3 independent experiments performed in triplicate. A 95% confidence level was used, assuming a normal distribution of values. Unpaired Student's t-test was used to compare the means of two conditions for independent data, statistically. The one-way analysis of variance (ANOVA) for any single independent variable was used to compare the differences between more than 2 groups and was followed by Tukey HSD or Bonferroni's post hoc adjustment. In electrophysiological experiments, 'n' indicates the number of cells analyzed. Values of $p < 0.05$ were considered statistically significant. Calculations were performed using GraphPad Prism software program (GraphPad, San Diego, CA, USA) and Microsoft Office Excel 2013 (Microsoft Corporation, Redmond, WA, USA).

3. Results

3.1. PRP Prevented TGF-β1- Induced Fibroblast to Myofibroblast Transition

Successful in vitro differentiation of NIH/3T3 fibroblasts towards myofibroblasts induced by the well-known pro-fibrotic factor TGF-β1 and the ability of PRP to prevent this transition were confirmed by morphological, biochemical and electrophysiological evaluations. Fibroblasts induced to differentiate by culturing in DM exhibited the typical features of myofibroblastic phenotype. Indeed, as judged by Western blotting analysis, they showed a significant increase of the expression of α-sma ($p < 0.05$), the most reliable marker of myofibroblasts, after 48 h and even more after 72 h of culture, as compared to control undifferentiated cells in PM (Figure 1A,B). Moreover, the immunocytochemical analysis at confocal microscopy, performed after 72 h of culture, confirmed the data of Western blotting and showed that this protein was well organized along filamentous structures (Figure 1C,D,I).

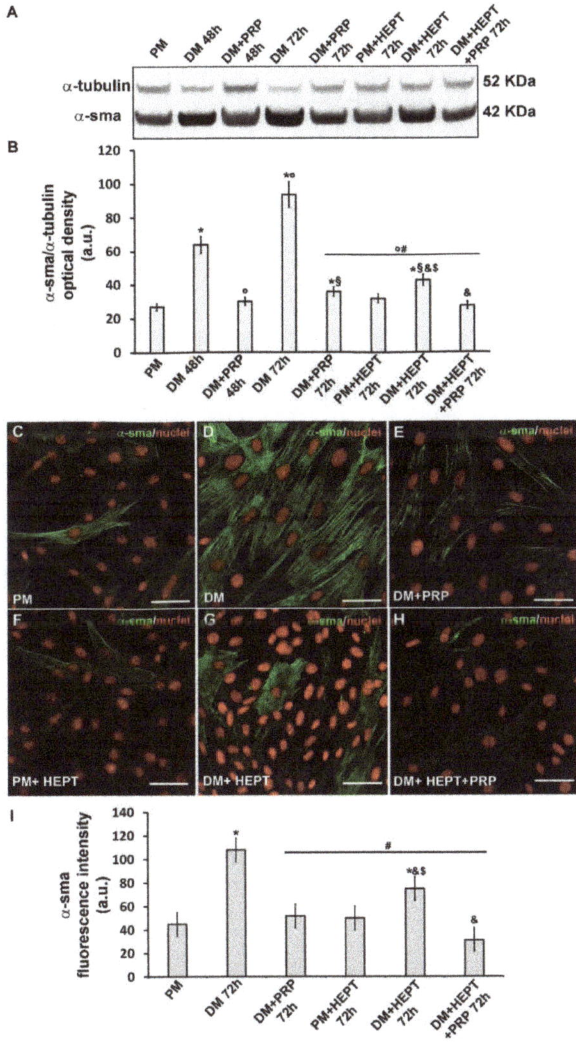

Figure 1. Evaluation of the effects PRP on fibroblast to myofibroblast transition and of the involvement of GJs: α-sma expression. Fibroblasts were induced to differentiate into myofibroblasts by culturing in differentiation medium (DM) in the presence or absence of PRP for 48 h and 72 h. Cells cultured in proliferation medium (PM) served as control undifferentiated cells. In parallel experiments, fibroblasts were cultured in PM or in DM in the presence of heptanol (HEPT), a common GJ channel blocker, in the presence or absence of PRP for 72 h. (**A**,**B**) Western Blotting analysis of α-sma expression. (**A**) Representative Blot. (**B**) Histogram showing the densitometric analysis of the bands normalized to α-tubulin. (**C–H**) Representative confocal fluorescence images of the cells immunostained with antibodies against α-sma (green) and counterstained with propidium iodide (PI) to detect nuclei. Scale bar: 50 μm. (**I**) Histogram showing the densitometric analysis of the intensity of the α-sma fluorescence signal performed on digitized images in 20 regions of interest (ROI) of 100 μm^2 for each confocal stack (10). Data shown are mean ± S.E.M. and represent the results of at least three independent experiments performed in triplicate. Significance of difference: * $p < 0.05$ versus PM; ° $p < 0.05$ versus DM 48 h; # $p < 0.05$ versus DM 72 h; § $p < 0.05$ versus DM + PRP 48 h; & $p < 0.05$ versus DM + PRP 72 h; $ $p < 0.05$ versus PM + HEPT 72 h (One-way ANOVA followed by the Tukey post hoc test).

Moreover, cells cultured in DM for 72 h, appeared much larger with a more polygonal shape as compared to the cells cultured in PM which, instead, were smaller and spindle-shaped as judged by the confocal fluorescence analysis after labeling with the membrane dye Alexa Fluor 488 conjugated WGA (Figure 2A,B). Differentiated cells also showed a robust increase ($p < 0.05$) in the expression of type-1 collagen at the cytoplasmic level and, in some cases, even outside the cells in a filamentous form (Figure 2D,E,G).

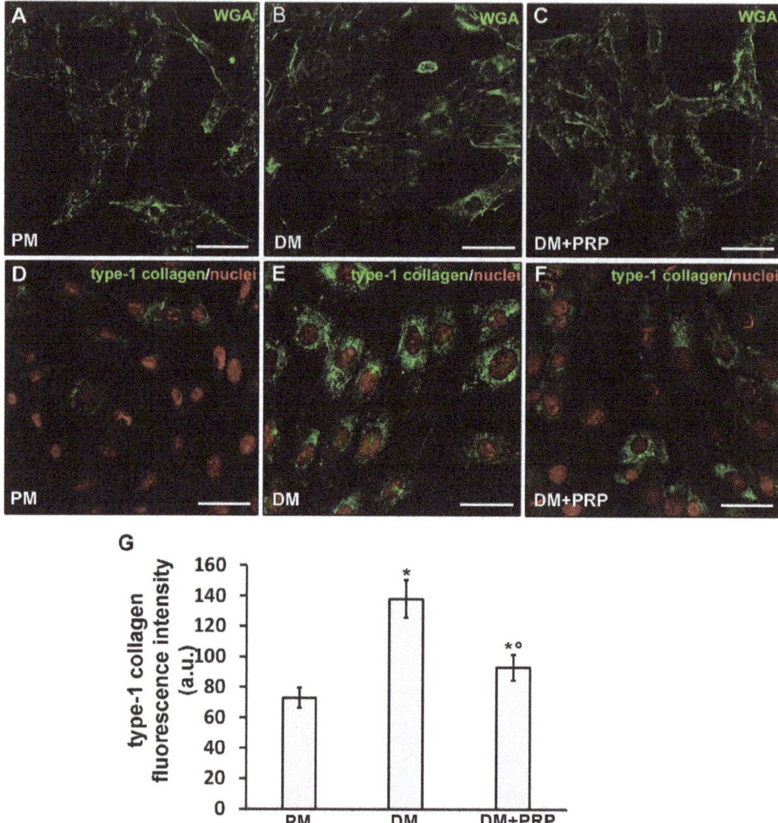

Figure 2. Effects of PRP on fibroblast to myofibroblast transition: Cell morphology and type-1 collagen expression. Fibroblasts were induced to differentiate into myofibroblasts by culturing in differentiation medium (DM) in the presence or absence of PRP for 72 h. The cells cultured in proliferation medium (PM) served as control undifferentiated cells. (**A–F**) Representative confocal fluorescence images of the cells (**A–C**) stained with Alexa Fluor 488-conjugated WGA (green) to reveal the plasma membrane and (**D–F**) immunostained with antibodies against type-1 collagen (green) and counterstained with propidium iodide (PI), to label nuclei. Scale bar: 50 μm. (**G**) Histogram showing the densitometric analysis of the intensity of type-1 collagen fluorescence signal performed on digitized images in 20 regions of interest (ROI) of 100 μm^2 for each confocal stack (10). Data are reported as mean ± S.E.M. and represent the results of at least three independent experiments performed in triplicate. Significance of difference: * $p < 0.05$ versus PM; ° $p < 0.05$ versus DM (One-way ANOVA followed by the Tukey post hoc test).

The electrophysiological analysis of the passive membrane properties achieved by the whole-cell patch-clamp technique confirmed that the cells cultured in DM acquired the myofibroblastic phenotype. First, the resting membrane potential, RMP, was recorded and it was found that the values recorded from the cells cultured in DM for 48 h tended to be more depolarized compared to control undifferentiated fibroblasts in PM, in accordance with previous observations [23,24]. The overall results from all of the experiments done are shown in Figure 3A and Table 1. The statistical analysis of the RMP values between the different conditions was achieved with one–way ANOVA that provided overall results for our data. Despite the observed tendency to depolarization, the differences between the means did not turned out to be statistically significant ($p = 0.26$; F = 1.45 < Fcrit = 3.15; df = 30).

We then analyzed the cell membrane resistance, R_m, in the voltage-clamp mode of our device. As shown in Figure 3B, R_m increased after 48 h and even more after 72 h of culture in DM. The one-way ANOVA analysis of R_m indicated statistical significance ($p = 0.00010$; F = 8.99 > Fcrit = 2.83; df = 50). To know which groups were significantly different from another, we used the Bonferroni post-hoc test. The resulting significance is indicated by the symbols depicted in Figure 3B and Table 1.

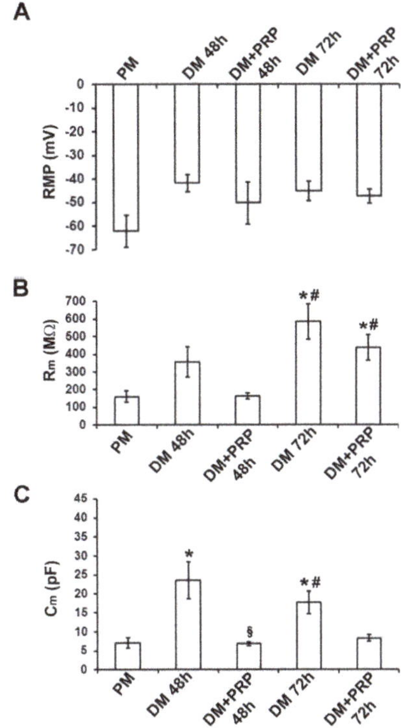

Figure 3. Effects of PRP on fibroblast to myofibroblast transition: Electrophysiological analysis of biophysical properties. (**A**) Resting membrane potential (RMP, in mV) recorded in the different conditions. Myofibroblasts have a tendency to be more depolarized. (**B**) Membrane resistance (R_m, in MΩ) shows higher values in myofibroblasts grown in differentiation medium (DM) compared to fibroblasts grown in proliferation medium (PM). (**C**) Membrane capacitance (C_m, in pF): Myofibroblasts show higher values compared to the undifferentiated cells in PM. All values are reported as mean ± S.E.M. and are listed in Table 1. * $p < 0.05$ versus PM; # $p < 0.05$ versus DM + PRP 48 h; § $p < 0.05$ versus DM 48 h (one-way ANOVA, followed by Bonferroni's post hoc test).

Table 1. Electrophysiological analysis of the membrane passive properties.

	PM	DM 48 h	DM + PRP 48 h	DM 72 h	DM + PRP 72 h
RMP (mV)	-62.0 ± 6.8 ($n = 6$)	-41.7 ± 3.75 ($n = 5$)	-50.2 ± 8.9 ($n = 7$)	-45.16 ± 4.06 ($n = 7$)	-47.37 ± 3.1 ($n = 6$)
R_m (M´Ω)	159.2 ± 31.0 (n = 6)	353.8 ± 85.9 (n = 5)	164.7 ± 17.3 (n = 14)	585.3 ± 100.9 [*,#] (n=11)	436.2 ± 71.3 [*,#] (n=15)
C_m (pF)	7.05 ± 1.2 ($n = 6$)	23.5 ± 4.7 [*] ($n = 9$)	6.82 ± 0.5 [§] ($n = 9$)	17.6 ± 2.9 [*,#] ($n = 8$)	8.3 ± 0.9 ($n = 14$)

Data are reported as mean ± S.E.M. [*] $p < 0.05$ versus PM; [#] $p < 0.05$ versus DM + PRP 48 h; [§] $p < 0.05$ versus DM 48 h (one-way ANOVA, followed by Bonferroni's post hoc test). The number of investigated cells is indicated by "n" in brackets for each condition.

Similarly, the cell capacitance, C_m, of cells cultured in DM, usually assumed as an index of cell surface, changed significantly ($p = 0.0085$; F = 4.52 > Fcrit = 2.86; df = 45; one way ANOVA). It tended to increase compared to that estimated in PM (Figure 3C; Table 1), being ($p < 0.05$) higher for cells in DM, especially after 48 h. These results were consistent with the observed cell morphology (Figure 2A,B).

The treatment with PRP actually counteracted the TGF-β1-induced fibroblast to myofibroblast transition. Indeed, the cells cultured in DM + PRP exhibited a significant ($p < 0.05$) reduction of α-sma (Figure 1A,B,E,I) with respect to differentiated cells in DM, together with different morphology, more similar to that of cells cultured in PM (Figure 2C) and a significantly reduced expression of type-1 collagen ($p < 0.05$) (Figure 2F,G).

Of interest, the electrophysiological analyses performed on the cells cultured in DM + PRP for the first time, revealed that R_m values tended to decrease compared to those measured in DM ($p > 0.05$) both after 48 h and 72 h of culture (Figure 3B; Table 1). As well, C_m values evaluated from cells cultured in DM + PRP were significantly reduced ($p < 0.05$) compared to those in DM, consistent with the observed changed morphology of these cells (Figure 3C; Table 1).

3.2. PRP Modifies Transjunctional Currents (I_j) and Gap Junctional Conductance (G_j) in Myofibroblast Pairs

Next, the transjunctional currents (I_j) were analyzed in cell pairs in different experimental conditions by the dual whole-cell technique. Most of the fibroblast pairs cultured in PM exhibited families of I_j current traces with a nearly heterogeneous time course. Typical tracings obtained from a not differentiated cell pair cultured in PM are depicted in Figure 4A.

A minority of the cell pairs cultured in PM (about 20%) showed current records with a symmetrical time course for negative and positive V_j (not shown), indicating the involvement of homotypic GJs. Moreover, they also showed a linear I_j-V_j plot, suggesting the presence of not-voltage-dependent connexons. By contrast, the remaining 80% of the cell pairs investigated showed a non-linear time course. In particular, only 40% of these cells showed a symmetrical voltage dependence for positive and negative V_j, suggesting the involvement of homotypic GJs and voltage-dependent connexons; 60% of the cells showed asymmetrical voltage dependence (Figure 4A). This may suggest the dominant presence of heterotypic GJs in control not differentiating fibroblasts.

We then analyzed the time course of the I_j evoked in myofibroblast pairs. Typical tracings of the I_j evoked in cell pairs cultured in DM for 48 h or 72 h are shown in Figure 4B,C, respectively. About 25% of these cells showed linear not-voltage-dependent I_j. Notably, about 75% of the cell pairs cultured in DM for 48 h exhibited a non-linear time course, suggesting the prevalent expression of voltage-dependent connexons. Only 33% of this kind of cell pairs showed an asymmetrical voltage-dependence suggesting a minor presence of heterotypic voltage-dependent GJs in differentiating myofibroblasts. In contrast, 67% of this kind of response was symmetrical for negative and positive V_j, suggesting that the majority of the GJs involved in this myofibroblastic population were voltage-dependent and homotypic.

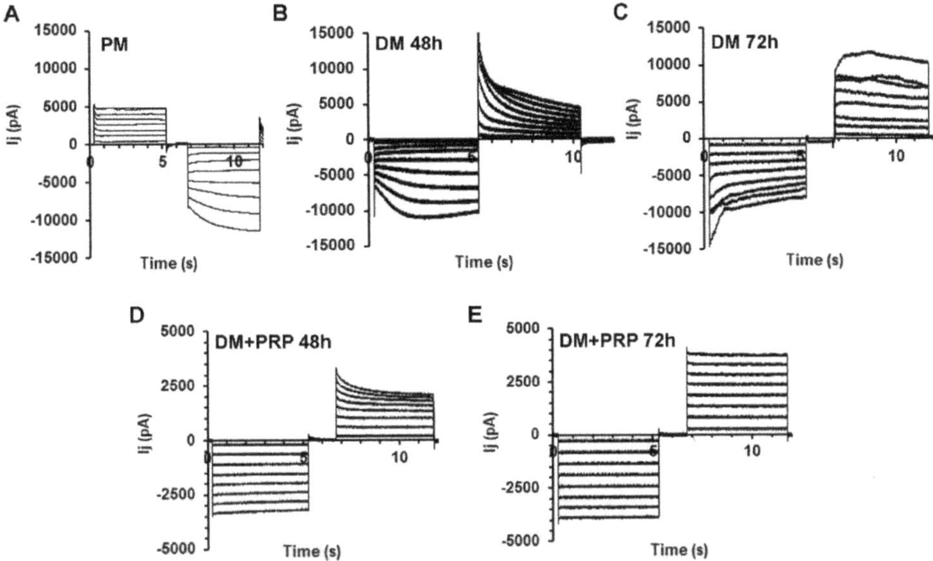

Figure 4. Time course of the transjunctional currents I_j, recorded from fibroblast and myofibroblast pairs in the absence or presence of PRP. (**A**) Representative I_j tracings (in pA) recorded in response to a bipolar pulse protocol applied to a fibroblast pair cultured in proliferation medium (PM). Note the asymmetrical time course between the two voltage polarities with a linear response for positive V_j. (**B,C**) Typical asymmetrical and almost voltage-dependent I_j tracings recorded from (**B**) a myofibroblast pair cultured in differentiation medium (DM) for 48 h and from (**C**) a myofibroblast pair in DM for 72 h. (**D,E**) Representative I_j recorded from a cell pair grown in (**D**) DM with PRP for 48 h (DM + PRP 48 h) and (**E**) for 72 h (DM + PRP 72 h). Note the completely linear and symmetrical responses in the latter condition.

After 72 h of culture in DM, we could observe an increase of the percentage of cell pairs with symmetrical voltage-dependent responses (about 80% in 72 h versus 67% in 48 h), indicating a progressive increase in the number of myofibroblasts exhibiting voltage-dependent and homotypic GJs.

The voltage dependence of I_j was evaluated by plotting the mean values of instantaneous currents, $I_{j,inst}$, (Figure 5A,C,E) and the steady-state currents, $I_{j,ss}$, (Figure 5B,D,F) as a function of V_j (I_j-V_j plot).

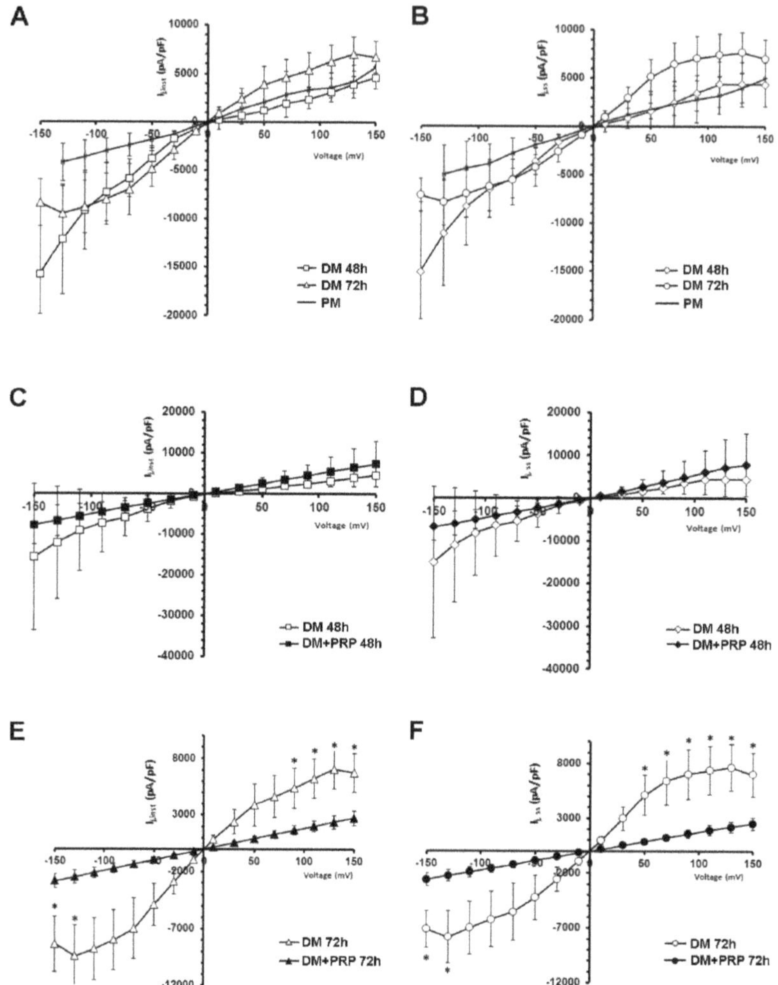

Figure 5. Voltage dependence of the transjunctional currents I_j, recorded from fibroblast and myofibroblast pairs in the absence or presence of PRP. (**A–F**) Transjunctional current values normalized for cell capacitance (in pA/pF), recorded from all of the fibroblast pairs cultured in proliferation medium (PM, continuous line, $n = 6$), and in differentiation medium (DM) for 48 h ($n = 9$) and 72 h ($n = 8$) plotted versus V_j. Panels A, C, and E show the $I_{j,inst}$ values whereas panels B, D, and F show the $I_{j,ss}$ values. Note that the plots related to DM show different slopes. (**C,D**) Comparison between I_j-V_j plot obtained from cell pairs cultured in DM for 48 h (open symbols, $n = 9$) and from cell pairs cultured in DM + PRP for 48 h (filled symbols, $n = 9$). Adding PRP to the culture medium for 48 h, altered the I_j voltage dependence, causing an almost complete linearity with voltage both for $I_{j,inst}$ and $I_{j,ss}$. (**E,F**) I_j-V_j plots related to 72 h treatments. The presence of PRP in DM for 72 h (DM + PRP 72 h, filled symbols, $n = 14$), strongly reduced the mean normalized current amplitude observed in DM 72 h (open symbols, $n = 8$) and altered the I_j voltage dependence, causing an almost complete linearity with voltage both for $I_{j,inst}$ and $I_{j,ss}$. All values represent mean $\pm$ S.E.M. * $p < 0.05$ (unpaired Student's t-test).

Notably, from a qualitative point of view, the I_j-V_j plots showed a different shape according to the different culture conditions, namely PM and DM 48 h and 72 h (Figure 5A,B). Ij currents recorded in

PM showed the smallest amplitude and a scarce deviation from linearity, especially for negative V_j, showing a similar slope for both the V_j polarities (Figure 5A,B). In contrast, the plot related to DM 48 h was almost linear and smoother for positive V_j (Figure 5A,B). The plot related to DM 72 h showed a kind of shoulder becoming S-shaped (Figure 5A,B). For negative V_j the resulting I_j-V_j plots showed a different steepness compared to that observed for positive V_j, and it was similar at any time in culture in DM for 48 h and 72 h (Figure 5A,B). Again, I_j data showing this asymmetrical V_j -dependence are indicative of heterotypic GJ channels. The I_j evaluated both at the peak ($I_{j,inst}$) and at the steady-state ($I_{j,ss}$) showed a progressively more marked voltage-dependence as the time in culture in DM increased. Based on this observation, we suggest a major involvement of voltage-dependent connexons during the differentiation time. Of note, when cells were cultured in DM + PRP for 48 h, about 75% of the cell pairs showed a linear time course (Figure 4D), even if this kind of response was not perfectly symmetrical for negative and positive V_j for all of the cell pairs investigated. When the mean values of all the $I_{j,inst}$ and $I_{j,ss}$ recorded were plotted versus V_j, the relation resulted approximately linear and symmetrical over the entire voltage range, clearly indicating the prevalence of not voltage-dependent homotypic GJ channels (Figure 5C,D) in this culture condition. On the other hand, the cells cultured in DM + PRP for 72 h exhibited only not-voltage dependent Ij (100%) (Figure 4E). Indeed, they showed a linear response and a perfectly symmetrical time course for positive and negative V_j. Again, the I_j-V_j plot analysis showed the I_j linearity with V_j and a marked symmetry, strongly indicating the presence of not-voltage-dependent homotypic GJ channels (Figure 5E,F). Remarkably, for the largest voltage steps applied, the mean current amplitudes recorded from cells cultured in DM for 72 h were statistically different ($p < 0.05$; multiple unpaired Student's t-test) to those estimated in DM + PRP at the same time. Of note, the comparison of I_j-V_j plots in Figure 5E,F with those in Figure 5C,D indicated that the maximal recorded normalized mean amplitude both of $I_{j,inst}$ and $I_{j,ss}$ resulted smaller in DM + PRP 72 h than in DM + PRP 48 h. For instance, the mean $I_{j,ss}$ value estimated for the +150 mV step pulse was 2508 ± 566 pA/pF in DM + PRP 72 h and 7801 ± 7006 pA/pF in DM + PRP 48 h (Figure 5C,D).

Therefore, the features of I_j observed in the cells cultured in DM + PRP, such as the symmetry of the time course and the linearity of $I_{j,inst}$ and $I_{j,ss}$ versus voltage plots lead us to suggest a lessened contribution of voltage-dependent connexons in this condition.

To test for the current really flowing through GJs, we added heptanol (1 mM), a regularly used GJ channel blocker [59], to the bath solution during the recordings. The I_j was evoked from a cell pair cultured in DM, and then records were acquired from the same cell pair. After that, heptanol was acutely added to the bath solution. The recorded currents were significantly reduced compared to those elicited without heptanol. The mathematical subtraction of these two sets of traces gave the heptanol-sensitive current that is the one flowing through the GJs. In contrast, heptanol added during the recordings to cell pairs cultured in DM in the presence of PRP usually caused only a slight reduction of the current amplitude, suggesting a minor number of functional GJs allowing the current flow. A typical experiment related to cell pairs cultured in DM or in DM + PRP for 72 h is shown in Figure 6A. Only the current amplitude obtained by applying two representative voltage pulses (+130 and −30 mV) is shown as an example. Similar results were systematically observed for the bulk of cell pairs investigated. Since the heptanol sensitive current had a very small value in DM + PRP 72 h, we suggest a very small amount of functional GJs in the cells cultured in this experimental condition (Figure 6B–D).

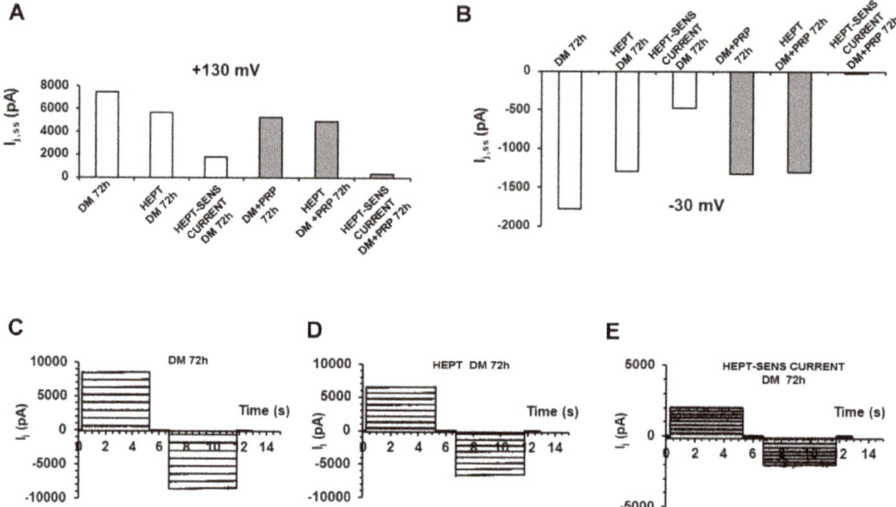

Figure 6. Evaluation of the current flowing through the GJs by acute addition of heptanol under different experimental conditions. (**A,B**) Evaluation of $I_{j,ss}$ (pA) related to a typical cell pair cultured for 72 h in differentiation medium (DM, white bars on the left side of each graph) or to a cell pair cultured for 72 h in DM + PRP (grey bars on the right). For clarity, only the current amplitude values obtained in response to a representative voltage pulse to +130 mV is reported in A, and to −30 mV in B. The current value recorded from the myofibroblast pair (DM 72 h) resulted clearly reduced after the acute addition of heptanol (HEPT, 1 mM) to the bath solution (HEPT DM 72 h). The current flowing through the GJs at the steady-state (HEPT-SENS CURRENT DM 72 h) is obtained by subtracting the current values recorded in the presence of heptanol from those recorded in DM alone. In contrast, the current amplitude recorded from the cell pair cultured in DM + PRP after acute addition of heptanol (HEPT DM + PRP 72 h) showed a slight reduction compared to DM + PRP 72h, having the heptanol-sensitive current a very small value (HEPT-SENS CURRENT DM + PRP 72h). Similar results were systematically observed for any voltage step applied in the bulk of the cell pairs investigated, suggesting a very small amount of functional GJs expressed under PRP treatment. (**C**) Characteristic time course of I_j (pA) recorded from a cell pair cultured in DM (DM 72 h) that exceptionally showed voltage independent features. (**D**) The same current traces recorded after acute heptanol addition (HEPT DM 72 h). (**E**) Resulting current traces (HEPT-SENS CURRENT DM 72 h) obtained by subtracting currents in D from currents in C, and representing the small heptanol-sensitive flux through the GJs. Note the different ordinate scale in E.

Noteworthy, even in those cell pairs cultured in DM for 72 h that exhibited not-voltage-dependent I_j, we could still measure a heptanol-sensitive current, suggesting that also the not-voltage dependent GJ functionality was hampered by the GJ blocker.

Finally, we analyzed the conductive properties of GJs. Intercellular current flow in cell pairs of fibroblasts and myofibroblasts were also used to study the dependency of the gap junctional conductance, G_j, on V_j by means of the G_j-V_j plot analysis. The related results are shown in Figure 7.

Data points obtained from cell pairs cultured in PM (Figure 7A) showed a horizontal distribution for positive V_j, suggesting the involvement of not-voltage-dependent GJs, in contrast to the not-linear distribution observed for negative V_j values. This asymmetrical voltage dependence of the G_j can suggest the prevalence of heterotypic GJs in proliferating fibroblastic cell pairs. These data points obtained from the cell pairs cultured in DM for 48 h (Figure 7B) did not follow a merely symmetrical relationship, being almost linear for negative V_j and more bell-shaped for positive V_j. Again, this may be consistent with the expression of more than one Cx isoform in myofibroblasts (possibly assembling in different combinations compared to those observed in PM) and hence confirms the involvement of heterotypic GJ channels in this cell population. After 72 h in DM, the G_j-V_j plot showed more or less the same shape as 48 h, but the G_j values for positive V_j resulted higher, suggesting a major contribution of the voltage-dependent component. In contrast, G_j-V_j plots related to the cells cultured in DM + PRP both at 48 h and 72 h were symmetrical and linear in any case, suggesting a lack of voltage-dependent GJs. This result suggests the involvement of homotypic GJ channels in this cell population.

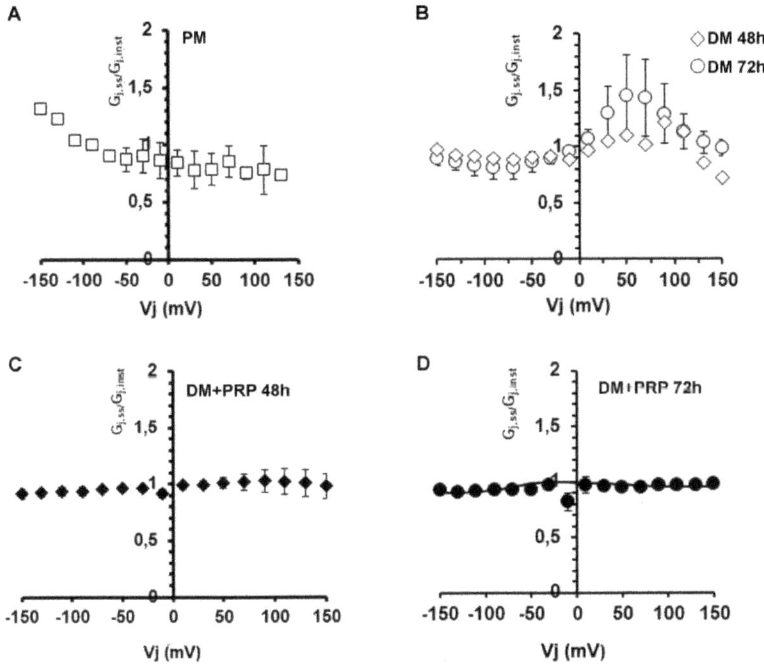

Figure 7. Voltage dependence of the transjunctional conductance G_j. (**A,B**) Voltage dependence of the transjunctional conductance obtained by plotting $G_{j,ss}/G_{j,inst}$ versus V_j related to (**A**) proliferation medium (PM) condition (open squares, $n = 6$), (**B**) differentiation medium (DM) condition at 48 h (open squares, $n = 9$) and 72 h (open circles, $n = 9$). These data obtained in DM show an asymmetrical distribution that becomes more bell-shaped for positive V_j. The treatment for 72 h gave higher values compared to 48 h, although not statistically significant ($p > 0.05$, multiple unpaired Student's t-test). (**C,D**) Symmetrical linear distribution observed under the concomitant treatment in DM + PRP at (**C**) 48 h (filled squares, $n = 9$), and (**D**) 72 h (filled circles, $n = 14$). All values represent mean ± S.E.M. Error bars are visible if they exceed the symbol size.

3.3. PRP Reduces Cx43 Expression and Increases Cx26 Expression in Differentiated Myofibroblasts

Since the electrophysiological experiments showed an increase of I_j and G_j functionality during myo-differentiation of fibroblasts, we cultured the cells in DM in the presence of the GJ channel blocker heptanol (1 mM) for 72 h, to test the effective involvement of GJs in myofibroblast generation. We first analyzed the expression of α-sma in this experimental condition. As assessed by Western blotting and confocal immunofluorescence analyses, the cells exposed to DM + heptanol exhibited a clear reduction of α-sma expression (Figure 1A,B,F,G,I) compared to control differentiated myofibroblasts cultured in DM, supporting a key role of the GJs in the acquisition of myofibroblastic phenotype. Of note, the cells cultured in DM + PRP showed a more robust reduction of α-sma then those cultured in DM + heptanol, likely suggesting that PRP-induced prevention of fibroblast myofibroblast differentiation involves the activation of multiple molecular mechanisms. The cells cultured in DM + heptanol + PRP showed a significantly reduced ($p < 0.05$) expression of α-sma as compared to that observed in the cells exposed to single treatment (i.e., DM + heptanol or DM + PRP).

Then, taking into consideration the increasingly marked voltage dependence of the I_j recorded during differentiation time, we performed experiments aimed to evaluate the expression of the typical Cx types forming voltage- dependent connexins, namely Cx43.

We found that Cx43 expression both at mRNA and protein levels significantly increased ($p < 0.05$) with time in the cells cultured in DM as compared to control cells in PM as judged by RT-PCR (Figure 8A) and Western blotting analyses (Figure 8B), respectively.

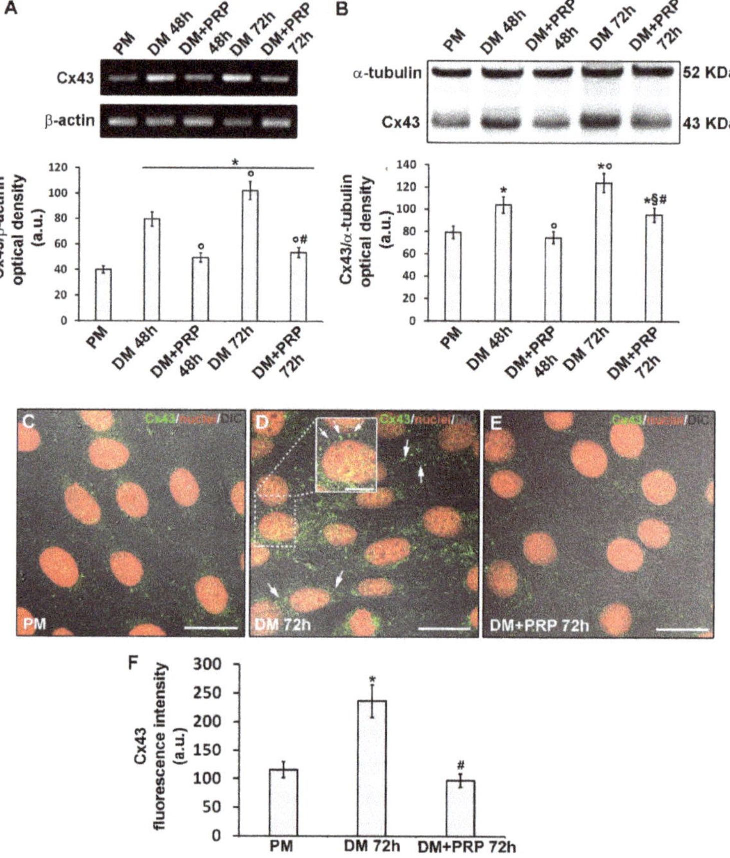

Figure 8. Cx43 expression and localization during fibroblast to myofibroblast transition and related PRP effects. Fibroblasts were induced to differentiate into myofibroblasts by culturing in differentiation medium (DM) in the presence or absence of PRP for 48 h and 72 h. The cells cultured in proliferation medium (PM) were used as control undifferentiated cells. (**A**) RT-PCR analysis of Cx43 expression in the indicated experimental conditions. Representative agarose gel is shown. The densitometric analysis of the bands normalized to β-actin is reported in the histogram. (**B**) Western Blotting analysis of Cx43 expression. Histogram showing the densitometric analysis of the bands normalized to α-tubulin. (**C–E**) Representative superimposed differential interference contrast (DIC, grey) and confocal fluorescence images of the cells immunostained with antibodies against Cx43 (green) and

counterstained with propidium iodide (PI, red) to label nuclei. Scale bar: 30 μm. Scale bar in the inset in D: 15 μm. Arrows indicate the localization of Cx43 at the membrane level of two adjacent cells. (**F**) Histogram showing the densitometric analysis of the intensity of the Cx43 fluorescence signal performed on digitized images in 20 regions of interest (ROI) of 100 μm^2 for each confocal stack (12). Data shown are mean ± S.E.M. and represent the results of at least three independent experiments performed in triplicate. Significance of difference: * $p < 0.05$ versus PM; ° $p < 0.05$ versus DM 48 h; # $p < 0.05$ versus DM 72 h; § $p < 0.05$ versus DM + PRP 48 h (One-way ANOVA followed by the Tukey post hoc test).

Confocal immunofluorescence analysis confirmed the increase of Cx43 expression in differentiated cells after 48 h (data not shown) and even more after 72 h of culture in DM as compared to control undifferentiated cells in PM (Figure 8C,D,F). Moreover, we demonstrated the protein localization either at the cytoplasmic level or at the cell membrane level of adjacent cells (Figure 8C,D). To confirm the key role of this Cx isoform in fibroblast to myofibroblast transition, we silenced the cells for the expression of Cx43 by specific siRNA before culturing them in DM for 72 h (Figure 9A).

These cells exhibited a significant reduction ($p < 0.05$) of α-sma expression (Figure 9B,C,D,E,I) compared to cell cultured in DM, suggesting that Cx43 was required for the differentiation process. Of note, the cells cultured in DM + PRP concomitantly to reduced α-sma, showed a significant reduction of Cx43 expression (Figure 8A,B,E,F). This outcome was consistent with the electrophysiological data showing the reduction of voltage-dependent responses in these cells. Notably, according to the results of the experiments achieved in cells cultured with heptanol (Figure 1), the cells cultured in DM + PRP showed reduced expression of α-sma with respect to the cells silenced for Cx43 cultured in DM (Figure 9 B–I). Cells silenced for Cx43 expression and exposed to DM + PRP exhibited reduced α-sma expression levels as compared to those observed in the cells exposed to single treatment (i.e., DM + siRNA or DM + PRP) (Figure 9 B–I).

Finally, even the occurrence of a not-voltage-dependent response in myofibroblast pairs (although in a minority) as well as of the increase of this type of response after treatment with PRP, we analyzed the expression of a typical Cx type forming not/scarcely voltage-dependent connexons, namely Cx26. Western blotting (Figure 10A,B) and confocal immunofluorescence (Figure 10C–H) analyses demonstrated that Cx26 expression significantly increased in the cells after culture in DM for 48 h with respect to undifferentiated cells cultured in PM ($p < 0.05$). By contrast, the cells cultured in DM for 72 h exhibited Cx26 expression levels comparable to those of undifferentiated cells. The cells cultured in DM for 48 h in the absence or presence of PRP exhibited comparable levels of Cx26 expression ($p > 0.05$). Of note, the cells cultured in DM + PRP for 72 h exhibited a slight but significant increase ($p > 0.05$) of Cx26 as compared to cells cultured in the absence of PRP (Figure 10A,B,F,G,H).

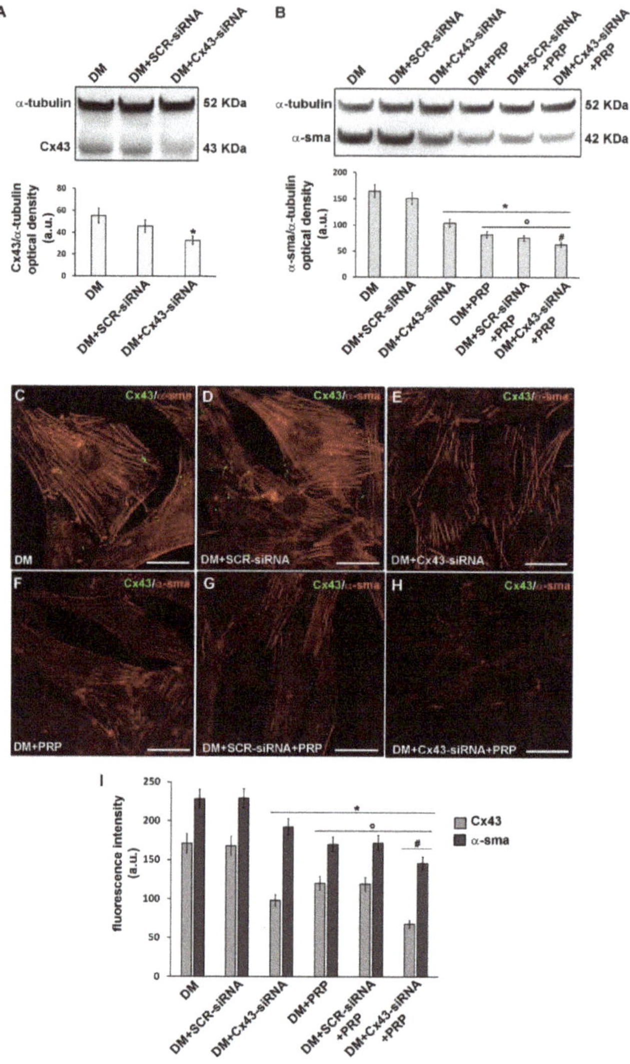

Figure 9. Effect of inhibition of Cx43 expression on fibroblast to myofibroblast transition and related PRP effects. Fibroblasts were silenced by specific Cx43-siRNA duplexes and cultured in differentiation medium (DM) for 48 h in the absence or presence of PRP. SCR-siRNA duplexes were used as an internal control. (**A,B**) Western Blotting analysis of (**A**) Cx43 and (**B**) α-sma expression. The densitometric analysis of the bands normalized to α-tubulin is reported in the histograms. (**C–H**) Representative confocal fluorescence images of the cells double immunostained with antibodies against Cx43 (green) and α-sma (red). Scale bar: 25 μm. (**I**) Histogram showing the densitometric analysis of the intensity of Cx43 and α-sma fluorescence signal performed on digitized images in 20 regions of interest (ROI) of 100 μm^2 for each confocal stack (12). Data shown are mean ± S.E.M. and represent the results of at least three independent experiments performed in triplicate. Significance of difference: * $p < 0.05$ versus DM; ° $p < 0.05$ versus DM + CX43 − siRNA; # $p < 0.05$ versus DM + PRP (One-way ANOVA followed by the Tukey post hoc test).

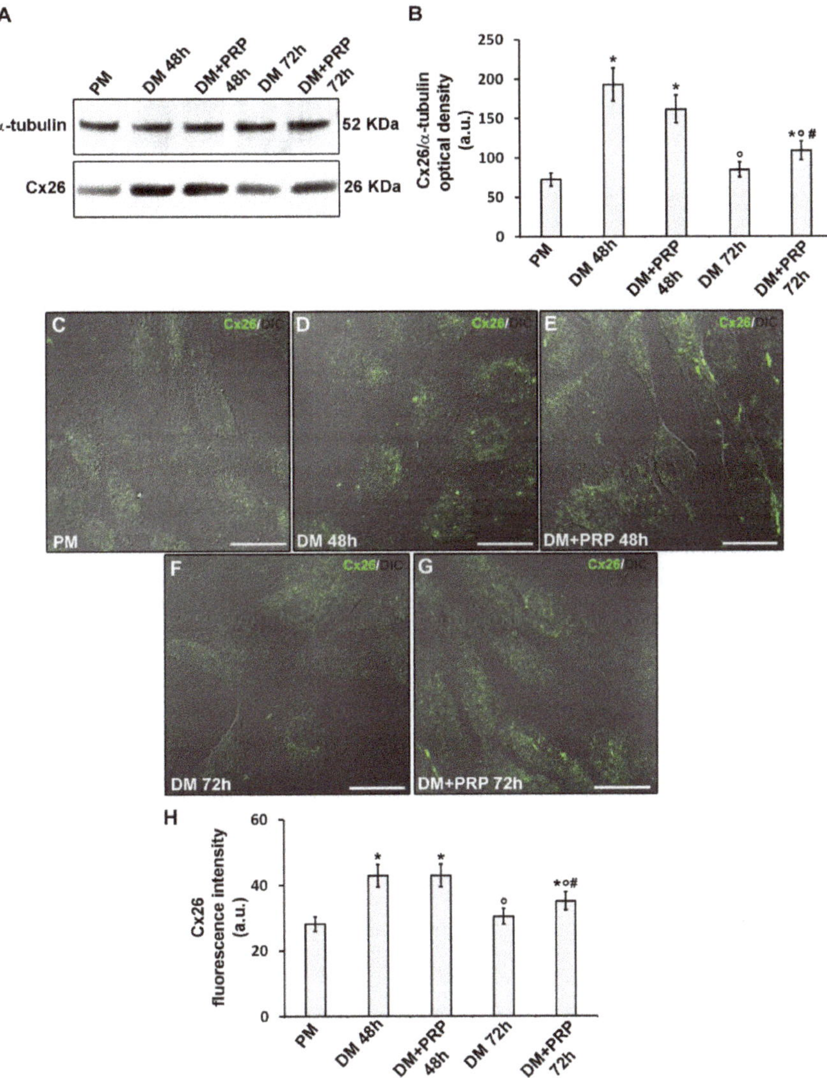

Figure 10. Cx26 expression during fibroblast to myofibroblast transition and related PRP effects. Fibroblasts were induced to differentiate into myofibroblasts by culturing in differentiation medium (DM) in the presence or absence of PRP for 48 h and 72 h. The cells cultured in proliferation medium (PM) served as control undifferentiated cells. (**A,B**) Western Blotting analysis of Cx26 expression. (**A**) Representative blot. (**B**) Histogram showing the densitometric analysis of the bands normalized to α-tubulin. (**C–G**) Representative superimposed differential interference contrast (DIC, grey) and confocal fluorescence images of the cells immunostained with antibodies against Cx26 (green) showing the cellular localization of the protein. Scale bar: 25 μm. (**H**) Histogram showing the densitometric analysis of the intensity of the Cx26 fluorescence signal performed on digitized images in 20 regions of interest (ROI) of 100 μm^2 for each confocal stack (12). Data shown are mean ± S.E.M. and represent the results of at least three independent experiments performed in triplicate. Significance of difference: * $p < 0.05$ versus PM; ° $p < 0.05$ versus DM 48 h; # $p < 0.05$ versus DM 72 h (One-way ANOVA followed by the Tukey post hoc test).

4. Discussion

In recent years, great attention has been paid to the identification of new therapeutic agents and treatments that may promote the repair/regeneration of damaged skeletal muscle. In such a context, several in vitro and in vivo studies provided evidence supporting the advantage of the use of PRP for muscle regenerative purpose [30,66]. In this line, we have recently demonstrated the capability of PRP to either stimulate proliferation and differentiation of myogenic progenitors, including satellite cells [58], or prevent the TGF-β1 induced differentiation of fibroblasts towards myofibroblasts [48,57].

These data led us to suggest that PRP, if properly administered along the cascade of events through which skeletal muscle repair/regeneration proceeds (which also includes the physiological fibrotic reparative response), could exert a double beneficial effect on the healing of injured muscle. This may consist in the direct activation of the resident cells effectors of muscle regeneration, responsible for the formation of new muscle fibers and, in parallel, in the modulation/prevention of an excessive fibrotic response, thus contributing to the recreation of a more hospitable and conducive microenvironment for muscle progenitor functionality and thus the promotion of tissue regeneration. Experiments are ongoing in our lab aimed to assess the effects of PRP on differentiated myofibroblasts, by evaluating the capability of this blood product to modulate their fate. The results should be of interest to support the anti-fibrotic action of PRP. However, the ability of PRP in antagonizing fibrotic signaling pathways is still an issue of debate. Some reports show limited effectiveness or even inefficacy of this blood-derived product in counteracting the skeletal muscle fibrotic response [30,42,44,67–75]. The great heterogeneity of the available PRP formulation, PRP dosage and application timing represent critical points that may account for the reported conflicting results concerning the effects of this blood product in the modulation of skeletal muscle tissue fibrosis.

Based on these considerations, studies aimed to support the anti-fibrotic effect of this blood product are needed, as well as researches focused on the identification of the cellular and molecular target mechanisms of PRP, underpinning its action.

4.1. PRP Counteracts Myofibroblast Generation

According to findings from our previous studies and other research groups [30,47,48,76,77], here we have confirmed the ability of PRP to counteract the core cellular process of the fibrotic response, namely differentiation of fibroblasts towards myofibroblasts induced by the pro-fibrotic agent TGF-β1, based on: i) morphological and biochemical analyses showing that the cells treated with TGF-β1 in the presence of PRP did not acquire a mature myofibroblastic phenotype; indeed they rather appeared more spindle-shaped and showed either a reduction of type-1 collagen expression and a lower expression of α-sma, that was also less organized in filamentous structure as compared to differentiated cells; ii) the novel electrophysiological recordings of the membrane passive properties and gap junctional functionality, showing the ability of PRP to modify such parameters with respect to those recorded in differentiated myofibroblasts. Particularly, in the present experiments we observed that differentiated myofibroblasts tended to have a more positive RMP and PRP treatment counteracted this occurrence. The RMP is always critical for cell function since any small alteration of its value can substantially change cell excitability, contractility, and other properties, such as cell migration [21]. The less positive membrane potential registered in the cells induced to differentiate in the presence of PRP may counteract the depolarization of myofibroblasts and hamper their contractility, leading to an altered functionality. In this regard, it was shown that depolarization causes enhancement of ventricular myofibroblast contractility [21]. In this view, the present findings may suggest that PRP can revert myofibroblast RMP towards a more 'dormant' condition, counteracting their full differentiation.

In addition, PRP action opposed the increase of R_m value observed in differentiating conditions, showing its ability to revert the effect on the resting conductive properties induced by TGF-β1. The R_m parameter, corresponding to the reciprocal value of the membrane conductance, G_m, gives an idea of the total resting ionic fluxes across the membrane; thus, its physiological relevance is strictly linked to cell excitability. Moreover, the C_m value assumed as an index of membrane surface area increased in the cells induced to differentiate as compared to proliferating cells. This observation was in agreement with the morphological analysis showing that the cells tended to increase their size upon TGF-β1-induced differentiation. Both phenomena were counteracted by PRP. Notably, the latter data are in accordance with the electrophysiological results described in our previous report dealing with the anti-fibrotic potential of relaxin [23].

4.2. Role of GJIC and Cx43 in Myofibroblast Generation

The main relevance of our study is the contribution toward defining the molecular and functional mechanisms regulating TGF-β1 induced fibroblast-myofibroblast transition, highlighting the role of GJs in this process as well as the involvement of voltage-dependent connexin isoform, namely Cx43. In particular, we found that the majority of differentiated myofibroblast pairs exhibited an enhancement of I_j amplitude in the course of differentiation, suggesting an increased functionality of GJs, especially of the voltage-dependent ones, with increasing exposure time to TGF-β1. The role of GJs in this differentiation process was confirmed by the use of heptanol, a common GJ blocker. When the cells were induced to differentiate in the presence of heptanol, they actually failed to acquire a myofibroblast phenotype, indicating an essential role of functional GJs in the promotion of fibroblasts differentiation towards myofibroblasts. These data are in good accordance with previous studies showing that the selective blockade of GJs downregulated myofibroblastic phenotype [78,79]. Therefore, it can be stated that GJIC is of crucial importance in our cell model to regulate the fibroblast transformation towards myofibroblast. However, a possible role of such intercellular communications in the functional coupling of mature myofibroblasts to coordinate their activity can also be speculated [18,51–54]. In fact, while it is well accepted that myofibroblasts are responsible for the reparative scar formation and contraction, it is not clear yet whether they act individually or behave synchronically [78]. In this regard, we can propose that myofibroblasts can, at least in part, act as a coordinated functional syncytium, thus that hindering intercellular communication may represent a therapeutic target in diseases characterized by an overabundance of these contractile cells. Another important point is the ability of myofibroblasts to interact, by GJs, with other resident cell types of tissue globally affecting the organ functionality [80–82]. The analysis of the transjunctional conductance, G_j, gave some interesting information. Usually, the higher the G_j, the faster the current flows from a cell to the adjacent one, resulting in faster propagation speed [83]. The estimated G_j is the overall result of the total number of GJ channels docked between cells, the single-channel conductance of each GJ channel, and their functional states (fully open, sub conductance, or closed states). The GJ functional states can be dynamically modulated by chemicals and transjunctional voltage. The transjunctional voltage-dependent gating is an intrinsic property in all characterized GJs. In this study, we found that the G_j in differentiated myofibroblast pairs showed a progressively more marked voltage-dependence, suggesting a prevalent expression of voltage-dependent Cxs in myofibroblasts. From a functional point of view, this may reflect the need for myofibroblasts to be coupled in response to stimuli that cause membrane potential alterations. However, myofibroblasts likely need to express a kind of Cx, such as Cx43, whose trafficking, half-life, and regulation (by phosphorylation) can be maximally modulated during differentiation to myofibroblast [84]. Corroborating this suggestion coming from electrophysiological records, here we found that myofibroblasts showed an increased expression of Cx43. Furthermore, we found that cells silenced for the expression of Cx43 did not exhibit a mature myofibroblastic phenotype when cultured in differentiation condition in the presence of TGF-β1, suggesting the requirement of Cx43 for fibroblast-myofibroblast transition. Collectively these data are in accordance with the findings of our recent study [48] demonstrating an upregulation of mRNA expression of Cx43 in TGF-β1 treated cells (i.e., myofibroblasts). As well they agree with the studies by Asazuma-Nakamura and co-workers (2009) [85] and by Paw and co-workers (2017) [86] showing that Cx43 positively regulated myofibroblastic differentiation of cardiac and bronchial fibroblasts, respectively. In parallel, other studies showed that Cx43 expression is largely modulated during wound repair, and the modulation of Cx43 expression and gap junctional communication can be beneficial to wound healing [87]. This process can be definitely altered by modulating Cx43 expression: wound closure can be delayed when Cx43 is overexpressed or accelerated when the levels of epidermal Cx43 are reduced [88–91]. In line with this, the transient blockade of Cx43 functions has been shown to reduce fibrosis as well as to promote experimental wound healing [89], and the normal GJ functions of Cx43 seems to be important for normal fibroblast function [92]. In this regard, multiple clinical trials are investigating Cx43 modulators and specific peptides targeting the intracellular loop, and the C-terminal tail region of this protein appear promising [93]. Based on our present findings, showing an increase of Cx43 expression in differentiated cells not only at the plasma membrane level but also in the cytoplasm, it is worth mentioning that a GJ independent function of Cx43 during fibroblast to myofibroblast transition cannot be excluded [59,94]. Taking into account the reported channel-independent influence of Cx43 on cytoskeleton remodeling and cell migration, we may speculate a similar role [94] for stress fiber assembly during fibroblast- myofibroblast transition.

4.3. GJs, Cx43, and Cx26 as Molecular Candidate Targets of the Potential Anti-fibrotic Action of PRP

Another main finding of this study is the compelling experimental evidence indicating, for the first time, that GJs and Cx43 are molecular candidate targets of the potential anti-fibrotic action of PRP. Indeed, PRP treatment affected the occurrence of voltage-dependent I_j, which was reduced at 48 h and even abolished after 72 h of culture; concomitantly, it prevented the TGF-β1 induced increase of Cx43 expression. Interestingly, the cells induced to differentiate in the presence of PRP mostly exhibited scarcely/not voltage-dependent I_j along with an observed upregulation of Cx26, especially after 48 h. We may suppose that these events reflect a compensatory mechanism to maintain a sort of cell-to-cell communication, upon PRP treatment. In addition, we may speculate a major role of Cx26 in this condition that may be linked to the ability of this Cx type to form hemichannels rather than highly regulated GJs. We should point out the ability of hemichannels themselves to act as membrane channels able to mediate cell communication with surrounding extracellular environment: recent studies confirm a link between hemichannel-mediated ATP release and the progression and development of fibrosis in different tissue types [95–98]. In this line, the role of Cx26 in forming scarcely voltage-dependent GJs and/or hemichannels and the possible relation with ATP content in our cell model is an interesting topic deserving further investigation.

4.4. Factors Released by PRP Possibly Modulating Cx Expression and GJ Functionality During Myofibroblast Generation

Finally, it is worth mentioning that the factors released by PRP possibly modulating Cx expression and GJ functionality during myo-differentiation process remain to be identified as well as their potential modes of action. We have previously demonstrated that PRP contains vascular endothelial growth factor (VEGF)-A and that, through this factor linking its receptor-1 (VEGFR-1 or Flt-1), PRP antagonizes TGF-β1/Smad3, thus preventing fibroblast myo-differentiation [30,48]. Taking into consideration that VEGF may modulate the expression of Cx43 and/or GJ functionality in different cell types [99–101] we may speculate that GJs/Cx43 might represent a downstream target of VEGF-A/VEGFR-1 mediated signaling in our experimental cell model. Studies are ongoing in our lab to assess this hypothesis. Moreover, it is known that GJ assembly and disassembly are events highly regulated by a sequence of protein kinase activation and phosphorylation events. This kind of regulation, as extensively reported for Cx43, has a net effect of reducing GJ communication [102–104]. Therefore, we may also postulate that VEGF-A or other factors released by PRP may reduce GJ/Cx43 functionality affecting such phosphorylation events. In such a view it has been reported that insulin-like growth factor (IGF)-1 is able to decrease gap junctional communication by inducing activation of PKCγ, enhancing the interaction between PKCγ and Cx43 and the phosphorylation of Cx43 by PKCγ [105] in epithelial cells. Since IGF-1 has been reported to be contained in PRP [106] and has been supposed as a potential modulator of fibrogenic events and pathways [14,107,108], it is tempting to speculate that similar interactions may also occur in our cell system. Therefore, the capability of PRP to counteract fibroblast differentiation towards myofibroblasts and its ability to modulate GJIC could also involve the IGF-1 signaling pathway. A full characterization of the releasing profile of likely cross-talking factors present in PRP are required to understand the molecular mechanisms underpinning the action of this plasma product.

5. Conclusions

In conclusion, the results of the present in vitro study provide the first experimental evidence that upregulation of Cx43 and the parallel increase of voltage-dependent GJ functionality are important mechanisms by which TGF-β1 endorses fibroblast differentiation towards myofibroblast, and that PRP treatment hampers this effect.

The main limitations of this study rely on the in vitro experimentation on the NIH/3T3 cell line. Obviously, the in vitro experimentation eliminates many paracrine/juxtacrine mechanisms, possibly regulating in situ intercellular interactions and cell functionality as well as the mechanical forces exerted by the surrounding microenvironment, including ECM stiffness, affecting cell behavior [16,18,109]. NIH/3T3 cells represent a widely used, reliable model to study fibroblast biology. We previously demonstrated that these cells show similar behavior, in terms of differentiation marker expression and electrophysiological parameters, to primary fibroblasts, such as human dermal fibroblasts, skeletal, and cardiac fibroblasts [23,48,55,56]. Nevertheless, the growing evidence on functional heterogeneity and the origin-linked response of fibroblasts to stimuli must be taken into account [110–112]. Therefore, we acknowledge that a different experimental set using primary cultures of skeletal muscle-derived fibroblasts could have offered in vitro findings possibly more closely related to in vivo conditions of skeletal muscle disease and fibrosis.

Another limitation is represented by the lack of a full characterization of the growth factors released by PRP. This should be relevant to understand better PRP mechanisms of action in the modulation of fibroblast-myofibroblast transition and to achieve a therapeutic translation of this approach. Furthermore, standardization of PRP preparation techniques as well as application protocols would allow performing meaningful comparative analyses.

However, despite these aspects, this research contributes to add further insights into molecular and functional mechanisms regulating fibroblast-myofibroblast transition. It likewise supports the anti-fibrotic action of PRP by means of its ability to hamper myofibroblast generation, targeting GJs, thus providing cues to novel therapeutic targets.

Author Contributions: Conceptualization, R.S., F.C., E.I., A.T., and C.S.; formal analysis, R.S., F.C., E.I., A.T., and C.S.; investigation, R.S., F.C., E.I., A.T., R.G., S.P., and C.S.; resources, R.S., F.C., P.P., F.B., S.Z.-O., and C.S.; data curation, R.S., F.C., E.I., A.T., and C.S.; writing—original draft preparation, R.S., and C.S.; writing—review and editing, R.S., F.C., E.I., A.T., S.Z.-O., and C.S.; visualization, R.S., F.C., E.I., A.T., and C.S.; funding acquisition, R.S., F.C., S.Z.-O., and C.S. All authors have read and agreed to the published version of the manuscript."

Funding: This research was supported by the annual financing fund by MIUR (Ministry of Education, University and Research, Italy)—University of Florence to R.S., F.C., S.Z.-O., and C.S. and by FFABR-MIUR 2017 (Financing Fund for Basic Research Activities) granted to R.S. and C.S.

Conflicts of Interest: The authors declare no conflict of interest.

References

1. Yin, H.; Price, F.; Rudnicki, M.A. Satellite cells and the muscle stem cell niche. *Physiol. Rev.* **2013**, *93*, 23–67. [CrossRef]
2. Doyle, M.J.; Zhou, S.; Tanaka, K.K.; Pisconti, A.; Farina, N.H.; Sorrentino, B.P.; Olwin, B.B. Abcg2 labels multiple cell types in skeletal muscle and participates in muscle regeneration. *J. Cell. Biol.* **2011**, *195*, 147–163. [CrossRef] [PubMed]
3. Judson, R.N.; Zhang, R.H.; Rossi, F.M. Tissue-resident mesenchymal stem/progenitor cells in skeletal muscle: Collaborators or saboteurs? *FEBS J.* **2013**, *280*, 4100–4108. [CrossRef]

4. Meng, J.; Chun, S.; Asfahani, R.; Lochmüller, H.; Muntoni, F.; Morgan, J. Human skeletal muscle-derived CD133(+) cells form functional satellite cells after intramuscular transplantation in immunodeficient host mice. *Mol. Ther.* **2014**, *22*, 1008–1017. [CrossRef] [PubMed]

5. Scicchitano, B.M.; Sica, G.; Musarò, A. Stem Cells and Tissue Niche: Two Faces of the Same Coin of Muscle Regeneration. *Eur. J. Transl. Myol.* **2016**, *26*, 6125. [CrossRef] [PubMed]

6. Tonlorenzi, R.; Rossi, G.; Messina, G. Isolation and Characterization of Vessel-Associated Stem/Progenitor Cells from Skeletal Muscle. *Methods Mol. Biol.* **2017**, *1556*, 149–177. [CrossRef] [PubMed]

7. Dumont, N.A.; Wang, Y.X.; Rudnicki, M.A. Intrinsic and extrinsic mechanisms regulating satellite cell function. *Development* **2015**, *142*, 1572–1581. [CrossRef]

8. Forcina, L.; Miano, C.; Pelosi, L.; Musarò, A. An Overview about the Biology of Skeletal Muscle Satellite Cells. *Curr. Genomics* **2019**, *20*, 24–37. [CrossRef]

9. Sacco, A.; Puri, P.L. Regulation of Muscle Satellite Cell Function in Tissue Homeostasis and Aging. *Cell Stem Cell.* **2015**, *16*, 585–587. [CrossRef]

10. Mann, C.J.; Perdiguero, E.; Kharraz, Y.; Aguilar, S.; Serrano, A.L.; Muñoz-Cánoves, P. Aberrant repair and fibrosis development in skeletal muscle. *Skelet. Muscle* **2011**, *1*, 21. [CrossRef]

11. Farup, J.; Madaro, L.; Puri, P.L.; Mikkelsen, U.R. Interactions between muscle stem cells, mesenchymal-derived cells and immune cells in muscle homeostasis, regeneration and disease. *Cell Death Dis.* **2015**, *6*, e1830. [CrossRef] [PubMed]

12. Ceafalan, L.C.; Fertig, T.E.; Popescu, A.C.; Popescu, B.O.; Hinescu, M.E.; Gherghiceanu, M. Skeletal muscle regeneration involves macrophage-myoblast bonding. *Cell Adhes. Migr.* **2018**, *12*, 228–235. [CrossRef] [PubMed]

13. Malecova, B.; Gatto, S.; Etxaniz, U.; Passafaro, M.; Cortez, A.; Nicoletti, C.; Giordani, L.; Torcinaro, A.; De Bardi, M.; Bicciato, S.; et al. Dynamics of cellular states of fibro-adipogenic progenitors during myogenesis and muscular dystrophy. *Nat. Commun.* **2018**, *9*, 3670. [CrossRef] [PubMed]

14. Forcina, L.; Miano, C.; Scicchitano, B.M.; Musarò, A. Signals from the Niche: Insights into the Role of IGF-1 and IL-6 in Modulating Skeletal Muscle Fibrosis. *Cells* **2019**, *8*, 232. [CrossRef] [PubMed]

15. Manetti, M.; Tani, A.; Rosa, I.; Chellini, F.; Squecco, R.; Idrizaj, E.; Zecchi-Orlandini, S.; Ibba-Manneschi, L.; Sassoli, C. Morphological evidence for telocytes as stromal cells supporting satellite cell activation in eccentric contraction-induced skeletal muscle injury. *Sci. Rep.* **2019**, *9*, 14515. [CrossRef] [PubMed]

16. Hinz, B.; McCulloch, C.A.; Coelho, N.M. Mechanical regulation of myofibroblast phenoconversion and collagen contraction. *Exp. Cell Res.* **2019**, *379*, 119–128. [CrossRef]

17. Weiskirchen, R.; Weiskirchen, S.; Tacke, F. Organ and tissue fibrosis: Molecular signals, cellular mechanisms and translational implications. *Mol. Aspects Med.* **2019**, *65*, 2–15. [CrossRef]

18. Tomasek, J.J.; Gabbiani, G.; Hinz, B.; Chaponnier, C.; Brown, R.A. Myofibroblasts and mechano-regulation of connective tissue remodelling. *Nat. Rev. Mol. Cell. Biol.* **2002**, *3*, 349–363. [CrossRef]

19. Koivumäki, J.T.; Clark, R.B.; Belke, D.; Kondo, C.; Fedak, P.W.; Maleckar, M.M.; Giles, W.R. Na(+) current expression in human atrial myofibroblasts: Identity and functional roles. *Front. Physiol.* **2014**, *5*, 275. [CrossRef]

20. Zhan, H.; Zhang, J.; Jiao, A.; Wang, Q. Stretch-activated current in human atrial myocytes and Na+ current and mechano-gated channels' current in myofibroblasts alter myocyte mechanical behavior: A computational study. *Biomed. Eng. Online* **2019**, *18*, 104. [CrossRef]

21. Chilton, L.; Ohya, S.; Freed, D.; George, E.; Drobic, V.; Shibukawa, Y.; Maccannell, K.A.; Imaizumi, Y.; Clark, R.B.; Dixon, I.M.; et al. K+ currents regulate the resting membrane potential, proliferation, and contractile responses in ventricular fibroblasts and myofibroblasts. *Am. J. Physiol. Heart. Circ. Physiol.* **2005**, *288*, 2931–2939. [CrossRef] [PubMed]

22. Kaur, K.; Zarzoso, M.; Ponce-Balbuena, D.; Guerrero-Serna, G.; Hou, L.; Musa, H.; Jalife, J. TGF-β1, released by myofibroblasts, differentially regulates transcription and function of sodium and potassium channels in adult rat ventricular myocytes. *PLoS ONE* **2013**, *8*, e55391. [CrossRef] [PubMed]

23. Squecco, R.; Sassoli, C.; Garella, R.; Chellini, F.; Idrizaj, E.; Nistri, S.; Formigli, L.; Bani, D.; Francini, F. Inhibitory effects of relaxin on cardiac fibroblast-to-myofibroblast transition: An electrophysiological study. *Exp. Physiol.* **2015**, *100*, 652–666. [CrossRef] [PubMed]

24. Sassoli, C.; Chellini, F.; Squecco, R.; Tani, A.; Idrizaj, E.; Nosi, D.; Giannelli, M.; Zecchi-Orlandini, S. Low intensity 635 nm diode laser irradiation inhibits fibroblast-myofibroblast transition reducing TRPC1 channel expression/activity: New perspectives for tissue fibrosis treatment. *Lasers Surg. Med.* **2016**, *48*, 318–332. [CrossRef]

25. Xue, M.; Jackson, C.J. Extracellular Matrix Reorganization During Wound Healing and Its Impact on Abnormal Scarring. *Adv. Wound Care (New Rochelle).* **2015**, *4*, 119–136. [CrossRef] [PubMed]

26. Jun, J.I.; Lau, L.F. Resolution of organ fibrosis. *J. Clin. Investig.* **2018**, *128*, 97–107. [CrossRef] [PubMed]

27. Horowitz, J.C.; Thannickal, V.J. Mechanisms for the Resolution of Organ Fibrosis. *Physiology (Bethesda)* **2019**, *34*, 43–55. [CrossRef]

28. Bochaton-Piallat, M.L.; Gabbiani, G.; Hinz, B. The myofibroblast in wound healing and fibrosis: Answered and unanswered questions. *F1000Research* **2016**, *5*, F1000 Faculty Rev-752. [CrossRef]

29. Rosenbloom, J.; Macarak, E.; Piera-Velazquez, S.; Jimenez, S.A. Human Fibrotic Diseases: Current Challenges in Fibrosis Research. *Methods Mol. Biol.* **2017**, *1627*, 1–23. [CrossRef] [PubMed]

30. Chellini, F.; Tani, A.; Zecchi-Orlandini, S.; Sassoli, C. Influence of Platelet-Rich and Platelet-Poor Plasma on Endogenous Mechanisms of Skeletal Muscle Repair/Regeneration. *Int. J. Mol. Sci.* **2019**, *20*, 683. [CrossRef]

31. Vu, T.D.; Pal, S.N.; Ti, L.K.; Martinez, E.C.; Rufaihah, A.J.; Ling, L.H.; Lee, C.N.; Richards, A.M.; Kofidis, T. An autologous platelet-rich plasma hydrogel compound restores left ventricular structure, function and ameliorates adverse remodeling in a minimally invasive large animal myocardial restoration model: A translational approach: Vu and Pal "Myocardial Repair: PRP, Hydrogel and Supplements". *Biomaterials* **2015**, *45*, 27–35. [CrossRef] [PubMed]

32. Jang, H.Y.; Myoung, S.M.; Choe, J.M.; Kim, T.; Cheon, Y.P.; Kim, Y.M.; Park, H. Effects of autologous platelet-rich plasma on regeneration of damaged endometrium in female rats. *Yonsei Med. J.* **2017**, *58*, 1195–1203. [CrossRef] [PubMed]

33. Moghadam, A.; Khozani, T.T.; Mafi, A.; Namavar, M.R.; Dehghani, F. Effects of platelet-rich plasma on kidney regeneration in gentamicin-induced nephrotoxicity. *J. Korean Med. Sci.* **2017**, *32*, 13–21. [CrossRef] [PubMed]

34. Andia, I.; Martin, J.I.; Maffulli, N. Advances with platelet rich plasma therapies for tendon regeneration. *Expert. Opin. Biol. Ther.* **2018**, *18*, 389–398. [CrossRef] [PubMed]

35. Sanchez-Avila, R.M.; Merayo-Lloves, J.; Riestra, A.C.; Berisa, S.; Lisa, C.; Sánchez, J.A.; Muruzabal, F.; Orive, G.; Anitua, E. Plasma rich in growth factors membrane as coadjuvant treatment in the surgery of ocular surface disorders. *Medicine (Baltimore)* **2018**, *97*, e0242. [CrossRef] [PubMed]

36. Sayadi, L.R.; Obagi, Z.; Banyard, D.A.; Ziegler, M.E.; Prussak, J.; Tomlinson, L.; Evans, G.R.D.; Widgerow, A.D. Platelet-rich plasma, adipose tissue, and scar modulation. *Aesthet. Surg. J.* **2018**, *38*, 1351–1362. [CrossRef]

37. Shoeib, H.M.; Keshk, W.A.; Foda, A.M.; Abo El Noeman, S.E. A study on the regenerative effect of platelet-rich plasma on experimentally induced hepatic damage in albino rats. *Can. J. Physiol. Pharmacol.* **2018**, *96*, 630–636. [CrossRef]

38. Tavukcu, H.H.; Aytaç, Ö.; Atuğ, F.; Alev, B.; Çevik, Ö.; Bülbül, N.; Yarat, A.; Çetinel, S.; Sener, G.; Kulaksızoğlu, H. Protective effect of platelet-rich plasma on urethral injury model of male rats. *Neurourol. Urodyn.* **2018**, *37*, 1286–1293. [CrossRef]

39. Terada, S.; Ota, S.; Kobayashi, M.; Kobayashi, T.; Mifune, Y.; Takayama, K.; Witt, M.; Vadalà, G.; Oyster, N.; Otsuka, T.; et al. Use of an anti-fibrotic agent improves the effect of platelet-rich plasma on muscle healing after injury. *J. Bone Jt. Surg. Am.* **2013**, *95*, 980–988. [CrossRef]

40. Cunha, R.C.; Francisco, J.C.; Cardoso, M.A.; Matos, L.F.; Lino, D.; Simeoni, R.B.; Pereira, G.; Irioda, A.C.; Simeoni, P.R.; Guarita-Souza, L.C.; et al. Effect of platelet-rich plasma therapy associated with exercise training in musculoskeletal healing in rats. *Transpl. Proc.* **2014**, *46*, 1879–1881. [CrossRef]

41. Anitua, E.; Pelacho, B.; Prado, R.; Aguirre, J.J.; Sánchez, M.; Padilla, S.; Aranguren, X.L.; Abizanda, G.; Collantes, M.; Hernandez, M.; et al. Infiltration of plasma rich in growth factors enhances in vivo angiogenesis and improves reperfusion and tissue remodeling after severe hind limb ischemia. *J. Control. Release* **2015**, *202*, 31–39. [CrossRef] [PubMed]

42. Cianforlini, M.; Mattioli-Belmonte, M.; Manzotti, S.; Chiurazzi, E.; Piani, M.; Orlando, F.; Provinciali, M.; Gigante, A. Effect of platelet rich plasma concentration on skeletal muscle regeneration: An experimental study. *J. Biol. Regul. Homeost. Agents* **2015**, *29*, 47–55. [PubMed]

43. Denapoli, P.M.; Stilhano, R.S.; Ingham, S.J.; Han, S.W.; Abdalla, R.J. Platelet-Rich Plasma in a Murine Model: Leukocytes, Growth Factors, Flt-1, and Muscle Healing. *Am. J. Sports Med.* **2016**, *44*, 1962–1971. [CrossRef] [PubMed]

44. Li, H.; Hicks, J.J.; Wang, L.; Oyster, N.; Philippon, M.J.; Hurwitz, S.; Hogan, M.V.; Huard, J. Customized platelet-rich plasma with transforming growth factor β1 neutralization antibody to reduce fibrosis in skeletal muscle. *Biomaterials* **2016**, *87*, 147–156. [CrossRef] [PubMed]

45. Zanon, G.; Combi, F.; Combi, A.; Perticarini, L.; Sammarchi, L.; Benazzo, F. Platelet-rich plasma in the treatment of acute hamstring injuries in professional football players. *Joints* **2016**, *4*, 17–23. [CrossRef]

46. Contreras-Muñoz, P.; Torrella, J.R.; Serres, X.; Rizo-Roca, D.; De la Varga, M.; Viscor, G.; Martínez-Ibáñez, V.; Peiró, J.L.; Järvinen, T.A.H.; Rodas, G.; et al. Postinjury Exercise and Platelet-Rich Plasma Therapies Improve Skeletal Muscle Healing in Rats But Are Not Synergistic When Combined. *Am. J. Sports Med.* **2017**, *45*, 2131–2141. [CrossRef]

47. Anitua, E.; Troya, M.; Orive, G. Plasma rich in growth factors promote gingival tissue regeneration by stimulating fibroblast proliferation and migration and by blocking transforming growth factor-β1-induced myodifferentiation. *J. Periodontol.* **2012**, *83*, 1028–1037. [CrossRef]

48. Chellini, F.; Tani, A.; Vallone, L.; Nosi, D.; Pavan, P.; Bambi, F.; Zecchi Orlandini, S.; Sassoli, C. Platelet-Rich Plasma Prevents In Vitro Transforming Growth Factor-β1- Induced Fibroblast to Myofibroblast Transition: Involvement of Vascular Endothelial Growth Factor (VEGF)-A/VEGF Receptor- 1-Mediated Signaling. *Cells* **2018**, *7*, 142. [CrossRef]

49. Willebrords, J.; Crespo Yanguas, S.; Maes, M.; Decrock, E.; Wang, N.; Leybaert, L.; da Silva, T.C.; Veloso Alves Pereira, I.; Jaeschke, H.; Cogliati, B.; et al. Structure, Regulation and Function of Gap Junctions in Liver. *Cell. Commun. Adhes.* **2015**, *22*, 29–37. [CrossRef]

50. Valiunas, V.; Gemel, J.; Brink, P.R.; Beyer, E.C. Gap junction channels formed by coexpressed connexin40 and connexin43. *Am. J. Physiol. Heart. Circ. Physiol.* **2001**, *281*, H1675–H1689. [CrossRef]

51. Gaudesius, G.; Miragoli, M.; Thomas, S.P.; Rohr, S. Coupling of cardiac electrical activity over extended distances by fibroblasts of cardiac origin. *Circ. Res.* **2003**, *93*, 421–428. [CrossRef] [PubMed]

52. Chilton, L.; Giles, W.R.; Smith, G.L. Evidence of intercellular coupling between co-cultured adult rabbit ventricular myocytes and myofibroblasts. *J. Physiol.* **2007**, *583*, 225–236. [CrossRef] [PubMed]

53. Zlochiver, S.; Muñoz, V.; Vikstrom, K.L.; Taffet, S.M.; Berenfeld, O.; Jalife, J. Electrotonic myofibroblast-to-myocyte coupling increases propensity to reentrant arrhythmias in two-dimensional cardiac monolayers. *Biophys. J.* **2008**, *95*, 4469–4480. [CrossRef]

54. Kakkar, R.; Lee, R.T. Intramyocardial fibroblast myocyte communication. *Circ. Res.* **2010**, *106*, 47–57. [CrossRef] [PubMed]

55. Sassoli, C.; Chellini, F.; Pini, A.; Tani, A.; Nistri, S.; Nosi, D.; Zecchi-Orlandini, S.; Bani, D.; Formigli, L. Relaxin prevents cardiac fibroblast-myofibroblast transition via notch-1-mediated inhibition of TGF-β/Smad3 signaling. *PLoS ONE* **2013**, *8*, e63896. [CrossRef] [PubMed]

56. Sassoli, C.; Nosi, D.; Tani, A.; Chellini, F.; Mazzanti, B.; Quercioli, F.; Zecchi-Orlandini, S.; Formigli, L. Defining the role of mesenchymal stromal cells on the regulation of matrix metalloproteinases in skeletal muscle cells. *Exp. Cell Res.* **2014**, *323*, 297–313. [CrossRef]

57. Chellini, F.; Tani, A.; Vallone, L.; Nosi, D.; Pavan, P.; Bambi, F.; Zecchi-Orlandini, S.; Sassoli, C. Platelet-Rich Plasma and Bone Marrow-Derived Mesenchymal Stromal Cells Prevent TGF-β1-Induced Myofibroblast Generation but Are Not Synergistic when Combined: Morphological in vitro Analysis. *Cells Tissues Organs* **2018**, *206*, 283–295. [CrossRef]

58. Sassoli, C.; Vallone, L.; Tani, A.; Chellini, F.; Nosi, D.; Zecchi-Orlandini, S. Combined use of bone marrow-derived mesenchymal stromal cells (BM-MSCs) and platelet rich plasma (PRP) stimulates proliferation and differentiation of myoblasts in vitro: New therapeutic perspectives for skeletal muscle repair/regeneration. *Cell Tissue Res.* **2018**, *372*, 549–570. [CrossRef]

59. Squecco, R.; Sassoli, C.; Nuti, F.; Martinesi, M.; Chellini, F.; Nosi, D.; Zecchi-Orlandini, S.; Francini, F.; Formigli, L.; Meacci, E. Sphingosine 1-phosphate induces myoblast differentiation through Cx43 protein expression: A role for a gap junction-dependent and -independent function. *Mol. Biol. Cell.* **2006**, *17*, 4896–4910. [CrossRef]

60. Formigli, L.; Sassoli, C.; Squecco, R.; Bini, F.; Martinesi, M.; Chellini, F.; Luciani, G.; Sbrana, F.; Zecchi-Orlandini, S.; Francini, F.; et al. Regulation of transient receptor potential canonical channel 1 (TRPC1) by sphingosine 1-phosphate in C2C12 myoblasts and its relevance for a role of mechanotransduction in skeletal muscle differentiation. *J. Cell. Sci.* **2009**, *122*, 1322–1333. [CrossRef]

61. Pierucci, F.; Frati, A.; Squecco, R.; Lenci, E.; Vicenti, C.; Slavik, J.; Francini, F.; Machala, M. Meacci, E. Non-dioxin-like organic toxicant PCB153 modulates sphingolipid metabolism in liver progenitor cells: Its role in Cx43-formed gap junction impairment. *Arch. Toxicol.* **2017**, *91*, 749–760. [CrossRef] [PubMed]

62. Sassoli, C.; Formigli, L.; Bini, F.; Tani, A.; Squecco, R.; Battistini, C.; Zecchi-Orlandini, S.; Francini, F.; Meacci, E. Effects of S1P on skeletal muscle repair/regeneration during eccentric contraction. *J. Cell. Mol. Med.* **2011**, *15*, 2498–2511. [CrossRef] [PubMed]

63. Pappone, P.A.; Lee, S.C. Purinergic receptor stimulation increases membrane trafficking in brown adipocytes. *J. Gen. Physiol.* **1996**, *108*, 393–404. [CrossRef] [PubMed]

64. Formigli, L.; Meacci, E.; Sassoli, C.; Chellini, F.; Giannini, R.; Quercioli, F.; Tiribilli, B.; Squecco, R.; Bruni, P.; Francini, F.; et al. Sphingosine 1-phosphate induces cytoskeletal reorganization in C2C12 myoblasts: Physiological relevance for stress fibres in the modulation of ion current through stretch-activated channels. *J. Cell. Sci.* **2005**, *118*, 1161–1171. [CrossRef] [PubMed]

65. Meacci, E.; Bini, F.; Sassoli, C.; Martinesi, M.; Squecco, R.; Chellini, F.; Zecchi-Orlandini, S.; Francini, F.; Formigli, L. Functional interaction between TRPC1 channel and connexin-43 protein: A novel pathway underlying S1P action on skeletal myogenesis. *Cell. Mol. Life Sci.* **2010**, *67*, 4269–4285. [CrossRef] [PubMed]

66. Andia, I.; Abate, M. Platelet-rich plasma in the treatment of skeletal muscle injuries. *Expert. Opin. Biol.* **2015**, *15*, 987–999. [CrossRef]

67. Cole, B.J.; Seroyer, S.T.; Filardo, G.; Bajaj, S.; Fortier, L.A. Platelet-rich plasma: Where are we now and where are we going? *Sports Health* **2010**, *2*, 203–210. [CrossRef]

68. Delos, D.; Leineweber, M.J.; Chaudhury, S.; Alzoobaee, S.; Gao, Y.; Rodeo, S.A. The effect of platelet-rich plasma on muscle contusion healing in a rat model. *Am. J. Sports Med.* **2014**, *42*, 2067–2074. [CrossRef]

69. Dimauro, I.; Grasso, L.; Fittipaldi, S.; Fantini, C.; Mercatelli, N.; Racca, S.; Geuna, S.; Di Gianfrancesco, A.; Caporossi, D.; Pigozzi, F.; et al. Platelet-rich plasma and skeletal muscle healing: A molecular analysis of the early phases of the regeneration process in an experimental animal model. *PLoS ONE* **2014**, *9*, e102993. [CrossRef]

70. Reurink, G.; Goudswaard, G.J.; Moen, M.H.; Weir, A.; Verhaar, J.A.; Bierma-Zeinstra, S.M.; Maas, M.; Tol, J.L. Dutch HIT-study Investigators. Rationale, secondary outcome scores and 1-year follow-up of a randomised trial of platelet-rich plasma injections in acute hamstring muscle injury: The Dutch Hamstring Injection Therapy study. *Br. J. Sports Med.* **2015**, *49*, 1206–1212. [CrossRef]

71. Kelc, R.; Vogrin, M. Concerns about fibrosis development after scaffolded PRP therapy of muscle injuries: Commentary on an article by Sanchez et al.: Muscle repair: Platelet-rich plasma derivates as a bridge from spontaneity to intervention. *Injury* **2015**, *46*, 428. [CrossRef]

72. Guillodo, Y.; Madouas, G.; Simon, T.; Le Dauphin, H.; Saraux, A. Platelet-rich plasma (PRP) treatment of sports-related severe acute hamstring injuries. *Muscles Ligaments Tendons J.* **2016**, *5*, 284–288. [CrossRef]

73. Grassi, A.; Napoli, F.; Romandini, I.; Samuelsson, K.; Zaffagnini, S.; Candrian, C.; Filardo, G. Is Platelet-Rich Plasma (PRP) Effective in the Treatment of Acute Muscle Injuries? A Systematic Review and Meta-Analysis. *Sports Med.* **2018**, *48*, 971–989. [CrossRef] [PubMed]

74. Rossi, L.A.; Molina Rómoli, A.R.; Bertona Altieri, B.A.; Burgos Flor, J.A.; Scordo, W.E.; Elizondo, C.M. Does platelet-rich plasma decrease time to return to sports in acute muscle tear? A randomized controlled trial. *Knee Surg. Sports Traumatol. Arthrosc.* **2017**, *25*, 3319–3325. [CrossRef] [PubMed]

75. Tonogai, I.; Hayashi, F.; Iwame, T.; Takasago, T.; Matsuura, T.; Sairyo, K. Platelet-rich plasma does not reduce skeletal muscle fibrosis after distraction osteogenesis. *J Exp Orthop.* **2018**, *5*, 26. [CrossRef] [PubMed]

76. Anitua, E.; de la Fuente, M.; Muruzabal, F.; Riestra, A.; Merayo-Lloves, J.; Orive, G. Plasma rich in growth factors (PRGF) eye drops stimulates scarless regeneration compared to autologous serum in the ocular surface stromal fibroblasts. *Exp. Eye Res.* **2015**, *135*, 118–126. [CrossRef] [PubMed]

77. Van der Bijl, I.; Vlig, M.; Middelkoop, E.; de Korte, D. Allogeneic platelet-rich plasma (**PRP**) is superior to platelets or plasma alone in stimulating fibroblast proliferation and migration, angiogenesis, and chemotaxis as relevant processes for wound healing. *Transfusion* **2019**, *59*, 3492–3500. [CrossRef]

78. Verhoekx, J.S.N.; Verjee, L.S.; Izadi, D.; Chan, J.K.K.; Nicolaidou, V.; Davidson, D.; Midwood, K.S.; Nanchahal, J. Isometric contraction of Dupuytren's myofibroblasts is inhibited by blocking intercellular junctions. *J. Investig. Dermatol.* **2013**, *133*, 2664–2671. [CrossRef]

79. Tarzemany, R.; Jiang, G.; Jiang, J.X.; Larjava, H.; Häkkinen, L. Connexin 43 Hemichannels Regulate the Expression of Wound Healing-Associated Genes in Human Gingival Fibroblasts. *Sci. Rep.* **2017**, *7*, 14157. [CrossRef]

80. Au, S.R.; Au, K.; Saggers, G.C.; Karne, N.; Ehrlich, H.P. Rat mast cells communicate with fibroblasts via gap junction intercellular communications. *J. Cell. Biochem.* **2007**, *100*, 1170–1177. [CrossRef]

81. Pistorio, A.L.; Ehrlich, H.P. Modulatory effects of connexin-43 expression on gap junction intercellular communications with mast cells and fibroblasts. *J. Cell. Biochem.* **2011**, *112*, 1441–1449. [CrossRef] [PubMed]

82. Foley, T.T.; Ehrlich, H.P. Through gap junction communications, co-cultured mast cells and fibroblasts generate fibroblast activities allied with hypertrophic scarring. *Plast. Reconstr. Surg.* **2013**, *131*, 1036–1044. [CrossRef] [PubMed]

83. Santos-Miranda, A.; Noureldin, M.; Bai, D. Effects of temperature on transjunctional voltage-dependent gating kinetics in Cx45 and Cx40 gap junction channels. *J. Mol. Cell. Cardiol.* **2019**, *127*, 185–193. [CrossRef]

84. Lastwika, K.J.; Dunn, C.A.; Solan, J.L.; Lampe, P.D. Phosphorylation of connexin 43 at MAPK, PKC or CK1 sites each distinctly alter the kinetics of epidermal wound repair. *J. Cell. Sci.* **2019**, *132*, jcs234633. [CrossRef]

85. Asazuma-Nakamura, Y.; Dai, P.; Harada, Y.; Jiang, Y.; Hamaoka, K.; Takamatsu, T. Cx43 contributes to TGF-beta signaling to regulate differentiation of cardiac fibroblasts into myofibroblasts. *Exp. Cell. Res.* **2009**, *315*, 1190–1199. [CrossRef] [PubMed]

86. Paw, M.; Borek, I.; Wnuk, D.; Ryszawy, D.; Piwowarczyk, K.; Kmiotek, K.; Wójcik-Pszczoła, K.A.; Pierzchalska, M.; Madeja, Z.; Sanak, M.; et al. Connexin43 Controls the Myofibroblastic Differentiation of Bronchial Fibroblasts from Patients with Asthma. *Am. J. Respir. Cell. Mol. Biol.* **2017**, *57*, 100–110. [CrossRef]

87. Coutinho, P.; Qiu, C.; Frank, S.; Tamber, K.; Becker, D. Dynamic changes in connexin expression correlate with key events in the wound healing process. *Cell. Biol. Int.* **2003**, *27*, 525–541. [CrossRef]

88. Kretz, M.; Euwens, C.; Hombach, S.; Eckardt, D.; Teubner, B.; Traub, O.; Willecke, K.; Ott, T. Altered connexin expression and wound healing in the epidermis of connexin-deficient mice. *J. Cell. Sci.* **2003**, *116*, 3443–3452. [CrossRef]

89. Qiu, C.; Coutinho, P.; Frank, S.; Franke, S.; Law, L.Y.; Martin, P.; Green, C.R.; Becker, D.L. Targeting connexin43 expression accelerates the rate of wound repair. *Curr. Biol.* **2003**, *13*, 1697–1703. [CrossRef]

90. Nakano, Y.; Oyamada, M.; Dai, P.; Nakagami, T.; Kinoshita, S.; Takamatsu, T. Connexin43 knockdown accelerates wound healing but inhibits mesenchymal transition after corneal endothelial injury in vivo. *Investig. Ophthalmol. Vis. Sci.* **2008**, *49*, 93–104. [CrossRef]

91. Montgomery, J.; Ghatnekar, G.S.; Grek, C.L.; Moyer, K.E.; Gourdie, R.G. Connexin 43-Based Therapeutics for Dermal Wound Healing. *Int. J. Mol. Sci.* **2018**, *19*, 1778. [CrossRef]

92. Cogliati, B.; Mennecier, G.; Willebrords, J.; Da Silva, T.C.; Maes, M.; Pereira, I.V.A.; Crespo-Yanguas, S.; Hernandez-Blazquez, F.J.; Dagli, M.L.Z.; Vinken, M. Connexins, Pannexins, and Their Channels in Fibroproliferative Diseases. *J. Membr. Biol.* **2016**, *249*, 199–213. [CrossRef] [PubMed]

93. Laird, D.W.; Lampe, P.D. Therapeutic strategies targeting connexins. *Nat. Rev. Drug. Discov.* **2018**, *17*, 905–921. [CrossRef] [PubMed]

94. Kameritsch, P.; Pogoda, K.; Pohl, U. Channel-independent influence of connexin 43 on cell migration. *Biochim. Biophys. Acta.* **2012**, *1818*, 1993–2001. [CrossRef] [PubMed]

95. Riteau, N.; Gasse, P.; Fauconnier, L.; Gombault, A.; Couegnat, M.; Fick, L.; Kanellopoulos, J.; Quesniaux, V.F.; Marchand-Adam, S.; Crestani, B.; et al. Extracellular ATP Is a Danger Signal Activating P2X7 Receptor in Lung Inflammation and Fibrosis. *Am. J. Respir. Crit. Care Med.* **2010**, *182*, 774–783. [CrossRef] [PubMed]

96. Lu, D.; Soleymani, S.; Madakshire, R.; Insel, P.A. ATP released from cardiac fibroblasts via connexin hemichannels activates profibrotic P2Y2 receptor. *FASEB* **2012**, *26*, 2580–2591. [CrossRef]

97. Ferrari, D.; Gambari, R.; Idzko, M.; Müller, T.; Albanesi, C.; Pastore, S.; La Manna, G.; Robson, S.C.; Cronstein, B. Purinergic signaling in scarring. *FASEB J.* **2016**, *30*, 3–12. [CrossRef]

98. Hills, C.; Price, G.W.; Wall, M.J.; Kaufmann, T.J.; Chi-Wai Tang, S.; Yiu, W.H.; Squires, P.E. Transforming Growth Factor Beta 1 Drives a Switch in Connexin Mediated Cell-to-Cell Communication in Tubular Cells of the Diabetic Kidney. *Cell. Physiol. Biochem.* **2018**, *45*, 2369–2388. [CrossRef]

99. Iyer, R.K.; Odedra, D.; Chiu, L.L.; Vunjak-Novakovic, G.; Radisic, M. Vascular endothelial growth factor secretion by nonmyocytes modulates Connexin-43 levels in cardiac organoids. *Tissue Eng. Part. A.* **2012**, *18*, 1771–1783. [CrossRef]

100. Wuestefeld, R.; Chen, J.; Meller, K.; Brand-Saberi, B.; Theiss, C. Impact of vegf on astrocytes: Analysis of gap junctional intercellular communication, proliferation, and motility. *Glia* **2012**, *60*, 936–947. [CrossRef]

101. Li, L.; Liu, H.; Xu, C.; Deng, M.; Song, M.; Yu, X.; Xu, S.; Zhao, X. VEGF promotes endothelial progenitor cell differentiation and vascular repair through connexin 43. *Stem. Cell. Res. Ther.* **2017**, *8*, 237. [CrossRef] [PubMed]

102. Boeldt, D.S.; Grummer, M.A.; Yi, F.; Magness, R.R.; Bird, I.M. Phosphorylation of Ser-279/282 and Tyr-265 positions on Cx43 as possible mediators of VEGF-165 inhibition of pregnancy-adapted Ca2+ burst function in ovine uterine artery endothelial cells. *Mol. Cell. Endocrinol.* **2015**, *412*, 73–84. [CrossRef] [PubMed]

103. Nimlamool, W.; Andrews, R.M.; Falk, M.M. Connexin43 phosphorylation by PKC and MAPK signals VEGF-mediated gap junction internalization. *Mol. Biol. Cell.* **2015**, *26*, 2755–2768. [CrossRef]

104. Solan, J.L.; Lampe, P.D. Spatio-temporal regulation of connexin43 phosphorylation and gap junction dynamics. *Biochim. Biophys. Acta Biomembr.* **2018**, *1860*, 83–90. [CrossRef] [PubMed]

105. Lin, D.; Boyle, D.L.; Takemoto, D.J. IGF-I-induced phosphorylation of connexin 43 by PKCgamma: Regulation of gap junctions in rabbit lens epithelial cells. *Investig. Ophthalmol. Vis. Sci.* **2003**, *44*, 1160–1168. [CrossRef] [PubMed]

106. Qiao, J.; An, N.; Ouyang, X. Quantification of growth factors in different platelet concentrates. *Platelets* **2017**, *28*, 774–778. [CrossRef]

107. Dong, Y.; Lakhia, R.; Thomas, S.S.; Dong, Y.; Wang, X.H.; Silva, K.A.; Zhang, L. Interactions between p-Akt and Smad3 in injured muscles initiate myogenesis or fibrogenesis. *Am. J. Physiol. Endocrinol. Metab.* **2013**, *305*, E367–E375. [CrossRef]

108. Andrade, D.; Oliveira, G.; Menezes, L.; Nascimento, A.L.; Carvalho, S.; Stumbo, A.C.; Thole, A.; Garcia-Souza, É.; Moura, A.; Carvalho, L.; et al. Insulin-like growth factor-1 short-period therapy improves cardiomyopathy stimulating cardiac progenitor cells survival in obese mice. *Nutr. Metab. Cardiovasc. Dis.* **2020**, *30*, 151–161. [CrossRef]

109. Sassoli, C.; Pierucci, F.; Zecchi-Orlandini, S.; Meacci, E. Sphingosine 1-Phosphate (S1P)/ S1P Receptor Signaling and Mechanotransduction: Implications for Intrinsic Tissue Repair/Regeneration. *Int. J. Mol. Sci.* **2019**, *20*, 5545. [CrossRef]

110. Lynch, M.D.; Watt, F.M. Fibroblast heterogeneity: Implications for human disease. *J. Clin. Investig.* **2018**, *128*, 26–35. [CrossRef]

111. Foote, A.G.; Wang, Z.; Kendziorski, C.; Thibeault, S.L. Tissue specific human fibroblast differential expression based on RNA sequencing analysis. *BMC Genomics* **2019**, *20*, 308. [CrossRef] [PubMed]

112. LeBleu, V.S.; Neilson, E.G. Origin and functional heterogeneity of fibroblasts. *FASEB J.* **2020**, *34*, 3519–3536. [CrossRef] [PubMed]

 *cells*

Article

Parkin Overexpression Attenuates Sepsis-Induced Muscle Wasting

Jean-Philippe Leduc-Gaudet [1,2,3,4,5], Dominique Mayaki [1], Olivier Reynaud [2,3,4,5],
Felipe E. Broering [1,2], Tomer J. Chaffer [1], Sabah N. A. Hussain [1,2,*,†]
and Gilles Gouspillou [2,3,4,5,6,*,†]

1   Meakins-Christie Laboratories and Translational Research in Respiratory Diseases Program,
    Research Institute of the McGill University Health Centre, Department of Critical Care,
    McGill University Health Centre, Montréal, QC H4A 3J1, Canada;
    jean-philippe.leduc-gaudet@mail.mcgill.ca (J.-P.L.-G.); dominique.mayaki@muhc.mcgill.ca (D.M.);
    felipe.broering@mail.mcgill.ca (F.E.B.); jordichaffer@gmail.com (T.J.C.)
2   Division of Experimental Medicine, McGill University, Montréal, QC H4A 3J1, Canada;
    oreynaud26@gmail.com
3   Département des sciences de l'activité physique, Faculté des Sciences, UQAM, Montréal,
    QC H2X 1Y4, Canada
4   Groupe de recherche en Activité Physique Adaptée, Montréal, QC H2X 1Y4, Canada
5   Département des sciences biologiques, Faculté des Sciences, UQAM, Montréal, QC H2X 1Y4, Canada
6   Centre de Recherche de l'Institut Universitaire de Gériatrie de Montréal, Montréal, QC H3W 1W5, Canada
*   Correspondence: sabah.hussain@muhc.mcgill.ca (S.N.A.H.); gouspillou.gilles@uqam.ca (G.G.);
    Tel.: +1-514-934-1934 (ext. 76222) (S.N.A.H.); +1-514-987-3000 (ext. 5322) (G.G.)
†   These authors contributed equally to this work as senior authors.

Received: 30 April 2020; Accepted: 8 June 2020; Published: 11 June 2020

Abstract: Sepsis elicits skeletal muscle weakness and fiber atrophy. The accumulation of injured mitochondria and depressed mitochondrial functions are considered as important triggers of sepsis-induced muscle atrophy. It is unclear whether mitochondrial dysfunctions in septic muscles are due to the inadequate activation of quality control processes. We hypothesized that overexpressing Parkin, a protein responsible for the recycling of dysfunctional mitochondria by the autophagy pathway (mitophagy), would confer protection against sepsis-induced muscle atrophy by improving mitochondrial quality and content. Parkin was overexpressed for four weeks in the limb muscles of four-week old mice using intramuscular injections of adeno-associated viruses (AAVs). The cecal ligation and perforation (CLP) procedure was used to induce sepsis. Sham operated animals were used as controls. All animals were studied for 48 h post CLP. Sepsis resulted in major body weight loss and myofiber atrophy. Parkin overexpression prevented myofiber atrophy in CLP mice. Quantitative two-dimensional transmission electron microscopy revealed that sepsis is associated with the accumulation of enlarged and complex mitochondria, an effect which was attenuated by Parkin overexpression. Parkin overexpression also prevented a sepsis-induced decrease in the content of mitochondrial subunits of NADH dehydrogenase and cytochrome C oxidase. We conclude that Parkin overexpression prevents sepsis-induced skeletal muscle atrophy, likely by improving mitochondrial quality and contents.

Keywords: muscle atrophy; septicemia; mitochondria; mitochondrial fusion; mitochondrial fission

---

1. Introduction

Sepsis is a complex syndrome characterized by an overwhelming infection that results in a severe systemic inflammatory response. Sepsis causes diverse vascular, metabolic and endocrine abnormalities that lead to multiple organ failure, and often result in death [1]. Amongst the very deleterious effects of sepsis is severe weakness, which involves both respiratory and limb skeletal muscles [2–5]. In the short term, sepsis-induced respiratory muscle weakness leads to difficulty removing patients from mechanical ventilation, increases the risk of the recurrence of respiratory failure, prolonged hospitalization and increased mortality [6]. In sepsis survivors discharged from the intensive care unit, the long-term ramifications of sepsis-induced limb muscle weakness included functional impairment, limited physical activity and poor quality of life [7].

There is currently a lack of effective therapies to either prevent or treat sepsis-induced skeletal muscle weakness, due in large part to the fact that its molecular and cellular bases are poorly understood. However, one clue lies at the ultrastructural level, where significant accumulations of damaged and dysfunctional mitochondria are characteristic of sepsis-induced muscle dysfunction [8,9]. Indeed, Bready et al. showed that, in human skeletal muscle, sepsis results in decreased complex I activity (a key enzyme of the mitochondrial electron transfer system) and declined the ATP/ADP ratio in skeletal muscles [10]. These defects in muscle bioenergetics were also observed in a rat model of sepsis [11]. By studying biopsies obtained from septic patients, Fredriksson K et al. described a 30% decrease in complex IV activity in limb skeletal muscles [12]. Several studies on experimental animals also reported that sepsis results in decreased mitochondrial respiration [13–16] and an increase in the mitochondrial production of reactive oxygen species (ROS) in skeletal muscle [17,18]. Sepsis has also been shown to increase the levels of morphologically abnormal mitochondria, such as those with disorganized cristae, translucent vacuoles and even myelin-like structures [13,19–21]. Recently, Owen et al. showed that persistent muscle weakness in mice that have survived sepsis is associated with abnormal mitochondrial ultrastructure, decreased respiration, decreased activity of complexes of the mitochondrial electron transfer system and persistent oxidative damage to muscle proteins [21].

In healthy muscles, damaged or dysfunctional mitochondria are selectively recycled in a process, known as mitophagy (selective autophagy of mitochondria), which is primarily regulated through the PINK1-Parkin pathway. Parkin, an E3 ubiquitin ligase encoded by the *Park2* gene, is a 465 amino acid protein that translocates to depolarized mitochondria to initiate mitophagy. Parkin-dependent mitophagy is regulated by PTEN-induced kinase 1 (PINK1), which acts upstream from Parkin. In healthy mitochondria, PINK1 is imported into the inner mitochondrial membrane and cleaved by PARL [22]. Cleaved PINK1 is then released into the cytosol where it is degraded by the proteasome system. In depolarized mitochondria, the importation of PINK1 into the inner mitochondrial membrane is blocked. PINK1 is no longer degraded and becomes phosphorylated and stabilized on the outer mitochondrial membrane [23–26]. Phosphorylated PINK1 triggers the recruitment of Parkin to the mitochondria. Parkin then ubiquitinates outer mitochondrial membrane proteins, including the fusion proteins MFN1, MFN2, MIRO and TOMM20 [27]. The degradation of MFN1 and MFN2 triggers mitochondrial fission and fragmentation, both of which are important to the recycling of mitochondria by the mitophagy pathway [28]. The functional importance of the PINK1-Parkin mitophagy pathway in regulating skeletal muscle mitochondrial function and quality in sepsis remains unknown. Recently, we reported that the genetic deletion of Parkin leads to the poor recovery of cardiac function in septic mice and increased sepsis-induced mitochondrial dysfunction in the heart [29]. We also demonstrated that autophagy is significantly induced in the skeletal muscles of septic mice and that the induction of autophagy is associated with increased muscle Parkin levels, suggesting that mitophagy was induced [20,30]. However, several morphologically and functionally abnormal mitochondria were observed in the electron micrographs of septic muscles, indicating that the mitophagy that was induced was likely insufficient to the task of completely recycling defective mitochondria [20,30]. Based on this reasoning, we hypothesized that enhancing mitophagy through Parkin overexpression would attenuate the impact of sepsis on skeletal muscles and their mitochondria. To test this hypothesis, Parkin was overexpressed for four weeks in the skeletal muscles of young mice using intramuscular injections of adeno-associated viruses (AAVs). The cecal ligation and perforation (CLP) procedure, a widely used model of sepsis [31], was used to induce sepsis. Sham-operated animals served as controls. We found that Parkin overexpression prevents sepsis-induced mitochondrial morphological injury and reverses the decline in mitochondrial protein content. We also found that Parkin overexpression protects against sepsis-induced myofiber atrophy. These findings indicate that defective mitophagy in sepsis can be therapeutically manipulated as a means of counteracting sepsis-induced muscle dysfunction.

2. Materials and Methods

2.1. Animal Procedures

All experiments were approved (#2014-7549) by the Research Ethics Board of the Research Institute of the McGill University Health Centre (MUHC-RI) and are in accordance with the principles outlined by the Canadian Council of Animal Care. Three-week-old male wild-type C57BL/6J mice (Charles River Laboratories, Saint-Constant, QC, Canada) were used for our experiments. All mice were group-housed under a standard 12:12 h light/dark cycle with food and water available ad libitum.

2.2. AAV Injections in Skeletal Muscle

All of the adeno-associated viruses (AAVs) used in our experiments were purchased from Vector Biolabs (Malvern, PA, USA) and were of Serotype 1, a serotype highly effective in transducing skeletal muscle cells [32]. Four-week-old mice were first anesthetized with an isoflurane (2.5 to 3.5%), and AAV1s containing a muscle specific promoter (muscle creatine kinase), a sequence coding for the reporter protein GFP and a sequence coding for Parkin (details on the AAV1 construction are available in Supplementary Figure S1) were then intramuscularly injected (25 μL per site; 1.5×10^{11} gc) into the gastrocnemius (GAS) muscles in the right leg. In this AAV1 construction, the sequences coding for Parkin and GFP were separated by a sequence coding for the auto-cleavable 2A peptide, allowing for the separation of the Parkin and GFP proteins once translated. Control AAV1s containing only the GFP sequence under the control of the MCK promoter were injected into the contralateral leg. Because the AAV1 recombination site in the wild-type AAV1s was deleted in these recombined AAV1s, both GFP and Parkin expression comprised episomal expression without integration into the host DNA.

2.3. Cecal Ligation and Perforation

After four weeks of AAV1 injection, the mice were subjected to cecal ligation and perforation or sham surgery. The cecal ligation and puncture (CLP) model, which closely mimics the clinical features of human sepsis [31], was performed to induce polymicrobial sepsis as described previously [30,33] with minor modifications. Briefly, the mice were first anesthetized with isoflurane (~3%; Piramal Critical Care). A midline abdominal incision (~2 cm) was then performed. The cecum was carefully ligated at ~1 cm from its distal portion. The ligated cecum was perforated by a through-and-through puncture performed with $25^{1/2}$ gauge needle in a sterile environment. Next, the ligated cecum was gently compressed to extrude a small amount of the cecal contents through the punctured holes. The cecum was then replaced in the abdominal cavity. The peritoneum was then closed in two separate layers using 3–0 absorbable polyfilament interrupted sutures. The skin was finally closed with a surgical staple (9 mm AutoClip® System, Fine Scientific tools, North Vancouver, BC, Canada). All of the animals received subcutaneous injections of buprenorphine (0.05 to 0.2 mg/kg in 1 mL of 0.9% saline) immediately after surgery. To minimize pain, buprenorphine was administered every 12 h (0.05 mg/kg in ~100 μL of 0.9% saline). The sham-operated mice were subjected to identical procedures with the exception of the cecum ligation and puncture. All of the animals were closely monitored for signs of excessive pain or distress, such as lack of movement, agonal breathing or excessive body loss (20%), by investigators and by the vivarium staff from the IR-MUHC. Any mouse reaching endpoint criteria was immediately euthanized.

2.4. Tissue Collection

Mice were anesthetized with isoflurane and subsequently euthanized by cervical dislocation 48 h after sham or CLP procedures. The gastrocnemius (GAS) muscles were carefully removed from both legs and cut in half; one half was mounted for histology and small strips were prepared for transmission electron microscope (TEM) analyses, as previously described [34]. The rest of the GAS was quickly frozen in liquid nitrogen and stored −80 °C until use for immunoblotting and qPCR experiments.

2.5. Fiber Size Determination

Muscles samples were mounted on plastic blocks in tragacanth gum and frozen in liquid isopentane cooled in liquid nitrogen. The samples were then stored until use at −80 °C. The samples were cut into 10 μm cross-sections using a cryostat (Leica Biosystem Inc., Concord, ON, Canada) at −20 °C and then mounted on lysine coated slides (Superfrost) to assess muscle fiber size, as described in [32,34]. To this end, the muscle cross-sections were first allowed to reach room temperature and were rehydrated with phosphate buffered saline (PBS, pH 7.2) and then blocked with goat serum (10% in PBS). The sections were then incubated with primary rabbit IgG polyclonal anti-laminin antibody (MilliporeSigma, Oakville, ON, Canada, L9393, 1:750) for 1 h at room temperature. The sections were then washed three times in PBS before being incubated for 1 h at room temperature with an Alexa Fluor 594 goat anti-rabbit IgG antibody (Invitrogen, Burlington, ON, Canada A-11037, 1:500). The sections were then washed three times in PBS and the slides were cover-slipped using Prolong Gold (Invitrogen, P36930) as mounting medium. The slides were imaged with a Zeiss Axio Imager 2 fluorescence microscope (Zeiss, Dorval, QC, Canada). The median minimum Feret's diameter of the muscle fibers, a reliable marker of myofiber size [35], was determined for each muscle sample using at least 200 fibers per muscle sample (average number ± SD of fiber analyzed for each group: sham AAV-GFP, 317 ± 62; sham AAV-Parkin, 300 ± 15; CLP AAV-GFP, 345 ± 30; CLP AAV-Parkin, 304 ± 46). Analyses were performed using ImageJ (NIH, Bethesda, MD, USA, https://imagej.nih.gov/ij/).

2.6. Transmission Electron Microscopy (TEM)

The samples for TEM were prepared as described in [34,36,37]. Briefly, small strips prepared from GAS were incubated in 2% glutaraldehyde buffer solution in 0.1 M cacodylate (pH 7.4) and were subsequently post-fixed in 1% osmium tetroxide in 0.1 M cacodylate buffer. Tissues were then dehydrated via increasing the concentrations of methanol to propylene oxide and infiltrated and embedded in EPON^{TM} resins at the Facility for Electron Microscopy Research (FEMR) of McGill University. Ultrathin sections (60 nm) were cut longitudinally using an ultramicrotome (Ultracut III, Reichert-Jung, Leica Biosystem Inc., Concord, ON, Canada) and mounted on nickel carbon-formvar-coated grids for electron microscopy. Uranyl acetate and lead citrate stained sections were then imaged using a FEI Tecnai 12 transmission electron microscope at 120 kV, and images were digitally captured using a XR80C CCD camera system (AMT, Woburn, MA, USA) at a magnification of 1400×. Individual intermyofibrillar (IMF) mitochondria from all groups were manually traced in longitudinal orientations using ImageJ 2.0.0 software (NIH, Bethesda, MD, USA, https://imagej.nih.gov/ij/) to measure the following morphological characteristics: area (in μm^2), perimeter (μm), circularity (4π·(surface area/perimeter^2)), Feret's diameter (longest distance (μm) between any two points within a given mitochondrion), aspect ratio (major axis/minor axis)—a measure of the "length to width ratio" and form factor (perimeter/4π·surface area)—a measure sensitive to the complexity and branching aspect of mitochondria [34,36,37]. An index of mitochondrial morphological complexity was finally calculated as follows: Mitochondrial complexity index = Aspect ratio × Form Factor. Details on the number of IMF mitochondria that were traced are available in the figure legends.

2.7. Immunoblotting

Frozen skeletal muscle tissues (15–30 mg) were homogenized in an ice-cold lysis buffer (50 mM Hepes, 150 mM NaCl, 100mM NaF, 5 mM EDTA, 0.5% Triton X-100, 0.1 mM DTT, 2 µg/mL leupeptin, 100 µg/mL PMSF, 2 µg/mL aprotinin, and 1 mg/100 mL pepstatin A, pH 7.2) using Mini-beadbeater (BioSpec Products) with a ceramic bead at 60 Hz. The muscle homogenates were kept on ice for 30 min with periodic agitation and were then centrifuged at 5000 g for 15 min at 4 °C, the supernatants were collected, and the pellets were discarded. The protein contents in each sample were determined using the Bradford method. The aliquots of crude muscle homogenates were mixed with Laemmli buffer (6×, reducing buffer, # BP111R, Boston BioProducts, Ashland, MA, USA) and subsequently denatured for 5 min at 95 °C. Equal amounts of protein extracts (30 µg per lanes) were separated by SDS-PAGE, and then transferred onto polyvinylidene difluoride (PVDF) membranes (Bio-Rad Laboratories, Saint-Laurent, QC, Canada) using a wet transfer technique. The total proteins on the membranes were detected with Ponceau-S solution (MilliporeSigma #P3504). The membranes were blocked in PBS + 1% Tween® 20 + 5% bovine serum albumin (BSA) for 1 h at room temperature and then incubated with the specific primary antibodies overnight at 4 °C. The complete list of antibodies used for immunoblots analysis can be found in Supplementary Table S1. Membranes were washed in PBST (3 × 5 min) and incubated with HRP-conjugated secondary anti-rabbit or anti-mouse secondary antibodies (Abcam, Toronto, ON, Canada, cat# Ab6728, Ab6721) for 1 h at room temperature, before further washing in PBST (3 × 5 min). Immunoreactivity was detected using an enhanced chemiluminescence substrate (Pierce™, Thermo Fisher Scientific, Saint-Laurent, QC, Canada) with the ChemiDoc™ XRS+ Imaging System. The optical densities (OD) of the protein bands were quantified using ImageLab software (Bio-Rad Laboratories) and normalized to loading control (Ponceau-stained PVDF membranes). Immunoblotting data are expressed as relative to Sham AAV-GFP.

2.8. Quantitative Real-Time PCR

Total RNA was extracted from frozen muscle samples using a PureLink™ RNA Mini Kit (Invitrogen Canada, Burlington, ON, Canada). The quantification and purity of RNA was assessed using the A260/A280 absorption method. Total RNA (2 µg) was reverse transcribed using a Superscript II® Reverse Transcriptase Kit and random primers (Invitrogen, Burlington, ON, Canada). The reactions were incubated at 42 °C for 50 min and at 90 °C for 5 min. The real-time PCR detection of mRNA expression was performed using a Prism® (Graphpad, San Diego, CA, USA) 7000 Sequence Detection System (Applied Biosystems, Foster, CA, USA). The cycle threshold (C_T) values were obtained for each target gene. The ΔC_T values (normalized gene expression) were calculated as C_T of target gene minus C_T of the geometric means of three housekeeping genes (*Cyclophilin B*, *β-Actin* and *18S*). The relative mRNA level quantifications of target genes were determined using the threshold cycle ($\Delta\Delta C_T$) method, as compared to sham AAV-GFP. The primer sequences for all genes are found in Supplementary Table S2.

2.9. Data Analysis and Statistics

All statistical analyses were performed using GraphPad Prism 8 (GraphPad, San Diego, CA, USA). Comparisons of initial body weight and body weight loss between sham-operated and CLP mice were performed using unpaired bilateral student *t*-tests (*p*-values < 0.05 were considered statistically significant). Comparisons of the effects of Parkin overexpression on parameters of interest were performed using two-way repeated measures analysis of variance (ANOVA) (except for comparisons of mitochondrial shape descriptors, as detailed below). Corrections for the multiple comparisons following two-way repeated measures ANOVA were performed with the two-stage step-up method of Benjamini and Krieger and Yekutieli (*q* < 0.1 was considered statistically significant). One-way ANOVA followed by the two-stage step-up method of Benjamini and Krieger and Yekutieli were used for the following comparisons: sham AAV-GFP vs. CLP AAV-GFP; sham AAV-GFP vs. CLP

AAV-Parkin; sham AAV-Parkin vs. CLP AAV-GFP; sham AAV-Parkin vs. CLP AAV-Parkin (except for comparisons of mitochondrial shape descriptors, as detailed below) ($q < 0.1$ was considered statistically significant). Differences for the median values of shape descriptors to assess mitochondrial morphology were assessed using a Kruskal–Wallis test followed by a Dunn's multiple comparisons test (adjusted p-values < 0.05 were considered statistically significant). The exact numbers of animals within each group in all figures are indicated in the figure legends.

3. Results

3.1. Successful Overexpression of Parkin in Skeletal Muscles of Sham and CLP Operated Mice

Four weeks after the intramuscular injections of AAVs, mice were subjected to cecal ligation and perforation (CLP) to induced polymicrobial sepsis. Sham-operated mice were used as control. At baseline (prior to sham and CLP procedures), body weight values were similar in the sham and CLP groups, as shown in Figure 1A. Body weight loss was more pronounced in the CLP group relative to the sham group ($-13.8 \pm 1.4\%$ vs. $-4.7\pm1.1\%$, respectively, $p < 0.05$), as shown in Figure 1B. As shown in Figure 1C,D, the intramuscular injection of AAV-Parkin significantly increased *Park2* mRNA expression and Parkin protein content, in the skeletal muscles of both Sham-operated and CLP mice. These results demonstrate that our approach was successful in overexpressing Parkin in mouse skeletal muscle.

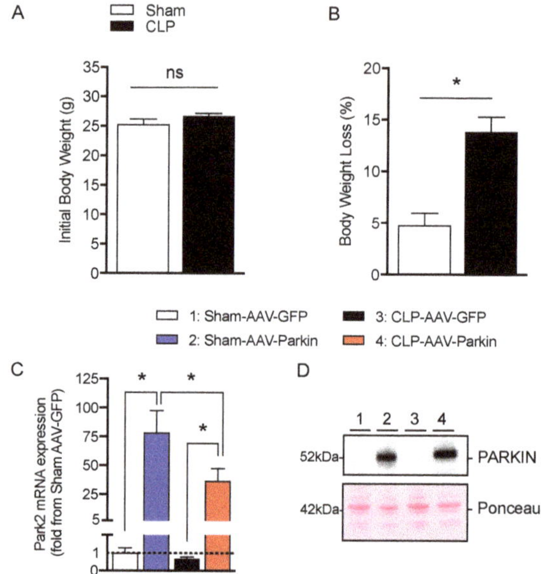

Figure 1. Effective Parkin overexpression in skeletal muscles of Sham and CLP operated mice. (**A**) Initial body weight and (**B**) percent of body weight loss in Sham-operated and or CLP mice. (**C**) qPCR analysis of *Park2* expression levels in the gastrocnemius muscles injected with either AAV-GFP or AAV-Parkin in Sham and CLP mice. (**D**) Representative Parkin immunoblots and its corresponding ponceau S stain performed on gastrocnemius samples of Sham and CLP mice injected with either AAV-GFP or AAV-Parkin. 1 = Sham-AAV-GFP; 2 = Sham-AAV-Parkin; 3 = CLP-AAV-GFP; 4 = CLP-AAV-Parkin. Data are presented as mean $\pm$ SEM ($n = 7$–9/group; * = statistically significant; ns = not statistically significant).

3.2. Parkin Overexpression Attenuates Sepsis-Induced Skeletal Muscle Atrophy

The effect of Parkin overexpression on muscle fiber size was evaluated 48 h after CLP, based on our previous reports which revealed that limb muscle atrophy develops at this time point [30,33]. In the sham group, Parkin expressing muscles had larger myofiber diameters relative to those expressing GFP, as shown in Figure 2B,C. This observation is in line with our previous report [32]. In the CLP group, GFP expressing muscles displayed a trend towards smaller myofiber diameters and a decreased proportion of large fibers relative to those expressing GFP in the sham group, as shown in Figure 2B,D, all of which are indicative of myofiber atrophy. As shown Figure 2B,E, no sign of atrophy was detected in the Parkin overexpressing muscles of CLP mice when compared to the Parkin expressing muscles of Sham-operated mice. In addition, the Parkin overexpressing muscles of CLP mice displayed larger myofibers vs. the GFP expressing muscles of CLP mice, as shown in Figure 2B,F. These results indicate that Parkin overexpression prevented the development of muscle atrophy in the CLP group and increased muscle fiber diameter in the sham group.

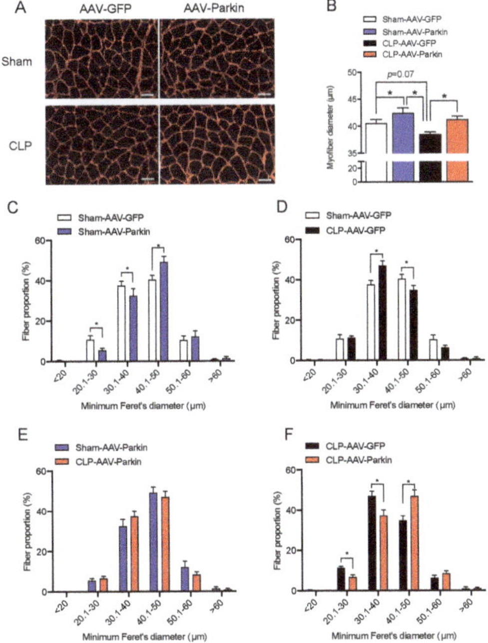

Figure 2. The impact of Parkin overexpression and sepsis on skeletal muscle fiber size. (**A**) Representative gastrocnemius (GAS) cryosections stained for laminin in all experimental groups. Scale bar: 50μm. (**B**) Quantification of minimum Ferret diameter of GAS myofibers of Sham and CLP animals injected with either AAV-GFP or AAV-Parkin. (**C**) Minimum Ferret distribution of the GAS myofibers of Sham AAV-GFP (n = 8 mice; 316 ± 21 fibers per GAS were traced) vs. Sham AAV-Parkin (n = 8 mice; 300 ± 5 fibers per GAS were traced). (**D**) Minimum Ferret distribution of the GAS myofibers of Sham AAV-GFP (n = 8 mice; 316 ± 21 fibers per GAS were traced) vs. CLP AAV-GFP (n = 6 mice; 345 ± fibers per GAS were traced). (**E**) Minimum Ferret distribution of the GAS myofibers of Sham AAV-Parkin (n = 8 mice; 300 ± 5 fibers per GAS were traced) vs. CLP AAV-Parkin (n = 6 mice; 304 ± 18 fibers per GAS were traced). (**F**) Minimum Ferret distribution of the GAS myofibers of CLP AAV-GFP (n = 6 mice; 345 ± fibers per GAS were traced) vs. CLP AAV-Parkin (n = 6 mice; 304 ± 18 fibers per GAS were traced). Data are presented as mean ± SEM. (n = 6–8/group; * = statistically significant).

3.3. The Impact of Parkin Overexpression and Sepsis on Skeletal Muscle Catabolic Signaling

We then investigated whether Parkin overexpression and sepsis affect the expression levels of apoptotic and autophagy-related genes. Neither Parkin overexpression nor sepsis affected the mRNA levels of pro-apoptotic *Bax*, *Bid*, *Bim*, and anti-apoptotic *Bcl2*, depicted in Figure 3A. As shown in Figure 3A, the expression level of BclXL was higher in septic animals. qPCR analyses revealed a significant increase in the mRNA levels of *Lc3b*, *Sqstm1*, *Gabarapl* and *Bnip3* in septic mice, shown in Figure 3B. The expression of *Gabarapl* was significantly higher in the Parkin overexpressing muscles of septic animals. No other impacts of Parkin overexpression on apoptotic and autophagy-related genes were observed. In line with our gene expression data, the protein contents of SQSTM1 (also known as p62) and BNIP3, two proteins regulating autophagy and mitophagy, were increased in septic animals, as shown in Figure 3C–E. The ratio of LC3-II to LC3-I was significantly increased in septic mice, suggesting an induction of autophagy, shown in Figure 3F. No impact of Parkin overexpression on the content of SQSTM1 and BNIP3 and the LC3-II to LC3-I ratio could be evidenced. We then assessed the expression levels of two key E3 ligases known to contribute to skeletal muscle atrophy [38,39], Fbxo32 (Atrogin-1) and Trim63 (MuRF1). The expression of these two E3 ligases was significantly increased in the skeletal muscle of septic animals, shown in Figure 3G. Parkin overexpression did not impact Fbxo32 and Trim63 expression. It is worth mentioning that neither Parkin overexpression nor sepsis had an impact on the content or phosphorylation levels of AKT and 4EBP1, two key proteins involved in the regulation of protein synthesis, as shown in Supplementary Figure S2. Taken altogether, these data indicate that Parkin overexpression did not attenuate sepsis-induced increases in catabolic signaling.

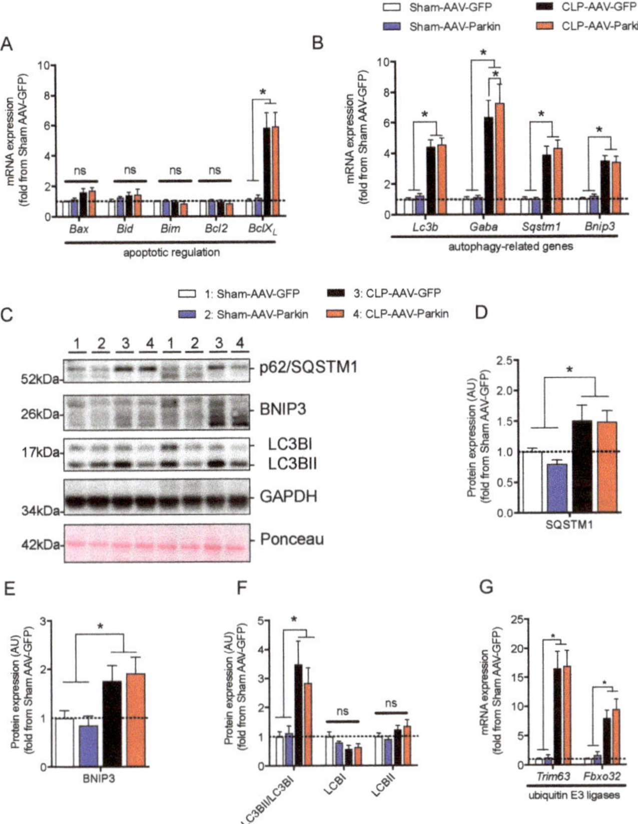

Figure 3. The impact of Parkin overexpression and sepsis on skeletal muscle catabolic signaling. (**A**) qPCR analysis of the mRNA expression of genes regulating apoptosis in the gastrocnemius (GAS) muscles of Sham and CLP animals injected with either AAV-GFP or AAV-Parkin. (**B**) qPCR analysis of autophagy-related gene expression in the gastrocnemius (GAS) muscles of Sham and CLP animals injected with either AAV-GFP or AAV-Parkin. *Gaba.* refers to *Gabarapl1*. (**C**) Immunoblot detection of SQSMT1(p62), BNIP3, LC3I/LC3II and GAPDH. (**D**) Quantification of SQSMT1 (p62) content. (**E**) Quantification of BNIP3 protein content. (**F**) Quantification of LC3I and LC3II protein content, as well as the LC3II to LC3I ratio. (**G**) qPCR analysis of *Fbxo32* (Atrogin-1) and *Trim63* (MuRF1) gene expression levels in the GAS muscles of Sham and CLP animals injected with either AAV-GFP or AAV-Parkin. 1 = Sham-AAV-GFP; 2 = Sham-AAV-Parkin; 3 = CLP-AAV-GFP; 4 = CLP-AAV-Parkin. Data are presented as mean ± SEM. (n = 6–9/group, * = statistically significant; ns = not statistically significant).

3.4. The Impact of Parkin Overexpression and Sepsis on the Expression of Genes and Proteins Regulating Mitochondrial Biology

Since Parkin plays a key role in mitochondrial quality control [23–26,40], and because sepsis is well known to impair mitochondrial function, we investigated whether Parkin overexpression could attenuate the impact of sepsis on skeletal muscle mitochondria. To this end, we first quantified the expression levels of the key transcriptional regulators of mitochondrial biology. As shown in Figure 4A,B, sepsis resulted in an increase in *Nrf1*, *Nrf2*, and *Sirt1* mRNA expression levels. In contrast, sepsis resulted in a decrease in the expression of *Pgc1-α, Tfam and Sirt3*, as shown in Figure 4A,B. In the skeletal muscles of both Sham-operated and CLP mice, Parkin overexpression resulted in a significant increase in *Nrf2* mRNA expression, depicted in Figure 4A. Parkin overexpression also led to an increased expression of *Sirt1* in the muscles of Sham-operated mice and an increase in *Tfam* expression in the muscles of CLP mice, as shown in Figure 4B.

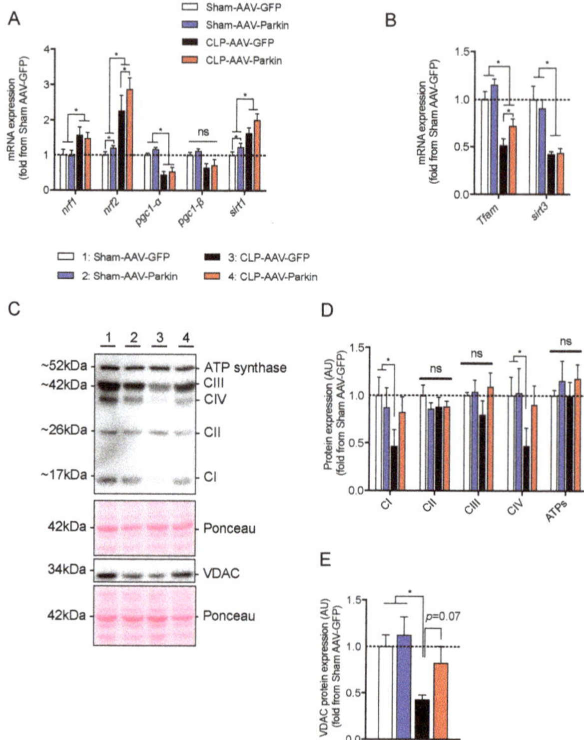

Figure 4. The impact of Parkin overexpression and sepsis in skeletal muscle on genes regulating mitochondrial biogenesis and on mitochondrial protein contents. (**A,B**) qPCR analysis of genes involved in mitochondrial biology. (**C**) Representative immunoblots performed with primary antibodies against representative subunits of the OXPHOS complexes and VDAC. Ponceau stains were used as loading controls. (**D,E**) Quantification of the contents of (**D**) representative subunits of the OXPHOS complexes and (**E**) VDAC. 1 = Sham-AAV-GFP; 2 = Sham-AAV-Parkin; 3 = CLP-AAV-GFP; 4 = CLP-AAV-Parkin. Data are presented as mean ± SEM. (n = 6–9/group, * = statistically significant; ns = not statistically significant).

We next assessed the impact of sepsis and Parkin overexpression on the content of proteins of the mitochondrial oxidative phosphorylation (OXPHOS) system. As shown in Figure 4D, sepsis significantly decreased the content of the representative subunits of Complex I and Complex IV. This finding is consistent with previous reports, which documented decreased mitochondrial contents in septic skeletal muscles [10,12–14]. Similarly, sepsis lowered VDAC protein content by 58% in the GFP expressing skeletal muscles, as shown in Figure 4E. Importantly, no impact of sepsis was observed on Complex I, Complex IV and VDAC contents in the Parkin overexpressing muscles, as shown in Figure 4D,E. Taken together, these data strongly suggest that Parkin overexpression prevented the inhibitory effect of sepsis on muscle mitochondrial content.

3.5. Effects of Parkin Overexpression and Sepsis on Mitochondrial Morphology and Dynamics

To analyze the impact of sepsis and Parkin overexpression on skeletal muscle mitochondrial morphology, we used TEM to evaluate the morphology of intermyofibrillar (IMF) mitochondria of the GAS of sham and CLP mice. Representative TEM images obtained from the GAS of sham and CLP mice are shown in Figure 5A–D and Supplemental Figure S3. In the CLP group, IMF mitochondria of GFP expressing muscles were larger, less circular and more complex (i.e., higher values of aspect ratio and form factor) than IMF mitochondria of GFP expressing muscles of the sham group, shown in Figure 5E–J. In the sham group, Parkin overexpressing muscles had larger, less circular and more complex IMF mitochondria compared to GFP expressing muscles, as shown in Figure 5E–J. In the CLP group, Parkin overexpressing muscles had smaller, more circular and simpler IMF mitochondria compared to GFP expressing counterparts, shown in Figure 5 E–J. Taken together, these results indicate that sepsis results in enlarged and more complex mitochondria, an impact that is abolished by Parkin overexpression.

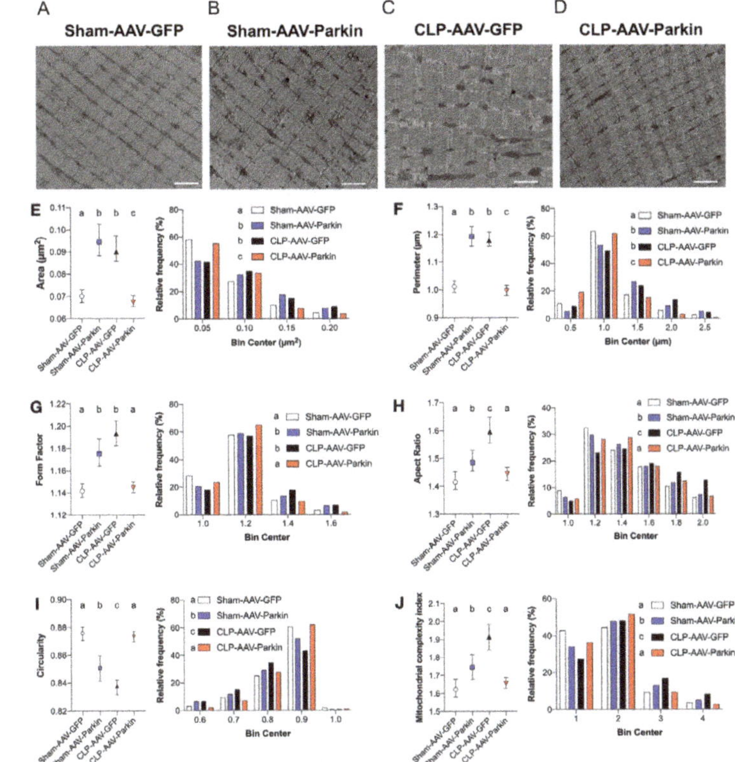

Figure 5. The impact of sepsis and Parkin overexpression on mitochondrial morphology in skeletal muscle. (**A–D**) Representative longitudinal TEM images from all groups that were used to assess mitochondrial morphology. Scale bar: 2µm. (**E–J**) Median values with 95% confidence interval (left) and relative frequencies (right) of multiple mitochondrial shape descriptors (Sham-AAV-GFP: n = 1246; Sham-AAV-Parkin: n = 728; CLP-AAV-GFP: n = 1149; CLP-AAV-Parkin: n = 1206). Groups not sharing a letter are significantly different (differences were tested using a Kruskal–Wallis test followed by a Dunn's multiple comparisons test; $p < 0.05$).

To gain better insights into the mechanisms underlying the impact of sepsis and Parkin overexpression on mitochondrial morphology, we next assessed the expression and content of major genes and proteins regulating mitochondrial dynamics. In the Sham-operated mice, Parkin overexpression had no impact on the mRNA expression and protein levels of *Mfn2*, *Opa1* and *Drp1*, as shown in Figure 6A–G. As shown in Figure 6A, sepsis in GFP expressing muscles resulted in a decrease in the mRNA levels of pro-fusion *Mfn2* and *Opa1* and pro-fission *Drp1*. In Parkin overexpressing muscles, CLP resulted in a significant decrease in the mRNA levels of Mfn2 and Drp1, while Opa1 expression remained unaffected, shown in Figure 6A. At the protein level, no impact of sepsis or Parkin overexpression could be evidenced on MFN2 and OPA1 protein content. Interestingly, DRP1 protein levels were lower in the GFP and Parkin expressing muscles of CLP mice, relative to the sham group, as shown in Figure 6D,E. Similarly, DRP1 phosphorylation on Ser^{616}, an activation site which triggers DRP1 translocation from the cytoplasm to mitochondria to promote mitochondrial fission [41], was also decreased in the GFP and Parkin expressing muscles of the CLP group, relative to the sham group, as shown in Figure 6D–G. These results indicate that sepsis seems to result in an inhibition of mitochondrial fission and that this effect was not influenced by Parkin overexpression.

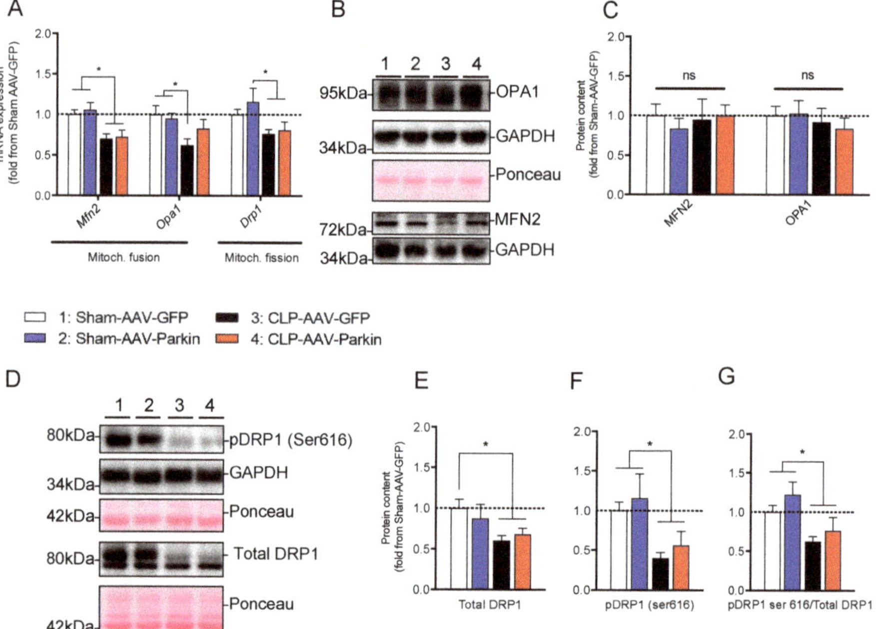

Figure 6. The impact of sepsis and Parkin overexpression on mitochondrial dynamics in skeletal muscle. (**A**) qPCR analysis of mitochondrial dynamic-related gene expression in the GAS muscles of Sham and CLP animals injected with either AAV-GFP or AAV-Parkin. (**B**) Representative immunoblots of OPA1, GADPH and MFN2. Ponceau stains or GAPDH immunoblots were used as loading controls. (**C**) Quantification of OPA1, GADPH and MFN2 content. (**D**) Representative immunoblots performed with primary antibodies against pDRP1(ser616) and total DRP1. Ponceau stains or GAPDH immunoblots were used as loading controls. (**E**) Quantification of DRP1 content. (**F**) Quantification of the contents of pDRP1(ser 616) content. (**G**) Quantification of the pDRP1(ser 616) to total DRP1 ratio. 1 = Sham-AAV-GFP; 2 = Sham-AAV-Parkin; 3 = CLP-AAV-GFP; 4 = CLP-AAV-Parkin. Data are presented as mean ± SEM. (*n* = 6–9/group; * = statistically significant; ns = not statistically significant).

4. Discussion

The accumulation of dysfunctional and injured mitochondria in skeletal muscles is believed to play a key role in the development of muscle weakness during sepsis [8,9]. In the current study, we investigated whether overexpressing Parkin, a key component of the PINK1-Parkin mitophagy pathway, could attenuate the negative impact of sepsis on skeletal muscles and their mitochondria. The current study indicates that Parkin overexpression prevented sepsis-induced accumulation of enlarged and complex mitochondria in the limb muscles of mice. Parkin overexpression also attenuated the sepsis-induced decrease in the content of complexes I and IV of the mitochondrial electron transfer system and prevented the development of limb muscle atrophy in septic mice. These results expand recent studies demonstrating that Parkin exerts protective effects on skeletal muscle health. Indeed, our group has recently reported that $Park2^{-/-}$ mice have decreased limb muscle contractility, depressed muscle mitochondrial respiration, increased mitochondrial uncoupling and enhanced susceptibility to the opening of mitochondrial permeability transition pore compared to wild-type (WT) mice [42]. $Park2^{-/-}$ mice also exhibit the impaired recovery of cardiac contractility and depressed cardiac mitochondrial functions in sepsis [29]. More recently, Peker et al. have reported that Parkin knockdown in C2C12 cells results in myotubular atrophy and that $Park2^{-/-}$ mice have decreased muscle mitochondrial respiration and increased levels of reactive oxygen species and fiber atrophy [43]. Parkin overexpression in the muscles of *Drosophila melanogaster* increased mitochondrial content, decreased proteotoxicity and extended lifespan [44]. Our finding that Parkin overexpression in the Sham group increased limb muscle fiber diameters is in accordance with our recent study documenting that Parkin overexpression for several months in young mice causes muscle hypertrophy, while in old mice, Parkin overexpression attenuates ageing-related loss of muscle mass and strength, increases mitochondrial content and enzymatic activities and protects from ageing-related oxidative stress, fibrosis and apoptosis [32]. Taken together, our current findings and published studies highlight the protective role that Parkin plays in skeletal muscle health.

Our findings that sepsis elicits distinct changes in skeletal muscle mitochondria, such as decreased VDAC level (a marker of mitochondrial content [45–47]), the downregulation of three mitochondrial biogenesis genes (*Pgc1-α*, *Tfam* and *Sirt3*) and decreased complexes I and IV levels, are in agreement with published studies on septic humans and experimental animals [10,12–14,20]. We report for the first time that Parkin overexpression in skeletal muscle prevents the inhibitory effects of sepsis on the expression of *Tfam* and on the content of complexes I and IV, as well as VDAC. Based on these results, we speculate that Parkin overexpression might have improved mitochondrial functions in septic muscles. This speculation is supported by the observation that Parkin overexpression increases mitochondrial content and enzymatic activities in normal skeletal muscles [32,44]. We should emphasize that, in the current study, Parkin overexpression increased *Nrf2* expression in the skeletal muscles of septic animals. Considering the role that this transcription factor has in the regulation of the expression of several anti-oxidant enzymes [48], we anticipate that increased *Nrf2* levels in muscles overexpressing Parkin might have contributed to the protection of mitochondrial morphology and contents in septic animals.

Mitochondria form a dynamic network constantly undergoing fusion and fission events that tightly regulate the shape (i.e., morphology), size and number of mitochondria [41,49]. In the present study, we show that sepsis significantly alters mitochondrial morphology by increasing the proportion of enlarged and more complex IMF mitochondria. These results extend previous observations, showing that sepsis causes major alterations of the mitochondrial ultrastructure in skeletal muscle [13,19–21]. This increase in mitochondrial size and complexity in septic muscles might have been caused by decreased DRP1 contents and activation [41], which are expected to alter the fusion/fission balance towards increased mitochondrial fusion. Since mitochondrial fission is required for mitochondrial degradation through mitophagy [50], it is possible that decreased DRP1 content and activation may play a role in the accumulation of damaged and dysfunctional mitochondria in septic muscles by impairing muscle capacity to recycle dysfunctional mitochondria through the mitophagy pathway.

It should also be noted that a decrease in DRP1 content per se might have also played a role in myofiber atrophy. Indeed, a recent study showed that muscle-specific DRP1 deletion results in severe muscle dysfunction, characterized by atrophy, weakness, fiber degeneration and mitochondrial dysfunction [51]. Importantly, we found that Parkin overexpression attenuated sepsis-induced changes in mitochondrial morphology and rendered muscle mitochondria to be smaller, more circular and simpler, relative to muscles expressing GFP. These findings are in agreement with previous reports, indicating that Parkin overexpression in skeletal muscles and neurons stimulates mitochondrial fragmentation [44,52]. We speculate that the decrease in mitochondrial size and complexity in septic muscles overexpressing Parkin might have facilitated the recycling of damaged/dysfunctional mitochondria.

A puzzling result of the present study is the increase in the proportion of enlarged and more complex mitochondria in the Parkin overexpressing muscles of sham-operated mice. The mechanisms behind the differences in the effects of Parkin on mitochondrial morphology in the sham and CLP groups remain unclear. We should point out, however, that although Parkin overexpression altered the mitochondrial morphology in skeletal muscles, none of the parameters related to mitochondrial dynamics that we investigated were affected by Parkin overexpression. Indeed, no significant impact of Parkin overexpression on the expression and protein content of MFN2, OPA1 and DRP1 was evident. Furthermore, Parkin overexpression had no effect on DRP1 phosphorylation on Ser^{616}, suggesting that there was no change in DRP1 activity. Further studies are therefore required to identify the mechanisms underlying the differential impact of Parkin overexpression on skeletal muscle mitochondrial morphology in healthy and septic animals.

In the current study, we report data indicating that autophagy was induced in septic skeletal muscles, as evidenced by the increased expression of several autophagy-related genes, including *Lc3b*, *Gabarapl1*, and *Sqstm1*, and by the increase in the LC3B-II/LC3B-I ratio in the muscles of CLP mice. These findings are in agreement with previous reports, which documented increased muscle autophagy in various models of sepsis [20,30,53]. An interesting finding in our study is that Parkin overexpression had no effects on the expression of autophagy-related genes and the LC3B-II/LC3B-I ratios in the sham and CLP groups. Given the key role that Parkin plays in the recycling of dysfunctional mitochondria by autophagosomes [23,50], our results indicate that basal and activated autophagy levels in normal and septic muscles, respectively, were sufficient to deal with increased mitophagy in muscles overexpressing Parkin. We also observed that BNIP3 mRNA and protein levels increased significantly in septic skeletal muscles, and that this induction was not influenced by Parkin overexpression. The BNIP3 protein localizes to the mitochondria and promotes PINK1-Parkin-independent mitophagy by interacting with the LC3 protein, resulting in the recruitment of autophagosomes to damaged mitochondria [54]. The lack of changes in BNIP3 levels in response to Parkin overexpression suggests that BNIP3-mediated mitophagy functions in an independent fashion to that of the PINK1-Parkin pathway. It is worth mentioning that the present study suffers from several limitations. First, we did not directly assess whether Parkin overexpression actually translates into increased mitophagic flux. Although it was recently reported that Parkin overexpression is sufficient to trigger higher mitochondrial clearance in cardiomyocytes [55], further studies should investigate whether Parkin overexpression is sufficient to increase mitophagy in healthy and septic skeletal muscles. Another important limitation arises from the fact that muscle contractility was not assessed in the present study. Further studies are therefore required to define whether Parkin overexpression can attenuate sepsis-induced skeletal muscle weakness.

5. Conclusions

The present study provides evidence that Parkin overexpression attenuates sepsis-induced myofiber atrophy and prevents sepsis-induced changes in mitochondrial morphology and protein contents. These findings suggest that targeting mitophagy may represent a promising therapeutic strategy to attenuate sepsis-induced skeletal muscle wasting.

Supplementary Materials: The following are available online at http://www.mdpi.com/2073-4409/9/6/1454/s1: Figure S1, Construction of the AAV1 designed to overexpress Parkin; Figure S2, The impact of sepsis and Parkin overexpression on the content and phosphorylation levels of proteins regulating protein synthesis; Figure S3, Additional TEM images; Table S1, List of antibodies; Table S2, List of qPCR primers.

Author Contributions: Designed and Conceived this Study, J.-P.L.-G., G.G. and S.N.A.H.; Collected, Analyzed and Interpreted the Data, Prepared all figures and tables and Wrote the first draft of the manuscript, J.-P.L.-G.; O.R., T.J.C., D.M. and F.E.B. were involved in data collection and analyses; G.G. and S.N.A.H. supervised the research, contributed to data analysis and interpretation and wrote the final version of the manuscript with J.-P.L.-G. Funding Acquisition: G.G. and S.N.A.H. All authors have read and approved the final version of the manuscript.

Funding: This work was funded by grants from the Natural Sciences and Engineering Council of Canada (NSERC, #RGPIN-2014-04668 awarded to Gilles Gouspillou) and from the Canadian Institute of Health Research (CIHR; MOP-93760 awarded to Sabah N. A. Hussain and MOV-409262 awarded to Sabah N. A. Hussain and Gilles Gouspillou). Gilles Gouspillou is also supported by a Chercheur Boursier Junior 1 salary award from the Fonds de Recherche du Québec en Santé (FRQS-35184). Tomer J Chaffer was supported by a Natural Sciences and Engineering Council of Canada Undergraduate Student Research Award (NSERC USRA). Jean-Philippe Leduc-Gaudet was supported by a CIHR Vanier Fellowship and currently holds a RI-MUHC Fellowship. The funders had no role in the design of the study; in the collection, analyses, or interpretation of data; in the writing of the manuscript, or in the decision to publish the results.

Acknowledgments: We thank Jeannie Mui from the Facility for Electron Microscopy Research (FEMR, McGill University, Montreal, QC, Canada) for her support and expertise. We are grateful for the technical support provided by Laurent Huck and the staff of the Meakins-Christie Laboratories at the Research Institute of the McGill University Health Center. We would like to thank Basil Petrof (McGill University and RI-MUHC) for his thoughtful discussions related to this study.

Conflicts of Interest: The authors declare no conflict of interest.

References

1. Angus, D.C.; van der Poll, T. Severe sepsis and septic shock. *N. Engl. J. Med.* **2013**, *369*, 840–851. [CrossRef] [PubMed]

2. Khan, J.; Harrison, T.B.; Rich, M.M.; Moss, M. Early development of critical illness myopathy and neuropathy in patients with severe sepsis. *Neurology* **2006**, *67*, 1421–1425. [CrossRef] [PubMed]

3. Tennila, A.; Salmi, T.; Pettila, V.; Roine, R.O.; Varpula, T.; Takkunen, O. Early signs of critical illness polyneuropathy in icu patients with systemic inflammatory response syndrome or sepsis. *Intensive Care Med.* **2000**, *26*, 1360–1363. [CrossRef] [PubMed]

4. Supinski, G.S.; Callahan, L.A. Diaphragm weakness in mechanically ventilated critically ill patients. *Crit. Care* **2013**, *17*, R120. [CrossRef] [PubMed]

5. De Jonghe, B.; Bastuji-Garin, S.; Durand, M.C.; Malissin, I.; Rodrigues, P.; Cerf, C.; Outin, H.; Sharshar, T. Respiratory weakness is associated with limb weakness and delayed weaning in critical illness. *Crit. Care Med.* **2007**, *35*, 2007–2015. [CrossRef]

6. Ali, N.A.; O'Brien, J.M., Jr.; Hoffmann, S.P.; Phillips, G.; Garland, A.; Finley, J.C.; Almoosa, K.; Hejal, R.; Wolf, K.M.; Lemeshow, S.; et al. Acquired weakness, handgrip strength, and mortality in critically ill patients. *Am. J. Respir. Crit. Care Med.* **2008**, *178*, 261–268. [CrossRef]

7. Iwashyna, T.J.; Ely, E.W.; Smith, D.M.; Langa, K.M. Long-term cognitive impairment and functional disability among survivors of severe sepsis. *JAMA* **2010**, *304*, 1787–1794. [CrossRef]

8. Friedrich, O.; Reid, M.; Van den Berghe, G.; Vanhorebeek, I.; Hermans, G.; Rich, M.; Larsson, L. The sick and the weak: Neuropathies/myopathies in the critically ill. *Physiol. Rev.* **2015**, *95*, 1025–1109. [CrossRef]

9. Callahan, L.A.; Supinski, G.S. Sepsis-induced myopathy. *Crit. Care Med.* **2009**, *37*, S354–S367. [CrossRef]

10. Brealey, D.; Brand, M.; Hargreaves, I.; Heales, S.; Land, J.; Smolenski, R.; Davies, N.A.; Cooper, C.E.; Singer, M. Association between mitochondrial dysfunction and severity and outcome of septic shock. *Lancet* **2002**, *360*, 219–223. [CrossRef]

11. Brealey, D.; Karyampudi, S.; Jacques, T.S.; Novelli, M.; Stidwill, R.; Taylor, V.; Smolenski, R.T.; Singer, M. Mitochondrial dysfunction in a long-term rodent model of sepsis and organ failure. *Am. J. Physiol. Regul. Integr. Comp. Physiol.* **2004**, *286*, R491–R497. [CrossRef] [PubMed]

12. Fredriksson, K.; Hammarqvist, F.; Strigard, K.; Hultenby, K.; Ljungqvist, O.; Wernerman, J.; Rooyackers, O. Derangements in mitochondrial metabolism in intercostal and leg muscle of critically ill patients with sepsis-induced multiple organ failure. *Am. J. Physiol. Endocrinol. Metab.* **2006**, *291*, E1044–E1050. [CrossRef] [PubMed]

13. Rooyackers, O.E.; Kersten, A.H.; Wagenmakers, A.J. Mitochondrial protein content and in vivo synthesis rates in skeletal muscle from critically ill rats. *Clin. Sci. (Lond.)* **1996**, *91*, 475–481. [CrossRef] [PubMed]

14. Callahan, L.A.; Supinski, G.S. Sepsis induces diaphragm electron transport chain dysfunction and protein depletion. *Am. J. Respir Crit. Care Med.* **2005**, *172*, 861–868. [CrossRef] [PubMed]

15. Crouser, E.D.; Julian, M.W.; Blaho, D.V.; Pfeiffer, D.R. Endotoxin-induced mitochondrial damage correlates with impaired respiratory activity. *Crit. Care Med.* **2002**, *30*, 276–284. [CrossRef]

16. Protti, A.; Carre, J.; Frost, M.T.; Taylor, V.; Stidwill, R.; Rudiger, A.; Singer, M. Succinate recovers mitochondrial oxygen consumption in septic rat skeletal muscle. *Crit. Care Med.* **2007**, *35*, 2150–2155. [CrossRef]

17. Alvarez, S.; Boveris, A. Mitochondrial nitric oxide metabolism in rat muscle during endotoxemia. *Free Radic Biol. Med.* **2004**, *37*, 1472–1478. [CrossRef]

18. Vanasco, V.; Cimolai, M.C.; Evelson, P.; Alvarez, S. The oxidative stress and the mitochondrial dysfunction caused by endotoxemia are prevented by alpha-lipoic acid. *Free Radic. Res.* **2008**, *42*, 815–823. [CrossRef]

19. Welty-Wolf, K.E.; Simonson, S.G.; Huang, Y.C.; Fracica, P.J.; Patterson, J.W.; Piantadosi, C.A. Ultrastructural changes in skeletal muscle mitochondria in gram-negative sepsis. *Shock* **1996**, *5*, 378–384. [CrossRef]

20. Mofarrahi, M.; Sigala, I.; Guo, Y.; Godin, R.; Davis, E.C.; Petrof, B.; Sandri, M.; Burelle, Y.; Hussain, S.N. Autophagy and skeletal muscles in sepsis. *PLoS ONE* **2012**, *7*, e47265. [CrossRef]

21. Owen, A.M.; Patel, S.P.; Smith, J.D.; Balasuriya, B.K.; Mori, S.F.; Hawk, G.S.; Stromberg, A.J.; Kuriyama, N.; Kaneki, M.; Rabchevsky, A.G.; et al. Chronic muscle weakness and mitochondrial dysfunction in the absence of sustained atrophy in a preclinical sepsis model. *eLife* **2019**, *8*. [CrossRef] [PubMed]

22. Jin, S.M.; Lazarou, M.; Wang, C.; Kane, L.A.; Narendra, D.P.; Youle, R.J. Mitochondrial membrane potential regulates pink1 import and proteolytic destabilization by parl. *J. Cell Biol.* **2010**, *191*, 933–942. [CrossRef] [PubMed]

23. Narendra, D.; Tanaka, A.; Suen, D.-F.; Youle, R.J. Parkin is recruited selectively to impaired mitochondria and promotes their autophagy. *J. Cell Biol.* **2008**, *183*, 795–803. [CrossRef] [PubMed]

24. Narendra, D.P.; Jin, S.M.; Tanaka, A.; Suen, D.-F.; Gautier, C.A.; Shen, J.; Cookson, M.R.; Youle, R.J. Pink1 is selectively stabilized on impaired mitochondria to activate parkin. *PLoS Biol.* **2010**, *8*, e1000298. [CrossRef]

25. Vives-Bauza, C.; Zhou, C.; Huang, Y.; Cui, M.; de Vries, R.L.; Kim, J.; May, J.; Tocilescu, M.A.; Liu, W.; Ko, H.S.; et al. Pink1-dependent recruitment of parkin to mitochondria in mitophagy. *Proc. Natl. Acad. Sci. USA* **2010**, *107*, 378–383. [CrossRef]

26. Matsuda, N.; Sato, S.; Shiba, K.; Okatsu, K.; Saisho, K.; Gautier, C.A.; Sou, Y.-S.; Saiki, S.; Kawajiri, S.; Sato, F. Pink1 stabilized by mitochondrial depolarization recruits parkin to damaged mitochondria and activates latent parkin for mitophagy. *J. Cell Biol.* **2010**, *189*, 211–221. [CrossRef]

27. Ni, H.-M.; Williams, J.A.; Ding, W.-X. Mitochondrial dynamics and mitochondrial quality control. *Redox Biol.* **2015**, *4*, 6–13. [CrossRef]

28. Tilokani, L.; Nagashima, S.; Paupe, V.; Prudent, J. Mitochondrial dynamics: Overview of molecular mechanisms. *Essays Biochem.* **2018**, *62*, 341–360.

29. Piquereau, J.; Godin, R.; Deschenes, S.; Bessi, V.L.; Mofarrahi, M.; Hussain, S.N.; Burelle, Y. Protective role of PARK2/Parkin in sepsis-induced cardiac contractile and mitochondrial dysfunction. *Autophagy* **2013**, *9*, 1837–1851. [CrossRef]

30. Stana, F.; Vujovic, M.; Mayaki, D.; Leduc-Gaudet, J.P.; Leblanc, P.; Huck, L.; Hussain, S.N.A. Differential regulation of the autophagy and proteasome pathways in skeletal muscles in sepsis. *Crit. Care Med.* **2017**, *45*, e971–e979. [CrossRef]

31. Buras, J.A.; Holzmann, B.; Sitkovsky, M. Animal models of sepsis: Setting the stage. *Nat. Rev. Drug Discov.* **2005**, *4*, 854–865. [CrossRef] [PubMed]

32. Leduc-Gaudet, J.P.; Reynaud, O.; Hussain, S.N.; Gouspillou, G. Parkin overexpression protects from ageing-related loss of muscle mass and strength. *J. Physiol.* **2019**, *597*, 1975–1991. [CrossRef] [PubMed]

33. Moarbes, V.; Mayaki, D.; Huck, L.; Leblanc, P.; Vassilakopoulos, T.; Petrof, B.J.; Hussain, S.N.A. Differential regulation of myofibrillar proteins in skeletal muscles of septic mice. *Physiol. Rep.* **2019**, *7*, e14248. [CrossRef] [PubMed]

34. Leduc-Gaudet, J.P.; Picard, M.; St-Jean Pelletier, F.; Sgarioto, N.; Auger, M.J.; Vallee, J.; Robitaille, R.; St-Pierre, D.H.; Gouspillou, G. Mitochondrial morphology is altered in atrophied skeletal muscle of aged mice. *Oncotarget* **2015**, *6*, 17923–17937. [CrossRef]

35. Briguet, A.; Courdier-Fruh, I.; Foster, M.; Meier, T.; Magyar, J.P. Histological parameters for the quantitative assessment of muscular dystrophy in the mdx-mouse. *Neuromuscul. Disord.* **2004**, *14*, 675–682. [CrossRef]

36. Picard, M.; Gentil, B.J.; McManus, M.J.; White, K.; St Louis, K.; Gartside, S.E.; Wallace, D.C.; Turnbull, D.M. Acute exercise remodels mitochondrial membrane interactions in mouse skeletal muscle. *J. Appl. Physiol.* **2013**, *115*, 1562–1571. [CrossRef]

37. Picard, M.; White, K.; Turnbull, D.M. Mitochondrial morphology, topology, and membrane interactions in skeletal muscle: A quantitative three-dimensional electron microscopy study. *J. Appl. Physiol.* **2013**, *114*, 161–171. [CrossRef]

38. Gomes, M.D.; Lecker, S.H.; Jagoe, R.T.; Navon, A.; Goldberg, A.L. Atrogin-1, a muscle-specific F-box protein highly expressed during muscle atrophy. *Proc. Natl. Acad. Sci. USA* **2001**, *98*, 14440–14445. [CrossRef]

39. Bodine, S.C.; Latres, E.; Baumhueter, S.; Lai, V.K.; Nunez, L.; Clarke, B.A.; Poueymirou, W.T.; Panaro, F.J.; Na, E.; Dharmarajan, K.; et al. Identification of ubiquitin ligases required for skeletal muscle atrophy. *Science* **2001**, *294*, 1704–1708. [CrossRef]

40. Kane, L.A.; Lazarou, M.; Fogel, A.I.; Li, Y.; Yamano, K.; Sarraf, S.A.; Banerjee, S.; Youle, R.J. Pink1 phosphorylates ubiquitin to activate parkin e3 ubiquitin ligase activity. *J. Cell Biol.* **2014**, *205*, 143–153. [CrossRef]

41. Chan, D.C. Fusion and fission: Interlinked processes critical for mitochondrial health. *Annu. Rev. Genet.* **2012**, *46*, 265–287. [CrossRef]

42. Gouspillou, G.; Godin, R.; Piquereau, J.; Picard, M.; Mofarrahi, M.; Mathew, J.; Purves-Smith, F.M.; Sgarioto, N.; Hepple, R.T.; Burelle, Y.; et al. Protective role of parkin in skeletal muscle contractile and mitochondrial function. *J. Physiol.* **2018**, *596*, 2565–2579. [CrossRef] [PubMed]

43. Peker, N.; Donipadi, V.; Sharma, M.; McFarlane, C.; Kambadur, R. Loss of parkin impairs mitochondrial function and leads to muscle atrophy. *Am. J. Physiol. Cell Physiol.* **2018**, *315*, C164–C185. [CrossRef]

44. Rana, A.; Rera, M.; Walker, D.W. Parkin overexpression during aging reduces proteotoxicity, alters mitochondrial dynamics, and extends lifespan. *Proc. Natl. Acad. Sci. USA* **2013**, *110*, 8638–8643. [CrossRef] [PubMed]

45. Gouspillou, G.; Sgarioto, N.; Norris, B.; Barbat-Artigas, S.; Aubertin-Leheudre, M.; Morais, J.A.; Burelle, Y.; Taivassalo, T.; Hepple, R.T. The relationship between muscle fiber type-specific pgc-1alpha content and mitochondrial content varies between rodent models and humans. *PLoS ONE* **2014**, *9*, e103044. [CrossRef] [PubMed]

46. Hernandez-Alvarez, M.I.; Thabit, H.; Burns, N.; Shah, S.; Brema, I.; Hatunic, M.; Finucane, F.; Liesa, M.; Chiellini, C.; Naon, D.; et al. Subjects with early-onset type 2 diabetes show defective activation of the skeletal muscle pgc-1{alpha}/mitofusin-2 regulatory pathway in response to physical activity. *Diabetes Care* **2010**, *33*, 645–651. [CrossRef] [PubMed]

47. Hanson, B.J.; Capaldi, R.A.; Marusich, M.F.; Sherwood, S.W. An immunocytochemical approach to detection of mitochondrial disorders. *J. Histochem. Cytochem.* **2002**, *50*, 1281–1288. [CrossRef]

48. Ma, Q. Role of nrf2 in oxidative stress and toxicity. *Annu. Rev. Pharm. Toxicol.* **2013**, *53*, 401–426. [CrossRef]

49. Suen, D.F.; Norris, K.L.; Youle, R.J. Mitochondrial dynamics and apoptosis. *Genes Dev.* **2008**, *22*, 1577–1590. [CrossRef]

50. Twig, G.; Shirihai, O.S. The interplay between mitochondrial dynamics and mitophagy. *Antioxid. Redox Signal.* **2011**, *14*, 1939–1951. [CrossRef]

51. Favaro, G.; Romanello, V.; Varanita, T.; Andrea Desbats, M.; Morbidoni, V.; Tezze, C.; Albiero, M.; Canato, M.; Gherardi, G.; De Stefani, D.; et al. Drp1-mediated mitochondrial shape controls calcium homeostasis and muscle mass. *Nat. Commun.* **2019**, *10*, 2576. [CrossRef] [PubMed]

52. Yu, W.; Sun, Y.; Guo, S.; Lu, B. The pink1/parkin pathway regulates mitochondrial dynamics and function in mammalian hippocampal and dopaminergic neurons. *Hum. Mol. Genet.* **2011**, *20*, 3227–3240. [CrossRef] [PubMed]

53. Kishta, O.A.; Guo, Y.; Mofarrahi, M.; Stana, F.; Lands, L.C.; Hussain, S.N.A. Pulmonary pseudomonas aeruginosa infection induces autophagy and proteasome proteolytic pathways in skeletal muscles: Effects of a pressurized whey protein-based diet in mice. *Food Nutr. Res.* **2017**, *61*, 1325309. [CrossRef] [PubMed]

54. Bellot, G.; Garcia-Medina, R.; Gounon, P.; Chiche, J.; Roux, D.; Pouysségur, J.; Mazure, N.M. Hypoxia-induced autophagy is mediated through hypoxia-inducible factor induction of bnip3 and bnip3l via their bh3 domains. *Mol. Cell. Biol.* **2009**, *29*, 2570–2581. [CrossRef]

55. Song, M.; Gong, G.; Burelle, Y.; Gustafsson, A.B.; Kitsis, R.N.; Matkovich, S.J.; Dorn, G.W., 2nd. Interdependence of parkin-mediated mitophagy and mitochondrial fission in adult mouse hearts. *Circ. Res.* **2015**, *117*, 346–351. [CrossRef]

Article

Key Components of Human Myofibre Denervation and Neuromuscular Junction Stability are Modulated by Age and Exercise

Casper Soendenbroe [1,2], Cecilie J. L. Bechshøft [1,3], Mette F. Heisterberg [1], Simon M. Jensen [1], Emma Bomme [1], Peter Schjerling [1,3], Anders Karlsen [1,3], Michael Kjaer [1,3], Jesper L. Andersen [1,3] and Abigail L. Mackey [1,2,3,*]

[1]   Institute of Sports Medicine Copenhagen, Department of Orthopedic Surgery M, Bispebjerg Hospital, Building 8, Nielsine Nielsens vej 11, 2400 Copenhagen NV, Denmark; caspersoendenbroe@outlook.dk (C.S.); cjleidersdorff@gmail.com (C.J.L.B.); metteflindt@hotmail.com (M.F.H.); simonmarqvard@gmail.com (S.M.J.); emmabomme@gmail.com (E.B.); Peter@mRNA.dk (P.S.); ak@anderskarlsen.dk (A.K.); michaelkjaer@sund.ku.dk (M.K.); Jesper.Loevind.Andersen@regionh.dk (J.L.A.)

[2]   Xlab, Department of Biomedical Sciences, Faculty of Health and Medical Sciences, University of Copenhagen, Blegdamsvej 3, 2200 Copenhagen N, Denmark

[3]   Center for Healthy Aging, Faculty of Health and Medical Sciences, University of Copenhagen, Blegdamsvej 3, 2200 Copenhagen N, Denmark

*   Correspondence: abigailmac@sund.ku.dk; Tel.: +45-3863-5366

Received: 28 February 2020; Accepted: 3 April 2020; Published: 6 April 2020

Abstract: The decline in muscle mass and function with age is partly caused by a loss of muscle fibres through denervation. The purpose of this study was to investigate the potential of exercise to influence molecular targets involved in neuromuscular junction (NMJ) stability in healthy elderly individuals. Participants from two studies (one group of 12 young and 12 elderly females and another group of 25 elderly males) performed a unilateral bout of resistance exercise. Muscle biopsies were collected at 4.5 h and up to 7 days post exercise for tissue analysis and cell culture. Molecular targets related to denervation and NMJ stability were analysed by immunohistochemistry and real-time reverse transcription polymerase chain reaction. In addition to a greater presence of denervated fibres, the muscle samples and cultured myotubes from the elderly individuals displayed altered gene expression levels of acetylcholine receptor (AChR) subunits. A single bout of exercise induced general changes in AChR subunit gene expression within the biopsy sampling timeframe, suggesting a sustained plasticity of the NMJ in elderly individuals. These data support the role of exercise in maintaining NMJ stability, even in elderly inactive individuals. Furthermore, the cell culture findings suggest that the transcriptional capacity of satellite cells for AChR subunit genes is negatively affected by ageing.

Keywords: sarcopenia; denervation; neuromuscular junction; heavy resistance exercise; acetylcholine receptor; cell culture; myogenesis; neonatal myosin; neural cell adhesion molecule

---

1. Introduction

The rate of loss of muscle mass increases with advancing age [1], and ultimately leads to impaired physical function in elderly individuals [2–4]. This age-dependent decline in muscle mass is partly due to a loss of individual muscle fibres [5] as a result of muscle fibre denervation [6–8]. While physical exercise is recognized as a strong countermeasure against the loss of muscle mass and has consistently been shown to maintain physical function and health in the last ten years of life [9,10], it is currently unclear whether denervation can be ameliorated or reversed by exercise.

It has been shown in animals that exercise causes positive adaptations to the neuromuscular junction (NMJ) that to some extent can attenuate the age-related degeneration of the NMJ [11]. Changes in expression of acetylcholine receptors (AChRs) with acute exercise have been suggested to indicate NMJ remodelling in animals [12,13] and represent a potential target for studying this in humans [14]. AChR are present in abundance at the NMJ [15] and are almost non-existent in the extra-synaptic region of the muscle fibre [16]. Upon experimental denervation, however, the $\alpha1$, $\beta1$, γ, and δ subunits increase extra synaptically [16–19], raising the possibility that these AChR subunits can be used as indicators of denervation associated with ageing. We recently observed a correlation between age and gene expression levels of the foetal γ AChR subunit in a large group ($n = 70$) of healthy elderly men ranging in age from 65 to 94 years, in conjunction with tissue markers of muscle fibre denervation, neural cell adhesion molecule (NCAM) and neonatal myosin (MHCn), at the protein level [20]. Direct comparisons with a younger cohort as well as the potential for exercise to influence AChR expression patterns are however lacking.

One of the challenges for ageing skeletal muscle is related to the decline in satellite cell function with age. Not only is satellite cell function important for tissue repair and maintenance, but it also has potential implications for maintenance of the NMJ, where myonuclei at this site must be capable of carrying out the specialization necessary to complete the formation of the NMJ. This includes producing a high concentration of AChRs at the membrane and a clustering of myonuclei, which become transcriptionally specialized and distinct from adjacent extra-synaptic myonuclei [21,22]. Whether this capacity declines with age is currently unknown. Satellite cells have been shown to play a vital role in maintaining the post-synaptic region in mice, both in terms of myonuclear clusters of AChRs and re-innervation of the regenerating NMJ [23,24]. In this context it is interesting that we have recently observed a poorer fusion capacity of satellite cells derived from old women compared to young women, accompanied by a distinctly different molecular profile throughout the myogenic program [25]. It remains unknown, however, to what extent this dysfunction in human satellite cells has implications for NMJ maintenance with increasing age.

Based on the above, the main purpose of this study was to investigate the influence of age and exercise on molecular markers of NMJ stability and muscle fibre denervation in healthy elderly individuals. An additional focus was to determine how ageing would alter the capacity of myonuclei in cell culture to produce key transcriptional elements for NMJ formation.

2. Materials and Methods

2.1. Experimental Design

This study is based on muscle biopsies collected from two studies, on 12 young and 12 elderly women [25], and on 25 elderly men [26], respectively. Both studies were approved by The Committees on Health Research Ethics for The Capital Region of Denmark (Ref: H-15017223, H-3-2012-081). All procedures conformed to the Declaration of Helsinki of 1975, revised in 2013, and the subjects gave written informed consent before participation. All participants were healthy, non-smokers, non-obese, and did not perform strenuous physical exercise on a regular basis. The men were part of a randomized controlled trial investigating the effect of the blood pressure-lowering medication losartan on the muscle response to exercise, where half of the participants received losartan and the other half placebo. Given the general lack of drug effect, the two groups were merged in the present study (separate group data are also provided for reference in online Supplementary Figure S1).

All participants performed a maximal strength test in a Leg Extension machine (M52, TechnoGym, Cesena, Italy) to determine the one-repetition maximum (1 RM), which was used to determine the load lifted during the subsequent bout of heavy resistance exercise. The Leg Extension exercise protocols consisted of both concentric and eccentric contractions. First, 4–5 sets of 12 concentric contractions at 70% of 1 RM were performed, followed by four sets of 4–6 eccentric contractions at 110% of 1 RM, as previously described [25,26]. The exercise was performed with one leg only, leaving the contralateral leg as a control. No other exercise was allowed during the study period.

2.2. Muscle Biopsies

For all participants, muscle biopsies were obtained from the vastus lateralis muscle, under local anaesthetic (1% lidocaine), using the percutaneous needle biopsy technique of Bergström [27], with five 6-mm needles and manual suction. Pieces of muscle tissue were aligned, embedded in Tissue-Tek, and then frozen in isopentane, pre-cooled in liquid nitrogen, and stored at −80 °C. The men had six muscle biopsies taken over 17 days, at the following time points: −10 and −3 days before exercise from the control, non-exercised leg, and from the exercised leg at +4.5 h and on days +1, +4, and +7 post exercise. The day −3 sample was excluded from the current study since its purpose was to investigate a potential effect of losartan in the rested state and is therefore superfluous in the current context. The young and elderly women had muscle biopsies collected from each leg five days after exercise, from which a part was embedded as described above and a part was used for cell culture, where myoblasts were plated in 12-well plates for three days of proliferation (12,000 cells per well), or three days of proliferation followed by four days of differentiation (20,000 cells per well), as previously described in detail [25].

2.3. RNA Extraction

100 cryo sections, 10 μm thick, from the embedded muscle tissue were homogenized in 1 mL of TriReagent (Molecular Research Center, Cincinnati, OH, USA) containing five stainless steel balls of 2.3 mm in diameter (BioSpec Products, Bartlesville, OK, USA), and one silicon-carbide sharp particle of 1 mm (BioSpec Products), by shaking in a FastPrep®-24 instrument (MP Biomedicals, Illkirch, France) at speed level four for 15 s. Cell culture cells were dissolved directly in the Trireagent. Bromo-chloropropane was added in order to separate the samples into an aqueous and an organic phase. Following isolation of the aqueous phase, RNA was precipitated using isopropanol. The RNA pellet was then washed in ethanol and subsequently dissolved in 20 μL RNAse-free water. Total RNA concentrations and purity were determined by spectroscopy at 260, 280, and 240 nm. Good RNA integrity was ensured by gel electrophoresis.

2.4. Real-Time RT-PCR

mRNA targets related to innervation were analysed for the current study. The specific primers are given in Table 1. Total RNA (500 ng for muscle and 150 ng for cell culture) was converted into cDNA in 20 μL using OmniScript reverse transcriptase (Qiagen, Redwood City, CA, USA) and 1 μM poly-dT (Invitrogen, Naerum, Denmark) according to the manufacturer's protocol (Qiagen). The same pool of cDNA used previously for the cells in culture [25] and the male muscle tissue [26] was used here. For each target mRNA, 0.25 μL cDNA were amplified in a 25-μL SYBR Green polymerase chain reaction (PCR) containing 1 × Quantitect SYBR Green Master Mix (Qiagen) and 100 nM of each primer (Table 1). The amplification was monitored real time using the MX3005P Real-time PCR machine (Stratagene, San Diego, CA, USA). The Ct values were related to a standard curve made with known concentrations of cloned PCR products or DNA oligonucleotides (Ultramer™ oligos, Integrated DNA Technologies, Inc., Leuven, Belgium) with a DNA sequence corresponding to the sequence of the expected PCR product. The specificity of the PCR products was confirmed by melting curve analysis after amplification. Ribosomal Protein Lateral Stalk Subunit P0 (RPLP0) mRNA was chosen as internal control. To validate this use, another unrelated "constitutive" mRNA, Glyceraldehyde-3-Phosphate Dehydrogenase (GAPDH), was measured and normalized with RPLP0. In the cell culture experiment GAPDH mRNA normalized to RPLP0 mRNA was constant, indicating that RPLP0 (and GAPDH) was indeed constant and suitable for normalization. However, in tissue the GAPDH/RPLP0 ratio was lower in the elderly female subjects and one and four days after exercise in the males, showing either a GAPDH decrease or a RPLP0 increase. However, the decrease in GAPDH was not reflected in the general pattern of the other mRNA when normalized to RPLP0, arguing against a general normalization error. We therefore chose to use retain RPLP0 for normalization. The GAPDH mRNA data from cell culture of the females and tissue of the males have been used as internal control in already published papers [25,26].

Table 1. Primers used for PCR. RPLP0: Ribosomal Protein Lateral Stalk Subunit P0; GAPDH: Glyceraldehyde-3-Phosphate Dehydrogenase; AChR: acetylcholine receptor; MuSK: muscle-specific-kinase; MHCn: neonatal myosin; MHCe: embryonic myosin heavy chain.

mRNA	Genbank	Sense	Antisense
RPLP0	NM_053275.3	GGAAACTCTGCATTCTCGCTTCCT	CCAGGACTCGTTTGTACCCGTTG
AchRα1	NM_000079.3	GCAGAGACCATGAAGTCAGACCAGGAG	CCGATGATGCAAACAAGCATGAA
AchRβ1	NM_000747.2	TTCATCCGGAAGCCGCCAAG	CCGCAGATCAGGGGCAGACA
AchRδ	NM_000751.2	CAGCTGTGGATGGGGCAAAC	GCCACTCGGTTCCAGCTGTCTT
AchRε	NM_000080.4	TGGCAGAACTGTTCGCTTATTTTCC	TTGATGGTCTTGCCGTCGTTGT
AchRγ	NM_005199.4	GCCTGCAACCTCATTGCCTGT	ACTCGGCCCACCAGGAACCAC
MuSK	NM_005592.3	TCATGGCAGAATTTGACAACCCTAAC	GGCTTCCCGACAGCACACAC
MHCe	NM_002470.3	CGGATATCGCAGAATCTCAAGTCAA	CTCCAGAAGGGCTGGCTCACTC
MHCn	NM_002472.2	CGGAAACATGAGCGACGAGTAAAA	CAGCCTGAGAACATTCTTGCGATCTT
GAPDH	NM_002046.6	GAGGGGCCATCCACAGTCTTCT	GACATGCCCAAGACCCAGAAGGA

2.5. Immunohistochemistry

For the female participants, cross sections (10 μm) from the biopsies of the exercised and control legs were cut at −20 °C in a cryostat. Sections from both legs of one individual were placed on the same glass slide (Thermo Scientific, Waltham, MA, USA) and stored at −80 °C until staining. For staining, two primary antibodies were diluted in 1% bovine serum albumin (BSA) in Tris-buffered saline (TBS) and applied to the sections (see Table 2), and then incubated in the refrigerator overnight. Afterwards two secondary antibodies (see Table 2) diluted in 1% BSA in TBS were applied for 45 min. At this point, the sections were fixed in 5% formaldehyde (Histofix, Histolab, Gothenburg, Sweden) for 12 min and then mounted with Prolong-Gold-Antifade (Invitrogen, Molecular Probes, OR, USA, catalogue #P36931), containing 4′,6-Di-amidino-2-phenylindole (DAPI). Slides were washed with TBS twice between all steps. Slides were kept in darkness at room temperature for 48 h and then moved to a −20 °C freezer. Two sections were also stained with NCAM and collagen XXII (made by Manuel Koch) [28], as previously described [29], since it was suspected that the NCAM staining in these sections was due to the presence of myotendinous junction and not denervated muscle fibres.

Table 2. Antibodies used for immunohistochemistry and immunocytochemistry. MHCn: neonatal myosin; MHCe: embryonic myosin heavy chain; NCAM: neural cell adhesion molecule.

Host	Antibody	Primary Antibody Company	Cat. no.	Concentration
Mouse	Dystrophin, IgG2b	Sigma-Aldrich	D8168	1:500
Mouse	Myosin 1, IgG1	Hybridoma Bank	A4.951	1:200
Mouse	MHCe, IgG1	Hybridoma Bank	F1.652	1:100
Mouse	MHCn, IgG1	Novocastra	NCL-MHCn	1:100
Mouse	NCAM, IgG1	Becton Dickinson	347740	1:50
Rabbit	Desmin, IgG	Abcam	AB32362	1:1000
Mouse	Myogenin, IgG1	Hybridoma Bank	F5D-s	1:50
Host	**Antibody**	**Secondary Antibody Company**	**Cat. no.**	**Concentration**
Goat	488, green, IgG1	Invitrogen	A-21121	1:500
Goat	568, red, IgG2b	Invitrogen	A-21144	1:200
Goat	568, red, IgG	Invitrogen	A-11036	1:500
Goat	488, green, IgG	Invitrogen	A-11029	1:500

2.6. Microscopy

All imaging was performed with a ×10/0.30NA objective and a 0.5× camera (Olympus DP71, Olympus Deutschland GmbH, Hamburg, Germany) mounted on a BX51 Olympus microscope, using the Olympus cellSens software (v.1.14). For all analyses, 1.7 × 1.3 mm greyscale images were captured.

Muscle fibre size and muscle fibre type composition analysis was only performed on the control leg. Non-overlapping images of high resolution (4080 × 3072 pixels) were captured to accommodate a semi-automated macro [30], run in ImageJ (v.1.51, U.S. National Institutes of Health, Bethesda, MD, USA). All analyses were conducted by the same person blinded to the age group. All included muscle fibres were manually checked, and fibres were excluded if the dystrophin staining was incomplete or if an area of the biopsy was longitudinally oriented. Fibres at the edge and around holes and folds in the biopsies were always excluded. After delineation of the muscle fibre cross-sectional area (CSA), fibre type was determined based upon the median light intensity. Fibres were classified as type I (positive for myosin type I staining) or type II (negative for myosin type I staining). Hybrid muscle fibres (low levels of type I myosin staining) were excluded from the analysis (a total of 131 fibres from all sections).

For the analysis of embryonic myosin heavy chain (MHCe)-, MHCn-, and NCAM-positive fibres, images at a resolution of 2040 × 1536 pixels were captured. For MHCe, only areas with positive staining were imaged, while for MHCn and NCAM the entire biopsy section was imaged (due to the relatively higher prevalence of positive fibres). Positively stained muscle fibres were determined as fibres with a complete dystrophin staining and a clear staining of one of the three markers. We extended the method used in our previous study [20] by also measuring the CSA of all transversely cut positive muscle fibres in the present study. All analyses were conducted by the same person, blinded to age group and leg of the sample. All values are expressed relative to the total number of fibres in the section. In a sub-analysis, four consecutive sections from two elderly subjects (both the exercised and the control leg) were additionally analysed for MHCn-positive fibres to determine whether small fibres could be found on consecutive sections. Overview images of the sections were initially used to identify areas of the biopsy that were present on all four consecutive sections. Peripherally positioned (at edges or holes) muscle fibres were not included. In total, 31 MHCn-positive muscle fibres were included across the two subjects and followed through the four consecutive sections (see online Supplementary Figure S2 for images).

2.7. Immunocytochemistry

For the cells cultured to differentiate, the fusion index was determined as reported earlier [25]. Briefly, coverslips were stained with the primary antibodies desmin and myogenin (see Table 2 for details) followed by the secondary antibodies goat anti-rabbit 568 (catalogue #A11036) and goat anti-mouse 488 (catalogue # A11029), and mounted with Prolong-Gold-Antifade containing DAPI (catalogue #P36931, Invitrogen), as described [25]. Fusion index was calculated as the percentage of desmin-positive nuclei within myotubes (containing three or more nuclei) divided by the total number of desmin-positive nuclei.

2.8. Statistics

All figures were prepared in GraphPad Prism (v.7.04, GraphPad Software, Inc., La Jolla, CA, USA) and all statistical analyses were conducted in SigmaPlot (v. 13.0, Systat Software Inc, San Jose, CA, USA), except subject characteristics and gene expression of the female subjects, which were analysed using Microsoft Excel 2016 (Microsoft Corporation, Redmond, Washington). p-Values below 0.05 were considered significant, and trends of $p < 0.1$ are also reported. mRNA data were normalized to RPLP0 and log-transformed before statistical analysis. For the female participants, unpaired t-tests (two-tailed) were performed between young and old for subject characteristics, fibre size, fibre type composition, and mRNA data. Paired t-tests (two-tailed) were conducted for the analysis of the exercise response (exercised leg vs. control leg). The Bonferroni correction was applied (multiplying the *p*-values ×3) to the *t*-test analyses on the mRNA data to correct for multiple testing. For correlation analyses, mRNA data were log-transformed and then subjected to Pearson's correlation. The number of MHCe-, MHCn-, and NCAM-positive fibres, which was not normally distributed, was subjected to the Mann–Whitney Rank Sum Test and Wilcoxon Signed Rank Test to compare differences between young and old subjects, and control versus exercised leg, respectively. For the male participants, data were analysed by one-way repeated measures analysis of variance, using Dunnett's method for multiple comparisons to compare each time point with baseline, where an overall main effect of time was found. The subject characteristics are presented as means with standard deviation and range, while muscle fibre size and composition are shown as individual values. MHCn- and NCAM-positive muscle fibres are presented as median and individual values.

3. Results

3.1. Subject Characteristics

Age, height, weight, BMI, and Leg Extension 1 RM for all subjects included in the analyses are provided in Table 3. The control muscle biopsy from one elderly woman was found to show irregularities (one fascicle filled with unusually large and small muscle fibres positive for NCAM, MHCn, and MHCe), and this subject was therefore excluded from all analyses.

Table 3. Subject characteristics. Average and standard deviations with ranges (superscript). Abbreviations: BMI, body mass index; 1 RM, one-repetition maximum; yr: years, kg: kilogram.

	Young Women			Old Women			Old Men		
	n = 12			*n* = 11			*n* = 25		
Age (yr)	23	± 3	20–28	74	± 3	71–78	70	± 7	64–90
Height (cm)	168	± 7	157–177	166	± 3	162–169	180	± 5	172–189
Weight (kg)	64	± 8	53–75	69	± 10	57–84	82	± 10	67–98
BMI (kg/m^2)	23	± 2	19–26	25	± 4	20–30	26	± 3	21–31
Knee extension 1RM (kg)	39	± 8	30–50	23	± 5	12–28	56	± 14	23–82

3.2. Tissue Immunohistochemistry at Baseline—Young and Elderly Women

On average, the numbers of fibres included in the fibre type and size analysis were 212 (129–352) for type I and 151 (68–247) for type II fibres in the young participants. The corresponding values for the elderly were 169 (85–267) type I and 143 (45–487) type II fibres. The type I fibre percentage was 59 ± 11% (35%–74%) for the young and 58 ± 15% (22%–75%) for the elderly, with no difference between them. As seen in Figure 1, the elderly had significantly smaller type II fibres compared to both their own type I fibres (−38%, *p* < 0.001) and the type II fibres in the young (−36%, *p* < 0.001).

On average, the number of fibres included in the immunohistochemical analysis of denervated fibres was 1080 [401–2270]. MHCe-positive fibres were only found in the excluded subject and are therefore not presented. The elderly had significantly more MHCn- and NCAM-positive fibres compared to the young (Figure 2).

No significant differences between the previously exercised and the control leg were found in either the young or the elderly for MHCn or NCAM (online Supplementary Figure S3). We evaluated the fibre size of all transversely cut MHCn- and NCAM-positive fibres from the control leg. A clear majority of the MHCn- and NCAM-positive muscle fibres were smaller than 150 μm^2 (online Supplementary Figure S3).

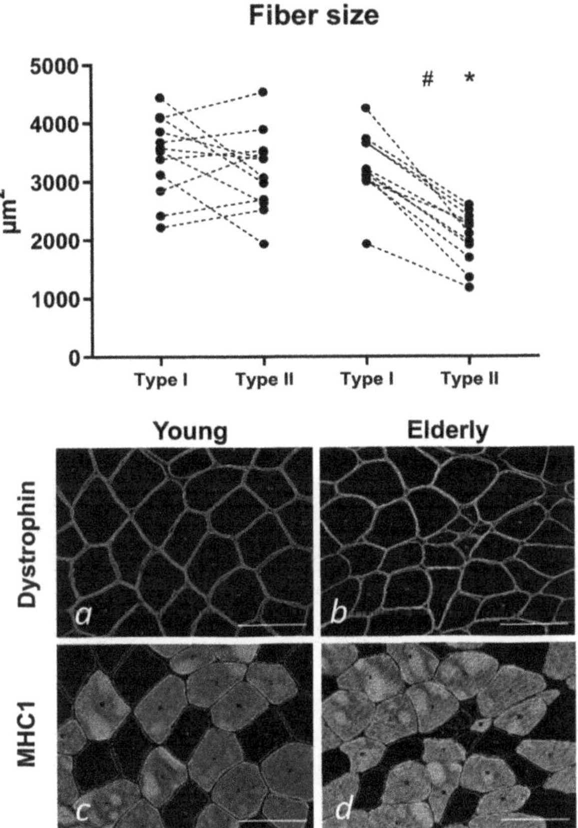

Figure 1. Muscle fibre size analysis in biopsy cross-sections from the control legs of 12 young and 11 elderly women. Individual values are displayed and with type I and II values for an individual connected by a dashed line. The type II fibres of the elderly individuals were significantly smaller than their own type I fibres and the type II fibres of the younger individuals. * $p < 0.001$ vs. young type II, # $p < 0.001$ between fibre types in elderly. Images (**a**–**d**) illustrate the analysis process. (b,d) show representative images of the same area, which has been delineated by the macro in ImageJ. This is an elderly subject with a mean fibre size of 3025 μm^2 and 1688 μm^2 for type I and II fibres, respectively. Similarly, a and c show representative images of the same area in a young subject with a mean fibre size of 3574 μm^2 and 3378 μm^2 for type I and II fibres, respectively. MHC1, myosin heavy chain 1. Scale bars = 100 μm.

One biopsy from the exercised leg of a young subject showed 13 NCAM-positive fibres (1.3% of total fibre count) all located adjacent to a thick band of connective tissue, reminiscent of the myotendinous junction (MTJ). Collagen XXII staining confirmed that this was in fact MTJ, so these fibres were not included in the analysis of this biopsy (see online Supplementary Figure S4 for image). One young subject had 13 (1.45% of total fibre count) NCAM-positive fibres, all of which were located at the edge of the biopsy. This area was not stained by collagen XXII and remained NCAM-positive on additional sections and was therefore not excluded from the analysis. In all other samples MHCn- and NCAM-positive fibres were randomly scattered in between normal muscle fibres.

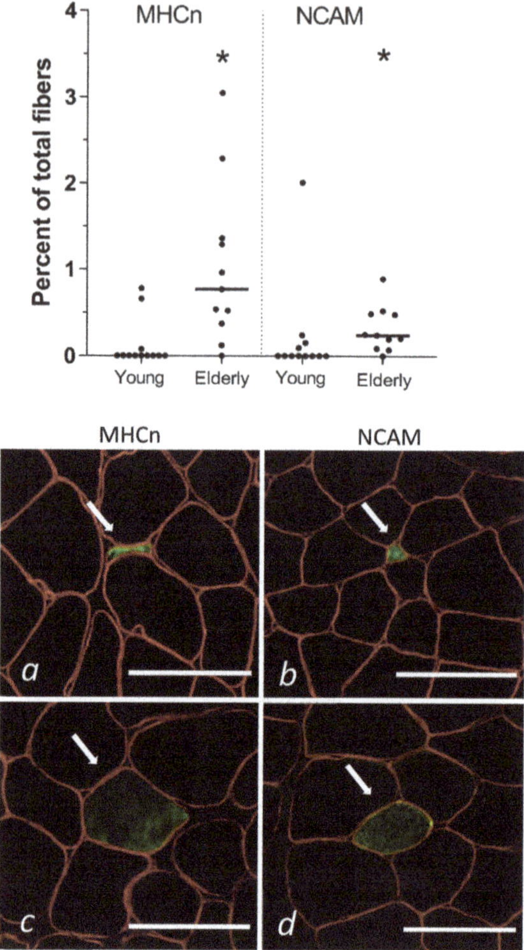

Figure 2. Muscle fibres positive for MHCn or NCAM in biopsy cross-sections from 12 young and 11 elderly women. Only the control leg is shown. Individual values are presented with the median (horizontal line). Panels show examples of small MHCn (**a**) and NCAM (**b**), and large MHCn (**c**) and NCAM (**d**) fibres (arrows). Positive fibres are green, dystrophin, red. * $p < 0.05$ vs. young. MHCn, neonatal myosin heavy chain; NCAM, neural cell adhesion molecule. Scale bars = 100 μm.

3.3. Tissue mRNA at Baseline and in Response to Exercise—Young and Elderly Women

The muscle tissue of the elderly women had significantly lower levels of AChR β1 mRNA compared to the young women, whereas levels of both AChR γ and MHCn mRNA were higher in the elderly compared to the young (Figure 3). Tendencies for differences were seen for gene expression levels of AChR α1 and muscle-specific-kinase (MuSK).

Both the elderly and the young women had a significant upregulation of AChR α1 mRNA in the previously exercised leg compared to the control leg (Figure 3). The exercise response of AChR δ mRNA only reached statistical significance in the elderly. AChR ε mRNA were detected in less than half of the samples at levels very close to detection limit of one molecule and with no preference for any group (data not shown).

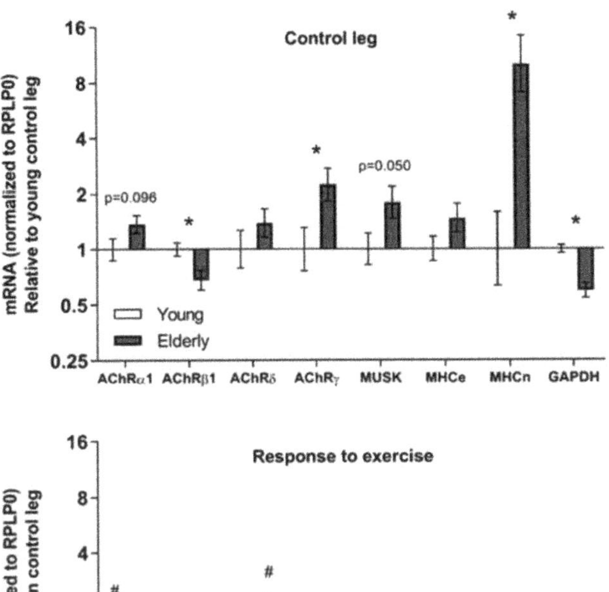

Figure 3. Gene expression in muscle biopsies of healthy young ($n = 12$) and elderly ($n = 11$) women, at rest (control) and five days after a single bout of one-legged exercise. mRNA data were normalized to RPLP0 and are shown as geometric means ± back-transformed SEM, relative to young control legs (control leg) and own control leg (response to exercise). * $p < 0.05$ elderly vs. young. # $p < 0.05$ vs. control leg. Tendencies are written. AChR: acetylcholine receptor; MuSK: muscle-specific-kinase; MHCe: embryonic myosin heavy chain; MHCn, neonatal myosin heavy chain; NCAM, neural cell adhesion molecule; GAPDH: Glyceraldehyde-3-Phosphate Dehydrogenase; RPLP0: Ribosomal Protein Lateral Stalk Subunit P0.

3.4. Cell Culture at Baseline and in Response to Exercise—Young and Elderly Women

The fusion index of the cell cultures from the rested and exercised legs of the elderly women was 36.3 ± 4.2% and 36.1 ± 5.0%, respectively, with the corresponding values for the young group being 52.2 ± 1.8% and 49.8 ± 2.2%, respectively (main effect of age, two-way repeated measures ANOVA).

All gene expression targets were more strongly expressed in differentiating compared to proliferating cells (see online Supplementary Figure S5). In the proliferating condition, the cells from the elderly had lower gene expression levels of MHCe and MHCn compared to young (Figure 4). Similarly, we also found a significantly lower level of MHCn gene expression in the differentiating cells in the control leg in the elderly compared to the young. AChR β1, δ, and γ all showed age-related tendencies.

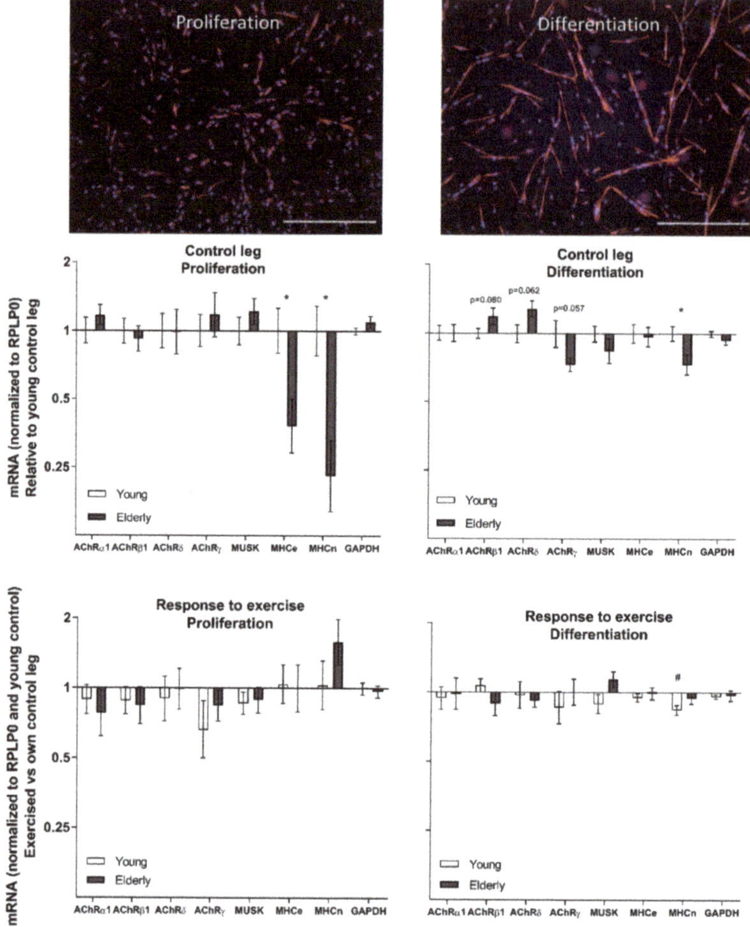

Figure 4. Images display cells in the proliferation condition (Desmin, red, and DAPI, blue) and in the differentiation condition (Desmin, red, Myogenin, green, and DAPI, blue), scale bars = 500 μm. Gene expression in cell cultures from the control and exercised legs of healthy young ($n = 12$) and elderly ($n = 11$) women. mRNA data were normalized to RPLP0 and are shown as geometric means ± back-transformed SEM, relative to young control leg (control leg) and own control leg (response to exercise). * $p < 0.05$ elderly vs. young. # $p < 0.05$ vs. control leg. Tendencies are written. AChR: acetylcholine receptor; MuSK: muscle-specific-kinase; MHCe: embryonic myosin heavy chain; MHCn, neonatal myosin heavy chain; NCAM, neural cell adhesion molecule; GAPDH: Glyceraldehyde-3-Phosphate Dehydrogenase; RPLP0: Ribosomal Protein Lateral Stalk Subunit P0.

Differentiating cells from the previously exercised leg from the young subjects demonstrated a lower gene expression of MHCn versus the control leg (Figure 4).

3.5. Tissue mRNA in Response to Exercise—Elderly Men

In general, gene expression in four out of the five AChR measured demonstrated a response to exercise. AChR α1 mRNA was downregulated 4.5 h and one day after the exercise and returned to baseline in four days (Figure 5). AChR β1 mRNA was downregulated at 1, 4, and 7 days. AChR δ mRNA showed a tendency for a decline 4.5 h after exercise and was upregulated seven days after the exercise. AChR γ mRNA decreased 4.5 h after the exercise bout. No significant exercise-induced changes in gene expression of the AChR ε subunit, MuSK, MHCe, or MHCn were observed.

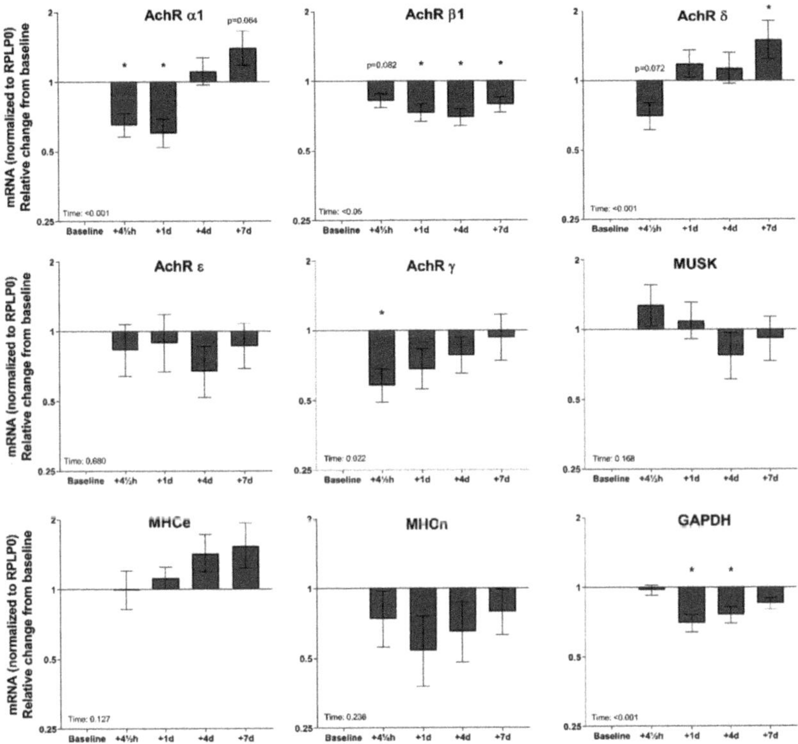

Figure 5. Gene expression in muscle biopsies of 25 healthy elderly men ten days before (baseline) and 4.5 h, one day, four days, and seven days after a single bout of exercise. mRNA data were normalized to RPLP0 and are shown as geometric means ± back-transformed SEM, relative to baseline. * $p < 0.05$ vs. baseline. Tendencies are written. AChR: acetylcholine receptor; MuSK: muscle-specific-kinase; MHCe: embryonic myosin heavy chain; MHCn, neonatal myosin heavy chain; NCAM, neural cell adhesion molecule; GAPDH: Glyceraldehyde-3-Phosphate Dehydrogenase; RPLP0: Ribosomal Protein Lateral Stalk Subunit P0.

4. Discussions

The most notable findings of the present study was that the skeletal muscle of elderly individuals, with morphological signs of ageing as demonstrated by a reduced type II muscle fibre CSA and a heightened number of denervated muscle fibres, has a significantly elevated gene expression level of the denervation-responsive AChR γ subunit and MHCn as compared to young healthy individuals. Our data also suggest an age effect on the capacity of satellite cell-derived myotubes to transcribe AChR genes, which is fundamental for NMJ maintenance. Furthermore, we provide novel insight into the transient changes in gene expression of all five muscle AChR subunits following heavy resistance exercise in healthy elderly human skeletal muscle. Together these data support the role of exercise in stimulating the stability of the NMJ, but also indicate age-related changes, even in healthy elderly individuals.

4.1. Muscle Fibre Denervation in Elderly Humans

Healthy elderly women with clear signs of ageing (lower type II muscle fibre CSA and lower muscle strength), also show a significantly heightened number of denervated muscle fibres compared to young healthy women, as evidenced by a greater proportion of fibres positive for NCAM and MHCn. When a muscle fibre loses its neural input, the plasticity of the peripheral nervous systems allows for adjacent nerve sprouts to attempt to re-innervate denervated muscle fibres through nerve sprouting [31]. It is believed that the increased synthesis of NCAM in denervated muscle fibres facilitates this innervation process [32,33]. Denervated muscle fibres will also revert into an immature myosin heavy chain configuration, as we found more MHCn-positive fibres in old subjects compared to the young. Furthermore, we also observed a substantial 10-fold higher gene expression level of MHCn in the muscle tissue of the elderly compared to the young females, reflecting a persisting synthesis of this distinct myosin isoform. Importantly, it should be noted that there is not a complete overlap between MHCn- and NCAM-positive stained fibres, which suggests that the rate at which these proteins aggregate in the muscle fibres following denervation might differ. In terms of denervated muscle fibre morphology, we observed a persisting MHCn and NCAM protein presence in even the smallest of muscle fibres (<75 μm^2). These miniature fibres are easily missed during regular biopsy assessments and could represent long-term denervated fibres that had atrophied over time [34] and undergone deterioration of muscle proteins [35], but maintained an increased and long-lasting cytoplasmic expression of MHCn [36] and NCAM [32]. The length of these miniature muscle fibres is a matter of uncertainty. We have previously been able to follow such fibres through 400 μm of consecutive biopsy sections in a selected subject [20]. In a sub analysis in the present study, we searched for MHCn-positive fibres through four consecutive sections and found that 13, 32, and 39% of the fibres had disappeared after 1, 2, and 3 sections, respectively. This implies a substantial number of miniature fibres ends, which could indicate that long-term denervated fibres are gradually degraded both transversally and longitudinally.

4.2. Ageing and Exercise Alter Acetylcholine Receptor Gene Expression

One of our most marked findings is that the gene expression of the AChR γ subunit is robustly elevated in the skeletal muscle of elderly compared to young females. This coincides with this subunit being a functionally distinct foetal subunit [37–39], which is increasingly expressed following both denervation [39,40] and neurotransmitter blocking [41]. Interestingly, in our group of male participants the muscle homogenate gene expression of the γ subunit was acutely downregulated after the exercise bout but had already returned to baseline after one day. We were also able to detect this subunit, as well as $\alpha 1$, $\beta 1$, and δ subunits, in both proliferating and differentiating cell cultures that were devoid of neural presence, meaning that satellite cell derived myonuclei can upregulate AChR gene expression without the presence of a nerve. As satellite cells have been shown to be crucial for maintaining the specialized post-synaptic region of the muscle fibre [23], it is interesting that we observed trends for an age effect in three out of the four subunits. It is worth noting that this is the case even with the conservative Bonferroni correction, but given that the age difference was not always in the same direction, it is possible that this is a real effect of age (and not an effect of general cell culture conditions), potentially reflecting an age-related satellite cell dysfunction that could negatively impact the maintenance of the NMJ. However, it should be noted that the cell cultures derived from satellite cells of the young subjects showed a significantly higher fusion index compared to the old subjects [25], indicative of a higher level of myotube maturity. Furthermore, we also observed a positive correlation between cell fusion index and gene expression levels of AChRγ (R = 0.74), MuSK (R = 0.75), and MHCe (R = 0.66) in the old group (Supplementary Figure S6). This would suggest that AChRγ gene expression in aneural cell cultures is increased concordant with myotube maturity and raises the possibility that the molecular differences we observed between the cell cultures from young and elderly muscle are determined by the extent of fusion. However we cannot rule out the opposite, i.e., that the lower gene expression levels contribute to the lower fusion index values.

Generally, our data show age and exercise effects on AChR γ subunit gene expression, in line with its suggested use in evaluating muscle fibre denervation in healthy individuals. In our previous study, we found a negative correlation between age and the AChR γ subunit in a large group of elderly men [20], which initially might seem to contradict our finding in the present study. However, it is important to acknowledge that a denervated muscle fibre is not in a "stable state", meaning that without the neural input the proteins will be degraded and the structure is gradually lost [34,35]. Ultimately, the muscle fibre completely disappears or is only present as a fraction of its former size and as such its contribution to the whole muscle gene expression profile will also decline.

Since our study includes one data set from males and the other from females, it is worth considering similarities in the pattern of the exercise response between the elderly male and female subjects, given that the day five timepoint of the females can be compared with the four- and seven-day timepoints of the male. In this way it seems that the α, δ, and γ subunits follow a similar pattern between the genders, with the first two subunits being upregulated in both male (α only a tendency) and female subjects after seven and five days, respectively, and the γ subunit being unaffected in both male and female subjects at these timepoints. The β subunit is consistently downregulated in the male subjects whereas this subunit is not affected in the elderly women five days after exercise. Whether this represents a true gender difference and what the functional significance might be however is unknown.

This study is to our knowledge the first to outline the gene expression time course for all AChR subunits following acute heavy resistance exercise and the first to analyse the expression of four out of five AChR subunits in both young and elderly individuals at rest and following acute exercise. The NMJ of humans is challenging to study molecularly since it is difficult to obtain actual human NMJs. Hence, we rely on extra-synaptic expression of various genes that are related to the NMJ. With this approach we observed that most subunits were found to be responsive to exercise, which would suggest that despite having reached an advanced age, there is a sustained tissue plasticity in terms of synthesizing new AChRs following an exercise stimulus. The subunit-specific responses also appear to be time-dependent, as some subunits were acutely reduced after exercise followed by a recovery phase, whereas others were downregulated for longer periods. The root of these widely diverging AChR subunit time courses is puzzling and, given evidence from animal studies that long-term exercise increases the size of the NMJ [11], it would be of interest to investigate the potential of lifelong exercise on NMJ adaptations in humans.

5. Conclusions

Taken together, these data support the concept that the loss of neural signal reverts certain muscle fibre proteins to an embryonic configuration (NCAM/MHCn/AChR γ) and that these markers are useful in evaluating the effectiveness of interventions to counteract the denervation-induced loss of muscle fibres in humans. Gene expression levels of the AChR γ subunit in particular repeatedly demonstrated sensitivity to age and exercise. The trends for age-related differences in the gene expression of AChR subunits in myotubes in cell culture were related to myogenic fusion index and potentially suggest a loss of satellite cell function in relation to the capacity to transcribe key molecules for NMJ stability. Finally, it can be speculated that the temporal manner of the AChR subunit gene expression response following exercise represents a beneficial stimulus for muscle mass preservation through strengthening of the NMJ.

Supplementary Materials: The following are available online at http://www.mdpi.com/2073-4409/9/4/893/s1, Figure S1: Gene expression in muscle biopsies of elderly men receiving losartan ($n = 13$) or placebo ($n = 12$). mRNA data were normalized to RPLP0, log-transformed, and are shown as geometric means ± back-transformed SEM, relative to baseline (−10d). Data were analysed with a two-way repeated measures ANOVA (treatment/time). * $p < 0.05$ vs. baseline. Tendencies are written; Figure S2: Panel showing four consecutive biopsy sections of an area with small MHCn-positive muscle fibres (a–d). The dotted squares (b,d) highlight the areas of the inserts (b_{1+2} and d_{1+2}), in which a distinct dystrophin membrane is visible around the small fibre. Note that one of the positive muscle fibres is no longer visible in (c) and (d). MHCn-positive fibres are red, dystrophin, green, and nuclei, blue. MHCn, neonatal myosin heavy chain. Scale bars = 20 µm; Figure S3: Muscle fibre size of MHCn- and NCAM-positive fibres in biopsy cross-sections from the control leg in 12 young and 11 elderly women, pooled and plotted on a logarithmic scale (a). The majority of the positive fibres were <100 μm^2.Muscle fibres positive for MHCn and NCAM for young and elderly women in control and exercised legs (b). No differences between control and exercised leg was observed for any variable. MHCn, neonatal myosin heavy chain; NCAM, neural cell adhesion molecule; Figure S4: Cross section of a muscle biopsy from one subject stained with NCAM (green) and collagen XXII (red). Nuclei are blue. NCAM-positive fibres are found in close proximity of a tendon-like structure and collagen XXII positivity confirms this is a myotendinous junction. These fibres were excluded from the analysis. Scalebar is 500 μm; Figure S5: Differentiating cells relative to proliferating cells in control leg of young women. mRNA data were normalized to RPLP0 and are shown as geometric means ± SEM. * $p < 0.05$ vs. proliferation; Figure S6: Myogenic fusion index correlates with cell culture mRNA levels of AChRγ, MuSK, and MHCe in rested leg of elderly ($n = 10$) but not young ($n = 11$) subjects. All mRNA data were log transformed and analysed with Pearson's correlation, with R and P values displayed.

Author Contributions: A.L.M., P.S., C.S., J.L.A., C.J.L.B., M.F.H., and M.K. contributed to the design of the project, while A.L.M., P.S., C.S., J.L.A., C.J.L.B., M.F.H., and M.K., S.M.J., E.B., A.K. acquired, analysed or interpreted the data of the project. C.S. and A.L.M. drafted the manuscript, and all authors gave intellectual feedback to the draft. All authors approve the final version of the manuscript to be published in *Cells* and are to be held accountable for all aspects of the work in ensuring that questions related to the accuracy or integrity of any part of the work are appropriately investigated and resolved. All authors have read and agreed to the published version of the manuscript.

Funding: Funding is gratefully acknowledged from The Nordea Foundation (Healthy Aging grant), The Danish Agency for Culture (FPK.2018-0036, FPK.2015-0020), The Lundbeck Foundation, The A.P. Møller Foundation for the Advancement of Medical Science, and Bispebjerg Hospital Research Funding.

Acknowledgments: The authors thank Anja Jokipii-Utzon and Camilla Brink Sørensen for excellent technical assistance with preparation of the muscle biopsies and the mRNA analysis. The monoclonal antibodies F1.652 (developmental MHC) and A4.951 (myosin heavy chain, human slow fibres), developed by Blau, H.M., were obtained from the Developmental Studies Hybridoma Bank, created by the NICHD of the NIH, and maintained at The University of Iowa, Department of Biology, Iowa City, IA 52242.

Conflicts of Interest: The authors declare no conflict of interest.

Abbreviations

MHCe	Embryonic myosin heavy chain
MHCn	Neonatal myosin heavy chain
NCAM	Neural cell adhesion molecule
MHC1	Myosin heavy chain 1
DYST	Dystrophin
AChR	Acetylcholine receptor
DAPI	4′,6-Di-amidino-2-phenylindole
TBS	Tris-buffered saline
BSA	Bovine serum albumin
mTOR	Mammalian target of rapamycin

References

1. Lexell, J.; Taylor, C.C.; Sjöström, M. What is the cause of the ageing atrophy? Total number, size and proportion of different fiber types studied in whole vastus lateralis muscle from 15- to 83-year-old men. *J. Neurol. Sci.* **1988**, *84*, 275–294. [CrossRef]

2. Bean, J.F.; Kiely, D.K.; Herman, S.; Leveille, S.G.; Mizer, K.; Frontera, W.R.; Fielding, R.A. The Relationship Between Leg Power and Physical Performance in Mobility-Limited Older People. *J. Am. Geriatr. Soc.* **2002**, *50*, 461–467. [PubMed]

3. Janssen, I.; Heymsfield, S.B.; Ross, R. Low Relative Skeletal Muscle Mass (Sarcopenia) in Older Persons Is Associated with Functional Impairment and Physical Disability. *J. Am. Geriatr. Soc.* **2002**, *50*, 889–896. [CrossRef] [PubMed]

4. Reid, K.F.; Naumova, E.N.; Carabello, R.J.; Phillips, E.M.; Fielding, R.A. Lower extremity muscle mass predicts functional performance in mobility-limited elders. *J. Nutr. Health Aging* **2008**, *12*, 493. [CrossRef] [PubMed]

5. Porter, M.M.; Vandervoort, A.A.; Lexell, J. Aging of human muscle: Structure, function and adaptability. *Scand. J. Med. Sci. Sports* **1995**, *5*, 129–142. [CrossRef]

6. Campbell, M.J.; McComas, A.J.; Petito, F. Physiological changes in ageing muscles. *J. Neurol. Neurosurg. Psychiatry* **1973**, *36*, 174–182. [CrossRef]
7. Tomlinson, B.E.; Irving, D. The numbers of limb motor neurons in the human lumbosacral cord throughout life. *J. Neurol. Sci.* **1977**, *34*, 213–219. [CrossRef]
8. Hepple, R.T.; Rice, C.L. Innervation and neuromuscular control in ageing skeletal muscle. *J. Physiol. (Lond.)* **2016**, *594*, 1965–1978. [CrossRef]
9. Snijders, T.; Leenders, M.; de Groot, L.C.P.G.M.; van Loon, L.J.C.; Verdijk, L.B. Muscle mass and strength gains following 6 months of resistance type exercise training are only partly preserved within one year with autonomous exercise continuation in older adults. *Exp. Gerontol.* **2019**, *121*, 71–78. [CrossRef]
10. Bechshøft, R.L.; Malmgaard-Clausen, N.M.; Gliese, B.; Beyer, N.; Mackey, A.L.; Andersen, J.L.; Kjær, M.; Holm, L. Improved skeletal muscle mass and strength after heavy strength training in very old individuals. *Exp. Gerontol.* **2017**, *92*, 96–105. [CrossRef]
11. Nishimune, H.; Stanford, J.A.; Mori, Y. ROLE of exercise in maintaining the integrity of the neuromuscular junction: Invited Review: Exercise and NMJ. *Muscle Nerve* **2014**, *49*, 315–324. [CrossRef]
12. Baehr, L.M.; West, D.W.D.; Marcotte, G.; Marshall, A.G.; De Sousa, L.G.; Baar, K.; Bodine, S.C. Age-related deficits in skeletal muscle recovery following disuse are associated with neuromuscular junction instability and ER stress, not impaired protein synthesis. *Aging* **2016**, *8*, 127–146. [CrossRef] [PubMed]
13. Hughes, D.C.; Marcotte, G.R.; Marshall, A.G.; West, D.W.D.; Baehr, L.M.; Wallace, M.A.; Saleh, P.M.; Bodine, S.C.; Baar, K. Age-related Differences in Dystrophin: Impact on Force Transfer Proteins, Membrane Integrity, and Neuromuscular Junction Stability. *J. Gerontol. Ser. A Biol. Sci. Med. Sci.* **2016**, *72*, 640–648. [CrossRef] [PubMed]
14. Sonjak, V.; Jacob, K.; Morais, J.A.; Rivera-Zengotita, M.; Spendiff, S.; Spake, C.; Taivassalo, T.; Chevalier, S.; Hepple, R.T. Fidelity of muscle fibre reinnervation modulates ageing muscle impact in elderly women. *J. Physiol.* **2019**, *597*, 5009–5023. [CrossRef] [PubMed]
15. Fambrough, D.M.; Drachman, D.B.; Satyamurti, S. Neuromuscular junction in myasthenia gravis: Decreased acetylcholine receptors. *Science* **1973**, *182*, 293–295. [CrossRef] [PubMed]
16. Merlie, J.P.; Sanes, J.R. Concentration of acetylcholine receptor mRNA in synaptic regions of adult muscle fibres. *Nature* **1985**, *317*, 66–68. [CrossRef] [PubMed]
17. Fambrough, D.M. Control of acetylcholine receptors in skeletal muscle. *Physiol. Rev.* **1979**, *59*, 165–227. [CrossRef] [PubMed]
18. Gundersen, K.; Rabben, I.; Klocke, B.J.; Merlie, J.P. Overexpression of myogenin in muscles of transgenic mice: Interaction with Id-1, negative crossregulation of myogenic factors, and induction of extrasynaptic acetylcholine receptor expression. *Mol. Cell. Biol.* **1995**, *15*, 7127–7134. [CrossRef]
19. Pestronk, A.; Drachman, D.B. Motor Nerve Sprouting and Acetylcholine Receptors. *Science* **1978**, *199*, 1223–1225. [CrossRef]
20. Soendenbroe, C.; Heisterberg, M.F.; Schjerling, P.; Karlsen, A.; Kjaer, M.; Andersen, J.L.; Mackey, A.L. Molecular indicators of denervation in aging human skeletal muscle. *Muscle Nerve* **2019**, *60*, 453–463. [CrossRef]
21. Sanes, J.R.; Lichtman, J.W. Induction, assembly, maturation and maintenance of a postsynaptic apparatus. *Nat. Rev. Neurosci.* **2001**, *2*, 791–805. [CrossRef] [PubMed]
22. Reist, N.E.; Werle, M.J.; McMahan, U.J. Agrin released by motor neurons induces the aggregation of acetylcholine receptors at neuromuscular junctions. *Neuron* **1992**, *8*, 865–868. [CrossRef]
23. Liu, W.; Klose, A.; Forman, S.; Paris, N.D.; Wei-LaPierre, L.; Cortés-Lopéz, M.; Tan, A.; Flaherty, M.; Miura, P.; Dirksen, R.T.; et al. Loss of adult skeletal muscle stem cells drives age-related neuromuscular junction degeneration. *Elife* **2017**, *6*, e26464. [CrossRef] [PubMed]
24. Liu, W.; Wei-LaPierre, L.; Klose, A.; Dirksen, R.T.; Chakkalakal, J.V. Inducible depletion of adult skeletal muscle stem cells impairs the regeneration of neuromuscular junctions. *Elife* **2015**, *4*, e09221. [CrossRef] [PubMed]
25. Bechshøft, C.J.L.; Jensen, S.M.; Schjerling, P.; Andersen, J.L.; Svensson, R.B.; Eriksen, C.S.; Mkumbuzi, N.S.; Kjaer, M.; Mackey, A.L. Age and prior exercise in vivo determine the subsequent in vitro molecular profile of myoblasts and nonmyogenic cells derived from human skeletal muscle. *Am. J. Physiol. Cell Physiol.* **2019**, *316*, C898–C912. [CrossRef] [PubMed]

26. Heisterberg, M.F.; Andersen, J.L.; Schjerling, P.; Bülow, J.; Lauersen, J.B.; Roeber, H.L.; Kjaer, M.; Mackey, A.L. Effect of Losartan on the Acute Response of Human Elderly Skeletal Muscle to Exercise. *Med. Sci. Sports Exerc.* **2018**, *50*, 225–235. [CrossRef]

27. Bergstrom, J. Percutaneous needle biopsy of skeletal muscle in physiological and clinical research. *Scand. J. Clin. Lab. Investig.* **1975**, *35*, 609–616. [CrossRef]

28. Koch, M.; Schulze, J.; Hansen, U.; Ashwodt, T.; Keene, D.R.; Brunken, W.J.; Burgeson, R.E.; Bruckner, P.; Bruckner-Tuderman, L. A novel marker of tissue junctions, collagen XXII. *J. Biol. Chem.* **2004**, *279*, 22514–22521. [CrossRef]

29. Jakobsen, J.R.; Mackey, A.L.; Knudsen, A.B.; Koch, M.; Kjaer, M.; Krogsgaard, M.R. Composition and adaptation of human myotendinous junction and neighboring muscle fibers to heavy resistance training. *Scand. J. Med. Sci. Sports* **2017**, *27*, 1547–1559. [CrossRef]

30. Karlsen, A.; Bechshøft, R.L.; Malmgaard-Clausen, N.M.; Andersen, J.L.; Schjerling, P.; Kjaer, M.; Mackey, A.L. Lack of muscle fibre hypertrophy, myonuclear addition, and satellite cell pool expansion with resistance training in 83-94-year-old men and women. *Acta Physiol. (Oxf.)* **2019**, *227*, e13271. [CrossRef]

31. Brown, M.C.; Holland, R.L.; Hopkins, W.G. Motor Nerve Sprouting. *Annu. Rev. Neurosci.* **1981**, *4*, 17–42. [CrossRef] [PubMed]

32. Covault, J.; Sanes, J.R. Neural cell adhesion molecule (N-CAM) accumulates in denervated and paralyzed skeletal muscles. *Proc. Natl. Acad. Sci. USA* **1985**, *82*, 4544–4548. [CrossRef] [PubMed]

33. Gillon, A.; Sheard, P. Elderly mouse skeletal muscle fibres have a diminished capacity to upregulate NCAM production in response to denervation. *Biogerontology* **2015**, *16*, 811–823. [CrossRef] [PubMed]

34. Viguie, C.A.; Lu, D.-X.; Huang, S.-K.; Rengen, H.; Carlson, B.M. Quantitative study of the effects of long-term denervation on the extensor digitorum longus muscle of the rat. *Anat. Rec.* **1997**, *248*, 346–354. [CrossRef]

35. Gosztonyi, G.; Naschold, U.; Grozdanovic, Z.; Stoltenburg-Didinger, G.; Gossrau, R. Expression of Leu-19 (CD56, N-CAM) and nitric oxide synthase (NOS) I in denervated and reinnervated human skeletal muscle. *Microsc. Res. Tech.* **2001**, *55*, 187–197. [CrossRef] [PubMed]

36. Doppler, K.; Mittelbronn, M.; Bornemann, A. Myogenesis in human denervated muscle biopsies. *Muscle Nerve* **2008**, *37*, 79–83. [CrossRef]

37. Mishina, M.; Takai, T.; Imoto, K.; Noda, M.; Takahashi, T.; Numa, S.; Methfessel, C.; Sakmann, B. Molecular distinction between fetal and adult forms of muscle acetylcholine receptor. *Nature* **1986**, *321*, 406–411. [CrossRef]

38. Gu, Y.; Hall, Z.W. Immunological evidence for a change in subunits of the acetylcholine receptor in developing and denervated rat muscle. *Neuron* **1988**, *1*, 117–125. [CrossRef]

39. Missias, A.C.; Chu, G.C.; Klocke, B.J.; Sanes, J.R.; Merlie, J.P. Maturation of the acetylcholine receptor in skeletal muscle: Regulation of the AChR gamma-to-epsilon switch. *Dev. Biol.* **1996**, *179*, 223–238. [CrossRef]

40. Goldman, D.; Staple, J. Spatial and temporal expression of acetylcholine receptor RNAs in innervated and denervated rat soleus muscle. *Neuron* **1989**, *3*, 219–228. [CrossRef]

41. Witzemann, V.; Brenner, H.R.; Sakmann, B. Neural factors regulate AChR subunit mRNAs at rat neuromuscular synapses. *J. Cell Biol.* **1991**, *114*, 125–141. [CrossRef] [PubMed]

cells

Article

mTORC1 Mediates Lysine-Induced Satellite Cell Activation to Promote Skeletal Muscle Growth

Cheng-long Jin [1], Jin-ling Ye [2], Jinzeng Yang [3], Chun-qi Gao [1], Hui-chao Yan [1], Hai-chang Li [4] and Xiu-qi Wang [1,*]

[1] College of Animal Science, South China Agricultural University/Guangdong Provincial Key Laboratory of Animal Nutrition Control/National Engineering Research Center for Breeding Swine Industry, Guangzhou 510642, China; jinchenglong1992@163.com (C.-l.J.); cqgao@scau.edu.cn (C.-q.G.); yanhc@scau.edu.cn (H.-c.Y.)

[2] Institute of Animal Science, Guangdong Academy of Agricultural Sciences, Guangzhou 510642, China; YEJL2014@163.com

[3] Department of Human Nutrition, Food and Animal Sciences, University of Hawaii, Honolulu, HI 96822, USA; jinzeng@hawaii.edu

[4] Department of Surgery, Davis Heart and Lung Research Institute, The Ohio State University, Columbus, OH 43210, USA; Haichang.Li@osumc.edu

* Correspondence: xqwang@scau.edu.cn; Tel./Fax: +86-20-38295462

Received: 25 September 2019; Accepted: 25 November 2019; Published: 30 November 2019

Abstract: As the first limiting amino acid, lysine (Lys) has been thought to promote muscle fiber hypertrophy by increasing protein synthesis. However, the functions of Lys seem far more complex than that. Despite the fact that satellite cells (SCs) play an important role in skeletal muscle growth, the communication between Lys and SCs remains unclear. In this study, we investigated whether SCs participate directly in Lys-induced skeletal muscle growth and whether the mammalian target of rapamycin complex 1 (mTORC1) pathway was activated both in vivo and in vitro to mediate SC functions in response to Lys supplementation. Subsequently, the skeletal muscle growth of piglets was controlled by dietary Lys supplementation. Isobaric tag for relative and absolute quantitation (iTRAQ) analysis showed activated SCs were required for longissimus dorsi muscle growth, and this effect was accompanied by mTORC1 pathway upregulation. Furthermore, SC proliferation was governed by medium Lys concentrations, and the mTORC1 pathway was significantly enhanced in vitro. After verifying that rapamycin inhibits the mTORC1 pathway and suppresses SC proliferation, we conclude that Lys is not only a molecular building block for protein synthesis but also a signal that activates SCs to manipulate muscle growth via the mTORC1 pathway.

Keywords: lysine; mTORC1; satellite cells; proliferation; skeletal muscle growth

---

1. Introduction

Lysine (Lys) is the first limiting essential amino acid for mammals consuming a predominantly cereal-based diet [1,2]. The important role of Lys in promoting skeletal muscle growth has already been demonstrated in animal husbandry, and this effect was attributed to increased protein synthesis [3,4]. Moreover, the functions of Lys in preventing human illnesses, such as osteoporosis and maldevelopment, have been intensively studied to protect human health [5,6]. In contrast, a low Lys diet has been used to treat glutaric aciduria type I and pyridoxine-dependent epilepsy [7,8], despite the fact that Lys deficiency causes severe body growth restriction and a reduction in body weight [9]. Furthermore, to study the mechanism of Lys in governing skeletal muscle growth, it has been reported that the mammalian target of rapamycin complex 1 (mTORC1) pathway is activated by Lys in the skeletal muscle of rats [10]. Additionally, Lys suppresses protein degradation in C2C12 myotubes via greater mTORC1 pathway phosphorylation [11]. However, protein synthesis is controlled by DNA in the nucleus, such that a higher number of cell nuclei in myofibers means greater protein synthesis efficiency [12].

As muscle stem cells are involved in skeletal muscle growth, satellite cells (SCs) are distributed in the basal lamina and sarcolemma of skeletal muscle fibers [13,14]. It has already been established that through proliferation [14], migration [15] and fusion into myotubes to form new nuclei, SCs contribute considerably to muscle fiber hypertrophy [16]. Moreover, in the study of SCs, the mTORC1 pathway is an invaluable index [17,18]. First, mTORC1 is critical for SC participation in skeletal muscle regeneration [19]. Another study showed that mTORC1 is also necessary for RNA-induced mitochondrial restoration in SC activation [17]. Furthermore, the addition of leucine (Leu) could promote proliferation in rat SCs via increasing mammalian target of rapamycin (mTOR) and ribosomal protein S6 kinase 1 (S6K1) phosphorylation [20]. Alway et al. found that a metabolite of Leu, β-hydroxy-β-methylbutyrate (HMB), promotes SC proliferation but does not activate the mTORC1 pathway [21]. Thus, investigating whether Lys could function as a signal regulatory factor that regulates SC proliferation through the mTORC1 pathway to promote skeletal muscle growth is an important endeavor.

In the current work, we aimed to expand our understanding of the role of Lys in governing skeletal muscle growth. Our research was designed to determine the specific skeletal muscle growth of piglets with dietary Lys supplementation in greater detail than a previous study [22]. Importantly, isobaric tag for relative and absolute quantitation (iTRAQ) analysis of the longissimus dorsi muscle displayed differentially expressed proteins related to SCs and the mTORC1 pathway, indicating the potential communication between Lys, the mTORC1 pathway and SCs in skeletal muscle growth. Then, we investigated the changes in proliferation and protein synthesis by accurately controlling Lys supplementation in medium to demonstrate that SC proliferation relies on mTORC1 pathway activation. Moreover, rapamycin was used to confirm the indispensable role of the mTORC1 pathway in the proliferation of SCs with Lys re-supplementation.

2. Materials and Methods

2.1. Ethics Statement

All animal procedures were performed in accordance with the Guidelines for the Care and Use of Laboratory Animals of South China Agricultural University (Guangzhou, China), and the experiments were approved by the Animal Ethics Committee (SCAU#0158Ethic Committee Approval Number) of South China Agricultural University (Guangzhou, China).

2.2. Animals and Sample Collection

The design for the feeding experiment is shown in Table S1. Briefly, a total of 30 Duroc × Landrace × Large White, male, weaned piglets with similar weights were divided randomly into 2 groups from days 0 to 14: the control group was fed a diet containing 1.31% Lys (n = 12), and the Lys deficiency group was fed a diet containing 0.83% Lys (n = 18). On day 15, six piglets closest to the average weight of each group were selected to determine skeletal muscle growth. Then, the remaining piglets in the Lys deficiency group were divided randomly into two groups from days 15 to 28: the Lys deficiency group was fed a diet containing 0.83% Lys (n = 6), and the Lys rescue group was fed a diet containing 1.31% Lys (n = 6). In addition, the remaining piglets in the control group were fed a diet containing 1.31% Lys between days 15 and 28 (n = 6). On day 29, all piglets were slaughtered, and the weight of their skeletal muscle was measured. Longissimus dorsi muscle samples were collected from all of the piglets at days 15 and 29, flash-frozen with liquid nitrogen and stored at − 80 °C.

2.3. Amino Acid Detection

To determine the content and concentration of amino acids in longissimus dorsi muscle, samples containing 20 mg protein were weighed and hydrolyzed with 6 mol/L hydrochloric acid (HCL) at 110 °C for 22 h. Then, the hydrolyzed liquid was transferred into a 50 mL volumetric flask with ultrapure water. Then, 1 mL of hydrolyzed liquid was dried by distillation and re-dissolved in 0.02 mol/L HCL. Finally, the amino acid composition was analyzed by an amino acid analyzer (Hitachi L-8900, Tokyo, Japan).

2.4. Protein Extraction

Tissue samples (n = 3) were excised and transferred into new tubes containing tissue lysis buffer (1% SDS, 8 mol/L urea) and 1 mmol/L phenylmethanesulfonyl fluoride (PMSF, Sigma-Aldrich, St. Louis, MO, USA). Then, the lysates were homogenized for 4 min using a TissueLyser (CK1000, Thmorgan, Beijing, China) and incubated on ice for 30 min. The lysates were centrifuged at 12,000× *g* and 4 °C for 15 min, and the supernatants were collected. The concentration of proteins was quantified using a micro-bicinchoninic acid assay (BCA) kit (Thermo-Fisher, Waltham, MA, USA) and separated on sodium dodecyl sulfate polyacrylamide gel electrophoresis (SDS-PAGE) gels.

2.5. iTRAQ Proteome Analysis

Proteins were treated with tris-(2-carboxyethyl)-phosphine (TECP, Sigma-Aldrich, St. Louis, MO, USA) and iodoacetamide and digested with trypsin. Then, the peptide mixture was labeled using the 8 plex iTRAQ reagent according to the manufacturer's instructions (Applied Biosystems, Foster City, CA, USA). Because there were eight samples, the peptides were divided into two parts for subsequent detection. For the first peptide group, the control group samples were labeled 115/116, the Lys deficiency group samples were labeled 117, the Lys rescue group samples were labeled 118/119, and the mixture (total of nine samples) was labeled 121. For the second peptide group, the control group samples were labeled 115, the Lys deficiency group samples were labeled 116/118, the Lys rescue group samples were labeled 119, and the mixture (total of nine samples) was labeled 121. Then, equal amounts of peptides from each peptide group were mixed together and vacuum dried.

Then, the peptides were separated by ultra-performance liquid chromatography (UPLC) with a Nano Aquity UPLC system (Waters, Milford, MA, USA) and analyzed in combination with a quadrupole-orbitrap mass spectrometer (Q-Exactive, Thermo-Fisher, Waltham, MA, USA) and an Easy-nLC 1200 (Thermo-Fisher, Waltham, MA, USA) for Nano LC-MS/MS analysis. Finally, the MS/MS data were searched using Protein Discoverer Software 2.1 against the Sus scrofa musculus database (UniProt, https://www.UniProt.org). The false discovery rate (FDR) applied to the control peptide level was defined as lower than 1%. For quantitative analysis, the $0.66 <$ fold change < 1.5 and *p*-value < 0.05 were the threshold values used to identify the differentially expressed proteins.

All identified proteins were annotated and classified by Gene Ontology (GO, http://www.geneontology.org), and the differentially expressed proteins were then analyzed by GOATOOLS 0.6.5 (https://pypi.org/project/goatools/) for the GO enrichment analysis. Data are available via ProteomeXchange with identifier PXD016396.

2.6. Immunohistochemical Analysis

First, the muscle samples were dehydrated with a 20% sucrose solution for 24 h and embedded in Tissue Tek to prepare the cryosections (5 μm, with at least six sections collected from each sample). Then, the tissue slides were incubated with Pax7 (MAB1675, R&D, Minneapolis, MN, USA) and Ki67 (NB500-170, Novus, Miami, FL, USA) at 4 °C overnight. After the slides were washed three times with phosphate-buffered saline (PBS), they were incubated with Alexa Fluor® 488 AffiniPure goat anti-mouse IgG (115-545-003, Jackson, West Grove, PA, USA) and Cy3-AffiniPure Goat anti-rabbit IgG (111-165-045, Jackson, West Grove, PA, USA) at room temperature for 90 min. Next, the slides were washed 3 times with PBS and incubated with 4′,6-diamidino-2-phenylindole (DAPI, Sigma-Aldrich, St. Louis, MO, USA) at room temperature for 5 min. Images were obtained using an immunofluorescence microscope (Ti2-U, Nikon, Tokyo, Japan TYPE, COMPANY, CITY, COUNTRY).

2.7. Western Blotting

Protein was extracted from the longissimus dorsi muscle or SCs with lysis buffer (RIPA, BioTeke, Beijing, China) and PMSF (Sigma-Aldrich, St. Louis, MO, USA). Next, the samples were centrifuged at 12,000× g and 4 °C for 15 min, and the protein concentration was determined using a micro BCA protein assay kit (Thermo-Fisher, Waltham, MA, USA). A total of 10 μg of protein was separated on 8–10% sodium dodecyl sulfate polyacrylamide gel electrophoresis (SDS-PAGE) gels and then transferred onto polyvinylidene fluoride membranes (PVDF, Millipore, Darmstadt, Germany). After blocking, the membranes were incubated with specific primary and second antibodies (Table S2). Immunoreactivity was detected using an electrochemiluminescence (ECL) Plus chemiluminescence detection kit (Millipore, Darmstadt, Germany) and a Fluor Chem M system (Protein Simple, Santa Clara, CA, USA). The band density was analyzed using ImageJ Analysis Software (https://imagej.nih.gov) after excluding the background density (n = 3). The results were confirmed by three independent experiments with three samples per treatment.

2.8. Isolation and Culture of SCs

The method used to isolate, purify and identify the SCs was performed as described previously with modification [23]. In this study, SCs were isolated from the longissimus dorsi muscle of 5-day-old Landrace piglets and cultured in Dulbecco's modified Eagle's Medium/Nutrient Mixture F-12 (DMEM/F-12, Thermo-fisher, Waltham, MA, USA) supplemented with 10% fetal bovine serum (FBS, Thermo-fisher, Waltham, MA, USA) at 37 °C and 5% CO_2. The medium was changed every 48 h.

2.9. Lys Depletion and Supplementation

After a 24 h period to allow adhesion, cells were starved for 6 h in FBS- and Lys-free DMEM/F12 medium. Then, the cells were cultured in 500 μmol/L Lys (control) and 0 μmol/L Lys (Lys deficiency) DMEM/F12 medium with 10% FBS for 24, 48 and 72 h to investigate cell proliferation. For proliferation rescue, due to the extreme decrease in proliferation after Lys deficiency for 48 h, we added sufficient Lys for another 72 h at this point. Lys concentrations in DMEM/F12, FBS and culture medium are displayed in Table S3.

2.10. Cell Proliferation Assay

For the 3-(4,5-dimethylthiazol-2-yl)-2, 5-diphenyltetrazolium bromide (MTT) assay, 20 μL 5 mg/mL MTT solution (Sigma-Aldrich, St. Louis, MO, USA) was added to each well and incubated for 4 h. Then, the plates were centrifuged at 1400× g for 15 min at 25 °C. A total of 150 μL dimethylsulfoxide (DMSO) working solution was added to each well after the supernatants were carefully discarded. The OD value of the product was evaluated using a microplate reader (Bio-Rad, Hercules, CA, USA) at a wavelength of 490 nm (n = 20). For the cell count assay [14,24], SCs were trypsinized and washed with PBS 3 times, and viable cells were counted using a hemocytometer under an automated cell counter (Count Star, Shanghai, China, n = 10).

2.11. Flow Cytometry

SCs were seeded at a density of 5×10^5 cells/well in 6-well culture plates (Corning, Corning, NY, USA) to detect the cell cycle distribution. The cultivation process was carried out as described above and according to a method described previously [25]. After harvesting at 24, 48 and 72 h, the cells were fixed in 70% ice-cold ethanol at −20 °C for cell cycle analysis. Before the samples were analyzed by flow cytometry using a Becton Dickinson Fluorescence Activating Cell Sorter Aria (BD Biosciences,

San Diego, CA, USA), the cells were centrifuged at 200× g and 4 °C for 5 min, re-suspended in 1 mL PBS, treated with 100 μL 200 mg/mL DNase-free RNase, incubated at 37 °C for 30 min, and treated with 100 μL 1 mg/mL propidium iodide (PI, Sigma-Aldrich, St. Louis, MO, USA) at room temperature (25 °C) for 10 min (n = 6).

2.12. Protein Synthesis

To measure protein synthesis, a nonradioactive technique called surface sensing of translation (SUnSET) was used [26,27]. In this study, 1 μg/mL puromycin (Millipore, Waltham, MA, USA) was added to all wells for an additional 30 min of culture, and puromycin was detected by western blotting with an anti-puromycin antibody (Millipore, Waltham, MA, USA, Table S2). The total protein concentration was determined by BCA (Thermo-fisher, Waltham, MA, USA).

2.13. Immunofluorescence Staining

SCs were cultured for 96 h for the proliferation rescue assay and differentiation rescue assay. First, SCs were fixed in 4% paraformaldehyde for 30 min and then permeabilized with 0.1% Triton-X-100 for 10 min. After blocking with 1% bull serum albumin (BSA) and 10% goat serum for 30 min, the SCs were stained with primary antibodies for 90 min and then probed with goat anti-rabbit IgG (Table S2). In addition, the nuclei were labeled with 4′,6-diamidino-2-phenylindole (DAPI, Sigma-Aldrich, St. Louis, MO, USA) for 5 min at room temperature. Images were obtained using immunofluorescence microscopy.

2.14. Rapamycin Inhibition

After Lys deficiency for 48 h, Lys rescue medium was added alone or combined with 20 or 50 nmol/L (nM) rapamycin for another 48 h. After a total of 96 h, cell viability was measured by MTT assay, and cell proliferation was measured by cell count assay. In addition, cell samples were collected to detect protein synthesis and the mTORC1 pathway by western blotting.

2.15. Statistical Analysis

The data were analyzed using Statistical Analysis System software (SAS, Version 9.2; SAS Institute, Cary, NC, USA). For control group and Lys deficiency group comparisons, the results were analyzed by t-test. For control group, Lys deficiency group and Lys rescue group comparisons, the mean data were assessed for significance using Tukey's test. The data are expressed as the mean ± S.E.M. Differences between treatments were considered statistically significant when $p < 0.05$ and extremely significant when $p < 0.01$.

3. Results

3.1. Skeletal Muscle Growth in Piglets Relies on Dietary Lys Supplementation

To determine the effects of dietary Lys supplementation on the skeletal muscle growth of weaned piglets, we developed the experimental design shown in Supplemental Table 1. After the piglets (initial body weight: control = 8.42 ± 0.11 kg versus Lys deficiency = 8.42 ± 0.08 kg) were fed the Lys-restricted diet for 14 d, we found that the growth of the piglets (final body weight: control = 11.91 ± 0.18 kg versus Lys deficiency = 11.33 ± 0.18 kg) was significantly suppressed (Table S4). In detail, compared with those of the control group, the relative weights of the longissimus dorsi muscle, extensor carpi radialis muscle, semimembranosus muscle, total forequarters muscle and total hindquarters muscle were significantly decreased by dietary Lys deficiency for 14 d (Table S4).

After dietary Lys deficiency for 14 d, the piglets received a diet we supplemented to match the level in the control diet for another 14 d. Obviously, the final weight of the piglets in the Lys rescue group was significantly increased compared with that of the piglets in the Lys deficiency group (Table 1). Moreover, compared with those after Lys deficiency for 28 d, the relative weights of the longissimus dorsi muscle, lateral head of triceps of brachii muscle, extensor carpi radialis muscle, biceps femoris muscle, semimembranosus muscle, semitendinosus muscle, cranial tibial muscle, soleus muscle, lateral head of gastrocnemius muscle and total hindquarters muscle were all increased in the Lys rescue group, which was subjected to dietary Lys deficiency for 14 d and re-supplemented for another 14 d (Table 1). Collectively, these findings suggest that skeletal muscle growth in piglets relies on dietary Lys supplementation.

Table 1. Effect of dietary Lys re-supplementation on the skeletal muscle growth of weaned piglets on 28 d (n = 5, %) [1].

Item	Control	Lys Deficiency	Lys Rescue	*p*-Value
Initial Weight (kg)	12.02 ± 0.29	11.32 ± 0.44	11.34 ± 0.35	0.336
Final Weight (kg)	17.95 ± 0.49 [a]	15.99 ± 0.30 [b]	17.44 ± 0.46 [a]	0.018
Longissimus Dorsi Muscle	2.00 ± 0.06 [ab]	1.86 ± 0.07 [b]	2.19 ± 0.06 [a]	0.018
Psoas Major Muscle	0.34 ± 0.01 [a]	0.28 ± 0.01 [b]	0.30 ± 0.01 [b]	0.014
Forequarters muscles				
Infraspinatus Muscle	0.24 ± 0.02	0.22 ± 0.01	0.21 ± 0.01	0.221
Supraspinatus Muscle	0.41 ± 0.02	0.40 ± 0.01	0.42 ± 0.02	0.538
Subclavius Muscle	0.23 ± 0.02	0.25 ± 0.01	0.22 ± 0.01	0.227
Latissimus Dorsi Muscle	0.26 ± 0.03	0.26 ± 0.04	0.27 ± 0.03	0.925
Long Head of Triceps of Brachii Muscle	0.63 ± 0.02	0.63 ± 0.01	0.67 ± 0.02	0.201
Lateral Head of Triceps of Brachii Muscle	0.17 ± 0.01 [a]	0.14 ± 0.004 [b]	0.18 ± 0.01 [a]	0.003
Extensor Carpi Radialis Muscle	0.13 ± 0.003 [a]	0.12 ± 0.002 [b]	0.13 ± 0.003 [a]	0.011
Extensor Muscle of Second- and Third-Digits	0.02 ± 0.001	0.02 ± 0.002	0.02 ± 0.002	0.684
Lateral Digital Extensor Muscle	0.02 ± 0.001	0.02 ± 0.001	0.02 ± 0.002	0.441
Total Forequarters Muscles	2.15 ± 0.07	2.02 ± 0.03	2.15 ± 0.03	0.111
Hindquarters muscles				
Middle Gluteus Medius Muscle	0.54 ± 0.02	0.48 ± 0.02	0.55 ± 0.03	0.102
Superficial Gluteal Muscle	0.17 ± 0.01	0.14 ± 0.01	0.15 ± 0.02	0.457
Biceps Femoris Muscle	1.15 ± 0.01 [a]	1.04 ± 0.04 [b]	1.21 ± 0.02 [a]	0.008
Semimembranosus Muscle	1.36 ± 0.06 [ab]	1.23 ± 0.05 [b]	1.46 ± 0.03 [a]	0.013
Semitendinosus Muscle	0.39 ± 0.02 [ab]	0.36 ± 0.01 [b]	0.41 ± 0.01 [a]	0.043
Tensor fascia Lata Muscle	0.18 ± 0.02	0.17 ± 0.01	0.20 ± 0.01	0.313
Cranial Tibial Muscle	0.04 ± 0.003 [a]	0.03 ± 0.001 [b]	0.04 ± 0.002 [a]	0.045
Long Peroneal Muscle	0.04 ± 0.003 [a]	0.03 ± 0.001 [b]	0.03 ± 0.002 [b]	0.035
Peroneus Tertius Muscle	0.08 ± 0.003	0.08 ± 0.002	0.08 ± 0.002	0.191
Gemelli Muscle	0.27 ± 0.01 [a]	0.23 ± 0.01 [b]	0.25 ± 0.01 [ab]	0.018
Soleus Muscle	0.23 ± 0.005 [a]	0.19 ± 0.01 [b]	0.23 ± 0.01 [a]	0.020
Lateral Head of Gastrocnemius Muscle	0.37 ± 0.005 [a]	0.31 ± 0.01 [b]	0.36 ± 0.02 [a]	0.027
Adductor Muscle	0.18 ± 0.02	0.19 ± 0.01	0.21 ± 0.01	0.301
Total Hindquarters Muscles	4.98 ± 0.12 [a]	4.49 ± 0.12 [b]	5.17 ± 0.11 [a]	0.004

[1] Values without the same small letters within the same line indicate a significant difference ($p < 0.05$).

3.2. Lys-induced Skeletal Muscle Growth in Relation to SC Activation Level

In addition to the great change in longissimus dorsi muscle mass with dietary Lys supplementation (Tables S1 and S4), we also found that the Lys concentration in the longissimus dorsi muscle was significantly reduced by dietary Lys deficiency for 14 d (Table S5), whereas it could be rescued by dietary Lys re-supplementation for an additional 14 d (Table 2). Furthermore, the concentrations of threonine (Thr), serine (Ser), glutamate (Glu) and arginine (Arg) showed the same changes as Lys (Tables S2 and S5).

Considering these findings, we collected muscle samples for iTRAQ analysis to learn more about the role of Lys in governing skeletal muscle growth (Figure 1). After GO enrichment analysis, we found that proteins related to the transition between slow and fast fibers, filamin binding, mitogen-activated protein kinase binding, cytoskeletal protein binding, actin cytoskeleton, microtubule binding, cytoskeleton organization, tubulin binding and muscle myosin complexes were enriched in the Lys deficiency versus control downregulated proteins (Figure 1A), indicating that the muscle structure was changed.

Table 2. Effect of dietary Lys re-supplementation on the concentrations of amino acids in the longissimus dorsi muscle on day 28 (freeze-dried basis, %) [1].

Item	Control	Lys Deficiency	Lys Rescue	p-Value
Aspartate	0.11 ± 0.003 [ab]	0.10 ± 0.006 [b]	0.12 ± 0.003 [a]	0.038
Threonine	0.06 ± 0.003 [a]	0.05 ± 0.003 [b]	0.06 ± 0.002 [a]	0.015
Serine	0.05 ± 0.000 [a]	0.04 ± 0.003 [b]	0.05 ± 0.002 [a]	0.002
Glutamate	0.21 ± 0.003 [a]	0.18 ± 0.012 [b]	0.21 ± 0.006 [a]	0.024
Glycine	0.07 ± 0.003	0.06 ± 0.000	0.07 ± 0.006	0.423
Alanine	0.07 ± 0.003	0.07 ± 0.006	0.08 ± 0.003	0.671
Valine	0.06 ± 0.003 [ab]	0.06 ± 0.003 [b]	0.07 ± 0.002 [a]	0.017
Isoleucine	0.07 ± 0.003 [ab]	0.06 ± 0.003 [b]	0.07 ± 0.002 [a]	0.025
Leucine	0.12 ± 0.003 [ab]	0.11 ± 0.009 [b]	0.13 ± 0.004 [a]	0.059
Tyrosine	0.05 ± 0.000	0.05 ± 0.003	0.05 ± 0.003	0.354
Phenylalanine	0.07 ± 0.000 [ab]	0.06 ± 0.006 [b]	0.07 ± 0.003 [a]	0.103
Lysine	0.11 ± 0.003 [a]	0.09 ± 0.006 [b]	0.11 ± 0.003 [a]	0.006
Histidine	0.05 ± 0.003	0.04 ± 0.003	0.05 ± 0.003	0.547
Arginine	0.09 ± 0.000 [a]	0.08 ± 0.006 [b]	0.10 ± 0.002 [a]	0.007

[1] Values without the same small letters within the same line indicate a significant difference ($p < 0.05$).

In addition, in the Lys deficiency versus control upregulated proteins, the enrichment of lipase inhibitor activity, negative regulation of homotypic cell-cell adhesion, negative regulation of cell activation, response to nutrients and negative regulation of wound healing were observed (Figure 1B), and these results suggested that the functions of the muscle cells were disturbed.

Furthermore, proteins related to protein kinase C inhibitor activity, protein Ser/Thr kinase inhibitor activity and cellular response to amino acid starvation were enriched in the Lys rescue versus control downregulated proteins (Figure 1C). Thus, the mTORC1 pathway may play a crucial role in Lys-controlled skeletal muscle growth.

More importantly, the results for the Lys rescue versus control upregulated proteins showed that the proteins related to the negative regulation of fat cell differentiation, regulation of fibroblast proliferation, regulation of lens fiber cell differentiation, positive regulation of cell proliferation, sarcolemma, basal plasma membrane, basolateral plasma membrane and cell-cell adhesion were enriched (Figure 1D). These data suggest that SCs may be activated.

Moreover, the results for the Lys rescue versus Lys deficiency downregulated proteins showed that the positive regulation of fibril organization, regulation of fibril organization and positive regulation of gap junction assembly were enriched (Figure 1E). The results for the Lys rescue versus Lys deficiency upregulated proteins showed that the muscle system process, negative regulation of fat cell differentiation, muscle hypertrophy, positive regulation of myoblast differentiation, regulation of myoblast differentiation, regulation of fibroblast proliferation, regulation of cell differentiation, positive regulation of muscle cell differentiation, regulation of muscle cell differentiation and regulation of developmental process were enriched (Figure 1F). In summary, these results indicate that skeletal muscle growth regulation by dietary Lys supplementation is possibly connected with the mTORC1 pathway and SCs.

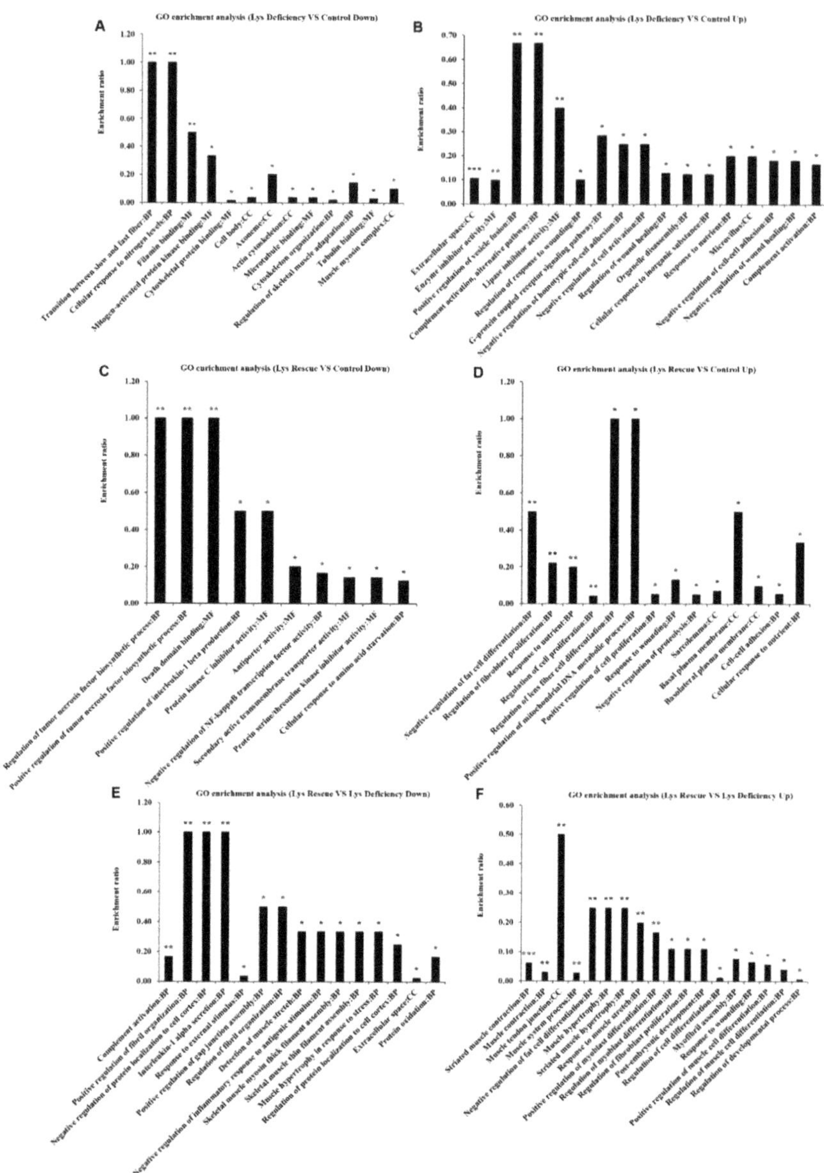

Figure 1. GO enrichment analysis of differentially expressed proteins in the longissimus dorsi muscle on day 28 according to iTRAQ analysis. (**A**) GO enrichment analysis results (Lys deficiency versus control) for downregulated proteins. (**B**) GO enrichment analysis results (Lys deficiency versus control) for upregulated proteins. (**C**) GO enrichment analysis results (Lys rescue versus control) for downregulated proteins. (**D**) GO enrichment analysis results (Lys rescue versus control) for upregulated proteins. (**E**) GO enrichment analysis results (Lys rescue versus Lys deficiency) for downregulated proteins. (**F**) GO enrichment analysis results (Lys rescue versus Lys deficiency) for upregulated proteins. The *x*-axis represents the different GO terms, and the *y*-axis represents the enrichment ratio (the ratio between the protein number enriched in the GO term and the protein number annotated to the GO term; the greater the ratio was, the greater the enrichment found was), * $p < 0.05$, ** $p < 0.01$, *** $p < 0.001$.

3.3. SCs and mTORC1 Activity Are Enhanced in Lys-induced Skeletal Muscle Growth

The specific marker Pax7 and the proliferation marker Ki67 were detected in SCs in the longissimus dorsi muscle on days 14 and 28 by immunohistochemical analysis (Figures S1A and S2A). Based on the total cells (stained with DAPI), dietary Lys deficiency for 14 d or 28 d significantly decreased the numbers of Pax7-positive cells and Pax7 + Ki67-positive cells compared with the control cells (Figures S1B and S2B). Interestingly, compared with those in the Lys deficiency group, the numbers of Pax7-positive cells, Ki67-positive cells and Pax7 + Ki67-positive cells were all increased in the Lys rescue group and were even higher than those in the control group (Figure 2B). In addition, compared with the control group and Lys rescue group, the Lys deficiency group had a decreased ratio of Pax7 + Ki67-positive cells to Pax7-positive cells, regardless of deficiency for 14 d or 28 d (Figures 1C and 2C). Taken together, these observations suggest that the status of SCs in terms of proliferation in the longissimus dorsi muscle is regulated by dietary Lys supplementation.

In addition, we observed that the key proteins in the mTORC1 pathway, such as p-mTOR (Ser2448), p-S6K1 (Thr389), p-S6 (Ser235), p-4EBP1 (Thr470) and eIF4E ($p = 0.083$), were all inhibited by dietary Lys deficiency for 14 d (Figure 1D,E), and this reduction was observed for samples from the group fed Lys-deficient diets for 28 d (Figure 2D,E). Fortunately, when dietary Lys deficiency was re-established to the control level at 14 d and sustained for another 14 d, the restricted key protein levels of the mTORC1 pathway were all increased (Figure 2D,E). In general, these data indicate that SCs, along with the mTORC1 pathway, are required for Lys-induced skeletal muscle growth.

3.4. Cell Proliferation Was Rescued with Increased mTORC1 by Lys Re-supplementation.

To gain further insight into the role of Lys in SCs, Lys supplementation treatment was designed as shown in Figure 3A. In addition, the Lys concentrations in cell culture are shown in Table S3. The MTT assay results showed that cell viability was significantly reduced under Lys deficiency for 48 h (Figures S2A and S3B), whereas cell viability was rescued by Lys re-supplementation for 48 h compared with Lys deficiency for 96 h (Figure 3B). The number of cells detected by an automated cell counter showed that SC proliferation was significantly decreased under Lys restriction conditions for 24 to 120 h (Figures S2B and S3C). In contrast, the number of SCs was increased after Lys was re-supplemented for 24 h after Lys deficiency (Figure 3C). However, cell viability and the number of SCs continued to increase at a slow rate in the deficiency group. This phenomenon might be explained by the existence of Lys in FBS (Figure S3). Because proliferation is determined by mitosis and because cell cycle distribution is a typically evaluated endpoint [25], we further investigated cell cycle distribution using flow cytometry after Lys deficiency for 48 h (Figure 2C–F). Compared with the control conditions, Lys deficiency resulted in an increased percentage of G1 cells at 48 h and 72 h, while the number of cells in the S phase was decreased (Figure 2E,F).

Moreover, because Lys plays an important role in protein synthesis, the SUnSET assay [26,27] was used to measure changes in protein synthesis in SCs. After Lys deficiency for 48 h, SCs evaluated by puromycin analysis showed an obvious decrease in protein synthesis (Figure S4A). However, protein synthesis was obviously increased after Lys was sufficiently supplemented for another 48 h (Figure 3D). Coomassie blue staining was used to verify equal protein loading (Figures 3E and 4B) [27]. Furthermore, a BCA assay was used to confirm the protein synthesis rate [28], and the results showed that total protein lysis in SCs cultured in Lys-deficient medium was extremely restricted at 48 h (Figure S4C) and was then enhanced after 48 h of Lys supplementation (Figure 3F).

Importantly, to investigate whether Lys-stimulated cell proliferation and protein synthesis were mediated by the mTORC1 pathway, the related proteins were analyzed by western blotting (Figures S4D,E and S3G,H). Compared with the levels in the control group, the levels of phosphorylated mTOR (Ser2448) and its downstream targets, phosphorylated S6K1 (Thr389), phosphorylated ribosomal protein S6 (S6, Ser235) and phosphorylated 4EBP1 (Thr70), and the levels of eukaryotic translation initiation factor 4E (eIF4E) were significantly decreased in the Lys deficiency group at 48 h and 96 h (Supplemental Figures 4E and 3H). However, compared with those under Lys deficiency for

96 h, the phosphorylated protein levels of mTOR and its downstream targets, such as 4EBP1 and S6K1, were restored by Lys supplementation for 48 h after Lys deficiency for 48 h (Figure 3H). Moreover, immunofluorescence staining further demonstrated that p-mTOR (Ser2448) expression was also increased after Lys supplementation (Figure S5). These data indicate that Lys-dependent SC proliferation and protein synthesis are related to mTORC1 pathways.

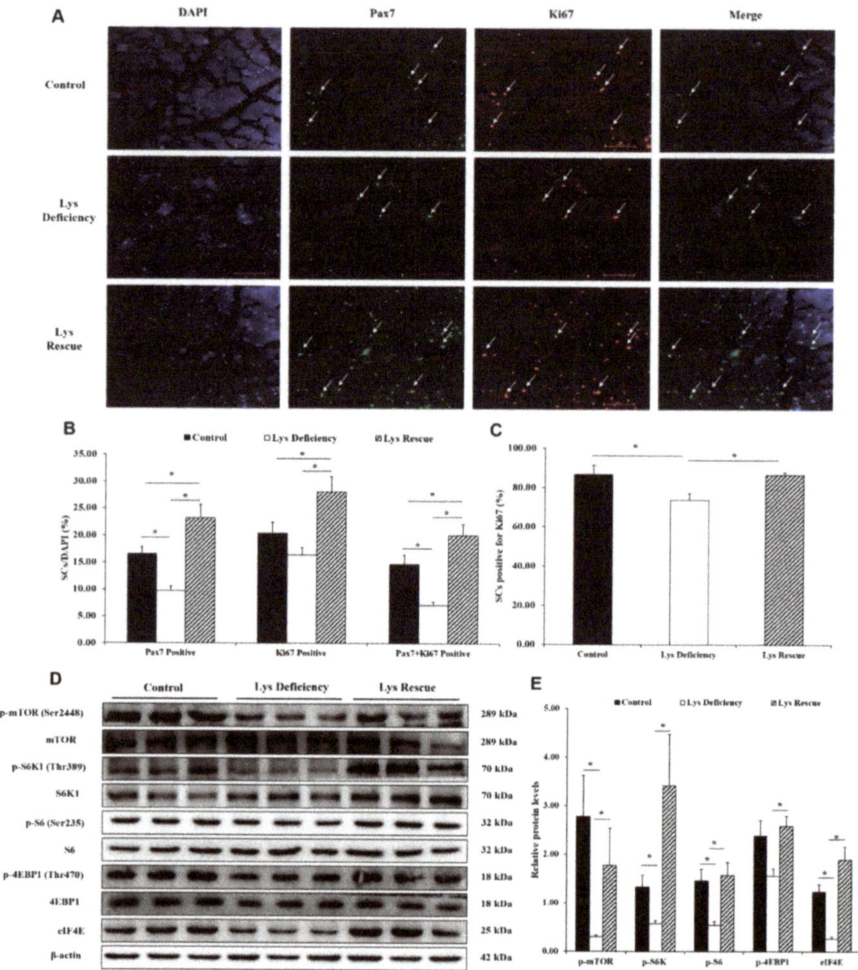

Figure 2. Activation of SCs and protein level of the mTORC1 pathway in the longissimus dorsi muscle on day 28. (**A**) Ki67 (red) and Pax7 (green) staining represents activated SCs during the proliferation period. Bar: 200×. (**B**) Percentage of cells positively stained for Ki67 (red), Pax7 (green) and Ki67 (red) + Pax7 (green) of the total cells (blue, DAPI). (**C**) Percentage of SCs positively stained for Ki67 (red) + Pax7 (green) to Pax7 (green). (**D**) Representative images of key proteins in the mTORC1 pathway detected by western blotting. (**E**) The values represent the ratio of the protein levels of p-mTOR (Ser2448), p-S6K1 (Thr389), p-S6 (Ser235) and p-4EBP1 (Thr470) to the total protein levels and the protein level of eIF4E to that of β-actin, n = 3. The results are shown as the means ± S.E.M. of three independent preparations. Statistical significance was assessed by ANOVA with Tukey's test, * $p < 0.05$.

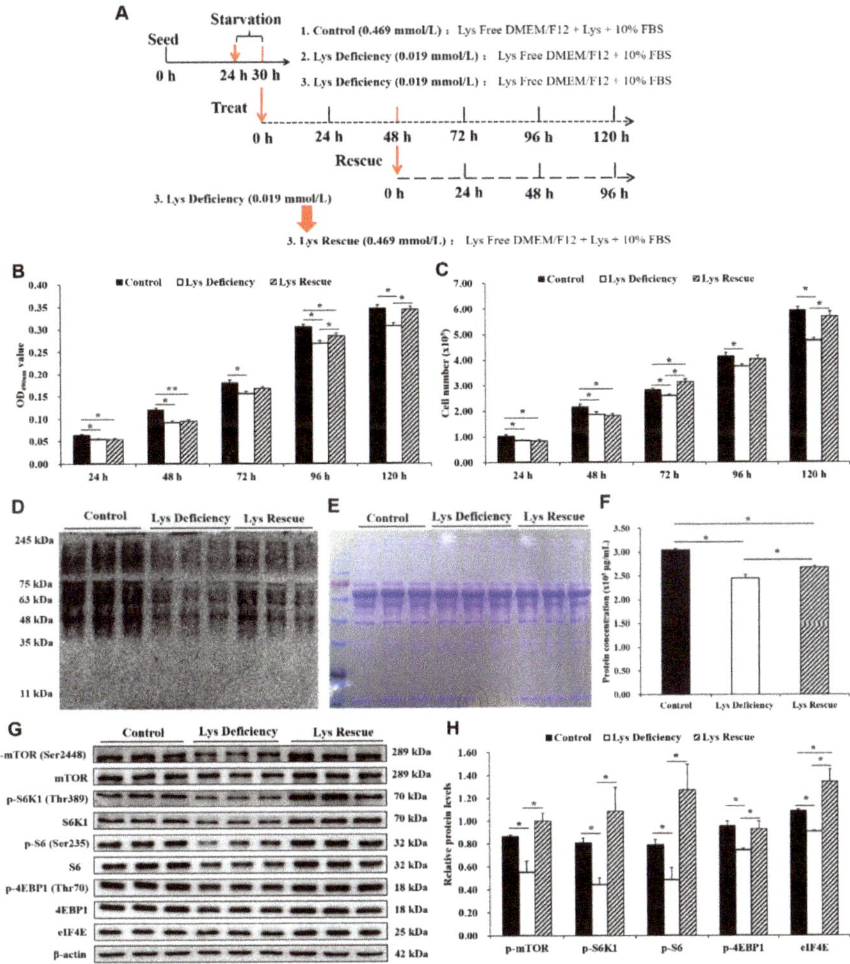

Figure 3. Changes in cell viability, proliferation, protein synthesis and mTORC1 pathway activation after Lys supplementation to sufficient levels. (**A**) Lys supplementation was changed from deficient to sufficient at 48 h, and the cells cultured for another 72 h. (**B**) MTT assays were used to measure cell viability, n = 20. (**C**) Cell proliferation was measured by cell counting assays, n = 10. (**D**) Representative image of the western blotting analyses for puromycin at 96 h, n = 3. (**E**) Coomassie blue staining was used to verify equal protein loading for puromycin measurements at 96 h, n = 3. (**F**) Total protein quantitation using bicinchoninic acid assays at 96 h, n = 3. (**G**) mTORC1 pathway-related proteins were measured by western blotting after Lys supplementation from deficiency for another 48 h (total 96 h). (**H**) The values represent the ratio of the phosphorylated protein levels to the total protein or β-actin level, n = 3. The bars are the means ± S.E.M. from the representative results of three independent experiments. Statistical significance was assessed by ANOVA and Tukey's test, * $p < 0.05$, ** $p < 0.01$.

3.5. Lys Rescue of SC Proliferation and mTORC1 Pathway Activation Were Inhibited by Rapamycin

To validate that mTORC1 pathway activity was crucial for Lys-regulated SC functions, SCs were treated with rapamycin under Lys rescue conditions for 48 h. As shown in Figure 4A, we found that the rescued cell viability by Lys re-supplementation was inhibited by the simultaneous addition of rapamycin and reduced to the Lys deficiency level. Moreover, Lys re-supplementation with rapamycin suppressed SC proliferation, and 50 nM rapamycin showed greater inhibition than 20 nM (Figure 4B). Furthermore, the SUnSET assay showed that 20 and 50 nM rapamycin restricted protein synthesis, which was rescued by Lys re-supplementation (Figure 4C). Protein amounts were also verified by Coomassie blue staining, and the results showed equal sample loading (Figure 4D). Apart from that, compared with the control group, all four other groups showed reductions in total protein concentrations (Figure 4E). Importantly, compared with Lys deficiency, Lys re-supplementation increased the total protein concentrations, whereas the increased protein concentrations were decreased by 20 ($p = 0.055$) and 50 nM ($p < 0.05$) rapamycin supplementation (Figure 4E). In addition, compared with control and Lys rescue conditions, Lys re-supplemented with rapamycin significantly decreased the protein levels of phosphorylated mTOR (Ser2448), phosphorylated S6K1 (Thr389), phosphorylated S6 (Ser235) and eIF4E (Figure 4F, G). Although there was no significant difference, the protein level of phosphorylated 4EBP1 (Thr70) was also decreased by 33.18% and 35.17% in the 20 and 50 nM rapamycin groups, respectively, compared to the Lys rescue group. Collectively, these results indicate that Lys-regulated SC functions are mediated by the mTORC1 pathway.

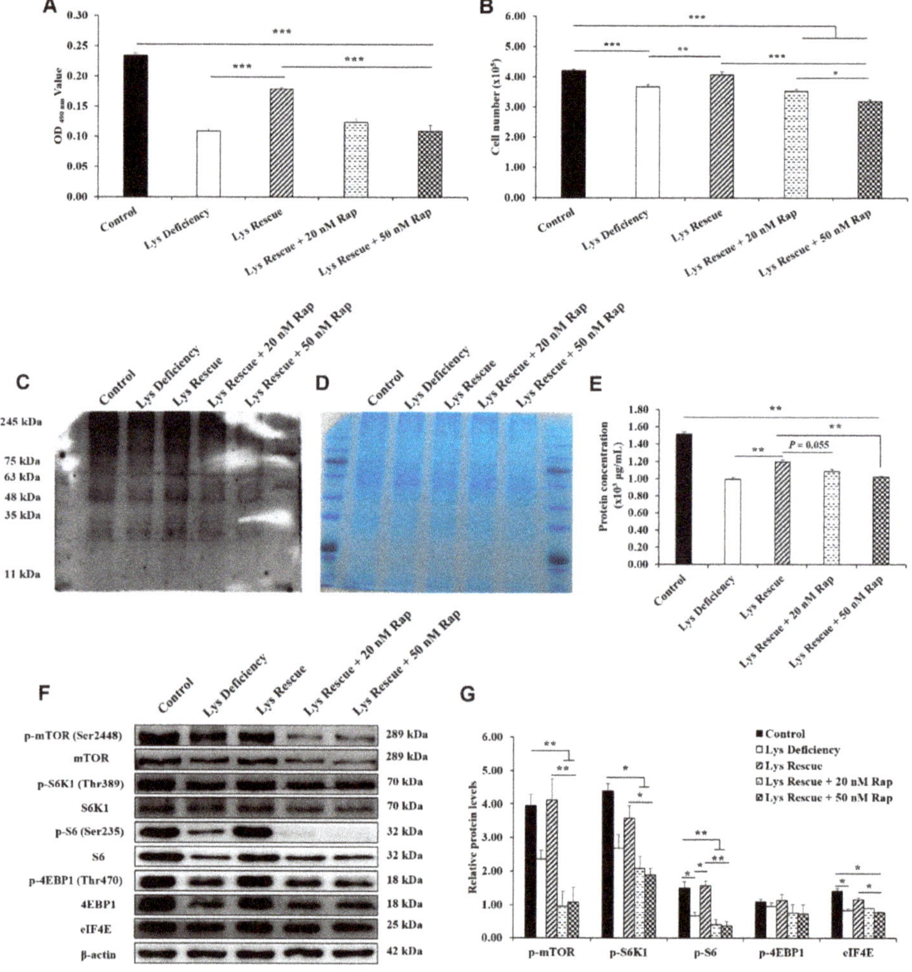

Figure 4. The increased cell viability, proliferation and protein synthesis induced by Lys rescue were inhibited by rapamycin, along with mTORC1 pathway downregulation. After Lys deficiency for

48 h, Lys was added to the medium alone or in combination with 20 or 50 nM rapamycin for another 48 h. (**A**) Cell viability was measured by MTT assays at 96 h, n = 20. (**B**) Cell counting assays were used to measure cell proliferation at 96 h, n = 10. (**C**) SUnSET measurements of protein synthesis were performed by incubating SCs in medium containing puromycin at 96 h. A representative image from the western blotting analyses for puromycin is shown, n = 3. (**D**) Coomassie blue staining was used to verify equal protein loading for puromycin measurements at 96 h, n = 3. (**E**) Total protein quantitation using bicinchoninic acid assays at 96 h is shown, n = 3. (**F**) Western blotting was used to detect the key proteins in the mTORC1 pathway at 96 h. (**G**) The values represent the ratio of the phosphorylated protein levels to the total protein or β-actin level, n = 3. The bars are the means $\pm$ S.E.M. from the representative results of three independent experiments. Statistical significance was assessed by ANOVA and Tukey's test, * $p < 0.05$, ** $p < 0.01$. *** $p < 0.001$.

4. Discussion

Lys is well known to be one of the most important essential amino acids for body growth [1,3,10,22,29]. However, the mechanisms through which Lys governs muscle mass are still debated. In addition, traditionally recognized muscle mass maintenance has been expanded from protein turnover to cell turnover [12,16]. Thus, the balance between myonuclear accretion and reduction is also an important factor in determining muscle mass, and the function of SC fusion into myotubes is important to study [16,30].

Early studies suggested that changes in whole-body weight were the main responses to dietary Lys supplementation or restriction [3,22,31]. Few studies have focused specifically on skeletal muscle growth. In our study, we found that the growth of almost all separated skeletal muscle was restricted during dietary Lys deficiency, while compensatory growth was shown after Lys supplementation was changed from deficient to sufficient [3,32]. In addition, some weights of specific skeletal muscles were unchanged, which could be caused by differences in myofiber type composition.

In previous studies, skeletal muscle mass accumulation was attributed to the relative efficiency between muscle protein synthesis and degradation [3,4,33]. Nevertheless, our GO enrichment analysis results obtained from the iTRAQ analysis showed that there were great changes in skeletal muscle structure and muscle cell function. Furthermore, SCs were found to participate in Lys-induced skeletal muscle growth. This is consistent with what was mentioned in a previous study, which did not provide compelling evidence [34].

To confirm the implications of the GO enrichment analysis for the longissimus dorsi muscle, we found that the SC proliferation ratios (indicated by Pax7 and Ki67 [35–37]) were accurately controlled by dietary Lys supplementation. Consistent with the compensatory growth of skeletal muscle, proliferation also showed the same tendency, especially Pax7 + Ki67-positive SCs. Taking these results together, we believe that SC turnover is required for Lys-induced skeletal muscle growth.

In the case of cell turnover, proliferation is the typical process by which cell numbers are increased by mitosis [14,24]. In this study, the suppressed Lys supply led to reduced cell numbers via changed cell cycle distribution such that there was an increased percentage of G1 phase cells and a decreased percentage of S phase cells. Similarly, a reduction in SC proliferation has also been detected in methionine (Met)- and cysteine (Cys)-restricted cell culture medium [38]. Furthermore, protein synthesis was suppressed directly by Lys deficiency and may thus ultimately cause muscle mass loss [31,32].

In addition to what we found under Lys deficiency conditions, the decrease in cell proliferation was suppressed by changing Lys supplementation from deficiency to sufficiency. These effects may be attributed to the function of Lys as an activator of cell mitotic activity [34]. Furthermore, protein synthesis was rescued by supplementing Lys to Lys-deficient cells. Despite the rescue growth effects of Lys supplementation found in our research, glycine seemed to suppress protein degradation weakly in cells in Lys-deficient medium [11]. These results could explain why Lys seems to play some special functions in life maintenance that have not been previously illustrated, that is, Lys is not used for only protein synthesis [29].

To obtain further insight from our findings, the mTORC1 pathway was measured to investigate the regulatory mechanisms of the Lys relationship with SCs and skeletal muscle growth. Notably, studies have shown that the mTORC1 pathway can be activated by energy, growth factors or nutrients, especially amino acids [38,39]. However, the function of Lys is not as defined as it is for Leu or Arg, as diets supplemented with Leu [40] or Arg [41] promote muscle growth and increase mTORC1 pathway activation. In fact, there has been little research on Lys interactions with mTORC1 in promoting muscle growth. In contrast, the key protein levels in the mTORC1 pathway were not altered after oral Lys administration in rats [33]. More importantly, previous studies showed that the mTORC1 pathway was activated by Lys to suppress protein degradation in vivo and in vitro [10,11]. In the present study, we found that the mTORC1 pathway was inhibited by reducing dietary Lys, and SC proliferation was likewise inhibited in Lys-deficient medium. Furthermore, the related proteins in the mTORC1 pathway were reactivated with complete Lys supplementation. In addition, the indispensable role of the mTORC1 pathway in Lys-governed SC function was verified by rapamycin [20]. We found that the inhibited proliferation and protein synthesis in the Lys deficiency group could be rescued by Lys re-supplementation, whereas these increases were suppressed by the simultaneous addition of rapamycin. Similar to our study, previous studies demonstrated that mTORC1 plays a crucial role in SC function [14,17–19]. Therefore, the mTORC1 pathway is necessary for Lys-induced SC activation in vivo and in vitro. However, the molecular mechanisms, such as extracellular Lys sensing in SCs and intracellular mTORC1 activation, need to be further studied.

5. Conclusions

In conclusion (Figure 5), our findings demonstrate that Lys supplementation exerts compensatory growth effects and that the functions of Lys in muscle mass accumulation are mediated by SCs and the mTORC1 pathway. Thus, Lys is not only a molecular building block for protein synthesis but also a signal that activates SCs to regulate muscle growth via the mTORC1 signaling pathway. These findings can provide us with a new target and therapeutic strategy for skeletal muscle regeneration and disease.

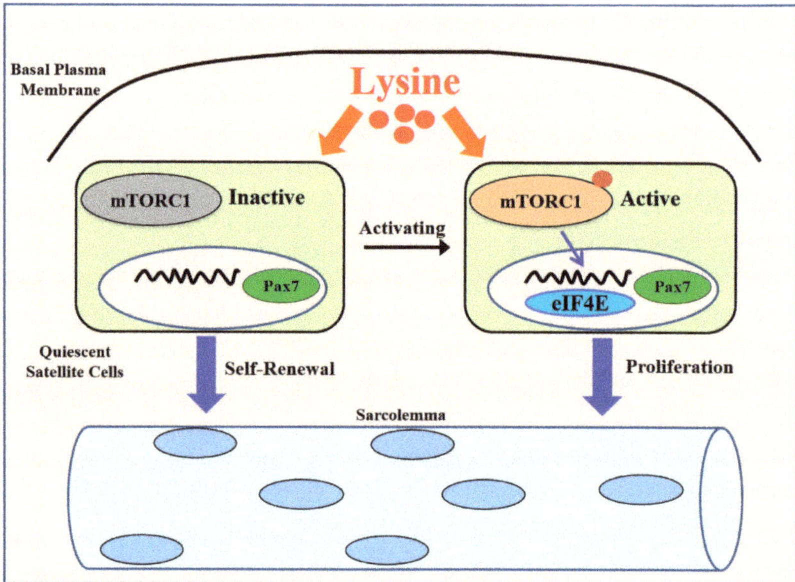

Figure 5. mTORC1 mediates Lys-induced SC activation to promote skeletal muscle growth. Lys supplementation activates the mTORC1 pathway to increase SC proliferation to enhance myogenic potential to promote skeletal muscle growth.

Supplementary Materials: The supplementary materials are available online at http://www.mdpi.com/2073-4409/8/12/1549/s1.

Author Contributions: Data curation, C.-l.J.; Formal analysis, C.-l.J. and J.-l.Y.; Funding acquisition, H.-c.Y. and X.-q.W.; Investigation, C.-l.J.; Methodology, C.-l.J., J.Y. and H.-c.L.; Project administration, C.-l.J. and C.-q.G.; Resources, X.-q.W.; Writing—original draft, C.-l.J.; Writing—review & editing, J.-l.Y., J.Y., C.-q.G., H.-c.L. and X.-q.W.

Funding: This research was funded by the National Key R&D Program of China (2018YFD0500403) and the Science and Technology Planning Project of Guangzhou, China (201704030005).

Conflicts of Interest: The authors declare no conflict of interest.

References

1. Li, P.F.; Zeng, Z.K.; Wang, D.; Xue, L.F.; Zhang, R.F.; Piao, X.S. Effects of the standardized ileal digestible lysine to metabolizable energy ratio on performance and carcass characteristics of growing-finishing pigs. *J. Anim. Sci. Biotechno.* **2012**, *3*, 9. [CrossRef]

2. Zeng, P.L.; Yan, H.C.; Wang, X.Q.; Zhang, C.M.; Zhu, C.; Shu, G.; Jiang, Q.Y. Effects of dietary lysine levels on apparent nutrient digestibility and serum amino acid absorption mode in growing pigs. *J. Anim. Sci.* **2013**, *26*, 1003–1011. [CrossRef]

3. Ishida, A.; Kyoya, T.; Nakashima, K.; Katsumata, M. Muscle protein metabolism during compensatory growth with changing dietary lysine levels from deficient to sufficient in growing rats. *J. Nutr. Sci. Vitaminol.* **2011**, *57*, 401–408. [CrossRef] [PubMed]

4. Mastellar, S.L.; Coleman, R.J.; Urschel, K.L. Controlled trial of whole body protein synthesis and plasma amino acid concentrations in yearling horses fed graded amounts of lysine. *Vet. J.* **2016**, *216*, 93–100. [CrossRef] [PubMed]

5. FAO/WHO/UNU Expert Consultation. Protein and amino acid requirements in human nutrition. World Health Organ. *Tech. Rep. Ser.* **2007**, *935*, 1–265.

6. Jennings, A.; MacGregor, A.; Spector, T.; Cassidy, A. Amino acid intakes are associated with bone mineral density and prevalence of low bone mass in women: Evidence from discordant monozygotic twins. *J. Bone Miner. Res.* **2016**, *31*, 326–335. [CrossRef]

7. Karnebeek, C.D.; Hartmann, H.; Jaggumantri, S.; Bok, L.A.; Cheng, B.; Connolly, M.; Coughlin, C.R.; Das, A.M.; Gospe, S.M.; Jakobs, C.; et al. Lysine restricted diet for pyridoxine-dependent epilepsy: First evidence and future trials. *Mol. Genet. Metab.* **2012**, *107*, 335–344. [CrossRef]

8. Boy, N.; Haege, G.; Heringer, J.; Assmann, B.; Mühlhausen, C.; Ensenauer, R.; Maier, E.M.; Lücke, T.; Hoffmann, G.F.; Müller, E.; et al. Low lysine diet in glutaric aciduria type I—effect on anthropometric and biochemical follow-up parameters. *J. Inherit. Metab. Dis.* **2013**, *36*, 525–533. [CrossRef]

9. Rodriguez, J.; Sanz, M.; Blanco, M.; Serrano, M.P.; Joy, M.; Latorre, M.A. The influence of dietary lysine restriction during the finishing period on growth performance and carcass, meat, and fat characteristics of barrows and gilts intended for dry-cured ham production. *J. Anim. Sci.* **2011**, *89*, 3651–3662. [CrossRef]

10. Sato, T.; Ito, Y.; Nagasawa, T. Dietary L-lysine suppresses autophagic proteolysis and stimulates Akt/mTOR signaling in the skeletal muscle of rats fed a low-protein diet. *J. Agric. Food. Chem.* **2015**, *63*, 8192–8198. [CrossRef]

11. Sato, T.; Ito, Y.; Nedachi, T.; Nagasawa, T. Lysine suppresses protein degradation through autophagic-lysosomal system in C2C12 myotubes. *Mol. Cell Biochem.* **2014**, *391*, 37–46. [CrossRef] [PubMed]

12. Gundersen, K. Muscle memory and a new cellular model for muscle atrophy and Hypertrophy. *J. Exp. Biol.* **2016**, *219*, 235–242. [CrossRef] [PubMed]

13. Ono, Y.; Calhabeu, F.; Morgan, J.E.; Katagiri, T.; Amthor, H.; Zammit, P.S. BMP signalling permits population expansion by preventing premature myogenic differentiation in muscle satellite cells. *Cell Death. Differ.* **2011**, *18*, 222–234. [CrossRef] [PubMed]

14. Wang, X.Q.; Yang, W.J.; Yang, Z.; Shu, G.; Wang, S.B.; Jiang, Q.Y.; Yuan, L.; Wu, T.S. The differential proliferative ability of satellite cells in Lantang and Landrace pigs. *PLoS ONE* **2012**, *7*, e32537. [CrossRef] [PubMed]

15. Wang, D.; Gao, C.Q.; Chen, R.Q.; Jin, C.L.; Li, H.C.; Yan, H.C.; Wang, X.Q. Focal adhesion kinase and paxillin promote migration and adhesion to fibronectin by swine skeletal muscle satellite cells. *Oncotarget* **2016**, *7*, 30845–30854. [CrossRef]

16. Pallafacchina, G.; Blaauw, B.; Schiaffino, S. Role of satellite cells in muscle growth and maintenance of muscle mass. *Nutr. Metab. Cardiovasc. Dis.* **2013**, *23*, 12–18. [CrossRef]

17. Jash, S.; Dhar, G.; Ghosh, U.; Adhya, S. Role of the mTORC1 complex in satellite cell activation by RNA-induced mitochondrial restoration: Dual control of cyclin D1 through microRNAs. *Mol. Cell Biol.* **2014**, *34*, 3594–3606. [CrossRef]

18. Gao, C.Q.; Zhi, R.; Yang, Z.; Li, H.C.; Yan, H.C.; Wang, X.Q. Low dose of IGF-I increases cell size of skeletal muscle satellite cells via Akt/S6K signaling pathway. *J. Cell Biochem.* **2015**, *116*, 2637–2648. [CrossRef]

19. Zhang, P.; Liang, X.; Shan, T.; Jiang, Q.; Deng, C.; Zheng, R.; Kuang, S. mTOR is necessary for proper satellite cell activity and skeletal muscle regeneration. *Biochem. Biophys. Res. Commun.* **2015**, *463*, 102–108. [CrossRef]

20. Dai, J.M.; Yu, M.X.; Shen, Z.Y.; Guo, C.Y.; Zhuang, S.Q.; Qiu, X.S. Leucine promotes proliferation and differentiation of primary preterm rat satellite cells in part through mTORC1 signaling pathway. *Nutrients* **2015**, *7*, 3387–3400. [CrossRef]

21. Alway, S.E.; Pereira, S.L.; Edens, N.K.; Hao, Y.; Bennett, B.T. β-Hydroxy-beta-methylbutyrate (HMB) enhances the proliferation of satellite cells in fast muscles of aged rats during recovery from disuse atrophy. *Exp. Gerontol.* **2013**, *48*, 973–984. [CrossRef] [PubMed]

22. Morales, A.; Garcia, H.; Arce, N.; Cota, M.; Zijlstra, R.T.; Araiza, B.A.; Cervantes, M. Effect of L-lysine on expression of selected genes, serum concentration of amino acids, muscle growth and performance of growing pigs. *J. Anim. Physiol. Anim. Nutr.* **2015**, *99*, 701–709. [CrossRef] [PubMed]

23. Jin, C.L.; Zhang, Z.M.; Ye, J.L.; Gao, C.Q.; Yan, H.C.; Li, H.C.; Yang, J.Z.; Wang, X.Q. Lysine-induced swine satellite cell migration is mediated by the FAK pathway. *Food Funct.* **2019**, *10*, 583–591. [CrossRef] [PubMed]

24. Ye, J.L.; Gao, C.Q.; Li, X.G.; Jin, C.L.; Wang, D.; Shu, G.; Wang, W.C.; Kong, X.F.; Yao, K.; Yan, H.C.; et al. EAAT3 promotes amino acid transport and proliferation of porcine intestinal epithelial cells. *Oncotarget* **2016**, *7*, 38681–38692. [CrossRef] [PubMed]

25. Li, X.G.; Sui, W.G.; Gao, C.Q.; Yan, H.C.; Yin, Y.L.; Li, H.C.; Wang, X.Q. L-Glutamate deficiency can trigger proliferation inhibition via down regulation of the mTOR/S6K1 pathway in pig intestinal epithelial cells. *J. Anim. Sci.* **2016**, *94*, 1541–1549. [CrossRef]

26. Schmidt, E.K.; Clavarino, G.; Ceppi, M.; Pierre, P. SUnSET, a nonradioactive method to monitor protein synthesis. *Nat. Methods.* **2009**, *6*, 275–277. [CrossRef]

27. Goodman, C.A.; Mabrey, D.M.; Frey, J.W.; Miu, M.H.; Schmidt, E.K.; Pierre, P.; Hornberger, T.A. Novel insights into the regulation of skeletal muscle protein synthesis as revealed by a new nonradioactive in vivo technique. *FASEB J.* **2011**, *25*, 1028–1039. [CrossRef]

28. McMillan, J.D.; Jennings, E.W.; Mohagheghi, A.; Zuccarello, M. Comparative performance of precommercial cellulases hydrolyzing pretreated corn stover. *Biotechnol. Biofuels.* **2011**, *4*, 29. [CrossRef]

29. Liao, S.F.; Wang, T.; Regmi, N. Lysine nutrition in swine and the related monogastric animals: Muscle protein biosynthesis and beyond. *Springerplus* **2015**, *4*, 147. [CrossRef]

30. Egner, I.M.; Bruusgaard, J.C.; Gundersen, K. Satellite cell depletion prevents fiber hypertrophy in skeletal muscle. *Development* **2016**, *143*, 2898–2906. [CrossRef]

31. Kim, J.; Lee, K.S.; Kwon, D.H.; Bong, J.J.; Jeong, J.Y.; Nam, Y.S.; Lee, M.S.; Liu, X.; Baik, M. Severe dietary lysine restriction affects growth and body composition and hepatic gene expression for nitrogen metabolism in growing rats. *J. Anim. Physiol. Anim. Nutr.* **2014**, *98*, 149–157. [CrossRef] [PubMed]

32. Yang, Y.X.; Guo, J.; Jin, Z.; Yoon, S.Y.; Choi, J.Y.; Wang, M.H.; Piao, X.S.; Kim, B.W.; Chae, B.J. Lysine restriction and realimentation affected growth, blood profiles and expression of genes related to protein and fat metabolism in weaned pigs. *J. Anim. Physiol. Anim. Nutr.* **2009**, *93*, 732–743. [CrossRef] [PubMed]

33. Sato, T.; Ito, Y.; Nagasawa, T. Regulation of skeletal muscle protein degradation and synthesis by oral administration of lysine in rats. *J. Nutr. Sci. Vitaminol.* **2013**, *59*, 412–419. [CrossRef] [PubMed]

34. Popha, S.; Mo, P.E.; Vieira, S.L. Satellite cell mitotic activity of broilers fed differing levels of lysine. *Int. J. Poult. Sci.* **2004**, *3*, 758–763.

35. Snijders, T.; Verdijk, L.B.; Smeets, J.S.J.; McKay, B.R.; Senden, J.M.G.; Hartgens, F.; Parise, G.; Greenhaff, P.; van Loon, L.J.C. The skeletal muscle satellite cell response to a single bout of resistance-type exercise is delayed with aging in men. *Age* **2014**, *36*, 9699. [CrossRef] [PubMed]

36. Mackey, A.L.; Andersen, L.L.; Frandsen, U.; Sjogaard, G. Strength training increases the size of the satellite cell pool in type I and II fibres of chronically painful trapezius muscle in females. *J. Physiol.* **2011**, *589*, 5503–5515. [CrossRef] [PubMed]

37. Snijders, T.; Verdijk, L.B.; Beelen, M.; McKay, B.R.; Parise, G.; Kadi, F.; van Loon, L.J.C. A single bout of exercise activates skeletal muscle satellite cells during subsequent overnight recovery. *Exp. Physiol.* **2012**, *97*, 762–773. [CrossRef]

38. Jewell, J.L.; Kim, Y.C.; Russell, R.C.; Yu, F.X.; Park, H.W.; Plouffe, S.W.; Tagliabracci, V.S.; Guan, K.L. Differential regulation of mTORC1 by leucine and glutamine. *Science* **2015**, *347*, 194–198. [CrossRef]

39. Wolfson, R.L.; Chantranupong, L.; Saxton, R.A.; Shen, K.; Scaria, S.M.; Cantor, J.R.; Sabatini, D.M. Sestrin2 is a leucine sensor for the mTORC1 pathway. *Science* **2016**, *351*, 43–48. [CrossRef]

40. Li, F.; Yin, Y.; Tan, B.; Kong, X.; Wu, G. Leucine nutrition in animals and humans: mTOR signaling and beyond. *Amino Acids* **2011**, *41*, 1185–1193. [CrossRef]

41. Yao, K.; Yin, Y.L.; Chu, W.Y.; Liu, Z.Q.; Deng, D.; Li, T.J.; Huang, R.L.; Zhang, J.S.; Tan, B.; Wang, W.C.; et al. Dietary arginine supplementation increases mTOR signaling activity in skeletal muscle of neonatal pigs. *J. Nutr.* **2008**, *138*, 867–872. [CrossRef] [PubMed]

Review

Identifying the Structural Adaptations that Drive the Mechanical Load-Induced Growth of Skeletal Muscle: A Scoping Review

Kent W. Jorgenson [1], Stuart M. Phillips [2] and Troy A. Hornberger [1,*]

[1] School of Veterinary Medicine and the Department of Comparative Biosciences, University of Wisconsin, Madison, WI 53706, USA; kwjorgenson@wisc.edu

[2] Department of Kinesiology, McMaster University, Hamilton, ON L8S 4K1, Canada; phillis@mcmaster.ca

* Correspondence: troy.hornberger@wisc.edu

Received: 23 June 2020; Accepted: 7 July 2020; Published: 9 July 2020

Abstract: The maintenance of skeletal muscle mass plays a critical role in health and quality of life. One of the most potent regulators of skeletal muscle mass is mechanical loading, and numerous studies have led to a reasonably clear understanding of the macroscopic and microscopic changes that occur when the mechanical environment is altered. For instance, an increase in mechanical loading induces a growth response that is mediated, at least in part, by an increase in the cross-sectional area of the myofibers (i.e., myofiber hypertrophy). However, very little is known about the ultrastructural adaptations that drive this response. Even the most basic questions, such as whether mechanical load-induced myofiber hypertrophy is mediated by an increase in the size of the pre-existing myofibrils and/or an increase in the number myofibrils, have not been resolved. In this review, we thoroughly summarize what is currently known about the macroscopic, microscopic and ultrastructural changes that drive mechanical load-induced growth and highlight the critical gaps in knowledge that need to be filled.

Keywords: fascicle; myofiber; myofibril; sarcomere; hypertrophy; hyperplasia; splitting; radial growth; longitudinal growth; exercise

1. Introduction

Skeletal muscle comprises approximately 40% of body mass and plays a critical role in posture, breathing, motion, and metabolic regulation [1]. As we age, the occurrence of age-related diseases, such as the loss of muscle mass (i.e., sarcopenia), are expected to become more prevalent [2]. For instance, between the ages of 25–80 years, the average individual will lose approximately 25% of their muscle mass [3,4]. This age-associated loss of muscle mass leads to an increased risk of fall-related injury, institutionalization, loss of independence, and disease [5–7]. Indeed, in the United States alone, the healthcare costs for muscle wasting related illnesses were estimated to be $18.5 billion in 2000 [8]. Based on this figure, reducing the rate of muscle wasting related diseases by even 10% could save a striking $1.1 billion in annual healthcare costs. The number of people over the age of 60 is expected to double by 2050, and thus, the costs associated with sarcopenia will only continue to increase [9]. Accordingly, the development of therapies that can restore, maintain, and/or increase muscle mass will be of great clinical and fiscal significance. However, to develop such therapies, we will first need to establish a comprehensive understanding of the mechanisms that regulate the size of this vital tissue.

Mechanical load-induced signals are one of the most widely recognized regulators of skeletal muscle mass. Indeed, historical evidence suggests that the growth-promoting effects of mechanical loading has been recognized since at least the 7th century BC [10]. During the last century, a variety of human and animal models have been used to further establish this point. For instance, in humans, resistance exercise is the most commonly used model of mechanical load-induced growth and it typically induces a 5–20% increase in skeletal muscle volume/mass within 8–16 weeks [11–17]. Similar changes in muscle mass have also been observed in animal models that are intended to mimic human resistance exercise [18–20]. Furthermore, animal models that use extreme forms of mechanical loading, such as synergist ablation, can promote a doubling of muscle mass within as little as 2 weeks [21–23]. Collectively, these models have provided extensive insight into the macroscopic and microscopic changes that contribute to the mechanical load-induced growth response, but surprisingly, the ultrastructural changes that drive these changes remain poorly understood. In this review, we will thoroughly summarize what is currently known about the structural adaptations that drive mechanical load-induced growth and highlight the critical gaps in knowledge that need to be filled.

2. Overview of Skeletal Muscle Structure

Before considering the structural changes that drive mechanical load-induced growth, we want to ensure that the reader appreciates the basic structural design of skeletal muscle. One of the easiest ways to appreciate this design is to consider skeletal muscle as a hierarchy of contractile machinery that is visible at the macroscopic level (viewable without magnification), followed by the microscopic level (viewable with standard microscopy), and finally the ultrastructural level (viewable with high resolution microscopy). Below we will provide a brief overview of the primary components that are found at each of these levels. For excellent illustrations and more comprehensive discussions on this topic, the reader is referred to the following reviews [24–26].

At the macroscopic level, it can be noted that skeletal muscles are connected to bones via tendinous attachments and enact their contractile function by providing movement and articulation of the skeletal system. Moreover, as illustrated in Figure 1, skeletal muscles are surrounded by an outer layer of connective tissue called the epimysium, and underneath the epimysium are bundles of myofibers (i.e., fascicles) that are surrounded by another layer of connective tissue called the perimysium [24]. In most skeletal muscles, the fascicles, and their associated myofibers, are not directly aligned with the longitudinal axis of the muscle, but instead are offset at an angle called the pennation angle (Figure 2) [27].

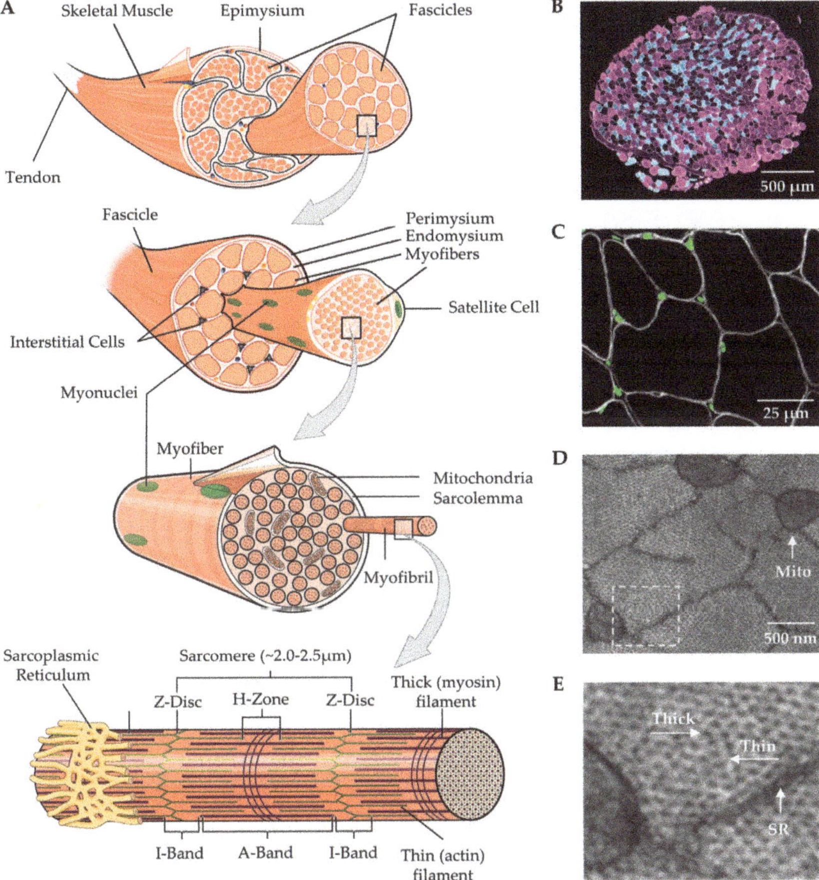

Figure 1. (**A**) Illustration of skeletal muscle structure copied with permission under the Creative Commons Attribution 4.0 International license and adapted for this review, available online: https://openstax.org/books/anatomy-and-physiology/pages/10-2-skeletal-muscle (accessed on 6 January 2020) [28]. (**B**) Cross-section of a mouse plantaris muscle that was subjected to immunohistochemistry for the identification of Type IIA (cyan), and Type IIB (magenta) myofibers as well as laminin to identify the endomysium (white). (**C**) Cross-section of a mouse plantaris muscle that was subjected to immunohistochemistry for the identification of dystrophin to identify the inner boundary of the sarcolemma (white) and nuclei (green). (**D**) Cross-section of a mouse plantaris muscle that was subjected to electron microscopy to highlight the sarcoplasmic reticulum (SR) that surrounds individual myofibrils as well as the mitochondria (Mito) that run between the myofibrils. (**E**) Higher magnification of the boxed region in D reveals the presence of the thick and thin myofilaments.

At the microscopic level, a cross-section of skeletal muscle will reveal the presence of individual myofibers (Figure 1A,B). The myofibers are multinucleated cells that are encased by a layer of connective tissue called the endomysium, and they are surrounded by interstitial cells such as fibroblasts, immune cells, pericytes and fibro-adipogenic progenitors (Figure 1A–C) [29,30]. Furthermore, another important class of cells, called satellite cells, resides between the endomysium and the plasma membrane of the myofibers (i.e., the sarcolemma) [31]. The endomysium is physically coupled to the sarcolemma, and everything the resides underneath the sarcolemma is typically referred to as the sarcoplasm.

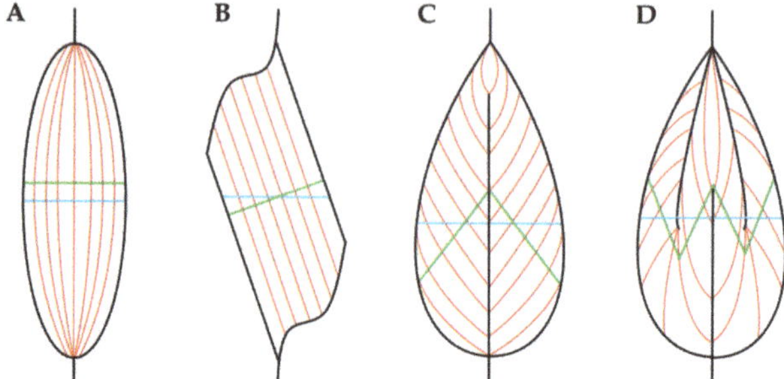

Figure 2. Various pennation angle arrangements of the fascicles/myofibers in skeletal muscle: (**A**) fusiform, (**B**) unipennate, (**C**) bipennante, (**D**) multipennate. Blue lines indicate the plane for the anatomical cross-sectional area (CSA) (i.e., the CSA that runs perpendicular to the longitudinal axis of the muscle), and green lines indicate the plane for physiological CSA (i.e., the CSA that runs perpendicular to the longitudinal axis of the fascicles/myofibers). Adapted under the Creative Commons Attribution-Share Alike 3.0 Unported license from original work by Uwe Gille (Available online: https://creativecommons.org/licenses/by-sa/3.0/deed.en (accessed on 6 January 2020).

The gelatinous sarcoplasm contains the primary ultrastructural elements of the myofiber, and as illustrated in Figure 1, an examination at the ultrastructural level reveals that ≈80% of the sarcoplasm is filled with an in-parallel array of rod-like structures called myofibrils [32–36]. The myofibrils are composed of a long in-series array of force-generating elements called sarcomeres and are surrounded by a mitochondrial reticulum and a membranous structure called the sarcoplasmic reticulum [37,38]. The sarcomeres within the myofibrils enact their function through the active sliding of thick and thin myofilaments, and in a longitudinal view, it can be seen that the sarcomeres consist of regions called the Z-disc, the I-band which contains the thin (actin) myofilaments, and the A-band which contains the thick (myosin) myofilaments [39,40]. It should also be noted that within a given species, the optimal/resting length of sarcomeres (2.0–2.5 μm) is highly conserved, and alterations in this length can profoundly influence force production [41–43].

When considering the structure of skeletal muscle, it is also essential to recognize that all myofibers are not created equal. For instance, some types of myofibers are heavily reliant on oxidative metabolism, exhibit a slow contractile speed and are resistant to fatigue. In contrast, other types of myofibers rely on anaerobic glycolytic metabolism, exhibit a fast contractile speed and rapidly fatigue when stimulated to contract [40]. Different fiber types are typically grouped according to the predominant isoform of the myosin heavy chain that they express, and these isoforms include Type I (slow oxidative), Type IIA (fast oxidative), and Type IIB (fast glycolytic) fibers [44]. It is also important to point out that humans do not express the Type IIB myosin isoform, but instead express a very similar (yet slightly slower) Type IIX myosin isoform [45]. However, since this differentiation was only solidified in 1990's [46–48], some older studies with human subjects used the Type IIB classification, while other studies have grouped Type IIB and Type IIX fibers together as a similar fiber type [49]. This use of IIB and IIX myosin labeling has led to some confusion when comparing earlier muscle growth studies to current studies, so it is important to keep this distinction in mind when relating fiber type-specific adaptations across the current body of literature.

3. Mechanical Load-Induced Growth of Skeletal Muscle at the Macroscopic Level

3.1. Whole Muscle

At the whole muscle level, mechanical load-induced growth can be mediated by an increase in the length and/or an increase in the cross-sectional area (CSA) of the muscle. Growth resulting from an increase in length is referred to as longitudinal growth, and it can occur in response to a variety of different perturbations. For instance, during development, the length of muscles can more than double from birth to the termination of bone growth [50–52]. Longitudinal growth can also be induced in adults by placing muscles in a chronically stretched state [53]. As a case in point, it has been shown that immobilizing a rat lower hindlimb in a fully dorsiflexed position can lead to a >20% increase in the length of the soleus muscle [54]. Likewise, limb-lengthening procedures can lead to a >20% increase in muscle length [55,56]. Indeed, even some of the more extreme models of mechanical load-induced growth can lead to an increase in muscle length [23,57]. For example, surgical removal of the gastrocnemius and soleus muscles (i.e., synergist ablation) is a commonly used rodent model for stimulating mechanical load-induced growth, and it has been reported that this can lead to a 13% increase length of the plantaris muscle [57,58]. Thus, it is clear that both adolescent and adult skeletal muscles can undergo longitudinal growth.

Although skeletal muscle is capable of undergoing longitudinal growth, most models of mechanical loading do not lead to notable alterations in whole muscle length [18,59]. Instead, mechanical load-induced growth is usually driven by an increase in the CSA of the muscle (also known as radial growth). For example, in humans, 8-16 weeks of resistance exercise will generally produce a 5–30% increase in whole muscle CSA but no change in muscle length [16,59–66]. Interestingly, the magnitude of increase in CSA is often greater than the increase that is observed for muscle volume/mass. An excellent example of this paradox was reported by Roman et al. (1993), whom reported that 12 weeks of resistance exercise led to a 14% increase in the volume of the elbow flexors, but the CSA at the mid-belly increased by 23% [60]. Importantly, however, the magnitude of the increase is CSA got progressively smaller towards the proximal and distal ends of the muscle, which explained why the muscle volume only increased by 14%. Simply put, the study by Roman et al. (1993) demonstrated that the radial growth response was not evenly distributed along the length of the muscle. Indeed, regional differences in the magnitude of radial growth have been reported in several animal and human-based studies, and in our opinion, this phenomenon represents an often overlooked aspect of the mechanical load-induced growth response [13,59,64,67–72].

3.2. Muscle Fascicles

Previous studies have shown that the initial mechanical load-induced increase in whole muscle CSA can be attributed, at least in part, to edema. However, the long-term changes are primarily caused by an expansion of the contractile elements [63,73,74]. For instance, using a rat model of resistance exercise, we have shown that 8 weeks of training resulted in a 24% increase in the CSA of the flexor hallucis longus muscle, and this was matched by a proportionate increase in total myofibrillar protein content and peak tetanic force production [18]. Thus, if the mechanical load-induced increase in whole muscle CSA was driven by an expansion of the contractile elements, then the increase should be reflective of the changes that occurred at the preceding level within the hierarchy of the contractile machinery (i.e., myofilaments → myofibrils → myofibers → fascicles → whole muscle).

Based on the aforementioned point, mechanical load-induced changes in whole muscle CSA should be driven by changes that happen at the level of the muscle fascicles, and there are effectively two predominant ways in which this is thought to occur: (1) longitudinal growth of the fascicles or (2) radial growth of the fascicles. It is also possible that mechanical loading could lead to an increase in the number of fascicles per muscle, but we are not aware of any studies that have attempted to answer this technically difficult question.

Upon first consideration, it can be challenging to appreciate how both longitudinal and radial growth of fascicles can lead to an increase in whole muscle CSA. Thus, to visualize these points, we have taken advantage of a geometric model that can be used to predict changes in the architectural properties of skeletal muscle [75,76]. Specifically, as shown in Figure 3, we used this model to illustrate how changes in either fascicle length (Lf), or fascicle diameter (Df), could produce a 30% increase in whole muscle CSA (the upper end of what is typically observed in humans after 8–16 weeks of resistance exercise). For simplicity, our model considers a hypothetical muscle that is composed of 50 fascicles aligned in parallel. Moreover, in the control (starting) state, the fascicle length to muscle length (Lm) ratio is 0.25, and the fascicles are offset at a pennation angle of 16° (similar to the properties of the vastus lateralis muscle in humans [77]). Based on these parameters, if the 30% increase in CSA was purely due to longitudinal growth of the fascicles, then fascicle length would have to increase by 11%, and the pennation angle would remain unaltered (Figure 3B). On the other hand, if the 30% increase in CSA was due exclusively to an increase in radial growth of the fascicles, then the fascicle diameter would have to increase by 14%, and this would result in a concomitant 15% increase in pennation angle (from 16°→18.4°) (Figure 3C). It is also worth noting that in the example of pure longitudinal growth, the number of fascicles visible in a cross-section of the mid-belly of the muscle would also increase by 30%, and could easily lead one to mistakenly conclude that the increase in whole muscle cross-sectional area was driven by new fascicle formation.

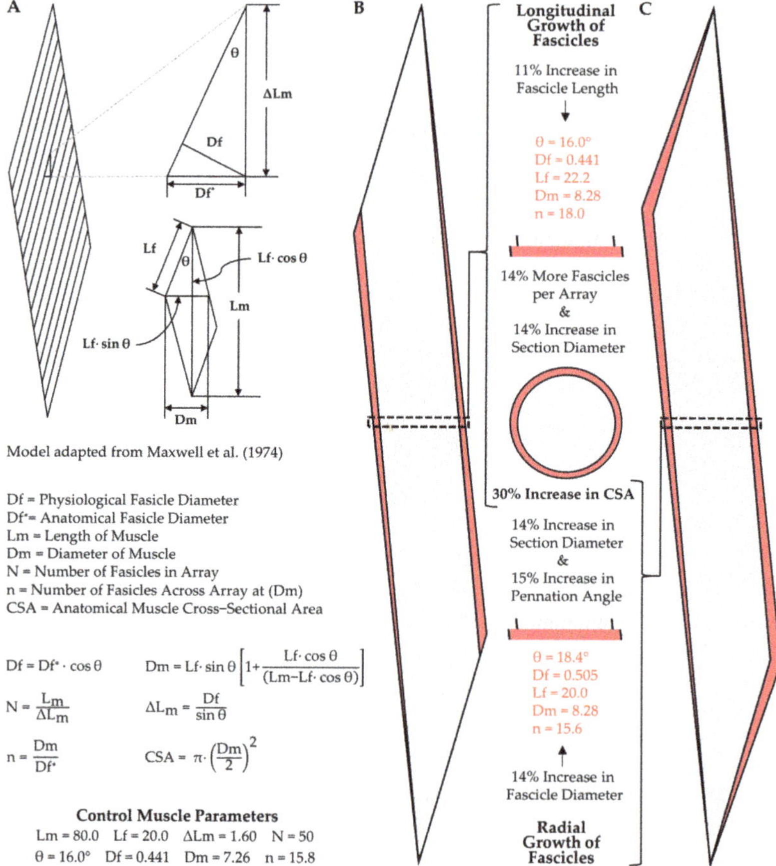

Figure 3. Illustration of how the longitudinal and radial growth of fascicles can lead to changes in muscle cross-sectional area (CSA). (**A**) Key elements of a geometric model that can be used to predict the architectural properties of skeletal muscle [75]. (**B**) Illustration of how an 11% increase in fascicle length would result in 30% increase in CSA, as well as a 30% increase in the number of fascicles per cross-section. (**C**) Illustration of how a 14% in fascicle diameter would lead to a 15% increase in the pennation angle and a 30% increase in the CSA, but essentially no change in the number of fascicles per cross-section.

Having illustrated how radial and longitudinal growth of fascicles can lead to an increase in whole muscle CSA, we will now consider the studies that have tested whether these types of adaptations occur. Specifically, we will first consider the studies that have examined whether mechanical load-induced alterations in whole muscle CSA are associated with changes in fascicle length, and fortunately, this has been a subject of extensive investigation [13,59,62,65,78–85]. For instance, Ema et al. (2016) recently compiled data from 38 different studies that addressed this topic and found that a significant positive relationship existed between the exercise-induced increases in muscle size and fascicle length [86]. Nonetheless, some of the studies that reported an increase in muscle size did not observe an increase in fascicle length [59,83–85], and there are even examples in which small but significant declines in fascicle length have been reported [78]. However, when Ema et al. (2016) compared the magnitude of change across all studies which showed a significant alteration in fascicle length versus those which did not, the average values from these studies still showed a 12.4% versus 7.7% increase in fascicle length, respectively [86]. Thus, there is a high level of support for the notion that mechanical load-induced alterations in whole muscle CSA can be driven, at least in part, by an increase in fascicle length.

As mentioned above, the radial growth of fascicles could also lead to an increase in whole muscle CSA. Importantly, as detailed by the work of Maxwell et al. (1974), and as illustrated in Figure 3C, a direct relationship exists between fascicle diameter and the pennation angle of the fascicles. Specifically, if the length of the muscle, the length of the fascicles, and the number of fascicles is held constant, then an increase in fascicle diameter will lead to an increase in the pennation angle. Hence, it is not surprising that most studies reporting a significant resistance exercise-induced increase in muscle size, but no change in fascicle length, instead find a significant increase in the pennation angle [59,83–85]. Indeed, just as with changes in fascicle length, Ema et al. (2016) determined that a significant positive relationship exists between the resistance exercise-induced increase in muscle size and pennation angle. On average, the studies that reported a significant change in pennation angle showed a 13.5% increase, while those that did not detect a significant change still found an average increase of 7.7% [86]. Accordingly, just as with changes in fascicle length, there is a high level of support for the notion that mechanical load-induced alterations in whole muscle CSA can be driven by an increase in fascicle diameter/pennation angle. Indeed, a collective view of the literature suggests that mechanical loading can lead to both longitudinal and radial growth of fascicles, and the exact contribution of these components is probably determined by a variety of different factors, such as the type of mechanical loads that are placed on the muscle (e.g., concentric vs. eccentric contractions) and the architectural properties of the muscle that is being considered (e.g., fusiform, unipennate, bipennate, etc.) [86–89].

4. Mechanical Load-Induced Growth of Skeletal Muscle at the Microscopic Level

As previously noted, mechanical load-induced alterations at each level of the skeletal muscle structure should be reflective of the changes that occurred at the preceding level within the hierarchy of the contractile machinery. Thus, having established that mechanical loading can lead to both longitudinal and/or radial growth of the fascicles, we will now consider how these changes can be mediated by alterations at the level of the myofibers.

4.1. Longitudinal Growth of Fascicles

Fascicles are composed of bundles of myofibers, and the myofibers can either run the entire length of the fascicle, or only part of the length of the fascicle and exhibit an intrafascicular termination [90–92]. For fascicles that are composed of myofibers that run the entire length of the fascicle, longitudinal growth of the fascicle would be exclusively dependent on the longitudinal growth of the individual myofibers. Alternatively, longitudinal growth of the fascicles with myofibers that exhibit intrafascicular terminations could result from longitudinal growth of the myofibers and/or the addition of new myofibers in-series. Although we are not aware of any studies that have addressed whether mechanical loading can lead to the formation of new myofibers in-series, a consistent body of literature has shown that myofibers are capable of undergoing longitudinal growth [57,72,93–95]. For instance, Alway et al. (1989) subjected the anterior latissimus dorsi (ALD) muscle of quails to chronic mechanical loading by securing a weight (10% of body mass) to one of the wings. In response to this perturbation, the mass of the ALD increased by 182%, which was associated with a 24% increase in the average length of the myofibers [72]. Similarly, Roy et al. (1982) demonstrated that in rats, chronic mechanical loading of the plantaris via synergist ablation resulted in doubling of its mass and a concomitant 19% increase in the myofiber to muscle length ratio [94]. Based on these, and related studies, it is clear that extreme models of mechanical loading can induce longitudinal growth of the myofibers.

Although a compelling body of evidence indicates that extreme models of mechanical loading can promote longitudinal growth of myofibers, only a handful of studies have directly addressed this topic within the confines of more physiologically relevant models. For instance, it has been shown that the eccentric contractions induced by downhill walking [96], and downhill running [97], can lead to an increase in the number of sarcomeres per myofiber; however, neither of these studies reported measurements of myofiber length. Indeed, we could only find one study that reported measurements of myofiber length within the context of a physiologically relevant model of mechanical load-induced growth [18]. In this case, rats were subjected to 8 weeks of resistance exercise which led to a 24% increase in whole muscle CSA, but myofiber length was not altered. Importantly; however, this study did not indicate whether the increase in muscle CSA was mediated by longitudinal vs. radial growth of the fascicles, and hence, it is difficult to extrapolate any meaningful insights from the data.

Given the paucity of data on this topic, we believe that it is worthwhile to mention unpublished results that we recently obtained from mice that had their plantaris muscles subjected to 16 days of myotenectomy (a much milder form of synergist ablation [98]). Specifically, we determined that myotenectomy led to 72% increase in the mass of the plantaris along with an 8% increase in length of the myofibers ($p < 0.01$). Likewise, Goh et al. (2019) recently described a high intensity interval training (HIIT) for mice that leads to a 17% increase in the mass of the extensor digitorum longus muscle [99], and this was associated with a 9% increase in the length of the myofibers ($p < 0.05$, personal communication from Dr Doug Millay). Thus, it appears that even physiologically relevant models of mechanical load-induced growth can induce longitudinal growth of myofibers; however, additional studies on this topic will need to be published before a clear consensus can be reached.

4.2. Radial Growth of Fascicles

As illustrated in Figure 4, radial growth of fascicles could result from an increase in the CSA of the existing myofibers (i.e., myofiber hypertrophy, Figure 4A) and/or an increase in the number of myofibers per cross-section (from myofiber splitting and/or hyperplasia, Figure 4B). These concepts have been widely studied within the context of mechanical load-induced growth, and in the following sections we will summarize the body of literature that exists on these topics. Before moving into these sections, we also want to point out that the radial growth of fascicles could result from the longitudinal growth of myofibers with intrafascicular terminations (Figure 4C) [91]. However, as mentioned above, very few studies have examined whether physiologically relevant models of mechanical loading can induce longitudinal growth of myofibers. Accordingly, this mechanism will not be subjected to further discussion.

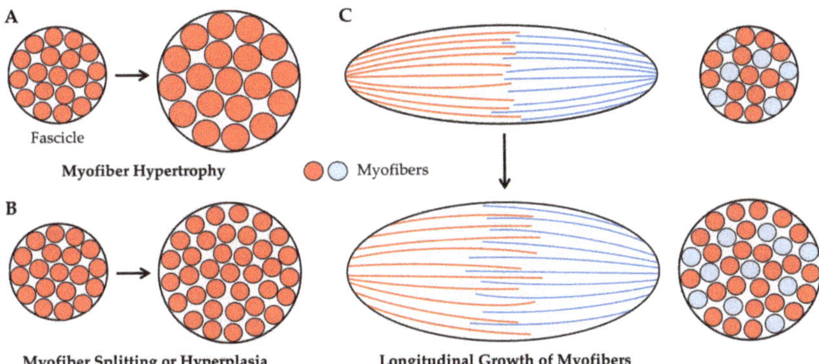

Figure 4. Illustration of how the radial growth of muscles fascicles could result from (**A**) myofiber hypertrophy, (**B**) myofiber splitting or hyperplasia, or (**C**) longitudinal growth of myofibers that exhibit intrafascicular terminations, such as those observed in the long sartorius and gracilis muscles of humans [90].

4.2.1. Myofiber Hypertrophy

Radial growth of myofibers leads to an increase in the CSA, and such a change is typically referred to as myofiber hypertrophy. Myofiber hypertrophy is, by far, the longest-standing and most widely acknowledged contributor to the mechanical load-induced growth of skeletal muscle. Indeed, the ability of mechanical loads to induce myofiber hypertrophy has been recognized since the late 1800's [100]. As summarized by Huan et al. [101], most of the early research on this topic used animals such as dogs [100], cats [102], mice [103], rats [104], hamsters [35], and birds [105]. Some of these animal-based studies employed rather extreme forms of chronic mechanical loading (e.g., synergist ablation, wing-weighting, etc.), whereas others used interventions that were intended to mimic human resistance exercise. A notable example was described by Goldspink (1964) in which young mice were trained to pull on a weighted cord so that they could gain access to their food, and it was determined that 25 days of this training resulted in a ≈30% increase in the CSA of myofibers within the biceps brachii [103]. Another classic example involves the model described by Gonyea and Ericson (1976) [102]. In this model, cats were operantly conditioned to move a weighted bar with their paw in exchange for a food reward, and it was found that the CSA of the myofibers within the flexor carpi radialis increased by 21–32% after 41 weeks of this type of training [102]. The magnitude of change in myofiber CSA observed in the above examples is similar to the 10–35% that is typically observed in humans after 8-16 weeks of resistance exercise [16,17,44,60,66,101,106–108]. However, this magnitude of change pales in comparison to what has been observed with some of the more extreme models of mechanical loading. For instance, Antonio and Gonyea (1993) observed an astonishing 142% increase in CSA of the myofibers of the ALD muscle after just 16 days of wing-weighting [109]. Simply stated, an extremely high level of evidence supports the notion that mechanical loading can induce myofiber hypertrophy and the capacity for this type of growth appear to be quite large.

4.2.2. Myofiber Splitting

As recently reviewed by Murach et al. (2019), split myofibers are characterized by the presence of "branching", "fragmentation", or "splitting" along the length of the myofiber [110]. Split myofibers can be found in healthy muscles, and an increased frequency of split myofibers is commonly observed in muscular dystrophy and various neurogenic myopathies [111,112]. An increased frequency of split myofibers has also been observed in muscles subjected to mechanical loading. For instance, the most extraordinary example of this was published by Antonio and Gonyea (1994) who reported that the frequency of split myofibers in the quail ALD muscle increased from 0.25% to 5.25% after 28 days of wing-weighting [113]. Tamaki et al. (1996) also found that the frequency of split myofibers in the rat plantaris muscle increased from 0.6% to 1.8% after 6 weeks of synergist ablation [114]. An increase in the occurrence of split myofibers (1.4% of all myofibers) has also been observed in powerlifters that used anabolic steroids [115]. Based on these reports, it would appear mechanical loading can result in an increased prevalence of split myofibers. However, it is important to point out that many of the studies that are frequently cited as providing support for this concept never actually quantified the number of split myofibers [116–118]. Moreover, there are multiple examples in which the number of split myofibers was quantified, and it was concluded that mechanical loading did not alter the frequency of their appearance [119–122]. Even the study by Antonio and Gonyea (1994) found that 16 days of wing-weighting resulted in an 88% increase in the mass of the ALD muscle, yet the frequency of split myofibers at this time point was still only 0.28% [113]. One potential explanation for this observation is that splitting along the entire length of the myofiber rapidly runs to completion, and thus, only a small fraction of the myofibers that split are effectively detected. However, if this were the case, then the total number of myofibers per muscle should increase. To test this, Antonio and Gonyea (1994) directly counted all of the fibers in the ALD muscles and found that the total number did not change after 16 days of wing-weighting [113]. Thus, it is our conviction that although mechanical loading may be capable of inducing myofiber splitting, the frequency of this event is low and thus does not typically make a major contribution to the overall growth process.

4.2.3. Hyperplasia

Hyperplasia refers to the generation of new myofibers, and as illustrated in Figure 4B, hyperplasia could lead to the radial growth of muscle fascicles. Indeed, numerous studies have shown that the number of myofibers per muscle rapidly increases during the early stages of developmental growth [123–125]. Although it is well accepted that hyperplasia occurs during developmental growth, whether hyperplasia can occur in adult skeletal muscles remains a subject of debate.

Part of the debate over whether hyperplasia occurs in adult skeletal muscle results from the types of measurements that have been used to address this question [126,127]. Specifically, two primary methods have been employed: (1) counting the number of myofibers per cross-section of the muscle, and (2) digestion of the muscle's connective tissue followed by a direct count of all myofibers present in the muscle. The direct counting method is ideal, but this approach requires the manual dissociation and counting of thousands of myofibers. Accordingly, most studies that describe measurements of hyperplasia are based on counts of the myofibers per cross-section. With this point in mind, it is imperative to recognize that the number of myofibers that appear in a cross-section can be highly influenced by changes in the architectural properties of the muscle (e.g., fiber length and/or pennation angle) [75,76,91,127]. Such effects have been thoroughly described by Maxwell et al. (1974), and can be appreciated by considering the illustrations presented in Figures 3 and 4 [75]. For instance, Figure 3B shows how an 11% increase in fascicle length would lead to a 30% in the number of fascicles per cross-section, and the same principles would hold at the level of the myofibers. Furthermore, as illustrated in Figure 4C, longitudinal growth of myofibers with intrafascicular terminations could also lead to an increase in the number of myofibers per cross-section. Hence, extreme caution needs to be exercised when interpreting the results from studies that rely on myofiber per cross-section counts to make conclusions about hyperplasia.

Unfortunately, the majority of studies that have examined whether mechanical loading induces hyperplasia have relied on counts of the myofibers per cross-section [22,35,109,116–118,128–130]; however, there are a handful of studies that have reported direct myofiber counts. For instance, investigators with ties to Dr Gonyea reported that 7–30 days of wing-weighting resulted in a 60–294% increase in muscle mass and a 30–50% increase in the number of myofibers per muscle [72,93,113,131]. In stark contrast, investigators with links to Dr Gollnick reported that 6–65 days of wing-weighting resulted in a 22–225% increase in muscle mass but no change in the number of myofibers per muscle [120]. The Gollnick group also reported that other extreme forms of mechanical loading such as synergist ablation does not alter the number of myofibers per muscle [119,121]. The conflicting conclusions from these groups has existed for over 25 years, and surprisingly, the controversy has still not been resolved [126,127,132,133]. Thus, in our opinion, the notion that extreme forms of mechanical loading can induce hyperplasia remains controversial.

Due to a minimal number of studies, a similar controversy exists with regards to whether more physiologically relevant models of mechanical loading can induce hyperplasia. Indeed, we are only aware of two studies that have directly addressed whether a resistance exercise-like stimulus can alter the number of myofibers per muscle. The first of these studies was performed by Gonyea et al. (1986) whom painstakingly counted the number of myofibers in the flexor carpi radialis of cats that had been subjected to 60–129 weeks of weight training, and the results indicated that the training stimulus led to a 9% increase in the number of myofibers per muscle (39,759 vs. 36,550 myofibers per muscle) [122]. Likewise, Tamaki et al. (1992) subjected rats to weight-lifting exercise and found that the number of myofibers in the plantaris muscle increased by 14% after 12 weeks of training; however, in this case, the absolute mass of the plantaris was not significantly altered by the training, and thus, the basis for the increase in fiber number is difficult to interpret [134]. Mixed and cautious interpretations can also be drawn from studies that have used myofiber per cross-section counts as a readout for hyperplasia, with some studies showing an increase in the number of myofibers per section [110,116,128], while other have reported no change [35,100,110]. Accordingly, a firm conclusion with regards to whether physiologically relevant forms of mechanical loading can induce hyperplasia remains elusive.

5. Mechanical Load-Induced Growth of Skeletal Muscle at the Ultrastructural Level

5.1. Longitudinal Growth of Myofibers

As summarized in the previous section, a compelling body of literature has shown that extreme models of mechanical loading can promote the longitudinal growth of myofibers. Furthermore, several lines of evidence suggest that longitudinal growth of myofibers can also be induced by physiologically relevant forms of mechanical loading. Since myofibers are composed of an in-series connection of sarcomeres, it follows that an increase in myofiber length would be mediated by an increase in the length of the sarcomeres and/or the serial addition of new sarcomeres. When considering these options, it is essential to bear in mind that the optimal length of sarcomeres ($\approx$2.5 µm) is highly conserved, and most muscles operate within a narrow range of the sarcomeres optimal length (94 $\pm$ 13%) [43]. Hence, it can be inferred that a mechanical load-induced increase in myofiber length would most likely be driven by the serial addition of new sarcomeres, as this would allow for the optimal length of the sarcomeres to be maintained in the elongated myofiber.

In support of the above rationale, Williams and Goldspink (1971) demonstrated that the increase in myofiber length that occurs during development is highly correlated with the serial addition of new sarcomeres [51], and a similar relationship is observed during the myofiber lengthening that occurs in response to increased mechanical loading. For instance, Williams and Goldpink (1973) demonstrated that the number of sarcomeres along the length of mouse soleus myofibers increases by 23% after $\approx$ 7 days of tenotomy (a milder form of the synergist ablation model) [135]. Likewise, Aoki et al. (2009) have shown that the number of sarcomeres along the length of rat soleus myofibers increases by 27% after just 4 days of chronic stretch [54]. Collectively, these, and many other studies [51,54,95–97,135–139], have not only indicated that mechanical loading could lead to the serial addition of new sarcomeres but also suggest that this type of growth can occur in a very rapid manner.

If mechanical loading leads to the serial addition of new sarcomeres, then it raises the question of where along the length of the myofibers the new sarcomeres are added. According to Goldspink (1983) "The point or points at which the sarcomeres are added has been rather uncertain until recently. With radioactively labeled amino acids and radioactively labeled adenosine the site of longitudinal growth was shown to be at the ends of the myofibrils" [140]. Although this is a fundamentally important conclusion, its validity remains highly contestable.

The first study that Goldspink cited as providing support for his conclusion was published by Griffin et al. (1971) and used ^{3}H-adenosine as a means for labeling where newly synthesized actin was deposited during the postnatal growth of myofibers [141]. Specifically, young mice were injected ^{3}H-adenosine, and then single myofibers were imaged with autoradiography. Based on the results, Griffin et al. concluded that the ^{3}H-adenosine was primarily deposited at the ends of the myofibers. Importantly, however, this conclusion was not supported by quantitative data, and the images included in the manuscript were far from persuasive [141].

The second study that Goldspink cited as support for his conclusion used ^{3}H-adenosine in an effort to identify where new sarcomeres were added in adult soleus muscles that were recovering from being immobilized in a shortened position [135]. The study began with a clear demonstration that serial sarcomere addition occurred during the recovery period. After establishing this point, the muscles were cut into 5 separate regions along the longitudinal axis and then analyzed for ^{3}H-adenosine. As shown in Figure 5, the outcomes revealed that the amount of ^{3}H-adenosine in the two most distal regions of the muscle was significantly elevated in muscles that were undergoing recovery. Importantly, however, whether the enhanced ^{3}H-adenosine deposition was due to formation of new sarcomeres at the ends of the myofibrils was not directly tested. Indeed, it could be argued that the results from this study simply reflect the type of regional differences in the mechanical load-induced growth that we described in Section 3.1.

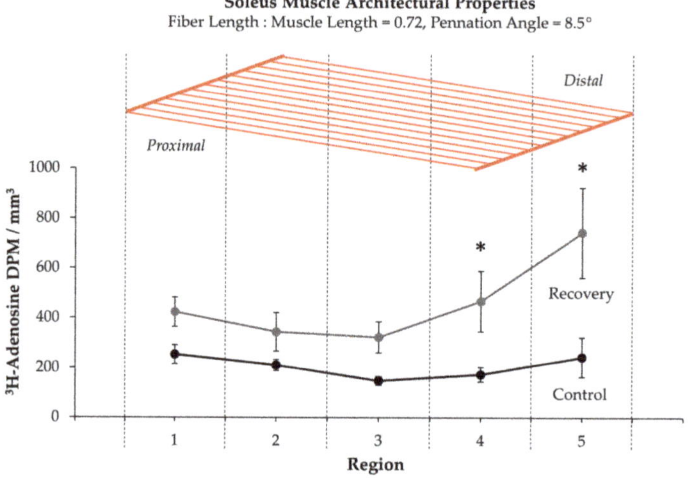

Figure 5. (Top) Schematic illustration of the soleus muscle and its basic architectural properties [142]. (Bottom) Summary of the data provided by Williams and Goldspink (1973) [135]. Values are presented as the means ± SEM and were analyzed with 2-way repeated measures ANOVA. *p*-values for the main effects (i.e., Treatment and Region) and interaction are provided. * Significantly different from the region-matched control condition.

In contrast to the notion that new sarcomeres are added at the ends of the myofibrils, others have provided evidence which suggests that new sarcomeres can be inserted throughout the length of the myofibrils [118,143–150]. For instance, when studying the developmental growth of single myofibers that possess two separate motor endplates, Bennett et al. (1985) discovered that the distance between the motor endplates increased in a manner that was directly proportional to the increase in myofiber length [146]. Similar evidence was obtained by Mackay and Harrop (1969) whom inserted wire markers at various points along the length of the sternomastoid and anterior gracilis muscles of 4 week old rats and then tracked their position with x-ray images during the subsequent 8 weeks of developmental growth [145]. In this case, a proportionate increase in the distance between wires occurred as the muscles grew in length, and this led the authors to conclude that the myofibers "must be adding new material at points all along their length as they grow". Indeed, Jahromi and Charlton (1979) obtained support for this concept when they found evidence of a longitudinal growth process that appears to involve the transverse splitting of sarcomeres that are embedded within the midst of the myofibrils (Figure 6A) [143].

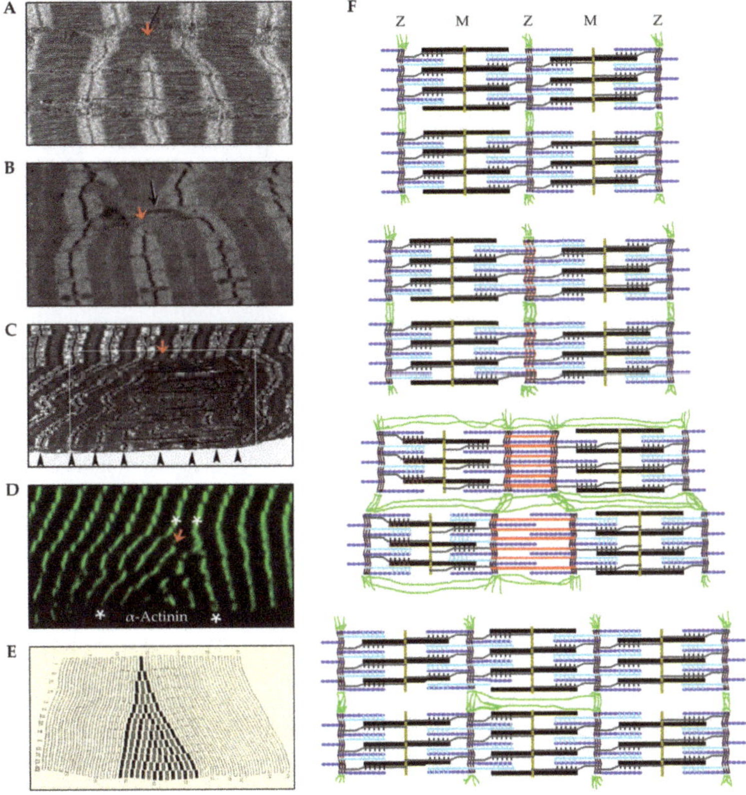

Figure 6. (**A**) Electron micrograph (EM) image that shows a group of myofibrils along with sarcomeres embedded within these myofibrils that possess a transverse split (red arrow) at the H-zone. The image was copied with permission under a Creative Commons License Attribution–Noncommercial–Share Alike 4.0 Unported license, and is available online at https://www.ncbi.nlm.nih.gov/pmc/articles/PMC2110374/ (accessed on 6/20/2020) [143]. (**B**) EM image of myofibrils from a muscle that was subjected to synergist ablation and appears to possess a transverse split at the H-zone (copied with permission from [118]). (**C,D**) EM image (**C**) and immunohistochemical image (**D**) of regions with "supernumerary sarcomeres" that are found in human skeletal muscles several days after being subjected to a bout of eccentric contractions (copied with permission from [147,148]). (**E**) Depiction of a "sphenode" region as detailed by Heidenhain (1919) [151]. (**F**) Illustration describing a mechanism for the in-series addition of new sarcomeres via transverse splitting at the Z-disc (copied with permission from [150]).

As described by Jahromi and Charlton (1979), the transverse splitting of sarcomeres appears to occur through an ordered sequence of events which include: (1) splitting of the thick filaments at the H-zone, (2) elongation of the two halves of the thick filaments along with the formation of new thin filaments in the previous H-zone, and (3) formation of a new Z-disc in the center of the newly formed thin filaments [143]. Although this process was originally described in crab skeletal muscles, there is evidence to suggest that the same process takes place in vertebrates. For instance, as shown in Figure 6B, Vaughan and Goldspink (1979) observed a similar phenomenon in soleus muscles of mice that had been subjected synergist ablation; however, in this instance, it was thought that the splitting was reflective of damage to the sarcomeres [118]. In fact, focal disruptions of the sarcomere, such as lesions, Z-disc streaming, and Z-disc smearing have long been viewed as markers of damage [152–156]. However, as detailed in a series of publications by Yu et al., these regions might simply be areas of remodeling that result in new sarcomere formation [147–149]. For instance, when examining soleus muscles from humans that had engaged in a bout of intense eccentric contractions, Yu et al. detected a 5-fold increase in the appearance of regions with "supernumerary sarcomeres" (Figure 6C,D) [147]. Such regions are remarkably similar to the "sphenode" regions that were described by Heidenhain over 100 years ago, which are characterized by the presence of additional sarcomeres that are out of register with the surrounding sarcomeres [151]. Interestingly, these regions appear to include areas that resemble H-zone transverse sarcomere splitting, as well as another potential type of transverse sarcomere splitting that occurs at the Z-disc (Figure 6F) [150]. Thus, when considering the studies that have been highlighted in this section, it is fair to conclude that mechanical loading can lead to the longitudinal growth of myofibers and this process is primarily driven by the serial addition of new sarcomeres. However, exactly how and where new sarcomeres get added along the length of the myofibrils remains to be resolved.

5.2. Radial Growth of Myofibers

In Section 4.2.1 we reviewed the evidence which indicates that the radial growth of myofibers (i.e., myofiber hypertrophy) is one of, if not the, primary contributor to the growth that occurs in response to increased mechanical loading. We will now examine what is known about the ultrastructural adaptations that drive this process. However, before going deeper into this topic, it is important to consider the concept of specific tension, which is defined as the maximal isometric force produced per CSA. At the myofiber level, the underlying premise for this concept is that the maximal isometric force is directly dependent on the number of the force-generating elements that act in parallel with the line of force production, and that the number of these elements is directly dependent on the CSA of the myofiber [157,158]. This thesis becomes particularly important when formulating hypotheses about the mechanisms that potentially contribute to the radial growth of the myofibers. For instance, if the CSA of a myofiber increases and specific tension remains constant, then it can be inferred that the radial growth was due to a proportionate addition of both force-generating elements (e.g., myofilaments/sarcomeres/myofibrils) and non-force-generating elements (e.g., mitochondria, sarcoplasmic reticulum, intracellular fluid, connective tissue, etc.). Alternatively, if the CSA of a myofiber increases and specific tension decreases, then it can be inferred that the radial growth was due to a disproportionately greater increase in the amount of non-force-generating elements. Thus, through measurements of specific tension, one can obtain fundamental insight into the mechanisms that drive the radial growth of the myofibers.

As summarized in a recent meta-analysis by Dankel et al. (2019), at least 15 different studies have assessed whether the specific tension of individual myofibers is impacted by resistance exercise. Importantly, the overwhelming majority of these studies have concluded that specific tension is either not significantly altered, or slightly increases in hypertrophied myofibers [159–167]. Similar observations have also been made in myofibers that were isolated from muscles that have adapted to extreme forms of mechanical loading, such as synergist ablation [168]. Thus, it would appear that the radial growth of myofibers is driven by a proportional increase in the force-generating and non-force-generating elements. However, despite this evidence, some have argued that a disproportionate increase in the non-force-generating elements can make a substantive contribution to radial growth. This type of radial growth has generically been referred to as sarcoplasmic hypertrophy, and in the following sections we will address in greater detail whether radial myofiber growth is driven by sarcoplasmic hypertrophy and/or the expansion of the force-generating elements that act in parallel with the line of force production.

5.2.1. Sarcoplasmic Hypertrophy

Anecdotal observations suggest that although bodybuilders have bigger muscles than powerlifters, they are not as strong. Such observations have led many to contend that myofiber hypertrophy in bodybuilders is due to a disproportionately larger increase in non-force-generating elements (i.e., sarcoplasmic hypertrophy). It has also been hypothesized that these non-force-generating elements could include osmotically active metabolites (e.g., creatine and glycogen) that would draw water into the myofiber, and/or organelles such as the sarcoplasmic reticulum and mitochondria [101,169,170]. However, the relevance of these hypotheses is dependent on whether sarcoplasmic hypertrophy makes a substantive contribution to the mechanical load-induced growth of myofibers. Thus, in this section, we will critically evaluate the evidence that surrounds this concept.

Several studies have been commonly cited as providing support for the existence of sarcoplasmic hypertrophy [32,36,163,171–174]. For instance, D'Antona et al. (2006) measured specific tension in single myofibers from recreationally active subjects, and from subjects that had engaged in bodybuilding for at least 2 years. With regards to providing support for sarcoplasmic hypertrophy, the often-cited outcome is that specific tension was lower in the Type I fibers of bodybuilders [163]. However, it is important to point out that the same study also observed an increase in specific tension of the Type IIA and IIX myofibers from the same bodybuilders [163]. The work of Meijer et al. (2015) is another frequently cited study that measured specific tension in single myofibers. In this case, specific tension was measured in myofibers from control subjects, bodybuilders, and powerlifters. Importantly, it was concluded that specific tension was lower in the myofibers obtained from bodybuilders [172]. At first glance it would appear that this study provides clear support for the notion that bodybuilders experience sarcoplasmic hypertrophy; however, 9 of the 12 bodybuilders in the study admitted to recent use of anabolic steroids [172]. This is noteworthy because the use of anabolic steroids has been associated with alterations in protein composition and the morphological properties of myofibers [175,176]. Indeed, MacDougall et al. (1982) reported a 9.8% decrease in the proportion of the myofiber CSA that is occupied by the myofibrils in elite bodybuilders and powerlifters (6 of 7 of whom admitted to the use of anabolic steroids), whereas only a 1.6% difference was observed after 6 months of resistance exercise in subjects that denied the use of anabolic steroids [32]. In addition to the aforementioned concerns, it also bears mentioning that the studies by D'Antona et al. and Meijer et al. were both cross-sectional in nature. This is important because it is well known that cross-sectional studies cannot be used to infer cause and effect relationships [177–180]. Thus, caution needs to be used when considering whether the outcomes of D'Antona et al. and Meijer et al. provide support for the presence of sarcoplasmic hypertrophy.

Other studies that have been cited as providing support for the existence of sarcoplasmic hypertrophy include the work of Penman (1969) who subjected participants to 8 weeks of an exercise intervention that included either progressive resistance exercise, isometric contractions, or stair running [171]. The frequently cited outcome from this study is that exercise led to a decrease in the "myosin concentration" (defined as number of myofibrils in a 5 μm^2 area) [171]. However, this study only included 2 subjects per group, there was a substantial amount of variance in the data, and no statistical analyses were performed.

Another commonly cited study involves the work of Toth et al. (2012) whom subjected older subjects ($\approx$73 years of age) to 18 weeks of resistance exercise and observed a significant decrease in the proportion of the myofiber CSA that was occupied by the myofibrils [36]. Importantly, however, the resistance exercise program employed in this study did not lead to a significant increase in myofiber CSA. Thus, if anything, the observed decrease in the proportion of the CSA that was occupied by the myofibrils would suggest that the resistance exercise program led to the selective loss of the myofibrils rather than a disproportionately large increase in non-force-generating elements (i.e., sarcoplasmic hypertrophy).

More recently, Haun et al. (2019) concluded that the myofiber hypertrophy that occurs after 6 weeks of high-volume resistance training can be largely attributed to sarcoplasmic hypertrophy [173]. Specifically, the key piece of evidence in this study was the observed trend for a decrease in the concentrations of myosin and actin after the 6 weeks of training (P = 0.052 and P = 0.055, respectively) [173]. Although these results are interesting, it should be noted that 15 subjects were analyzed in this study, and they only represented a subset of the 31 subjects that participated in the original training intervention [180]. More importantly, the 15 subjects that were examined only included the subjects who showed an "increase" in myofiber CSA (responders by the authors' definition) [173]. This is important because when all 31 subjects from the original training intervention were considered, it was determined that the 6 weeks of training did not induce myofiber hypertrophy [180]. Accordingly, the results of Haun et al. (2019) cannot be viewed as being representative of the whole population and, are therefore, difficult to interpret within the context of whether sarcoplasmic hypertrophy normally makes a substantive contribution to the mechanical load-induced growth of myofibers.

In summary, we remind the reader that as summarized by Dankel et al. (2019), a large number of longitudinal studies have shown that specific tension is preserved in myofibers that have experienced radial growth as a result of increased mechanical loading [159–168]. This consistent body of evidence strongly suggests that the radial growth of myofibers is not driven by sarcoplasmic hypertrophy, but rather is due to a proportionate increase in the force-generating and non-force-generating elements that act in parallel with the line of force production.

5.2.2. Expansion of the Force-Generating Elements

In myofibers from vertebrates, the force-generating myofilaments are contained within the sarcomere and organized into a hexagonal array of thick and thin myofilaments [181]. The overall geometry and spacing between the myofilaments is highly conserved, and thus, any changes in the number of force-generating myofilaments that are aligned in parallel would likely be matched by a proportionate alteration in the CSA that is occupied by the sarcomeres/myofibrils [182,183]. Given that specific tension is preserved in myofibers that have experienced radial growth as a result of increased mechanical loading, and that specific tension is dependent on the number of in parallel force-generating elements, it would follow that the radial growth is mediated by a propionate increase in the CSA that is occupied by the sarcomeres/myofibrils. Indeed, a handful of studies have directly tested this thesis, and all of them reported that induction of myofiber hypertrophy was associated with minimal changes (≤4%) in the relative proportion of the CSA that was occupied by the myofibrils [32–35,184]. For instance, MacDougall et al. (1982) reported that 6 months of resistance exercise in humans led to a 22–25% increase in the CSA of myofibers along with almost no change in the proportion of the CSA that was occupied by the myofibrils (84.2% vs. 82.6% in the pre- and post-trained states, respectively) [32]. Put differently, the data from MacDougall et al. indicated that the total area occupied by the myofibrils increased by ≈23%, but whether this was due to radial growth of the pre-existing myofibrils (myofibril hypertrophy) and/or an increase the number of myofibrils (myofibril hyperplasia) was not determined (Figure 7A) [32]. In fact, we are not aware of any studies that have systematically addressed whether mechanical load-induced myofiber hypertrophy is mediated by myofibril hypertrophy and/or myofibril hyperplasia. In our opinion, it is easy to envision how the induction of myofibril hypertrophy and/or myofibril hyperplasia could serve as the foundational events by which mechanical loading drives the radial growth of myofibers, thus the lack of knowledge on this topic is quite surprising.

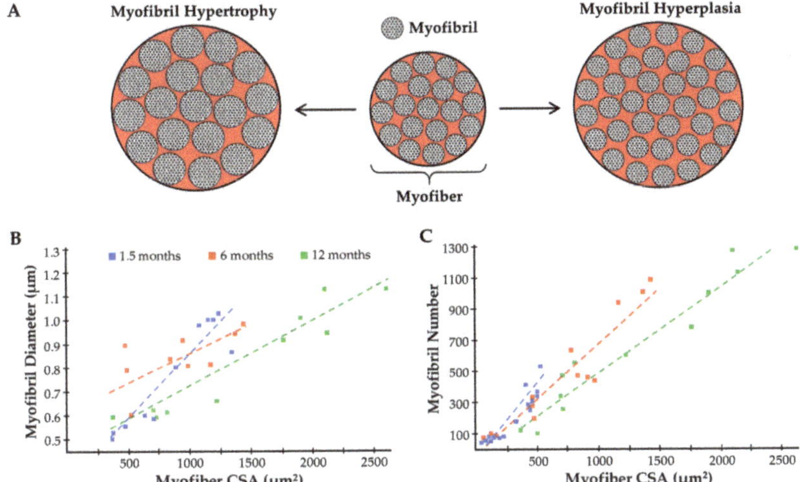

Figure 7. (**A**) Illustration of how an increase in the CSA of the pre-existing myofibrils (myofibril hypertrophy) and an increase in the number of myofibrils (myofibril hyperplasia) can contribute to the radial growth of myofibers. (**B,C**) Summary of the data from Goldspink (1970) which highlights the relationship that exists between myofiber CSA and myofibril diameter (**B**), as well as myofiber CSA and myofibril number (**C**), in mice of various ages [185].

Even though the concepts of myofibril hypertrophy and myofibril hyperplasia have not been thoroughly examined within the confines of mechanical load-induced skeletal muscle growth, there is still much that can be learned from related fields of study (e.g., developmental growth of skeletal muscle, mechanical load-induced growth of the heart, etc.). For instance, seminal work by Goldspink (1970) used mice of various ages to establish that a positive linear relationship exists between myofibril diameter and myofiber CSA, and a similar relationship was also found to exist between myofibril number and myofiber CSA (Figure 7B,C) [185]. Collectively, the results of this study provided some of the first evidence that both myofibril hypertrophy and myofibril hyperplasia could contribute to the radial growth of myofibers. Moreover, these observations provided the basis for Dr Goldspink's intriguing model of radial growth which involves a process he called myofibril splitting [141,185–187]. Specifically, Dr Goldspink proposed that the increase in myofibril number that occurs during the radial growth of myofibers could be explained by the longitudinal splitting of pre-existing myofibrils. In support of his hypothesis, he published numerous longitudinal images of single myofibrils that appeared to split into two smaller daughter myofibrils (Figure 8A) [185–187]. Moreover, he demonstrated that the splits usually occurred in the middle of Z-disc, and were typically found in myofibrils that are twice as large as myofibrils that did not contain splits [185].

In addition to his observations on longitudinal splitting, Dr Goldspink also noted that the thin myofilaments in sarcomeres do not run directly perpendicular to the Z-disc, but instead are offset at a slightly oblique angle ($\approx$6–10°) [186]. This was an important observation because it suggested that the thin myofilaments could exert outward radial forces on the Z-disc when the sarcomeres contract. Indeed, this became a key part of his myofibril splitting model in which it was proposed that myofibrils initially undergo hypertrophy and, as their diameter increases, the outward radial forces that they exert on the Z-disc also increases. The outward radial forces place a strain on the center of the Z-disc, and when these forces reach a critical threshold, it causes the Z-disc to break (Figure 8B). The break begins at the center of the Z-disc and forms a split which then propagates through the remainder of the myofibril and ultimately forms two smaller daughter myofibrils.

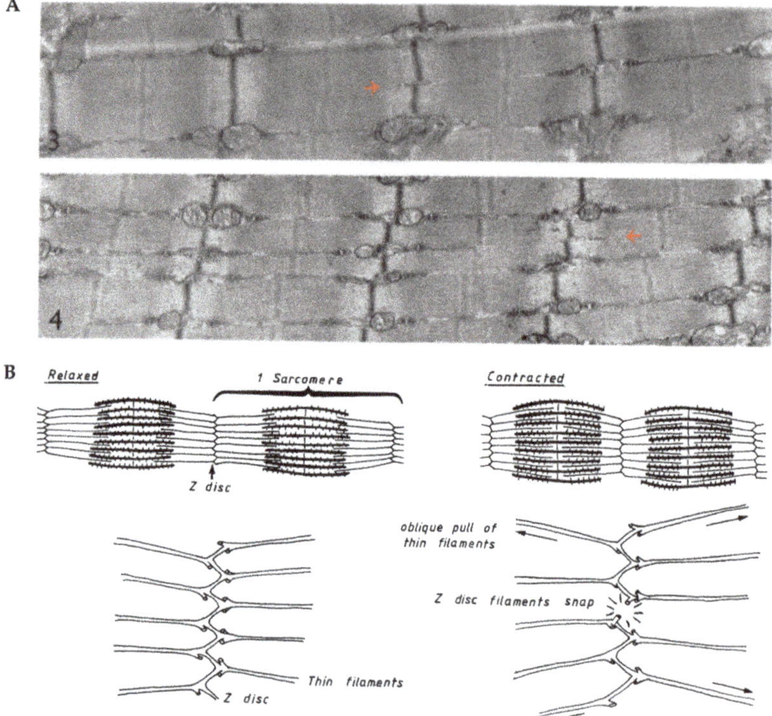

Figure 8. (**A**) Electron micrographs of longitudinal sections from mouse skeletal muscle. Red arrows highlight myofibrils that appear to split into two smaller daughter myofibrils (copied with permission from [186]). (**B**) Illustration from Goldspink (1983) which describes how the oblique angle of the thin myofilaments could exert outward radial forces on the Z-disc when the sarcomeres contract (copied with permission from [140]).

Dr Goldspink's model of myofibril splitting was developed over 40 years ago, and it has frequently served as the textbook explanation of how myofibril number could increase during the radial growth of myofibers [188–191]. However, despite being widely accepted, the validity of the model has not been rigorously tested. For instance, we are not aware of any direct evidence that a single myofibril can split into daughter myofibrils. Furthermore, we are not aware of any studies that have established whether the outward radial forces generated by the obliquely aligned myofilaments would be physically capable of "breaking" the Z-disc. In addition to limited evidence, there are also parts of myofibril splitting model that seem to be incomplete. For instance, as shown in Figure 8B, it has been shown that the diameter of the myofibrils is directly related to the size of the myofibers, and from our point of view, Dr Goldspink's model is not capable of explaining this relationship [185]. Nevertheless, the general concepts of the myofibril splitting model are well reasoned and, as such, it will serve as framework for remainder of our discussions on myofibril hypertrophy and myofibril hyperplasia.

5.2.3. Myofibril Hypertrophy

If we assume that the basic concepts of the myofibril splitting model are correct, and that they can be applied to the radial growth of myofibers that occurs in response to increased mechanical loading, then the first part of the overall growth process would involve myofibril hypertrophy. This initial hypertrophic response would continue until the myofibrils reached the critical size that induces splitting. The splitting would result in the formation of daughter myofibrils that would then undergo hypertrophy until they split, and the cycle would repeat until the radial growth of the myofiber commenced. We will now examine the limited body of literature that surrounds this thesis.

To the best of our knowledge, only one study has addressed whether mechanical load-induced myofiber hypertrophy is associated with myofibril hypertrophy [184]. This study was performed by Ashmore and Summers (1981) and was focused on defining the changes that occur in the patigialis muscle of young chickens after 1–7 days of wing-weighting [184]. Importantly, the same group had previously demonstrated their model of wing-weighting leads to an ≈55% in myofiber CSA after 7 days [136], and not surprisingly, their 1981 publication revealed that the increase in myofiber CSA was matched by a proportionate increase in the CSA that was occupied by the myofibrils [185]. In this study, they also found that the average CSA of the individual myofibrils increased by 36% after 7 days, and this was associated with a 2.6-fold increase in the proportion of myofibrils that presented with signs of splitting [136]. When taken together these results are very noteworthy because they provide critical support for the notion that mechanical loading can induce myofibril hypertrophy, and that this effect is associated with an increase in myofibril splitting.

The results of Ashmore and Summers (1981) provided support for the notion that mechanical loading can induce myofibril hypertrophy, and therefore raise questions about the processes that drive this response [184]. When considering these processes it is important to remember that the force-generating myofilaments within the myofibrils are organized into a hexagonal array and the spacing between the myofilaments is highly conserved [181–183]. Thus, it can be predicted that an increase in the CSA of the myofibril would be met by a proportionate increase in the number of force-generating myofilaments per CSA. If this is correct, then one is left with the question of where the new myofilaments get deposited.

As illustrated in Figure 9, some possible locations of new myofilament deposition include but are not limited to: (A) the periphery of the pre-existing myofibril, (B) the center of the pre-existing myofibril, or (C) throughout the pre-existing myofibril. All these options seem plausible, but options B and C would likely require extensive remodeling of the pre-existing myofilament lattice, whereas option A presumably would not. Thus, from a resources/energetic standpoint, the deposition of new myofilaments at the periphery of the pre-existing myofibril would appear to be the most cost-effective and least disruptive option.

The work of Morkin (1970) is often cited as providing support for the notion that new myofilaments are added to the periphery of myofibrils [192]. Specifically, in this study, rat diaphragm muscles were incubated with ^{3}H-leucine to label newly synthesized proteins, and then electron microscope autoradiography was used to identify the location of the newly synthesized proteins [192]. As shown in Figure 10A, the location of the newly synthesized proteins was indicated by the presence of relatively large (≈300 nm) electron dense grains. The quantitative results from this study are shown in Figure 10B with the bars indicating how frequently the center of the grains appeared at various distances from the periphery of the myofibril, and the green highlighted curve illustrating the theoretical distribution of the grains that would be expected if the myofibrils were labeled exclusively at the periphery. At first glance, the close match between the theoretical and observed values appears to provide compelling support for the conclusion that new myofilaments are added to the periphery of the myofibrils [192]. However, this evidence becomes less persuasive when one considers that ribosomes are typically localized in the intermyofibrillar space and many of these ribosomes appear in polysomal configurations which is indicative of active protein synthesis (Figure 10C) [193,194]. This point leads us to question how well the data from Morkin (1970) would fit with a different hypothesis. In this case, the hypothesis was that the ribosomes in the intermyofibrillar space are actively engaged in the synthesis of new proteins. In Figure 10D,E, we have illustrated how well the data from Morkin (1970) fit with the theoretical distribution of the grains that would be expected if the myofibrils were labeled exclusively at the periphery, and compared that with the theoretical distribution of the grains that would be expected if newly synthesized proteins were located exclusively within the intermyofibrillar space. The key point from this illustration is that the data appears to be consistent with both theoretical distributions, and this is because the resolution provided by autoradiography simply does not allow for a clear distinction between the two possibilities.

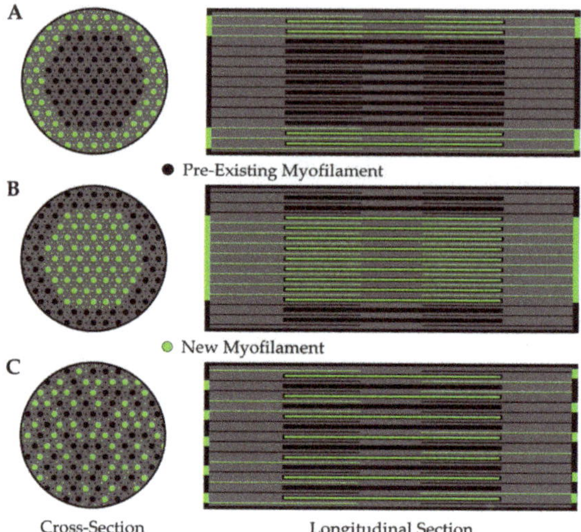

Figure 9. Illustration of where new myofilaments might be added during myofibril hypertrophy. The described possibilities include: (**A**) the periphery of the pre-existing myofibril, (**B**) the center of the pre-existing myofibril, or (**C**) throughout the pre-existing myofibril.

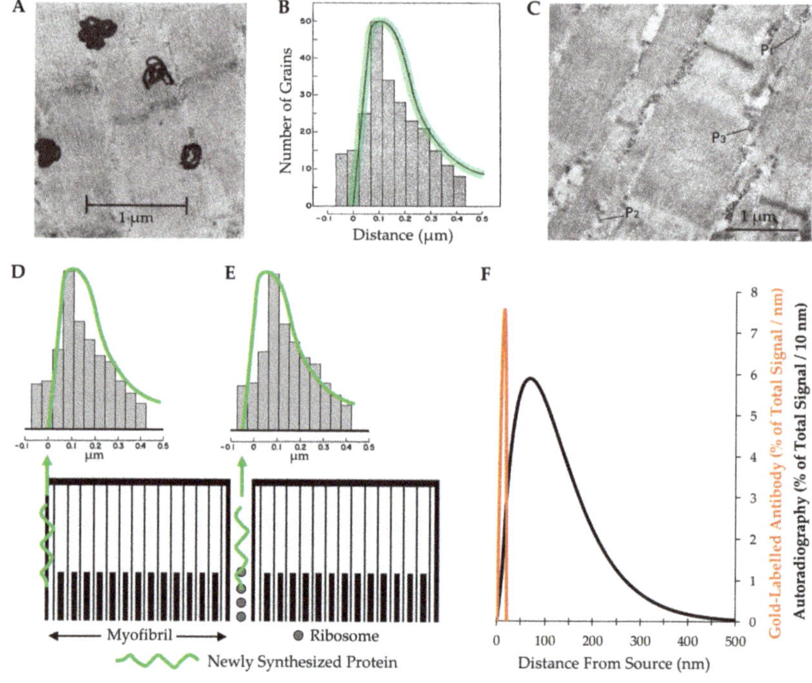

Figure 10. (**A**) Electron microscope autoradiograph from Morkin (1970) which shows the large electron dense grains that were used to identify the location of newly synthesized proteins in the rat diaphragm.

(**B**) Bars represent the frequency distribution of the grains relative to the periphery of the myofibril, and the green shaded curve illustrates the theoretical distribution that would be expected if the newly synthesized proteins were located exclusively at the periphery of the myofibril. The images in both A and B were copied with permission from [192]. (**C**) Electron micrograph of the levator ani muscle from an adult rat which reveals the presence of ribosomes in the intermyofibrillar space. Please note that many of the ribosomes appear in different polysomal configurations (P1, P2 and P3) (copied with permission from [193]). (**D,E**) Illustration of how well the data from Morkin 1970 fit with the theoretical distribution that would be expected if the newly synthesized proteins were located exclusively at the periphery of the myofibril (**D**) versus being located exclusively within the intermyofibrillar space (**E**). (**F**) A graph illustrating the theoretical radial distribution of the signal obtained with electron microscope autoradiography versus with immunoelectron microscopy that employed a primary antibody (15 nm diameter) conjugated to a 10 nm gold-particle [195–197].

The limitations of the resolution that can be obtained with electron microscope autoradiography have been thoroughly described by Caro (1962) and Salpeter et al. (1969) [195,198]. Importantly, both of these studies demonstrate that under typical conditions, 50% of the grains will develop within ≈130 nm of the source and 95% of the grains will develop within ≈300 nm [195,198] (Figure 10F). This level of resolution would be outstanding if the goal was to identify the location of newly synthesized proteins within a myofiber (typical diameter of 25,000 nm), but it is far from ideal when the goal is to identify the location of newly synthesized proteins within a myofibril (typical diameter 850 nm). To effectively accomplish this goal, technologies that offer a much higher level of resolution are needed, and fortunately, such technologies are now available. For instance, it is now possible to identify the location of newly synthesized proteins with immunological and click-chemistry-based technologies [199,200]. This is noteworthy because, as illustrated in Figure 10F, a typical immunoelectron microscopy-based approach will result in 100% of the signal appearing within 20 nm of the source, and the use of more advanced approaches (e.g., 1 nm gold conjugated Fab antibody fragments, or click-chemistry-based linkers) can allow for a resolution of less than 7 nm [197,201–203]. Thus, although we still do not know whether mechanical load-induced hypertrophy of myofibers is driven by myofibril hypertrophy, or where new myofilaments get added during the process of myofibril hypertrophy, the technologies that are need to answer these fundamental questions are now within our reach.

5.2.4. Myofibril Hyperplasia

As mentioned in the previous section, the study by Ashmore and Summers (1981) provided support for the notion that the mechanical load-induced radial growth of myofibers is associated with myofibril hypertrophy, but unfortunately, the study did not address the concept of myofibril hyperplasia [184]. In fact, we are not aware of any studies that have directly addressed this concept, and the only study we could find that even came remotely close was performed by Holmes and Rasch (1958) [204]. Specifically, this study involved 7 weeks of training rats with progressively more intense running and concluded that the number of myofibrils per myofiber in the sartorius muscle was not significantly altered by the training regime [204]. However, it was also determined that the training regime did not lead to a significant increase in mass of the sartorius muscle and, thus, it is difficult to extrapolate any meaningful insights from the data.

Although we are not aware of any studies in skeletal muscle that have directly addressed whether mechanical load-induced myofiber hypertrophy is associated with myofibril hyperplasia, there are a few studies that have addressed this topic in the heart. For instance, Toffolo and Ianuzzo (1994) used aortic constriction to subject rat hearts to mechanical overload and found that after 30 days, the cardiomyocyte area had increased by ≈50% and this was associated with an ≈70% increase in the number of myofibrils per cardiomyocyte [205]. An increase in the number of myofibrils per cardiomyocyte has also been observed in hypertrophied human hearts that were examined postmortem [206]. Furthermore, Anversa et al. (1980) examined heart papillary muscles after 8 days of mechanical overload and observed a 55%

increase in the CSA of the cardiomyocytes that was occupied by the myofibrils, but no change in the CSA of the individual myofibrils, thus implying an increase in myofibril number [207]. Taken together, these studies consistently suggest that an increase in mechanical loading can lead to an increase in myofibril number in the heart, but whether the same effect occurs in skeletal muscles remains to be determined.

5.2.5. The Radial Growth of Myofibers—Closing Remarks

As we have discussed, a substantial body of evidence indicates that the mechanical load-induced radial growth of myofibers is mediated by a proportional increase in the force-generating and non-force-generating elements. The force-generating elements are contained within the myofibrils, and the myofibrils account for ≈80% of the myofiber CSA. Thus, it can be argued that the bulk of the radial growth is driven by an expansion of the myofibrils. However, whether this expansion is due to hypertrophy of the individual myofibrils and/or myofibril hyperplasia remains to be established. Based on our collective view of the literature, we propose that both processes are involved, and can be explained by a model that we have defined as the "myofibril expansion cycle". Specifically, as illustrated in Figure 11, the myofibril expansion cycle begins with the deposition of new myofilaments around the periphery of the pre-existing myofibrils, and results in myofibril hypertrophy. Once the myofibrils reach a critical size, they split and subsequently form two smaller daughter myofibrils. The daughter myofibrils are then able to enter another round of the cycle, and the cycle repeats until the radial growth of the myofiber has commenced. Clearly, our model is based on an integration of hypotheses that were proposed more than 40 years ago, and as emphasized throughout this section, the validity of these hypotheses have not been rigorously tested. Fortunately, the technologies that are needed to test these hypotheses are now available. Thus, we hope that this section will help to inspire new investigations into this seemingly forgotten, yet critically important aspect of skeletal muscle biology.

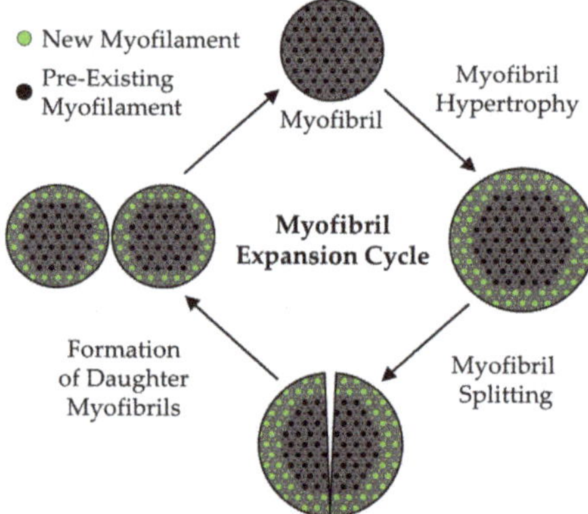

Figure 11. Illustration of the proposed "Myofibril Expansion Cycle".

6. Take Home Messages

Mechanical loads are one of the most potent regulators of muscle mass and the maintenance of muscle mass plays a critical role in health and quality of life. In Table 1 we have summarized the major structural adaptations that have been implicated in the mechanical load-induced growth of skeletal muscle. Based on our review, we have also considered whether each of these adaptations makes a substantive contribution to the overall growth process, as well as the level of evidence that is available to support that conclusion. The table also lists some of the major gaps in knowledge that we identified during our review of the literature. Importantly, this is not meant to be an exhaustive summary, and exclusion from the table does not indicate that a given adaptation or gap in knowledge is unimportant (e.g., satellite cell fusion, are satellite cells necessary for mechanical load-induced growth, etc.).

Table 1. The Structural Adaptations that Drive the Mechanical Load-Induced Growth of Skeletal Muscle.

Adaptation	Evidence	Gaps in Knowledge
Longitudinal Growth of Fascicles	High	Does mechanical loading alter the number of fascicles? Can mechanical loading lead to the addition of new myofibers in-series?
Radial Growth of Fascicles	High	To what extent does myofiber hyperplasia, myofiber splitting, and the lengthening of myofibers with intrafascicular terminations contribute to the radial growth of fascicles?
Myofiber Splitting	Low	Do physiologically relevant models of mechanical loading induce myofiber splitting?
Myofiber Hyperplasia	Low & Controversial	To what extent, if any, does myofiber hyperplasia contribute to the radial growth of fascicles?
Longitudinal Growth of Myofibers	Mixed - Model Dependent	Do physiologically relevant forms of mechanical loading induce the longitudinal growth of myofibers? Where, and how, are new sarcomeres added during the longitudinal growth of myofibers?
Radial Growth of Myofibers	Extremely High	Is mechanical load-induced myofiber hypertrophy driven by myofibril hypertrophy and/or myofibril hyperplasia?
Sarcoplasmic Hypertrophy	Low & Controversial	Are there specific conditions during which sarcoplasmic hypertrophy might make substantive contribution to the mechanical load-induced growth of myofibers?
Myofibril Hypertrophy	Low	Does mechanical loading lead to myofibril hypertrophy? Where are new myofilaments deposited during myofibril hypertrophy?
Myofibril Hyperplasia	Very Low	Does mechanical loading lead to myofibril hyperplasia? Are new myofibrils generated via the process of myofibril splitting?

As documented in this review, several of the adaptations that we consider as having weak supporting evidence have been engrained in the literature as "textbook" mechanisms (e.g., the longitudinal growth of myofibers is driven by the addition of new sarcomeres at the ends of myofibers, new myofibrils are formed via myofibril splitting, etc.). We hope that after reading this review, the reader appreciates how little we actually know about the structural adaptations that drive skeletal muscle growth, and the number of extremely fundamental gaps in knowledge that remain to be filled.

Author Contributions: K.W.J., S.M.P. and T.A.H.; all contributed to the writing of this manuscript. All authors have read and agreed to the published version of the manuscript.

Funding: The research reported in this publication was supported by the National Institute of Arthritis and Musculoskeletal and Skin Diseases of the National Institutes of Health under Award Number AR074932 to SMP and TAH. The content is solely the responsibility of the authors and does not necessarily represent the official views of the National Institutes of Health.

Conflicts of Interest: The authors declare no conflict of interest.

References

1. McLeod, M.; Breen, L.; Hamilton, D.L.; Philp, A. Live strong and prosper: The importance of skeletal muscle strength for healthy ageing. *Biogerontology* **2016**, *17*, 497–510. [CrossRef]
2. Doherty, T.J. Invited Review: Aging and sarcopenia. *J. Appl. Physiol.* **2003**, *95*, 1717–1727. [CrossRef]
3. Mitchell, W.K.; Williams, J.; Atherton, P.; Larvin, M.; Lund, J.; Narici, M. Sarcopenia, dynapenia, and the impact of advancing age on human skeletal muscle size and strength; A quantitative review. *Front. Physiol.* **2012**, *3*, 260. [CrossRef]
4. Kalyani, R.R.; Corriere, M.; Ferrucci, L. Age-related and disease-related muscle loss: The effect of diabetes, obesity, and other diseases. *Lancet Diabetes Endocrinol.* **2014**, *2*, 819–829. [CrossRef]
5. Shou, J.; Chen, P.-J.; Xiao, W.-H. Mechanism of increased risk of insulin resistance in aging skeletal muscle. *Diabetol. Metab. Syndr.* **2020**, *12*, 14. [CrossRef] [PubMed]
6. Beaudart, C.; Zaaria, M.; Pasleau, F.; Reginster, J.-Y.; Bruyère, O. Health Outcomes of Sarcopenia: A Systematic Review and Meta-Analysis. *PLoS ONE* **2017**, *12*, e0169548. [CrossRef] [PubMed]

7. Chen, H.; Ma, J.; Liu, A.; Cui, Y.; Ma, X. The association between sarcopenia and fracture in middle-aged and elderly people: A systematic review and meta-analysis of cohort studies. *Injury* **2020**, *51*, 804–811. [CrossRef]

8. Janssen, I.; Shepard, D.S.; Katzmarzyk, P.T.; Roubenoff, R. The Healthcare Costs of Sarcopenia in the United States. *J. Am. Geriatr. Soc.* **2004**, *52*, 80–85. [CrossRef] [PubMed]

9. UN. *World Population Prospects 2019: Highlights*; ST/ESA/SER.A/423; Department of Economic and Social Affairs, Population Division: New York, NY, USA, 2019.

10. Todd, J. From Milo to Milo: A History of Barbells, Dumbells, and Indian Clubs. *Iron Game History* **1995**, *3*, 4–16.

11. Abe, T.; Kojima, K.; Kearns, C.F.; Yohena, H.; Fukuda, J. Whole body muscle hypertrophy from resistance training: Distribution and total mass. *Br. J. Sports Med.* **2003**, *37*, 543–545. [CrossRef]

12. Trappe, T.A.; Carroll, C.C.; Dickinson, J.M.; LeMoine, J.K.; Haus, J.M.; Sullivan, B.E.; Lee, J.D.; Jemiolo, B.; Weinheimer, E.M.; Hollon, C.J. Influence of acetaminophen and ibuprofen on skeletal muscle adaptations to resistance exercise in older adults. *Am. J. Physiol. Regul. Integr. Comp. Physiol.* **2011**, *300*, R655–R662. [CrossRef] [PubMed]

13. Franchi, M.V.; Atherton, P.J.; Reeves, N.D.; Fluck, M.; Williams, J.; Mitchell, W.K.; Selby, A.; Beltran Valls, R.M.; Narici, M.V. Architectural, functional and molecular responses to concentric and eccentric loading in human skeletal muscle. *Acta Physiol.* **2014**, *210*, 642–654. [CrossRef] [PubMed]

14. Mandić, M.; Rullman, E.; Widholm, P.; Lilja, M.; Dahlqvist Leinhard, O.; Gustafsson, T.; Lundberg, T.R. Automated assessment of regional muscle volume and hypertrophy using MRI. *Sci. Rep.* **2020**, *10*, 2239. [CrossRef] [PubMed]

15. Mitchell, C.J.; Churchward-Venne, T.A.; Parise, G.; Bellamy, L.; Baker, S.K.; Smith, K.; Atherton, P.J.; Phillips, S.M. Acute post-exercise myofibrillar protein synthesis is not correlated with resistance training-induced muscle hypertrophy in young men. *PLoS ONE* **2014**, *9*, e89431. [CrossRef]

16. Aagaard, P.; Andersen, J.L.; Dyhre-Poulsen, P.; Leffers, A.M.; Wagner, A.; Magnusson, S.P.; Halkjaer-Kristensen, J.; Simonsen, E.B. A mechanism for increased contractile strength of human pennate muscle in response to strength training: Changes in muscle architecture. *J. Physiol.* **2001**, *534*, 613–623. [CrossRef] [PubMed]

17. Mitchell, C.J.; Churchward-Venne, T.A.; West, D.W.; Burd, N.A.; Breen, L.; Baker, S.K.; Phillips, S.M. Resistance exercise load does not determine training-mediated hypertrophic gains in young men. *J. Appl. Physiol.* **2012**, *113*, 71–77. [CrossRef] [PubMed]

18. Hornberger, T.A., Jr.; Farrar, R.P. Physiological hypertrophy of the FHL muscle following 8 weeks of progressive resistance exercise in the rat. *Can. J. Appl. Physiol.* **2004**, *29*, 16–31. [CrossRef]

19. Wong, T.S.; Booth, F.W. Skeletal muscle enlargement with weight-lifting exercise by rats. *J. Appl. Physiol.* **1988**, *65*, 950–954. [CrossRef]

20. Ogasawara, R.; Fujita, S.; Hornberger, T.A.; Kitaoka, Y.; Makanae, Y.; Nakazato, K.; Naokata, I. The role of mTOR signalling in the regulation of skeletal muscle mass in a rodent model of resistance exercise. *Sci. Rep.* **2016**, *6*, 31142. [CrossRef]

21. Lowe, D.A.; Alway, S.E. Animal models for inducing muscle hypertrophy: Are they relevant for clinical applications in humans? *J. Orthop. Sports Phys. Ther.* **2002**, *32*, 36–43. [CrossRef]

22. Goodman, C.A.; Frey, J.W.; Mabrey, D.M.; Jacobs, B.L.; Lincoln, H.C.; You, J.S.; Hornberger, T.A. The role of skeletal muscle mTOR in the regulation of mechanical load-induced growth. *J. Physiol.* **2011**, *589*, 5485–5501. [CrossRef]

23. Holly, R.G.; Barnett, J.G.; Ashmore, C.R.; Taylor, R.G.; Mole, P.A. Stretch-induced growth in chicken wing muscles: A new model of stretch hypertrophy. *Am. J. Physiol.* **1980**, *238*, C62–C71. [CrossRef]

24. Frontera, W.R.; Ochala, J. Skeletal Muscle: A Brief Review of Structure and Function. *Calcif. Tissue Int.* **2015**, *96*, 183–195. [CrossRef]

25. Lieber, R.L. Skeletal muscle adaptability. I: Review of basic properties. *Dev. Med. Child. Neurol.* **1986**, *28*, 390–397. [CrossRef] [PubMed]

26. Henderson, C.A.; Gomez, C.G.; Novak, S.M.; Mi-Mi, L.; Gregorio, C.C. Overview of the Muscle Cytoskeleton. *Comprehensive Physiol.* **2017**, *7*, 891–944. [CrossRef]

27. Narici, M.; Franchi, M.; Maganaris, C. Muscle structural assembly and functional consequences. *J. Exp. Biol.* **2016**, *219*, 276–284. [CrossRef]

28. Betts, G.J.; Kelly, A.Y.; Wise, J.A.; Johnson, E.; Poe, B.; Kruse, D.H.; Korol, O.; Johnson, J.E.; Womble, M.; DeSaix, P. *Anatomy and Physiology*; OpenStax: Houston, TX, USA, 2013.

29. Biferali, B.; Proietti, D.; Mozzetta, C.; Madaro, L. Fibro-Adipogenic Progenitors Cross-Talk in Skeletal Muscle: The Social Network. *Front. Physiol.* **2019**, *10*, 1074. [CrossRef]

30. Tedesco, F.S.; Moyle, L.A.; Perdiguero, E. Muscle Interstitial Cells: A Brief Field Guide to Non-satellite Cell Populations in Skeletal Muscle. *Methods Mol. Biol.* **2017**, *1556*, 129–147. [CrossRef] [PubMed]

31. Murach, K.A.; Fry, C.S.; Kirby, T.J.; Jackson, J.R.; Lee, J.D.; White, S.H.; Dupont-Versteegden, E.E.; McCarthy, J.J.; Peterson, C.A. Starring or Supporting Role? Satellite Cells and Skeletal Muscle Fiber Size Regulation. *Physiology* **2018**, *33*, 26–38. [CrossRef]

32. MacDougall, J.D.; Sale, D.G.; Elder, G.C.; Sutton, J.R. Muscle ultrastructural characteristics of elite powerlifters and bodybuilders. *Eur. J. Appl. Physiol. Occup. Physiol.* **1982**, *48*, 117–126. [CrossRef]

33. Seiden, D. Quantitative analysis of muscle cell changes in compensatory hypertrophy and work-induced hypertrophy. *Am. J. Anat.* **1976**, *145*, 459–465. [CrossRef] [PubMed]

34. Luthi, J.M.; Howald, H.; Claassen, H.; Rosler, K.; Vock, P.; Hoppeler, H. Structural changes in skeletal muscle tissue with heavy-resistance exercise. *Int. J. Sports Med.* **1986**, *7*, 123–127. [CrossRef] [PubMed]

35. Goldspink, G.; Howells, K.F. Work-induced hypertrophy in exercised normal muscles of different ages and the reversibility of hypertrophy after cessation of exercise. *J. Physiol.* **1974**, *239*, 179–193. [CrossRef]

36. Toth, M.J.; Miller, M.S.; VanBuren, P.; Bedrin, N.G.; LeWinter, M.M.; Ades, P.A.; Palmer, B.M. Resistance training alters skeletal muscle structure and function in human heart failure: Effects at the tissue, cellular and molecular levels. *J. Physiol.* **2012**, *590*, 1243–1259. [CrossRef]

37. Lavorato, M.; Loro, E.; Debattisti, V.; Khurana, T.S.; Franzini-Armstrong, C. Elongated mitochondrial constrictions and fission in muscle fatigue. *J. Cell Sci.* **2018**, *131*. [CrossRef] [PubMed]

38. Dirksen, R.T. Sarcoplasmic reticulum-mitochondrial through-space coupling in skeletal muscle. *Appl. Physiol. Nutr. Metab.* **2009**, *34*, 389–395. [CrossRef] [PubMed]

39. Gordon, A.M.; Homsher, E.; Regnier, M. Regulation of contraction in striated muscle. *Physiol. Rev.* **2000**, *80*, 853–924. [CrossRef]

40. Billeter, R.; Hoppeler, H. Muscular Basis of Strength. In *Strength and Power in Sport*; Blackwell Science Ltd.: Oxford, UK, 2003.

41. Edman, K.A. The velocity of unloaded shortening and its relation to sarcomere length and isometric force in vertebrate muscle fibres. *J. Physiol.* **1979**, *291*, 143–159. [CrossRef]

42. Gordon, A.M.; Huxley, A.F.; Julian, F.J. The variation in isometric tension with sarcomere length in vertebrate muscle fibres. *J. Physiol.* **1966**, *184*, 170–192. [CrossRef] [PubMed]

43. Burkholder, T.J.; Lieber, R.L. Sarcomere length operating range of vertebrate muscles during movement. *J. Exp. Biol.* **2001**, *204*, 1529–1536.

44. Fry, A.C. The role of resistance exercise intensity on muscle fibre adaptations. *Sports Med.* **2004**, *34*, 663–679. [CrossRef]

45. Scott, W.; Stevens, J.; Binder–Macleod, S.A. Human Skeletal Muscle Fiber Type Classifications. *Phys. Ther.* **2001**, *81*, 1810–1816. [CrossRef] [PubMed]

46. Hilber, K.; Galler, S.; Gohlsch, B.; Pette, D. Kinetic properties of myosin heavy chain isoforms in single fibers from human skeletal muscle. *FEBS Lett.* **1999**, *455*, 267–270. [CrossRef]

47. Smerdu, V.; Karsch-Mizrachi, I.; Campione, M.; Leinwand, L.; Schiaffino, S. Type IIx myosin heavy chain transcripts are expressed in type IIb fibers of human skeletal muscle. *Am. J. Physiol. Cell Physiol.* **1994**, *267*, C1723–C1728. [CrossRef]

48. Ennion, S.; Sant'ana Pereira, J.; Sargeant, A.J.; Young, A.; Goldspink, G. Characterization of human skeletal muscle fibres according to the myosin heavy chains they express. *J. Muscle Res. Cell Motil.* **1995**, *16*, 35–43. [CrossRef]

49. Pette, D.; Peuker, H.; Staron, R.S. The impact of biochemical methods for single muscle fibre analysis. *Acta Physiol. Scand.* **1999**, *166*, 261–277. [CrossRef]

50. Goldspink, G. Increase in Length of Skeletal Muscle during Normal Growth. *Nature* **1964**, *204*, 1095–1096. [CrossRef]

51. Williams, P.E.; Goldspink, G. Longitudinal growth of striated muscle fibres. *J. Cell Sci.* **1971**, *9*, 751–767. [PubMed]

52. Beaucage, K.L.; Pollmann, S.I.; Sims, S.M.; Dixon, S.J.; Holdsworth, D.W. Quantitative in vivo micro-computed tomography for assessment of age-dependent changes in murine whole-body composition. *Bone Rep.* **2016**, *5*, 70–80. [CrossRef]

53. Zöllner, A.M.; Abilez, O.J.; Böl, M.; Kuhl, E. Stretching skeletal muscle: Chronic muscle lengthening through sarcomerogenesis. *PLoS ONE* **2012**, *7*, e45661. [CrossRef]

54. Aoki, M.S.; Soares, A.G.; Miyabara, E.H.; Baptista, I.L.; Moriscot, A.S. Expression of genes related to myostatin signaling during rat skeletal muscle longitudinal growth. *Muscle Nerve* **2009**, *40*, 992–999. [CrossRef]

55. Simpson, A.H.; Williams, P.E.; Kyberd, P.; Goldspink, G.; Kenwright, J. The response of muscle to leg lengthening. *J. Bone Joint Surg. Br.* **1995**, *77*, 630–636. [CrossRef] [PubMed]

56. Caiozzo, V.J.; Utkan, A.; Chou, R.; Khalafi, A.; Chandra, H.; Baker, M.; Rourke, B.; Adams, G.; Baldwin, K.; Green, S. Effects of distraction on muscle length: Mechanisms involved in sarcomerogenesis. *Clin. Orthop. Relat. Res.* **2002**, S133–S145. [CrossRef] [PubMed]

57. Roy, R.R.; Edgerton, V.R. Response of mouse plantaris muscle to functional overload: Comparison with rat and cat. *Comp. Biochem. Physiol. Physiol.* **1995**, *111*, 569–575. [CrossRef]

58. Terena, S.M.; Fernandes, K.P.; Bussadori, S.K.; Deana, A.M.; Mesquita-Ferrari, R.A. Systematic review of the synergist muscle ablation model for compensatory hypertrophy. *Rev. Assoc. Med. Bras. (1992)* **2017**, *63*, 164–172. [CrossRef] [PubMed]

59. Kawakami, Y.; Abe, T.; Kuno, S.Y.; Fukunaga, T. Training-induced changes in muscle architecture and specific tension. *Eur. J. Appl. Physiol. Occup. Physiol.* **1995**, *72*, 37–43. [CrossRef] [PubMed]

60. Roman, W.J.; Fleckenstein, J.; Stray-Gundersen, J.; Alway, S.E.; Peshock, R.; Gonyea, W.J. Adaptations in the elbow flexors of elderly males after heavy-resistance training. *J. Appl. Physiol. (1985)* **1993**, *74*, 750–754. [CrossRef] [PubMed]

61. Welle, S.; Totterman, S.; Thornton, C. Effect of age on muscle hypertrophy induced by resistance training. *J. Gerontol. Biol. Sci. Med. Sci.* **1996**, *51*, M270–M275. [CrossRef]

62. Angleri, V.; Ugrinowitsch, C.; Libardi, C.A. Crescent pyramid and drop-set systems do not promote greater strength gains, muscle hypertrophy, and changes on muscle architecture compared with traditional resistance training in well-trained men. *Eur. J. Appl. Physiol.* **2017**, *117*, 359–369. [CrossRef]

63. Damas, F.; Libardi, C.A.; Ugrinowitsch, C. The development of skeletal muscle hypertrophy through resistance training: The role of muscle damage and muscle protein synthesis. *Eur. J. Appl. Physiol.* **2018**, *118*, 485–500. [CrossRef]

64. Housh, D.J.; Housh, T.J.; Johnson, G.O.; Chu, W.K. Hypertrophic response to unilateral concentric isokinetic resistance training. *J. Appl. Physiol. (1985)* **1992**, *73*, 65–70. [CrossRef] [PubMed]

65. McMahon, G.E.; Morse, C.I.; Burden, A.; Winwood, K.; Onambele, G.L. Impact of range of motion during ecologically valid resistance training protocols on muscle size, subcutaneous fat, and strength. *J. Strength Cond. Res.* **2014**, *28*, 245–255. [CrossRef] [PubMed]

66. Bellamy, L.M.; Joanisse, S.; Grubb, A.; Mitchell, C.J.; McKay, B.R.; Phillips, S.M.; Baker, S.; Parise, G. The acute satellite cell response and skeletal muscle hypertrophy following resistance training. *PLoS ONE* **2014**, *9*, e109739. [CrossRef] [PubMed]

67. Blazevich, A.J.; Cannavan, D.; Coleman, D.R.; Horne, S. Influence of concentric and eccentric resistance training on architectural adaptation in human quadriceps muscles. *J. Appl. Physiol. (1985)* **2007**, *103*, 1565–1575. [CrossRef] [PubMed]

68. Narici, M.V.; Hoppeler, H.; Kayser, B.; Landoni, L.; Claassen, H.; Gavardi, C.; Conti, M.; Cerretelli, P. Human quadriceps cross-sectional area, torque and neural activation during 6 months strength training. *Acta Physiol. Scand.* **1996**, *157*, 175–186. [CrossRef] [PubMed]

69. Sakuma, K.; Yamaguchi, A.; Katsuta, S. Are region-specific changes in fibre types attributable to nonuniform muscle hypertrophy by overloading? *Eur. J. Appl. Physiol. Occup. Physiol.* **1995**, *71*, 499–504. [CrossRef]

70. Gardiner, P.F.; Jasmin, B.J.; Corriveau, P. Rostrocaudal pattern of fiber-type changes in an overloaded rat ankle extensor. *J. Appl. Physiol. (1985)* **1991**, *71*, 558–564. [CrossRef]

71. Antonio, J. Nonuniform Response of Skeletal Muscle to Heavy Resistance Training: Can Bodybuilders Induce Regional Muscle Hypertrophy? *J. Strength Cond. Res.* **2000**, *14*, 102–113. [CrossRef]

72. Alway, S.E.; Winchester, P.K.; Davis, M.E.; Gonyea, W.J. Regionalized adaptations and muscle fiber proliferation in stretch-induced enlargement. *J. Appl. Physiol. (1985)* **1989**, *66*, 771–781. [CrossRef]

73. Counts, B.R.; Buckner, S.L.; Mouser, J.G.; Dankel, S.J.; Jessee, M.B.; Mattocks, K.T.; Loenneke, J.P. Muscle growth: To infinity and beyond? *Muscle Nerve* **2017**, *56*, 1022–1030. [CrossRef]

74. Damas, F.; Phillips, S.M.; Lixandrao, M.E.; Vechin, F.C.; Libardi, C.A.; Roschel, H.; Tricoli, V.; Ugrinowitsch, C. Early resistance training-induced increases in muscle cross-sectional area are concomitant with edema-induced muscle swelling. *Eur. J. Appl. Physiol.* **2016**, *116*, 49–56. [CrossRef] [PubMed]

75. Maxwell, L.C.; Faulkner, J.A.; Hyatt, G.J. Estimation of number of fibers in guinea pig skeletal muscles. *J. Appl. Physiol.* **1974**, *37*, 259–264. [CrossRef] [PubMed]

76. Jorgenson, K.W.; Hornberger, T.A. The Overlooked Role of Fiber Length in Mechanical Load-Induced Growth of Skeletal Muscle. *Exerc. Sport Sci. Rev.* **2019**, *47*, 258–259. [CrossRef] [PubMed]

77. O'Brien, T.D.; Reeves, N.D.; Baltzopoulos, V.; Jones, D.A.; Maganaris, C.N. Muscle-tendon structure and dimensions in adults and children. *J. Anat.* **2010**, *216*, 631–642. [CrossRef]

78. Timmins, R.G.; Ruddy, J.D.; Presland, J.; Maniar, N.; Shield, A.J.; Williams, M.D.; Opar, D.A. Architectural Changes of the Biceps Femoris Long Head after Concentric or Eccentric Training. *Med. Sci. Sports Exerc.* **2016**, *48*, 499–508. [CrossRef]

79. Sharifnezhad, A.; Marzilger, R.; Arampatzis, A. Effects of load magnitude, muscle length and velocity during eccentric chronic loading on the longitudinal growth of the vastus lateralis muscle. *J. Exp. Biol.* **2014**, *217*, 2726–2733. [CrossRef]

80. Reeves, N.D.; Maganaris, C.N.; Longo, S.; Narici, M.V. Differential adaptations to eccentric versus conventional resistance training in older humans. *Exp. Physiol.* **2009**, *94*, 825–833. [CrossRef]

81. Baroni, B.M.; Geremia, J.M.; Rodrigues, R.; De Azevedo Franke, R.; Karamanidis, K.; Vaz, M.A. Muscle architecture adaptations to knee extensor eccentric training: Rectus femoris vs. vastus lateralis. *Muscle Nerve* **2013**, *48*, 498–506. [CrossRef]

82. Ullrich, B.; Holzinger, S.; Soleimani, M.; Pelzer, T.; Stening, J.; Pfeitter, M. Neuromuscular Responses to 14 Weeks of Traditional and Daily Undulating Resistance Training. *Int. J. Sports Med.* **2015**, *36*, 554–562. [CrossRef]

83. Ema, R.; Wakahara, T.; Miyamoto, N.; Kanehisa, H.; Kawakami, Y. Inhomogeneous architectural changes of the quadriceps femoris induced by resistance training. *Eur. J. Appl. Physiol.* **2013**, *113*, 2691–2703. [CrossRef]

84. Erskine, R.M.; Jones, D.A.; Williams, A.G.; Stewart, C.E.; Degens, H. Inter-individual variability in the adaptation of human muscle specific tension to progressive resistance training. *Eur. J. Appl. Physiol.* **2010**, *110*, 1117–1125. [CrossRef] [PubMed]

85. Alegre, L.M.; Ferri-Morales, A.; Rodriguez-Casares, R.; Aguado, X. Effects of isometric training on the knee extensor moment-angle relationship and vastus lateralis muscle architecture. *Eur. J. Appl. Physiol.* **2014**, *114*, 2437–2446. [CrossRef] [PubMed]

86. Ema, R.; Akagi, R.; Wakahara, T.; Kawakami, Y. Training-induced changes in architecture of human skeletal muscles: Current evidence and unresolved issues. *J. Phys. Fitness Sports Med.* **2016**, *5*, 37–46. [CrossRef]

87. Franchi, M.V.; Atherton, P.J.; Maganaris, C.N.; Narici, M.V. Fascicle length does increase in response to longitudinal resistance training and in a contraction-mode specific manner. *Springerplus* **2016**, *5*, 94. [CrossRef] [PubMed]

88. Timmins, R.G.; Shield, A.J.; Williams, M.D.; Lorenzen, C.; Opar, D.A. Architectural adaptations of muscle to training and injury: A narrative review outlining the contributions by fascicle length, pennation angle and muscle thickness. *Br. J. Sports Med.* **2016**, *50*, 1467–1472. [CrossRef]

89. Franchi, M.V.; Reeves, N.D.; Narici, M.V. Skeletal Muscle Remodeling in Response to Eccentric vs. Concentric Loading: Morphological, Molecular, and Metabolic Adaptations. *Front. Physiol.* **2017**, *8*, 447. [CrossRef]

90. Young, M.; Paul, A.; Rodda, J.; Duxson, M.; Sheard, P. Examination of intrafascicular muscle fiber terminations: Implications for tension delivery in series-fibered muscles. *J. Morphol.* **2000**, *245*, 130–145. [CrossRef]

91. Paul, A.C.; Rosenthal, N. Different modes of hypertrophy in skeletal muscle fibers. *J. Cell Biol.* **2002**, *156*, 751–760. [CrossRef]

92. Swatland, H.J.; Cassens, R.G. Muscle growth: The problem of muscle fibers with an intrafascicular termination. *J. Anim. Sci.* **1972**, *35*, 336–344. [CrossRef]

93. Alway, S.E.; Gonyea, W.J.; Davis, M.E. Muscle fiber formation and fiber hypertrophy during the onset of stretch-overload. *Am. J. Physiol.* **1990**, *259*, C92–C102. [CrossRef]

94. Roy, R.R.; Meadows, I.D.; Baldwin, K.M.; Edgerton, V.R. Functional significance of compensatory overloaded rat fast muscle. *J. Appl. Physiol. Respir. Environ. Exerc. Physiol.* **1982**, *52*, 473–478. [CrossRef] [PubMed]

95. Lindsey, C.A.; Makarov, M.R.; Shoemaker, S.; Birch, J.G.; Buschang, P.H.; Cherkashin, A.M.; Welch, R.D.; Samchukov, M.L. The effect of the amount of limb lengthening on skeletal muscle. *Clin. Orthop. Relat. Res.* 2002. [CrossRef] [PubMed]

96. Butterfield, T.A.; Leonard, T.R.; Herzog, W. Differential serial sarcomere number adaptations in knee extensor muscles of rats is contraction type dependent. *J. Appl. Physiol. (1985)* **2005**, *99*, 1352–1358. [CrossRef] [PubMed]

97. Lynn, R.; Morgan, D.L. Decline running produces more sarcomeres in rat vastus intermedius muscle fibers than does incline running. *J. Appl. Physiol. (1985)* **1994**, *77*, 1439–1444. [CrossRef]

98. You, J.S.; McNally, R.M.; Jacobs, B.L.; Privett, R.E.; Gundermann, D.M.; Lin, K.H.; Steinert, N.D.; Goodman, C.A.; Hornberger, T.A. The role of raptor in the mechanical load-induced regulation of mTOR signaling, protein synthesis, and skeletal muscle hypertrophy. *FASEB J.* **2019**, *33*, 4021–4034. [CrossRef]

99. Goh, Q.; Song, T.; Petrany, M.J.; Cramer, A.A.; Sun, C.; Sadayappan, S.; Lee, S.J.; Millay, D.P. Myonuclear accretion is a determinant of exercise-induced remodeling in skeletal muscle. *Elife* **2019**, *8*. [CrossRef]

100. Morpurgo, B. Eine experimentelle Studie. *Arch. Pathol. Anat. Physiol. Klin. Med.* **1897**, *150*, 522–554. [CrossRef]

101. Haun, C.T.; Vann, C.G.; Roberts, B.M.; Vigotsky, A.D.; Schoenfeld, B.J.; Roberts, M.D. A Critical Evaluation of the Biological Construct Skeletal Muscle Hypertrophy: Size Matters but So Does the Measurement. *Front. Physiol.* **2019**, *10*, 247. [CrossRef] [PubMed]

102. Gonyea, W.J.; Ericson, G.C. An experimental model for the study of exercise-induced skeletal muscle hypertrophy. *J. Appl. Physiol.* **1976**, *40*, 630–633. [CrossRef]

103. Goldspink, G. The Combined Effects of Exercise and Reduced Food Intake on Skeletal Muscle Fibers. *J. Cell Comp. Physiol* **1964**, *63*, 209–216. [CrossRef] [PubMed]

104. Schiaffino, S.; Pierobon Bormioli, S.; Aloisi, M. Cell proliferation in rat skeletal muscle during early stages of compensatory hypertrophy. *Virchows Archiv B* **1972**, *11*, 268–273. [CrossRef]

105. Sola, O.M.; Christensen, D.L.; Martin, A.W. Hypertrophy and hyperplasia of adult chicken anterior latissimus dorsi muscles following stretch with and without denervation. *Exp. Neurol.* **1973**, *41*, 76–100. [CrossRef]

106. Conceicao, M.S.; Vechin, F.C.; Lixandrao, M.; Damas, F.; Libardi, C.A.; Tricoli, V.; Roschel, H.; Camera, D.; Ugrinowitsch, C. Muscle Fiber Hypertrophy and Myonuclei Addition: A Systematic Review and Meta-analysis. *Med. Sci. Sports Exerc.* **2018**, *50*, 1385–1393. [CrossRef] [PubMed]

107. Kosek, D.J.; Kim, J.S.; Petrella, J.K.; Cross, J.M.; Bamman, M.M. Efficacy of 3 days/wk resistance training on myofiber hypertrophy and myogenic mechanisms in young vs. older adults. *J. Appl. Physiol. (1985)* **2006**, *101*, 531–544. [CrossRef] [PubMed]

108. Campos, G.E.; Luecke, T.J.; Wendeln, H.K.; Toma, K.; Hagerman, F.C.; Murray, T.F.; Ragg, K.E.; Ratamess, N.A.; Kraemer, W.J.; Staron, R.S. Muscular adaptations in response to three different resistance-training regimens: Specificity of repetition maximum training zones. *Eur. J. Appl. Physiol.* **2002**, *88*, 50–60. [CrossRef]

109. Antonio, J.; Gonyea, W.J. Progressive stretch overload of skeletal muscle results in hypertrophy before hyperplasia. *J. Appl. Physiol. (1985)* **1993**, *75*, 1263–1271. [CrossRef]

110. Murach, K.A.; Dungan, C.M.; Peterson, C.A.; McCarthy, J.J. Muscle Fiber Splitting Is a Physiological Response to Extreme Loading in Animals. *Exerc. Sport Sci. Rev.* **2019**, *47*, 108–115. [CrossRef]

111. Schwartz, M.S.; Sargeant, M.; Swash, M. Longitudinal fibre splitting in neurogenic muscular disorders–its relation to the pathogenesis of "myopathic" change. *Brain* **1976**, *99*, 617–636. [CrossRef]

112. Pichavant, C.; Pavlath, G.K. Incidence and severity of myofiber branching with regeneration and aging. *Skeletal Muscle* **2014**, *4*, 9. [CrossRef]

113. Antonio, J.; Gonyea, W.J. Muscle fiber splitting in stretch-enlarged avian muscle. *Med. Sci. Sports Exerc.* **1994**, *26*, 973–977. [CrossRef]

114. Tamaki, T.; Akatsuka, A.; Tokunaga, M.; Uchiyama, S.; Shiraishi, T. Characteristics of compensatory hypertrophied muscle in the rat: I. Electron microscopic and immunohistochemical studies. *Anat. Rec.* **1996**, *246*, 325–334. [CrossRef]

115. Eriksson, A.; Lindström, M.; Carlsson, L.; Thornell, L.-E. Hypertrophic muscle fibers with fissures in power-lifters; fiber splitting or defect regeneration? *Histochem. Cell Biol.* **2006**, *126*, 409–417. [CrossRef] [PubMed]

116. Ho, K.W.; Roy, R.R.; Tweedle, C.D.; Heusner, W.W.; Van Huss, W.D.; Carrow, R.E. Skeletal muscle fiber splitting with weight-lifting exercise in rats. *Am. J. Anat.* **1980**, *157*, 433–440. [CrossRef]

117. Gonyea, W.; Ericson, G.C.; Bonde-Petersen, F. Skeletal muscle fiber splitting induced by weight-lifting exercise in cats. *Acta Physiol. Scand.* **1977**, *99*, 105–109. [CrossRef] [PubMed]

118. Vaughan, H.S.; Goldspink, G. Fibre number and fibre size in a surgically overloaded muscle. *J. Anat.* **1979**, *129*, 293–303. [PubMed]

119. Gollnick, P.D.; Timson, B.F.; Moore, R.L.; Riedy, M. Muscular enlargement and number of fibers in skeletal muscles of rats. *J. Appl. Physiol. Respir. Environ. Exerc. Physiol.* **1981**, *50*, 936–943. [CrossRef] [PubMed]

120. Gollnick, P.D.; Parsons, D.; Riedy, M.; Moore, R.L. Fiber number and size in overloaded chicken anterior latissimus dorsi muscle. *J. Appl. Physiol. Respir. Environ. Exerc. Physiol.* **1983**, *54*, 1292–1297. [CrossRef]

121. Timson, B.F.; Bowlin, B.K.; Dudenhoeffer, G.A.; George, J.B. Fiber number, area, and composition of mouse soleus muscle following enlargement. *J. Appl. Physiol.* **1985**, *58*, 619–624. [CrossRef]

122. Gonyea, W.J.; Sale, D.G.; Gonyea, F.B.; Mikesky, A. Exercise induced increases in muscle fiber number. *Eur. J. Appl. Physiol. Occup. Physiol.* **1986**, *55*, 137–141. [CrossRef]

123. Li, M.; Zhou, X.; Chen, Y.; Nie, Y.; Huang, H.; Chen, H.; Mo, D. Not all the number of skeletal muscle fibers is determined prenatally. *BMC Dev. Biol.* **2015**, *15*, 42. [CrossRef]

124. Timson, B.F.; Dudenhoeffer, G.A. Skeletal muscle fibre number in the rat from youth to adulthood. *J. Anat.* **1990**, *173*, 33–36. [PubMed]

125. Rehfeldt, C.; Stickland, N.C.; Fiedler, I.; Wegner, J. Environmental and Genetic Factors as Sources of Variation in Skeletal Muscle Fibre Number. *Basic Appl. Myol.* **1999**, *9*, 235–253.

126. Kelley, G. Mechanical overload and skeletal muscle fiber hyperplasia: A meta-analysis. *J. Appl. Physiol.* **1996**, *81*, 1584–1588. [CrossRef] [PubMed]

127. Taylor, N.A.; Wilkinson, J.G. Exercise-induced skeletal muscle growth. Hypertrophy or hyperplasia? *Sports Med.* **1986**, *3*, 190–200. [CrossRef] [PubMed]

128. Gonyea, W.J. Role of exercise in inducing increases in skeletal muscle fiber number. *J. Appl. Physiol. Respir. Environ. Exerc. Physiol.* **1980**, *48*, 421–426. [CrossRef]

129. Alway, S.E. Stretch induces non-uniform isomyosin expression in the quail anterior latissimus dorsi muscle. *Anat. Rec.* **1993**, *237*, 1–7. [CrossRef]

130. Abruzzo, P.M.; Esposito, F.; Marchionni, C.; di Tullio, S.; Belia, S.; Fulle, S.; Veicsteinas, A.; Marini, M. Moderate exercise training induces ROS-related adaptations to skeletal muscles. *Int. J. Sports Med.* **2013**, *34*, 676–687. [CrossRef] [PubMed]

131. Alway, S.E. Perpetuation of muscle fibers after removal of stretch in the Japanese quail. *Am. J. Physiol.* **1991**, *260*, C400–C408. [CrossRef] [PubMed]

132. Antonio, J.; Gonyea, W.J. Skeletal muscle fiber hyperplasia. *Med. Sci. Sports Exerc.* **1993**, *25*, 1333–1345. [CrossRef] [PubMed]

133. Brown, L.E.; Incledon, T. Skeletal Muscle Fiber Hyperplasia: Why It Can or Cannot Occur in Humans. *Strength Cond. J.* **2000**, *22*, 28. [CrossRef]

134. Tamaki, T.; Uchiyama, S.; Nakano, S. A weight-lifting exercise model for inducing hypertrophy in the hindlimb muscles of rats. *Med. Sci. Sports Exerc.* **1992**, *24*, 881–886. [CrossRef]

135. Williams, P.E.; Goldspink, G. The effect of immobilization on the longitudinal growth of striated muscle fibres. *J. Anat.* **1973**, *116*, 45–55. [PubMed]

136. Barnett, J.G.; Holly, R.G.; Ashmore, C.R. Stretch-induced growth in chicken wing muscles: Biochemical and morphological characterization. *Am. J. Physiol.* **1980**, *239*, C39–C46. [CrossRef]

137. Koh, T.J. Do adaptations in serial sarcomere number occur with strength training? *Hum. Mov. Sci.* **1995**, *14*, 61–77. [CrossRef]

138. Matano, T.; Tamai, K.; Kurokawa, T. Adaptation of skeletal muscle in limb lengthening: A light diffraction study on the sarcomere length in situ. *J. Orthop. Res.* **1994**, *12*, 193–196. [CrossRef]

139. Kinney, M.C.; Dayanidhi, S.; Dykstra, P.B.; McCarthy, J.J.; Peterson, C.A.; Lieber, R.L. Reduced skeletal muscle satellite cell number alters muscle morphology after chronic stretch but allows limited serial sarcomere addition. *Muscle Nerve* **2017**, *55*, 384–392. [CrossRef]

140. Goldspink, G. Alterations in myofibril size and structure during growth, exercise and change in environmental temperature. In *Handbook of Physiology*; American Physiological Society, Bethesda: Rockville, Maryland, USA, 1983; Volume 10, pp. 539–554.

141. Griffin, G.E.; Williams, P.E.; Goldspink, G. Region of longitudinal growth in striated muscle fibres. *Nat. New Biol.* **1971**, *232*, 28–29. [CrossRef]

142. Burkholder, T.J.; Fingado, B.; Baron, S.; Lieber, R.L. Relationship between muscle fiber types and sizes and muscle architectural properties in the mouse hindlimb. *J. Morphol.* **1994**, *221*, 177–190. [CrossRef] [PubMed]

143. Jahromi, S.S.; Charlton, M.P. Transverse sarcomere splitting. A possible means of longitudinal growth in crab muscles. *J. Cell Biol.* **1979**, *80*, 736–742. [CrossRef] [PubMed]

144. Crawford, G.N. An experimental study of muscle growth in the rabbit. *J. Bone Jt. Surg. Br.* **1954**, *36-B*, 294–303. [CrossRef]

145. Mackay, B.; Harrop, T.J. An experimental study of the longitudinal growth of skeletal muscle in the rat. *Acta Anat.* **1969**, *72*, 38–49. [CrossRef] [PubMed]

146. Bennett, M.R.; Fernandez, H.; Lavidis, N.A. Development of the mature distribution of synapses on fibres in the frog sartorius muscle. *J. Neurocytol.* **1985**, *14*, 981–995. [CrossRef] [PubMed]

147. Yu, J.G.; Carlsson, L.; Thornell, L.E. Evidence for myofibril remodeling as opposed to myofibril damage in human muscles with DOMS: An ultrastructural and immunoelectron microscopic study. *Histochem. Cell Biol.* **2004**, *121*, 219–227. [CrossRef] [PubMed]

148. Yu, J.G.; Furst, D.O.; Thornell, L.E. The mode of myofibril remodelling in human skeletal muscle affected by DOMS induced by eccentric contractions. *Histochem. Cell Biol.* **2003**, *119*, 383–393. [CrossRef]

149. Yu, J.G.; Thornell, L.E. Desmin and actin alterations in human muscles affected by delayed onset muscle soreness: A high resolution immunocytochemical study. *Histochem. Cell Biol.* **2002**, *118*, 171–179. [CrossRef]

150. Carlsson, L.; Yu, J.G.; Moza, M.; Carpen, O.; Thornell, L.E. Myotilin: A prominent marker of myofibrillar remodelling. *Neuromuscul. Disord.* **2007**, *17*, 61–68. [CrossRef]

151. Heidenhain, M. Uber die Noniusfelder der Muskelfaser. *Anat. Hefte* **1919**, *56*, 321–402.

152. Friden, J. Changes in human skeletal muscle induced by long-term eccentric exercise. *Cell Tissue Res.* **1984**, *236*, 365–372. [CrossRef]

153. Orfanos, Z.; Godderz, M.P.; Soroka, E.; Godderz, T.; Rumyantseva, A.; van der Ven, P.F.; Hawke, T.J.; Furst, D.O. Breaking sarcomeres by in vitro exercise. *Sci. Rep.* **2016**, *6*, 19614. [CrossRef]

154. Friden, J.; Sjostrom, M.; Ekblom, B. Myofibrillar damage following intense eccentric exercise in man. *Int. J. Sports Med.* **1983**, *4*, 170–176. [CrossRef]

155. Gibala, M.J.; MacDougall, J.D.; Tarnopolsky, M.A.; Stauber, W.T.; Elorriaga, A. Changes in human skeletal muscle ultrastructure and force production after acute resistance exercise. *J. Appl. Physiol.* **1995**, *78*, 702–708. [CrossRef]

156. Clarkson, P.M.; Hubal, M.J. Exercise-induced muscle damage in humans. *Am. J. Phys. Med. Rehabil.* **2002**, *81*, S52–S69. [CrossRef] [PubMed]

157. Close, R.I. Dynamic properties of mammalian skeletal muscles. *Physiol. Rev.* **1972**, *52*, 129–197. [CrossRef] [PubMed]

158. Fitts, R.H.; McDonald, K.S.; Schluter, J.M. The determinants of skeletal muscle force and power: Their adaptability with changes in activity pattern. *J. Biomech.* **1991**, *24* (Suppl. S1), 111–122. [CrossRef]

159. Erskine, R.M.; Jones, D.A.; Maffulli, N.; Williams, A.G.; Stewart, C.E.; Degens, H. What causes in vivo muscle specific tension to increase following resistance training? *Exp. Physiol.* **2011**, *96*, 145–155. [CrossRef]

160. Trappe, S.; Williamson, D.; Godard, M.; Porter, D.; Rowden, G.; Costill, D. Effect of resistance training on single muscle fiber contractile function in older men. *J. Appl. Physiol.* **2000**, *89*, 143–152. [CrossRef]

161. Trappe, S.; Godard, M.; Gallagher, P.; Carroll, C.; Rowden, G.; Porter, D. Resistance training improves single muscle fiber contractile function in older women. *Am. J. Physiol. Cell Physiol.* **2001**, *281*, C398–C406. [CrossRef]

162. Widrick, J.J.; Stelzer, J.E.; Shoepe, T.C.; Garner, D.P. Functional properties of human muscle fibers after short-term resistance exercise training. *Am. J. Physiol. Regul. Integr. Comp. Physiol.* **2002**, *283*, R408–R416. [CrossRef]

163. D'Antona, G.; Lanfranconi, F.; Pellegrino, M.A.; Brocca, L.; Adami, R.; Rossi, R.; Moro, G.; Miotti, D.; Canepari, M.; Bottinelli, R. Skeletal muscle hypertrophy and structure and function of skeletal muscle fibres in male body builders. *J. Physiol.* **2006**, *570*, 611–627. [CrossRef]

164. Paoli, A.; Pacelli, Q.F.; Cancellara, P.; Toniolo, L.; Moro, T.; Canato, M.; Miotti, D.; Neri, M.; Morra, A.; Quadrelli, M.; et al. Protein Supplementation Does Not Further Increase Latissimus Dorsi Muscle Fiber Hypertrophy after Eight Weeks of Resistance Training in Novice Subjects, but Partially Counteracts the Fast-to-Slow Muscle Fiber Transition. *Nutrients* **2016**, *8*, 331. [CrossRef]

165. Claflin, D.R.; Larkin, L.M.; Cederna, P.S.; Horowitz, J.F.; Alexander, N.B.; Cole, N.M.; Galecki, A.T.; Chen, S.; Nyquist, L.V.; Carlson, B.M.; et al. Effects of high- and low-velocity resistance training on the contractile properties of skeletal muscle fibers from young and older humans. *J. Appl. Physiol.* **2011**, *111*, 1021–1030. [CrossRef]

166. Pansarasa, O.; Rinaldi, C.; Parente, V.; Miotti, D.; Capodaglio, P.; Bottinelli, R. Resistance training of long duration modulates force and unloaded shortening velocity of single muscle fibres of young women. *J. Electromyogr. Kinesiol.* **2009**, *19*, e290–e300. [CrossRef]

167. Dankel, S.J.; Kang, M.; Abe, T.; Loenneke, J.P. Resistance training induced changes in strength and specific force at the fiber and whole muscle level: A meta-analysis. *Eur. J. Appl. Physiol.* **2019**, *119*, 265–278. [CrossRef]

168. Mendias, C.L.; Schwartz, A.J.; Grekin, J.A.; Gumucio, J.P.; Sugg, K.B. Changes in muscle fiber contractility and extracellular matrix production during skeletal muscle hypertrophy. *J. Appl. Physiol.* **2017**, *122*, 571–579. [CrossRef]

169. Vann, C.G.; Roberson, P.A.; Osburn, S.C.; Mumford, P.W.; Romero, M.A.; Fox, C.D.; Moore, J.H.; Haun, C.T.; Beck, D.T.; Moon, J.R.; et al. Skeletal Muscle Myofibrillar Protein Abundance Is Higher in Resistance-Trained Men, and Aging in the Absence of Training May Have an Opposite Effect. *Sports* **2020**, *8*, 7. [CrossRef] [PubMed]

170. Aguiar, A.F.; de Souza, R.W.; Aguiar, D.H.; Aguiar, R.C.; Vechetti, I.J., Jr.; Dal-Pai-Silva, M. Creatine does not promote hypertrophy in skeletal muscle in supplemented compared with nonsupplemented rats subjected to a similar workload. *Nutr. Res.* **2011**, *31*, 652–657. [CrossRef] [PubMed]

171. Penman, K.A. Ultrastructural changes in human striated muscle using three methods of training. *Res. Q* **1969**, *40*, 764–772. [CrossRef]

172. Meijer, J.P.; Jaspers, R.T.; Rittweger, J.; Seynnes, O.R.; Kamandulis, S.; Brazaitis, M.; Skurvydas, A.; Pisot, R.; Simunic, B.; Narici, M.V.; et al. Single muscle fibre contractile properties differ between body-builders, power athletes and control subjects. *Exp. Physiol.* **2015**, *100*, 1331–1341. [CrossRef] [PubMed]

173. Haun, C.T.; Vann, C.G.; Osburn, S.C.; Mumford, P.W.; Roberson, P.A.; Romero, M.A.; Fox, C.D.; Johnson, C.A.; Parry, H.A.; Kavazis, A.N.; et al. Muscle fiber hypertrophy in response to 6 weeks of high-volume resistance training in trained young men is largely attributed to sarcoplasmic hypertrophy. *PLoS ONE* **2019**, *14*, e0215267. [CrossRef] [PubMed]

174. Roberts, M.D.; Romero, M.A.; Mobley, C.B.; Mumford, P.W.; Roberson, P.A.; Haun, C.T.; Vann, C.G.; Osburn, S.C.; Holmes, H.H.; Greer, R.A.; et al. Skeletal muscle mitochondrial volume and myozenin-1 protein differences exist between high versus low anabolic responders to resistance training. *PeerJ* **2018**, *6*, e5338. [CrossRef] [PubMed]

175. Kadi, F.; Eriksson, A.; Holmner, S.; Thornell, L.E. Effects of anabolic steroids on the muscle cells of strength-trained athletes. *Med. Sci. Sports Exerc.* **1999**, *31*, 1528–1534. [CrossRef] [PubMed]

176. Eriksson, A.; Kadi, F.; Malm, C.; Thornell, L.E. Skeletal muscle morphology in power-lifters with and without anabolic steroids. *Histochem. Cell Biol.* **2005**, *124*, 167–175. [CrossRef] [PubMed]

177. Kesmodel, U.S. Cross-sectional studies - what are they good for? *Acta Obs. Gynecol. Scand.* **2018**, *97*, 388–393. [CrossRef] [PubMed]

178. Setia, M.S. Methodology Series Module 3: Cross-sectional Studies. *Indian J. Derm.* **2016**, *61*, 261–264. [CrossRef]

179. Mann, C.J. Observational research methods. Research design II: Cohort, cross sectional, and case-control studies. *Emerg. Med. J.* **2003**, *20*, 54–60. [CrossRef]

180. Haun, C.T.; Vann, C.G.; Mobley, C.B.; Roberson, P.A.; Osburn, S.C.; Holmes, H.M.; Mumford, P.M.; Romero, M.A.; Young, K.C.; Moon, J.R.; et al. Effects of Graded Whey Supplementation During Extreme-Volume Resistance Training. *Front. Nutr.* **2018**, *5*, 84. [CrossRef]

181. Huxley, H.E. Electron microscope studies of the organisation of the filaments in striated muscle. *Biochim. Biophys. Acta* **1953**, *12*, 387–394. [CrossRef]

182. Millman, B.M. The filament lattice of striated muscle. *Physiol. Rev.* **1998**, *78*, 359–391. [CrossRef]

183. Luther, P.K.; Squire, J.M. The intriguing dual lattices of the Myosin filaments in vertebrate striated muscles: Evolution and advantage. *Biology* **2014**, *3*, 846–865. [CrossRef] [PubMed]

184. Ashmore, C.R.; Summers, P.J. Stretch-induced growth in chicken wing muscles: Myofibrillar proliferation. *Am. J. Physiol.* **1981**, *241*, C93–C97. [CrossRef]

185. Goldspink, G. The proliferation of myofibrils during muscle fibre growth. *J. Cell Sci.* **1970**, *6*, 593–603. [PubMed]

186. Goldspink, G. Changes in striated muscle fibres during contraction and growth with particular reference to myofibril splitting. *J. Cell Sci.* **1971**, *9*, 123–137. [PubMed]

187. Patterson, S.; Goldspink, G. Mechanism of myofibril growth and proliferation in fish muscle. *J. Cell Sci.* **1976**, *22*, 607–616. [PubMed]

188. MacInstosh, B.; Gardiner, P.; McComas, A. *Skeletal Muscle: Form and Function*; Human Kinetics: Champaign, IL, USA, 2006.

189. Komi, P.V. *Strength and Power in Sport*; John Wiley & Sons: Hoboken, NJ, USA, 2008.

190. Bourne, G.H. *The Structure and Function of Muscle*; Elsevier: Amsterdam, The Netherlands, 2013.

191. McComas, A.J. *Neuromuscular Function and Disorders*; Butterworth-Heinemann: Oxford, UK, 2013.

192. Morkin, E. Postnatal muscle fiber assembly: Localization of newly synthesized myofibrillar proteins. *Science* **1970**, *167*, 1499–1501. [CrossRef]

193. Galavazi, G.; Szirmai, J.A. The influence of age and testosterone on the ribosomal population in the m. levator ani and a thigh muscle of the rat. *Z Zellforsch Mikrosk Anat.* **1971**, *121*, 548–560. [CrossRef]

194. Galavazi, G. Identification of helical polyribosomes in sections of mature skeletal muscle fibers. *Z Zellforsch Mikrosk Anat.* **1971**, *121*, 531–547. [CrossRef]

195. Salpeter, M.M.; Bachmann, L.; Salpeter, E.E. Resolution in electron microscope radioautography. *J. Cell Biol.* **1969**, *41*, 1–32. [CrossRef]

196. Caro, L.G. High-resolution autoradiogaphy. II. The problem of resolution. *J. Cell Biol.* **1962**, *15*, 189–199. [CrossRef]

197. Goodman, C.A.; Mabrey, D.M.; Frey, J.W.; Miu, M.H.; Schmidt, E.K.; Pierre, P.; Hornberger, T.A. Novel insights into the regulation of skeletal muscle protein synthesis as revealed by a new nonradioactive in vivo technique. *FASEB J.* **2011**, *25*, 1028–1039. [CrossRef]

198. Tom Dieck, S.; Muller, A.; Nehring, A.; Hinz, F.I.; Bartnik, I.; Schuman, E.M.; Dieterich, D.C. Metabolic labeling with noncanonical amino acids and visualization by chemoselective fluorescent tagging. *Curr. Protoc. Cell Biol.* **2012**, *56*, 7.11.1–7.11.29. [CrossRef]

199. Hermann, R.; Walther, P.; Muller, M. Immunogold labeling in scanning electron microscopy. *Histochem. Cell Biol.* **1996**, *106*, 31–39. [CrossRef] [PubMed]

200. Huang, B.; Babcock, H.; Zhuang, X. Breaking the diffraction barrier: Super-resolution imaging of cells. *Cell* **2010**, *143*, 1047–1058. [CrossRef] [PubMed]

201. Raulf, A.; Spahn, C.K.; Zessin, P.J.; Finan, K.; Bernhardt, S.; Heckel, A.; Heilemann, M. Click chemistry facilitates direct labelling and super-resolution imaging of nucleic acids and proteinsdagger. *Rsc Adv.* **2014**, *4*, 30462–30466. [CrossRef] [PubMed]

202. Mobius, W.; Posthuma, G. Sugar and ice: Immunoelectron microscopy using cryosections according to the Tokuyasu method. *Tissue Cell* **2019**, *57*, 90–102. [CrossRef]

203. Klein, J.S.; Gnanapragasam, P.N.; Galimidi, R.P.; Foglesong, C.P.; West, A.P., Jr.; Bjorkman, P.J. Examination of the contributions of size and avidity to the neutralization mechanisms of the anti-HIV antibodies b12 and 4E10. *Proc. Natl. Acad. Sci. USA* **2009**, *106*, 7385–7390. [CrossRef]

204. Holmes, R.; Rasch, P.J. Effect of exercise on number of myofibrils per fiber in sartorius muscle of the rat. *Am. J. Physiol.* **1958**, *195*, 50–52. [CrossRef]

205. Toffolo, R.L.; Ianuzzo, C.D. Myofibrillar adaptations during cardiac hypertrophy. *Mol. Cell Biochem.* **1994**, *131*, 141–149. [CrossRef]

206. Richter, G.W.; Kellner, A. Hypertrophy of the human heart at the level of fine structure. An analysis and two postulates. *J. Cell Biol.* **1963**, *18*, 195–206. [CrossRef]

207. Anversa, P.; Olivetti, G.; Melissari, M.; Loud, A.V. Stereological measurement of cellular and subcellular hypertrophy and hyperplasia in the papillary muscle of adult rat. *J. Mol. Cell. Cardiol.* **1980**, *12*, 781–795. [CrossRef]

Article

Priority Strategy of Intracellular Ca^{2+} Homeostasis in Skeletal Muscle Fibers during the Multiple Stresses of Hibernation

Jie Zhang [1,2,†], Xiaoyu Li [3,†], Fazeela Ismail [1,2], Shenhui Xu [1,2], Zhe Wang [1,2], Xin Peng [1,2], Chenxi Yang [4], Hui Chang [1,2,*], Huiping Wang [1,2] and Yunfang Gao [1,2,*]

[1]   Key Laboratory of Resource Biology and Biotechnology in Western China, College of Life Sciences, Northwest University, Ministry of Education, Xi'an 710069, China; zhangjie1@stumail.nwu.edu.cn (J.Z.); rjfazila123@gmail.com (F.I.); xushenhui@stumail.nwu.edu.cn (S.X.); wangzhe754778887@126.com (Z.W.); e1084875064@163.com (X.P.); wanghp@nwu.edu.cn (H.W.)
[2]   Shaanxi Key Laboratory for Animal Conservation, Northwest University, Xi'an 710069, China
[3]   Human Functional Genomics Laboratory, Northwest University, Xi'an 710069, China; lixiaoyualian@gmail.com
[4]   College of Biological Science and Engineering, North Minzu University, Yinchuan 750021, China; rain300861@163.com
*   Correspondence: changhui@nwu.edu.cn (H.C.); gaoyunf@nwu.edu.cn (Y.G.)
†   J.Z., X.L. contributed equally to the study.

Received: 19 November 2019; Accepted: 19 December 2019; Published: 22 December 2019

Abstract: Intracellular calcium (Ca^{2+}) homeostasis plays a vital role in the preservation of skeletal muscle. In view of the well-maintained skeletal muscle found in Daurian ground squirrels (*Spermophilus dauricus*) during hibernation, we hypothesized that hibernators possess unique strategies of intracellular Ca^{2+} homeostasis. Here, cytoplasmic, sarcoplasmic reticulum (SR), and mitochondrial Ca^{2+} levels, as well as the potential Ca^{2+} regulatory mechanisms, were investigated in skeletal muscle fibers of Daurian ground squirrels at different stages of hibernation. The results showed that cytoplasmic Ca^{2+} levels increased in the skeletal muscle fibers during late torpor (LT) and inter-bout arousal (IBA), and partially recovered when the animals re-entered torpor (early torpor, ET). Furthermore, compared with levels in the summer active or pre-hibernation state, the activity and protein expression levels of six major Ca^{2+} channels/proteins were up-regulated during hibernation, including the store-operated Ca^{2+} entry (SOCE), ryanodine receptor 1 (RyR1), leucine zipper-EF-hand containing transmembrane protein 1 (LETM1), SR Ca^{2+} ATPase 1 (SERCA1), mitochondrial calcium uniporter complex (MCU complex), and calmodulin (CALM). Among these, the increased extracellular Ca^{2+} influx mediated by SOCE, SR Ca^{2+} release mediated by RyR1, and mitochondrial Ca^{2+} extrusion mediated by LETM1 may be triggers for the periodic elevation in cytoplasmic Ca^{2+} levels observed during hibernation. Furthermore, the increased SR Ca^{2+} uptake through SERCA1, mitochondrial Ca^{2+} uptake induced by MCU, and elevated free Ca^{2+} binding capacity mediated by CALM may be vital strategies in hibernating ground squirrels to attenuate cytoplasmic Ca^{2+} levels and restore Ca^{2+} homeostasis during hibernation. Compared with that in LT or IBA, the decreased extracellular Ca^{2+} influx mediated by SOCE and elevated mitochondrial Ca^{2+} uptake induced by MCU may be important mechanisms for the partial cytoplasmic Ca^{2+} recovery in ET. Overall, under extreme conditions, hibernating ground squirrels still possess the ability to maintain intracellular Ca^{2+} homeostasis.

Keywords: calcium homeostasis; hibernation; mitochondria; sarcoplasmic reticulum; skeletal muscle

---

1. Introduction

The maintenance of cytoplasmic calcium (Ca^{2+}) homeostasis is important for the preservation of a normal structure and function of skeletal muscle fibers. Skeletal muscle inactivity can trigger Ca^{2+} homeostasis disturbance, often characterized by cytoplasmic Ca^{2+} overload [1]. A direct consequence of this overload is the activation of calpain system-mediated protein degradation [2]. In addition, an increased cytoplasmic Ca^{2+} concentration can promote cell apoptosis [3]. Increased protein degradation and cell apoptosis are both involved in skeletal muscle loss.

Hibernation is a unique survival strategy exhibited by various mammals in order to cope with adverse environments in winter, during which hibernators not only face the challenge of prolonged skeletal muscle inactivity, but also deal with other stresses, including hypoxia, fasting, and repeated ischemia-reperfusion during the torpor-arousal cycle. However, various studies have reported that skeletal muscle is well-maintained in hibernators during hibernation [4,5]. Therefore, hibernators can be considered typical anti-atrophy models, with their unique skeletal muscle preservation mechanism undoubtedly an attractive and valuable research topic.

Previous findings from our laboratory showed that, under adverse conditions over several months of hibernation, the cytoplasmic Ca^{2+} concentration in skeletal muscle fibers of Daurian ground squirrels increased transiently during inter-bout arousal, partially recovered after re-entering torpor, and almost recovered to pre-hibernation levels in the post-hibernation stage, thus exhibiting good Ca^{2+} homeostasis during the entire hibernation cycle [6]. During long-term hibernation, the torpor-arousal cycle likely plays an important role in protecting skeletal muscle from atrophy by avoiding or alleviating persistent and excessive cytoplasmic Ca^{2+} overload-induced protein degradation. Therefore, exploring the potential mechanisms involved in Ca^{2+} homeostasis during hibernation could help reveal the mechanisms against disuse-induced skeletal muscle atrophy of hibernators. To date, however, only one study (from our lab) has reported on sarcoplasmic reticulum Ca^{2+} pump (SERCA) expression in skeletal muscles during hibernation [7]. As such, the regulatory mechanisms involved in intracellular Ca^{2+} homeostasis in skeletal muscle fibers are far from having been clarified.

The level of intracellular Ca^{2+} is closely related to the expression level and activity of Ca^{2+} transport proteins or channels located in the plasma membrane and intracellular Ca^{2+} storage membrane (mainly sarcoplasmic reticulum (SR) and mitochondria), as well as intracellular Ca^{2+} binding proteins. Increased extracellular Ca^{2+} influx and intracellular Ca^{2+} storage/release (especially in the SR) both contribute to an increase in the intracellular Ca^{2+} concentration. Store-operated Ca^{2+} entry (SOCE) is the most important channel transporting extracellular Ca^{2+} into the cytosol. Stromal interaction molecule-1 (STIM1) located in the endoplasmic reticulum (ER) and Orai1 (also known as calcium-release-activated calcium-modulator, CRACM1) located in the cell membrane are two essential components required for SOCE [8–10]. With external stimulation, Ca^{2+} is released from the STIM1 EF-hand domain, which triggers the aggregation and movement of STIM1 to ER/plasma membrane (PM) binding sites, as well as the Orai1 aggregation of STIM1, and leads to the activation of SOCE and Ca^{2+} influx [11–14]. The ryanodine receptor (RyR) is a major SR Ca^{2+} release channel. Specifically, when sensing cell membrane depolarization, exterior membrane L-type calcium channels (surface membrane and T tubules) and dihydropyridine receptors (DHPR) combine to activate RyR, resulting in substantial SR Ca^{2+} release [15,16]. The RyR family is comprised of three isoforms (i.e., RyR1–3), with RyR1 exclusively expressed and particularly enriched in skeletal muscle [17]. Leucine zipper-EF-hand-containing transmembrane protein 1 (LETM1) is a Ca^{2+}-H^+ exchanger located in the mitochondrial membrane. When the mitochondrial Ca^{2+} concentration is high, LETM1 will extrude excess Ca^{2+} from the mitochondria into the cytoplasm [18]. Therefore, LETM1 is another possible contributor to elevated cytoplasmic Ca^{2+} levels.

In contrast to the above mechanisms, however, the increase in Ca^{2+} efflux, intracellular Ca^{2+} uptake of the Ca^{2+} pool, and binding capacity of free Ca^{2+} binding protein in the cytoplasm all effectively decrease cytoplasmic Ca^{2+}. Plasma membrane Ca^{2+} ATPase (PMCA) can eject Ca^{2+} from the cytosol into the external medium, thereby attenuating the cytoplasmic Ca^{2+} concentration. PMCA3 is the major isoform expressed in skeletal muscle [19]. As a primary active transporter located in the SR membrane, SR/ER Ca^{2+} ATPase (SERCA) can decrease cytoplasmic Ca^{2+} levels by pumping Ca^{2+} from the cytosol into the SR, which is one of the key factors attenuating cytoplasmic Ca^{2+} overload in skeletal muscle fibers [20]. The mitochondrial calcium uniporter (MCU) complex is considered a major channel for the transportation of Ca^{2+} into mitochondria [21]. Mitochondrial calcium uptake 1 and 2 (MICU1 and 2) are two regulatory subunits of MCU [22]. When the Ca^{2+} concentration in the intermembrane space is low, the heterodimers of MICU1 and MICU2 block the MCU channel and inhibit the entry of Ca^{2+} into the mitochondria. In contrast, when the Ca^{2+} level is high upon stimulation, the binding of Ca^{2+} to the MICU protein elicits a conformational change, resulting in the opening of the channel and the transportation of Ca^{2+} into the mitochondria [21,23]. Calmodulin (CALM), a Ca^{2+} binding protein located in the cytoplasm, can directly reduce the concentration of cytoplasmic free Ca^{2+} by combining with four Ca^{2+} ions [24]. Overall, Ca^{2+} uptake channels, extrusion mechanisms, and free Ca^{2+} binding proteins all contribute to intracellular Ca^{2+} homeostasis.

What, then, is the role of Ca^{2+} channels in Ca^{2+} fluctuations during the torpor-arousal cycle? To answer this question, we investigated the cytoplasmic, SR, and mitochondrial Ca^{2+} levels in the plantaris (PL, calf muscle) and adductor magnus (AM, thigh muscle) muscles of Daurian ground squirrels during different hibernation states (i.e., summer active, pre-hibernation, late torpor (entering a new bout after more than 5 d), inter-bout arousal (arousing spontaneously for less than 12 h), early torpor (entering a new bout for less than 48 h), and post-hibernation). Furthermore, a comprehensive and time-course investigation was carried out to explore the roles of the above major Ca^{2+} transport proteins/channels, including SOCE, RyR1, LETM1, PMCA3, SERCA1, and MCU, as well as the major Ca^{2+} binding protein CALM, in the fluctuations of Ca^{2+} concentration throughout hibernation.

2. Materials and Methods

2.1. Animals and Groups

All animal procedures and care and handling protocols were in accordance with the approval granted by the Laboratory Animal Care Committee of the China Ministry of Health (approval No. MH-55). The Daurian ground squirrels used in the experimental procedures were captured from the Weinan region, Shaanxi Province, China. Upon return to the laboratory, all squirrels were maintained in an animal room under a temperature range of 18–25 °C and modified daily light conditions (coincident with local sunrise and sunset). After one month of adaptation, the adult individuals were weight-matched and divided into six groups (n = 6–8): (i) Summer active group (SA): samples were collected in mid-June; (ii) pre-hibernation group (PRE): samples were collected in mid-September; (iii) late torpor group (LT): after two months hibernation, animals entered into a new hibernation bout and were in continuous torpor for at least 5 d, with a stable body temperature (Tb) of 5–8 °C; (iv) inter-bout arousal group (IBA): after two months hibernation, animals entered into a new hibernation bout and were fully aroused, with the Tb returned to 34–37 °C for less than 12 h; (v) early torpor group (ET): after two months hibernation, animals entered into a new hibernation bout, with Tb maintained at 5–8 °C for less than 24 h; (vi) post-hibernation group in spring (POST): animals awaking from hibernation and maintaining a Tb of 36–38 °C for more than 3 d in March of the following year. Animals in the SA and PRE groups were maintained in an environment with a natural light:dark photoperiod until sacrifice. When the ground squirrels gradually entered torpor in early November, the animals were transferred to a 4–6 °C dark hibernaculum. Due to observations occurring twice a day under weak light, these animals were housed under a 2:22 light-dark cycle. The Tb of animals was measured using a visual thermometer with thermal imaging (Fluke, VT04, Everett, Washington, DC, USA). The different states of the animals used here are shown in Figure 1.

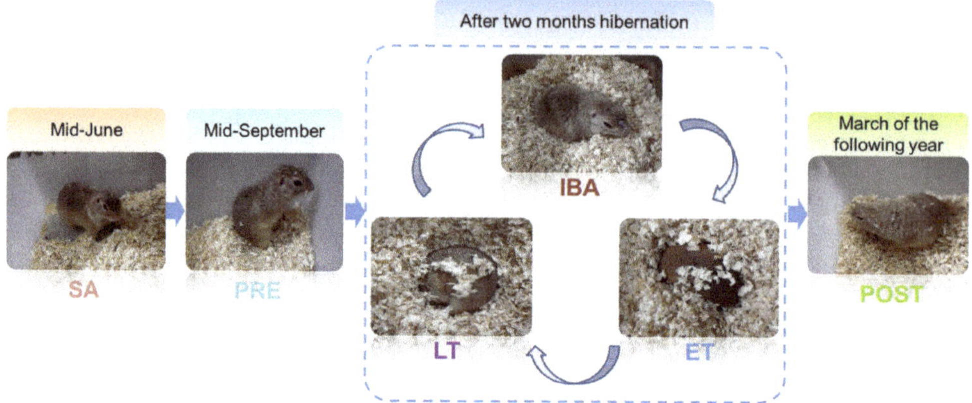

Figure 1. Representative images and sampling time of Daurian ground squirrels in different periods. SA, summer active; PRE, pre-hibernation; LT, late torpor; IBA, inter-bout arousal; ET, early torpor; POST, post-hibernation.

2.2. Muscle Sample Collection and Preparation

Muscle sample collection was carried out at 9:00 am for the SA, PRE, LT, and POST groups. Due to the specific sampling procedures in the IBA and ET groups, it could not be guaranteed that sample collection in these two groups always occurred at 9:00 am. Animals were anesthetized with sodium pentobarbital (90 mg/kg). The two distinct skeletal muscles (PL and AM) were carefully isolated and surgically removed. Subsequently, left leg muscles were treated with embedding medium for frozen section cutting and staining and right leg muscles were used for all other experiments. Following surgical intrusion, the squirrels were euthanized via sodium pentobarbital overdose injection.

2.3. Skeletal Muscle Fiber Cross-Sectional Area (CSA) Determination

As described previously [25], immunofluorescence and confocal analyses were used to measure the muscle fiber cross-sectional area (CSA) in frozen sections. Briefly, 10-μm thick frozen cross-sections were cut from the muscle mid-belly at −20 °C with a CM1850 cryostat (Leica, Wetzlar, Germany) and stored at −80 °C until further staining. After fixing in 4% paraformaldehyde for 30 min, slices were permeabilized in 0.1% Triton X-100 for 30 min, blocked with 1% bovine serum albumin (BSA) in phosphate-buffered saline (PBS) at room temperature for 60 min, and then incubated at 4 °C overnight with an anti-laminin antibody (1:500, Boster, BA1761-1, Wuhan, China) to visualize muscle fiber CSA. Subsequently, after washing them three times with PBS (10 min/time), the sections were incubated at 37 °C for 2 h with a 647-labeled IgG secondary antibody (1:400, Thermo Fisher Scientific, A-21235, Eugene, OR, USA). Finally, the slices were treated with anti-fade mounting medium (Life Technologies, 1427588, Eugene, OR, USA). Images were visualized via confocal laser scanning microscopy (Olympus, FV1000, Tokyo, Japan) at a 40× objective magnification. Image-Pro Plus 6.0 was used to measure muscle fiber CSA. In detail, eight images were captured from each sample, and the CSA of all complete muscle fibers (about 30) within each picture was then analyzed. Therefore, the CSA of ~250 muscle fibers per skeletal muscle sample was determined.

2.4. Single Muscle Fiber Isolation

We anaesthetized the squirrels with 90 mg/kg sodium pentobarbital, after which muscle samples (including the tendons) were carefully removed from the neighboring tissues and sarcolemma, ensuring that the blood and nerve supply remained intact. The muscle samples were subsequently separated into two full-length strips along the longitudinal axis using a pair of tweezers. The muscle strips, which were obtained from the same middle region, were then washed with 20 mL of PBS (137 mM sodium chloride, 2.7 mM potassium chloride, 4.3 mM disodium chloride, 1.4 mM monopotassium phosphate, pH 7.4) and digested with 3 mL of enzymatic digestion solution containing 0.35% collagenase I (Sigma-Aldrich, C0130-1G, Saint Quentin Fallavier, France), followed by orbital shaker incubation at 37 °C for 2 h and saturation with 95% O_2 and 5% CO_2 to ensure complete digestion of the muscle fiber samples. Finally, the digestion solution was removed with a PBS rinse and the muscle samples were added to Dulbecco's modified Eagle's medium (DMEM) (Hyclone, AC10221937, Pittsburgh, PA, USA) containing 10% fetal bovine serum (FBS) (Everygreen, 11011-8611, Hangzhou, China), 25 μM of N-benzyl-p-toluenesulfonamide (BTS) (TCI, B3082, Shanghai, China), and 0.1 M HEPES (Guoan, H0082, Xi'an, China), and were carefully stirred with a pipette. The digested single muscle fiber samples were finally plated on culture chamber slides and viewed via inverted microscopy (Olympus, IX2-ILL100, Tokyo, Japan).

2.5. Measurement of Cytoplasmic Ca^{2+}

We used fluo-3-acetoxymethylester (Fluo-3/AM) (Invitrogen, Carlsbad, CA, USA), which demonstrates increased fluorescence upon Ca^{2+} binding, to determine cytoplasmic free Ca^{2+}. Briefly, after washing the samples three times with fresh PBS, dye (5 mM Fluo-3/AM) was slowly added along the sides of the single muscle fibers, followed by incubation in the dark at 37 °C for 30 min. After incubation, the glass slide-mounted Fluo-3/AM-loaded fibers were washed with fresh PBS three times (20 s/time, 1-min process). The slide was quickly placed on the microscope stage, with the fibers focused in the bright field (20-s process) and scanned via laser confocal microscopy in combination with an Olympus FV10-ASW system (Tokyo, Japan) under 488-nm krypton/argon laser illumination, with fluorescence detected at 526 nm. According to their length, three to five pictures were captured at 10× objective magnification for each muscle fiber (10-s capture process for each picture). In consideration of the influence of muscle fiber size on the fluorescence intensity, the average fluorescence intensity (total fluorescence intensity/total area of selected region) was used to measure the Ca^{2+} levels. Specifically, the average fluorescence intensity of 10 different regions in each picture was measured using Olympus Fluoview v4.2 software. All pictures (3–5 pictures, depending on the muscle fiber length) of each fiber, with 10 muscle fibers per sample, were used for statistical analysis.

2.6. Measurement of Sarcoplasmic Reticulum Ca^{2+}

We used magnesium-Fluo-4-acetoxymethylester (mag-Fluo-4/AM) (M14206, Thermo Fisher Scientific, Eugene, OR, USA), which demonstrates increased fluorescence upon Ca^{2+} binding, to indicate SR free Ca^{2+}, as per Park et al. (2000) [26]. Briefly, after washing samples twice with fresh PBS, dye (5 mM mag-Fluo-4/AM) was slowly added along the sides of the single muscle fibers, followed by incubation in the dark at 37 °C for 30 min. After incubation, the glass slide-mounted mag-Fluo-4/AM-loaded fibers were washed with fresh PBS three times (20 s/time, 1-min process). The slide was then quickly placed on the microscope stage, with the fibers focused in the bright field (20-s process) and scanned via laser confocal microscopy in combination with an Olympus FV10-ASW system (Japan) under 488-nm krypton/argon laser illumination, with fluorescence detected at 526 nm. Analysis and statistical methods were similar to those used for the measurement of cytoplasmic Ca^{2+} mentioned above.

2.7. Measurement of Mitochondrial Ca^{2+}

We used Rhod-2/AM (R1244, Thermo Fisher Scientific, USA), which demonstrates increased fluorescence upon Ca^{2+} binding in the mitochondria, to determine mitochondrial free Ca^{2+} [27]. Briefly, after washing samples twice with fresh PBS, dye (5 μM Rhod-2/AM) was slowly added along the sides of the single muscle fibers, followed by incubation in the dark at 37 °C for 30 min. After incubation, the glass slide-mounted Rhod-2/AM-loaded fibers were washed with fresh PBS three times (20 s/time, 1-min process). The slide was then quickly placed on the microscope stage, and the fibers were focused in the bright field (20-s process) and scanned via laser confocal microscopy in combination with an Olympus FV10-ASW system (Japan) under 594-nm krypton/argon laser illumination, with fluorescence detected at 618 nm. Analysis and statistical methods were similar to those used for the measurement of cytoplasmic Ca^{2+} mentioned above.

2.8. Total RNA Extraction and Quantitative Real-Time Polymerase Chain Reaction (RT-PCR)

As per Fu et al. (2016) [6] and in accordance with the manufacturer's protocols, we extracted total RNA from the muscle samples using an RNAiso Plus kit (TaKaRa Biotechnology, 9109, Dalian, China). RNA quality was characterized using the OD260/OD280 ratio, after which selected samples (those exhibiting OD260/OD280 > 1.8) were reverse transcribed into cDNA using an appropriate reagent (TaKaRa Biotechnology, RR036A China) and stored (−20 °C) for the following analyses. Here, qRT-PCR was undertaken using a SYBR Premix Ex Taq II kit (TaKaRa Biotechnology, RR820A, China), following the protocols stated by the manufacturer. The resultant dissolution and amplification curves were observed and selected, with the α-tubulin reference gene and $2^{-\Delta\Delta ct}$ method then being applied to analyze the relative mRNA concentrations of STIM1, ORAI1, RyR1, LETM1, PMCA3, SERCA1, MCU, MICU1, MICU2, CALM, and α-tubulin. The primers used for the above genes (Sangon, Nanjing, China) are listed in Table 1.

Table 1. Primers used for quantitative real-time PCR experiments.

Genes	Primer Sequence
STIM1	forward: 5′-CAGTTCTCATGGCCCGAGTT-3′
	reverse: 5′-GTGGGGAATGCGTGTGTTTC-3′
ORAI1	forward: 5′-CGCAAGCTCTACTTGAGCCG-3′
	reverse: 5′-CATCGCTACCATGGCGAAGC-3′
RYR1	forward: 5′-GGTACTGGTCGGGATACCCT-3′
	reverse: 5′- GACCTCGGGACTCTCAATCA-3′
Letm1	forward: 5′-ACTGGTCCCTTTCCTGGTCT-3′
	reverse: 5′-CTTCAGCCTCTCCTCCTTGA-3′
PMCA3	forward: 5′-CGGCGGTCTTCGGTCCTCAG-3′
	reverse: 5′-TGGGCTTGGCGGCAGAGAG-3′
SERCA1	forward: 5′-GGTACTGGTCGGGATACCCT-3′
	reverse: 5′-GCTGGATAGAGCCTGTGACC-3′
MCU	forward: 5′-TGGTGTGTTTTTACGGCAAC-3′
	reverse: 5′-TCATCAAGGAGGAGGAGGTC-3′
MICU1	forward: 5′-TGGGTATGCGTCACAGAGAT-3′
	reverse: 5′-GATGGTCAGTTTCCCCTTGA-3′
MICU2	forward: 5′-TGACACCACGAGACTTCCTCT-3′
	reverse: 5′-GATTCCTGCCAATACCTCCTC-3′
CALM	forward: 5′-GGCACCATTGACTTCCCAGA-3′
	reverse: 5′-TCTGCCGCACTGATGTAACC-3′
α-tubulin	forward: 5′-AATGCCTGCTGGGAGCTCTA-3′
	reverse: 5′-CAGCGCCTGTCTCACTGAAG-3′

2.9. Protein Extraction and Western Blotting Analysis

Muscle samples (~0.1 g) were weighed and fully homogenized with 1 mM RIPA Lysis Buffer (Heart, WB053A, Xi'an, China), 1% protease inhibitor cocktail (Heart, WB053B, Xi'an, China), and 1% phenylmethylsulfonyl fluoride (PMSF, Heart, WB053C, Xi'an, China). After 15 min of centrifugation at 4 °C and 15,000 rpm, the supernatants were removed and placed into new tubes, with soluble protein concentrations then detected using a Pierce™ BCA Protein Quantitation kit (Thermo Fisher Scientific, 23227, USA). The supernatants were mixed with 1 × SDS loading buffer (100 mM Tris, 5% glycerol, 5% 2-β-mercaptoethanol, 4% SDS, and bromophenol blue, pH 6.8) at a 1:4 *v/v* ratio, followed by boiling and then storage at −20 °C for further analysis.

Western blotting procedures were as described by Zhang et al. (2017) [28]. In brief, we first separated the muscle protein extracts using SDS-PAGE on 10% Laemmli gels (acrylamide/bisacrylamide ratio of 37.5:1 for STIM1, ORAI1, LETM1, PMCA3, SERCA1, MCU, MICU1, CALM) and on 6% Laemmli gels (acrylamide/bisacrylamide ratio of 37.5:1 for RyR1), respectively. Following electrophoresis (for 60 min at 120 V), the proteins were electrically transferred to 0.45-μm pore polyvinylidene difluoride (PVDF) membranes (Millipore, IPVH00010, Merck kGaA, Darmstadt, Germany) using the Bio-Rad (1703930) semidry transfer apparatus (Hercules, CA, USA) at 15 V for 30–40 min. We then blocked the membranes at room temperature for 2 h using 5% skim milk in TBST (containing 10 mM Tris-HCl, 150 mM NaCl, 0.05% Tween-20, pH 7.6), followed by overnight incubation at 4 °C with primary STIM1 (1:1000, CST, 5668S, Danvers, MA, USA), ORAI1 (1:1000, Thermo, MA5-15776, Eugene, OR, USA), RyR1 (1:1000, CST, 8153S, USA), LETM1 (1:1000, CST, 14997S, USA), PMCA3 (1:1000, Abcam, ab3530, Cambridge, UK), SERCA1 (1:1000, CST, 4219S, USA), MCU (1:1000, CST, 14997S, USA), MICU1 (1:750, CST, 12524S, USA), and CALM (1:1000, CST, 4830S, USA) antibodies in TBST containing 0.1% BSA. The membranes where then washed three times with TBST (10 min/time), followed by 1.5-h incubation at room temperature with horseradish peroxidase (HRP)-conjugated anti-rabbit or anti-mouse secondary antibodies (Thermo Fisher Scientific, A27014, USA). The membranes were again washed with TBST (four times × 10 min), with the resulting immunoblots being visualized using enhanced chemiluminescence reagents (Thermo Fisher Scientific, NCI5079, USA), in accordance with the manufacturer's instructions. Blot quantification was conducted using Image-Pro Plus 6.0 software. Total protein staining of the gel was used as the normalization control for all blots. In detail, as described previously, 0.5% 2,2,2-trichloroethanol (TCE) was first added to the gel [29–31]. After electrophoresis, the gel was irradiated on the UV platform of the electrophoresis gel imaging analysis system (G: box, GBOX Cambridge, UK) for 5 min, with the signal then being collected. As described previously [32,33], the original images captured with no gain were stored. After that, the fluorescence intensity of each lane (after removal of the background fluorescence intensity) was determined with Image-Pro Plus 6.0, with the internal reference being used to correct the fluorescence intensity of the target protein. Specificity detection of the complete SDS-PAGE lane for each antibody used in the present study is shown in Figure S1.

2.10. Co-Localization Analysis of ORAI1/STIM1

Briefly, 10-μm thick frozen cross-sections were cut from the muscle mid-belly at −20 °C with a CM1850 cryostat (Leica, Wetzlar, Germany). After fixing samples in 4% paraformaldehyde for 30 min, slices were permeabilized in 0.1% Triton X-100 for 30 min, blocked with 1% BSA in PBS at room temperature for 60 min, and then incubated at 4 °C overnight with an anti-ORAI1 antibody (1:50, Thermo, MA5-15776, USA). On the second day, after washing samples three times with PBS (10 min/time), the sections were incubated at 37 °C for 2 h with an Alexa Fluor FITC-conjugated secondary antibody (1:300, Thermo Fisher Scientific, Rockford, IL, USA). After again washing samples three times with PBS (10 min/time), the slices were incubated at 4 °C overnight with an anti-STIM1 antibody (1:300, CST, 5668S, USA). On the third day, after washing samples three times with PBS (10 min/time), the sections were incubated at 37 °C for 2 h with a 647-labeled IgG secondary antibody (1:200, Thermo Fisher Scientific, A-21235, USA). The slices were then washed three times with PBS (10 min/time) and dried and treated with anti-fade mounting medium (Life Technologies, 1427588, USA). Images were visualized and captured via confocal laser scanning microscopy (Olympus, FV1000, Japan) at a 40× objective magnification with krypton/argon laser illumination at 488 and 647 nm and captured at 526 and 665 nm. As previously described [34,35], the co-localization of ORAI1/STIM1 was calculated by Pearson's correlation coefficients using Image-Pro Plus 6.0.

2.11. Statistical Analysis

Data are presented as means ± SEM. SPSS Statistics 17.0 was used for all statistical tests. Group differences were determined via one-way analysis of variance (ANOVA) with Fisher's least significant difference (LSD) post hoc test. When no homogeneity was detected, ANOVA-Dunnett's T3 method was applied. A value of $p < 0.05$ was considered statistically significant.

3. Results

3.1. Skeletal Muscle Mass and Single Muscle Fiber CSA of PL and AM during Different Hibernation Periods

Changes in skeletal muscle morphology were observed by analyzing the muscle mass (MM) (Figure 2B) and muscle fiber CSA (Figure 2A,C) in different groups. The results showed that, compared with the SA group, slight decreases in muscle mass and single muscle fiber CSA (15–20%) were observed during hibernation.

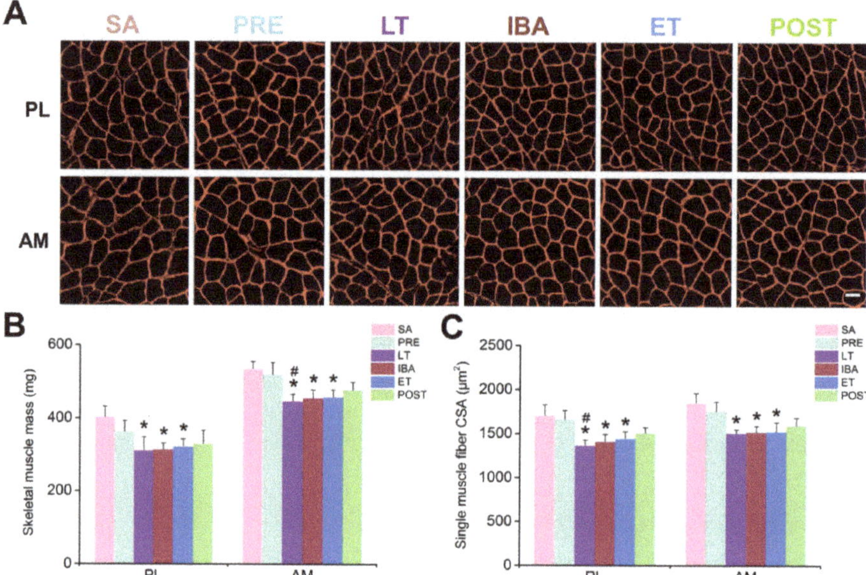

Figure 2. Changes in muscle mass and single muscle fiber CSA in PL and AM muscles during different periods. (**A**) Representative fluorescence images of single muscle fiber CSA in PL and AM muscles. 400× magnification, scale bar = 100 µm. (**B**) Histogram depicting muscle mass of PL and AM muscles during different periods. (**C**) Histogram depicting single muscle fiber CSA in PL and AM muscles during different periods. CSA, cross-sectional area; PL, plantaris; AM, adductor magnus. SA, summer active group; PRE, pre-hibernation group; LT, late torpor group; IBA, inter-bout arousal group; ET, early torpor group; POST, post-hibernation group. Values are means ± SEM, n = 6–8. * $p < 0.05$ compared with SA; # $p < 0.05$ compared with PRE.

3.2. Cytoplasmic Ca²⁺ Level in Single Skeletal Muscle Fibers during Different Hibernation Periods

In comparison with that in the SA and PRE groups, the cytoplasmic Ca^{2+} level in the single PL and AM muscle fibers increased significantly during hibernation, and almost recovered to SA levels post-hibernation. During the torpor-arousal cycle, the cytoplasmic Ca^{2+} level partially recovered when animals re-entered the torpor state. Compared with the results in the LT group, significant decreases in cytoplasmic Ca^{2+} levels were observed in the PL (24%) and AM muscles (32%) of the ET group (Figure 3).

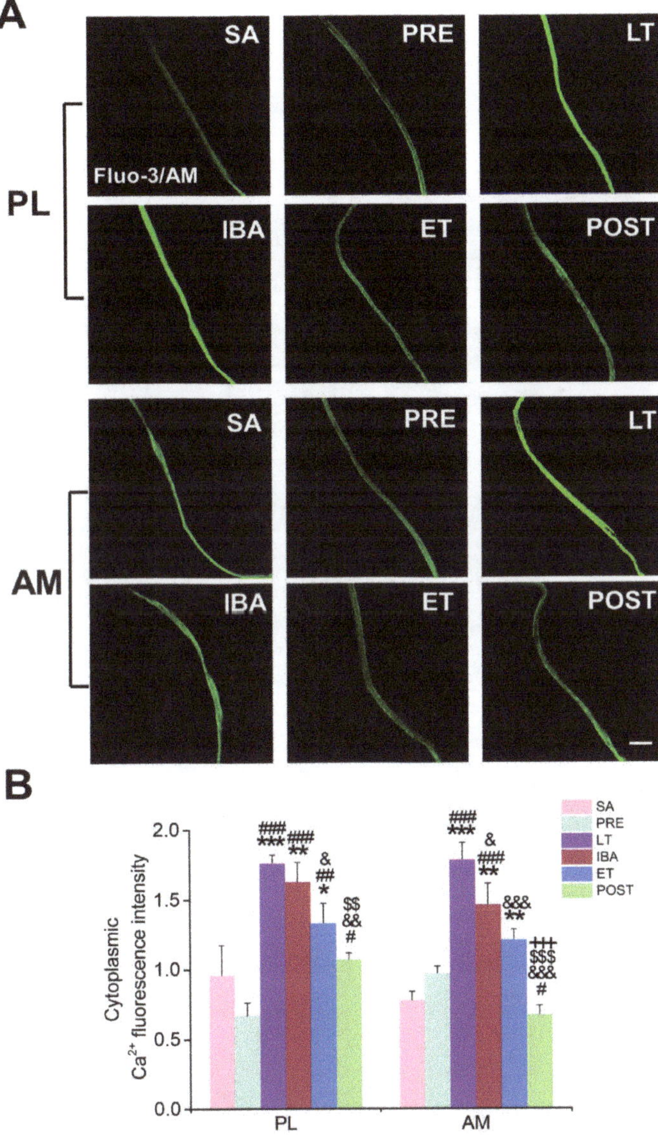

Figure 3. Changes in cytoplasmic Ca^{2+} concentration in PL and AM muscles during different periods. (**A**) Representative fluorescence images of single PL and AM muscle fibers. Scale bar = 100 μm. (**B**) Histogram depicting the cytoplasmic Ca^{2+} fluorescence intensity in PL and AM muscles during different periods. PL, plantaris; AM, adductor magnus. SA, summer active group; PRE, pre-hibernation group; LT, late torpor group; IBA, inter-bout arousal group; ET, early torpor group; POST, post-hibernation group. Values are means ± SEM, n = 6–8. * $p < 0.05$, ** $p < 0.01$, and *** $p < 0.001$ compared with SA; # $p < 0.05$, ## $p < 0.01$, and ### $p < 0.001$ compared with PRE; & $p < 0.05$, && $p < 0.01$, and &&& $p < 0.001$, compared with LT; \$\$ $p < 0.01$ and \$\$\$ $p < 0.001$ compared with IBA; +++ $p < 0.001$ compared with ET.

3.3. SR Ca²⁺ Level in Single Skeletal Muscle Fibers during Different Hibernation Periods

In contrast to the results produced for cytoplasmic Ca^{2+}, the SR Ca^{2+} level in single PL and AM muscle fibers decreased significantly during hibernation and almost recovered to SA levels post-hibernation (Figure 4). During the torpor-arousal cycle, compared with that in the LT and IBA groups, the SR Ca^{2+} level in the PL muscle was significantly elevated by 27–34% when animals re-entered the torpor state, indicating that the SR Ca^{2+} level partially recovered in the ET group.

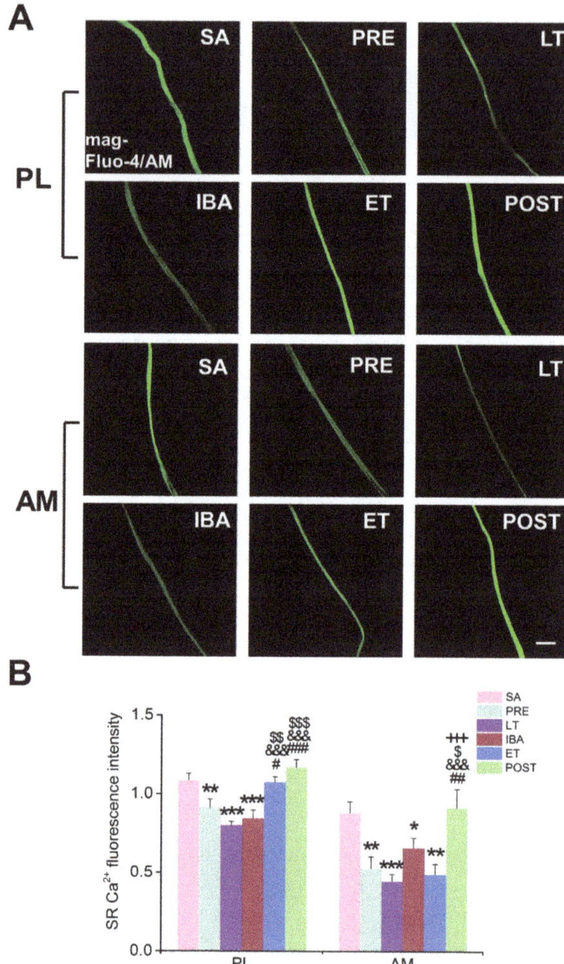

Figure 4. Changes in the SR Ca^{2+} concentration in PL and AM muscles during different periods. (**A**) Representative fluorescence images of single PL and AM muscle fibers. Scale bar = 100 μm. (**B**) Histogram depicting the cytoplasmic Ca^{2+} fluorescence intensity in PL and AM muscles during different periods. SR, sarcoplasmic reticulum; PL, plantaris; AM, adductor magnus. SA, summer active group; PRE, pre-hibernation group; LT, late torpor group; IBA, inter-bout arousal group; ET, early torpor group; POST, post-hibernation group. Values are means ± SEM, n = 6–8. * $p < 0.05$, ** $p < 0.01$, and *** $p < 0.001$ compared with SA; # $p < 0.05$, ## $p < 0.01$, and ### $p < 0.001$ compared with PRE; &&& $p < 0.001$ compared with LT; \$ $p < 0.05$, \$\$ $p < 0.01$, and \$\$\$ $p < 0.001$ compared with IBA; +++ $p < 0.001$ compared with ET.

3.4. Mitochondrial Ca^{2+} Level in Single Skeletal Muscle Fibers during Different Hibernation Periods

Compared with that in the SA or PRE groups, the mitochondrial Ca^{2+} level in the single muscle fibers was elevated to varying degrees during hibernation (in PL muscle of the IBA group and in AM muscle of the LT and ET groups) (Figure 5).

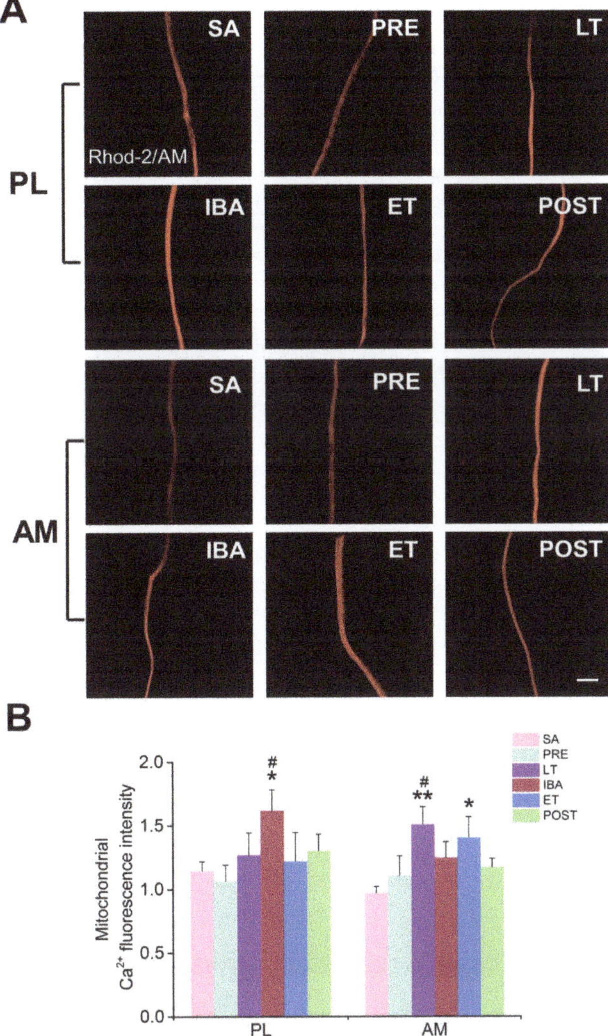

Figure 5. Changes in the mitochondrial Ca^{2+} concentration in PL and AM muscles during different periods. (**A**) Representative fluorescence images of single PL and AM muscle fibers. Scale bar = 100 µm. (**B**) Histogram depicting the mitochondrial Ca^{2+} fluorescence intensity in PL and AM muscles during different periods. PL, plantaris; AM, adductor magnus. SA, summer active group; PRE, pre-hibernation group; LT, late torpor group; IBA, inter-bout arousal group; ET, early torpor group; POST, post-hibernation group. Values are means ± SEM, n = 6–8. $*\ p < 0.05$ and $**\ p < 0.01$ compared with SA; $\#\ p < 0.05$ compared with PRE.

We comprehensively analyzed the changes in cytoplasmic, SR, and mitochondrial Ca^{2+} levels during different periods. Firstly, the opposite changes in cytoplasmic and SR Ca^{2+} suggest that SR Ca^{2+} participates in fluctuation of the cytoplasmic Ca^{2+} level during hibernation. In addition, the slight elevation in mitochondrial Ca^{2+} during hibernation may result from increased cytoplasmic Ca^{2+} or SR Ca^{2+} leakage. Further studies were subsequently carried out to explore the mechanisms involved in intracellular Ca^{2+} fluctuations during hibernation.

3.5. Relative mRNA and Protein Levels of Key Ca^{2+} Transport Proteins/Channels in Skeletal Muscle Fibers during Different Hibernation Periods

The mRNA and protein expression levels of several major Ca^{2+} transport proteins/channels located in the cytoplasm, SR, and mitochondria, including STIM1, ORAI1, RyR1, LETM1, PMCA3, SERCA1, MCU, MICU1, MICU2, and the free Ca^{2+} binding protein CALM, as well as the Pearson correlation coefficients for the co-localization of ORAI1 and STIM1, were detected to explore the potential mechanisms involved in intracellular Ca^{2+} fluctuation during hibernation. It should be clarified that, as mRNA expression is very sensitive to both internal and external environmental factors, large variability in the size of the error bars occurred in the mRNA statistical results. Therefore, we set a fold-change of $\geq$2-fold as the threshold for the biological significance of mRNA expression.

The mRNA and protein expression levels, as well as the Pearson correlation coefficients for the co-localization of ORAI1 and STIM1, were detected to explore the role of the SOCE channel in intracellular Ca^{2+} level fluctuation during hibernation. In the PL muscle, compared with that in the SA group, the mRNA expression levels of both STIM1 and ORAI1 increased during IBA (Figure 6A, B). Their protein expression levels showed an increasing (though non-significant) trend (Figure 7B, C). In addition, the Pearson correlation coefficients for the co-localization of ORAI1 and STIM1 were significantly elevated in the IBA group (Figure 8). In the AM muscle, the mRNA and protein expression levels of STIM1 and ORAI1 showed no significant change in the IBA group. However, Pearson's correlation coefficients for the co-localization of ORAI1 and STIM1 exhibited slight increases in the IBA group. Overall, the Pearson's correlation coefficients for the co-localization of ORAI1 and STIM1 in PL and AM muscle increased when ground squirrels aroused from torpor.

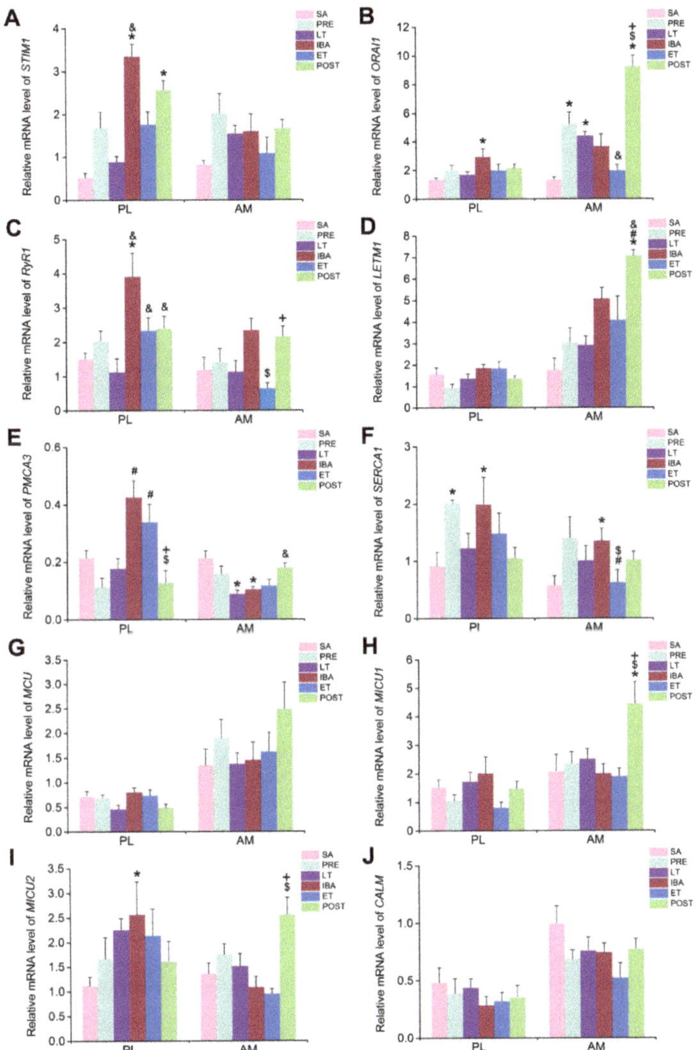

Figure 6. Changes in the mRNA expression of distinct Ca^{2+} transport and binding proteins in PL and AM muscles during different periods. Histograms depicting (**A**) stromal interaction molecule-1 (*STIM1*) mRNA expression, (**B**) *ORAI1* mRNA expression, (**C**) ryanodine receptor 1 (*RyR1*) mRNA expression, (**D**) leucine zipper-EF-hand containing transmembrane protein 1 (*LETM1*) mRNA expression, (**E**) plasma membrane Ca^{2+} ATPase (PMCA)3 mRNA expression, (**F**) SR Ca^{2+} ATPase 1 (*SERCA1*) mRNA expression, (**G**) mitochondrial calcium uniporter (*MCU*) mRNA expression, (**H**) mitochondrial calcium uptake 1 (*MICU1*) mRNA expression, (**I**) mitochondrial calcium uptake 2 (*MICU2*) mRNA expression, and (**J**) calmodulin (*CALM*) mRNA expression in PL and AM muscles during different periods. PL, plantaris; AM, adductor magnus. SA, summer active group; PRE, pre-hibernation group; LT, late torpor group; IBA, inter-bout arousal group; ET, early torpor group; POST, post-hibernation group. Values are means ± SEM, n = 6–8. * ($p < 0.05$ and fold change ≥ 2-fold), compared with SA; # ($p < 0.05$ and fold change ≥ 2-fold) compared with PRE; & ($p < 0.05$ and fold change ≥ 2-fold) compared with LT; $ ($p < 0.05$ and fold change ≥ 2-fold) compared with IBA; + ($p < 0.05$ and fold change ≥ 2-fold) compared with ET.

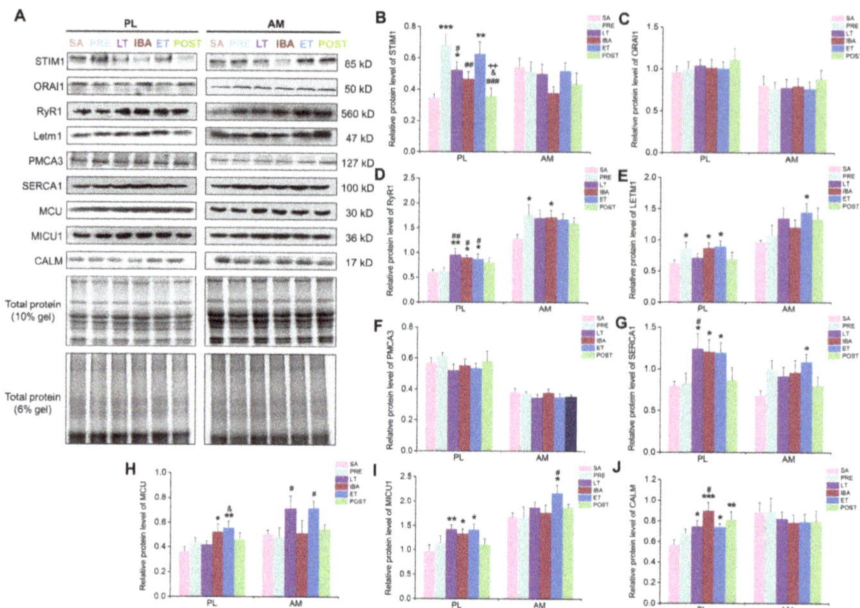

Figure 7. Changes in the protein expression of distinct Ca^{2+} transport and binding proteins in PL and AM muscles during different periods. Histograms depicting (**A**) representative Western blot images of STIM1, ORAI1, RyR1, LETM1, PMCA3, SERCA1, MCU, MICU1, and CALM in PL and AM muscles during different periods. Histograms depicting (**B**) STIM1 protein expression, (**C**) ORAI1 protein expression, (**D**) RyR1 protein expression, (**E**) LETM1 protein expression, (**F**) PMCA3 protein expression, (**G**) SERCA1 protein expression, (**H**) MCU protein expression, (**I**) MICU1 protein expression, and (**J**) CALM protein expression in PL and AM muscles during different periods. PL, plantaris; AM, adductor magnus. SA, summer active group; PRE, pre-hibernation group; LT, late torpor group; IBA, inter-bout arousal group; ET, early torpor group; POST, post-hibernation group. Values are means ± SEM, n = 6–8. * $p < 0.05$, ** $p < 0.01$, and *** $p < 0.001$ compared with SA; # $p < 0.05$, ## $p < 0.01$, and ### $p < 0.001$ compared with PRE; & $p < 0.05$ compared with LT; $ $p < 0.05$ and $$ $p < 0.01$ compared with IBA; ++ $p < 0.01$ compared with ET.

In view of the opposite changes between cytoplasmic and SR Ca^{2+} during hibernation, and to further explore whether the increased cytoplasmic Ca^{2+} during hibernation resulted from SR Ca^{2+} release, the mRNA and protein expression levels of RyR1 were measured. As shown in Figure 6C, in the PL muscle, compared with that in the SA group, the mRNA expression of RyR1 was significantly elevated in the IBA group. Its protein expression also increased (47–60%) in the distinct hibernation groups (Figure 7D). In the AM muscle, compared with that in the SA group, no significant change in RyR1 mRNA expression occurred (Figure 6C); however, its protein expression was significantly increased by 32% in the IBA group (Figure 7D). Overall, the protein expression of RyR1 in the PL and AM muscles increased significantly during hibernation (especially when animals aroused from torpor).

To further explore whether increased cytoplasmic Ca^{2+} during hibernation resulted from mitochondrial Ca^{2+} extrusion, the mRNA and protein expression levels of LETM1 were detected. In the PL muscle, compared with that in the PRE group, the mRNA expression of LETM1 increased significantly in the IBA and ET groups. Its protein expression level also significantly increased (37–40%) in the IBA and ET groups. In the AM muscle, compared with that in the SA group, the mRNA expression of LETM1 was dramatically elevated in the POST group (Figure 6D). The protein expression level was also elevated by (36–48%) in the LT, ET, and POST groups (Figure 7E). Overall, the mRNA and protein expression levels of LETM1 in the PL and AM muscles were elevated during hibernation and after post-hibernation.

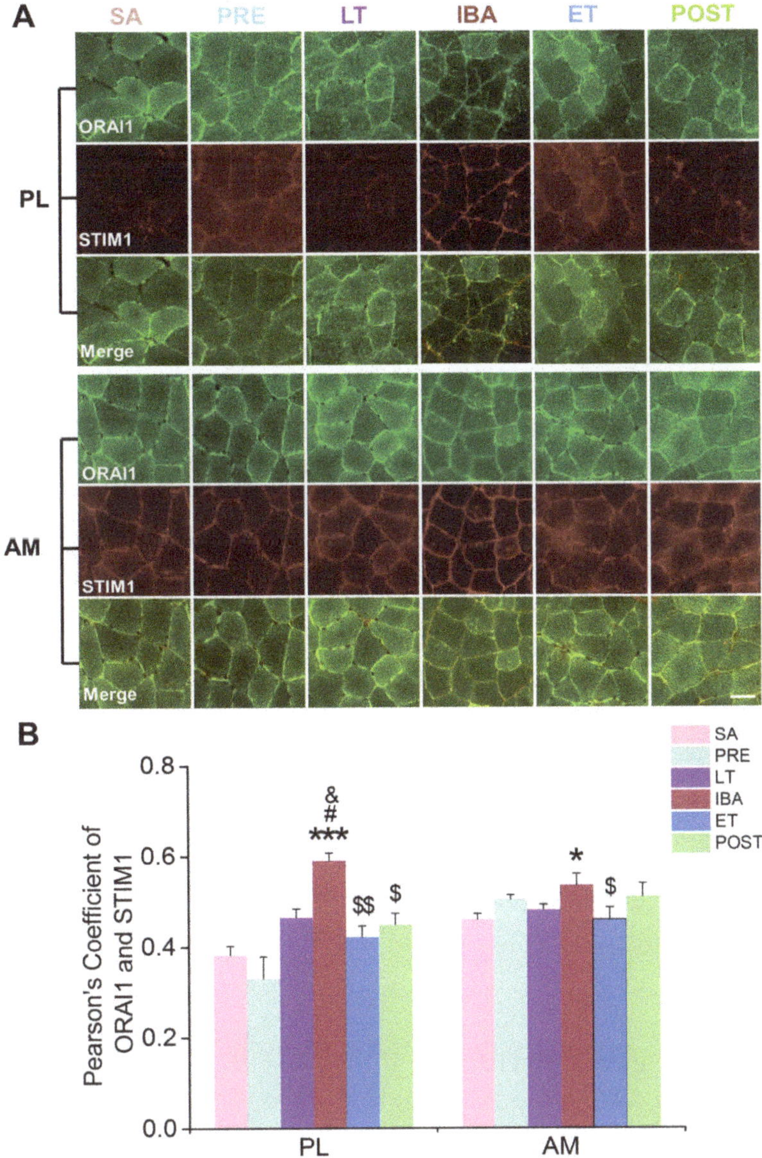

Figure 8. Changes in Pearson correlation coefficients of ORAI1/STIM1 in PL and AM muscles during different periods. (**A**) Representative fluorescence images in PL and AM muscles during different periods. 800× magnification, scale bar = 100 μm. (**B**) Histogram depicting Pearson correlation coefficients of ORAI1/STIM1 in PL and AM muscles during different periods. PL, plantaris; AM, adductor magnus. SA, summer active group; PRE, pre-hibernation group; LT, late torpor group; IBA, inter-bout arousal group; ET, early torpor group; POST, post-hibernation group. Values are means ± SEM, n = 6–8. * $p < 0.05$ and *** $p < 0.001$ compared with SA; # $p < 0.05$ compared with PRE; & $p < 0.05$ compared with LT; \$ $p < 0.05$ and \$\$ $p < 0.01$ compared with IBA.

The results showed that the cytoplasmic Ca^{2+} level partially recovered when the ground squirrels re-entered torpor after 12–24 h of IBA, which led us to consider the Ca^{2+} level recovery mechanisms involved in this process. The mRNA and protein expression levels of PMCA3 (a major plasma membrane Ca^{2+} ATPase in skeletal muscle) were first detected. The results showed that, compared with the SA and PRE groups, the mRNA level of PMCA3 in the PL muscle increased significantly in the IBA and ET groups (Figure 6E). A lower mRNA expression of PMCA3 was observed during hibernation in the AM muscle. In addition, the protein expression level of PMCA3 showed no significant differences among the different groups in either the PL or AM muscles (Figure 7F).

The mRNA and protein expression levels of SERCA1, a highly important Ca^{2+} pump located in the SR, were then detected. Compared with that in the SA group, the mRNA expression level of SERCA1 in the PL and AM muscles increased significantly in the IBA group (Figure 6F). Consistent with the changes in mRNA expression levels, the protein level of SERCA1 in the PL and AM muscles also increased significantly (43–58%) in the LT, IBA, and ET groups (Figure 7G). Overall, elevated SERCA1 mRNA and protein expression levels were observed in the PL and AM muscles during hibernation.

Mitochondria also plays a critical role in Ca^{2+} storage in skeletal muscle. Therefore, the mRNA and protein expression levels of the MCU complex, a major mitochondrial Ca^{2+} uptake channel, were determined. In the PL and AM muscles, compared with those in the SA group, no significant changes in mRNA expression levels of MCU, MICU1, or MICU2 were observed during hibernation (Figure 6G,H,I). However, the protein expression levels of MCU and MICU1 were increased significantly (29–51%) in the LT, IBA, and ET groups (Figure 7H,I).

In addition to the Ca^{2+} channels or pumps, free Ca^{2+} binding protein plays a major role in cytoplasmic Ca^{2+} levels. Therefore, we also detected the mRNA and protein expression levels of CALM. As shown in Figure 6J, in the PL and AM muscles, there were no significant differences in the mRNA expression levels of CALM among the different groups. However, the protein expression level was significantly increased during hibernation in the PL muscle, but showed no significant differences in the AM muscle (Figure 7J).

4. Discussion

A comprehensive time-course investigation was firstly carried out to measure the cytoplasmic, SR, and mitochondrial Ca^{2+} levels, as well as clarify the possible Ca^{2+} regulatory mechanisms, in the skeletal muscles of Daurian ground squirrels during different hibernation states. Fluctuations in intracellular Ca^{2+} levels were observed during the torpor-arousal cycle, with Ca^{2+} levels partially recovered when the animals aroused and then re-entered torpor (Figure 9A). Further investigation suggested that the Ca^{2+} proteins/channels and free Ca^{2+} binding protein located in the cytoplasm, SR, and mitochondria all participated in the fluctuation of intracellular Ca^{2+} (Figure 9B).

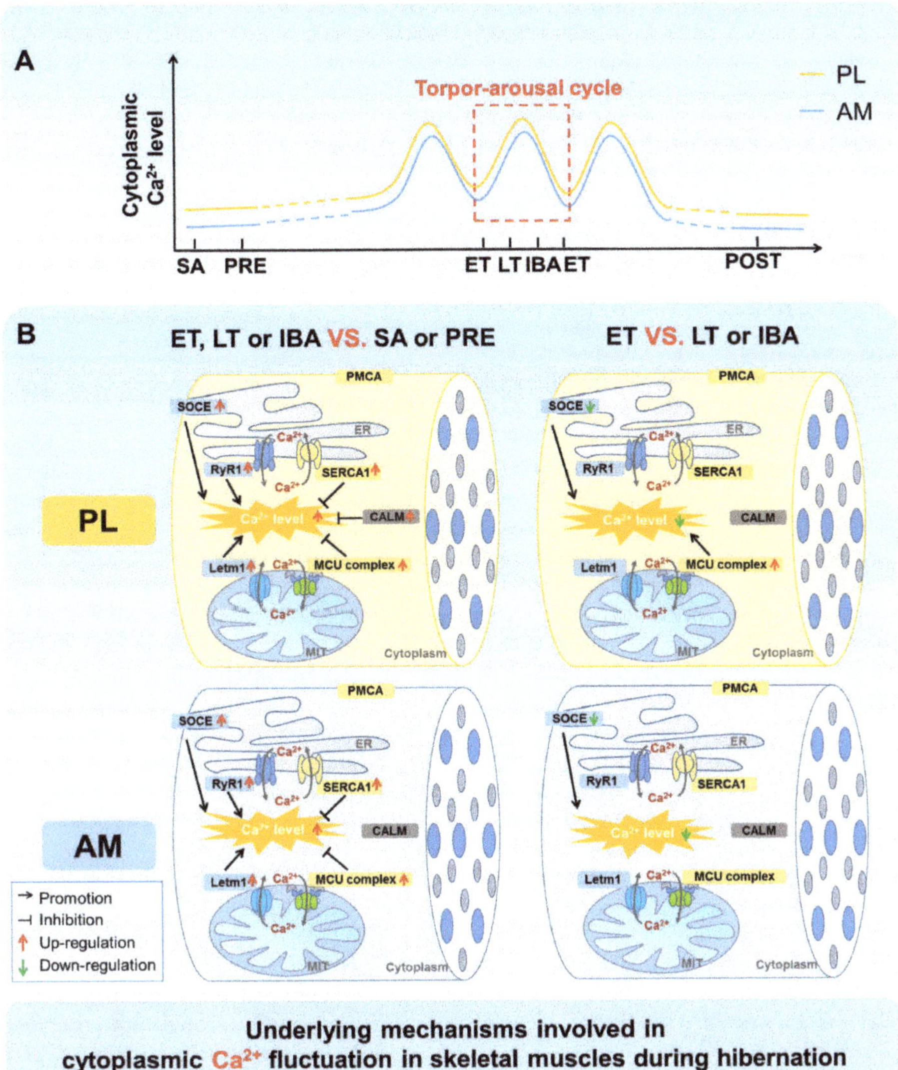

Figure 9. Graphical summary of the study. Our research focused on the potential roles of major Ca^{2+} channels and proteins in cytoplasmic Ca^{2+} fluctuations in skeletal muscles of Daurian ground squirrels during hibernation. (**A**) Fluctuation of cytoplasmic Ca^{2+} levels in skeletal muscle fibers during the torpor-arousal cycle. (**B**) Increased activation probability or protein expression of SOCE, RyR1, and LETM1 may participate in the periodic elevation of cytoplasmic Ca^{2+} during hibernation, whereas the increased expression of SERCA1, MCU, and CALM (only for PL) may be potential mechanisms through which hibernators can attenuate cytoplasmic Ca^{2+} and restore Ca^{2+} homeostasis during hibernation. Compared with that in LT or IBA groups, the decreased activation of SOCE and increased expression of MCU may participate in the partial down-regulation of cytoplasmic Ca^{2+}. PL, plantaris; AM, adductor magnus. SA, summer active group; PRE, pre-hibernation group; LT, late torpor group; IBA, inter-bout arousal group; ET, early torpor group; POST, post-hibernation group.

During prolonged hibernation, hibernators experience various stressful conditions, including prolonged inactivity, hypoxia, fasting, and repeated ischemia-reperfusion from the torpor-arousal cycle. Here, after several months of hindlimb inactivity during hibernation, slight decreases in muscle mass and single muscle fiber CSA (15–20%) were observed, suggesting that only limited skeletal muscle loss occurs in the PL and AM muscles, consistent with our previous report on PL and gastrocnemius muscles [36]. Obviously, hibernators can be considered good anti-atrophy models, and their unique skeletal muscle preservation mechanisms deserve further exploration. In view of the critical role of intracellular Ca^{2+} homeostasis in skeletal muscle maintenance, the present study focused on changes in Ca^{2+} levels in the different compartments (including cytoplasm, SR, and mitochondria) of skeletal muscle fibers during different hibernation stages. The results showed that the cytoplasmic Ca^{2+} levels in the PL and AM muscles were elevated to varying degrees during hibernation, suggesting that, similar to the phenomenon found in non-hibernators, prolonged skeletal muscle disuse during hibernation was accompanied by elevated cytoplasmic Ca^{2+} levels [37–40]. However, fluctuations in cytoplasmic Ca^{2+} levels were observed during the torpor-arousal cycle, with an increase during LT and partial recovery when the squirrels re-entered torpor after transient arousal. Therefore, periodic arousal from torpor may be a strategy employed by hibernators to adjust and stabilize cytoplasmic Ca^{2+} levels to effectively avoid excessive Ca^{2+}-induced skeletal muscle loss or damage. Our previous study found that, compared with pre-hibernation levels, serum Ca^{2+} concentrations in ground squirrels increased significantly during hibernation and recovered after post-hibernation [41]. This raises the question of whether increased serum Ca^{2+} concentrations during hibernation influence cytoplasmic Ca^{2+} levels. SOCE is the main channel of extracellular Ca^{2+} influx. Higher co-localization coefficients of its two major components, i.e., STIM1 and ORAI1, represent a higher activation probability of SOCE [35]. Our results showed that the co-localization coefficients of STIM1 and ORAI1 increased significantly during LT and IBA, suggesting that the activation probability of SOCE increased during these two stages. Furthermore, the extracellular Ca^{2+} influx mediated by SOCE may be one reason for the increased cytoplasmic Ca^{2+} level during LT and IBA. In addition to extracellular Ca^{2+} influx, Ca^{2+} release from intracellular Ca^{2+} storage can also cause elevated cytoplasmic Ca^{2+} levels. RyR1 is the main Ca^{2+} release channel located in the SR. The results showed that its protein expression level was up-regulated to varying degrees during different periods of hibernation. Combined with the phenomenon that the Ca^{2+} concentration in the SR showed the opposite change to that of cytoplasmic Ca^{2+} during hibernation, the increased release of SR Ca^{2+} was likely a major reason for the increased Ca^{2+} level in skeletal muscles of ground squirrels during hibernation. In addition to SR Ca^{2+} release, several proteins also mediated Ca^{2+} efflux from the mitochondria to the cytoplasm. The present study showed that the protein expression of LETM1, a major channel that mediates Ca^{2+} efflux in the mitochondrial membrane, was increased to varying degrees during different periods of hibernation, intimating that Ca^{2+} release mediated by LETM1 in the mitochondria may also be involved in the increase in cytoplasmic Ca^{2+} levels during hibernation.

As a Ca^{2+} pump located in the cell membrane, PMCA3 can pump intracellular Ca^{2+} out of the cell. In the current study, however, the PMCA3 protein expression levels showed no significant differences during hibernation. Therefore, PMCA3 did not appear to play a substantial role in the fluctuation of cytoplasmic Ca^{2+} in hibernation. In contrast, the protein expression levels of SERCA1, a major Ca^{2+} uptake channel in the SR, were significantly up-regulated during different stages of hibernation, contrary to the decrease in SERCA activity found in non-hibernating animals under disuse conditions [42]. Our previous study showed that the protein content of SERCA2 in the soleus and extensor digitorum longus muscles of Daurian ground squirrels increased significantly during hibernation and IBA [7]. Other studies have also reported that SERCA activity is more resistant to temperature reduction in hibernating cardiac muscle than that in non-hibernating rats [43,44]. Such evidence indicates that, although faced with various stresses during hibernation, including a low temperature, low metabolism, and prolonged skeletal muscle disuse, SERCA1 still maintains high activity and can pump more cytoplasmic Ca^{2+} into the SR, thus avoiding an excessive increase in cytoplasmic Ca^{2+} and related skeletal muscle injury. In addition, the protein expression of MCU, a major Ca^{2+} absorption channel located in the mitochondrial membrane, was elevated during hibernation, and may therefore be another important mechanism for the absorption of Ca^{2+} into the mitochondria and alleviation of the Ca^{2+} level in the cytoplasm. However, we know that when the concentration of mitochondrial Ca^{2+} reaches a certain threshold, mitochondrial depolarization is triggered and, as a consequence, the pro-apoptotic protein Bax is activated, translocated, and inserted into the outer membrane via Bax/Bax-homo-oligomerization [45]. This is followed by the formation and opening of a mitochondrial permeability transition pore (mPTP), through which cytochrome C (a mitochondria-residing apoptogenic factor) is released into the cytosol, leading to the cleavage of nuclear DNA and cell apoptosis [3,46]. In the present study, the level of mitochondrial Ca^{2+} only increased slightly during hibernation. Therefore, the simultaneously higher expression of MCU and LETM1 may be a strategy employed by hibernating ground squirrels to control the balance between cytoplasmic and mitochondrial Ca^{2+}, and thereby avoid mitochondrial Ca^{2+} concentration-induced skeletal muscle damage, such as cell apoptosis. As a free Ca^{2+} binding protein, CALM can relieve the elevation of cytoplasmic Ca^{2+} by binding to free Ca^{2+} in the cytoplasm. Our results showed that the protein expression of CALM in the PL muscle increased significantly, and the enhanced free Ca^{2+} binding capacity mediated by the elevated protein expression of CALM may thus be another important mechanism in hibernating ground squirrels to alleviate cytoplasmic Ca^{2+} levels during hibernation.

During the torpor-arousal cycle, the intracellular Ca^{2+} level showed increased-decreased fluctuation. Compared with that during LT and IBA, the cytoplasmic Ca^{2+} level in the skeletal muscles of the Daurian ground squirrels was down-regulated to varying degrees after re-entry into torpor (ET). Compared with levels in LT, the co-localization coefficients of STIM1 and ORAI1 decreased to varying degrees, representing decreased Ca^{2+} influx mediated by SOCE, which may be another critical mechanism used to avoid the intracellular Ca^{2+} increase caused by persistent Ca^{2+} influx during hibernation. In addition, we found that the higher protein expression of Ca^{2+} uptake channel MCU in ET may be another vital mechanism through which PL muscle can relieve cytoplasmic Ca^{2+} overload. It is worth noting that, after arousal from hibernation, the intracellular Ca^{2+} levels (i.e., cytoplasmic, SR, and mitochondrial) returned to the levels found in the SA and PRE groups, thus reflecting a strong and rapid Ca^{2+} recovery ability of these ground squirrels. We further found that the expression of Ca^{2+} transport proteins/channels or free Ca^{2+} binding protein also returned to the levels observed in the SA and PRE groups. This may be due to the sampling time of the POST group, which occurred 3 d after the squirrels aroused from torpor, and all physiological states had returned to normal.

By comparing and analyzing the mRNA and protein expression levels of each Ca^{2+} transporter protein, we found that the levels were not always consistent in the same stage. In particular, during LT, the mRNA levels of most Ca^{2+} transport proteins/channels were decreased or showed no significant change, whereas the corresponding protein expression levels were significantly increased. In view of these results, we checked the transcription and proteomic analysis results of the skeletal muscles during different hibernation stages (unpublished data) and found that the predicted results were consistent with our experimental findings. Previous studies have demonstrated that transcriptional elongation and initiation are essentially arrested during hibernation [47,48], which may be one of the main reasons for the lower mRNA expression of Ca^{2+} transport proteins/channels in skeletal muscles of Daurian ground squirrels during LT. In other words, the inconsistent changes in mRNA and protein expression levels may be the result of total inhibition at the transcriptional level during LT, although the corresponding protein expression was not affected during hibernation.

In summary, despite experiencing stresses such as a low temperature, low metabolism, and prolonged hindlimb inactivity during hibernation, hibernating ground squirrels still possess a strong Ca^{2+} operation ability. Here, by regulating the activity and protein expression of Ca^{2+} pumps, Ca^{2+} channels, and Ca^{2+} binding proteins in the cytoplasm, SR, and mitochondrial membrane, the dynamic balance of intracellular Ca^{2+} homeostasis was well-maintained during hibernation. Therefore, maintaining intracellular Ca^{2+} homeostasis and avoiding skeletal muscle injury caused by its disturbance appear to be priority strategies employed by hibernating squirrels to cope with the various stresses induced during the torpor-arousal cycle.

Supplementary Materials: The following are available online at http://www.mdpi.com/2073-4409/9/1/42/s1, Figure S1: Represent image of the complete SDS-PAGE lane for each antibody used in present study.

Author Contributions: J.Z. and X.L. participated in the study design and carried out the molecular laboratory work, acquisition of data, data analysis and interpretation, and drafting of the manuscript. F.I., S.X., Z.W., C.Y., and X.P. helped draft the manuscript. Y.G. contributed to the conception and design of the study and final approval of the submitted version. H.C. and H.W. critically revised the paper for important intellectual content. All authors have read and agreed to the published version of the manuscript.

Funding: This work was supported by funds from the National Natural Science Foundation of China (No. 31772459), Shaanxi Province Natural Science Basic Research Program (2018JM3015), and Key Project of North Minzu University (No. 2017KJ28, 2016skky02).

Conflicts of Interest: The authors declare no conflicts of interest.

References

1. Gao, Y.; Arfat, Y.; Wang, H.; Goswami, N. Muscle atrophy induced by mechanical unloading: Mechanisms and potential countermeasures. *Front. Physiol.* **2018**, *9*, 235. [CrossRef]

2. Goll, D.E.; Thompson, V.F.; Li, H.; Wei, W.; Cong, J. The calpain system. *Physiol. Rev.* **2003**, *83*, 731–801. [CrossRef]

3. Powers, S.K.; Wiggs, M.P.; Duarte, J.A.; Zergeroglu, A.M.; Demirel, H.A. Mitochondrial signaling contributes to disuse muscle atrophy. *Am. J. Physiol. Endocrinol. Metab.* **2012**, *303*, E31–E39. [CrossRef] [PubMed]

4. Gao, Y.F.; Wang, J.; Wang, H.P.; Feng, B.; Dang, K.; Wang, Q.; Hinghofer-Szalkay, H.G. Skeletal muscle is protected from disuse in hibernating dauria ground squirrels. *Comp. Biochem. Physiol. Part A Mol. Integr. Physiol.* **2012**, *161*, 296–300. [CrossRef] [PubMed]

5. Cotton, C.J. Skeletal muscle mass and composition during mammalian hibernation. *J. Exp. Biol.* **2016**, *219*, 226–234. [CrossRef] [PubMed]

6. Fu, W.; Hu, H.; Dang, K.; Chang, H.; Du, B.; Wu, X.; Gao, Y. Remarkable preservation of Ca^{2+} homeostasis and inhibition of apoptosis contribute to anti-muscle atrophy effect in hibernating daurian ground squirrels. *Sci. Rep.* **2016**, *6*, 27020. [CrossRef]

7. Guo, Q.L.; Mi, X.; Sun, X.Y.; Li, X.Y.; Fu, W.W.; Xu, S.H.; Wang, Q.; Arfat, Y.; Wang, H.P.; Chang, H.; et al. Remarkable plasticity of Na^+, K^+-atpase, Ca^{2+}-atpase and serca contributes to muscle disuse atrophy resistance in hibernating daurian ground squirrels. *Sci. Rep.* **2017**, *7*, 10509. [CrossRef]

8. Varnai, P.; Hunyady, L.; Balla, T. Stim and orai: The long-awaited constituents of store-operated calcium entry. *Trends Pharmacol. Sci.* **2009**, *30*, 118–128. [CrossRef]

9. Salido, G.M.; Sage, S.O.; Rosado, J.A. Biochemical and functional properties of the store-operated Ca^{2+} channels. *Cell. Signal.* **2009**, *21*, 457–461. [CrossRef]

10. Vig, M.; Peinelt, C.; Beck, A.; Koomoa, D.L.; Rabah, D.; Koblan-Huberson, M.; Kraft, S.; Turner, H.; Fleig, A.; Penner, R.; et al. Cracm1 is a plasma membrane protein essential for store-operated Ca^{2+} entry. *Science* **2006**, *312*, 1220–1223. [CrossRef]

11. Taylor, C.W. Store-operated Ca^{2+} entry: A stimulating storai. *Trends Biochem. Sci.* **2006**, *31*, 597–601. [CrossRef] [PubMed]

12. Xu, P.; Lu, J.; Li, Z.; Yu, X.; Chen, L.; Xu, T. Aggregation of stim1 underneath the plasma membrane induces clustering of orai1. *Biochem. Biophys. Res. Commun.* **2006**, *350*, 969–976. [CrossRef] [PubMed]

13. Luik, R.M.; Wang, B.; Prakriya, M.; Wu, M.M.; Lewis, R.S. Oligomerization of stim1 couples er calcium depletion to crac channel activation. *Nature* **2008**, *454*, 538–542. [CrossRef] [PubMed]

14. Avila-Medina, J.; Mayoral-Gonzalez, I.; Dominguez-Rodriguez, A.; Gallardo-Castillo, I.; Ribas, J.; Ordonez, A.; Rosado, J.A.; Smani, T. The complex role of store operated calcium entry pathways and related proteins in the function of cardiac, skeletal and vascular smooth muscle cells. *Front. Physiol.* **2018**, *9*, 257. [CrossRef] [PubMed]

15. Laver, D.R. Ca^{2+} stores regulate ryanodine receptor Ca^{2+} release channels via luminal and cytosolic Ca^{2+} sites. *Clin. Exp. Pharmacol. Physiol.* **2007**, *34*, 889–896. [CrossRef]

16. Felder, E.; Protasi, F.; Hirsch, R.; Franzini-Armstrong, C.; Allen, P.D. Morphology and molecular composition of sarcoplasmic reticulum surface junctions in the absence of dhpr and ryr in mouse skeletal muscle. *Biophys. J.* **2002**, *82*, 3144–3149. [CrossRef]
17. Raffaello, A.; Mammucari, C.; Gherardi, G.; Rizzuto, R. Calcium at the center of cell signaling: Interplay between endoplasmic reticulum, mitochondria, and lysosomes. *Trends Biochem. Sci.* **2016**, *41*, 1035–1049. [CrossRef]
18. Jiang, D.; Zhao, L.; Clapham, D.E. Genome-wide rnai screen identifies letm1 as a mitochondrial Ca^{2+}/H^+ antiporter. *Science* **2009**, *326*, 144–147. [CrossRef]
19. Brini, M.; Cali, T.; Ottolini, D.; Carafoli, E. The plasma membrane calcium pump in health and disease. *FEBS J.* **2013**, *280*, 5385–5397. [CrossRef]
20. Primeau, J.O.; Armanious, G.P.; Fisher, M.E.; Young, H.S. The sarcoendoplasmic reticulum calcium atpase. *Sub-Cell. Biochem.* **2018**, *87*, 229–258.
21. Wang, C.H.; Wei, Y.H. Role of mitochondrial dysfunction and dysregulation of Ca^{2+} homeostasis in the pathophysiology of insulin resistance and type 2 diabetes. *J. Biomed. Sci.* **2017**, *24*, 70. [CrossRef] [PubMed]
22. Perocchi, F.; Gohil, V.M.; Girgis, H.S.; Bao, X.R.; McCombs, J.E.; Palmer, A.E.; Mootha, V.K. Micu1 encodes a mitochondrial ef hand protein required for Ca^{2+} uptake. *Nature* **2010**, *467*, 291–296. [CrossRef] [PubMed]
23. Patron, M.; Checchetto, V.; Raffaello, A.; Teardo, E.; Vecellio Reane, D.; Mantoan, M.; Granatiero, V.; Szabo, I.; De Stefani, D.; Rizzuto, R. Micu1 and micu2 finely tune the mitochondrial Ca^{2+} uniporter by exerting opposite effects on mcu activity. *Mol. Cell* **2014**, *53*, 726–737. [CrossRef] [PubMed]
24. Klee, C.B.; Vanaman, T.C. Calmodulin. *Adv. Protein Chem.* **1982**, *35*, 213–321.
25. Zhang, J.; Wei, Y.; Qu, T.; Wang, Z.; Xu, S.; Peng, X.; Yan, X.; Chang, H.; Wang, H.; Gao, Y. Prosurvival roles mediated by the perk signaling pathway effectively prevent excessive endoplasmic reticulum stress-induced skeletal muscle loss during high-stress conditions of hibernation. *J. Cell. Physiol.* **2019**, *234*, 19728–19739. [CrossRef]
26. Park, M.K.; Petersen, O.H.; Tepikin, A.V. The endoplasmic reticulum as one continuous Ca^{2+} pool: Visualization of rapid Ca^{2+} movements and equilibration. *EMBO J.* **2000**, *19*, 5729–5739. [CrossRef]
27. Ainbinder, A.; Boncompagni, S.; Protasi, F.; Dirksen, R.T. Role of mitofusin-2 in mitochondrial localization and calcium uptake in skeletal muscle. *Cell Calcium* **2015**, *57*, 14–24. [CrossRef]
28. Zhang, J.; Li, Y.; Li, G.; Ma, X.; Wang, H.; Goswami, N.; Hinghofer-Szalkay, H.; Chang, H.; Gao, Y. Identification of the optimal dose and calpain system regulation of tetramethylpyrazine on the prevention of skeletal muscle atrophy in hindlimb unloading rats. *Biomed Pharm.* **2017**, *96*, 513–523. [CrossRef]
29. Kazmin, D.; Edwards, R.A.; Turner, R.J.; Larson, E.; Starkey, J. Visualization of proteins in acrylamide gels using ultraviolet illumination. *Anal. Biochem.* **2002**, *301*, 91–96. [CrossRef]
30. Edwards, R.A.; Jickling, G.; Turner, R.J. The light-induced reactions of tryptophan with halocompounds. *Photochem. Photobiol.* **2002**, *75*, 362–368. [CrossRef]
31. Ladner, C.L.; Yang, J.; Turner, R.J.; Edwards, R.A. Visible fluorescent detection of proteins in polyacrylamide gels without staining. *Anal. Biochem.* **2004**, *326*, 13–20. [CrossRef] [PubMed]
32. Li, R.; Shen, Y. An old method facing a new challenge: Re-visiting housekeeping proteins as internal reference control for neuroscience research. *Life Sci.* **2013**, *92*, 747–751. [CrossRef] [PubMed]
33. Posch, A.; Kohn, J.; Oh, K.; Hammond, M.; Liu, N. V3 stain-free workflow for a practical, convenient, and reliable total protein loading control in western blotting. *J. Vis. Exp.* **2013**, *82*, 50948. [CrossRef] [PubMed]
34. Butorac, C.; Muik, M.; Derler, I.; Stadlbauer, M.; Lunz, V.; Krizova, A.; Lindinger, S.; Schober, R.; Frischauf, I.; Bhardwaj, R.; et al. A novel stim1-orai1 gating interface essential for crac channel activation. *Cell Calcium* **2019**, *79*, 57–67. [CrossRef]
35. Ross, G.R.; Bajwa, T.; Edwards, S.; Emelyanova, L.; Rizvi, F.; Holmuhamedov, E.L.; Werner, P.; Downey, F.X.; Tajik, A.J.; Jahangir, A. Enhanced store-operated Ca^{2+} influx and orai1 expression in ventricular fibroblasts from human failing heart. *Biol. Open* **2017**, *6*, 326–332. [CrossRef]
36. Ma, X.; Chang, H.; Wang, Z.; Xu, S.; Peng, X.; Zhang, J.; Yan, X.; Lei, T.; Wang, H.; Gao, Y. Differential activation of the calpain system involved in individualized adaptation of different fast-twitch muscles in hibernating daurian ground squirrels. *J. Applied Physiol.* **2019**, *127*, 328–341. [CrossRef]
37. Wu, X.; Gao, Y.F.; Zhao, X.H.; Cui, J.H. Effects of tetramethylpyrazine on nitric oxide synthase activity and calcium ion concentration of skeletal muscle in hindlimb unloading rats. *Zhonghua Yi Xue Za Zhi* **2012**, *92*, 2075–2077.

38. Ingalls, C.P.; Warren, G.L.; Armstrong, R.B. Intracellular Ca^{2+} transients in mouse soleus muscle after hindlimb unloading and reloading. *J. Appl. Physiol.* **1999**, *87*, 386–390. [CrossRef]

39. Ingalls, C.P.; Wenke, J.C.; Armstrong, R.B. Time course changes in $[Ca^{2+}]i$, force, and protein content in hindlimb-suspended mouse soleus muscles. *Aviat. Space Environ. Med.* **2001**, *72*, 471–476.

40. Hu, N.F.; Chang, H.; Du, B.; Zhang, Q.W.; Arfat, Y.; Dang, K.; Gao, Y.F. Tetramethylpyrazine ameliorated disuse-induced gastrocnemius muscle atrophy in hindlimb unloading rats through suppression of Ca^{2+}/ros-mediated apoptosis. *Appl. Physiol. Nutr. Metab. Physiol. Appl. Nutr. Metab.* **2017**, *42*, 117–127. [CrossRef]

41. Hu, H.X.; Du, F.Y.; Fu, W.W.; Jiang, S.F.; Cao, J.; Xu, S.H.; Wang, H.P.; Chang, H.; Goswami, N.; Gao, Y.F. A dramatic blood plasticity in hibernating and 14-day hindlimb unloading daurian ground squirrels (spermophilus dauricus). *J. Comp. Physiol. B Biochem. Syst. Environ. Physiol.* **2017**, *187*, 869–879. [CrossRef] [PubMed]

42. Li, S.; Yang, Z.; Gao, Y.; Li, G.; Wang, H.; Hinghofer-Szalkay, H.G. Ligustrazine and the contractile properties of soleus muscle in hindlimb-unloaded rats. *Aviat. Space Environ. Med.* **2012**, *83*, 1049–1054. [CrossRef] [PubMed]

43. Liu, B.; Belke, D.; Wang, L. Ca^{2+} uptake by cardiac sarcoplasmic reticulum at low temperature in rat and ground squirrel. *Am. J. Physiol.* **1997**, *272*, 1121–1127. [CrossRef] [PubMed]

44. Wang, S.Q.; Lakatta, E.G.; Cheng, H.; Zhou, Z.Q. Adaptive mechanisms of intracellular calcium homeostasis in mammalian hibernators. *J. Exp. Biol.* **2002**, *205*, 2957–2962.

45. Zha, H.; Aime-Sempe, C.; Sato, T.; Reed, J.C. Proapoptotic protein bax heterodimerizes with Bcl-2 and homodimerizes with bax via a novel domain (BH3) distinct from BH1 and BH2. *J. Biol. Chem.* **1996**, *271*, 7440–7444. [CrossRef]

46. Green, D.R.; Kroemer, G. The pathophysiology of mitochondrial cell death. *Science* **2004**, *305*, 626–629. [CrossRef]

47. Pan, P.; Treat, M.D.; van Breukelen, F. A systems-level approach to understanding transcriptional regulation by p53 during mammalian hibernation. *J. Exp. Biol.* **2014**, *217*, 2489–2498. [CrossRef]

48. van Breukelen, F.; Martin, S.L. Reversible depression of transcription during hibernation. *J. Comp. Physiol. B Biochem. Syst. Environ. Physiol.* **2002**, *172*, 355–361. [CrossRef]

Article

Satellite Cells in Skeletal Muscle of the Hibernating Dormouse, a Natural Model of Quiescence and Re-Activation: Focus on the Cell Nucleus

Manuela Malatesta [1], Manuela Costanzo [1], Barbara Cisterna [1,*] and Carlo Zancanaro [1]

Anatomy and Histology Section, Department of Neurosciences, Biomedicine and Movement Sciences, University of Verona, Strada Le Grazie, 8 I-37134 Verona, Italy; manuela.malatesta@univr.it (M.M.); manuela.costanzo@univr.it (M.C.); carlo.zancanaro@univr.it (C.Z.)
* Correspondence: barbara.cisterna@univr.it

Received: 23 March 2020; Accepted: 21 April 2020; Published: 23 April 2020

Abstract: Satellite cells (SCs) participate in skeletal muscle plasticity/regeneration. Activation of SCs implies that nuclear changes underpin a new functional status. In hibernating mammals, periods of reduced metabolic activity alternate with arousals and resumption of bodily functions, thereby leading to repeated cell deactivation and reactivation. In hibernation, muscle fibers are preserved despite long periods of immobilization. The structural and functional characteristics of SC nuclei during hibernation have not been investigated yet. Using ultrastructural and immunocytochemical analysis, we found that the SCs of the hibernating edible dormouse, *Glis glis*, did not show apoptosis or necrosis. Moreover, their nuclei were typical of quiescent cells, showing similar amounts and distributions of heterochromatin, pre-mRNA transcription and processing factors, as well as paired box protein 7 (Pax7) and the myogenic differentiation transcription factor D (MyoD), as in euthermia. However, the finding of accumulated perichromatin granules (i.e., sites of storage/transport of spliced pre-mRNA) in SC nuclei of hibernating dormice suggested slowing down of the nucleus-to-cytoplasm transport. We conclude that during hibernation, SC nuclei maintain similar transcription and splicing activity as in euthermia, indicating an unmodified status during immobilization and hypometabolism. Skeletal muscle preservation during hibernation is presumably not due to SC activation, but rather to the maintenance of some functional activity in myofibers that is able to counteract muscle wasting.

Keywords: Hibernation; electron microscopy; immunocytochemistry

1. Introduction

Satellite cells (SCs) represent a population of postnatal mononucleated stem cells [1] that are located between the basal lamina and the sarcolemma of skeletal muscle fibers, and are clearly detectable by means of electron microscopy [2]. SCs are able to occasionally fuse with muscle fibers in order to compensate for the muscle turnover caused by daily wear and tear, or support muscle hypertrophy, thereby underpinning skeletal muscle plasticity [3,4]. SCs exert their physiological role in close interplay with their local environment (the SC niche; [5]). While several non-SC cell populations are involved in muscle plasticity [6], SCs represent a key factor in muscle growth and regeneration, and research thereupon is steadily increasing over time. SCs are typically in a quiescent state, showing a minimum amount of cytoplasm and organelles therein [7]. Upon stimulation (e.g., physical exercise or muscle damage), SCs activate and re-enter the cell cycle, leading to proliferation and/or differentiation. Activation of SCs implies that a transition takes place in the cell nucleus from a low/absent to a high transcriptional activity [8]. The most obvious morphological counterpart of such a transition is a reduction in the amount of heterochromatin, and it was shown that the activation and differentiation of SCs are characterized by an important shift from condensed to lightly packed chromatin [9].

Hibernation is an adaptation to adverse winter conditions adopted by several mammals [10], which is characterized by greatly reduced metabolic activity and lowered body temperature while maintaining homeostasis. Upon arousal, bodily functions are resumed in full. Accordingly, cells in the organs of hibernating animals undergo periodical cycles of deactivation and reactivation. Intriguingly, skeletal muscle mass and strength are preserved during hibernation as well as the fiber size despite long periods of immobilization, contrary to what happens in non-hibernating mammals [11–13]. Protective mechanisms apparently take place in skeletal muscle during hibernation, which most likely involve, among others, inhibition of proteolysis, a decrease in autophagy and increased oxidative capacity [14,15]. The preservation of muscle mass during hibernation could also involve SC activation; however, the role of SCs in the prevention of atrophy has barely been investigated [16,17]. In particular, to the best of our knowledge, the structural and functional characteristics of the cell nucleus in SCs during hibernation have not been investigated so far.

Over the last several years, the morphology of the cell nucleus during the hibernation/arousal cycle has been studied in several tissue types of different species in our laboratory [18–27]. In the skeletal muscle, we found that the fine structure of the muscle fiber is well preserved during hibernation in the edible dormouse (*Glis glis*, Gliridae), with myonuclei showing morphological evidence of transcriptional activity [28]. In the present work, we investigated whether hibernation affects the structural and functional features of the SC nucleus. Both the ultrastructural and immunocytochemical characteristics of the SC nuclei were analyzed in active and hibernating edible dormice, with a focus on the key nuclear constituents involved in RNA transcription and processing. The results show that the SC nuclei are similar in hibernating and euthermic dormice, suggesting that factors other than SC activation are operating during hibernation in the immobilized skeletal muscle to prevent atrophy.

2. Materials and Methods

This is a retrospective study conducted on specimens obtained from six male edible dormice captured in 1998/1999 for the purpose of multiple investigations upon permission from the Regione del Veneto (Decreto n. 76 del 20 Gennaio, 1998). Adult (approximately 1–2-year-old) wild-living animals were trapped and maintained in an outdoor animal house supplied with food and bedding material. Under such conditions, they spontaneously began to hibernate in November and awoke in March. Three animals were killed during the euthermic period (June–July), and three during deep hibernation (January, after at least three days of continuous torpor). Euthermic animals were decapitated under deep anesthesia; hibernating animals were taken from the cage and immediately decapitated.

Samples of the right quadriceps muscle were processed for transmission electron microscopy for either morphology or immunocytochemistry.

For morphological analysis, muscle samples were fixed by immersion in 2.5% (*v/v*) glutaraldehyde and 2% (*v/v*) paraformaldehyde in 0.1 M Sørensen phosphate buffer pH 7.4 at 4 °C for 2 h, post-fixed with 1% (*v/v*) OsO_4 at 4 °C for 1 h, dehydrated through graded acetone and embedded in Epon 812. Ultrathin sections (70–80 nm in thickness) were placed on copper grids coated with a Formvar layer and stained with Reynold's lead citrate prior to observation.

For immunocytochemistry, muscle samples were fixed by immersion in 4% (*v/v*) paraformaldehyde in 0.1 M Sørensen phosphate buffer at 4 °C for 2 h, washed, treated with 0.5 M NH_4Cl in PBS 0.1 M pH 7.4 to block free aldehydes, dehydrated with ethanol and embedded in LR White resin polymerized under UV light. Ultrathin sections (70–80 nm in thickness) were placed on nickel grids coated with a Formvar-carbon layer and treated with the following probes: mouse monoclonal antibodies directed against the active phosphorylated form of RNA polymerase II (Abcam, Cambridge, MA, USA; ab24759) and the DNA/RNA hybrid molecules [29], both occurring at pre-mRNA transcription sites [28,30]; the (Sm)snRNP (small nuclear RiboNucleoProtein) core protein (Abcam; ab3138), involved in the co-transcriptional splicing of pre-mRNA [31]; the myogenic differentiation transcription factor D (MyoD; [32] (Abcam; ab16148)); and rabbit polyclonal antibody directed against the SC-specific paired box protein 7 (Pax7) transcription factor [33] (Abcam; ab34360). In detail, the sections were floated for 3 min on normal goat serum (NGS) diluted 1:100 in PBS, incubated for 17 h at 4 °C with the primary antibody diluted in PBS containing 0.1% (*w/v*) bovine serum albumin (Fluka, Buchs, Switzerland) and 0.05% (*v/v*) Tween 20. After rinsing, sections were floated on NGS, and then reacted for 30 min at room temperature with the specific secondary 12 or 6 nm gold-conjugated antibody (Jackson ImmunoResearch Laboratories Inc., West Grove, PA, USA) diluted 1:10 in PBS. Finally, the sections were rinsed and air-dried. The control grids were treated as above, but the primary antibody was omitted from the incubation mixture, and then processed as described. After the immunocytochemical procedure, the sections were treated with Uranyl Acetate Replacement Stain (Electron Microscopy Sciences, Hatfield, PA, USA) for 30 min and lead citrate for 45 s in order to weakly stain the heterochromatin and visualize the structural constituents in the interchromatin space. All grids were observed in a Philips Morgagni TEM operating at 80 kV and equipped with a Megaview III camera for digital image acquisition.

Quantitative assessment of immunolabelling was carried out by estimating the gold particle density over the interchromatin space (i.e., the nucleoplasmic region devoid of heterocromatin clumps) in sections treated in the same run. Briefly, the surface areas of the nucleoplasmic region and the heterochromatin were measured in 10 randomly selected electron micrographs (×22,000) of SC nuclei from each animal using a computerized image analysis system (AnalySIS Image processing, Soft Imaging System GmbH, Münster, Germany). The interchromatin space area was calculated, the gold particles present over the interchromatin space were counted and their density was expressed as number/µm^2. Background evaluation was carried out on the resin (in the areas devoid of tissue) of the immunolabelled samples as well as on the tissue of control samples. The same procedure was used to assess the density of perichromatin granules (PG; representing sites of storage and/or transport of spliced pre-mRNA [34]) over the interchromatin space.

To estimate the amount of heterochromatin, the percentage of the heterochromatin area within the total nucleoplasm area was calculated.

For each analyzed variable, the Kolmogorov–Smirnov two-sample test was performed in order to verify the hypothesis of identical distribution among animals in each group, and then the mean ± standard error of the mean (SEM) was calculated. Comparisons of variables in the two groups (euthermic and hibernating) were performed with one-way ANOVA (significance set at $p \leq 0.05$).

3. Results

In all of the muscle samples, SCs were morphologically recognizable as small cells located between the sarcolemma and the myofiber basal lamina; they showed scarce cytoplasm and an ovoid nucleus with an irregular border and abundant heterochromatin clumps (Figure 1).

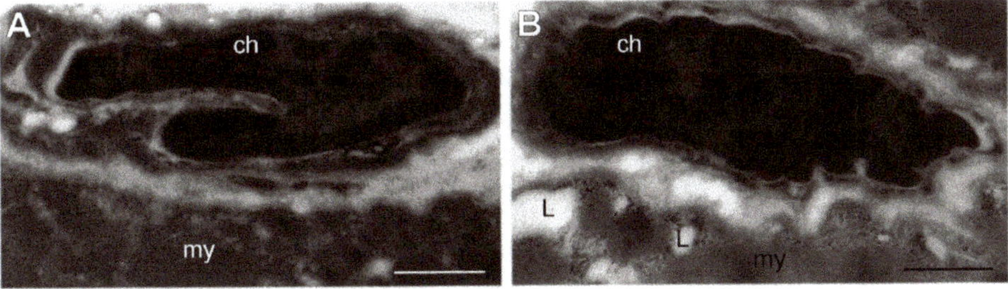

Figure 1. Transmission electron micrographs of satellite cells (SCs) bordering a myofiber (my) in skeletal muscles from euthermic (**A**) and hibernating (**B**) dormice. In both seasonal phases, SC nuclei contain large amounts of heterochromatin (ch). (**B**) The accumulation of lipid droplets (L) in the myofiber is a typical feature of hibernating edible dormice [28]. Bars: 1 µm.

In the LR White embedded samples, the usual RNP structural constituents involved in pre-mRNA transcription and processing were evident in the nucleoplasm (Figure 2): a few perichromatin fibrils (PFs; representing the in situ form of nascent transcripts, as well as of their splicing and 3′ end processing [34,35]) and PGs were mainly distributed at the periphery of the heterochromatin clumps, and small clusters of interchromatin granules (IG; representing the storage, assembly and phosphorylation sites for transcription and splicing factors [34]) occurred in the interchromatin space (not shown).

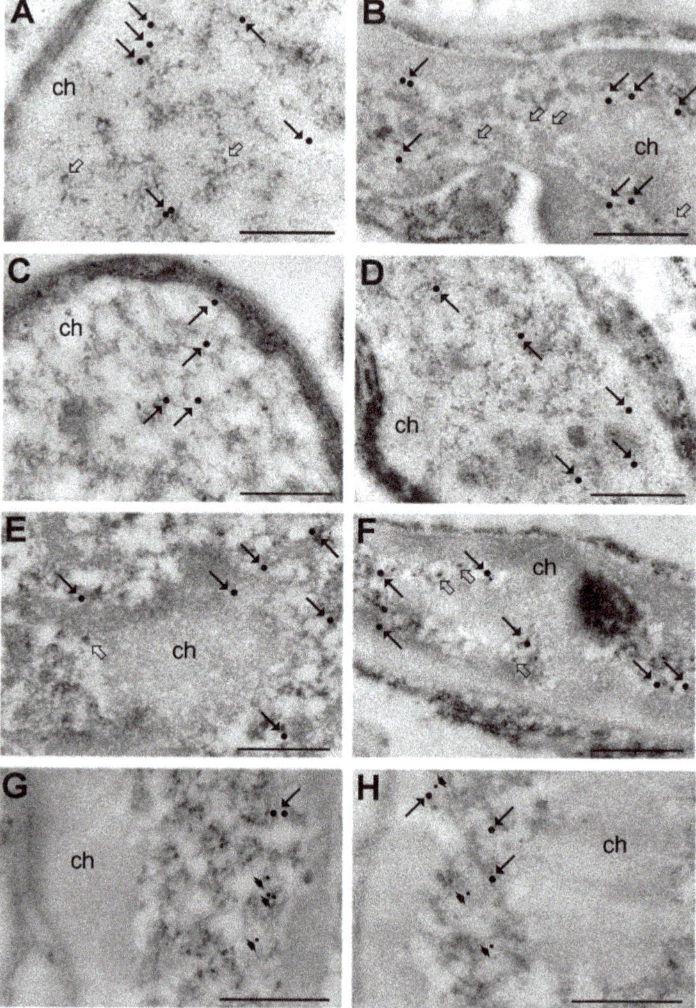

Figure 2. Immunoelectron microscopy. SC nuclei from euthermic (**A,C,E,G**) and hibernating (**B,D,F,H**) dormice; immunolabelling for RNA polymerase II (**A,B**; arrows), DNA/RNA hybrid molecules (**C,D**; arrows), small nuclear RiboNucleoProtein ((Sm)snRNP) core protein (**E,F**; arrows), paired box protein 7 (Pax7) (**G,H**; arrows) and the myogenic differentiation transcription factor D (MyoD) (**G,H**; arrowheads). All antibodies specifically label perichromatin fibrils (PFs) that mostly occur at the periphery of heterochromatin clumps (ch). Perichromatin granules (PGs) are indicated by open arrows (**A,B,E,F**). Gold particles were digitally enhanced to improve their visibility. Bars: 500 nm.

SC nuclei were structurally similar in euthermic and hibernating dormice, and morphological evidence of apoptosis or necrosis was never found in any of the muscle samples examined. No statistically significant difference in the percentage of heterochromatin was found between hibernating and euthermic dormice (Figure 3A). Conversely, PG density increased in hibernating vs. euthermic dormice (Figure 3B).

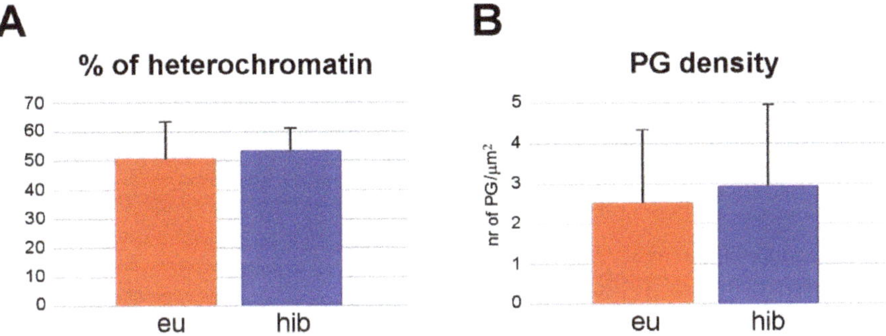

Figure 3. Quantitative evaluation of the percentage of heterochromatin (**A**) and PG density (**B**) (mean ± standard error of the mean (SEM)) in SC nuclei from skeletal muscles of euthermic (eu) and hibernating (hib) dormice. No significant difference was found between euthermia and hibernation for heterochromatin ($p = 0.091$), whereas PG density was significantly higher in hibernating dormice ($p = 0.002$).

The distribution of the immunolabelling for phosphorylated polymerase II, DNA/RNA hybrid molecules, (Sm)snRNP, PAX7 and MyoD was similar in SC nuclei from hibernating and euthermic dormice, being almost exclusively associated with PFs at the edge of the heterochromatin clumps (Figure 2). Quantitative evaluation of the immunolabelling revealed similar densities of all probes in SC nuclei of hibernating and euthermic dormice (Figure 4). Background values were negligible in all the immunolabelling experiments (not shown).

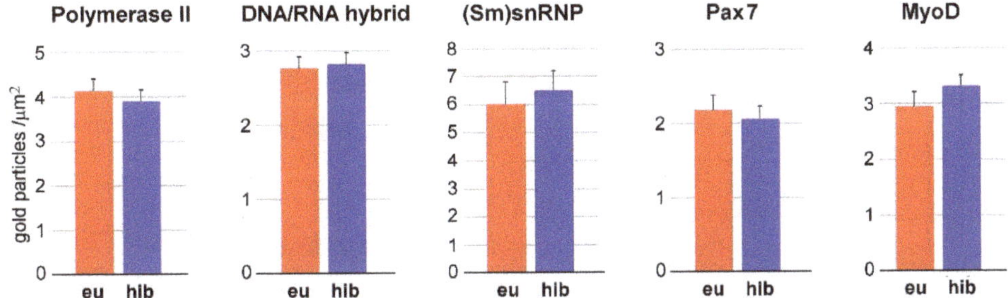

Figure 4. Quantitative immunoelectron microscopy. Labelling density (gold particles/µm²) of RNA processing factors in the interchromatin space (mean ± SE) of SC nuclei from skeletal muscles of euthermic (eu) and hibernating (hib) dormice. No significant difference was found between euthermia and hibernation.

4. Discussion

The absence of morphologically recognizable apoptotic or necrotic nuclei in SCs suggests that hibernation does not negatively affect the viability of the SC pool; this is consistent with findings in hindlimb skeletal muscles of late torpid thirteen-lined ground squirrels showing that the number of SCs does not decrease during deep hibernation in comparison with euthermia [16].

The SCs bordering the myofibers of hibernating dormice were morphologically similar to those found in euthermic animals; in particular, the structural features of their nuclei were typical of quiescent cells with low nuclear activity, i.e., showing abundant clumps of heterochromatin and a few PFs [34,35]. Qualitative analysis was confirmed by quantitative evaluation of heterochromatin, which was comparable in euthermic and hibernating dormice (Figure 3), consistent with previous observations in myonuclei of the same hibernating species [28].

Similarly, the in situ analysis of pre-mRNA transcription and processing factors did not reveal differences in both their intranuclear distribution and amount between euthermic and hibernating dormice.

Activated RNA polymerase II and DNA/RNA hybrid molecules as well as snRNPs were specifically located in PFs. Similar results were obtained in myonuclei of the same edible dormice [28] as well as in the liver and brown adipocytes of hazel dormice [20]. This finding suggests that in the SCs of hibernating dormice, the organization of the mRNA transcription and processing machinery is maintained.

Pax7 (a transcription factor marker of both quiescent and active SC) as well as MyoD (a transcription factor increasing in activated SCs [32,33]), were specifically associated with PFs, i.e., sites of transcription. Such an association was previously observed in murine myoblasts in vitro [36], and now, for the first time, in the SCs of the intact muscle.

The presence of similar amounts of immunolabelling for both Pax7 and MyoD in SC nuclei of euthermic and hibernating edible dormice indicates the absence of changes in their activation state along the hibernation cycle, in accordance with findings by Brooks et al. [16] in ground squirrels. Taken together, these results suggest that SC nuclei do not undergo modification in transcription and early splicing during hibernation. However, accumulation of PGs was found during hibernation in these nuclei. Since PGs are involved in storage/transport of spliced pre-mRNA [34], this finding indicates hibernation-associated changes in pre-mRNA processing and/or a slowdown of intranuclear or nucleus-to-cytoplasm transport of mRNAs [37]. Typically, PG accumulation due to the impairment of pre-mRNA processing is accompanied by PF clustering. For example, PF clustering has been found during ageing in different cell types, inclusive of SCs [30,38]. However, in SC nuclei of hibernating edible dormice, no PF clustering was observed. Indeed, an accumulation of PGs unaccompanied by PF clustering was observed in brown adipocytes of hibernating hazel dormice, and was interpreted as a consequence of continuing transcription and splicing activity paralleled by a reduced export of mature mRNA to the cytoplasm, likely to be promptly used upon arousal [39]. Accumulation of PGs during hibernation also suggests that, in spite of the maintenance of transcription and splicing rate, SC activity undergoes some decrease in hibernating edible dormice. This suggestion is supported by findings in hibernating ground squirrels, where an inhibition of both SC activation and myoblast differentiation were shown [17].

Interestingly, no nuclear bodies were observed during hibernation inside the SC nuclei of edible dormice, whereas myonuclei of the same animals showed some amorphous bodies [28]. Different types of nuclear bodies involved in the storage/assembly of RNA processing factors have been shown to form in various tissues during hibernation [23] and rapidly disappear upon arousal when massive nuclear reactivation takes place [21]. Their absence in SCs during hibernation could be related to the naturally quiescent state of these cells, which present a low metabolic activity even in euthermia.

In conclusion, the SC nuclei of hibernating edible dormice maintain similar transcription and splicing activity as in euthermia, although the nucleus-to-cytoplasm transport undergoes a slowing down. Therefore, the SC activation state is unmodified during hibernation, supporting the idea that skeletal muscle preservation during this seasonal phase, characterized by prolonged inactivity and starvation, is not due to SCs, but rather to the maintenance of some functional activity in myofibers that is able to counteract muscle wasting [17,28].

Author Contributions: Conceptualization, writing—review and editing, M.M. and C.Z.; investigation, M.C. and B.C.; writing—original draft preparation, M.M. and B.C.; supervision, C.Z. All authors have read and agreed to the published version of the manuscript.

Funding: This research received no external funding.

Conflicts of Interest: The authors declare no conflict of interest.

References

1. Dumont, N.A.; Bentzinger, C.F.; Sincennes, M.C.; Rudnicki, M.A. Satellite cells and skeletal muscle regeneration. *Compr. Physiol.* **2015**, *5*, 1027–1059. [PubMed]
2. Mauro, A. Satellite cell of skeletal muscle fibers. *J. Biophys. Biochem. Cytol.* **1961**, *9*, 493–495. [CrossRef] [PubMed]
3. Snijders, T.; Nederveen, J.P.; McKay, B.R.; Joanisse, S.; Verdijk, L.B.; van Loon, L.J.; Parise, G. Satellite cells in human skeletal muscle plasticity. *Front. Physiol.* **2015**, *6*, 283. [CrossRef] [PubMed]
4. Forcina, L.; Miano, C.; Pelosi, L.; Musarò, A. An Overview about the biology of skeletal muscle satellite cells. *Cur. Genomics.* **2019**, *20*, 24–37. [CrossRef] [PubMed]
5. Mashinchian, O.; Pisconti, A.; Le Moal, E.; Bentzinger, C.F. The muscle stem cell niche in health and disease. *Curr. Top. Dev. Biol.* **2018**, *126*, 23–65. [PubMed]
6. Yin, H.; Price, F.; Rudnicki, M.A. Satellite cells and the muscle stem cell niche. *Physiol. Rev.* **2013**, *93*, 23–67. [CrossRef]
7. Muir, A.R.; Kanji, A.H.; Allbrook, D. The structure of the satellite cells in skeletal muscle. *J. Anat.* **1965**, *99*, 435–444.
8. Schultz, E.; Gibson, M.C.; Champion, T. Satellite cells are mitotically quiescent in mature mouse muscle: An EM and radioautographic study. *J. Exp. Zool.* **1978**, *206*, 451–456. [CrossRef]
9. Shi, X.; Garry, D.J. Muscle stem cells in development, regeneration, and disease. *Genes Dev.* **2006**, *20*, 1692–1708. [CrossRef]
10. Geiser, F. Hibernation. *Curr. Biol.* **2013**, *23*, R188–R193. [CrossRef]
11. Cotton, C.J.; Harlow, H.J. Avoidance of skeletal muscle atrophy in spontaneous and facultative hibernators. *Physiol. Biochem. Zool.* **2010**, *83*, 551–560. [CrossRef] [PubMed]
12. Andres-Mateos, E.; Brinkmeier, H.; Burks, T.N.; Mejias, R.; Files, D.C.; Steinberger, M.; Soleimani, A.; Marx, R.; Simmers, J.L.; Lin, B.; et al. Activation of serum/glucocorticoid-induced kinase 1 (SGK1) is important to maintain skeletal muscle homeostasis and prevent atrophy. *EMBO Mol. Med.* **2013**, *5*, 80–91. [CrossRef] [PubMed]
13. Ivakine, E.A.; Cohn, R.D. Maintaining skeletal muscle mass: Lessons learned from hibernation. *Exp. Physiol.* **2014**, *99*, 632–637. [CrossRef] [PubMed]
14. Tessier, S.N.; Storey, K.B. Lessons from mammalian hibernators: Molecular insights into striated muscle plasticity and remodeling. *Biomol. Concepts.* **2016**, *7*, 69–92. [CrossRef]
15. Chang, H.; Peng, X.; Yan, X.; Zhang, J.; Xu, S.; Wang, H.; Wang, Z.; Ma, X.; Gao, Y. Autophagy and Akt-mTOR signaling display periodic oscillations during torpor-arousal cycles in oxidative skeletal muscle of Daurian ground squirrels (Spermophilus dauricus). *J. Comp. Physiol. B* **2020**, *190*, 113–123. [CrossRef]
16. Brooks, N.E.; Myburgh, K.H.; Storey, K.B. Muscle satellite cells increase during hibernation in ground squirrels. *Comp. Biochem. Physiol. B Biochem. Mol. Biol.* **2015**, *189*, 55–61. [CrossRef]
17. Andres-Mateos, E.; Mejias, R.; Soleimani, A.; Lin, B.M.; Burks, T.N.; Marx, R.; Lin, B.; Zellars, R.C.; Zhang, Y.; Huso, D.L.; et al. Impaired skeletal muscle regeneration in the absence of fibrosis during hibernation in 13-lined ground squirrels. *PLoS ONE* **2012**, *7*, e48884. [CrossRef]
18. Malatesta, M.; Zancanaro, C.; Biggiogera, M. Immunoelectron microscopic characterization of nucleolus-associated domains during hibernation. *Microsc. Res. Tech.* **2011**, *74*, 47–53. [CrossRef]
19. Suozzi, A.; Malatesta, M.; Zancanaro, C. Subcellular distribution of key enzymes of lipid metabolism during the euthermia-hibernation-arousal cycle. *J. Anat.* **2009**, *214*, 956–962. [CrossRef]
20. Malatesta, M.; Biggiogera, M.; Baldelli, B.; Barabino, S.M.; Martin, T.E.; Zancanaro, C. Hibernation as a far-reaching program for the modulation of RNA transcription. *Microsc. Res. Tech.* **2008**, *71*, 564–572. [CrossRef]

21. Malatesta, M.; Luchetti, F.; Marcheggiani, F.; Fakan, S.; Gazzanelli, G. Disassembly of nuclear bodies during arousal from hibernation: An in vitro study. *Chromosoma* **2001**, *110*, 471–477. [CrossRef] [PubMed]
22. Malatesta, M.; Gazzanelli, G.; Battistelli, S.; Martin, T.E.; Amalric, F.; Fakan, S. Nucleoli undergo structural and molecular modifications during hibernation. *Chromosoma* **2000**, *109*, 506–513. [CrossRef] [PubMed]
23. Malatesta, M.; Cardinali, A.; Battistelli, S.; Zancanaro, C.; Martin, T.E.; Fakan, S.; Gazzanelli, G. Nuclear bodies are usual constituents in tissues of hibernating dormice. *Anat. Rec.* **1999**, *254*, 389–395. [CrossRef]
24. Tamburini, M.; Malatesta, M.; Zancanaro, C.; Martin, T.E.; Fu, X.D.; Vogel, P.; Fakan, S. Dense granular bodies: A novel nucleoplasmic structure in hibernating dormice. *Histochem. Cell. Biol.* **1996**, *106*, 581–586. [CrossRef] [PubMed]
25. Malatesta, M.; Zancanaro, C.; Tamburini, M.; Martin, T.E.; Fu, X.D.; Vogel, P.; Fakan, S. Novel nuclear ribonucleoprotein structural components in the dormouse adrenal cortex during hibernation. *Chromosoma* **1995**, *104*, 121–128. [CrossRef]
26. Malatesta, M.; Zancanaro, C.; Martin, T.E.; Chan, E.K.; Amalric, F.; Lührmann, R.; Vogel, P.; Fakan, S. Cytochemical and immunocytochemical characterization of nuclear bodies during hibernation. *Eur. J. Cell. Biol.* **1994**, *65*, 82–93.
27. Malatesta, M.; Zancanaro, C.; Martin, T.E.; Chan, E.K.; Amalric, F.; Lührmann, R.; Vogel, P.; Fakan, S. Is the coiled body involved in nucleolar functions? *Exp. Cell. Res.* **1994**, *211*, 415–419. [CrossRef]
28. Malatesta, M.; Perdoni, F.; Battistelli, S.; Muller, S.; Zancanaro, C. The cell nuclei of skeletal muscle cells are transcriptionally active in hibernating edible dormice. *BMC Cell Biol.* **2009**, *10*, 19. [CrossRef]
29. Testillano, P.S.; Gorab, E.; Risueno, M.C. A new approach to map transcription sites at the ultrastructural level. *J. Histochem. Cytochem.* **1994**, *42*, 1–10. [CrossRef]
30. Malatesta, M.; Perdoni, F.; Muller, S.; Zancanaro, C.; Pellicciari, C. Nuclei of aged myofibres undergo structural and functional changes suggesting impairment in RNA processing. *Eur. J. Histochem.* **2009**, *53*, 97–106. [CrossRef]
31. Lührmann, R.; Kastner, B.; Bach, M. Structure of spliceosomal snRNPs and their role in pre-mRNA splicing. *Biochim. Biophys. Acta* **1990**, *1087*, 265–292. [CrossRef]
32. Legerlotz, K.; Smith, H.K. Role of MyoD in denervated, disused, and exercised muscle. *Muscle Nerve* **2008**, *38*, 1087–1100. [CrossRef] [PubMed]
33. Seale, P.; Sabourin, L.A.; Girgis-Gabardo, A.; Mansouri, A.; Gruss, P.; Rudnicki, M.A. Pax7 is required for the specification of myogenic satellite cells. *Cell* **2000**, *102*, 777–786. [CrossRef]
34. Fakan, S. Ultrastructural cytochemical analyses of nuclear functional architecture. *Eur. J. Histochem.* **2004**, *48*, 5–14. [PubMed]
35. Biggiogera, M.; Cisterna, B.; Spedito, A.; Vecchio, L.; Malatesta, M. Perichromatin fibrils as early markers of transcriptional alterations. *Differentiation* **2008**, *76*, 57–65. [CrossRef]
36. Cisterna, B.; Giagnacovo, M.; Costanzo, M.; Fattoretti, P.; Zancanaro, C.; Pellicciari, C.; Malatesta, M. Adapted physical exercise enhances activation and differentiation potential of satellite cells in the skeletal muscle of old mice. *J. Anat.* **2016**, *228*, 771–783. [CrossRef]
37. Puvion-Dutilleul, F.; Puvion, E. Relationship between chromatin and perichromatin granules in cadmium-treated isolated hepatocytes. *J. Ultrastruct. Res.* **1981**, *74*, 341–350. [CrossRef]
38. Malatesta, M.; Biggiogera, M.; Cisterna, B.; Balietti, M.; Bertoni-Freddari, C.; Fattoretti, P. Perichromatin fibrils accumulation in hepatocyte nuclei reveals alterations of pre-mRNA processing during aging. *Dna Cell Biol.* **2010**, *29*, 49–57. [CrossRef]
39. Zancanaro, C.; Malatesta, M.; Vogel, P.; Osculati, F.; Fakan, S. Ultrastructural and morphomentrical analyses of the brown adipocyte nucleus in a hibernating dormouse. *Biol. Cell* **1993**, *79*, 55–61. [CrossRef]

MDPI
St. Alban-Anlage 66
4052 Basel
Switzerland
Tel. +41 61 683 77 34
Fax +41 61 302 89 18
www.mdpi.com

Cells Editorial Office
E-mail: cells@mdpi.com
www.mdpi.com/journal/cells

9 783039 434367